Rice Blast Disease

RICE BLAST DISEASE

Edited by

R.S. Zeigler, S.A. Leong* and P.S. Teng

Entomology and Plant Pathology Division
International Rice Research Institute
PO Box 933
1099 Manila
The Philippines

and

**Department of Plant Pathology*
University of Wisconsin-Madison
Russell Laboratories
1630 Linden Drive
Madison, WI 53706
USA

CAB INTERNATIONAL

in association with the

International Rice Research Institute

CAB INTERNATIONAL
Wallingford
Oxon OX10 8DE
UK

Tel: Wallingford (0491) 832111
Telex: 847964 (COMAGG G)
E-mail: cabi@cabi.org
Fax: (0491) 833508

Published in association with:

International Rice Research Institute
P.O. Box 933
1099 Manila
The Philippines

A catalogue entry for this book is available from the British Library.

ISBN 0 85198 935 7

Typeset by Colset Pte. Ltd., Singapore
Printed and bound in the UK at the University Press, Cambridge

Contents

Foreword

Rice will continue to be the major human staple crop well into the next century. With the twin forces of population growth and economic expansion, particularly in Asia, world rice requirements are expected to increase by 1.7% annually between 1990 and 2025. Although seemingly small, this growth rate translates into an *additional 13 million tons of rough rice per year*. With less land available to expand rice-growing areas and with competing demands from urbanization and industrialization on existing rice lands and irrigation water, production increases will have to come from intensification of existing lands in favorable and less favorable areas. However, the recipe for this intensification – better nitrogen nutrition and less water per ton of rice – can result in serious increases in pest and disease pressure; especially rice blast disease. Thus food security for the next century demands that, among many other efforts, we eliminate the risk of catastrophic losses to this disease.

The rice blast disease has long been recognized as the most explosive and potentially damaging disease of the rice crop. The complexity of the disease and the formidable challenge of the diversity of the pathogen have dictated that no lone institution or group can expect to solve the problem in isolation. From the first international conference devoted to the disease held in 1965, 30 years ago, research on blast world wide has reflected strong commitment to collaboration among scientists from universities, international and national research centers. The conference on which this book is based is the fourth in an *ad hoc* series and continues to reflect the spirit of international collaboration and free exchange of ideas and results. Over 120 scientists from more than 20 countries openly shared their very latest findings, much as yet unpublished in the formal literature. From the contents of the proceedings it is clear that tremendous progress has been made in our understanding of, and our ability to manipulate, the rice-pathogen system in order to reduce losses.

There are three striking features of this conference and the proceedings. First there is a good mix of participation from developing countries, international institutes, and 'advanced' institutes from the industrialized world. Such partnership is a necessary precondition for successfully attacking any complex field problem. It is hoped that the 'North–South' collaboration evident in many of the papers will soon be joined by 'South–South' collaboration among scientists from principal rice-growing countries. Second, scientists dealing with equally difficult problems in other crops and disciplines were included to 'cross fertilize' the thinking of the group. Too often do small groups of specialists talk only among themselves, not availing themselves of relevant progress being made by others. Finally, a strong social science perspective has been included. This recognition that solutions to problems, especially complex problems, require full farmer participation is still lacking in a disturbing number of similar discussions.

The basic to applied research outputs presented here offer great promise of success with this up-to-now unsolved problem. I thank and congratulate the participants: first for devoting their energies to such an important disease; and, second, for taking such pains to summarize their considerable efforts in a forum and style that will be accessible to all the world's rice researchers and students.

Klaus Lampe
Director General
International Rice Research Institute

Preface

Blast continues to be the most destructive fungal disease of rice despite decades of research towards its control. Because of the importance of the disease, scientists from around the world have met periodically to share their experiences, discuss research strategies and develop collaborative projects. The conference upon which this book is based was held in Madison, Wisconsin, USA, August 9–13, 1993, and represents the fourth international meeting of blast researchers held since 1960. Although rice blast has been addressed in a number of fora over the past 30 years, not since the first international meeting in 1965 has a diverse group met expressly to assess thoroughly our understanding of the pathogen, its interaction with the plant, and the development of epidemics in the field. Indeed, the last several conferences focused primarily on the development of resistant cultivars. However, progress in many relevant areas of science has been made rapidly over the past decade. Clearly it was time to bring researchers together from around the world to explore how these new advances on many scientific fronts may help us relegate blast to the status of a minor disease of rice.

The principal focus of the conference was to explore approaches to manage rice blast by integrating basic knowledge about the disease with developments in rice technology, biological control, breeding for disease resistance, and indigenous agricultural knowledge. With recent progress in rice and fungal transformation and the mapping and cloning of rice genes, it is reasonable to consider that biotechnology will make major contributions towards controlling this disease. However, technology in itself can never provide a sole solution – particularly if it disrupts the fabric of a society or degrades the environment. Likewise, the application of biotechnology must also have sound ecological bases. It is with this in mind that we have broadened the expertise of the participants.

This was the first blast conference to integrate knowledge from traditional and emerging technologies to aid in developing management plans for rice blast disease. Over 100 scientists participated, presenting 34 papers and 39 posters. These proceedings bring together almost all of what was presented at the conference and represent a consolidated summary of our current understanding of this disease. They also bring together diverse and sometimes conflicting views. In preparing manuscripts the authors were asked to summarize their very latest published and unpublished data and knowledge. In particular they were encouraged to speculate on the implications of their knowledge and propose hypotheses for testing and further areas for research. Areas of concern are highlighted in the summary of the discussion sessions. This volume also brings together an enormous amount of the latest information in the form of poster abstracts. Through the combination of in-depth chapters and these abstracts, we hope that this will serve as a reference and guide for blast researchers for years to come.

As in past meetings speakers from outside the blast community of researchers have been invited to bring different perspectives and new approaches to our efforts. In broad terms these participants have brought experience in other pathosystems from both evolutionary and social perspectives. We hope that from their contributions and from the range of disciplines represented a synergy has been established from which new, multidisciplinary, and transnational approaches to manage the disease will evolve. Indeed, an atmosphere of teamwork developed during the conference that persists as of this writing. We could never have brought so many people together from so many countries without the assistance of many dedicated people. It is impossible to mention all who helped, but several should be highlighted. In addition to the editors, J.L. Notteghem of CIRAD and A.-A. Sy of WARDA participated actively in conceptualizing the conference and identifying speakers and participants. Logistical support was very ably provided by Maureen Sundell, Patricia Gaitan and others of the University of Wisconsin Extension Conferences office, Shirley Knickmeier and Allen Budde of the USDA, and numerous graduate students and postdoctoral fellows of the Department of Plant Pathology, University of Wisconsin. Robert Bright of the University of Wisconsin Cooperative Extension assisted in organizing the discussion sessions. Yolanda Mendoza and Veronica Lantican of IRRI, and Beverly Bagnes, Evangeline Sison, Cecilia Capillo, and Gemma Eduarte of the University of the Philippines, spent long hours checking and retyping the manuscripts and abstracts. Linda Gabriel of IRRI performed the heroic task of managing the very complex editing process and kept it on track to meet a very tight schedule.

Financial and logistical support was generously provided by Agracetus Company; Agrigenetics Company; American Cyanamid; Ciba-Geigy Corporation; Dow Elanco Export; Monsanto Company; Centre Tecnique de Cooperation Agricole et Rurale; The Biotechnology Center, University of Wisconsin; The College of Agriculture and Life Sciences, University of Wisconsin; The Department of Plant Pathology, University of Wisconsin; and The United States Department of Agriculture.

Finally, on behalf of blast researchers and rice farmers worldwide, we would like to thank the conference participants and the authors. It is through their interest, dedication and efforts that the conference and this volume were such successes.

R.S. Zeigler
P.S. Teng
International Rice Research Institute
Los Baños, The Philippines

and

S.A. Leong
United States Department of Agriculture
Madison, Wisconsin, USA

I Pathogen Biology

1 Cell Biology of Pathogenesis

R.J. HOWARD

DuPont Central Research and Development, Experimental Station, PO Box 80402, Wilmington, Delaware 19880–0402, USA

INTRODUCTION

The interactions between cells of *Magnaporthe grisea* and those of the host involve a complex of biological influences which can lead to blast disease. Pathogenesis in general, and the initial infection steps in particular, may be viewed as a sequence of discrete, critical events. Disruption of this process at any point will result in failure of the pathogen to colonize host tissue, and may disclose a new avenue for developing innovative crop protection strategies and tools. However, to recognize and capture such opportunities we must first understand the very bases of the disease process at the cellular and molecular level. This will require a far greater knowledge of the cell biology of both fungus and plant. Current efforts are being made to develop and apply numerous modern techniques for cytological analysis of *Magnaporthe* growth and differentiation. Together with physiological studies of appressorium function, this research is contributing significantly to our understanding of the early stages of host–pathogen interaction.

Cryo techniques, including specimen preservation by means of ultra-low temperatures rather than chemical fixatives, have been especially useful in this regard. In combination with scanning (SEM) and transmission (TEM) electron microscopy, cyro 'fixation' has greatly enhanced ultrastructural, immunological and cytochemical analyses. With this approach, we have begun to examine the morphogenesis of conidiophores and conidiogenesis from diseased tissues using low voltage, field emission *cryoSEM* (Jeffree and Read, 1991). Subsequent events including release of spore tip mucilage, germination, germ tube growth, and appressorium formation have been further characterized, in combination with TEM examination of specimens prepared by *freeze substitution* (Howard and O'Donnell, 1987). This chapter summarizes from a cytological perspective our current understanding of infection structure morphogenesis and function.

Conidiogenesis

The process of conidium formation begins with egress from infected tissues and production of conidiophores. Cells which give rise to these long, upright structures emerge from infected tissue through natural openings provided by stomata, or directly by eruption of the plant host cuticle from underlying cells of the pathogen (Fig. 1.1). The outer cuticle of rice leaves is adorned by a fairly even distribution of wax crystals (Zee, 1981), and the sites where the pathogen exits the host directly are marked by apparent tearing of the cuticle and separation of the wax covering. Hashioka and Nakai (1974) have reported that egress by *M. grisea* closely resembles the process of initial penetration into the host (Aist, 1981). TEM analysis did not provide evidence of melanized conidiophore mother cells, suggesting that the mechanism of egress from rice tissues is different from that leading to ingress from melanized appressoria (see below). However, tearing of the cuticle during egress points to a mechanism that may include mechanical force. It is noteworthy that following disruption of a gene for cutinase, *cut1*$^{-}$ mutants were found to be completely competent in pathogenesis, including the sporulation phase (Sweigard *et al.*, 1992). Thus, enzymatic weakening of the cuticle may not be required during fungal ingress or egress.

Conidiogenous cells are bottle-shaped and form successively and sympodially, in a characteristic pattern (Fig. 1.2). However, some variation in the shape and arrangement of conidiogenous cells has been noted among various strains of the fungus. Environmental and physiological factors may contribute to these differences (Asuyama, 1965; Ou, 1985). Further study of this important stage of development would be most appropriate, using the methods devised for the study of this process in other fungi (e.g. Hanau *et al.*, 1989), and would complement research on the genetics of sporulation (Leung and Shi, this volume).

Holoblastic conidiogenesis gives rise first to a more or less spherical conidium initial (Fig. 1.2) easily distinguished from the conidiogenous cell by the unique surface 'ornamentation' which is characteristic of mature conidia. The nature and function of this outer wall layer is unknown. Subsequently the conidium elongates from the spherical form, and tapers (Fig. 1.3). Mature conidia of *M. grisea* are generally three-celled, pyriform and exhibit a basal appendage at the point of attachment to the conidiophore. Mutations of the *SMO*$^{+}$ genetic locus have been reported to cause a number of gross deviations from the normal process of conidiogenesis, resulting in conidia which exhibit a wide variety of unusual morphologies (Hamer *et al.*, 1989).

Spore tip mucilage

During maturation, the conidium swells and releases a droplet of *spore tip mucilage* while still attached to the conidiophore (Fig. 1.4). This material is found only at the conidial apex, and serves to attach the conidium to nearly any surface even under water (Hamer *et al.*, 1988). The droplet of mucilage assumes the form of a thin, viscous pool upon contact with a surface, and firmly attaches the

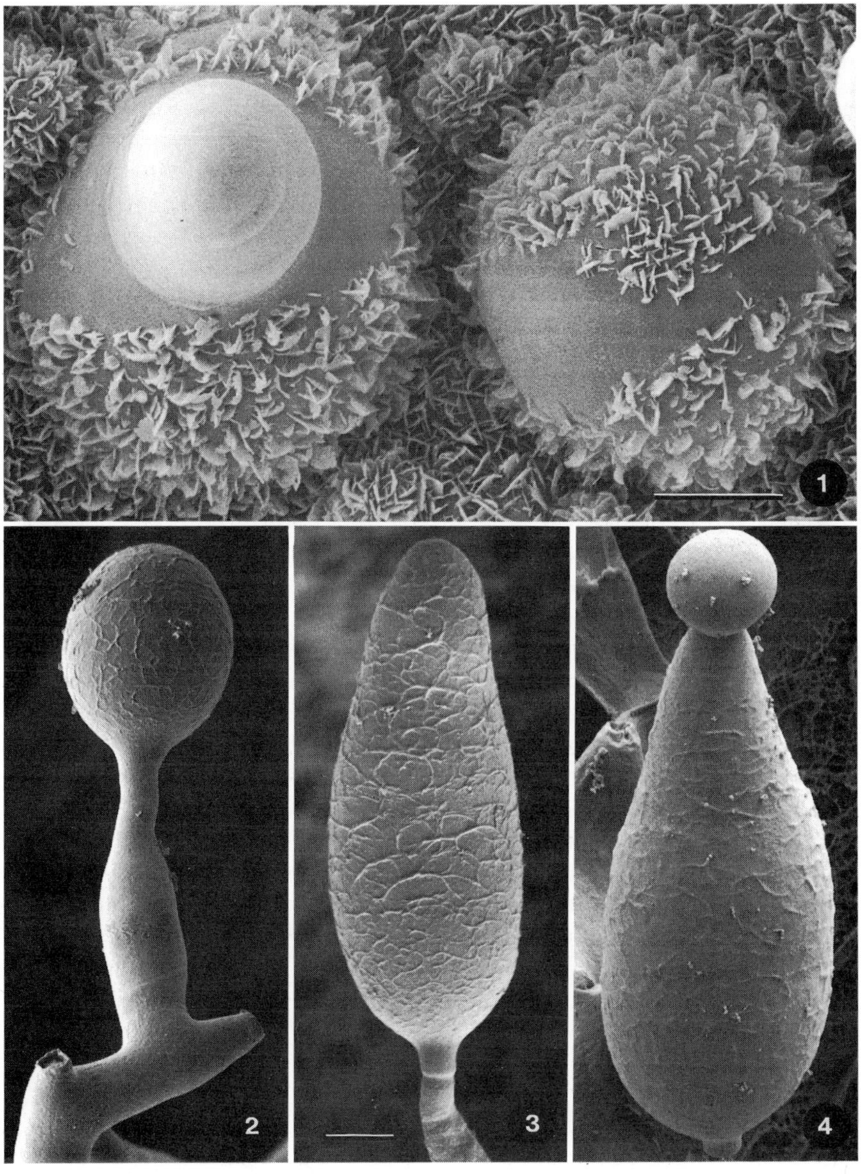

Figs 1.1–1.4. CryoSEM micrographs depicting various stages of conidiogenesis; scale bars = 2.5 μm, Figs 1.2–1.4 same magnification. **Fig. 1.1.** Eruption of conidiophore initials through a rice leaf cuticle. Apparent tearing of the cuticle is evident along the lines where leaf wax crystals have separated. **Fig. 1.2.** Conidiophore bearing an incipient conidium. Evident at this early stage of development, the young conidial surface exibits a unique reticular coating. Note also the near-right angle pattern by which conidiogenous cells are arranged. **Fig. 1.3.** Intermediate stage of conidal expansion where growth has been primarily in length. **Fig. 1.4.** Fully expanded conidium exibits an apical droplet of *spore tip mucilage*, while still attached to the conidiophore. [Fig. 1.1, unpublished micrograph by R.J. Howard and Dr N.D. Read. Fig. 1.4, from Braun and Howard (1994a), reprinted courtesy of Springer-Verlag.]

conidium, by the apex, to the underlying substrate (Fig. 1.5). Vigorous attempts to wash adherent conidia from surfaces are typically unsuccessful, showing the tenacity and strength of attachment (Hamer *et al.*, 1988).

Spore tip mucilage is believed to derive from an accumulation of material in the periplasmic space at the tips of conidia (Hamer *et al.*, 1988; Braun and Howard, 1994a), an observation made possible by the 'dry fixation' of conidia harvested in the absence of free water. An 'intact' conidial apex containing unreleased mucilage has never been found among conidia harvested using conventional aqueous washing techniques. Also elusive are the particular culture conditions under which spore tip mucilage can be produced reliably. Loss of spore tip mucilage during preparation of aqueous inoculum may partially explain the difficulties in obtaining uniform pathogenicity assays, and could reduce the efficacy of inocula.

Apparently, variation exists among strains with respect to the amount of, perhaps even the ability to produce, spore tip mucilage, a trait that might be expected to dissipate with serial culturing techniques (for proper methods of culture storage see Valent *et al.*, 1986). Optimum mucilage production seems related also to culture age and maturity of conidia.

Spore tip mucilage binds the plant lectin concanavalin A (Hamer *et al.*, 1988; Bourett *et al.*, 1993), indicating that it contains glucose and/or mannose residues. Pre-incubation of harvested conidia in buffered concanavalin A renders them incapable of adhesion. Isolation and partial characterization of the mucilage has demonstrated protein, carbohydrate and lipid components. The sole carbohydrate, 25% of the material by weight, was an α-1,2-mannose disaccharide, explaining the high affinity for concanavalin A (J.A. Sweigard and B. Valent, DuPont Company, 1989, personal communication).

Production, accumulation, release, chemical composition and physical properties of spore tip mucilage are subjects well deserving of further analysis. Increasingly, we are finding evidence for the importance and wide occurrence of similar mechanisms for adhesion among diverse plant pathogens (Braun and Howard, 1994a,b). In terms of epidemiology, an active mechanism for adhesion of conidia to the host surface may be far more important than is now appreciated. In addition, targeting such a mechanism could represent a potential site of action for a novel crop protectant strategy.

Appressorium Development

Until recently little consideration was given to stages of pathogenesis preceding penetration. There was no knowledge of mechanisms for adhesion to host surfaces, nor were there any careful studies of pathogen morphogenesis prior to the time of ingress. Rather, attention was focused on the end product, appressoria for example, and their role in penetration. Again, however, little attention was given to the mechanisms underlying these processes. Many recent reports clearly demonstrate the utility of conducting detailed studies of the events preceding ingress, and we have made considerable progress in expanding our knowledge of the pre-penetration stages of host–pathogen interactions (for a recent review see Hardham, 1992; Mendgen and Deising, 1993).

Germination, growth of germ tubes, formation of appressoria, and penetration of the underlying substrate can all occur in the absence of exogenous nutrients. Thus, much work on these stages of pathogenesis has been conducted *in vitro*, where growth and differentiation can be studied under carefully controlled conditions in sterile distilled water (Howard *et al.*, 1991a). An *in vitro* system is ideally suited for manipulating and observing the extremely delicate structures which are involved in the infection process. Obviously such conditions do not represent the natural state and we must keep this fact at the forefront of our thinking (see Valent, 1990).

Germ tubes and initiation of appressorium formation

Germination of conidia, which can occur even before release from conidiophores (unpublished observation), seems to be triggered by nothing more than the availability of free water (*cf.* Lee and Dean, 1993). Typically, a single germ tube emerges from the apical and/or basal cell of three-celled conidia (Fig. 1.6), but rarely from the middle cell (Bourett and Howard, 1990). In the presence of exogenous nutrients, such as potato dextrose broth, we have observed considerable deviation from this pattern of germination. Under nutrient-rich conditions a single conidium will give rise to many germ tubes from any or all cells. Weintraub *et al.* (1958) reported that germination can be accelerated by untreated guttation fluid from rice.

In water, germination begins after 30–90 min. Subsequent growth and differentiation of germ tubes may depend upon the surface over which it occurs. Lee and Dean (1993) report a correlation between the hydrophobicity of substrate and the induction of appressorium formation. We have examined appressorium formation in water on various inert substrates exhibiting a range of hydrophobicities, including Teflon, polystyrene, polyethylene and glass. We found considerable variation in the response to these surfaces among different strains of the pathogen, but observed no correlation between substrate hydrophobicity and appressorium formation (R. Howard and M. Ferrari, unpublished data). Uchiyama *et al.* (1979) have reported that certain rice leaf wax fractions stimulated appressorium formation while others were inhibitory. Clearly, certain physical features of a substrate can be very important in triggering the induction of appressorium formation in some fungi (Hoch and Staples, 1991; Read *et al.*, 1992; Terhune *et al.*, 1993 and references therein). Substrate features which influence infection structure development in *M. grisea* need further characterization.

During germ tube elongation a cluster of cytoplasmic vesicles can be found near the cell apex, lying next to the substratum rather than at the forefront of the apical dome as is characteristic of vegetative hyphae (Bourett and Howard, 1990; *cf.* Bourett and Howard, 1991). As the germ tube tips grow, the cluster of vesicles remains in the same location. Formation of a terminal 'hook' and swelling at the germ tube apex (Fig. 1.6) mark the onset of appressorium induction, usually 2–4 h after conidia are harvested.

The relationship of the nuclear cycle to appressorium development has been documented using DAPI staining (Bourett and Howard, 1990). Each cell of

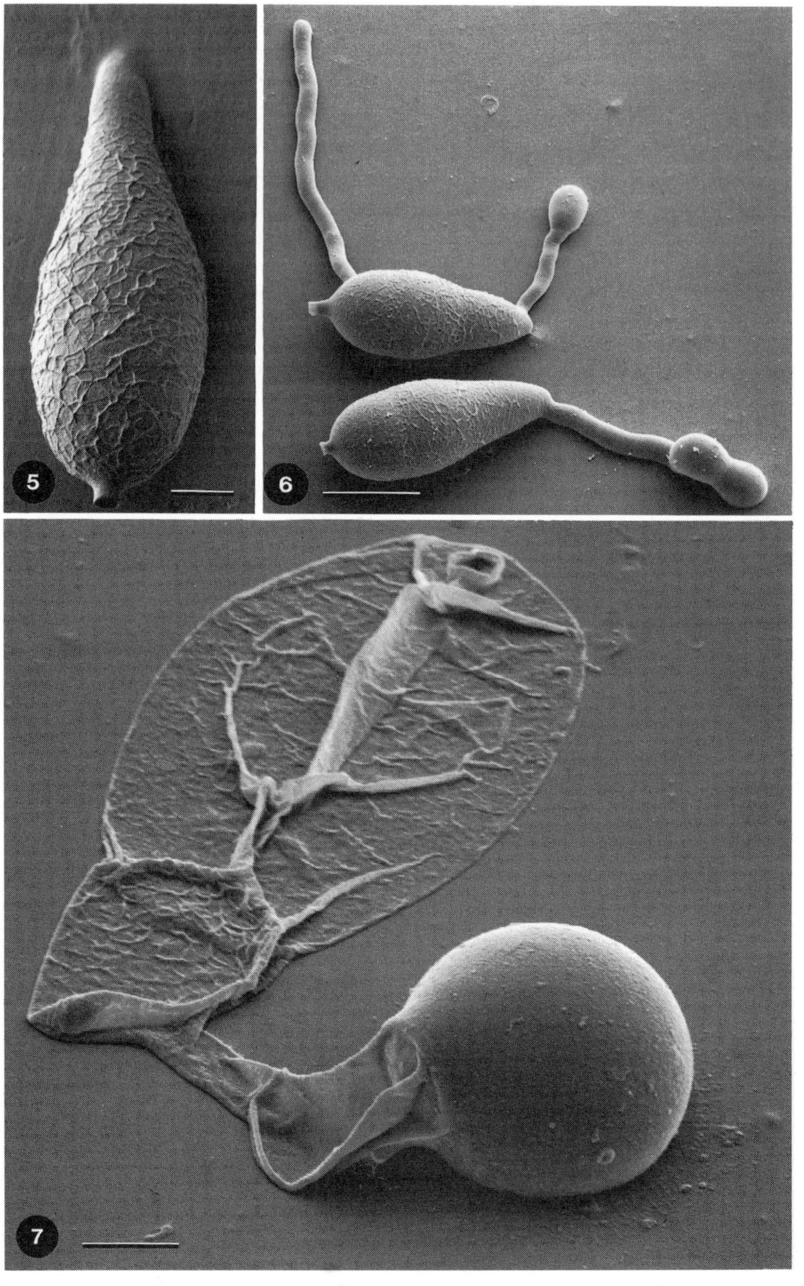

Figs 1.5–1.7. CryoSEM micrographs depicting early stages of pathogen interaction with a Mylar substrate. **Fig. 1.5.** *M. grisea* conidium attached at the apex to the substrate by way of spore tip mucilage. Surface reticulations near the conidial apex are obscured by a coating of the mucilage. Scale bar = 2.5 μm. **Fig. 1.6.** Early stages of appressorium development. Germination typically occurs from apical and/or basal conidial cell, giving rise to the filamentous germ tube (upper left). The onset of appressorium development is marked by swelling of the germ

the conidium contains a single nucleus. As the germ tube emerges from one of the cells, the nucleus in that conidial cell undergoes a mitotic division. One of the daughter nuclei remains in the conidium while the second daughter moves through the germ tube and enters the forming appressorium. A septum, visible in Fig. 1.7, then forms across the end of the germ tube. Modification of this pattern of nuclear behavior can be encountered when growing germlings on different substrates. Physical characteristics of surfaces seem to influence length of germ tubes, hence the nuclear cycle, and appressorial induction by unknown mechanisms.

Even during early stages of appressorium development, the highly specialized nature of cell wall lying against the substratum and around the future site of penetration can be recognized. This wall is juxtaposed with the vesicle cluster, is much thinner than the other walls, consists of only one layer, and apparently lacks chitin (Figs 1.8 and 1.9). In addition to the single nucleus and numerous vesicles, the expanding appressorium contains a high concentration of glycogen rosettes, many lipid bodies and smooth membrane cisternae, in addition to a normal complement of other organelles (Bourett and Howard, 1990). Under *in vitro* conditions, appressoria complete expansion, 4–8 h after conidia are harvested.

Melanization and the appressorium pore

Deposition of one final cell wall layer, just outside the plasma membrane, represents the final stage of appressorium formation. This wall layer, composed of melanin, covers the entire cell perimeter, except across the region in contact with the substratum (Fig. 1.10). The bottom zone lacking melanin is the same specialized area recognized earlier in development. After melanization, this specialized area of the cell surface appears to lack any cell wall, and is called the *appressorium pore* (Howard and Ferrari, 1989). High magnification TEM micrographs of the pore show the appressorial cytoplasm separated from the underlying substrate by only the plasma membrane.

The melanin layer appears homogeneous in the electron microscope (Fig. 1.11) and is responsible for imparting a dark pigmentation to the appressorium when observed by light microscopy (Fig. 1.12). This *M. grisea* melanin, like so many others among pathogenic fungi (Wheeler and Bell, 1988), is composed of polymerized dihydroxynaphthalene, one of the end products of the pentaketide biosynthetic pathway (Wheeler and Stipanovic, 1985). In mature appressoria the layer of melanin is about 100 nm thick, rendering the appressorium pore easily visible in the light microscope. By focusing the microscope and one's attention on the plane of contact with the substrate, a ring

---

tube apex (middle and lower right). Scale bar = 10 μm. (Micrograph by R.J. Howard and Dr N.D. Read.) **Fig. 1.7.** Fully formed, globose, highly turgid appressorium attached to collapsed conidium and germ tube. A septum separates the appressorium from the germ tube. Scale bar = 2.5 μm. [Reprinted from Braun and Howard (1994a), with permission from Springer-Verlag.]

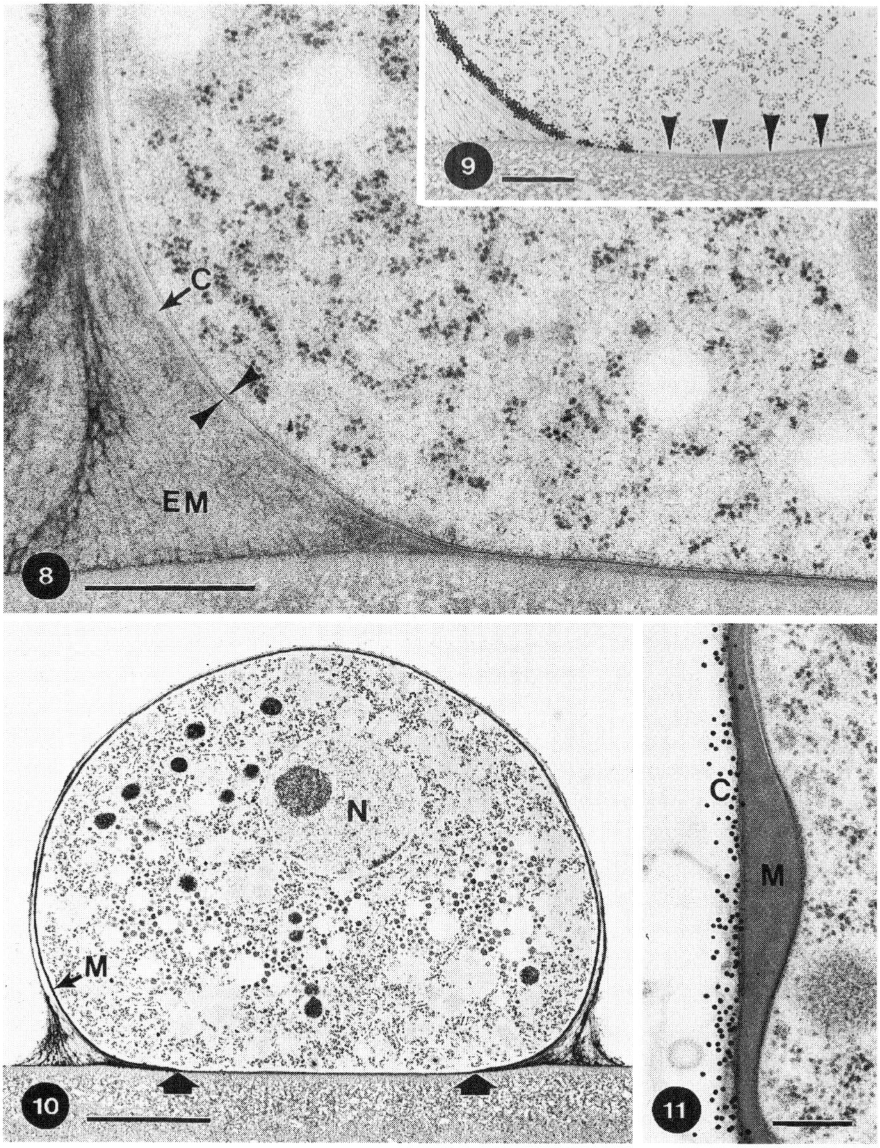

Figs 1.8–1.11. Transmission electron micrographs of appressorium structure before (Figs 1.8 and 1.9) and after (Figs 1.10 and 1.11) melanization. **Fig. 1.8.** The plasma membrane (between arrowheads) is clearly visible in this fully expanded appressorium, and lies adjacent to the electron lucent cell wall layer containing chitin (C). Two layers of extracellular matrix (EM) surround the appressorium and are most abundant near the subtratum. Scale bar = 500 nm. **Fig. 1.9.** An equivalent appressorial region as shown in the previous figure, decorated on thin sections by colloidal gold-wheat germ agglutinin. Distribution of the probe indicates that a surface region on the appressorium lacks chitin, lying against the substrate (arrowheads). Scale bar = 500 nm. **Fig. 1.10.** Fully formed, melanized appressorium. The wall layer of melanin (M) extends around the entire cell, except across the region in contact with the substrate (between arrows), comprising the appressorium pore. Scale bar = 2 μm.

of material can be observed surrounding the perimeter of the appressorium pore (Fig. 1.13). Howard and Ferrari (1989) have suggested that this *pore ring* might play a role in the mechanism by which the appressorium pore is sealed tightly to the surface of the substrate.

Appressorium Function

Many studies on the mode-of-action of the so called 'antipenetrant' crop protectant chemicals indicate their inhibition of melanin biosynthesis. The conclusions of these studies, however, must always be open to the possibility that for any given xenobiotic, there could be countless sites of action within a cell which remain unknown. The question of mode-of-action, then, must remain open as well. Fortunately, for those who work with *M. grisea*, we can rely upon genetic analysis to aid our inquiry.

Genetic or chemical perturbation of the pentaketide pathway of melanin biosynthesis has been shown to block pathogenesis by preventing successful penetration of the host surface (Woloshuk *et al.*, 1980; Yamaguchi *et al.*, 1982; Sisler, 1986; Chida and Sisler, 1987; Chumley and Valent, 1990). As we shall see, such perturbation destroys the molecular integrity of the melanin layer which renders the appressorium incapable of generating the tremendous force required to penetrate the rice leaf surface. Thus, by studying the role of melanin during penetration we can advance our understanding of infection structure function considerably.

Three classes of *M. grisea* pigment mutants have been identified and shown to have unlinked single gene defects (Chumley and Valent, 1990). The three mutant strains grow normally, form appressoria and, when introduced into host tissues through wounds, were able to cause typical symptoms of blast disease. The strains were deficient only in their ability to penetrate host surfaces and to synthesize dihydroxynaphthalene melanin. Ultrastructural analysis of appressoria formed by one of these genetically defined pigment mutants shows that the cell wall lacks melanin, the innermost layer observed in wild type appressoria (Howard and Ferrari, 1989). We may conclude, therefore, that the melanin layer of the appressorium cell wall is required for normal function, i.e. penetration of the host.

Appressorial turgor pressure

Loss of the ability to form elevated appressorial turgor pressure has been shown to result from either chemical or genetic perturbation of the pentaketide pathway of melanin biosynthesis (Howard and Ferrari, 1989). Melanized and unmelanized

---

Fig. 1.11. (opposite) Cell wall of the septum separating the germ tube (left) from the cytoplasm of the appressorium (right). The layer of melanin (M) is conspicuous, while the layer of chitin (C) has been demonstrated by the binding of wheat germ lectin-gold complex. Scale bar = 250 nm. [Figs 1.8–1.11 reprinted from Bourett and Howard (1990).]

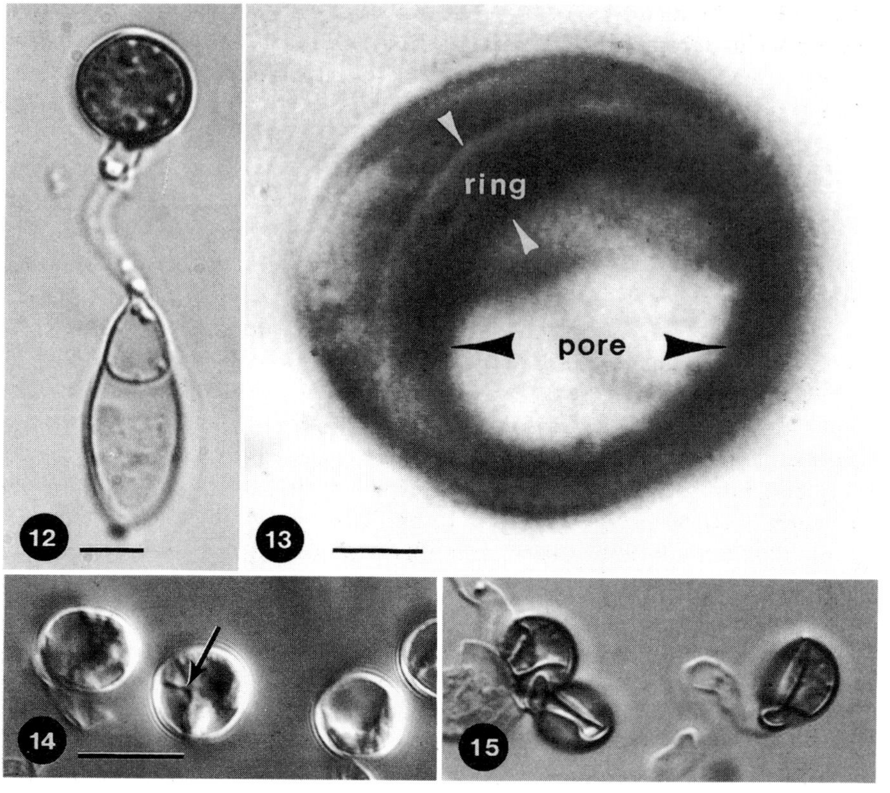

Figs 1.12–1.15. Differential interference contrast light micrographs. **Fig. 1.12.** Melanization of the appressorium wall imparts a dark pigmentation unique to that structure. Scale bar = 5 μm. **Fig. 1.13.** Video-enhanced contrast image of the plane of contact between an appressorium and substrate. The appressorium pore exists as an apparent hole in the cell wall, surrounded by a *ring* of thickened wall material. Scale bar = 2 μm. **Figs 1.14 and 1.15.** Responses of appressoria subjected to hyperosmotic stress depend upon the porosity of cell walls. Scale bar = 10 μm. **Fig. 1.14.** *Plasmolysis* of unmelanized appressoria incubated in 1 M KCl. **Fig. 1.15.** *Cytorrhysis* of melanized appressoria induced by 3 M KCl. [Figs 1.12–1.15 reprinted from Howard and Ferrari (1989), with the permission of Academic Press.]

appressoria respond differentially to hyperosmotic solutions: unmelanized appressoria show plasmolysis (Fig. 1.14) whereas melanized appressoria collapse, a response known as cytorrhysis (Fig. 1.15). Early experiments of this type were conducted by Woloshuk *et al.* (1983), but their findings were inconclusive. Subsequent work (Howard and Ferrari, 1989) demonstrated a dramatic difference in turgor pressure between the two groups of appressoria. Wild type appressoria, formed in the absence of a melanin biosynthesis inhibitor, exhibited highly elevated turgor. In addition, these cells were capable of elaborating an extension (penetration peg) into the surface of non-biodegradable substrates, such as plastic coverslips composed of polyvinyl chloride and polyvinyl acetate.

Attempts to measure appressorial turgor directly, using a pressure probe, have not been successful. However, estimates of normal appressorial turgor pressure have been obtained by plotting the number of cells which collapse as a function of extracellular osmotic pressure (Fig. 1.16a), analogous to classical methods which plot incipient plasmolysis. These experiments have demonstrated that appressoria can generate enormous turgor pressure, at least 8.0 MPa (80 bars or 1161 psi), during incubation on a Mylar substrate (Howard *et al.*, 1991b). This is the highest turgor pressure ever reported in a living organism.

Six different Mylar membranes exhibiting a range of surface hardnesses were used to prove that this high turgor pressure is required for penetration. After forming in water on the Mylar membranes, appressoria were incubated in various solutions of polyethylene glycol (PEG)-8000, effectively lowering appressorial turgor pressure. A correlation was found between the resulting turgor pressure and the ability to penetrate the different membranes (Fig. 1.16b). The turgor pressure that was necessary for penetration of a particular Mylar was found to be a function of the hardness of that substrate (Howard *et al.*, 1991b). Turgor pressure below the critical level for each surface resulted in failure of the appressorium to penetrate. In similar experiments, lowering the turgor pressure of appressoria on rice leaves, again using PEG, also inhibited penetration of the host. We can conclude that highly elevated turgor pressure is necessary for a mechanical component of penetration during pathogenesis. In addition, the rice leaf apparently presents a formidable barrier to ingress in the form of a rather hard surface.

The concentration of cytoplasmic solutes necessary to generate appressorial pressure of this magnitude would vary depending upon the solute molecule. If the solute were sucrose, for example, the cytoplasmic concentration would need to be 1.65 M (Money, 1989). Money (1994) has argued that this is quite feasible. The concentration of trehalose in spores of *Phycomyces* reportedly exceeds 1 M, which could generate a pressure of roughly 8.0 MPa when placed in pure water. Bourett and Howard (1990) have reported finding a huge supply of glycogen within appressoria that dissipates during maturation. They suggest that a glycogen catabolite might serve as the osmolyte responsible for generating turgor. A comparison using NMR of extracellular incubation fluids from melanin-deficient mutants and wild type appressoria has indicated the presence of a large quantity of a glycol which apparently leaked from the unmelanized cells (R.J. Howard and M.A. Ferrari, unpublished data). This result would support the hypothesis that glycogen is metabolized to the form of an unknown sugar alcohol. Further work is clearly needed to address this important stage of pathogenesis.

Role of melanin in appressorium function

Experiments using genetic mutants, or melanin biosynthesis inhibiting chemicals have established two requirements for the generation of enormous turgor in appressoria. First, the melanin layer must form in the appressorial cell wall and second, the appressorium must adhere tightly to the underlying substrate effectively sealing the appressorial pore. These parameters suggest that the melanin layer is acting as a permeability barrier, allowing for an increase in the concentration of some cytoplasmic solute (Howard and Ferrari, 1989).

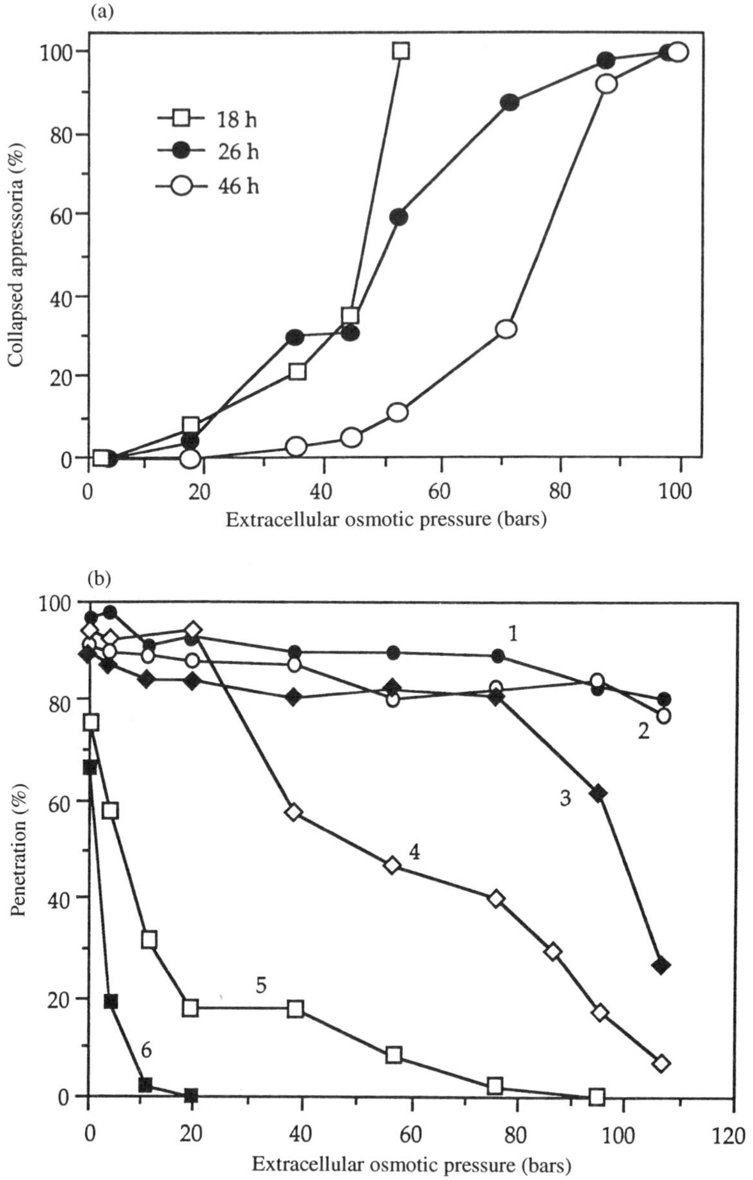

Fig. 1.16. Turgor pressure in melanized appressoria. (a) Increase of appressorial turgor during incubation on a Mylar substrate. After various periods of incubation in water, appressoria were tested in different solutions of PEG-8000. Collapse of appressoria indicates a hyperosmotic condition. (b) Six different Mylars, designated by numbers 1–6 (softest to hardest), were used to test the ability of appressoria to penetrate as a function of cell turgor. Each data point represents the observation of 100 appressoria. Following formation and an 18-h incubation in water, appressoria were placed in various concentrations of PEG-8000 and monitored over a 10-day period. An obvious direct correlation exists between penetration, cell turgor, and hardness of the substratum. [Reprinted from Howard *et al.* (1991b).]

Appressoria that form on a permeable surface, such as dialysis membrane, do not generate highly elevated osmotic pressure as they are not sealed from the extracellular environment. This can be demonstrated using hyperosmotic solutions. In such a case, extracellular solute can enter melanized cells by way of the appressorium pore. This path of entry is not possible for appressoria formed on an impermeable surface because the cell adheres tightly to the substrate.

Another method has been used to demonstrate that the layer of melanin in appressorial cell wall acts as a permeability barrier. Howard and Ferrari (1989) killed melanized appressoria, formed on a solid surface, by boiling them in water. They found that such appressoria could still respond by collapsing when placed in a hyperosmotic environment. Heat-killed melanin deficient appressoria did not respond at all to a hyperosmotic shock, nor did any appressoria formed on dialysis membrane. These results confirm that melanin in the cell wall was acting as a differentially permeable barrier in a manner typical of the plasma membrane in other cells. If the appressorial cell membrane was responsible for this function, the barrier would have been destroyed by boiling, and would not be dependent upon the presence of melanin. If the appressorial cell membrane was acting as the permeability barrier, then extracellular solute would have entered all cells, causing plasmolysis, rather than only those cells formed on dialysis membrane or those lacking melanin.

How could an appressorial cell wall layer of melanin act as a differentially permeable barrier? In the absence of an active transport mechanism, differential permeability could be imposed physically: molecules larger than a specific size would be excluded by the pores through which flux of smaller molecules could occur. In fact, the existence of restrictive wall porosity is well known for plants and fungi (see Money, 1990) and represents the basis for the cytorrhysis response of walled cells to hyperosmotic shock (Carpita *et al.*, 1979). Pore sizes in these walls have been estimated with a *solute exclusion technique*, where pore diameter is estimated by the size of the smallest excluded molecules. Determination of exclusion thresholds in appressoria has shown that melanized walls have a significantly smaller porosity than unmelanized walls (Howard *et al.*, 1991b). This evidence provided further support for the concept that melanin acts as the differentially permeable layer that is responsible for enormous appressorium turgor pressures.

Different explanations have been proposed for the role of melanin in the penetration process which the reader is encouraged to consider (see Woloshuk *et al.*, 1983; Kubo and Furusawa, 1991). However, there are no experimental data to support these alternatives.

Mechanism of Penetration: Force and Enzymes

Penetration from appressoria in many plant pathogens occurs soon after formation of infection structures. There are several advantages to a quick resolution of these early stages of pathogenesis. The longer the fungus remains on the surface of the host, the more vulnerable it will be to desiccation, competition from other phyloplane inhabitants, host defense mechanisms, depletion of stored nutrients, and failure. So it is with special attention that we must examine the time required for *M. grisea* to achieve ingress. Ou (1985) has reviewed much of the relevant information pertaining to this question. Environmental parameters such as air and soil temperatures, macro and micro climatic relative humidity, hours of continuous leaf wetness, degree of light intensity and duration and timing of dark periods have all been considered as important. But regardless of these influences, the time required for ingress by *M. girsea* appears longer than for many other foliar plant pathogens. *In vitro* studies have indicated a period of 24–31 h was required before penetration pegs formed on cellophane (Bourett and Howard, 1990), and progressively longer periods were required as the hardness of the substrate was increased (Howard *et al.*, 1991b). Despite the fact that appressoria were fully formed about 6 h after germination, a significant period of incubation seems necessary before the next stage of pathogenesis. For the hardest Mylar tested, a period of 8 days was required before penetration pegs began to appear. Seemingly quiescent, appressoria may be building turgor pressure during this period to a level sufficient for mechanical rupture of the underlying substrate.

For purposes of this discussion, Mylar can be considered non-biodegradable and we can conclude that enzymes did not play a role in altering Mylar surface structure in any way that would aid deformation by the penetration peg. On the hardest Mylar tested, the smallest decrease in appressorial turgor blocked penetration (Fig. 1.16b). A comparable decrease in appressorial turgor blocked penetration of rice leaves by 95% (Howard *et al.*, 1991b), indicating the significance of a mechanical component during penetration of the host. The rice leaf could be penetrated only when appressorial turgor was at its highest, suggesting that leaf hardness is in the range of the hardest Mylar tested. Apparently, the rice leaf surface is very hard indeed. In this regard, it is interesting to note: (i) the correlation reported between high levels of silicon available to the plant host and increased resistance to blast disease (e.g. Datnoff *et al.*, 1992; Winslow, 1992); and (ii) the suggestion that Si-rich layers or localized Si deposits in epidermal cell walls may slow or inhibit fungal ingress (Yoshida *et al.*, 1962; Heath and Stumpf, 1986; Carver *et al.*, 1987; Samuels *et al.*, 1991).

Let us now examine the time required for penetration of the hardest Mylar, about 8 days, and the host surface, less than 2 days. But we have just considered evidence suggesting these two surfaces are of similar hardness. The obvious and important difference is that the rice leaf may be subject to enzymatic modification. The quicker penetration of rice, compared to Mylar, can be interpreted as evidence of such enzymatic activity (Howard *et al.*, 1991b). Thus, it is likely that there is a necessary role for both a mechanical force and enzymes during penetration of the host. The elegant design of the appressorium pore may function to allow for this possibility (see below).

Penetration pegs

Marking the end of ostensible quiescence and just prior to substrate penetration, the appressorium pore is blanketed by the deposition of new cell wall in the form of the *pore wall overlay* (Bourett and Howard, 1990). Up until this stage of

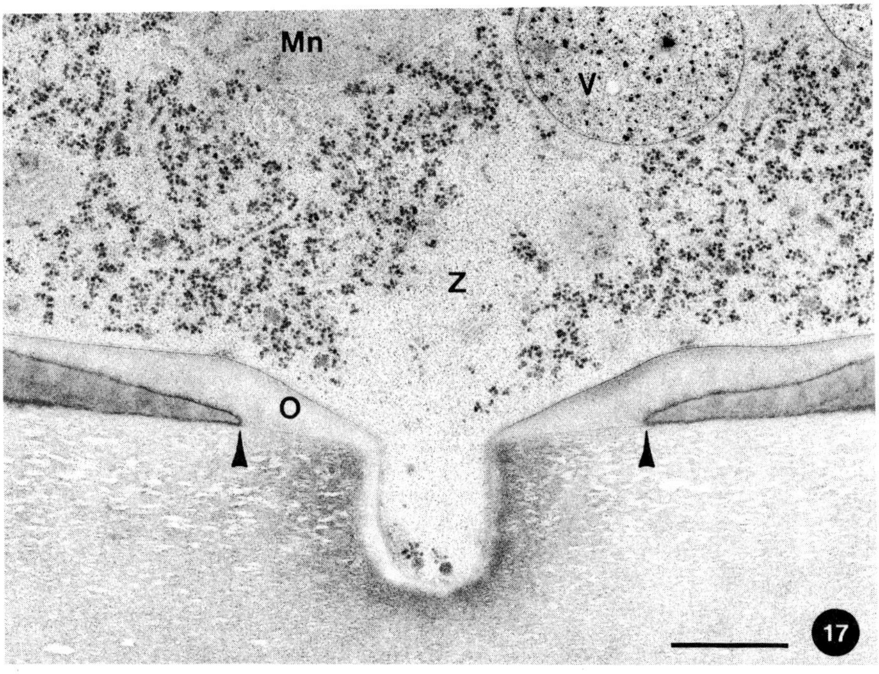

Fig. 1.17. Site of penetration peg formation by an appressorium on cellophane membrane. A zone of exclusion (Z) fills the peg and extends into the adjacent region of appressorial cytoplasm. This young peg has formed near the outer edge of the appressorium pore, through the nascent pore wall overlay (O). Mn, mitochondrion; arrowhead, perimeter of appressorium pore; V, vacuole. Scale bar = 500 nm. [Reprinted from Bourett and Howard (1992). Copyright, Springer-Verlag.]

development no cell wall could be discerned in the pore region, only naked plasma membrane. The pore wall overlay appeared initially as one-layered (Bourett and Howard, 1992), but clearly two-layered at a later stage of development (Bourett and Howard, 1990). During the latter stage, wall overlay closest to the cytoplasm bound both concanavalin A and wheat germ agglutinin indicating the presence of glucose or mannose as well as chitin. The outer layer showed no affinity for either probe. The chemical nature of the substantial outer wall layer covering the pore remains completely unknown but, because of the relevance to penetration peg wall composition, represents an important subject for study.

A tip growing penetration peg develops through the overlay, and is surrounded by a cell wall that is continuous with the overlay (Fig. 1.17). Nascent pegs exhibit vesicles near their apex, but contain few other organelles. Penetration peg cytoplasm, together with a small region of adjacent cytoplasm in the appressorium, has been characterized as a *zone-of-exclusion*, excluding most organelles, comprised in part of actin (Bourett and Howard, 1992). Immunofluorescence localization, as well as electron microscope observation (unpublished data), has shown the presence of numerous actin filaments lying against the pore wall overlay, radiating from the peg and perpendicular to the axis of penetration. In addition, fluorescence techniques have been used to document plaques of actin within the appressorium pore region, which most likely represent *filasomes* (Howard, 1981; Bourett and Howard, 1991). Abundant filasomes have been observed in the same region of appressoria in taxonomically diverse fungi (e.g. Mims and Richardson, 1989; Swann and Mims, 1991). Because filasomes have been found consistently near regions of wall synthesis in other fungal cell types, their presence at the appressorium–substrate interface might signify a high level of exo- and/or endocytosis across this area of plasma membrane. Hence, there may be a large flux of molecules passing through the appressorium pore, between the fungal cytoplasm and the host surface in either or both directions.

Significance of the appressorium pore

Sonication has been used to test the strength of appressorial adhesion to various surfaces (Howard and Ferrari, 1989), but as demonstrated by Uchiyama *et al.* (1979) it also can provide a unique vantage from which to study the pathogen–substrate interface (Fig. 1.18). Utilizing this unique perspective, one is unavoidably impressed by the apparent sophistication of the appressorium. The appressorium pore can be observed and appreciated as the critical zone of interaction between the fungus and the substrate. We have already established that the entire perimeter of appressoria is lined with a layer of melanin, impermeable to any molecule larger than water. Such a size restriction would certainly exclude molecules one might expect to play an important role during pathogenesis including enzymes, carbohydrate signal molecules or even ions in the hydration state. The appressorium pore is a solution to this potential fungal problem, as it represents the proverbial window unto the host. And, that signs of copious endo- or exocytosis have been found in the pore region supports this *window* concept of the pore. The elegance of appressorium design and function allow this window to remain fully *open* to the flux of diverse molecular species, while the force of an enormous osmotic pressure is generated and maintained.

Conclusion

We have seen how careful cytological analysis has allowed us to learn much about the cell biology of *M. grisea* as it relates to early pathogenesis. Thus far we have examined the *in vitro* growth and differentiation of *Magnaporthe*, apart from the host. This is only a beginning and there is so much more to learn. The application and development of new technologies must continue for our understanding of rice blast disease to grow. This is particularly important at the level of host–pathogen interaction. The next stages of fungal development also can be studied *in vitro*, as cells resembling infection hyphae have been reported to form in cellophane membrane following penetration of this artificial substrate. It will be important to compare the cell biology of these structures to those that form in host tissues. Such comparisons could provide important insight and new direction

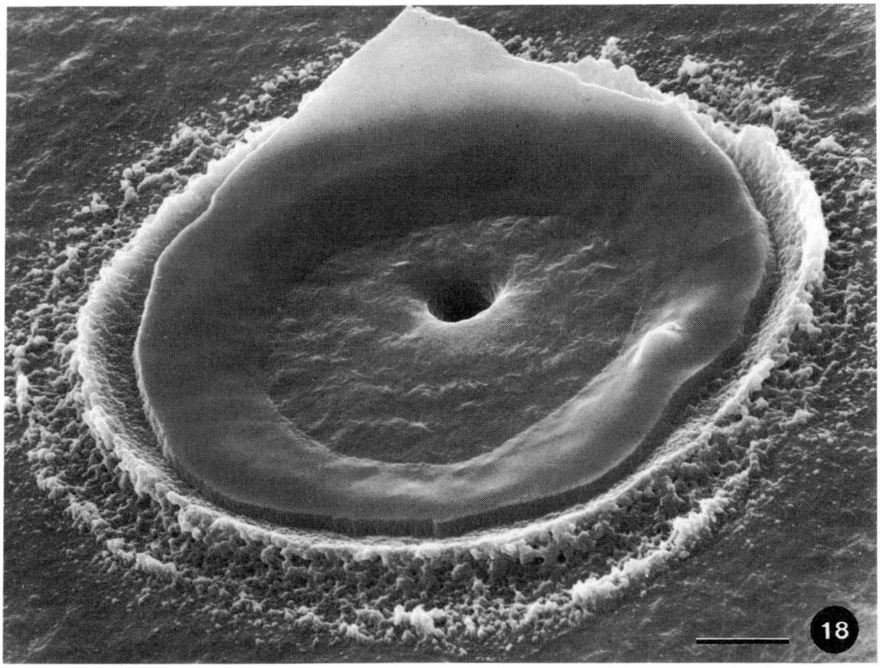

Fig. 1.18. Remnants of an appressorium formed on polyethylene, following sonication. The entire upper portion of the cell, including the cytoplasm and pore wall overlay, has been removed. The site where a penetration peg mechanically indented the substrate is clearly visible within the appressorium pore. The remaining smooth wall layer is melanin, surrounded on the outside by extracellular matrix material. Scale bar = 1.0 μm. [Reprinted from Braun and Howard (1994a). Copyright, Springer-Verlag.]

for subsequent work. But, of course, our primary interest must focus on the biology of the pathogen as it relates to that of the host. To apply cryo fixation to the study of plant tissues will require the use of freezing methods compatible with large specimens, such as high pressure freezing (Mendgen *et al.*, 1991). The application of *in situ* hybridization (McFadden, 1991), immunolocalization (Bourett and Howard, 1991), or other subcellular probes (Bourett *et al.*, 1993), will allow us to learn more about the gene expression of individual cells during pathogenesis as well as the ability to follow specific cellular components at the ultrastructural level. These techniques can provide information about individual pathogens at individual host encounter sites, or of individual cells of the host. Based on the advances in understanding resulting from previous techniques development, efforts to adapt these and other evolving methods to the study of rice blast disease, pathogen and host, will yield a handsome return. Clearly, we are very fortunate to have such an economically and socially important model system with which to ask basic biological questions.

Acknowledgement

I wish to thank sincerely Timothy M. Bourett, Margaret A. Ferrari and Michael

A. Picollelli for their invaluable collaboration and expert assistance in conducting our research efforts.

References

Aist, J.R. (1981) Development of parasitic conidial fungi in plants. In: Cole, G.T. and Kendrick, B. (eds), *Biology of Conidial Fungi, Vol. 2*. Academic Press, New York, pp. 75–110.

Asuyama, H. (1965) Morphology, taxonomy, host range, and life cycle of *Pyricularia oryzae*. In: *The Rice Blast Disease. Proceedings of a Symposium at the International Rice Research Institute, July, 1963*. The Johns Hopkins Press, Baltimore, Maryland, pp. 9–22.

Bourett, T.M. and Howard, R.J. (1990) *In vitro* development of penetration structures in the rice blast fungus *Magnaporthe grisea*. *Canadian Journal of Botany* 68, 329–342.

Bourett, T.M. and Howard, R.J. (1991) Ultrastructural immunolocalization of actin in a fungus. *Protoplasma* 163, 199–202.

Bourett, T.M. and Howard, R.J. (1992) Actin in penetration pegs of the fungal rice blast pathogen, *Magnaporthe grisea*. *Protoplasma* 168, 20–26.

Bourett, T.M., Picollelli, M.A. and Howard, R.J. (1993) Postembedment labeling of intracellular concanavalin A-binding sites in freeze-substituted fungal cells. *Experimental Mycology* 17, 223–235.

Braun, E.J. and Howard, R.J. (1994a) Adhesion of fungal spores and germlings to host plant surfaces. *Protoplasma* (in press).

Braun, E.J. and Howard, R.J. (1994b) Adhesion of *Cochliobolus heterostrophus* conidia and germlings to leaves and artificial surfaces. *Experimental Mycology* (in press).

Carpita, N.C., Sabularse, D., Montezinos, D. and Delmer, D.P. (1979) Determination of pore size of cell walls of living plant cells. *Science* 205, 1144–1147.

Carver, T.L.W., Zeyen, R.J. and Ahlstrand, G.G. (1987) The relationship between insoluble silicon and success or failure of attempted primary penetration by powdery mildew (*Erysiphe graminis*) germlings on barley. *Physiological and Molecular Plant Pathology* 31, 133–148.

Chida, T. and Sisler, H.D. (1987) Restoration of appressorial penetration ability by melanin precursors in *Pyricularia oryzae* treated with antipenetrants and in melanin-deficient mutants. *Journal of Pesticide Science* 12, 49–55.

Chumley, F.G. and Valent, B. (1990) Genetic analysis of melanin-deficient, non-pathogenic mutants of *Magnaporthe grisea*. *Molecular Plant-Microbe Interactions* 3, 135–143.

Datnoff, L.E., Snyder, G.H. and Deren, C.W. (1992) Influence of silicon fertilizer grades on blast and brown spot development and on rice yields. *Plant Disease* 76, 1011–1013.

Hamer, J.E., Howard, R.J., Chumley, F.G. and Valent, B. (1988) A mechanism for surface attachment in spores of a plant pathogenic fungus. *Science* 239, 288–290.

Hamer, J.E., Valent, B. and Chumley, F.G. (1989) Mutations at the *SMO* genetic locus affect the shape of diverse cell types in the rice blast fungus. *Genetics* 122, 351–361.

Hanau, R.M., Fanklin, D., Yang, Z. and Nicholson, R.L. (1989) Conidiogenous cell development in *Helminthosporium carbonum*. *Experimental Mycology* 13, 337–347.

Hardham, A.R. (1992) Cell biology of pathogenesis. *Annual Review of Plant Physiology and Plant Molecular Biology* 43, 491–526.

Hashioka, Y. and Nakai, Y. (1974) Fine structure of the rice blast. XI. Outthrust of conidiophore cells of *Pyricularia oryzae*, etc., through an epidermal outer wall of a host leaf. *Research Bulletin of the Faculty of Agriculture of Gifu University* 36, 9–18

Heath, M.C. and Stumpf, M.A. (1986) Ultrastructural observations of penetration sites of the cowpea rust fungus in untreated and silicon-depleted French bean cells. *Physiological and Molecular Plant Pathology* 29, 27–39.

Hoch, H.C. and Staples, R.C. (1991) Signaling for infection structure formation in fungi. In: Cole, G.T. and Hoch, H.C. (eds), *The Fungal Spore and Disease Initiation in Plants and Animals*. Plenum Press, New York, pp. 25–46.

Howard, R.J. (1981) Ultrastructural analysis of hyphal tip cell growth in fungi: Spitzenkörper, cytoskeleton and endomembranes after freeze-substitution. *Journal of Cell Science* 48, 89–103.

Howard, R.J. and Ferrari, M.A. (1989) Role of melanin in appressorium function. *Experimental Mycology* 13, 403–418.

Howard, R.J. and O'Donnell, K.L. (1987) Freeze substitution of fungi for cytological analysis. *Experimental Mycology* 11, 250–269.

Howard, R.J., Bourett, T.M. and Ferrari, M.A. (1991a) Infection by *Magnaporthe*: an *in vitro* analysis. In: Mendgen, K. and Lesemann, D.-E. (eds), *Electron Microscopy of Plant Pathogens*. Springer-Verlag, New York, pp. 251–264.

Howard, R.J., Ferrari, M.A., Roach, D.H. and Money, N.P. (1991b) Penetration of hard substrates by a fungus employing enormous turgor pressures. *Proceedings of the National Academy of Sciences of the United States of America* 88, 11281–11284.

Jeffree, C.E. and Read, N.D. (1991) Ambient- and low-temperature scanning electron microscopy. In: Hall, J.L. and Hawes, C. (eds), *Electron Microscopy of Plant Cells*. Academic Press, London, pp. 313–413.

Kubo, Y. and Furusawa, I. (1991) Melanin biosynthesis. Prerequisite for successful invasion of the plant host by appressoria of *Colletotrichum* and *Pyricularia*. In: Cole, G.T. and Hoch, H.C. (eds), *The Fungal Spore and Disease Initiation in Plants and Animals*. Plenum Press, New York, pp. 205–218.

Lee, Y.-H. and Dean, R.A. (1993) cAMP regulates infection structure formation in the plant pathogenic fungus *Magnaporthe grisea*. *The Plant Cell* 5, 693–700.

McFadden, G.I. (1991) *In situ* hybridization techniques: molecular cytology goes ultrastructural. In: Hall, J.L. and Hawes, C. (eds), *Electron Microscopy of Plant Cells*. Academic Press, London, pp. 219–255.

Mendgen, K. and Deising, H. (1993) Tansley Review No. 48. Infection structures of fungal plant pathogens – a cytological and physiological evaluation. *New Phytologist* 124, 193–213.

Mendgen, K., Welter, K., Scheffold, F. and Knauf-Beiter, G. (1991) High pressure freezing of rust infected plant leaves. In: Mendgen, K. and Lesemann, D.-E. (eds), *Electron Microscopy of Plant Pathogens*. Springer-Verlag, New York, pp. 31–42.

Mims, C.W. and Richardson, E.A. (1989) Ultrastructure of appressorium development by basidiospore germlings of the rust fungus *Gymnosporangium juniperi-virginianae*. *Protoplasma* 148, 111–119.

Money, N.P. (1989) Osmotic pressure of aqueous polyethylene glycols. Relationship between molecular weight and vapor pressure deficit. *Plant Physiology* 91, 766–769.

Money, N.P. (1990) Measurement of pore size in the hyphal cell wall of *Achlya bisexualis*. *Experimental Mycology* 14, 234–242.

Money, N.P. (1994) Osmotic adjustment and the role of turgor in filamentous fungi. In: Wessels, J.G.H. and Meinhardt, F. (eds), *The Mycota, Vol. 2. Growth, Differentiation and Sexuality*. Springer-Verlag, New York (in press)

Ou, S.H. (1985) *Rice Diseases*, 2nd edn. Commonwealth Mycological Institute, Kew, UK, 380 pp.

Read, N.D., Kellock, L.J., Knight, H. and Trewavas, A.J. (1992) Contact sensing during infection by fungal pathogens. In: Callow, J.A. and Green, J.R. (eds), *Perspectives in Plant Cell Recognition. Society for Experimental Biology Seminar Series 48.* Cambridge University Press, UK, pp. 137–172.

Samuels, A.L., Glass, A.D.M., Ehret, D.L. and Menzies, J.G. (1991) Distribution of silicon in cucumber leaves during infection by powdery mildew fungus (*Sphaerotheca fuliginea*). *Canadian Journal of Botany* 69, 140–146.

Sisler, H.D. (1986) Control of fungal diseases by compounds acting as antipenetrants. *Crop Protection* 5, 306–313.

Swann, E.C. and Mims, C.W. (1991) Ultrastructure of freeze-substituted appressoria produced by aeciospore germlings of the rust fungus *Arthuriomyces peckianus. Canadian Journal of Botany* 69, 1655–1665.

Sweigard, J.A., Chumley, F.G. and Valent, B. (1992) Disruption of a *Magnaporthe grisea* cutinase gene. *Molecular and General Genetics* 232, 183–190.

Terhune, B.T., Bojko, R.J. and Hoch, H.C. (1993) Deformation of stomatal guard cell lips and microfabricated artificial topographies during appressorium formation by *Uromyces. Experimental Mycology* 17, 70–78.

Uchiyama, T., Ogasawara, N., Nanba, Y. and Ito, H. (1979) Conidial germination and appressorial formation of the plant pathogenic fungi on the coverglass or cellophane coated with various lipid components of plant leaf waxes. *Agricultural and Biological Chemistry* 43, 383–384.

Valent, B. (1990) Rice blast as a model system for plant pathology. *Phytopathology* 80, 33–36.

Valent, B., Crawford, M.S., Weaver, C.G. and Chumley, F.G. (1986) Genetic studies of fertility and pathogenicity in *Magnaporthe grisea* (*Pyricularia oryzae*). *Iowa State Journal of Research* 60, 569–594.

Weintraub, R.L., Miller, W.E. and Schantz, E.J. (1958) Chemical stimulation of germination of spores of *Picricularia oryzae. Phytopathology* 48, 7–10.

Wheeler, M.H. and Bell, A.A. (1988) Melanins and their importance in pathogenic fungi. In: McGinnis, M.R. (ed.), *Current Topics in Medical Mycology*, Vol. 2. Springer-Verlag, New York, pp. 338–387.

Wheeler, M.H. and Stipanovic, R.D. (1985) Melanin biosynthesis and the metabolism of flaviolin and 2-hydroxyjuglone in *Wangiella dermatitidis. Archives of Microbiology* 142, 234–241.

Winslow, M.D. (1992) Silicon, disease resistance, and yield of rice genotypes under upland cultural conditions. *Crop Science* 32, 1208–1213.

Woloshuk, C.P., Sisler, H.D., Tokousbalides, M.C. and Dutky, S.R. (1980) Melanin biosynthesis in *Pyricularia oryzae*: site of tricyclazole inhibition and pathogenicity of melanin-deficient mutants. *Pesticide Biochemistry and Physiology* 14, 256–264.

Woloshuk, C.P., Sisler, H.D. and Vigil, E.L. (1983) Action of the antipenetrant, tricyclazole, on appressoria of *Pyricularia oryzae. Physiological Plant Pathology* 22, 245–259.

Yamaguchi, I., Sekido, S. and Misato, T. (1982) The effect of non-fungicidal anti-blast chemicals on the melanin biosythesis and infection by *Pyricularia oryzae. Journal of Pesticide Science* 7, 523–529.

Yoshida, S., Ohnishi, Y. and Kitagishi, K. (1962) Histochemistry of silicon in rice plant. III. The presence of cuticle-silica double layer in the epidermal tissue. *Soil Science and Plant Nutrition* 8, 1–5.

Zee, S.-Y. (1981) *Wheat and Rice Plants. A Scanning Electron Microscope Survey*. Cosmos Books Ltd., Hong Kong.

2 Signalling Systems and Gene Expression Regulating Appressorium Formation in *Magnaporthe grisea*

R.A. Dean, Y.-H. Lee, T.K. Mitchell and D.S. Whitehead

Department of Plant Pathology and Physiology, 120 Long Hall, Clemson University, Clemson, SC 29634–0377, USA

Introduction

Plant pathogenic fungi have evolved highly adapted cellular and morphological processes to infect plants. However, considering the agronomic significance of the infection process, relatively little is known about the differentiation of appressoria or the genes central to their development. This chapter outlines strategies to elucidate the mechanisms of appressorium initiation by *Magnaporthe grisea*.

In nature, infection by *M. grisea* is initiated when a conidium lands on a leaf surface. The hydrated conidium attaches firmly by releasing mucilage from the apex thereby preventing dislodgment by wind and rain (Hamer *et al.*, 1988). One or more germ tubes emerge from any of the three cells of the conidium, although germination from the middle cell is rare (Bourett and Howard, 1990). Apical growth ceases within a few hours when the germ-tube tip hooks to form an appressorium initial and begins to swell. During maturation, the appressorium becomes melanized, except at a well-defined pore between the appressorium and the substratum. A granular substance, believed to be an adhesive, accumulates at the substratum interface (Howard and Ferrari, 1989). As hydrostatic pressure increases within the appressorium, a narrow hypha emerges through the pore and penetrates directly into the plant (Howard, 1991). These specialized events undoubtedly require new gene products. The recognition of a plant surface and communication of this information to affect morphological events has much in common with other well-understood stimulus-response relationships (Kincaid, 1991; Pawson, 1991). We shall address this topic further here.

Acquiring knowledge of the mechanisms controlling the infection process provides opportunities for the development of novel disease control strategies. The identification of genes or processes essential for infection will facilitate design

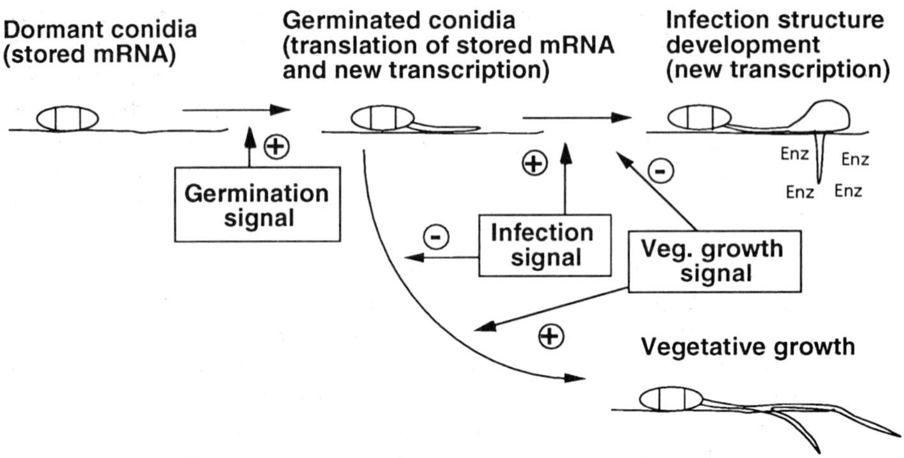

Fig. 2.1. Diagrammatic representation of signals and genes regulating the switch from vegetative growth to appressorium formation in *Magnaporthe grisea*. ⊕ indicates a positively acting signal and ⊖ indicates a negatively acting signal regulating the developmental process shown.

of chemical control agents. Such chemicals that specifically target critical steps in pathogenesis may not have general fungicidal effects that impact on non-target organisms. In this chapter, we discuss recognition by the blast fungus of a plant surface that stimulates appressorium formation, and strategies to identify genes that play a central role in the development of appressoria.

Critical Steps and Mechanisms Regulating Infection-related Morphogenic Events

Conidium germination and emergence of the germ tube are illustrated in Fig. 2.1. A number of environmental signals control both germination and the fate of the germ-tube tip. Conidia do not germinate when held in aqueous suspension unless nutrients are added or the conidia contact a solid surface (Lee and Dean, 1993a). Presumably, this prevents precocious germination. The destiny of the germ tube depends on additional signals – an infection signal, a vegetative growth signal, or both. Cellular response in *M. grisea* to the infection signal overrides signals for vegetative growth because appressorium formation is not greatly influenced by the presence of nutrients such as sugars. Upon sensing the infection signal, such as the hydrophobic leaf surface, a fundamental change occurs: apical extension ceases and the region immediately behind the hyphal tip buckles up and away from the surface and the tip begins to swell into the appressorium.

Research on *Neurospora* and *Aspergillus* supports the hypothesis that transcripts stored within the dormant conidium are translated during germination (Brambl *et al.*, 1987). Whether transcription is required for germination remains an open question. Contrary to conventional theory, Brambl *et al.* (1987) reported that RNA synthesis inhibitors inhibited germination of *Neurospora*. However, when we use transcriptional inhibitors, such as actinomycin D and proflavine sulfate, we are unable, at any concentration, to block germination of *M. grisea* in liquid media or on hard surfaces, whereas cycloheximide, a translation inhibitor, completely blocks germination at 3.55 μM (unpublished data).

If transcription is not required for germination, then it is conceivable that transcripts for appressorium formation, or factors specific for transcription of appressorium genes, could also be pre-formed. Results using transcription inhibitors were inconclusive. Both actinomycin D and proflavine sulfate slightly reduce appressorium formation when added to ungerminated conidia. Other data argue that conidia do not contain pre-formed factors for infection structure formation. In certain instances, germ tubes continue to extend up to 12 h before appressoria begin to develop. During this time the germ tube nucleus has undergone numerous mitotic divisions that are separated by septa, therefore pre-formed factors present in conidia would have been diluted greatly. Furthermore, when vegetative mycelium is forced to make contact with an inductive surface, hyphal tips that are in contact with the substratum differentiate into appressoria within 6 h. Hyphae that remain unattached to the surface continue to grow vegetatively (Lee and Dean, 1993a).

What then is involved in environment recognition by hyphal tips and how is this information relayed into the cell to induce transcription and the formation of the appressorium?

Intracellular Signalling Leading to Appressorium Formation

The mechanisms involved in recognition and adaptive response to stimuli are fundamentally similar in all eukaryotes investigated (Kincaid, 1991; Pawson, 1991). A series of common steps is generally involved. An environmental stimulus interacts with a primary sensor to initiate a response. The series of events leading to the response is described as signal transduction and is mediated through second-messenger pathways.

Hydrophobicity of the plant surface is a principal environmental stimulus

The surface on which conidia of *M. grisea* land has little effect on germination (Lee and Dean, 1993a). Suspensions of conidia germinate equally well and as rapidly on glass as they do on wax surfaces. However, for our wild type strains (designated 70–15 and 70–6, gift of Dr A. Ellingboe, University of Wisconsin), the frequency with which appressoria form correlates strongly with the degree of surface hydrophobicity. Wax paper, the most hydrophobic surface tested, is the most effective inductive surface; more than 93% of germinating conidia form appressoria. On hydrophilic surfaces such as clean glass or the hydrophilic side of Gelbond (FMC BioProducts, Maine) the germ tube develops into typical vegetative mycelium. Hamer *et al.* (1988) also reported that surfaces with similar properties to those of a rice leaf induced appressoria. In other phytopathogenic fungi, topographical features of the leaf surface are important in the induction of appressoria. *Uromyces appendiculatus* responds to artificial ridges of a height similar to those formed by guard cells (Hoch *et al.*, 1987). However, *M. grisea* does not penetrate via stomates and fails to form appressoria in response to ridges regardless of height tested (Lee and Dean, 1994).

Proteins of the extracellular matrix may act as primary sensors

Proteins associated with the extracellular matrix (ECM) play a crucial role in cell-substratum adhesion, differentiation and signal transduction in mammalian cells (Endelman and Buck, 1992). Two such ECM components are the adhesive glycoproteins vitronectin and fibronectin. They are linked to the cytoskeleton via membrane-bound, membrane-spanning receptors, known as integrins. It has been shown *in vitro* that vitronectin undergoes a conformational change depending upon various chemical and physical stimuli, including the hydrophobicity of the surface to which it is bound (Underwood *et al.*, 1993). We hypothesize that analogs of these proteins are present in *M. grisea* and are important for the triggering of appressorium formation by hydrophobic surfaces. We have several pieces of evidence to support this theory. Western blotting using polyclonal antisera raised to the human proteins reveals the presence of a vitronectin-like protein of approximately 95 kDa and a fibronectin-like protein of 60 kDa in protein extracts of *M. grisea* mycelium and spores. The addition of IgG purified from polyclonal antiserum raised to human vitronectin or fibronectin to conidial suspensions prevents both conidial attachment and appressorium formation, but does not affect germination. Furthermore, addition of the antibody after conidia have attached to a hydrophobic surface severely reduces appressorium formation. This suggests that adhesion is not a prerequisite for appressorium formation. As shown in Fig. 2.2, the vitronectin-like protein immunolocalizes predominantly in the spore-tip mucilage; it is detected only infrequently on developing appressoria, but the thick cell wall layers covering the maturing appressoria may mask antibody binding (D.S. Whitehead *et al.*, unpublished data).

Cyclic AMP acts as a second messenger in appressorium formation

The transduction of environmental information within a cell is controlled by a number of small molecules, such as cyclic nucleotides, Ca^{++}, and phosphoinositides. The transduced signal is relayed by a protein phosphorylation cascade which involves protein kinases and phosphatases. In fungi, cyclic adenosine monophosphate (cAMP) is an important regulator of development (Pall, 1981; Devreotes, 1983; van Haastert, 1991; Pereyra *et al.*, 1992; Thevelein, 1992). As shown in Fig. 2.3, the addition of cAMP to conidia germinated on non-inductive surfaces, such as glass, induces appressorium formation to levels

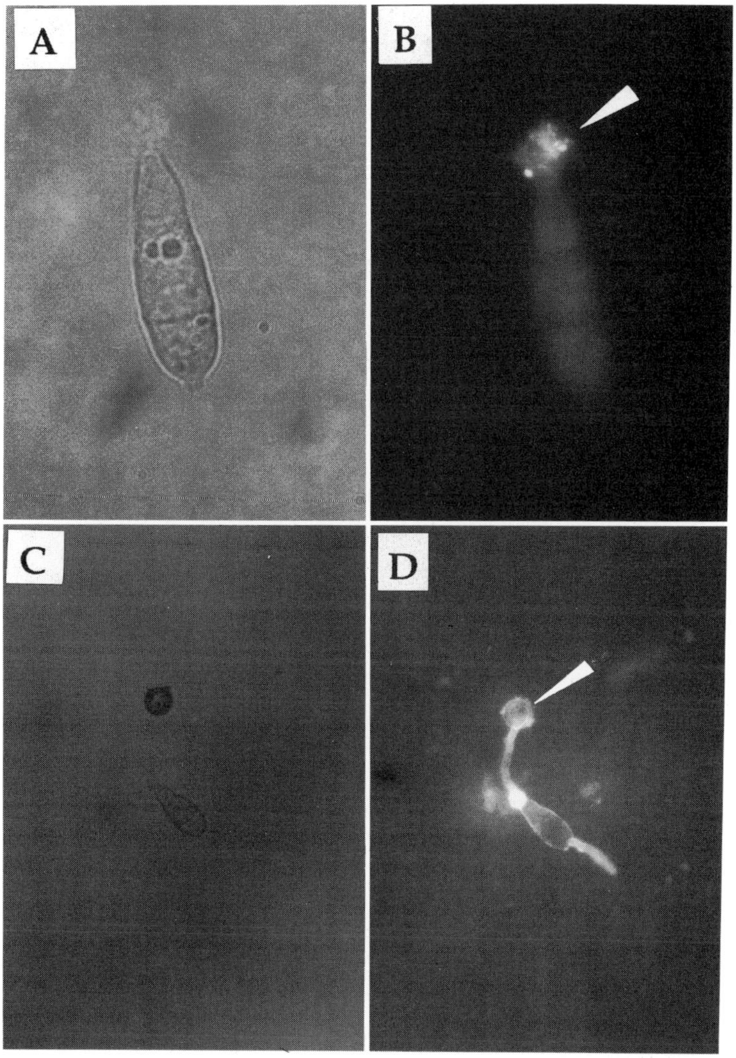

Fig. 2.2. Immunofluorescent localization of vitronectin-like protein in conidia and appressorium of *M. grisea*. (A) and (B) conidium attached to hydrophobic surface. (C) and (D) conidium, germ tube and developing appressorium at 6 h. (B) and (D) viewed with fluorescence optics. Antibody binds primarily to spore-tip mucilage (indicated by arrow in B) and occasionally to the developing appressorium (indicated by arrow in D).

similar to those induced by hydrophobic surfaces. Appressorium formation is cAMP dosage dependent, 10 mM being most effective. Concentrations in this range have been typically used to examine signal transduction in fungi. Presumably these levels, which are in the order of 10^4 times higher than normal endogenous levels, are required because of low permeability and rapid turnover of cAMP in fungal cells (Pall, 1981). Appressoria are induced specifically by cAMP; non-cyclic adenine nucleotides and adenosine were completely ineffective.

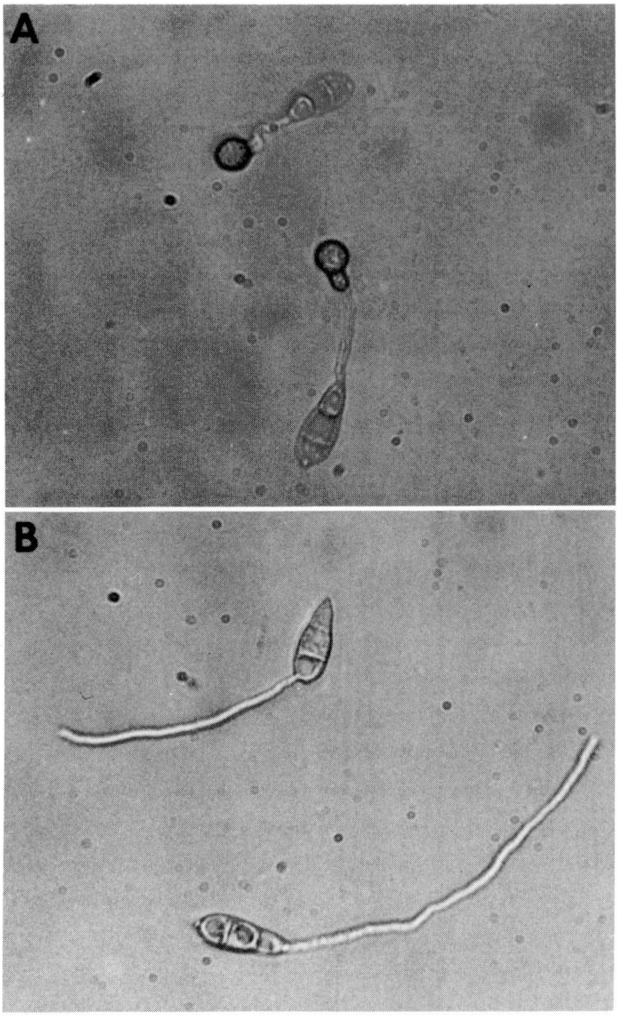

Fig. 2.3. Effect of cyclic AMP on conidia of *M. grisea* germinated on a hydrophilic surface. Conidia were germinated in presence of 10 mM cAMP (A) or its absence (B) on the hydrophilic surface of Gelbond. Photographs were taken at 10 h.

8-Bromocyclic AMP and monobutyryl-cAMP, commonly used effective analogs in fungi and animals, are also highly effective at inducing appressoria (Kay, 1989).

Under normal circumstances, cellular cAMP concentrations are tightly regulated by adenylate cyclase and phosphodiesterase – enzymes which are responsible for synthesis and degradation, respectively (Kincaid, 1991). The addition of 3-isobutyl-1-methyl-xanthine (IBMX), the most commonly used phosphodiesterase inhibitor (Pall, 1981), induces appressorium formation on hydrophilic surfaces to levels above those induced by hydrophobic surfaces (Lee and Dean, 1993a).

```
Primer:     1F                   2F              1B                         2B
MgAPK       YRDLKPENLLLDRHGHLKITDFGFAKRVPDKTWTLCGTPDYLAPEVVSNKGYNKSVDWWSLG

TPK1        ********I***KN**I**********Y***V*Y*********I*****T*P****I****F*
TPK2        ********I****N**I**********E*QTV***********I****ITT*P**********
TPK3        ********I***KN**I**********Y***V*Y*********I*****T*P*********F*
cAPKa       **********I*QQ*YIQV*********KGR*********E*****IILS*****A****A**
cAPKb       **********I*HQ*YIQV*********KGR*********E*****IILS*****A****A**
cAPKg       H*********I*QQ*Y*QV*********KGR*********E*****IILS*****A****A**
EcAPK       **********I*ST*Y**V*********KGR*********E*****IILS*****A****A**
AplC        **********I*SY*Y**V*********KGR*********E*****IILS***D*A****A**
SAK         ********I***HL*Y**V********KLR*H*********Q*****MIL*RS*ASQWTTGG**
DCO         **********I*SQ*Y**V*********KGR*********E*****IILS*****A****A**
DC1         ********I***QR*YI******ET***DGR*S*******E*****I*QLRP********AF*
DC2         **********IN*D*************KLR*R*********E*I***IIQS**H**A****A**
```

Fig. 2.4. Comparison of cAMP-dependent protein kinase catalytic-domain sequences from *Saccharomyces cerevisiae*, human, *Caenorhabditis elegans*, *Aplysia californica*, *Drosophila* and the predicted amino acid of the 189 bp PCR product from *M. grisea*. Identical matches are indicated with an asterisk. Primer sequences used to amplify this region were derived from regions designated 1F, 2F, 1B and 2B (shown in bold). TPK 1, 2, & 3 – *Saccharomyces cerevisiae*; cAPK a, b & g – human EcAPK - *Caenorhabditis elegans* Aplc & SAK – marine mollusk *Aplysia californica* DCO; 1 & 2 – *Drosophila*.

Recently, the gene for the principal target for cAMP, cAMP-dependent protein kinase, was found in *M. grisea* (unpublished data). Figure 2.4 shows the conserved regions of the catalytic-domain from 12 published cAMP-dependent protein kinases of other organisms. Cloning of the *M. grisea* gene was accomplished by DNA amplification of this region using degenerate oligonucleotides as primers for the polymerase chain reaction (PCR). The deduced amino acid sequence of the 189 bp PCR product is shown in Fig. 2.4. A comparison of this sequence with cAMP-dependent protein kinases from yeast showed 80% identity (Hanks *et al.*, 1988).

Gene Expression During Appressorium Formation

A number of strategies have been used to identify the genes having key roles in developmental processes. Two of the most versatile and powerful have been applied to appressorium development: the cloning of differentially expressed genes and the isolation of mutants (Lee and Dean, 1993b; D.S. Whitehead *et al.*, unpublished data).

Transcriptionally regulated genes

The strategy of differential hybridization is to identify clones in a cDNA (DNA synthesized from an mRNA template) library or genomic DNA library that hybridize more strongly to mRNA (cDNA) probes prepared from cells induced to form appressoria than to probes prepared from cells not stimulated. In a search for genes preferentially expressed during appressorium formation, a genomic cosmid library was constructed in the vector pUI (a gift of Dr B. Miller, University of Idaho) (Lee and Dean, 1993b). Approximately 1000 clones (10 microtiter plates × 96 wells) were transferred to nylon membranes supported in a dot blot apparatus. Two cosmids (28C6 and 29G11) that showed the greatest degree of differential expression and two others (17G8 and 20D7) that showed very little difference in hybridization intensity were digested with restriction endonucleases, electrophoresed through agarose and Southern blotted. ^{32}P-labeled cDNA prepared from infection structures hybridized to a single 8 kb *Eco*RI fragment in cosmid 29G11 and primarily to a 5 kb *Eco*RI fragment in cosmid 28C6. These fragments contain single copy genes because only fragments of a similar size were hybridized in Southern blots of genomic DNA of *M. grisea*. The 5 kb and 8 kb fragments hybridized to transcripts of ≈1.2 kb and ≈3.1 kb, respectively, isolated from conidia germinated to form infection structures. Little hybridization to RNA isolated from conidia or vegetative cells was detected, even after long exposure times to film. Initial DNA sequence analysis of these two genes, designated *mif*23 (*M*agnaporthe *inf*ection) from cosmid 28C6 and *mif*29 from cosmid 29G11, revealed little similarity with other known genes (Lee and Dean, 1993b). The advantage of differential screening of genomic DNA cosmid libraries is that relatively few clones need to be screened. Since the average size of the inserts in our library in pUI is *c.* 30 kb, the entire genome of *M. grisea*, estimated at 3.8×10^4 kb (Hamer *et al.*, 1989), is contained within a few thousand cosmid clones. However, the identification of many clones containing genes expressed during appressorium formation may be masked by the expression of other housekeeping genes, that is, genes expressed at all growth stages. Indeed, cosmids 28C6 and 29G11 appear to contain only *mif*23 and *mif*29, respectively, as no hybridization of insert DNA was observed with cDNA probes obtained from vegetative cells of conidia.

Appressorium-deficient mutants

Isolation of appressorium-deficient mutants

The use of mutants is a powerful and widely exploited technique to elucidate the genetic basis of a developmental pathway (Hartwell, 1974; Leung *et al.*, this volume; Shaw *et al.*, 1991; Timberlake, 1991). Mutants are especially valuable in systems such as *M. grisea* which are amenable to DNA-mediated transformation, because affected genes can be isolated by complementation.

Screening of 1500 UV survivors yielded 16 auxotrophs and 6 mutants blocked in germination or appressorium formation on artificial hydrophobic surfaces. The appressorium-deficient mutants, shown in Fig. 2.5, are similar in color and colony morphology to the wild type. All mutants germinate as well as the wild type (>90%), except for a single mutant which fails to germinate in liquid media, on agar media, or hard surfaces. The phenotypes of the mutants range from a complete inability to form appressoria on any surface tested, to the production of a greatly reduced number of appressoria that are frequently irregularly shaped. A few of the mutants also displayed a slightly elevated level of appressorium formation on a hydrophilic surface.

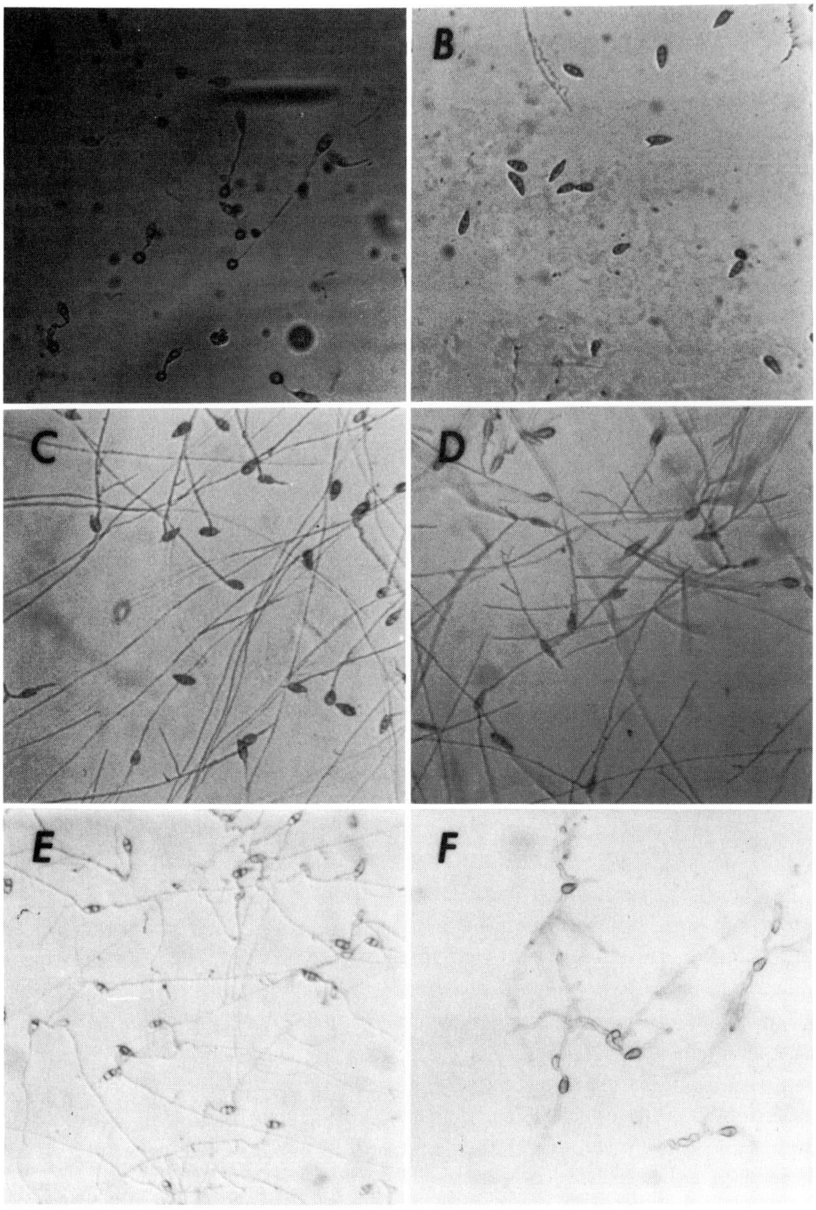

Fig. 2.5. Phenotypes of appressorium-deficient mutants of *M. grisea*. Conidia were germinated on the hydrophobic surface of Gelbond and photographed after 10 h. (A) wild type; (B) non-germinating mutants; (C) and (D) mutants that produce long, straight and unbranched germ tubes and no appressoria; (E) and (F) mutants that produce convoluted germ tubes – mutant in panel (F) forms misshapen appressoria at a low frequency.

Further evidence for a central role of cAMP in appressorium formation

To explore whether accumulation of cAMP is affected in any of the mutants, both cAMP and IBMX were tested for their potential to restore the capability to form appressoria. It was found that although some of the mutants were insensitive to cAMP and IBMX, several mutants could be restored to form appressoria on a hydrophobic surface, but not on a hydrophilic surface.

Several explanations may account for the ability of cAMP and IBMX to restore appressorium formation by these mutants. Mutants may be unable to sense their environment or to regulate cAMP within their cells. Overall, these results provide further evidence that cAMP is involved in regulating appressorium formation. However, the genetic analysis of these mutants is still in progress and it is not known with certainty that single gene mutations are involved.

Appressorium-deficient mutants are less virulent

The mutants are less virulent than the parental wild type. Most commonly, mutants formed far fewer appressoria than the wild type which resulted in smaller, generally irregularly shaped lesions.

The frequency of appressorium formation by several of the mutants on rice leaves is in contrast with results obtained using artificial hydrophobic surfaces, on which appressoria have never been observed. Perhaps another trigger other than, or in addition to, surface hydrophobicity is important. Other data indicate that a component of the waxy surface of rice leaves can induce appressorium formation in a manner not directly related to hydrophobicity (unpublished data).

Working Hypothesis for the Induction of Appressoria

Although the knowledge of the mechanisms regulating appressorium formation is incomplete, the available data for *M. grisea* can be formulated into a model. '*The interaction between cell-wall matrix proteins of the emerging germ tube and a hydrophobic surface results in the accumulation of cAMP in the fungus. Elevated levels of cAMP in turn trigger transductive pathways resulting in new gene expression leading to infection structure formation.*' This model will continue to evolve and be refined as additional information is uncovered. As it now stands, the model represents a reasonable framework to contemplate and formulate further experimentation.

Acknowledgements

This is paper No. 3474 of the South Carolina Agricultural Experiment Station, Clemson, SC 29634. We are indebted to our coworkers for helpful discussions during the course of this work. We also thank Dr E. Zehr for critical reading of this manuscript and members of the Department of Plant Pathology and Physiology for their support. The research summarized here was supported in part by grants from the US Department of Agriculture-NRICGP (No. 91-37303-6653), National Science Foundation (Nos STI-9108772 and IBN-9219561), and Clemson University.

References

Bourett, T.M. and Howard, R.J. (1990) *In vitro* development of penetration structures in the rice blast fungus *Magnaporthe grisea*. *Canadian Journal of Botany* 68, 329–342.

Brambl, R., Plesofsky-Vig, N., Hammett, J.R. and Russell, P.J. (1987) Preserved polyadenylated ribonucleic acid in dormant conidia of *Neurospora crassa* and new RNA synthesis during spore germination. *Experimental Mycology* 11, 317–330.

Devreotes, P.N. (1983) Cyclic nucleotides and cell–cell communication in *Dictyostelium discoideum*. *Advances in Cyclic Nucleotide Research* 15, 55–97.

Endelman, J.M. and Buck, C.A. (1992) The integrins: A general overview. In: Hook, M. and Switalski, L. (eds), *Microbial Adhesion and Invasion*. Springer-Verlag, New York, pp. 15–32.

Hamer, J.E., Howard, R.J., Chumley, F.G. and Valent, B. (1988) A mechanism for surface attachment in spores of a plant pathogenic fungus. *Science* 239, 288–290.

Hamer, J.E., Farrall, L., Orbach, M.J., Valent, B. and Chumley, F.G. (1989) Host species-specific conservation of a family of repeated DNA sequences in the genome of a fungal plant pathogen. *Proceedings of the National Academy of Sciences of the United States of America* 86, 9981–9985.

Hanks, S.K., Quinn, A.M. and Hunter, T. (1988) The protein kinase family: Conserved features and deduced phylogeny of the catalytic domains. *Science* 241, 42–52.

Hartwell, L.H. (1974) *Saccharomyces cerevisiae* cell cycle. *Bacteriological Reviews* 38, 164–198.

Hoch, H.C., Staples, R.C., Whitehead, B., Comeau, J. and Wolf, E.D. (1987) Signalling for growth orientation and cell differentiation by surface topology in *Uromyces*. *Science* 235, 1659–1662.

Howard, R.J. (1991) Penetration of hard substances by a fungus employing enormous turgor pressures. *Proceedings of the National Academy of Sciences of the United States of America* 88, 11281–11284.

Howard, R.J. and Ferrari, M.A. (1989) Role of melanin in appressorium function. *Experimental Mycology* 13, 403–418.

Kay, R.R. (1989) Evidence that elevated intracellular cyclic AMP triggers spore maturation in *Dictyostelium*. *Development* 105, 753–759.

Kincaid, R.L. (1991) Signalling mechanisms in microorganisms: Common themes in the evolution of signal transduction pathways. In: Greengard, P. and Robinson, G.A. (eds), *Advances in Second Messenger and Phosphoprotein Research*. Raven Press, New York, pp. 165–183.

Lee, Y.-H. and Dean, R.A. (1993a) Cyclic AMP regulates infection structure formation by the plant pathogenic fungus *Magnaporthe grisea*. *Plant Cell* 5, 693–700.

Lee, Y.-H. and Dean, R.A. (1993b) Genes specifically expressed during infection structure formation of *Magnaporthe grisea*. *Experimental Mycology* 17, 215–222.

Lee, Y.-H. and Dean, R.A. (1994) Hydrophobicity of contact surface induces appressorium formation of *Magnaporthe grisea*. *FEMS Letters* 115, 71–76.

Pall, M.L. (1981) Adenosine 3′,5′-phosphate in fungi. *Microbiological Reviews* 45, 462–480.

Pawson, T. (1991) Signal transduction in the control of cell growth and development. *Trends in Genetics* 7, 343–345.

Pereyra, E., Zaremberg, V. and Moreno, S. (1992) Effect of dibutyryl-cAMP on growth and morphology of germinating *Mucor rouxii* sporangiospores. *Experimental Mycology* 16, 93–101.

Shaw, C.H., Loake, G.J., Brown, A.P., Garrett, C.S., Deakin, W., Alton, G., Hall, M., Jones, S.A., O'Leary, M. and Primavesi, L. (1991) Isolation and characterization of behavioral mutants and genes of *Agrobacterium tumefaciens. Journal of General Microbiology* 137, 1939–1953.

Thevelein, J.M. (1992) The RAS-adenylate cyclase pathway and cell cycle control in *Saccharomyces cerevisiae. Antonie van Leeuwenhoek* 62, 109–130.

Timberlake, W.E. (1991) Temporal and spatial controls of *Aspergillus* development. *Current Opinion in Genetics and Development* 1, 351–357.

Underwood, P.A., Steele, J.G. and Dalton, B.A. (1993) Effects of polystyrene surface chemistry on the biological activity of solid phase fibronectin and vitronectin, analyzed with monoclonal antibodies. *Journal of Cell Science* 104, 793–803.

van Haastert, P.J.M. (1991) Transmembrane signal transduction pathways in *Dictyostelium. Advances in Second Messenger and Phosphoprotein Research* 23, 185–226.

3 Genetic Regulation of Sporulation in the Rice Blast Fungus

HEI LEUNG AND ZHIXIN SHI

Department of Plant Pathology, Washington State University, Pullman, WA 99164-6430, USA

INTRODUCTION

The disease cycle of rice blast involves three distinct phases: infection, colonization, and sporulation. Infection is initiated by a conidium landing on the leaf surface of a rice plant. The conidium produces a germ tube which forms an appressorium. In a compatible interaction, the appressorium produces an infection peg which penetrates the epidermis (see Chapter 1, this volume). Subsequent hyphal growth inside the host cells results in the development of a lesion. Under high relative humidity, a lesion on a susceptible plant produces conidia profusely for 3–4 days. These conidia are easily dislodged and dispersed, providing the inoculum for the next infection cycle. The polycyclic nature of the disease makes sporulation a key determinant of the severity of blast epidemics. Hence, interference with the sporulation pathway can be a powerful approach to reduce the epidemiological potential of the fungus.

Our strategy to develop stable resistance is to dissect the mechanisms controlling the disease cycle and then intervene at specific points in pathways that could reduce blast epidemics. We envision that by delineating the pathway of sporulation, we can select critical steps of the pathway against which specific plant components or chemicals can be incorporated into the rice plant by breeding and/or genetic engineering. Also, identification of gene products involved in sporulation may lead to the design of anti-sporulation compounds as potential fungicides.

In this chapter, we will discuss the genetic regulation of sporulation in *Magnaporthe grisea* (Barr, 1977) and the progress made towards the isolation of major genes controlling the production of conidia.

Genetic Analysis of Sporulation in Model Experimental Fungi

The most detailed genetic studies of sporulation have been performed in two experimental fungi, *Aspergillus nidulans* and *Neurospora crassa*. A brief overview of the work on these model organisms is given below to provide a framework for the study of sporulation in *M. grisea*.

In *A. nidulans*, extensive mutational analysis shows that there are two main classes of mutations: class I mutants lose the competence to be induced to sporulate (Butnick *et al.*, 1984) and class II mutants affect the morphogenesis of conidiophores and conidia. Genes determining the shift from mycelial growth to conidial development have also been referred to as strategic genes whereas genes determining the structural development of the conidiophores are called tactical genes (Clutterbuck, 1969).

The regulatory roles of the major sporulation genes in *A. nidulans* have been determined through molecular cloning of the genes (see Timberlake, 1990, for review). In *A. nidulans*, three loci, *bristle* (*brlA*), *abacus* (*abaA*), and *wet* (*wetA*), form the central pathway regulating the development and differentiation of conidiophores and conidia. Mutations at the *brlA* locus cause indeterminant growth of the conidiophore without subsequent differentiation. Mutations at *abaA* result in swellings at intervals of the phialide, which fails to produce conidia. *wetA* mutations give unpigmented conidia with defective walls that break down during conidial maturation. *brlA* activates genes that are required for the formation of conidiophore, and *abaA* further activates *brlA* and additional genes involved in conidiophore development. *abaA* also activates *wetA* which in turn activates genes involved in spore differentiation. A second genetic pathway called *STUNTED* consists of at least two genetic loci, *stunted* (*stuA*) and *medusa* (*medA*), which modulate the cell type and spatial organization of the conidiophore structure (Miller, 1990). The *stuA* mutation blocks the development of medulae and phialides, but produces a small number of conidia directly from the vesicle.

Similar mutations are found in *N. crassa* where specific stages in the sporulation pathway are blocked, resulting in unique morphogenetic phenotypes (Matsuyama *et al.*, 1974; Selirennikoff *et al.*, 1974; Springer and Yanofsky, 1989). Sporulation-specific genes in *N. crassa* have been isolated by screening a genomic library with a cDNA probe enriched in sequences expressed during sporulation (Berlin and Yanofsky, 1985). Although the exact phenotypes to which these cDNA correspond are not known, the temporal order of expression of these genes in the sporulation pathway has been determined by correlating gene expression with known sporulation-defective mutants (Roberts *et al.*, 1980; Hager and Yanofsky, 1990).

A common theme observed in *A. nidulans* and *N. crassa* is that relatively few genetic loci have a major morphogenetic effect on sporulation but a larger number of genes (estimated at about 1000 in *A. nidulans*; Timberlake, 1990) are involved in modifying the process. In the context of intervention of sporulation, inhibition of genes in the 'strategic' and 'tactical' categories would be more effective than targetting genes controlling later stages in spore morphogenesis.

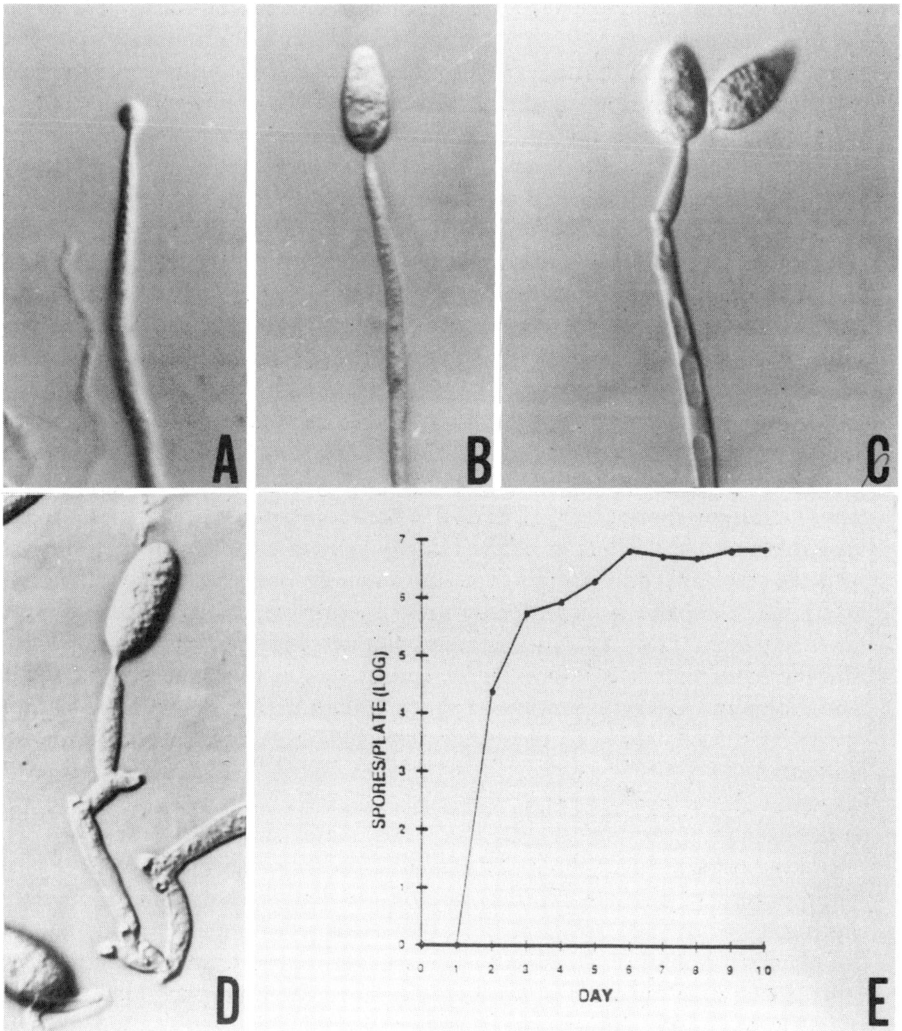

Fig. 3.1. (A)–(D) The normal sequence of conidium formation in *Magnaporthe grisea*. (A) Swelling of the apex of a conidiophore; (B) the first immature conidium not fully septate; (C) the second conidium begins to form while the first conidium becomes fully septate; (D) mature conidia dislodged, showing the sympodial branching pattern of the conidiophore; (E) the time course of sporulation on oatmeal agar. Sporulation is assayed by washing conidia with 1 ml of water from a 6-day-old culture on a 5-cm-diameter oatmeal plate.

Normal Pathway of Sporulation in *M. grisea*

Figures 3.1(A)–(D) depict the normal sequence of sporulation in *M. grisea*. Conidiogenesis in *M. grisea* is holoblastic (Cole, 1986), which means that conidia are formed by a 'blowing-out' of the apex of the conidiophore. A septum is then formed to delimit the conidium. The active apical tip moves to the side to produce the next conidium, resulting in 3–5 conidia borne sympodially on a mature conidiophore. Each conidium is a three-celled structure, and each cell contains an identical nucleus derived from a common mother nucleus. On a medium favoring sporulation, a single conidium produces conidiophores about 48 h after germination (Fig. 3.1E). The first crop of conidia are produced on the conidiophores about 60 h after germination of a conidium.

The critical factors for *in vitro* sporulation are the nutrients in the growth medium and light. Many complex nutrient media can support sporulation of *M. grisea*; the most commonly used media are oatmeal, prune juice, rice polish (Ou, 1985). However, some media only support mycelial growth but not sporulation. For example, a complete medium composed of yeast extract and sucrose supports normal vegetative growth but completely suppresses sporulation. In contrast to *A. nidulans* where light is essential for sporulation in wild type strains (Mooney and Yager, 1990), most wild type strains of *M. grisea* produce spores in the dark. However, sporulation can be stimulated by light particularly in the UVA (320–450 nm) range. Figure 3.2 shows the level of sporulation in representative genetic lineages from the Philippines (N. Bandian and H. Leung, unpublished data). There is considerable genetic variation in sporulation capacity and the responsiveness to light induction among isolates collected from nature.

Little is known about how host genotypes influence the amount of sporulation *in planta*. It is known that rice varieties with partial resistance to blast show a reduction in the number of lesions, lesion size, and sporulation (Yeh and Bonman, 1986), but the resistance components that specifically suppress sporulation have not been genetically resolved.

Genetic Regulation of Sporulation

Our approach to dissect the pathway of sporulation is mutational analysis. In most cases, Guy11 (MAT2), a hermaphroditic rice-infecting strain, is used for mutagenesis (Leung *et al.*, 1988). A series of near-isogenic lines of Guy11 of opposite mating type (MAT1) has been developed by backcrossing ascospore progeny to Guy11 eight times (H. Leung, unpublished data). The MAT1 near-isogenic Guy11 strains are used as a tester to facilitate the genetic analysis of primary mutants by reducing the heterogeneity expressed in segregating progeny.

Diepoxyoctane is used to generate mutations affecting the sporulation pathway (Chumley and Valent, 1990). Survivors are individually examined under a stereomicroscope for abnormal sporulation. We also make use of naturally occurring variants that show either quantitative or qualitative differences in sporulation. Induced or spontaneous mutants are first crossed to fertile strains to obtain F_1 ascospore progeny and the inheritance of the variant phenotypes confirmed by backcrosses and sib crosses. The epistatic relationship among genetic loci are tested by crossing two mutants to produce double mutants. Whenever possible, tetrad analysis is used to confirm single-gene inheritance and the epistatic relationship between two genes. Phenotypic characterizations (growth, sporulation, pathogenicity) are performed with the primary mutants and segregating progeny.

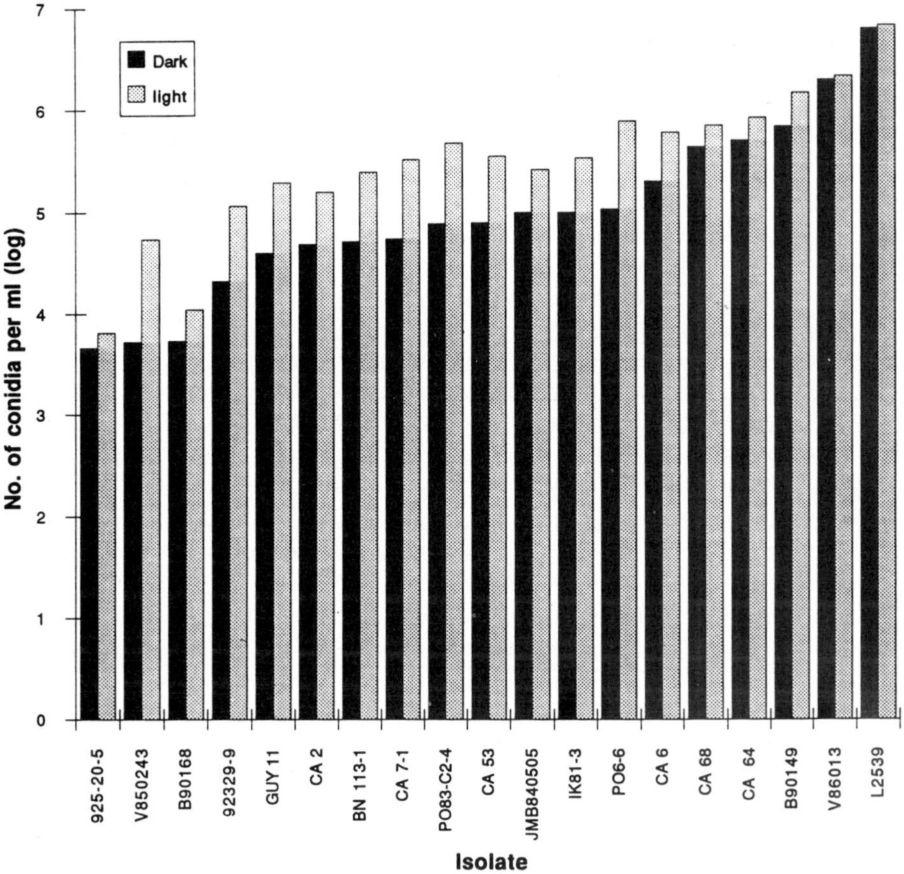

Fig. 3.2. Variation in sporulation capacity and the responsiveness to light induction among field isolates of *M. grisea* from the Philippines. Sporulation is assayed by washing conidia with 1 ml of water from a 6-day-old culture on a 5-cm-diameter oatmeal plate. Light induction is performed by exposing a 4-day-old culture to light (320–400 nm) for 8 h and returned to the dark for 24 h before assayed for sporulation.

Genes that block major morphogenetic steps

The first sporulation variant we found in *M. grisea* was a mutation causing abnormal conidium morphology as well as a 97% reduction in sporulation (Shi and Leung, 1994). This mutant, Guy11E46, was recovered fortuitously from survivors of an electroporation experiment in which germinated conidia of Guy11 were subjected to a 3000 V cm^{-1} electrical shock. In this experiment, transformation vector pAN7-2 (Punt *et al.*, 1987) was added to the conidial suspension. However, subsequent analysis showed that the mutant was not resistant to hygromycin B and did not contain vector DNA in the genome. Similar variants were not found in two subsequent electroporation experiments performed under identical conditions. We are therefore not certain whether the variant was caused by electric shock or simply a result of spontaneous mutation.

Figure 3.3A–B shows the morphology of the conidiophore and conidium of the mutant. Instead of producing a cluster of 4–5 conidia sympodially borne on a conidiophore, the mutant produces an aerial hypha bearing a terminal, elongated conidium (4 × 50 μm). Genetic analysis showed that the mutation is conditioned by a single gene, designated *con1*. Crosses between *con1* strains and a previously described spore morphology (*smo*) mutant (Hamer *et al.*, 1989) showed that the *con1* mutation is epistatic to *smo*.

An interesting feature of the *con1* mutation is the extent of pleiotropy. The growth rate of the *con1* mutant is about 73% of the wild type. Most importantly, conidia of the *con1* strains do not produce normal appressoria (Fig. 3.3C) and are non-pathogenic on rice lines susceptible to the wild type parent. The lack of appressorium formation is not the only cause of nonpathogenicity because the *con1* strains do not cause lesions on rice even when conidia are injected into the leaf sheath (i.e. despite wounding of the leaf). The *con1* mutation could be a regulatory gene that affects the morphogenesis of conidia and appressoria as well as other pathogenic attributes. Alternatively, the pleiotropy could result from a deletion of several genes on a chromosome.

A second mutation, designated *con2*, was recovered from survivors after mutagenesis of Guy11 using diepoxyoctane. When incubated in the dark, the *con2* mutant produces conidiophores with swollen hyphal tips that fail to develop conidia (Fig. 3.3D–G). Under continuous light, some abnormally shaped and immature conidia are formed (Fig. 3.3F, G). Although the *con2* mutation suppresses sporulation, it causes no detectable changes in colony morphology, growth rate, or sexual fertility. Allelism tests showed that *con1* and *con2* are independent loci, with *con2* being epistatic to *con1*. Furthermore, *con2* is genetically linked (19 centimorgan (cM)) to *con1*. Thus, two genes controlling the key steps in conidium and conidiophore development are on the same chromosome. It would be of interest to determine whether genes involved in sporulation tend to cluster, as has been found with the spore-specific genes (*SpoC1* region) in *A. nidulans* (Clutterbuck and Timberlake, 1992).

A mutant that restores the production of conidia was recovered from a *con2* strain. The reversion mutant produces almost the same quantity of conidia in the dark as the wild type parent but many conidia are non-septate and irregular in shape. Genetic analysis showed that the reversion is caused by mutation at a third locus (*con3*) independent of *con2*. It is possible that *con3* is a suppressor mutation specific to *con2*. Alternatively, there could be a secondary pathway which negatively regulates sporulation. Thus, in the presence of *con2* (i.e. when the main sporulation pathway is blocked), a mutated *con3* allele will restore some sporulation. About 40% of germinated conidia of *con2* and *con3* strains form appressoria. Preliminary data suggest that *con2* and *con3* strains are nonpathogenic on rice.

Genes that modify the spore morphology

While *con1* and *con2* block the normal development of conidiophores and conidia, a different class of sporulation mutation modifies the morphology of conidia but has little effect on the amount of spores produced. Hamer *et al.* (1989)

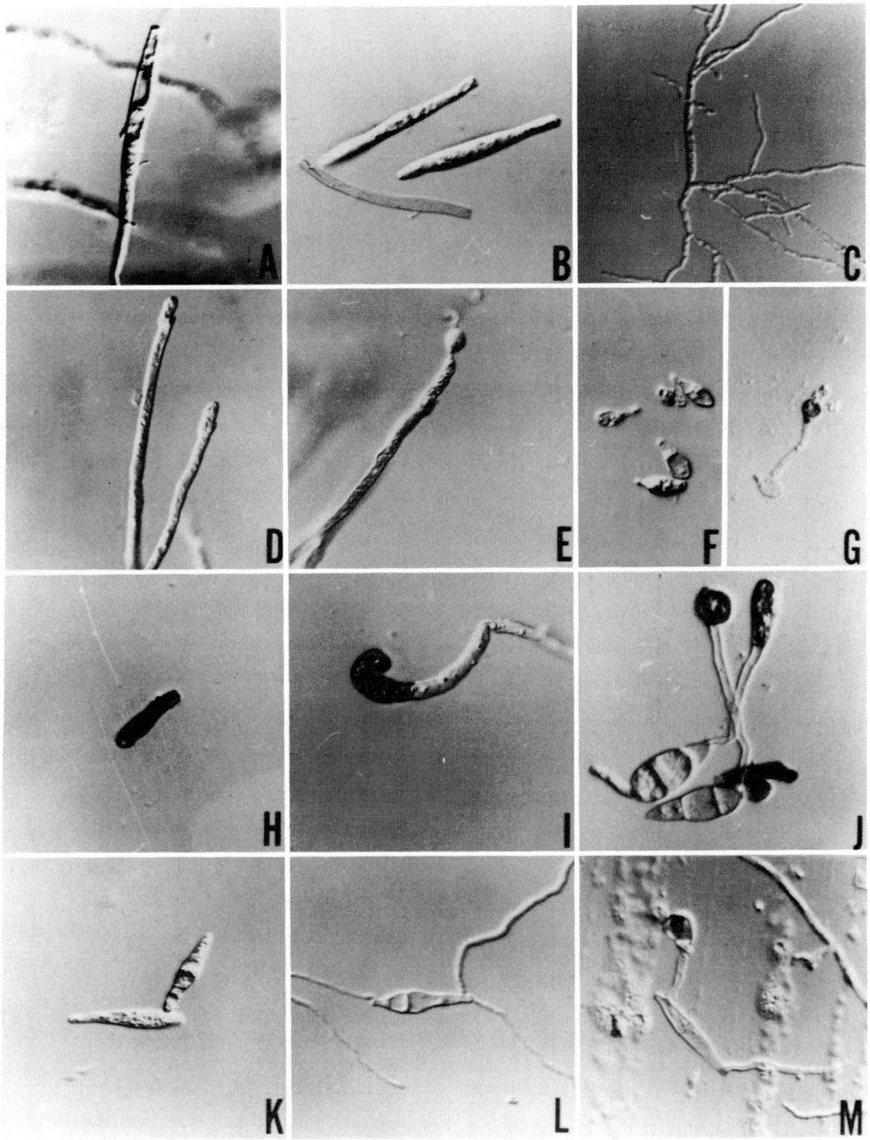

Fig. 3.3. Morphology of sporulation mutants of *M. grisea*. (A–C) *con1* mutation. (A) and (B) Morphology of a conidiphore and conidia of *con1* mutation. (C) Indeterminant hyphal growth from a *con1* conidium. No appressorium is formed. (D)–(G) *con2* mutation. (D) and (E) *con2* mutant produces a swelling at the apex of a conidiophore. (F) and (G) A few non-septate, abnormally-shaped conidia are produced. (H)–(J) Aleuriospore-forming mutation. (H) Pigmented, thick-walled aleuriospore. (I) An aleuriospore borne on the tip of a conidiophore. (J) Some proportion of the wild type conidia from a aleuriospore-forming strain produces an aleuriospore (arrow) instead of an appressorium at the tip of the germling. (K)–(M) 5SMA13 spore morphology mutation. (K) Elongated conidia. This mutation has 80% reduction in sporulation. (L) and (M) Only a low percentage of germinated conidia produce appressoria. Phenotypes were observed in primary mutants as well as in mutants derived from crosses.

reported a spore morphology mutation called *smo* that determines the cell shape of conidia and asci. Although cell shapes are affected, there is no reduction in the sporulation capacity of *smo* mutants. The abnormal spore shape, however, causes a reduction in pathogenicity on rice (Hamer and Givan, 1990).

By mutagenesis of Guy11 with diepoxyoctane, we have recovered a number of mutants (about 5% of survivors) which produce pigmented, thick-walled conidia called aleuriospores (Fig. 3.3H–J). Formation of aleuriospores does not replace normal sporulation; rather aleuriospores are produced in addition to wild type conidia. The relative abundance of aleuriospores is dependent on the genetic background of the strain. Unlike chlamydospores, which are usually formed from intercalary cells in aged cultures, aleuriospores are formed exclusively at the tips of actively growing hyphae. A proportion of wild type conidia from an aleuriospore-forming strain produce aleuriospores instead of appressoria at the end of the germ tube (Fig. 3.3J). Thus, aleuriospore formation seems to result from an abrupt change from active hyphal growth to a resting stage. The ability to form aleuriospores is controlled by two complementary genes, *aleu1* and *aleu2*. Mutant *aleu* alleles are expressed in a *smo* genetic background indicating that the *aleu* loci act independently in modifying spore morphology. Since it is unlikely that a double mutation occurs in a single mutagenic treatment of Guy11, one of the loci controlling aleuriospore formation in Guy11 probably has been mutated prior to mutagenesis. This is supported by the observation that a mutagenized population of strain 2539 did not yield any aleuriospore-forming mutants. Repeated attempts to test cross Guy11 with aleuriospore-forming strains have not been successful.

A sporulation mutant 5SMA13 (provided by J. Sweigard and B. Valent, E.I. Du Pont de Nemours & Company, Inc., Delaware, USA) was recovered after transformation with an integrative vector carrying hygromycin B resistance. This mutant suffers from a 80% reduction in sporulation and produces elongated conidia (Fig. 3.3K–M). Segregation analysis shows that the 5SMA13 mutation is controlled by a single gene but the spore mutation is not associated with hygromycin B resistance. Thus, the mutation is not due to gene inactivation caused by integration of the transforming DNA into the genome of *M. grisea*. The spore morphology and appressorium behavior of conidia of 5SMA13 are distinct from those of *con1* mutants. The width/length ratio of conidia of 5SMA13 is much greater than that of *con1* strains. Furthermore, appressoria are produced by some conidia of 5SMA13 (Fig. 3.3M) but not by conidia of *con1* mutants. These observations suggest that 5SMA13 is different from the *con1* mutation. Allelism test are in progress to determine the epistatic relationships between 5SMA13 and other sporulation mutations.

Genes that have quantitative effects on sporulation

As shown in Fig. 3.2, there is a high degree of genetic variation in sporulation among field isolates. The amount of sporulation in field isolates can differ by 1000-fold. The available variability allows us to examine the genetic difference between the high- and low-sporulating isolates.

A field isolate designated CH40-1 (MAT2) collected from Hunan, China,

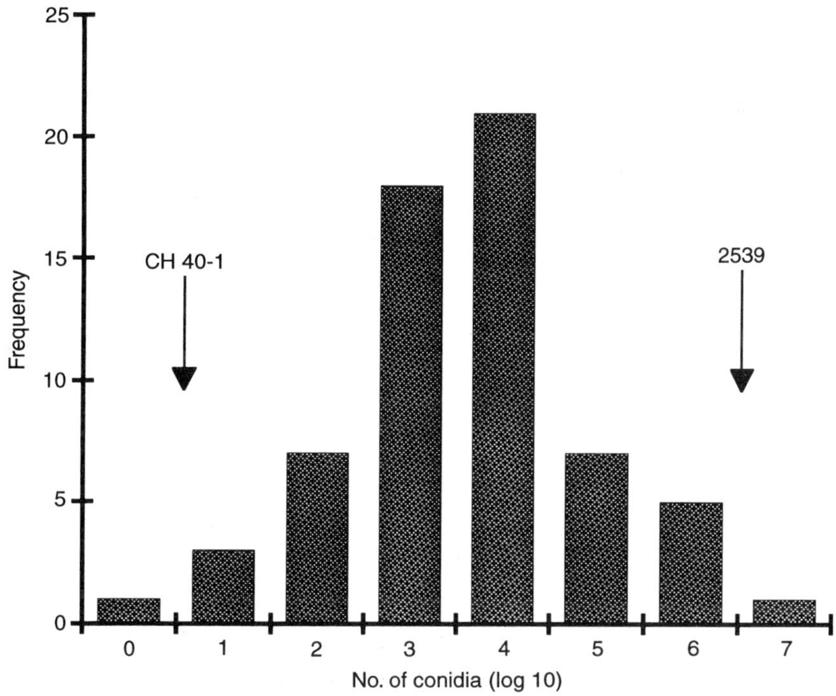

Fig. 3.4. Segregation in sporulation of F_1 progeny derived from a cross between a low (CH40-1) and high sporulating (2539) strain. The normal distribution of segregating progeny suggests that the amount of sporulation is under polygenic control.

produces very few spores on oatmeal agar. This strain was crossed to the high-sporulating strain 2539. Progeny derived from this cross were assayed for sporulation. A normal distribution of sporulation among the F_1 progeny is observed (Fig. 3.4), suggesting that the difference between the high and low sporulation strains is controlled by polygenes. This segregating population offers the possibility of locating quantitative trait loci (QTL) controlling sporulation in *M. grisea*.

Based on the above genetic data, a pathway of genetic regulation of sporulation in *M. grisea* is shown in Fig. 3.5. So far, we have defined a linear, central pathway determining the development of the conidiophores and conidia (*con1* and *con2*). This central pathway is modified by other genes that affect the morphology of spores (*aleu1* and *aleu2*) and the number of spores produced (polygenes, *con3*). Further mutational analysis will likely reveal a more complicated pattern of interactions between different developmental pathways. For the purpose of inhibiting sporulation, it will be critical to determine whether there are secondary pathways that can compensate sporulation when the primary pathway is blocked. The finding of the *con3* locus seems to support this possibility. Another significant observation is that most of the sporulation mutations have an strong pleiotropic effect on other developmental processes. The most notable feature is a reduction in appressorium formation and a loss of pathogenicity. The *con1* mutation is found to cosegregate with a loss of

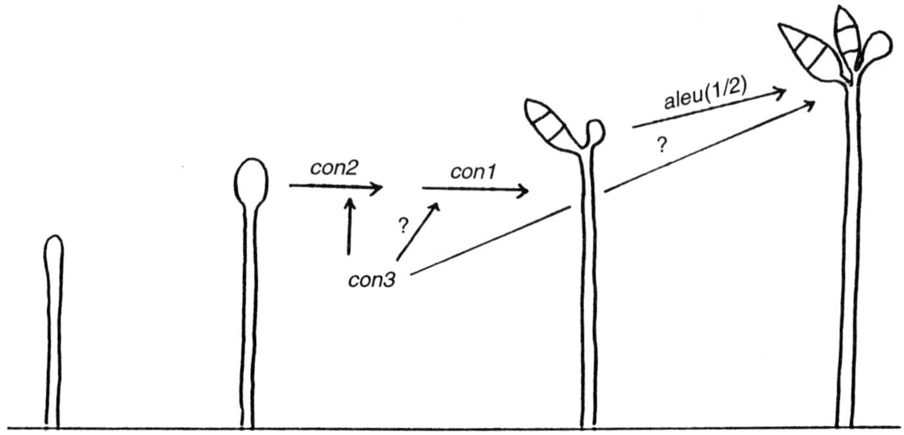

Fig. 3.5. A schematic diagram of the sporulation pathway in *M. grisea*.

pathogenicity. Preliminary data suggest that *con2* and *con3* are also non-pathogenic even though these two mutations have no adverse effect on vegetative growth. More extensive genetic and cytological analysis is in progress to determine whether genes controlling sporulation and spore morphogenesis also play an important role in pathogenesis.

Towards Cloning of Sporulation Genes

Since *con1* and *con2* represent the major genes controlling the development of conidia, they are the logical targets for molecular cloning. Although it is not known whether genes controlling sporulation tend to cluster on one chromosome, the genetic linkage between *con1* and *con2* makes chromosome walking an attractive cloning strategy.

Map-based cloning has been facilitated by the recent development of RAPD-bulked segregant analysis (Williams *et al.*, 1990; Michelmore *et al.*, 1991) (Fig. 3.6). This mapping technique involves the enzymatic amplification of bulked DNA with 10-base random primers in a polymerase chain reaction. By pooling DNA of progeny according to a specific genotype, one can generate two isogenic samples with almost identical 'genomes' except for the regions flanking the target gene. Amplification of DNA of two bulked samples reveals polymorphic markers around the target gene because most loci unlinked to the target locus are rendered monomorphic by bulking. Thus, RAPD-bulked segregant analysis offers both the efficiency and flexibility in tagging specific genes with DNA markers.

We have successfully applied RAPD-bulked segregant analysis to identify DNA markers linked to the *con1* and *con2* locus. About 65% of the more than 200 random primers tested gave DNA amplification with genomic DNA of *M. grisea*. By pooling DNA of 15 progeny of each phenotype, four out of 125 primers yielded genetic markers linked to the *con1* locus. Figure 3.7 shows

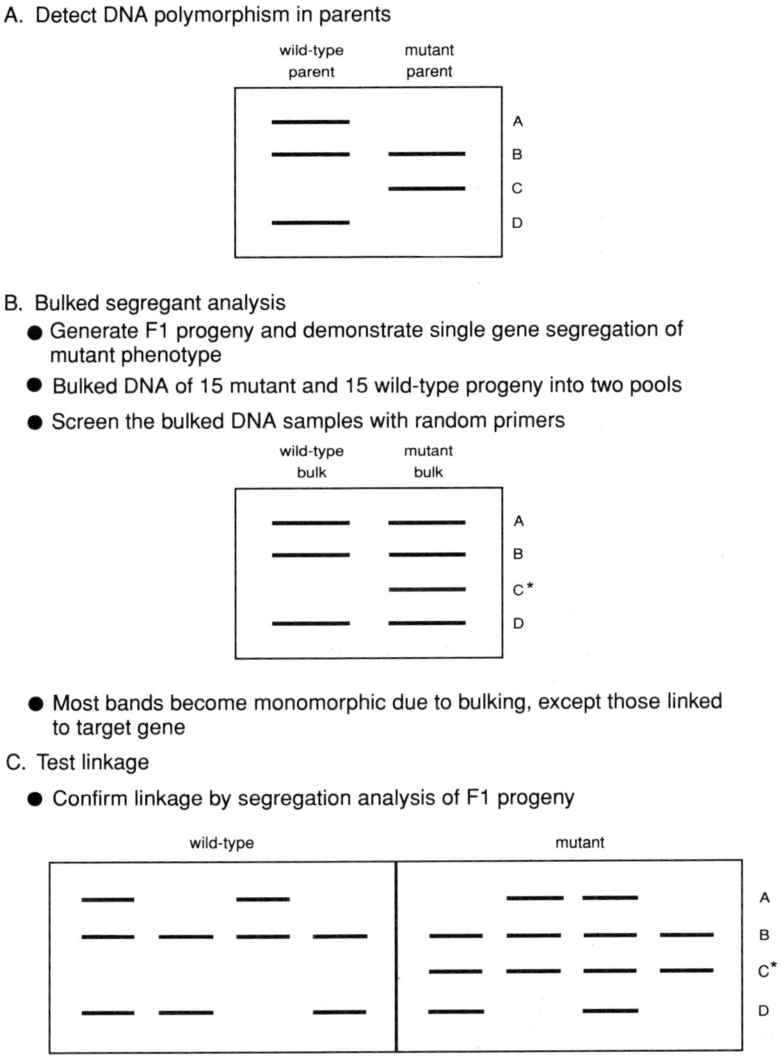

Fig. 3.6. A schematic diagram to illustrate the principle of bulked segregant analysis. Note that segregation in a haploid ascomycete occurs in the F_1 generation. In this example, band C (*) is the polymorphic locus linked to the mutated gene. This scheme is a modification of that of Michelmore *et al.* (1991).

a RAPD marker that cosegregates with the mutant phenotype. The overall percentage of useful primers in bulked segregant analysis is about 3% in *M. grisea*. The linkage map surrounding the *con1* and *con2* locus is shown in Fig. 3.8. The high efficiency of this mapping strategy can be exploited to obtain a high resolution map with markers 1–2 cM from the target gene before initiating chromosome walking in a genomic library.

A RAPD marker 1.3 cM from the *con2* locus was used as a hybridization probe to select cosmid clones in a genomic library constructed in pAN7-2

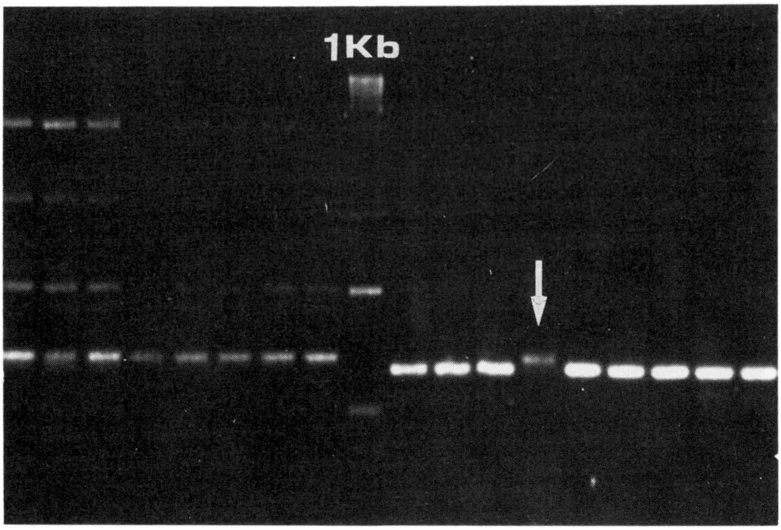

Fig. 3.7. An ethidium bromide-stained gel showing a RAPD marker linked to the *con1* locus. Each lane contains PCR-amplified products of a single segregant. Wild type and *con1* segregants are on the left and right side of the 1 kb size marker ladder, respectively. Note the single recombinant (arrow).

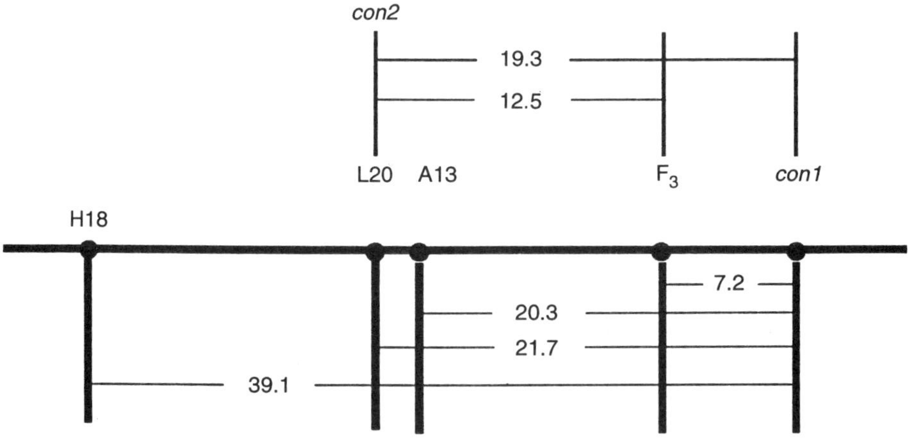

Fig. 3.8. Linkage map generated by RAPD-bulked segregant analysis around the region containing the *con1* and *con2* gene. Distance is expressed in percent of recombination.

(Punt *et al.*, 1987). From a cosmid library of 4000 clones, two overlapping cosmids were found to contain the RAPD marker. The end fragments of the overlapping clones were subcloned and used to select more overlapping cosmids. As of now, cosmid clones spanning about 70 kb on either side of the RAPD marker have been isolated and are being used for genetic complementation of the *con2* mutant.

For the cloning of the *con1* gene, RAPD markers close to *con1* will be selected by additional cycles of bulk segregant analysis. Once a RAPD marker less than

2 cM from the locus is found, it will be used to select cosmid clones from the library.

Sporulation in relation to disease control

Monogenic resistance has been widely used in rice to defend against multiple races of the blast fungus. Such forms of resistance, however, are often vulnerable to genetic adaptation by the pathogen, and become ineffective after 1–3 years (Bonman and MacKill, 1988). Due to this lack of durability, increasing emphasis has been placed on the development of partial or incomplete resistance that tolerates some degree of infection. In contrast to complete race-specific resistance, which prevents disease by a hypersensitive reaction to pathogen attack, partial resistance retards disease development by acting at multiple stages of the disease cycle. In varieties with partial resistance, there is a reduction in the lesion number, lesion size, and sporulation of the fungus on the plant (Yeh and Bonman, 1986; Wang *et al.*, 1994). It is postulated that by allowing some degree of infection the selective pressure for new virulent types can be reduced. Another argument is that it would be more difficult for the fungus to overcome resistance that acts at multiple stages of fungal development. The challenge ahead is to exploit bioengineering to develop the kinds of resistance that are consistent with the objectives of slowing the evolution of new virulences and maintaining disease below an economic threshold.

Cornelissen and Melchers (1993) reviewed the strategies for control of fungal diseases with transgenic plants. The strategies include the use of antifungal proteins, toxic compounds, and inhibitors of fungal enzymes. de Wit (1992) suggested a more elaborate system in which a cloned avirulence gene from a pathogen is placed under the control of a regulated and inducible promoter in the plant. Upon challenge by a non-specific elicitor, the avirulence gene will be induced in the plant and trigger defense mechanisms. With the exception of de Wit's proposal, most approaches do not take into account the developmental process of the pathogen. Without addressing how engineered resistance mechanisms act on the pathogen, the instability problems experienced in conventional breeding will likely continue.

The successful use of anti-penetrant fungicides in blast control provides a useful model for developing future control strategies (Sisler, 1986). The anti-penetrant fungicides are not toxic to the fungus but inhibit melanin synthesis which subsequently prevent penetration of the host by the fungus (Howard and Ferrari, 1989). Despite extensive use of these fungicides in many rice growing countries, no mutant resistant to melanin inhibition has yet been found, suggesting that inhibition of melanin formation apparently exerts little selection pressure on the fungal population. It is also possible that there are no easy developmental alternatives that can compensate the role of melanin.

Our rationale for targeting sporulation genes in disease control is that by specifically suppressing spore production, rice blast can be changed from an explosive disease to one that has limited epidemiological potential. Anti-sporulation compounds or resistance will allow the fungus to infect rice but be unable to produce multiple infection cycles. In epidemiological terms, anti-sporulation mechanisms will be equivalent to non-race specific, horizontal resistance by reducing the amount of inoculum in the field. An anti-sporulation mechanism may also exert less selection pressure on the fungus as compared to fungitoxic mechanisms. However, it remains to be tested whether a blockage in sporulation can be readily compensated by other developmental pathways.

Analysis of the genetic regulation of sporulation is a prerequisite to pursuing an anti-sporulation strategy for blast control. The developmental stages defined by the *con1* and *con2* mutations represent two potential intervention points. Although specific intervention tactics are not possible until the biochemical nature of sporulation genes are understood, some practical applications of cloned sporulation genes can be envisioned. First, sporulation genes can be used as targets to develop an efficient screen for anti-sporulation compounds. Once a sporulation gene is isolated, it should be feasible to develop a reporter gene system to assess the biological activities of chemicals or biocontrol agents *in vitro*. Recently, A.M. Rosales *et al.* (personal communication) have identified antifungal secondary metabolites produced by six strains of rice-associated *Pseudomonas* species. It would be of interest to determine the effects of these compounds on sporulation. If appropriate gene products or compounds can be isolated, it might be possible to incorporate such genes or compounds into the rice plant to express anti-sporulation activity.

The identification of sporulation genes can also help dissect the components of resistance in rice varieties. QTLs have been identified in a rice cultivar Moroberekan with durable resistance against blast (Wang *et al.*, 1994). Does one kind of QTL reduce lesion size and another kind of QTL reduce sporulation? Some of these QTLs have been bred into a common susceptible rice cultivar to produce near-isogenic QTL lines (R.J. Nelson, IRRI, personal communication). These genetic materials can be assayed for their effects on specific sporulation genes. This will eventually lead to a better understanding of the biological effects of QTL and help the selection and accumulation of QTL in a breeding program.

Acknowledgements

We thank Dave Christian for assistance in the preparation of photographs, D. Chen, M. Bernardo and R. Nelson for providing isolates of representative genetic lineages from the Philippines, and B. Valent and J. Sweigard for providing the 5SMA13 mutant series. This work was supported in part by Rockefeller Foundation grant RF89025 #98 and USDA grant 92-37303-7803. Contribution PPNS No. 0193 from Washington State Agricultural Research Center.

References

Barr, M.E. (1977) *Magnaporthe*, *Telimenella*, and *Hyponectria* (Physosporellaceae). *Mycologia* 69, 952–966.

Berlin, V. and Yanofsky, C. (1985) Isolation and characterization of genes differentially expressed during conidiation of *Neurospora crassa. Molecular and Cellular Biology* 5, 849–855.

Bonman, J.M. and MacKill, D.J. (1988) Durable resistance to rice blast disease. *Oryza* 25, 103–110.

Butnick, N.Z., Yager, L.N., Kurts, M.B. and Champe, S.P. (1984) Genetic analysis of mutants of *Aspergillus nidulans* blocked at an early stage of sporulation. *Journal of Bacteriology* 160, 541–545.

Chumley, F.G. and Valent, B. (1990) Genetic analysis of melanin deficient, non-pathogenic mutants of *Magnaporthe grisea. Molecular Plant-Microbe Interactions* 3, 135–143.

Clutterbuck, A.J. (1969) A mutational analysis of conidial development in *Aspergillus nidulans. Genetics* 63, 317–327.

Clutterbuck, A.J. and Timberlake, W.E. (1992) Genetic regulation of sporulation in the fungus *Aspergillus nidulans*. In: Russo, V.E.A., Brody, S., Cove, D. and Ottolenghi, S. (eds), *Development: The Molecular Genetic Approach*. Springer-Verlag, New York pp. 103–120.

Cole, G.T. (1986) Models of cell differentiation in conidial fungi. *Microbiological Review* 50, 95–132.

Cornelissen, B.J.C. and Melchers, L.S. (1993) Strategies for control of fungal diseases with transgenic plants. *Plant Physiology* 101, 709–712.

de Wit, P.J.G.M. (1992) Molecular characterization of gene-for-gene systems in plant–fungus interactions and the application of avirulence genes in control of plant pathogens. *Annual Review of Phytopathology* 30, 391–418.

Hager, K.M. and Yanofsky, C. (1990) Gene expressed during conidiation in *Neurospora crassa*: molecular characterization of con-13. *Gene* 96, 153–159.

Hamer, J.E. and Givan, S. (1990) Genetic mapping with dispersed repeated sequences in the rice blast fungus: Mapping the SMO locus. *Molecular and General Genetics* 223, 487–495.

Hamer, J.E., Valent, B. and Chumley, F.G. (1989) Mutations at the SMO genetic locus affect the shape of diverse cell types in the rice blast fungus. *Genetics* 122, 351–361.

Howard, R.J. and Ferrari, M.A. (1989) Role of melanin in appressorium formation. *Experimental Mycology* 13, 403–418.

Leung, H., Borromeo, E.S., Bernardo, M.A. and Notteghem, J.L. (1988) Genetic analysis of virulence in the rice blast fungus *Magnaporthe grisea. Phytopathology* 78, 1227–1233.

Matsuyama, S.S., Nelson, R.E. and Siegel, R.W. (1974) Mutations specifically blocking differentiation of macroconidia in *Neurospora crassa. Developmental Biology* 41, 278–287.

Michelmore, R.W., Paran, I. and Kesseli, R.V. (1991) Identification of markers linked to disease-resistance genes by bulked segregant analysis: a rapid method to detect markers in specific genomic regions by using segregating populations. *Proceedings of National Academy of Sciences of the United States of America* 88, 9828–9832.

Miller, B.L. (1990) The developmental genetics of asexual reproduction in *Aspergillus nidulans. Seminars in Developmental Biology* 1, 207–219.

Mooney, J.L. and Yager, L.N. (1990) Light is required for conidiation in *Aspergillus nidulans. Genes and Development* 4, 1473–1482.

Ou, S.H. (1985) *Rice Diseases*, 2nd edn. Commonwealth Mycological Institute, Kew, Surrey, UK, 380 pp.

Punt, P.J., Oliver, R.P., Dingemanse, M.A., Pouwels, P.H. and van den Hondel, C.A.M.J.J. (1987) Transformation of *Aspergillus* based on the hygromycin B resistance marker from *Escherichia coli. Gene* 56, 117–124.

Roberts, A.N., Berlin, V., Hager, K.M. and Yanofsky, C. (1988) Molecular analysis of a *Neurospora crassa* gene expressed during conidiation. *Molecular and Cellular Biology* 8, 2411–2418.

Selirennikoff, C.P., Nelson, R.E. and Siegel, R.W. (1974) Phase-specific genes for macroconidiation in *Neurospora crassa. Genetics* 78, 679–690.

Shi, Z. and Leung, H. (1994) Genetic analysis and rapid mapping of a sporulation mutation in *Magnaporthe grisea. Molecular Plant-Microbe Interactions* 7, 113–120.

Sisler, H.D. (1986) Control of fungal diseases by compounds acting as antipenetrants. *Crop Protection* 5, 306–313.

Springer, M.L. and Yanofsky, C. (1989) A morphological and genetic analysis of conidiophore development in *Neurospora crassa. Genes and Development* 3, 559–571.

Timberlake, W.E. (1990) Molecular genetics of *Aspergillus* development. *Annual Review of Genetics* 24, 5–36.

Wang, G., MacKill, D.J., Bonman, J.M., McCouch, S.R., Champoux, M. and Nelson, R.J. (1994) RFLP mapping of genes conferring complete and partial resistance to blast in a durably resistant rice cultivar. *Genetics* 136, 1421–1434.

Williams, J.G.K., Kubelik, J.R., Livak, K.J., Rafalski, J.A. and Tingey, S.V. (1990) DNA polymorphisms amplified by arbitrary primers are useful as genetic markers. *Nuclear Acids Research* 18, 6531–6535.

Yeh, W.H. and Bonman, J.M. (1986) Assessment of partial resistance to *Pyricularia oryzae* in six rice cultivars. *Plant Pathology* 35, 319–323.

4 Genetic Interactions in *Magnaporthe grisea* that Affect Cultivar Specific Avirulence/Virulence on Rice

A.H. Ellingboe and Chih-Cheng T. Chao

Departments of Plant Pathology and Genetics, University of Wisconsin, 1630 Linden Drive, Madison, WI 53706, USA

Introduction

The blast disease of rice, like most diseases, is controlled by some combination of host plant resistance, cultural practices, and fungicides. Host plant resistance appears to be the most promising method for control of blast, and there is a long history of the development of blast resistant cultivars of rice. It seems very likely that, in the future, there will be increased emphasis on methods such as genetic resistance that do not pollute the environment.

The ability of a plant to express resistance is also dependent on the genotype of the pathogen. There is a mutual interdependence of the genes in both rice and in *Magnaporthe grisea* for resistance and cultivar-specific avirulence/virulence, respectively. A rice plant cannot be resistant to an isolate of *M. grisea* unless the pathogen has the genes that make it avirulent on that rice plant. An isolate of *M. grisea* cannot be avirulent on a rice plant unless the rice plant has genes that make it resistant to that isolate. Thus it is possible to learn much about the genetics of host resistance by studies of the genetics of the pathogen.

Some of the first studies on the genetics of *M. grisea* involved crosses between isolates pathogenic on several different grasses with isolates pathogenic on rice (Yaegashi, 1977; Yaegashi and Asaga, 1981; Valent *et al.*, 1986; Kolmer and Ellingboe, 1988; Valent *et al.*, 1991). Isolates from grasses commonly functioned as hermaphrodites whereas the isolates from rice usually functioned only as males in crosses. Most isolates from rice were of only one mating type. However, one isolate, Guy11, obtained from J.L. Notteghem, was *MAT1–2* and has been found to be sexually compatible with many isolates from rice (Leung *et al.*, 1988).

Our laboratory has developed several populations of isolates that are all pathogenic on rice and in which the mating type genes are the only restrictions on the ability to cross siblings (Chao and Ellingboe, 1991). These populations were developed from an initial cross of an isolate pathogenic on weeping lovegrass (AR4) with an isolate pathogenic on rice (CH104–3) (Kolmer and Ellingboe, 1988). Sib selection for fertility and pathogenicity on rice for several generations was as follows:

AR4 × CH104-3
↓
Nine generations of sib selection for fertility and pathogenicity on rice ↓ ↙ Guy11
↓
↓
↓
↓
crosses 70 and 76

(All siblings from cross 76 will mate if of opposite mating type; all siblings from cross 76 are pathogenic on rice.)

Isolates were obtained that were pathogenic on rice, of both mating types, and hermaphroditic by the fifth generation. Guy11 was introduced into the pedigree at the fifth generation and again in later generations of the sib selection process (Chao and Ellingboe, 1991) for three reasons.

1. It was one of the few isolates from rice that would function as both male and female in crosses between rice isolates.
2. Isolates from the fifth and sixth generations of this sib selection procedure, when crossed with Guy11, gave progenies *all* of which were pathogenic on rice (Ellingboe *et al.*, 1988). (Prior to the crosses with Guy11, some progenies were nonpathogenic on rice even though both parents were pathogenic on rice.)
3. Guy11 brought virulence to many different rice cultivars. Within the descendents of crosses 70 and 76 were isolates that were pathogenic on all cultivars but one, as shown in Table 4.1.

Table 4.1. Pathogenic isolates of rice cultivars.

Pathogen isolates	Rice cultivars		
	Katy	Leah	Lemont
70–6	+	+	+
70–14	−	+	+

+ = compatible interaction; − = incompatible interaction.

These isolates have permitted the systematic analysis of avirulence/virulence on a single rice cultivar. Progenies of cross 70 were selected that differed in their avirulence/virulence on only one cultivar (e.g. 70–14 × 70–6).

Segregation for Avirulence/Virulence on Rice Cultivar Katy

The genetic segregation for avirulence/virulence on cultivar Katy will be used as an illustration of the segregation on one cultivar. Similar results were obtained

Table 4.2. Crosses between virulent and avirulent siblings.

	No. of crosses	No. of crosses segregating	Ratio A:V
Avirulent × avirulent	26	1	3:1
Avirulent × virulent	14	0	–
Avirulent × virulent	34	34	1:3
			7:9
			1:1
			5:3
			3:1

on other cultivars. The distinction between avirulent and virulent isolates on Katy is usually clear. Isolates either give small black flecks (avirulent) or large lesions with gray centers (virulent). Therefore, the segregation for avirulence/virulence among the progenies is easily scored.

Two progenies from cross 70 were selected because they differed only in their avirulence/virulence on cultivar Katy. Isolate 70–6 was virulent on 11 rice cultivars and isolate 70–14 was avirulent on Katy but virulent on the other ten cultivars (Lau *et al.*, 1993). A cross between these two isolates (cross 76) gave 54 progenies. All 54 progenies were virulent on ten cultivars. The segregation for reaction with Katy was 26 avirulent (small flecks) and 28 virulent (large lesions with gray centers). The two classes of progenies were clearly distinguishable. The segregation ratio suggested that one locus with two alleles was the proper interpretation of the data. To test this hypothesis, crosses were made between siblings that were both avirulent on Katy, between siblings that were both virulent on Katy, and between virulent and avirulent siblings (Lau *et al.*, 1993). A summary of the results is shown in Table 4.2.

The data show that segregation is occurring at more than one locus. One cross between two avirulent siblings gave approximately 1/4 virulent progenies. A cross of two avirulent isolates that yields virulent progenies must be segregating for at least two genes. The simplest interpretation is that the two isolates contained different avirulence genes (e.g. *P1p2* × *p1P2* → *P1P2*, *P1p2*, *p1P2* and *p1p2*). In this example, the first three progenies would be avirulent and only the last would be virulent. But if segregation was occurring at two loci in cross 76, why did parent cross 70 segregate 26 avirulent:28 virulent, which is consistent with avirulence/virulence on Katy being controlled by two alleles at one locus? Crosses between siblings, one of which was virulent and the other avirulent on Katy, gave a variety of avirulent:virulent segregation ratios. From the intercrosses among the siblings, backcrossing to the parent isolates, and test crosses with isolates of known genotype that are virulent on Katy, it was possible to show that cross 76 was segregating for four loci, two avirulence genes, *P11* and *P12*, and two suppressors, *S11* and *S12* (Lau *et al.*, 1993). The expected segregation in a cross heterozygous for two avirulence genes and two corresponding suppressors is 7 avirulent:9 virulent, which is also an acceptable fit to the observed segregation of 26 avirulent:28 virulent among the progenies.

The evidence for the suppressors was based on the recovery of avirulent progenies in certain crosses between virulent parents (Ellingboe, 1992). Lawrence *et al.* (1981) had presented evidence that the recovery of avirulent progenies from a cross of two isolates of *Melampsora lini* that were virulent on the flax cultivar Williston Brown was due to the segregation at two loci rather than an example of virulence being dominant to avirulence. (*M. lini* is dikaryotic, functionally like a diploid, so dominance can be determined.) They presented evidence that avirulence on *cv.* Williston Brown was controlled by a dominant avirulence gene, but that a dominant suppressor of the avirulence allele was also segregating in the crosses. The implications from the dominance of both loci was that avirulence was an active function, and that suppression of the avirulence allele was also an active function. While suppression of the avirulence allele was not always complete (intermediate phenotypes were common) with *M. lini*, suppression of avirulence has always been complete with *M. grisea*. Intermediate phenotypes on Katy were not observed. As *M. grisea* is a haploid organism, dominance cannot be used to determine whether an allele at the *S* locus is active for the suppression or required for the expression of the corresponding avirulence gene. Following the pattern established with *Melampsora lini*, the assumption with *M. grisea* is that the avirulence allele *P* represents the active allele and that the allele *S* suppresses the expression of *P*. However, the possibility exists that the allele designated by *s* may be the active allele and that the *s* allele may be necessary for the expression (not suppression) of the avirulence specified by the corresponding *P* gene.

The interactions between the *P* and *S* genes is illustrated in the following crosses. A cross between an avirulent and a virulent isolate may give several types of segregation ratios. If the cross is segregating at both loci, a ratio of 1 avirulent : 3 virulent is observed.

P12s12 × *p12S12*
avirulent ↓ virulent

P12S12 – virulent
P12s12 – avirulent
p12S12 – virulent
p12s12 – virulent

1 avirulent : 3 virulent segregation ratio

If the cross is heterozygous only for the *P* locus, the segregation among progenies is expected to be 1 avirulent : 1 virulent on Katy.

P12s12 × *p12s12*
avirulent ↓ virulent

P12s12 – avirulent
p12s12 – virulent

1 avirulent : 1 virulent segregation ratio

If the cross is heterozygous only for the *S* locus, then the segregation among progenies is also expected to be 1 avirulent : 1 virulent on Katy.

P12s12 × *P12S12*
avirulent ↓ virulent
P12S12 – virulent
P12s12 – avirulent

1 avirulent : 1 virulent segregation ratio

The 1 avirulent : 3 virulent ratio suggests that two genes are necessary to give avirulence. The second and third crosses each suggest the segregation of an 'avirulence' gene. The segregation at the *P* locus can be seen only in the presence of the *s* allele and the segregation at the *S* locus can be seen only the presence of the *P* allele. In the second cross, the *P* locus is the avirulence gene. In the third cross, the *S* locus is the avirulence gene. Which is the locus controlling the specificity? Only one genotype in this example gives a unique, cultivar-specific phenotype, namely, avirulence, and that is *P12s12*.

The phenomenon of two interacting loci being required for the expression of cultivar-specific avirulence is not unique to cultivar Katy. Of 12 cultivar-specific avirulence genes studied by us to date (Ellingboe, 1992; unpublished findings), all 12 have a specific second locus that determines whether the avirulence will be expressed as illustrated below:

P1 – *S1*
P2 – *S2*
P3 – *S3*
P4 – *S4*
:
:
:
P12 – *S12*

For 12 'avirulence' genes examined to date there is always a second locus involved. Each second locus affects only one avirulence gene. The specificity of the interactions between *P* and *S* genes suggests that both might be necessary for the expression (not suppression) of avirulence on host lines with a corresponding *R* gene. It is possible that the products of both genes are necessary for cultivar-specific avirulence and, therefore, the distinction between *P* and *S* genes may be erroneous. The distinction of which allele at a locus is the active allele can be tested by mutational analysis as the expectation of mutation of avirulence and suppressor genes would be different.

Induced Mutations to Increased Virulence on Katy

Isolate 70–14 has been shown to have the genotype *P11P12S11s12* (Lau *et al.*, 1993). This isolate is avirulent on cultivar Katy based on *P12s12*. *P11* does not give avirulence on Katy because of the presence of *S11*. Mutations to increased virulence on Katy have been obtained following mutagenesis with nitrosoguanidine (NTG) (Lau and Ellingboe, 1993). Four mutants to increased virulence have been analyzed to date.

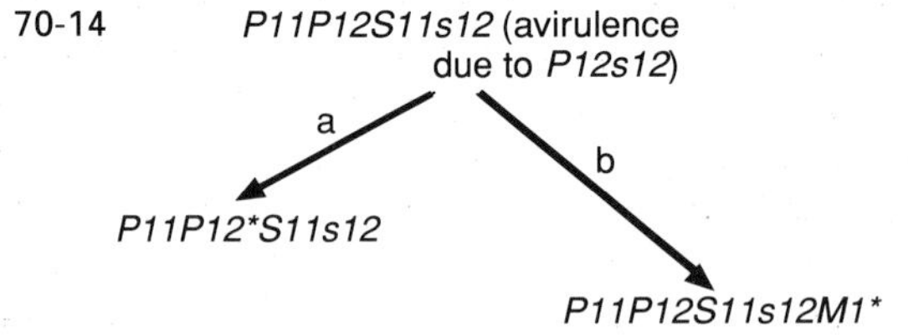

Fig. 4.1. Mutations induced in 70–14 to increased virulence on Katy. (a) Mutation at the *P12* locus; (b) mutation at a different locus (*M1*) that affects the expression of *P12*.

As stated earlier, the reactions of Katy to isolates 70–14 and 70–6 are very distinct. Isolate 70–14 produces minute flecks whereas 70–6 produces large lesions on Katy. Several mutants to increased virulence (from 70–14) gave an intermediate phenotype on Katy. The intermediate phenotypes produce lesions that range from minute flecks to large lesions with gray centers on individual leaves of Katy. The intermediate phenotypes are distinct from either parent. The mutants have been single spored several times and all subcultures give the same range in lesion types. Therefore, the intermediate phenotypes are not the result of a mixture of genotypes.

Mutations to increased virulence in 70–14 could occur by mutation of the *P12* locus. Increased virulence of 70–14 could also be obtained by mutation of genes other than *P12* that are required for the expression of *P12* (Fig. 4.1). The distinction between mutations at *P12* and at other loci is by different expectations in test crosses. If the mutation occurred at the *P12* locus, a cross of the mutant with Guy11 (*p11p12S11s12*) would be expected to yield progenies of only intermediate and virulent phenotypes in approximately a 1 : 1 ratio.

*P12*s12* × *p12s12* (Guy11)

↓

		Expected ratio
*P12*s12*	intermediate virulence	1
p12s12	virulent	1

Two mutants to intermediate virulence on Katy, when crossed with Guy11, gave only two types of progenies and in approximately equal frequencies. The conclusion drawn was that the two independent mutations occurred at or near the *P12* locus.

If the mutation to increased virulence on Katy occurred at a locus other than *P12*, and at a locus that is not tightly linked to *P12*, a cross of the mutant with Guy11 might be expected to give three types of progenies, including progenies that are avirulent on Katy.

*P12s12M1** × *p12s12* (Guy11)

↓

		Expected ratio
P12s12M1	avirulent	1
*P12s12M1**	intermediate	1
p12s12M1	virulent	1
*p12s12M1**	virulent	1

Two of the mutants to intermediate virulence, when crossed with Guy11, gave progenies that were avirulent, intermediate in virulence, and virulent on Katy, in a ratio of approximately 1:1:2 (Lau and Ellingboe, 1993). The recovery of avirulent progenies from a cross of an isolate with intermediate virulence with virulent isolate Guy11 indicates that loci other than *P12* and *S12* are required for the expression of cultivar-specific avirulence on Katy controlled by *P12* and *S12*. Therefore, at least three genes are necessary for the cultivar-specific avirulence tentatively ascribed to *P12*. None of the mutations affected the expression of avirulence/virulence on other cultivars.

If the two mutations do not map to the *P12* locus, do they map to the same locus? If the mutations occurred at the same locus (*M1*), crosses between isolates that contain these mutations should give progenies that are all intermediate in virulence on Katy.

P12s12 M1* × *P12s12M1***

↓

*P12s12M1** – intermediate
*P12s12M1*** – intermediate

If the mutations were at two loci, *M1* and *M2*, crosses between isolates of intermediate virulence that contain different mutations should yield some avirulent progenies.

*P12s12M1*M2* × *P12s12M1M2**

↓

P12s12M1M2	avirulent
*P12s12M1*M2*	intermediate
*P12s12M1M2**	intermediate
*P12s12M1*M2**	unique? Yes, virulent

The crosses revealed that the two mutations to intermediate virulence, that did not map to *P12*, were at different loci, tentatively designated *M1* and *M2* (A. Ellingboe, unpublished findings). Of greater interest is that isolates were recovered that gave full virulence on Katy. These progenies are postulated to be the double mutants *M1*M2**. These observations suggest that there is an interaction between the *M1* and *M2* genes to affect the expression of cultivar-specific avirulence on Katy.

The mutants with *M1** or *M2** have very similar phenotypes on Katy. Each are distinguishable from the parent isolate 70–14 from which the mutant was obtained, and the virulent isolate 70–6. The analyses of the intercrosses of *M1** and *M2** recorded above were made in crosses that were homozygous for *P11* and *S11*. A cross of a mutant of 70–14 (genotype *P11P12S11s12M1**) with 70–6 (genotype *p11p12s11S12M1*) should segregate for five loci. It is possible to predict the expected phenotypes among the progenies of this cross, and their expected ratios, if the *M1* locus affects only *P12* or if it also affects *P11*, *S11* and/or *S12*. The results obtained to date suggest, but do not prove, that *M1** affects the expression of only *P12*. The results with *M2** , however, suggests that *M2** affects the expression of both *P11* and *P12*, and possibly *S11* (Fig. 4.2). It is clear that *M1** and *M2** have different effects on the segregation in crosses with 70–6, but

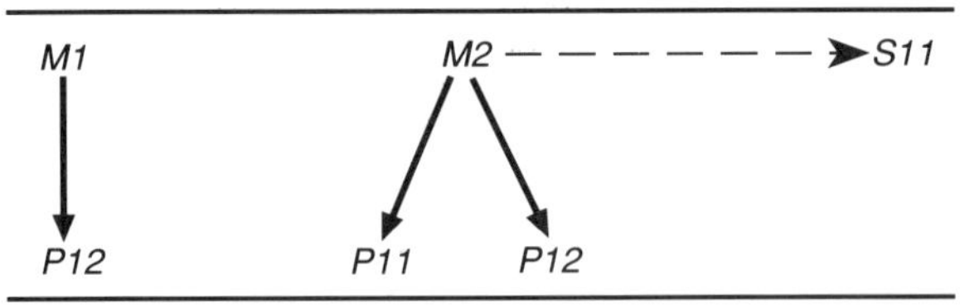

Fig. 4.2. The apparent interactions of *M1* and *M2* with *P12, P11* and *S11*. *M1* affects the expression of *P12* but not *P11* whereas *M2* affects the expression of *P12, P11* and possibly *S11*.

the specificity of each *M* mutation to different *P* and *S* alleles is not clear. Additional test crosses will be necessary to establish with some degree of certainty the interaction between *M*, *P* and *S* genes.

Possible Interpretations of the Role of *P*, *S* and *M* Genes in Cultivar-specific Avirulence

The assumptions in this study about functions of *P* and *S* genes have been that cultivar-specific avirulence/virulence in *M. grisea* is controlled by genes that function similarly to the genes in *Melampsora lini*. In *Melampsora lini*, avirulence on a particular host line is dominant to virulence. Suppressors of dominant avirulence genes are also dominant to lack of suppression. One suppressor in *Melampsora lini* appears to suppress avirulence controlled by one avirulence gene. The results also suggest that a second suppressor, possibly allelic to the first, suppresses avirulence controlled by four different avirulence genes (Lawrence *et al.*, 1981). The dominance relationships would suggest that avirulence is an active function and that suppression of avirulence is an active function. Following this pattern, the assumption with *M. grisea* is that the avirulence allele represents the active function and, thus, the *P* designation, and that suppression of *P* is an active function by the gene designated *S*. The determination that two of the mutations in 70–14 to increased virulence on Katy mapped to *P12* is consistent with *P12* having an active function for avirulence. The observation that for each gene designated as an avirulence gene there is a second locus (designated *S*) (Ellingboe, 1992) suggests that the *S* allele could function either as a suppressor of avirulence or that the *s* allele could be required for the expression of avirulence. The mutual interdependence of the *P* and *S* genes suggests that they may be indistinguishable. Both loci may be required for cultivar-specific avirulence. If suppression is the active function of the *S* locus, then a mutation of the *S* locus in the presence of the corresponding *P* gene [*PxSx* (virulent) → *PxSx** (avirulent)] should lead to cultivar-specific decreases in virulence. Mutations to cultivar-specific decreases in virulence have been obtained in *M. grisea*, but have not been mapped. The difficulty of mapping and distinguishing *P* and *S* genes based on segregation of avirulent and virulent progenies was illustrated by the cross of two avirulent progenies (76–25 × 76–39) from the cross 70–14 × 70–6 that yielded virulent progenies (Lau *et al.*, 1993). The cross was homozygous for two *P* genes

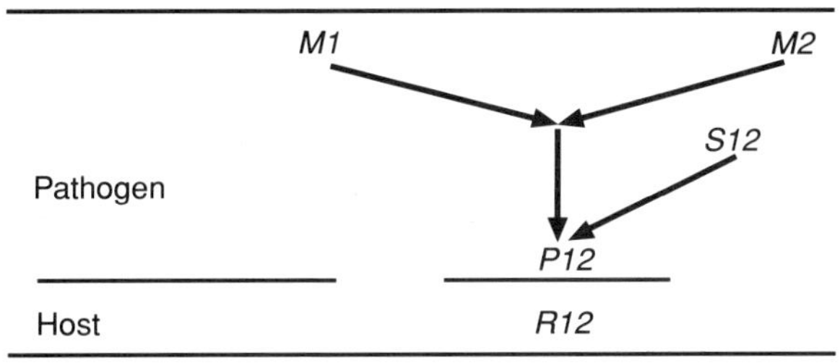

Fig. 4.3. The interactions between *P12*, *S12*, *M1* and *M2* if it is assumed that the *P12* locus determines the specificity of the interactions with Katy and the other three genes determine whether the *P12* gene is expressed.

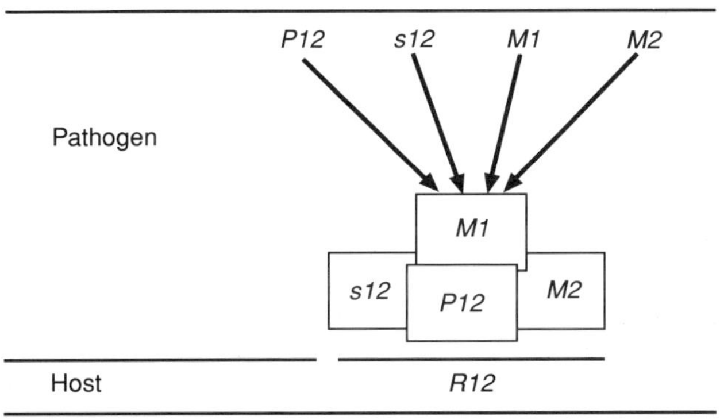

Fig. 4.4. The interactions between *P12*, *s12*, *M1* and *M2* if the products of all four genes are required for the presentation of the *P12* specificity to the host.

but segregating for two *S* genes (*P11P12S11s12* × *P11P12s11S12*). The segregation observed was 3 avirulent: 1 virulent, the same ratio expected for the segregation of two avirulence genes.

One model for the interaction of the pathogen genes is presented in Fig. 4.3. Segregation in crosses between rice cultivars that are resistant (Katy) and susceptible (Lemont), respectively, to 70–14 suggested that a single gene, *R12*, was segregating for resistance to isolate 70–14 of genotype *P11P12S11s12M1M2*. The locus designated *P12* is considered to determine the specificity that is corresponding to the host gene *R12* in Katy. The *S* locus is considered to determine whether *P12* is expressed or suppressed. The M loci, which interact, also affect the expression of *P12*. An alternative hypothesis is that the products of all four genes (*P12*, *S12*, *M1* and *M2*) are necessary for the presentation of the *P12* specificity in the interactions with plants containing *R12* (Fig. 4.4).

Which gene confers specificity is difficult to determine. In most studies the gene that segregates in a cross to give either virulent or avirulent progenies (terms that are relative to one another) on a given host cultivar is considered an avirulence gene. Consider the crosses presented in Table 4.3.

In each of these crosses the first listed parent is less virulent than the second. Relative to one another, the first is avirulent and the second is virulent. In cross 1

Table 4.3. Crosses showing 'avirulence' loci.

Cross	Ratio of avirulence:virulence	Locus showing segregation
1 *P12s12M1M2* × *p12s12M1M2*	1:1	*P12*
2 *P12s12M1M2* × *p12S12M1M2*	1:1	*S12*
3 *P12s12M1M2* × *P12s12M1*M2*	1:1	*M1*
4 *P12s12M1M2* × *P12s12M1M2**	1:1	*M2*

the *P12* locus would be identified as the 'avirulence' locus. In cross 2, the *S12* locus would be identified as the 'avirulence' locus. In crosses 3 and 4, the *M1* and *M2* loci, respectively, would be identified as the 'avirulence' loci. If there are batteries of genes, all of which are required for the expression of cultivar-specific avirulence, then it is important to distinguish which plays the principal role in determining the specificity. In Figs 4.3 and 4.4, the assumption is that the principal role in the specificity for plants with *R12* is determined by *P12*, and that the products of *s12*, *M1* and *M2* are necessary for conferring the specificity of *P12* to the plant with *R12*. The tentative conclusion that *P12* contains the specificity for determining avirulence on cultivar Katy is based on two observations. One is that it was segregating in crosses between wild type isolates of *M. grisea*. A second is that two of the mutants to increased virulence were at or near the *P12* locus. Neither argument is particularly compelling.

The precise mechanisms of interaction of the *P*, *S*, *M1* and *M2* loci are of great interest. The fact that the mechanisms of interaction remain to be uncovered does not negate the observations that at least four genes are involved in the expression of a single specificity, namely avirulence on Katy. Each of the four genes could, and would, be classified as an avirulence gene if segregation was occurring at only one of the four loci in a segregating generation. This observation raises the question of how many other reports of the segregation, and cloning, of 'avirulence' genes were of the genes that determine specificity, or of genes that regulate the expression of the genes that determine the specificity.

Implications for the Basic Biology of Host–Pathogen Interactions

Segregation for avirulence/virulence on a host line does not necessarily mean that an avirulence gene has been identified. Segregation could be at a locus that is involved in the expression of the gene for cultivar-specific avirulence.

Mutation to increased virulence on a host line does not necessarily mean the mutation is at the locus that determines cultivar-specific avirulence. The mutation could be at the locus that regulates or interacts with the gene for cultivar-specific avirulence.

Complementation of a mutant of a gene for avirulence (or allele for virulence) by genetic transformation does not prove it is the gene for specificity.

Complementation of *P*, *S*, *M1* and *M2* would all give the same result, namely, a change in the degree of avirulence/virulence on a given cultivar.

Implications for Breeding for Disease Resistance in Rice

There are several ways to increase virulence. Mutations at any one of several loci can give increased virulence on a host cultivar. Two of the four mutations described herein were at the *P12* locus, but the other two mutations were each at a unique locus. The cross of rice cultivars Katy × Lemont gave 3 resistant:1 susceptible F_2 progeny when tested with isolate 70–14 (C. Chao and A. Ellingboe, unpublished findings). Though only one gene in Katy conferred resistance to isolate 70–14, mutations in at least four genes in *M. grisea* can lead to increased virulence on Katy. It is easy to see why a cultivar with a single gene for resistance to one or more of the strains of a pathogen provides a tenuous, and usually temporary, resistance.

Analysis of a new race with increased virulence on a host line with one effective gene for resistance will almost always look like a one gene-for-one gene interaction. Whether the mutation to increased virulence occurred at the locus for determining the specificity, or at a locus that affects the expression of the gene that determines the specificity, may not be clear.

Practical Uses of Laboratory Strains of *M. grisea* with Sexual Competence that are Pathogenic on Rice

Sib selection for fertility and pathogenicity on rice has permitted the development of isolates whereby the only major limitation on crossing is the mating type locus. Crosses in which simple segregations of avirulence/virulence were observed on specific cultivars (e.g. cross 76) were subjected to intensive test crossing. The extensive test crossing has shown that instead of a simple one locus with two alleles controlling avirulence/virulence on Katy, for example, avirulence/virulence was controlled by four loci with definite patterns of interactions of *P* and *S* genes (Lau *et al.*, 1993). Without the highly fertile strains of *M. grisea* there would have been an erroneous interpretation of the inheritance of avirulence/virulence on cultivar Katy.

The highly fertile isolates that are virulent on many cultivars have also been crossed successfully with field isolates. For example, an isolate from Texas has been crossed with lab strains and shown to contain at least four genes that control avirulence on Katy, three genes for avirulence on cultivar Leah, and three genes that control avirulence/virulence on cultivars Lemont (C. Chao and A. Ellingboe, unpublished findings). Whereas the initial cross gave avirulent:virulent ratios of 33:2, 28:7, 29:6, on Katy, Leah, and Lemont, respectively, backcrosses of the avirulent progenies to 70–6 in each of several generations has led to isolates that are avirulent on one cultivar but virulent on the other two cultivars. Backcrosses of avirulent progenies to virulent lab strains has permitted the establishment of avirulent isolates that show single-gene segregation (e.g. ratios of 1 avirulent:1 virulent) when crossed with virulent siblings. Intercrosses among these presumptively 'single gene' isolates have established which genes are present in each isolate (C. Chao and A. Ellingboe, unpublished findings).

Crosses have also been made between our lab isolates and isolates from the fields of Arkansas. The detailed analyses of the genes controlling avirulence/ virulence from the Arkansas isolates can be compared with the analyses of genes controlling avirulence/virulence in the isolates from Texas. Does the isolate from Arkansas have the same genes controlling avirulence/virulence on Katy as the isolate from Texas? These comparisons can now be made. The isolates with single but different genes controlling avirulence/virulence can also be used to determine if the same gene or genes in Katy give resistance to the Texas and Arkansas isolates of *M. grisea*.

The laboratory strains with high fertility and virulence on many rice cultivars provide an opportunity to determine the genes for avirulence/virulence of most any field isolate on many different rice cultivars. The systematic genetic analysis of field isolates should permit comparisons between field isolates for genes in common and for genes that are different.

The establishment of *M. grisea* lines, each with single and different genes governing avirulence on a given rice cultivar, e.g. Katy, should permit a systematic analysis of the genes for resistance in rice.

Conclusions

1. Segregation ratios of avirulent:virulent progenies in crosses between isolates of *M. grisea* may be very misleading.
2. Cultivar-specific avirulence has been found to involve two loci tentatively identified as an avirulence gene and a corresponding suppressor gene in the pathogen in 12 examples to date.
3. Mutations to increased virulence on Katy have permitted the identification of two additional loci that are necessary for the expression of the *P12* locus for avirulence on Katy. These mutations do not affect virulence on other cultivars.
4. There are several genes that control avirulence/virulence on each of several cultivars tested.
5. Resistance in a cultivar of rice (e.g. Katy) can be negated by mutations at many different loci in *M. grisea*.
6. The interpretation of the gene-for-gene hypothesis as meaning one gene for avirulence in the pathogen with one corresponding gene for resistance in the host is not consistent with our data on genetic analysis of the *M. grisea*–rice interactions.
7. The implications of these results for breeding for blast resistance further emphasizes the versatility of the pathogen, the need to anticipate the options available to the pathogen, and the need to systematically analyze the genes in the host that affect interactions with the pathogen.

References

Chao, C.C.T. and Ellingboe, A.H. (1991) Selection for mating competence in *Magnaporthe grisea* pathogenic on rice. *Canadian Journal of Botany* 69, 2130–2134.

Ellingboe, A.H. (1992) Segregation of avirulence/virulence on three rice cultivars in sixteen crosses of *Magnaporthe grisea. Phytopathology* 82, 597–601.

Ellingboe, A.H., Wu, B.-C. and Robertson, W. (1988) Fertility and pathogenicity to rice in *Magnaporthe grisea*. In: Palacios, R. and Verma, D.P.S. (eds), *Molecular Genetics of Plant-Microbe Interactions*. APS Press, St. Paul, Minnesota, pp. 235–240.

Kolmer, J.A. and Ellingboe, A.H. (1988) Genetic relationships between fertility and pathogenicity and virulence to rice in *Magnaporthe grisea. Canadian Journal of Botany* 66, 891–897.

Lau, G.W. and Ellingboe, A.H. (1993) Genetic analysis of mutations to increased virulence in *Magnaporthe grisea. Phytopathology* 83, 1093–1096.

Lau, G.W., Chao, C.T. and Ellingboe, A.H. (1993) Interactions of genes controlling avirulence/virulence of *Magnaporthe grisea* on rice cultivars Katy. *Phytopathology* 83, 375–382.

Lawrence, G.J., Mayo, G.M.E. and Shepherd, K.W. (1981) Interactions between genes controlling pathogenicity in the flax rust fungus. *Phytopathology* 71, 12–19.

Leung, H., Borromeo, E.S., Bernardo, M.A. and Notteghem, J.L. (1988) Genetic analysis of virulence in the rice blast fungus *Magnaporthe grisea. Phytopathology* 78, 1227–1233.

Valent, B., Crawford, M.J., Weaver, G.G. and Chumley, F.G. (1986) Genetic studies of fertility and pathogenicity in *Magnaporthe grisea* (*Pyricularia oryzae*). *Iowa State Journal of Research* 60, 569–594.

Valent, B., Farrell, L. and Chumley, F.G. (1991) *Magnaporthe grisea* genes for pathogenicity and virulence identified through a series of backcrosses. *Genetics* 127, 87–101.

Yaegashi, H. (1977) On sexuality of blast fungi, *Pyricularia* spp. *Annals of the Phytopathological Society of Japan* 43, 432–439.

Yaegashi, H. and Asaga, K. (1981) Further studies on the inheritance of pathogenicity in crosses of *Pyricularia oryzae* with *Pyricularia* spp. from fingermillet. *Annals of the Phytopathological Society of Japan* 47, 677–679.

5 Genomic Structure and Variability in *Pyricularia grisea*

V. Shull and J.E. Hamer

Department of Biological Sciences, Purdue University, West Lafayette, Indiana 47907, USA

Introduction

Historically, *Pyricularia oryzae* (isolates pathogenic on the genus *Oryza*), and *Pyricularia grisea* (isolates pathogenic on monocots other than the genus *Oryza*), were considered distinct. However, the production of the sexual stage on defined media (termed *Magnaporthe grisea*; Hebert, 1971), and the subsequent crossing of rice pathogens with goosegrass pathogens (Yaegashi, 1977), as well as the cross infectivity of *P. oryzae* and *P. grisea* isolates (many of these isolates infect barley; Mackill and Bonman, 1986), has demonstrated that this classification scheme is ambiguous. A literature review has established *Pyricularia grisea* as the correct name for the anamorphic state producing both rice blast disease and gray leaf spot disease on other monocots (Rossman *et al.*, 1990).

In addition to host-specificity, early studies describing the extent of variation in this fungus relied on physical characteristics such as spore and appressorial morphologies, and germ tube branching (Suzuki, 1963). Many of these phenotypes are variable, and have contributed to the view that *P. grisea* is a genetically unstable species (Ou, 1985). More recent biochemical studies describing isozyme polymorphisms (Leung and Williams, 1986) and double-stranded RNA content (Hunst *et al.*, 1986) detected some variation, although it was not meaningfully correlated with pathotype. Current studies rely on advances in techniques used to investigate the molecular genetics of filamentous fungi (Timberlake and Marshall, 1989), and have focused on the nuclear and mitochondrial genomes. These molecular genetic studies demonstrate that although *P. grisea* is composed of groups of isolates that are genetically diverse, individual isolates are not extraordinarily genetically unstable (Levy *et al.*, 1991, 1993).

After the exhaustive effort to produce the sexual stage in the laboratory (Hebert, 1971; Yaegashi and Nishihara, 1976; Yaegashi, 1977; Kato and Yamaguchi, 1982; Valent *et al.*, 1986; Valent and Chumley, 1987; Valent *et al.*, 1991), it is hard to accept that the telomorph is sufficiently prevalent in the field that meiotic recombination plays a major role in the production of genetic variability. In this review, we propose that *P. grisea* currently exists as a number of genetically distinct, asexually reproducing populations. These distinct populations occur because of strong selection to maintain host specificity (Koch and Köhler, 1990), and perhaps geographic isolation. Four types of molecular genetic approaches (including mitochondrial restriction fragment length polymorphism (RFLP) analysis, ribosomal cistron RFLP analysis, the characterization of restriction site polymorphisms associated with single copy genomic regions, and the characterization of restriction site polymorphisms associated with transposable element sequences) support our contention.

Restriction sites within the mitochondrial genome, within the ribosomal DNA cistron, and associated with single copy regions of the nuclear genome are somewhat variable. These analyses broadly divide *P. grisea* into at least eight genetically distinct haplotypes (haploid RFLP types), and have the potential to describe the timing of host-specific evolution. For example, these haplotypic studies indicate that certain host-specific forms of the pathogen may have evolved more than once. This conclusion is supported by the observation that certain haplotypes have differentially acquired or selectively amplified repetitive DNA sequences.

Restriction sites associated with transposable element sequences are highly variable when compared to other regions of the nuclear and mitochondrial genomes of rice pathogen field isolates. These polymorphism profiles may be used to subdivide the rice pathogen haplotype into groups of isolates, with each group containing a distinct repetitive DNA genotype (Levy *et al.*, 1991, 1993). Intermediate genotypes are rarely described, strongly suggesting that sexuality and parasexuality do not currently occur in the field (Tibayrenc *et al.*, 1991), or that if they occur, viable progeny are rarely produced.

We begin this chapter by reviewing recent work describing the sequence and organization of the *P. grisea* nuclear and mitochondrial genomes, although we do not cover the construction of genetic maps (Leong *et al.*, Chapter 6, this volume). We continue by reviewing the range of variation that has been described in these genomes. The extent of this variation suggests that *P. grisea* is a complex species composed of groups of isolates with distinct genotypes. In addition to our laboratory, other laboratories studying genomic variation in *P. grisea* include: Dr M.-H. Lebrun, Université de Paris-Sud; Dr S.A. Leong, University of Wisconsin; Dr H. Leung, Washington State University; Dr R.J. Nelson, International Rice Research Institute; Dr M.J. Orbach, University of Arizona; and Dr B. Valent, E.I. Du Pont de Nemours and Co. We are grateful to these laboratories for providing us with unpublished data to include in this chapter.

Genomic Structure

Protein coding genes

Two genes from *P. grisea* have been cloned, and the sequence published. Analysis of these genes reveals a structure rather typical of *Ascomycetes*, including a strong

G/C bias in codon usage (Gurr *et al.*, 1988). Sequence analysis of a genetic locus encoding the cutinase enzyme (*CUT1*) shows that 195 out of 228 amino acid codons end in G or C (Sweigard *et al.*, 1992). The promoter region of *CUT1* contains a consensus TATAA element at position −117; however, a consensus CCAAT element is not present. A 3′-polyadenylation signal is also absent. *CUT1* is interrupted by two introns. Each intron contains the conserved 5′ donor (GT), and 3′ acceptor (AG) sequences required for correct splicing. Compared to the cutinase genes from other fungi, the second intron of *CUT1* is unique (Sweigard *et al.*, 1992). This intron interrupts a serine codon. This serine residue is contained within the active site of the mature enzyme.

The promoter region of *MPG1*, the first pathogenicity gene cloned from *P. grisea* (Talbot *et al.*, 1993b), is similar to *CUT1* except that it contains two pyrimidine-rich tracts at positions −18 and −40 that are homologous to functional promoter elements in *Aspergillus nidulans* (Hamer and Timberlake, 1987). *MPG1* also contains a TATAA element at position −52, and two mRNA transcription initiation sites have been mapped by primer extension. A consensus polyadenylation signal (AATAAA) is also found in the 3′ untranslated region of the gene. Finally, *MPG1* contains a 77 base pair (bp) intron with consensus splice signals as described above.

Ribosomal DNA (rDNA) genes

The 18*S*, 5.8*S*, and 28*S* ribosomal RNA genes are organized in a transcriptional unit of 8 kb (Lebrun *et al.*, 1990) that is tandemly repeated. The rDNA locus has been genetically mapped to a single position on a chromosome (Leong *et al.*, Chapter 6, this volume), and suppressed meiotic recombination has also been demonstrated at this locus (Romao and Hamer, 1992), consistent with results in *Saccharomyces cerevisiae* (Petes, 1979).

The 5*S* genes of *P. grisea* have also been studied to a limited extent (V. Shull and J.E. Hamer, unpublished data). Southern blot analysis using polymerase chain reaction (PCR) amplified probes, indicates that the 5*S* genes are dispersed throughout the genome. This result is consistent with studies in *Neurospora crassa* (Free, 1979).

Mitochondrial DNA (mtDNA) genome

Size estimates of the *P. grisea* mitochondrial genome range from 32–39 kb and are based on the analysis of isolates from the Philippines (Borromeo *et al.*, 1993). A restriction map of the mitochondrial genome has not yet been constructed. This map would be a useful reference to begin the characterization of specific RFLPs occurring in this genome (see p. 70).

Karyotype

Early light microscopy studies of mitotic nuclei reported a significant variation in the number of *P. grisea* chromosomes, and various aneuploidies were commonly observed (Suzuki, 1963). The discovery of the sexual stage prompted a study of nuclear division and chromosome behavior during ascus development (Leung and Williams, 1987). Rice pathogens and weeping lovegrass pathogens were crossed for at least nine generations to produce a highly fertile mating population with over 90% ascospore viability. Among the 20 cells examined at pachytene, the chromosome number was always six. This result was confirmed at diakinesis. Further, these authors reported that asynchronous disjunction at both meiotic and mitotic anaphase produces lagging chromosomes, giving the appearance of variable chromosome numbers.

Current karyotypic studies have relied on pulsed-field gel electrophoresis (PFGE) (Hamer *et al.*, 1989; Skinner *et al.*, 1991; Talbot *et al.*, 1993a; Skinner *et al.*, 1993). In general strains analyzed by this method contain six or seven chromosomes varying in size from 3 to 12 Mb. (The *P. grisea* genome has been estimated to be approximately 38 Mb in size; Hamer *et al.*, 1989). Chromosome length polymorphisms between isolates are frequently detected by PFGE, indicating that the overall genome size may be variable, and that chromosomes that have the same electrophoretic mobilities may not necessarily be homologs (Hamer *et al.*, 1989; Talbot *et al.*, 1993a). Further, studies using single copy hybridization probes have shown that chromosome length polymorphisms may arise from large deletions and translocations (Skinner *et al.*, 1993; M.J. Orbach, F.G. Chumley, and B. Valent, unpublished data).

Rice pathogens (particularly isolates collected in the US) contain up to four 'minichromosomes' that range in size from 0.47 to 2.2 Mb, and 'minichromosome' length polymorphisms are also frequently identified (Talbot *et al.*, 1993a). These 'minichromosomes' may be supernumerary because they display aberrant meiotic segregation behavior (Valent and Chumley, 1991). They also contain genomic repeated sequences suggesting that they may be derived from the larger chromosomes (Talbot *et al.*, 1993a).

The chromosomal and 'minichromosomal' electrophoretic karyotypes of rice pathogen field isolates are variable (i.e. groups of isolates that comprise a single pathotype, and groups of isolates that contain a distinct repetitive DNA genotype, often have different electrophoretic karyotypes (Talbot *et al.*, 1993a). Electrophoretic karyotypes can also change upon prolonged serial transfer (Talbot *et al.*, 1993a). Karyotypic variation in asexually reproducing phytopathogenic fungi appears to be common (Mills and McCluskey, 1990), and it has been suggested that Imperfect fungi are more likely than Perfect fungi to contain chromosome length polymorphisms because recurrent meioses select against chromosomal aberrations (Kistler and Miao, 1992).

Variation within the Ribosomal DNA Cistron, the Mitochondrial Genome, and Associated with Single Copy Regions of the Nuclear Genome

The collective host range of *P. grisea* is extensive; however, individual isolates are normally host specific. Although this suggests that *P. grisea* is organized into a number of genetically distinct forms, the low fertility of many field isolates has made it difficult to study the genetic basis of host-specificity. In a few genetic crosses, host specificity has been shown to segregate as a single gene (Yaegashi, 1978; Yaegashi and Asaga, 1981; Valent *et al.*, 1986; Valent and Chumley, 1987), or as a complex multi-locus trait (Valent and Chumley, 1989).

In addition to exhibiting host-specificity, most rice pathogens have restricted cultivar compatibility; however, these isolates are also female sterile, and cannot be crossed to determine the genetic basis of pathotype. The recent discovery of the hermaphroditic rice pathogen field isolate Guy11 (Leung *et al.*, 1988), and the development of the hermaphroditic weeping lovegrass (*Eragrostis curvula*) pathogen 4091-5-8 (Valent *et al.*, 1986), have made genetic crosses possible. In genetic crosses to either Guy11, 4091-5-8, or strains derived from these two strains, avirulence towards different rice cultivars has been shown to segregate as a single gene (Yaegashi, 1978; Valent *et al.*, 1991; Silué *et al.*, 1992a, 1992b; S.A. Leong *et al.*, Chapter 6, this volume), as two unlinked genes (Silué *et al.*, 1992b), or as a complex multi-locus trait involving the interaction of suppressor genes with avirulence genes (Ellingboe *et al.*, 1990; Lau *et al.*, 1993).

Molecular genetic techniques provide a complementary approach to describing the degree of relatedness among strains that differ by host specificity or avirulence genes (Michelmore and Hulbert, 1987). With these techniques, large numbers of isolates may be examined, allowing researchers to quantify the degree of genetic variation within and between host-specific forms.

RFLPs within the ribosomal DNA cistron

The rDNA cistron contains two regions where RFLPs are commonly found. The first is composed of two fully transcribed regions and is called the internal transcribed spacer (ITS1 and ITS2). The second is partially transcribed and is referred to as the intergenic spacer (IGS). Length polymorphisms within the IGS are common in *Ascomycetes* (Garber *et al.*, 1988), whereas restriction site polymorphisms in the ITS regions are less common, and are normally conserved at the species level (Bruns *et al.*, 1991). Initial studies using 13 restriction enzymes and the Guy11 rDNA cistron as a probe, failed to detect polymorphisms in the ITS regions of three rice pathogens from diverse geographies (Lebrun *et al.*, 1990). However, a *Dra*1 RFLP was identified that distinguished these rice pathogens from *Eleusine coracana* pathogens, suggesting that rice pathogens may possess a unique ITS rDNA haplotype. (This conclusion will be strengthened below when more isolates are considered.) Three classes of *Eco*R1 length polymorphisms were also identified in the IGS region of these rice pathogens. It may be meaningful to subdivide rice pathogen field isolates into groups with distinct IGS rDNA haplotypes.

In this same study, two domains of the 26*S* rRNA gene were sequenced in rice pathogen Guy11 and two pathogens of *Eleusine* spp., all from different geographic locations (Lebrun *et al.*, 1990). The rDNA Domain One sequence is usually conserved at the species level, but is often useful for comparing genera (Gaudet *et al.*, 1989). The Domain Two sequence is not normally conserved at the species level. The Domain One and Two sequences are identical in these three *P. grisea* pathogens, as would be expected for isolates comprising a single species.

In another study, four *Eco*RI RFLP profiles were identified in 54 Philippine isolates (18 host-specific forms; Borromeo *et al.*, 1993). Three of the four RFLP profiles are each uniquely represented by an isolate that infects *Cyperus* spp. (see pp. 71, 76). The fourth RFLP profile is shared by the remaining 15 host-specific forms, including rice pathogens. This sample set (which only includes Philippine isolates), does not differentiate between rice pathogens and other host-specific forms of *P. grisea*.

A second study used 84 isolates comprising 29 host-specific forms (M.-H. Lebrun *et al.*, unpublished data; see Table 5.1). The isolates analyzed in this study included pathogens of *Oryza sativa* (rice), *Eleusine* spp. (goosegrasses and finger millets), *Triticum aestivum* (wheat), *Digitaria* spp. (crabgrasses), and *Pennisetum* spp. (pearl millets); they were from broadly dispersed geographies (including Africa, Asia, North America, and South America), and were collected over a 17-year period. Restriction enzyme analysis of the rDNA cistron identified six haplotypes: 1, 2, 3, 4, 5, and 6. Surprisingly, strains pathogenic on *Eleusine* and *Pennisetum* spp. contain two different rDNA haplotypes (haplotypes 2 and 3, and 5 and 6, respectively), suggesting that these host specificities may have evolved at least twice, or that at least two distinct fungal populations developed after pathogenicity to these hosts evolved. Haplotype 2 also contains isolates that are pathogenic toward wheat, suggesting that wheat and certain *Eleusine* spp. pathogens may be derived from a common ancestor that was pathogenic on both hosts (or was pathogenic on a common ancestor to these hosts). Haplotype 1 is only composed of rice pathogens, in stark contrast to Borromeo *et al.* (1993) (see below). These results demonstrate that *P. grisea* is an exceedingly complex species; its populations have diverged to produce haplotypes that show a level of diversity more commonly found among species than among members of a single species (Bruns *et al.*, 1991).

RFLPs within the mitochondrial genome

RFLPs in the mitochondrial genome also describe the phylogenetic relatedness of *P. grisea* host-specific forms. Initially the study by Borromeo *et al.* (1993), reported on 18 Philippine isolates (17 host-specific forms), which contained six mtDNA haplotypes as judged by digestion with five different restriction enzymes. Each RFLP haplotype was distinct (no two haplotypes had more than 10% of their bands in common). In addition, excluding strains isolated from *Cyperus rotundus*, multiple isolates from a single host always had identical mtDNA haplotypes. Although this study was expanded to include 57 other Philippine isolates (including 20 rice pathogens), additional mtDNA haplotypes were not identified. The rice-infecting isolates analyzed in this study comprise a single haplotype that is shared with ten other host-specificities (akin to the rDNA haplotype results from this study). Of the six haplotypes identified, three were uniquely represented by an isolate that infects *Cyperus* spp. Two repetitive DNA probes originally cloned from rice pathogen field isolates failed to hybridize to the nuclear genomes of these three strains, suggesting that they may not belong to the *P. grisea* species. In addition, two of these three strains have an unusually large mitochondrial genome size of at least 55 kb. This further indicates that these *Cyperus* spp. pathogens do not belong to *P. grisea*.

A more diverse study (containing 84 isolates and 29 host-specific forms; see p. 70) has identified four major mtDNA haplotypes: I, II, III, and IV (A.R. Kubelik, J.A. Sweigard, F.G. Chumley, and B. Valent, unpublished data; see Table 5.1). Haplotype I isolates may differ by at least one polymorphic restriction fragment, and have therefore been subdivided into four groups: Ia, Ib, Ic, and Ie. By the same criterion, haplotype II has been subdivided into two groups: IIa, and IIb; overall this sample set contains eight mtDNA haplotypes. Isolates that infect *Digitaria* and *Pennisetum* spp. comprise mitochondrial DNA haplotypes II, III, and IV. These isolates lack restriction fragments that are found in haplotype I, and in some cases fail to hybridize to haplotype-I-specific probes (A.R. Kubelik, J.A. Sweigard, F.G. Chumley, and B. Valent, unpublished data). These results indicate that pathogens of *Digitaria* and *Pennisetum* spp. are very distinct from many host-specific forms of *P. grisea*, including rice pathogens.

RFLPs associated with single copy regions of the nuclear genome

Early studies of nine rice pathogens, using six restriction enzymes and five single copy probes cloned from rice pathogen Guy11, detected only eight polymorphic restriction fragments (out of 102 scored bands; Lebrun *et al.*, 1990). Therefore these isolates could not be organized into single copy nuclear DNA haplotypes. However, when Guy11 was compared to three isolates pathogenic on *Eleusine coracana*, *Triticum aestivum*, and *Zea mays*, each of the five probes detected some degree of polymorphism (overall 37% of the bands were polymorphic), demonstrating that the four host specificities may be genetically distinct.

Eighty additional isolates have now been analyzed, and Table 5.1 shows that *P. grisea* contains at least eight single copy nuclear DNA haplotypes: A, B, C, D, E, F, and G, and an eighth group consisting of pathogens of *Pennisetum* spp. that hybridize only weakly to a number of single copy probes including the *CUT1* gene, the *MPG1* gene, and the *ILV1* gene (acetolactate synthase; V. Shull and J.E. Hamer, unpublished data). This result was surprising because pathogens of *Pennisetum* spp. are clearly related to other *P. grisea* isolates because they contain diagnostic repetitive DNA sequences.

The rDNA, mtDNA, and single copy nuclear DNA haplotype data suggest that *P. grisea* is composed of at least eight composite haplotypes: a, b, c, d, e, f, g, and h. It should be noted that we have based our overall conclusions on a sample set containing 84 isolates. These isolates are from broadly dispersed geographies collected over a 17-year period and represent 29 host specificities. In

Table 5.1. *Pyricularia grisea* RFLP haplotypes. For each haplotype, at least seven isolates were examined by each of the three RFLP analysis methods. The rDNA haplotypes were generated using four restriction enzymes: *Eco*RI, *SSp*I, *Sac*II, and *Pst*I. A total of 13–17 fragments were analysed depending on the isolate (M.-H. Lebrun, unpublished data). The mtDNA haplotypes were generated using seven restriction enzymes, each having a 4 bp recognition sequence, and approximately 50 fragments were analyzed depending on the isolate (A.R. Kubelik, J.A. Sweigard, F.G. Chumley and B. Valent, unpublished data). The single copy DNA haplotypes were generated using three restriction enzymes and three single copy probes (two Guy11 genomic clones and the *ILV1* gene). Haplotype I, which contains pathogens of *Pennisetum* spp., only weakly hybridize to these single copy probes. Approximately 15–20 fragments were analyzed depending on the isolate (M.-H. Lebrun, unpublished data).

	RFLP haplotype methods				
Composite haplotype	rDNA	mtDNA	Single copy DNA	Plant host	Geographic location
a	1	Ia	A	*Oryza sativa*	Africa: CD, ML Asia: CH, IN, JP, PH South America: BR, GY
b	2	Ib	B	*Triticum aestivum*	South America: BR
c	2	Ie	C	*Eleusine* spp.	Africa: ML, RW Asia: IN, JP, NP
d	3	Ic	D	*Eleusine* spp.	Africa: CD, MD, RW, UG Asia: IN, JP South America: BR

Table 5.1. *(cont.)*

Composite haplotype	RFLP haplotype methods			Plant host	Geographic location
	rDNA	mtDNA	Single copy DNA		
e	4	IIa	E	*Digitaria* spp.	Africa: ML North America: US South America: BR
f	4	IIb	F	*Digitaria* spp.	Asia: IN, JP, PH
g	5	IV	G	*Pennisetum* spp.	Africa: BF, CD, ML Asia: IN
h	6	III	Weak hybridization	*Pennisetum* spp.	Africa: ML Asia: PH North America: US South America: BR

Abbreviations: BF, Burkina Faso; BR, Brazil; CH, China; GY, Guyana; IC, Ivory Coast; IN, India; JP, Japan; MD, Madagascar; ML, Mali; NP, Nepal; PH, the Philippines; RW, Rwanda; UG, Uganda; US, the United States of America.

other words, this sample set appears to maximize the genetic diversity of *P. grisea* in space and time. We recognize that a small sample size can lead to over-simplified conclusions; however, it is likely that these isolates describe much of the genetic variability in the *P. grisea* species. (A core collection of strains representing these eight haplotypes has been assembled by M.-H. Lebrun to be used as 'reference RFLP profiles' to readily determine the haplotypes of new isolates.)

This haplotypic analysis indicates that pathogens of *Eleusine*, *Digitaria*, and *Pennisetum* spp. are each represented by at least two genetically distinct populations. At this level of molecular genetic analysis, the rice pathogens comprise another genetically distinct group with a highly uniform genetic structure. However, rice pathogen-specific repetitive DNA probes (containing transposable element sequences) will provide a more sensitive tool to further differentiate the rice pathogen population.

Variation Associated with Transposable Element Sequences

Three transposable elements have been identified in *P. grisea*. These include a LINE-like retroposon (Valent and Chumley, 1991), and two *Gypsy* class retrotransposons called *fosbury* (M. Farman, Y. Tosa and S.A. Leong, unpublished data; M.-H. Lebrun, B. Valent, and F.G. Chumley, unpublished data; and V. Shull and J.E. Hamer, unpublished data), and *grasshopper* (Dobinson *et al.*, 1993). Figure 5.1 details the basic structure of these two classes of retroelements. A detailed discussion of retroposons and retrotransposons can be found in Hutchison *et al.* (1989), and Bingham and Zachar, (1989), respectively.

Rice pathogen-specific DNA elements

Magnaporthe grisea repetitive (MGR) DNA sequences

MGR sequences were isolated by screening a genomic DNA library constructed from rice pathogen field isolate O-137 (Hamer *et al.*, 1989). Clones that hybridized to radiolabeled O-137 genomic DNA, but did not hybridize to radiolabeled poly(A)$^-$ RNA, were designated MGR. Restriction mapping of these clones did not reveal a consensus structure. Rather, MGR appears to contain at least three different repetitive DNA sequences that have been subcloned and designated pCB583, pCB586 and pCB608.

MGR sequences are amplified and dispersed only throughout the genomes of rice pathogens, which contain approximately 100 and 50 restriction fragments that hybridize to subclones pCB583 and pCB586, respectively. (Hereafter, sequence homologs of pCB583 will be designated MGR583, and sequence homologs of pCB586 will be designated MGR586.) Generally non-rice pathogens contain only one or a few restriction fragments that hybridize to these MGR probes. Similar episodes of selective repetitive DNA sequence amplification have been documented in certain mammalian taxa (Pascale *et al.*, 1990). The appearance

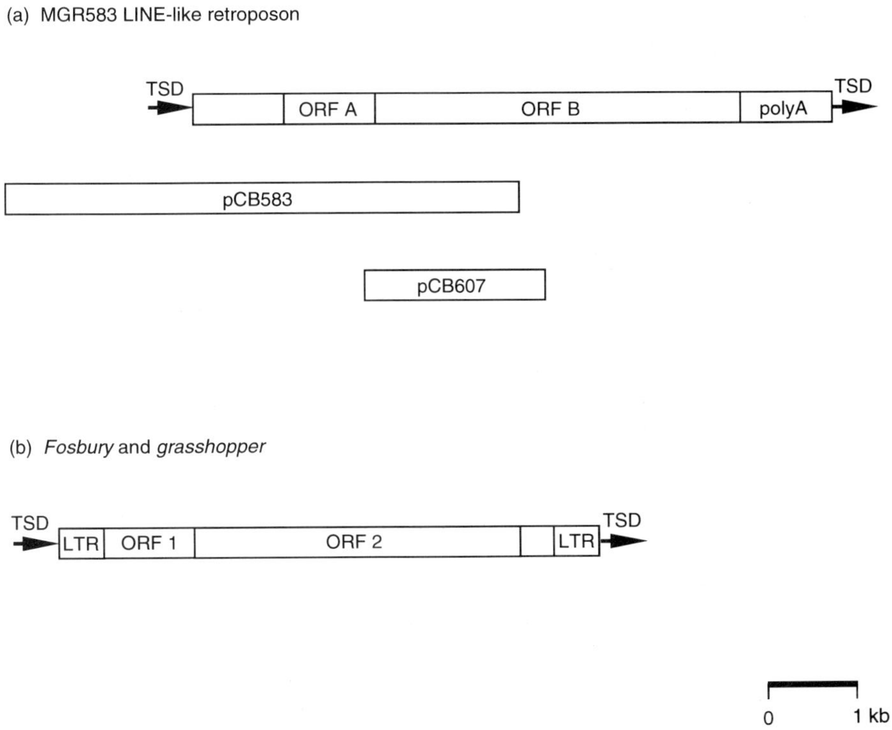

Fig. 5.1. The general structures of two classes of retroelements characterized in *P. grisea*. **(a)** The MGR583 LINE-like retroposon is approximately 7 kb in length (Valent and Chumley, 1991). This retroposon contains a polyadenylated tract (poly A) at its 3′ terminus, indicating that it was probably reverse transcribed from an mRNA template (Hutchison *et al*., 1989). The regions within the MGR583 LINE-like retroposon that hybridize to pCB583 and pCB607 are indicated below the element. **(b)** The *Gypsy* class retrotransposons *fosbury* (M. Farman, Y. Tosa and S.A. Leong, unpublished data; M.-H. Lebrun, B. Valent, and F.G. Chumley, unpublished data; and V. Shull, and J.E. Hamer, unpublished data), and *grasshopper* (Dobinson *et al*., 1993), are each approximately 6–8 kb in length. Both retrotransposons are flanked by long terminal repeats (LTRs) that contain transcription initiation and termination signals (Bingham and Zachar, 1989). The *fosbury* LTRs are 253 bp; the *grasshopper* LTRs are 198 bp. Open reading frame 1 (ORF1) encodes proteins homologous to retroviral core proteins.

Both the retroposon open reading frame B (ORF B), and the retrotransposon ORF 2 encode a number of enzymatic functions including an RNA directed DNA polymerase (reverse transcriptase), and an endonuclease (integrase). The integrase enzyme restricts the chromosomal DNA, allowing insertion, and generates a short (<10 bp) target site duplication (TSD) immediately flanking the retroelement. Since the integration activity is endogenously encoded, the presence of a TSD is strong evidence that the retroelement has inserted by a transposition mechanism.

of MGR583 and MGR586 in the *P. grisea* genome may predate the evolution of host specificity, because each host-specific form contains at least one copy of these sequences (Hamer *et al.*, 1989; Dobinson *et al.*, 1993; Borromeo *et al.*, 1993; see Fig. 5.2). Thus, homology to pCB583 and pCB586 appears to indicate that a given isolate belongs within the *P. grisea* species, and isolates (such as pathogens of *Cyperus* spp.) that lack MGR homology may not belong to this species.

Subclone pCB583 hybridizes intensely to a 2 kb *Bam*HI genomic DNA fragment in all strains examined. The 2 kb *Bam*HI fragment (pCB607) hybridizes to a 4.2 kb *Hin*dIII fragment. This 4.2 kb *Hin*dIII fragment has been sequenced and found to contain an open reading frame (ORF B; see Fig. 5.1), that is most similar to mammalian LINE (poly(A)-containing) retroposons (Valent and Chumley, 1991). Four RNA species of 7.5, 2.6, 2.4, and 0.5 kb are transcribed from this MGR583 LINE-like retroposon (Hamer *et al.*, 1989; Valent and Chumley, 1991).

Screening of a Guy11 genomic DNA library with radiolabeled pCB583, pCB586, and pCB607 has demonstrated that there is a physical association between MGR583 and MGR586, and confirmed that there is an association between MGR583 and MGR607 (V. Shull and J.E. Hamer, unpublished data). However, MGR586 and MGR607 are not associated in a statistically significant manner. This suggests that pCB583 contains two different sequence elements. One region of MGR583 is associated with MGR607 sequences. These sequences contain part of the LINE-like retroposon (Valent and Chumley, 1991; see Fig. 5.1). A second region of MGR583 is associated with MGR586 sequences. Although this MGR583/MGR586 element has not been characterized in detail, radiolabeled pCB586 does not hybridize to RNA isolated from rice pathogens suggesting that MGR586 is not abundantly transcribed (J.E. Hamer, unpublished data).

Restriction site polymorphisms associated with MGR586 sequences are sufficiently variable to uncover genetic divergence within the rice pathogen population. While all rice pathogens comprise one rDNA haplotype (1), one mtDNA haplotype (Ia), and one single copy nuclear DNA haplotype (A), the pCB586 probe detects variation within this haplotypic group that appears to be related to pathotype (Levy *et al.*, 1991, 1993; see Table 5.1).

Fosbury

MGR should represent all of the dispersed repetitive DNA sequences specific to rice pathogen field isolate O-137 (Hamer *et al.*, 1989). However, another rice pathogen-specific repetitive DNA sequence has been identified both by its ability to transpose (M.-H. Lebrun, B. Valent and F.G. Chumley, unpublished data; V. Shull and J.E. Hamer, unpublished data), and by sequence analysis (M. Farman, Y. Tosa and S.A. Leong, unpublished data).

Genetic analysis of MGR586 segregation identified a novel RFLP that was present in only one member of a tetrad derived from strain O-137 (V. Shull and J.E. Hamer, unpublished data). This aberrant segregation is consistent with a mitotic transposition mechanism. Restriction mapping and hybridization analysis of a genomic clone containing the MGR586 RFLP has shown that it was created

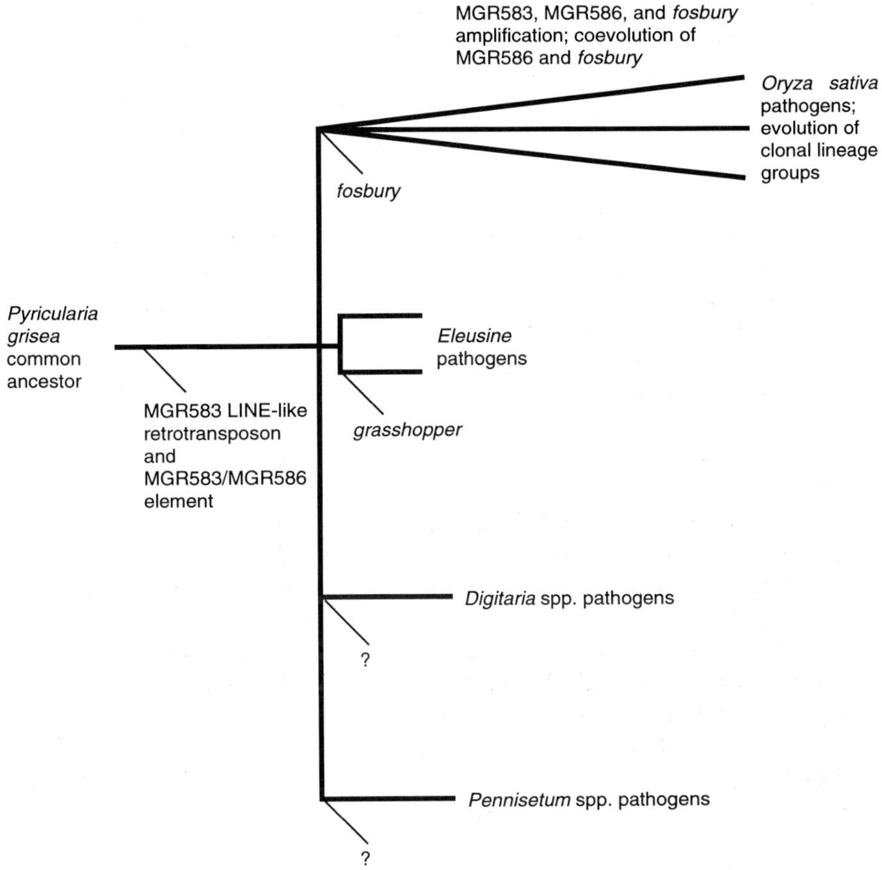

Fig. 5.2. Repetitive DNA elements that have been differentially acquired or selectively amplified in subpopulations of *P. grisea*. Each host specific form of *P. grisea* contains at least one restriction fragment that hybridizes to pCB583 and pCB586; therefore the MGR583 LINE-like retrotransposon and the MGR583/MGR586 element were probably acquired before the evolution of host specificity (Hamer *et al.*, 1989; Dobinson *et al.*, 1992; Borromeo *et al.*, 1993). The MGR sequences are highly amplified only in the genomes of rice pathogens. Restriction fragments with homology to a *fosbury* probe are found only in the genomes of rice pathogens (V. Shull and J.E. Hamer, unpublished data), indicating that this element was acquired concurrent with or after the evolution of pathogenicity towards rice. Further, pCB586 and a *fosbury* probe resolve rice pathogen field isolates into the same clonal lineage groups (V. Shull and J.E. Hamer, unpublished data), suggesting that MGR586 and *fosbury* have coevolved in the genomes of rice pathogens. A *grasshopper* probe hybridizes only to restriction fragments from the genomes of certain pathogens of *Eleusine* spp., indicating that this element was probably acquired concurrent with or after the divergence of this host specific form (Dobinson *et al.*, 1992).

The differential acquisition or selective amplification of these retroelements complements the mtDNA, rDNA, and single copy nuclear DNA RFLP analyses. Note that this figure is not intended to be an accurate phylogenetic tree.

by the insertion of a terminally redundant, >6.9 kb repetitive DNA element called *fosbury*. (Dick Fosbury invented the 'Fosbury Flop,' and went on to win the 1968 Olympic high jump gold medal with a record leap of 7 foot 4¼ inches). In this particular case, *fosbury* is inserted into a single copy genomic DNA sequence that is adjacent to an MGR583/MGR586 element (V. Shull and J.E. Hamer, unpublished data).

Fosbury was also identified through a transposon trapping protocol (M.-H. Lebrun, B. Valent and F.G. Chumley, unpublished data). In this scheme, a chlorate-resistant *P. grisea* strain that was unable to utilize nitrate was transformed with the *A. nidulans* nitrate reductase (*niaD*) gene (Daboussi *et al.*, 1992). This transformant was chlorate-sensitive and able to utilize nitrate. Spontaneous chlorate-resistant mutants were then selected. One mutant that has been characterized in detail resulted from the insertion of a 5.6 kb *fosbury* element (see Fig. 5.1), 850 bp upstream of the *niaD* transcription start site.

The two active *fosbury* elements differ in size; however, their restriction maps are similar. The *fosbury* element identified by tetrad analysis contains a duplication of internal sequences (V. Shull and J.E. Hamer, unpublished data). Figure 5.1 displays a *fosbury* element that does *not* contain this duplication.

Sequence analysis of the reverse transcriptase open reading frame has demonstrated that *fosbury* is a member of the *Gypsy* class family of retrotransposons (M. Farman, Y. Tosa and S.A. Leong, unpublished data; M.-H. Lebrun, B. Valent and F.G. Chumley, unpublished data). Members of this family contain long terminal repeats (LTRs). The 5′ LTR of each of the cloned elements is 253 bp in size, and at least 99% identical (M. Farman, Y. Tosa and S.A. Leong, unpublished data; M.-H. Lebrun, B. Valent and F.G. Chumley, unpublished data; V. Shull and J.E. Hamer, unpublished data). Each LTR is flanked by a six base pair inverted repeat, with the overall structure: $^{5'}$TGTCAC ... GTGACA$^{3'}$.

Sequencing of the genomic DNA flanking the *fosbury* element that transposed into the *niaD* locus identified a five base pair target site duplication with the structure $^{5'}$ATATT$^{3'}$ (M.-H. Lebrun, B. Valent and F.G. Chumley, unpublished data). Sequencing of the genomic DNA upstream of the *fosbury* insertion identified by tetrad analysis has shown that this region of the genome is 80% A/T (V. Shull and J.E. Hamer, unpublished data). Target sites that are A/T-rich are commonly reported for many *Gypsy* class retrotransposons (Bingham and Zachar, 1989).

A study using a *fosbury* probe and a worldwide collection of isolates has demonstrated that this element is present in approximately 100 copies in the genomes of rice pathogen field isolates, but is not present in other host-specific forms of *P. grisea* (V. Shull and J.E. Hamer, unpublished data). These data suggest that the introduction of *fosbury* into the genome was concurrent with the evolution of pathogenicity toward rice (see Fig. 5.2). Clearly, the acquisition of MGR583 and MGR586 pre-dates the introduction of *fosbury* into the *P. grisea* genome.

Restriction site polymorphisms surrounding *fosbury* elements are sufficiently divergent that a *fosbury* probe resolves the rice pathogen population into clonal lineage groups (V. Shull and J.E. Hamer, unpublished data), that are identical to the clonal lineage groups defined by the pCB586 probe (Levy *et al.*, 1991, 1993). Further, the screening of a Guy11 genomic DNA library with a *fosbury* probe and the pCB586 probe demonstrates that *fosbury* does not share a close physical association with MGR586 (V. Shull and J.E. Hamer, unpublished data). Although they reside in different genomic locations, the rate of accumulation of sequence polymorphisms surrounding *fosbury* and MGR586 are similar, suggesting these repetitive DNA elements have coevolved in the genomes of rice pathogens (see Fig. 5.2).

Grasshopper

Repetitive DNA sequences specific to any host-limited form may be identified by constructing a genomic DNA library from the pathogen of interest, screening this library with a genomic DNA probe from this pathogen, and counter-screening with a genomic DNA probe from another host-specific form. Clones that hybridize to both probes contain the rDNA cistron (Dobinson *et al*, 1993), or other repetitive DNA sequences common to both pathogens. Clones that only hybridize to the first probe contain repetitive sequences unique to this host-specific form (see question marks in Fig. 5.2).

Using this strategy, a repetitive DNA element originating in *Eleusine* pathogen G-22 has been cloned from the genome of laboratory strain 4091-5-8 (derived from a cross between G-22 and weeping lovegrass pathogen G-17), called *grasshopper* (Dobinson *et al.*, 1993; see Fig. 5.1). An internal 4.3 kb region of *grasshopper* contains a long open reading frame (ORF 2; see Fig. 5.1), encoding at least 1430 amino acids (Dobinson *et al.*, 1993; K.F. Dobinson, personal communication). Analysis of conserved domains and predicted amino acid motifs within this ORF indicates that *grasshopper* is related to members of the *Gypsy* class family of LTR-containing retrotransposons. Sequencing each end of *grasshopper* demonstrated that the element is flanked by identical 198 bp LTRs, and that each LTR is flanked by 5 bp inverted repeats yielding the overall structure $^{5'}$TGTTA ... TAACA$^{3'}$ (these inverted repeat sequences are identical to those of the *Cladosporium fulvum* retrotransposon *Cft-1*, which is a close relative of *grasshopper*). In addition, genomic DNA sequences flanking the element contain a five base-pair target site duplication ($^{5'}$ATAAA$^{3'}$), suggesting that this particular *grasshopper* element was generated by transposition.

A *grasshopper* probe hybridizes to all the chromosome-sized DNA molecules resolved in CHEF separations of 4091-5-8 and G-22 chromosomes, and genomic reconstruction experiments have shown that strain 4091-5-8 contains approximately 65 copies of *grasshopper*. This suggests that *grasshopper* is dispersed throughout the genome of 4091-5-8 (Dobinson *et al.*, 1993).

Isolates pathogenic on *Eleusine* spp. are broadly divided into two haplotypes (c and d; see Table 5.1). Haplotype c isolates contain many restriction fragments that hybridize to a *grasshopper* probe, while haplotype d isolates lack homology to *grasshopper* (Dobinson *et al.*, 1993; K.F. Dobinson, personal communication). Isolates containing intermediate copy numbers of *grasshopper* have not been identified. It appears that *grasshopper* has been acquired by a horizontal transfer event concurrent with or after the divergence of pathogens of *Eleusine* spp. (Dobinson *et al.*, 1993). The differential acquisition of *grasshopper* is consistent with the other RFLP analyses suggesting that *Eleusine* spp. pathogens can be subdivided into two haplotypes, and that pathogenicity towards this host may have evolved at least twice.

Summary and Implications

From an epidemiological standpoint it is important to define groups of genetically similar isolates and to understand their interrelatedness. To this end, the mtDNA, rDNA, and single copy nuclear DNA RFLP analyses have been useful in demonstrating that *P. grisea* is composed of a limited number of genetically distinct haplotypes, contrasting with the isozyme study of Leung and Williams (1986). In some cases these haplotypes correlate with host specificity. For example, composite haplotype a is only composed of rice pathogens. In other cases the haplotypes do not correlate with host specificity. Two distinct haplotypes appear to infect *Diditaria*, *Eleusine*, and *Pennisetum* spp. It is not known if the degree of genetic divergence described in *P. grisea* is idiosyncratic, or represents a general feature of fungal phytopathogens.

Haplotype distinctiveness is also reflected by differential repetitive DNA content, and transposable element probes may be useful indicators of recent common ancestry. For example, *grasshopper* is present only in a subset of *Eleusine* pathogens (haplotype c), while *fosbury*, and amplified copy numbers of MGR583 and MGR586, are only identified in the genomes of rice pathogens (haplotype a). It remains to be determined if similar retroelements exist in pathogens of *Digitaria* and *Pennisetum* spp. (see question marks in Fig. 5.2). Such sequences may define the degree of genetic differentiation within these haplotypes and may also be used to rapidly assess the degree of genetic hybridization between haplotypes.

It has been suggested that many microorganisms exist as groups of clonal populations, and that two hallmarks of an asexual population structure are the overrepresentation of alleles within each population, and the absence of recombinant genotypes (Tibayrenc *et al.*, 1991). RFLPs associated with MGR586 (Levy *et al.*, 1993), and *fosbury* are consistent with our proposal that the rice pathogen population is primarily clonal. These data suggest that although the telomorphic stage could exist in nature, it is not sufficiently prevalent that meiotic recombinants can be readily identified. Although there is considerable evidence suggesting that the rice pathogen population is clonal, the evidence in other host-specific forms remains meager. *Grasshopper* probes are now available to analyse isolates comprising haplotype c (many *Eleusine* spp. pathogens). These isolates are often the most fertile stains of *P. grisea*, and they could provide a rigorous test of widespread asexuality in this fungus.

Recently it has been suggested that certain populations which appear clonal may in fact rely on sex to produce genetic diversity (Smith *et al.*, 1993). In these 'epidemic populations', meiotic recombination occurs; however, only the most fit individuals survive. Due to the small sample sizes referred to in this paper, we cannot discern between the absolute clonality we argue for, and the possibility that sexuality occurs and with only a few individuals surviving to form the basis of lineages. To distinguish between these possibilities an analysis of sexuality and parasexuality within some rice pathogen lineages is currently underway in several laboratories. It is known that repetitive DNA probes detect a small amount of variation within each lineage (Levy *et al.*, 1993), and based on our studies of *fosbury* (V. Shull and J.E. Hamer, unpublished data), we understand at least one source for this variation is the insertion of a retrotransposon. Recently the analysis of another eukaryotic microbial population using dispersed repeated sequences (tRNA genes), suggested the possibility of meiotic recombination; however, the authors conclude that this result is artifactual (Francis and Eisenberg, 1993). Transposable elements are known to insert near tRNA genes in this species, and the insertion and excision of these sequences, not sexuality, is responsible for the 'shuffling' of RFLP bands. Thus transposable elements may be responsible for the generation of distinctive clones within clonal lineages of *P. grisea*.

If a sexual cycle is normally lacking in *P. grisea*, then definitions of this species that include the 'sharing of a common gene pool' are not germane. We suggest that hybridization to pCB583 and pCB586 reliably indicates that a given isolate belongs to this species, and isolates (such as pathogens of *Cyperus* spp.) that lack homology to these sequences do not belong to *P. grisea*. The caveat to this hypothesis is that repetitive DNA sequences may be transmitted horizontally, and strains that are not phylogenetically related may contain the same repetitive DNA sequence. In addition to these repetitive DNA probes, a set of single copy probes may also prove useful in establishing taxonomy. Lack of hybridization with these probes would exclude an isolate from the *P. grisea* species. By this method, isolates from *Pennisetum* spp. may not belong to *P. grisea*; however, more study is warranted on this group of interesting isolates.

Although we suggest that *P. grisea* is presently clonal, it is likely that the recent evolutionary history of the fungus included sex. After all, it is currently possible to cross certain isolates and produce fertile laboratory strains, suggesting that the sexual life cycle was once important. In this scenario, meiotic recombination may be largely responsible for the array of host specificities we currently observe. The current rice pathogen lineages that have been described may in fact represent 'fixed' polymorphisms inherited from a common sexual ancestor.

Current studies on genetic differentiation in *P. grisea* suggest two possible evolutionary scenarios. It is possible that *P. grisea* reached a high level of adaptation for pathogenicity towards different monocots early in its evolutionary history. In the ensuing time, genomic differentiation has had little consequence on this species' present morphology and physiology. Alternatively, *P. grisea* may be viewed as a recently evolved parasite with a genome currently undergoing a dramatic level of reshaping. In this case, it may be that insufficient time has elapsed for these reshaping events to affect morphology and physiology. To discriminate between these alternatives, a phylogenetic analysis of rDNA sequences is warranted. (Protein coding sequences may also be help answer this question; however, they are not the best source of phylogenetic data because they may arise by duplication and subsequently evolve under different selection pressures.) It may also be useful to re-examine the extent of isozyme variation in *P. grisea* using a core group of strains encompassing the eight composite haplotypes described in Table 5.1 (M.-H. Lebrun, unpublished data). Earlier studies suggested a paucity of isozyme variation within this species, a result that would support the recent emergence of *P. grisea* pathogens (Leung and Williams, 1986). However, this sample set was largely composed of rice pathogens, a group that is representative of a single haplotype and monophyletic in origin.

Recently, it has been suggested that the genomes of the Gramineae family (which includes, rice, wheat, maize, barley, millet, oats, rye, sorghum, sugarcane, and approximately 10,000 other species of grasses), contain different permutations of one tractable genome (Bennetzen and Freeling, 1993). The genetically distinct clonal populations of *P. grisea*, a pathogen of many Gramineae species, also possess different permutations of a single genome. Several parallels between the Gramineae and *P. grisea* genomes are intriguing. First, genomic rearrangements including large inversions and translocations have punctuated the evolution and domestication of the Gramineae, and grasses such as maize that contain active transposable elements can display a high degree of genomic instability. The result is that these species currently contain numerous tracts with collinear or inverted order of genetic markers at the 2–20 cM level. Genomic rearrangements are also common in *P. grisea*. These isolates contain active transposons, display a high degree of karyotypic variability (genomic instability). In addition, studies using single copy hybridization probes have shown that chromosome length polymorphisms may arise from large deletions and translocations (M.J. Orbach, F.G. Chumley and B. Valent, unpublished data) and genetic mapping studies have confirmed that translocations and duplications occur (Skinner *et al.*, 1993; Sweigard *et al.*, 1993; K.F. Dobinson and J.R. Stout, unpublished data). As a result, mapping strains can differ in their order of closely linked genetic markers (Leong *et al.*, chapter 6, this volume).

In the Gramineae discrepancies in genome size (for example, the maize genome is 35 times larger than that of rice), may be largely attributable to repetitive DNA content. *P. grisea* isolates may have a slightly variable genome size (as judged by CHEF), and some host-specific forms appear to have undergone selective repetitive DNA sequence amplification events (e.g. MGR in rice pathogens) or have differentially acquired repetitive DNA sequences by horizontal transfer (e.g. *grasshopper* in certain *Eleusine* spp. pathogens). Although these changes may not dramatically alter the total genome size, they nevertheless mirror the genomic reshaping events seen in the Gramineae.

Finally, allelic variation at shared genetic loci, or the acquisition of new genes (neomorphic mutations), account for the morphological and behavioral differences that exist between Gramineae species. Similarly, neomorphic mutations in *P. grisea* appear to be responsible for this species' spectrum of pathogenicity. For example, studies using cloned host specificity and avirulence genes demonstrate that host-limited forms of the fungus, and rice pathogen clonal lineages, differ by unique genetic loci confering host specificity and avirulence, respectively (Levy *et al.*, 1993; Valent and Chumley, chapter 7, this volume).

In conclusion, we hope that the future of rice blast research includes continued studies identifying genetic differentiation within and between *P. grisea* haplotypes. As a composite species, *P. grisea* infects a number of monocots, the most economically important group of plants. Thus, *P. grisea* contains the genetic potential to overcome many of the resistance mechanisms that are arrayed within the monocots. The genetic tractability of this pathogen may allow researchers to investigate disease resistance mechanisms in wheat, barley, millet, and perhaps even maize. The recent findings of a high degree of genetic similarity within the genomes of the Gramineae suggests that such studies would be extremely fruitful and may yield fundamental principles of host–pathogen interactions within the monocots.

Acknowledgments

Work in J.E. Hamer's laboratory is supported by the National Science Foundation, the United States Department of Agriculture, the Rockfeller Foundation, and the David and Lucille Packard Foundation. We wish to thank Dr K.F. Dobinson and Dr M.-H. Lebrun for critical reading of the manuscript. Finally, we wish to thank Dr M.-H. Lebrun for supplying, and allowing us to discuss, the data in Table 5.1.

References

Bennetzen, J.L. and Freeling, M. (1993) Grasses as a single genetic system: genome composition, collinearity and compatibility. *Trends in Genetics* 9(8), 259–260.

Bingham, P.M. and Zachar, Z. (1989) Retrotransposons and the FB transposon in *Drosophila melanogaster*. In: Berg, D.E. and Howe, M.M. (eds), *Mobile DNA*. American Society for Microbiology, Washington DC, pp. 485–502.

Borromeo, E.S., Nelson, R.J., Bonman, J.M. and Leung, H. (1993) Genetic differentiation among isolates of *Pyricularia* infecting rice and weed hosts. *Phytopathology* 83, 393–399.

Bruns, T.D., White, T.J. and Taylor, J.W. (1991) Fungal molecular systematics. *Annual Review of Ecological Systems* 22, 525–564.

Daboussi, M.-J., Langin, T. and Brygoo, Y. (1992) Fot1, a new family of fungal transposable elements. *Molecular and General Genetics* 232, 12–16.

Dobinson, K.F., Harris, R.E. and Hamer, J.E. (1993) *Grasshopper*, a long terminal repeat (LTR) retroelement in the phytopathogenic fungus *Magnaporthe grisea*. *Molecular Plant-Microbe Interactions* 6, 114–126.

Ellingboe, A.H., Wu, B.-C. and Robertson, W. (1990) Inheritance of avirulence/virulence in a cross of two isolates of *Magnaporthe grisea* pathogenic to rice. *Phytopathology* 80(1), 108–111.

Francis, D. and Eisenberg, R. (1993) Genetic structure of a natural population of *Dictyostellium discoideum*, a cellular slime mould. *Molecular Ecology* 2, 385–391.

Free, S. (1979) Arrangement of the genes coding for ribosomal ribonucleic acids in *Neurospora crassa*. *Journal of Bacteriology* 137, 1219–1226.

Garber, R.C., Turgeon, B.G., Selker, E.U. and Yoder, O.C. (1988) Organization of ribosomal RNA genes in the fungus *Cochliobolus heterostrophus*. *Current Genetics* 14, 573–582.

Gaudet, J., Julien, J., Lafay, J.F. and Brygoo, T. (1989) Phylogeny of some *Fusarium* species, as determined by large subunit rRNA sequence comparison. *Molecular Biology and Evolution* 6(3), 227–249.

Gurr, S.J., Unkles, S.E. and Kinghorn, J.R. (1988) The structure and organization of nuclear genes in filamentous fungi. In: Kinghorn, J.R. (ed.), *Gene Structure in Eukaryotic Microbes*. IRL Press, Oxford, pp. 93–109.

Hamer, J.E., Farrall, L., Orbach, M.J., Valent, B. and Chumley, F.G. (1989) Host species-specific conservation of a family of repeated DNA sequences in the genome of a fungal plant pathogen. *Proceedings of the National Academy of Sciences of the United States of America* 86, 9981–9985.

Hamer, J.E. and Timberlake, W.E. (1987) Functional organization of the *Aspergillus nidulans trpC* promoter. *Molecular and Cellular Biology* 7(7), 2352–2359.

Hebert, T.T. (1971) The perfect state of *Pyricularia grisea. Phytopathology* 61, 83–87.

Hunst, P.L., Latterell, F.M. and Rossi, A.E. (1986) Variation in double-stranded RNA from isolates of *Pyricularia oryzae. Phytopathology* 76, 674–678.

Hutchinson, C.A., III, Hardies, S.C., Loeb, D.D., Shehee, W.R. and Edgell, M.H. (1989) LINEs and related retroposons: long interspersed repeated sequences in the eucaryotic genome. In: Berg, D.E. and Howe, M.M. (eds), *Mobile DNA*. American Society for Microbiology, Washington DC, pp. 593–618.

Kato, H. and Yamaguchi, T. (1982) The perfect state of *Pyricularia oryzae* Cav. from rice plants in culture. *Annals of the Phytopathology Society of Japan* 48, 607–612.

Kistler, H.C. and Miao, V.P.W. (1992) New modes of genetic change in filamentous fungi. *Annual Reviews in Phytopathology* 30, 131–152.

Koch, G. and Köhler, W. (1990) Isozyme variation and genetic distances of *Erysiphe graminis* DC. formae speciales. *Journal of Phytopathology* 129, 89–101.

Lau, G.W., Chao, C.T. and Ellingboe, A.H. (1993) Interaction of genes controlling avirulence/virulence of *Magnaporthe grisea* on rice cultivar Katy. *Phytopathology* 83(4), 375–382.

Lebrun, M.-H., Capy, M.P., Garcia, N., Dutertre, M., Brygoo, Y., Notteghem, J.L. and Valés, M. (1990) Biology and genetics of *Pyricularia oryzae* and *Pyricularia grisea* populations: Current situation and development of RFLP markers. In: *Rice Genetics Conference*. International Rice Research Institute, Los Baños, The Philippines.

Leung, H. and Williams, P.H. (1986) Enzyme polymorphism and genetic differentiation among geographic isolates of the rice blast fungus. *Phytopathology* 76, 778–783.

Leung, H. and Williams, P.H. (1987) Nuclear division and chromosome behavior during meiosis and ascosporogenesis in *Pyricularia oryzae. Canadian Journal of Botany* 65, 112–123.

Leung, H., Borromeo, E.S., Bernardo, M.A. and Notteghem, J.L. (1988) Genetic analysis of virulence in the rice blast fungus *Magnaporthe grisea. Phytopathology* 78, 1227–1233.

Levy, M., Romao, J., Marchetti, M.A. and Hamer, J.E. (1991) DNA fingerprinting with a dispersed repeated sequence resolves pathotype diversity in the rice blast fungus. *Plant Cell* 3, 95–102.

Levy, M., Correa-Victoria, F.J., Zeigler, R.S., Xu, S. and Hamer, J. (1993) Genetic diversity of the rice blast fungus in a disease nursery in Colombia. *Phytopathology* 83, 1427–1433.

Mackill, A.O. and Bonman, J.M. (1986) New hosts of *Pyricularia oryzae. Plant Disease* 70, 125–127.

Michelmore, R.W. and Hulbert, S.H. (1987) Molecular markers for genetic analysis of phytopathogenic fungi. *Annual Reviews in Phytopathology* 25, 383–404.

Mills, D. and McCluskey, K. (1990) Electrophoretic karyotypes of fungi: the new cytology. *Molecular Plant-Microbe Interactions* 3, 351–357.

Ou, S.H. (1985) *Rice Diseases*, 2nd edn. Commonwealth Mycological Institute, Kew, UK, 380 pp.

Pascale, E., Valle, E. and Furano, A.V. (1990) Amplification of an ancestral mammalian L1 family of long interspersed repeated DNA occurred just before the murine radiation. *Proceedings of the National Academy of Sciences of the United States of America* 87, 9481–9485.

Petes, T. D. (1979) Meiotic mapping of yeast ribosomal deoxyribonucleic acid on chromosome XII. *Journal of Bacteriology* 138(1), 185–192.

Romao, J. and Hamer, J.E. (1992) Genetic organization of a repeated DNA sequence family in the rice blast fungus. *Proceedings of the National Academy of Sciences of the United States of America* 89, 5316–5320.

Rossman, A.Y., Howard, R.J. and Valent, B. (1990) *Pyricularia grisea*, the correct name for the rice blast disease fungus. *Mycologia* 82, 509–512.

Silué, D., Notteghem, J.L. and Tharreau, D. (1992a) Evidence for a gene-for-gene relationship in the *Oryza sativa–Magnaporthe grisea* pathosystem. *Phytopathology* 82(5), 577–580.

Silué, D., Tharreau, D. and Nottegham, J.L. (1992b) Identification of *Magnaporthe grisea* avirulence genes to seven rice cultivars. *Phytopathology* 82(12), 1462–1467.

Skinner, D.Z., Budde, A.D. and Leong, S.A. (1991) Molecular karyotype analysis of fungi. In: Bennet, J.W. and Lasure, L.L. (eds), *More Gene Manipulations in Fungi*. Academic Press, San Diego, California.

Skinner, D.Z., Budde, A.D., Farmen, M.L., Smith, J.R., Leung, H. and Leong, S.A. (1993) Genome organization of *Magnaporthe grisea*: genetic map, electrophoretic karyotype, and occurrence of repeated DNAs. *Theoretical and Applied Genetics* 87, 545–557.

Smith, J.M., Smith, N.H., O'Rourke, M. and Spratt, B.G. (1993) How clonal are bacteria? *Proceedings of the National Academy of Sciences of the United States of America* 90, 4384–4388.

Suzuki, H. (1963) Origin of variation in *Piricularia oryzae*. In: Ou, S.H. (ed.), *The Rice Blast Disease*. The Johns Hopkins Press, Baltimore, Maryland, pp. 111–149.

Sweigard, J.A., Chumley, F.G. and Valent, B. (1992) Cloning and analysis of *CUT1*, a cutinase gene from *Magnaporthe grisea*. *Molecular and General Genetics* 232, 174–182.

Sweigard, J.A., Valent, B., Orbach, M.J., Walter, A.M., Rafalski, A. and Chumley, F.G. (1993) Genetic map of the rice blast fungus *Magnaporthe grisea*. In: O'Brien, S.J. (ed.), *Genetic Maps*. Cold Spring Harbor Laboratory, New York, pp. 3.112–3.115.

Talbot, N.A., Salch, Y., Ma, M. and Hamer, J.E. (1993a) Karyotype variation within clonal lineages of the rice blast fungus *Magnaporthe grisea*. *Applied and Environmental Microbiology* 59, 585–593.

Talbot, N.J., Ebbole, D.J. and Hamer, J.E. (1993b) Identification and characterization of *MPG1*, a gene involved in pathogenicity from the rice blast fungus *Magnaporthe grisea*. *The Plant Cell* 5, 1575–1590.

Tibayrenc, M., Kjellberg, F., Arnaud, J., Oury, B., Breniére, S.F., Dardé, M.-L. and Ayala, F.J. (1991) Are eukaryotic microorganisms clonal or sexual? A population genetics vantage. *Proceedings of the National Academy of Sciences of the United States of America* 88, 5129–5133.

Timberlake, W.E. and Marshall, M.A. (1989) Genetic engineering in filamentous fungi. *Science* 244, 1313–1317.

Valent, B. and Chumley, F.G. (1987) Genetic analysis of host species specificity in *Magnaporthe grisea*. In: *Molecular Strategies for Crop Protection*. Alan R. Liss, Inc., New York, pp. 83–93.

Valent, B. and Chumley, F.G. (1989) Genes for cultivar specificity in the rice blast fungus, *Magnaporthe grisea*. In: *Advanced NATO workshop on Molecular Signals in Microbe-Plant Symbiotic Systems*. Biddinghuizen, The Netherlands.

Valent, B. and Chumley, F.G. (1991) Molecular genetic analysis of the rice blast fungus, *Magnaporthe grisea*. *Annual Review of Phytopathology* 29, 443–467.

Valent, B., Crawford, M.S., Weaver, C.G. and Chumley, F.G. (1986) Genetic studies of fertility and pathogenicity in *Magnaporthe grisea* (*Pyricularia oryzae*). *Iowa State Journal of Research* 60, 569–594.

Valent, B., Farrall, L. and Chumley, F.G. (1991) *Magnaporthe grisea* genes for pathogenicity and virulence identified through a series of backcrosses. *Genetics* 127, 87–101.

Yaegashi, H. (1977) On the sexuality of blast fungi, *Pyricularia* spp. *Annals of the Phytopathological Society of Japan* 43, 432–439.

Yaegashi, H. (1978) Inheritance of pathogenicity in crosses of *Pyricularia* isolates from weeping lovegrass and finger millet. *Annals of the Phytopathology Society of Japan* 44, 626–632.

Yaegashi, H. and Asaga, K. (1981) Further studies on the inheritance of pathogenicity in crosses of *Pyricularia oryzae* with *Pyricularia* sp. from finger millet. *Annals of the Phytopathology Society of Japan* 47, 677–679.

Yaegashi, H. and Nishihara, N. (1976) Production of the perfect stage in *Pyricularia* from cereals and grasses. *Annals of the Phytopathology Society of Japan* 42, 511–515.

6 Molecular Genetic Approach to the Study of Cultivar Specificity in the Rice Blast Fungus

S.A. Leong[1,2], M. Farman[2], J. Smith[1*], A. Budde[1], Y. Tosa[2,3] and N. Nitta[2]

*USDA-ARS Plant Disease Resistance Research Unit[1] and Department of Plant Pathology[2], University of Wisconsin, Madison, WI 53706, USA; [3]Faculty of Agriculture, Kochi University, B200 Monobe, Nankoku-shi, 783 Japan; *Present address: Pioneer Hi-Bred International Inc., Plant Breeding Division, Department of Corn Breeding, 1 Mile West RR50, PO Box 596, Kekaha, Kauai, HI 96752, USA*

Introduction

Magnaporthe grisea (Barr, 1977) is the teleomorph of a complex genus of Ascomycete fungi composed of interfertile anamorphs (*Pyricularia grisea* cav.) which cause blast of rice and/or gramineae other than rice (Rossman *et al.*, 1990). As indicated below, this complexity may contribute to the pathogenic variability attributed to this organism and the rapidity with which this fungus can overcome genetic resistance of rice hosts.

Host genetic resistance is the major method of disease control for rice blast disease (Bonman *et al.*, 1992). While much has been learned about the genetic basis of disease resistance in the host (McCouch *et al.*, this volume) and cultivar specificity in the pathogen (see below; Ellingboe and Chao, this volume; Notteghem *et al.*, this volume; Valent and Chumley, this volume), we still have little understanding of the molecular basis and evolution of cultivar-specific interactions of *M. grisea* with rice. This information is essential for the rational design of plants with durable resistance to blast and for the strategic deployment of disease-resistant cultivars.

Resistance to rice blast disease can be conferred by major genes, which offer complete resistance to certain pathotypes of the fungus, or minor genes, which act in a polygenic fashion to provide quantitative or partial resistance (Bonman *et al.*, 1992; McCouch *et al.*, this volume). Much attention is now being devoted to the mapping of genes conferring both complete and partial resistance to blast in order to rapidly introgress these genes into elite varieties and assess the relative contribution and stability of each type of resistance in a common genetic background (McCouch *et al.*, this volume; Zeigler *et al.*, this volume). This mapping information is also being used to facilitate the cloning of disease resistance genes by map-based cloning strategies as a first step toward understanding the molecular basis of disease resistance (Yu *et al.*, 1991; for approach see chapter 11, this volume).

Population Biology of *M. grisea*

Mapping efforts in the host are being complemented by studies of the population biology of *M. grisea* and the analysis of the molecular and genetic basis of cultivar/host specificity in *M. grisea*. The former studies (Hamer *et al.*, 1989; Hamer, 1991; Levy *et al.*, 1991, 1993; Borromeo *et al.*, 1993; Correa *et al.*, this volume) have shown by DNA fingerprinting methods that:

1. *M. grisea* isolates that are pathogenic to rice can be organized into many clonal lineages worldwide;
2. rice pathogen populations in each country are composed of a smaller and usually unique number of clonal lineages;
3. each lineage has a limited range of cultivar specificity toward rice;
4. rice pathogens and grass pathogens are generally host-limited; and
5. most lineages are indigenous to a country, although several lineages can be found throughout the world.

These population data are now being used to develop resistance breeding strategies based on lineage-exclusion (Correa *et al.*, this volume; Zeigler *et al.*, this volume).

While DNA fingerprinting studies of the pathogen can usually be used to distinguish rice and grass-specific forms of *M. grisea*, exceptional grass pathogenic isolates have been found that have rice pathogen-like fingerprints (Correa *et al.*, this volume; Skinner *et al.*, 1993). In one case where pathotype was evaluated (Correa *et al.*, this volume), these isolates were not found to be highly pathogenic on rice. By contrast, an exceptional grass isolate having a grass pathogen fingerprint, but which is pathogenic on 13 rice cultivars as well as fingermillet, has been found in Bangladesh (Shahjahan *et al.*, 1993). Mackill and Bonman (1986) have also reported isolates of *M. grisea* from grass weeds that were pathogenic to rice and A.S. Prabhu (Wisconsin, 1993, personal communication) has observed that some Brazilian rice pathogenic isolates of *M. grisea* can attack wheat. Interestingly, one of the rice-pathogenic grass isolates described by Mackill and Bonman (1986) was analyzed by Borromeo *et al.* (1993) and found to have fingerprint patterns consistent with grass pathogens. Thus DNA fingerprinting cannot always be used reliably to determine pathotype of this fungus even at the level based on host genus. These studies also raise questions about the evolution of pathotypes.

Molecular Genetic Analysis of Host/Cultivar Specificity in *M. grisea*

The second experimental approach with *M. grisea* has focused on the genetic analysis of cultivar/host specificity by several labs (Kolmer and Ellingboe, 1988; Leung *et al.*, 1988; Ellingboe *et al.*, 1990; Valent *et al.*, 1991; Ellingboe, 1992; Silue and Notteghem, 1992; Silue *et al.*, 1992a,b; Lau *et al.*, 1993; Smith and Leong, 1994), the generation of RFLP maps of the fungus (Romao and Hamer, 1992; Budde *et al.*, 1993; Hayashi and Naito, 1993; Skinner *et al.*, 1993; Sweigard *et al.*, 1993; M. Farman and S.A. Leong, unpublished findings), the genetic mapping of cultivar/host specificity genes (Hayashi and Naito, 1993; Sweigard *et al.*, 1993; Smith and Leong, 1994; Tharreau *et al.*, 1994), the map-based cloning of several of these genes and their preliminary characterization (chapter 7, this volume). These studies complement those being done on the host as they provide data on the molecular basis of cultivar/host specificity and provide insight about the mechanisms that lead to defeat of disease resistance genes of the host. For example, Barbara Valent's laboratory at Dupont has cloned two unstable cultivar/host specificity genes in *M. grisea* (see chapter 7, this volume). *Avr2-YAMO* has been shown to undergo frequent, spontaneous changes that include single base pair alterations in the protein coding sequence as well as deletions and insertions. The resultant mutant fungal isolate becomes virulent on rice cultivar Yashiro-mochi upon which it was previously avirulent. Likewise, frequent, spontaneous mutations in the gene *Pwl2*, a host specificity gene for weeping lovegrass, give rise to strains that are able to infect weeping lovegrass (see chapter 7, this volume). The implications of such phenotypic and genetic changes on the durability of disease resistance are clear.

Molecular and genetic studies (Leung *et al.*, 1988; Valent *et al.*, 1991; Smith and Leong, 1994; chapter 7, this volume) have shown that the ability to attack different cultivars of rice or other host species may be conferred in many cases by single host/cultivar specificity genes, although the degree of pathogenicity, e.g. number of lesions, lesion size, and spore yield, may be modulated by multiple genes in the pathogen. A more complex genetic relationship has been described by Ellingboe (1992; chapter 4, this volume) and Notteghem *et al.* (chapter 9, this volume) to describe cultivar specificity towards rice. In the studies of Notteghem and coworkers, rice pathogens were the source of the host/cultivar specificity genes, while in those reported by Chumley and Valent (1991), Valent *et al.* (1991) and chapter 7 (this volume), the genes originated primarily from grass pathogens identified in crosses between laboratory and field isolates. The origin of the mutiple genes required for expression of avirulence in isolates studied by Ellingboe and coworkers is not clear (A.H. Ellingboe, Wisconsin, 1994, personal communication). Alleles of *Avr2-YAMO*, which was cloned from a rice pathogen and conditions cultivar specificity toward rice cultivar Yashiro-mochi, have also been found in other *M. grisea* isolates that infect grasses belonging to *Pennisetum* spp., *Digitaria* spp., *Setaria* spp. and *Panicum* spp. (Valent and Chumley, this volume). Alleles from *Pennisetum* and *Digitaria* spp. prevent the infection of rice cultivar Yashiro-mochi when transformed into a normally virulent rice pathogenic isolate of *M. grisea*. Homologs of *Pwl2*, a gene controlling specificity for weeping lovegrass, are also found in *M. grisea* strains infecting *Pennisetum*, *Digitaria*, *Leersia* and *Paspalum* spp. as well as barley, maize, and wheat. Alleles of *Pwl2* cloned from *Pennisetum* spp. have been shown to be functional (chapter 7, this volume). Taken together, the fingerprinting and molecular genetic data suggest that *M. grisea* may have the basic ability to attack a wide range of grass hosts/rice cultivars and that this ability is restricted primarily by host/cultivar specificity genes. As these genes mutate, the host/cultivar range of *M. grisea* increases.

Similar observations have been made for bacterial pathogens of plants (for review see Long and Staskawicz, 1993). For example, an avirulence gene from a tomato strain of *Xanthomonas campestris* pv. *vesicatoria* confers cultivar specificity on alfalfa, *Phaseolus* bean, corn, cotton and cowpea when introduced into pathovars of *Xanthomonas campestris* that attack these crop plants (Whalen *et al.*, 1988). Mutation of this gene in *X. c.* pv. *vesicatoria* allowed the bacterium to grow better than the wild type strain in bean carrying the corresponding disease resistance gene, although a hypersensitive response was still elicited. Likewise avirulence genes of *Pseudomonas syringae* pv. *tomato* can confer cultivar specificity to *Pseudomonas syringae* pv. *glycinea* on soybean (Kobayashi *et al.*, 1989). One of these is homologous to the *avr*A gene of *P. s.* pv. *glycinea*, a gene which specifies cultivar specificity on soybean. *avr*Rpt2, another avirulence gene from *P. syringae* pv. *tomato* was found to determine avirulence on both arabidopsis and soybean (Whalen *et al.*, 1991). Finally, a family of avirulence genes from *Xanthomonas campestris* pv. *oryzae* has been shown to be structurally related to avirulence genes from Xanthomonads that attack dicot crop plants (Hopkins *et al.*, 1992). These data suggest that avirulence genes may play a role in non-host as well as host resistance and that similar resistance genes may be present in plant species that are non-hosts. Whether or not such conclusions can be made for the blast pathosystem awaits further molecular characterization of cultivar/host specificity genes of the pathogen and the corresponding resistance genes in the host.

A New Approach to Blast Resistance in Rice

Molecular characterization of a stable rice cultivar specificity gene from *M. grisea* has not yet been done. *Avr2-YAMO* appears to mutate readily (Valent and Chumley, this volume). This extreme mutability may be due to the chromosomal location of the gene. *Avr2-YAMO* maps to within 2 kb of a telomere and undergoes deletion, insertion and point mutations. Telomeric sequences in plants (Broun *et al.*, 1992) and yeast (Horowitz *et al.*, 1984) are known to undergo high frequency heritable changes. However, instability of genes in many organisms is not always related simply to chromosomal location. For example in humans, genetic instability of three different genes associated with neurological disorders was recently found to be caused by an expansion of a triple repeat sequence within these genes (Morell, 1993). Cultivar specificity in some Xanthomonads is attributed to the presence of a 102 bp sequence which is directly repeated within *avr* genes (Herbers *et al.*, 1992; Hopkins *et al.*, 1992;

De Feyter *et al.*, 1993). Intergenic recombination among this class of *avr* genes in *X. c.* pv. *malvacearum* appears to give rise to new races through deletion of all or portions of the gene (De Feyter *et al.*, 1993). Although not yet documented, intragenic recombination could also lead to deletion of selected 102 bp repeats yielding *avr* genes with new cultivar specificities. Thus the inherent stability of an *avr* gene may not be simply predicted from its genomic location.

The incorporation of a stable avirulence gene from *M. grisea* into rice which shows resistance to pathogens carrying this gene could provide a novel means of creating rice with disease resistance to other pathotypes of the fungus as such an engineered plant might express disease resistance in a controlled manner (de Wit, 1992). This approach has already met with some preliminary success in the *Cladosporium fulvum*/tomato system (Jones *et al.*, 1993). The laboratory of Jonathan Jones has engineered the *avr9* cultivar specificity gene of *Cladosporium fulvum* into tomato and tobacco under the control of the Cauliflower mosaic virus 35S promoter. Intercellular fluids of these plants containing the *avr9* gene product caused a resistant response when injected into tomato carrying the *cf9* resistance gene but not the *cf0* resistance gene. When the *avr9* transgenic tomato plant was crossed with a *cf9* tomato, the progeny that resulted grew very poorly, indicating that presence of the *avr* gene product was negatively affecting growth of the resistant plant as one would expect for a systemic necrotic response. These data also indicate that the regulated (wound-inducible and/or elicitor-inducible) and tissue-specific expression of this *avr* gene will likely be necessary to prevent a negative growth effect on the plant. Thus the isolation and characterization of stable, cultivar specificity genes from *M. grisea* may have practical merit in the development of transgenic rice plants with durable field resistance to a wide range of rice blast pathotypes. Requisite to such a strategy is the fundamental analysis of a stable *avr* gene, its expression, its product, and its function in the host – parasite interaction.

Our studies over the last seven years have focused on:

1. an analysis of the genome organization of *M. grisea* (RFLP map, molecular karyotype analysis, identification and characterization of repeated DNAs, the determination of the relationship of genetic to physical distance, the analysis of genome stability);
2. the genetic characterization of a stable cultivar specificity gene and an ongoing attempt to clone this gene by chromosome walking;
3. the successful application of dominant selectable markers (hygromycin and phleomycin resistance) and the GUS reporter gene to transformation of *M. grisea*; and
4. the study of stability of transforming DNA in the fungus. We describe here results in these four areas.

Genetic Map of *Magnaporthe grisea*

A genetic map was constructed for a cross between the *M. grisea* rice isolate Guy11 (*Mat 1–2*, hermaphrodite), a rice pathogenic field isolate from French Guyana, and the laboratory strain 2539 (*Mat 1–1*, hermaphrodite) (Fig. 6.1) (Skinner *et al.*,

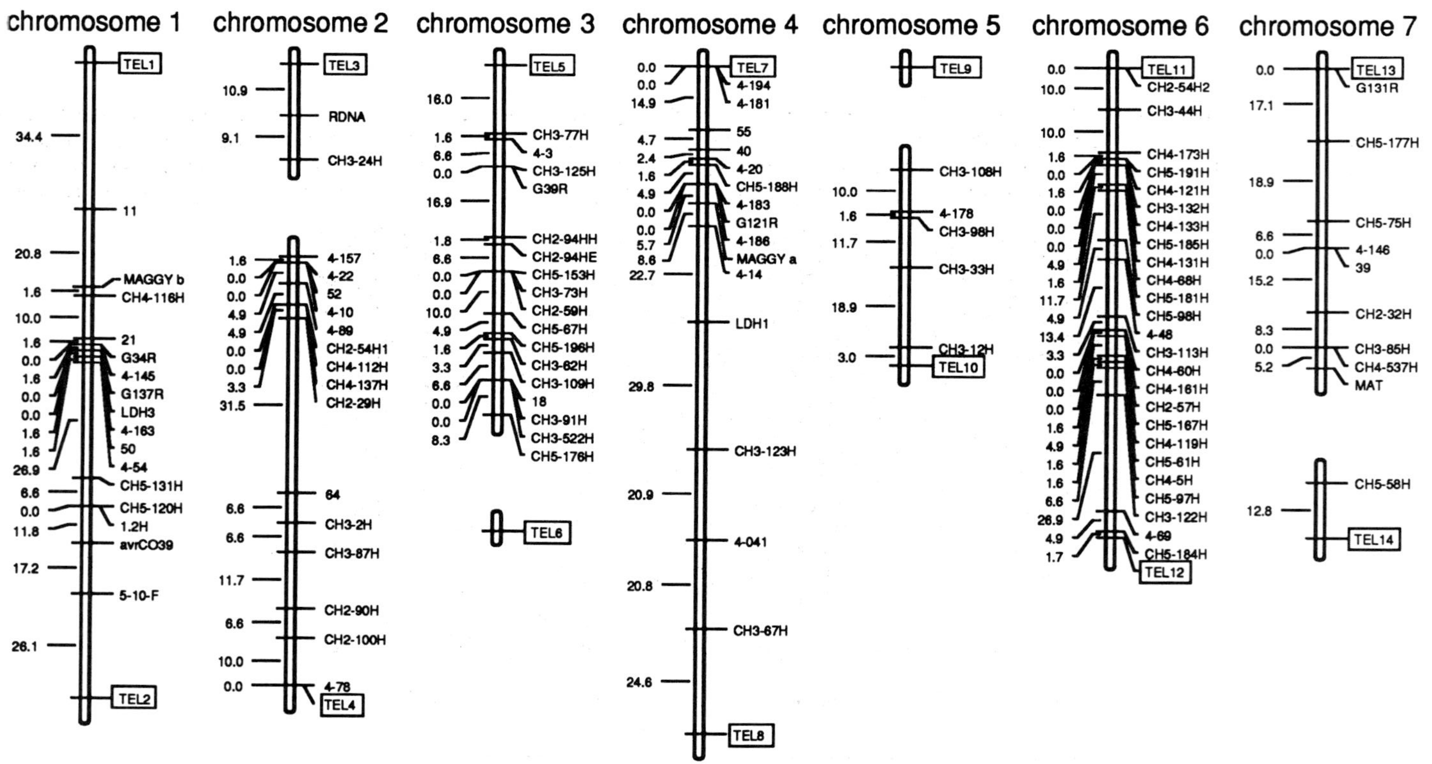

Fig. 6.1. Genetic map of *M. grisea* cross Guy11 × 2539. The map was constructed following segregation of markers in 61 random ascospore progeny (Leung *et al.*, 1988) using version 1.0 of the MAPMAKER program (Lander *et al.*, 1987). The Kosambi mapping function was employed and the distances reported are in cM. Parameters employed were maximum recombination fraction of 0.3 and minimum LOD number of 4.0.

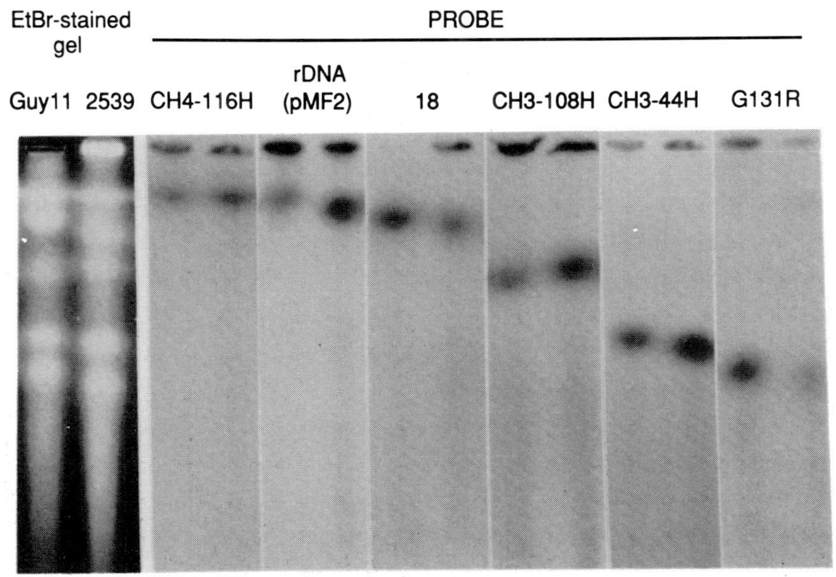

Fig. 6.2. Electrophoretic karyotype of *M. grisea* strains Guy11 and 2539. The chromosomes were isolated and separated by CHEF electrophoresis (Skinner *et al.*, 1991) as described in Skinner *et al.* (1993). The panel on the left represents an ethidium bromide gel showing the separation of chromosomes. Chromosomal DNA from the same gel was transferred to nylon using the Southern blot procedure (Sambrook *et al.*, 1989) and then probed sequentially with ^{32}P-labeled RFLP markers representing six of the seven chromosomes.

1993). Segregation data for 97 RFLP markers, two isoenzyme loci, 14 telomeres, two retroelements, and the mating-type locus were analyzed in 61 progeny of this cross using the MAPMAKER program (Lander *et al.*, 1987). Segregation of telomeric RFLPs allowed 12 telomeres to be genetically mapped and the map to be completely delimited (M. Farman and S.A. Leong, unpublished findings). As shown below, we now know that portions of the genome of our mapping population are not well represented by molecular or phenotypic markers. Based on the current map, we estimate the size of the *M. grisea* genome to be 917 cM. This estimation is comparable to those of Romao and Hamer (1992) and Sweigard *et al.* (1993) who estimated the genome size of *M. grisea* to be 802 cM and 840 cM, respectively, using different *M. grisea* isolates and markers.

Linkage groups were assigned to corresponding chromosomes by hybridization of selected markers to Southern blots of electrophoretically separated chromosomes (Fig. 6.2). From these physical and genetic analyses we determined that the *M. grisea* isolates studied here contain seven chromosomes. Similar results were obtained by Sweigard *et al.* (1993) for a cross involving strain 6043, an F_1 progeny of the cross we employed, and the laboratory strain 4224-7-8 and by Hayashi and Naito (1993) for a cross of rice pathogenic field isolates. Thus earlier cytological studies (Leung and Williams, 1987) may have underestimated the chromosome content of *M. grisea*. A small translocation involving chromosome 2 of 2539 and chromosome 6 of Guy11 was also confirmed by probing the electrophoretic karyotype of parent strains and of progeny representative of parental ditype and nonparental ditype asci (Skinner *et al.*, 1993).

The combined use of physical and stringent genetic criteria (log of the odds (LOD) > 4, Theta < 0.30 using MAPMAKER) to map telomeres was essential to the correct assignment of telomeric RFLPs to specific chromosome ends. For example, the assignment of telomeric RFLPs to chromosome ends was verified by second-dimensional electrophoresis of restriction endonuclease-treated, electrophoretically separated chromosomes (Fig. 6.3). We also found it necessary to employ several restriction endonucleases to confirm segregation data in the RFLP studies of telomeres as aberrantly sized telomeric restriction fragments occurred with significant frequency in meiotic progeny (M. Farman and S.A. Leong, unpublished findings).

Achilles' heel cleavage (AC) techniques (Koob and Szybalski, 1990; Koob *et al.*, 1993) were successfully applied to physical analysis of the *M. grisea* genome. As shown in Figs. 6.4 and 6.5, these methods allow one to cleave a complex genome at a predetermined site. RFLP marker 11 of chromosome 1 was incorporated into plasmid vector pSPlac7-2-2 (M. Farman and S.A. Leong, unpublished findings) which contains the hygromycin resistance cassette of pAN7-2 (P. Punt, unpublished vector described below) and a synthetic *lacO* sequence (Koob and Szybalski, 1990). The resultant plasmid was targeted to the marker 11 locus on chromosome 1 by transformation (see below). AC of the *lacO* integrated at RFLP marker 11 has shown that this marker is 1.8 mgbp from one end of chromosome 1 in both parent isolates (Fig. 6.6) (M. Farman and S.A. Leong, unpublished findings). The resulting fragment was shown to hybridize to pSPlac11 as well as the telomere oligonucleotide probe $(TTAGGG)_3$. RecA-assisted Achilles' cleavage at marker CH5-120H released a 1.2 mgbp fragment

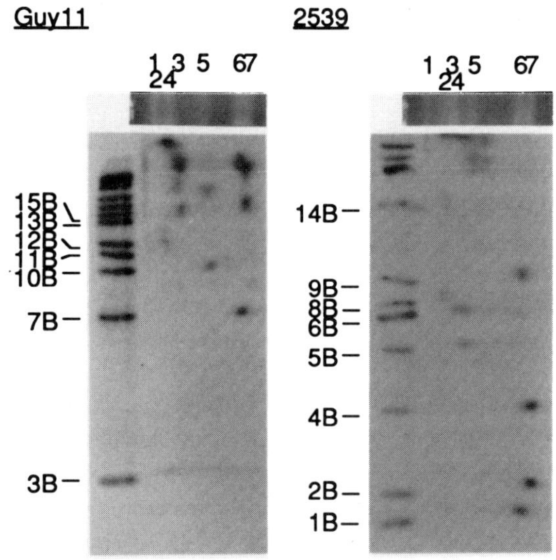

Fig. 6.3. Second dimensional electrophoresis of restricted chromosomes of Guy11 and 2539. Each panel shows two autoradiograms: an upper horizontal one showing hybridization of the telomere-specific oligonucleotide probe $(TTAGGG)_3$ (Schechtman, 1989) to whole chromosomes following the first dimensional electrophoresis by CHEF electrophoresis; beneath this autoradiogram is shown the electrophoretic pattern obtained after *Bam*HI digestion of these chromosomes and the separation of the fragments in 0.7% agarose using standard electrophoresis conditions. DNA in these gels was transferred to nylon and probed with the telomere probe.

from the other end of chromosome 1 (Fig. 6.7) (M. Farman and S.A. Leong, unpublished findings). These results indicate that the paucity of markers on chromosome 1 (Fig. 6.1) is due to a lack of markers at the distal ends of this chromosome. Whether this is due to a greater similarity in the DNA of the parent strains in these regions of the genome remains to be determined. Besides providing information about the physical organization of *M. grisea* chromosomes, the AC methods can be used to isolate additional RFLP markers as well as clone genes contained within the released DNA fragments (see below).

Mapping of a Gene Controlling Cultivar Specificity on Rice Cultivar CO39

Segregation analysis of virulence (the ability to produce conidiating lesions) on rice cultivar CO39 and five CO39 near isogenic lines carrying different, single resistance genes to blast (Mackill and Bonman, 1992; Table 6.1) was evaluated in the *M. grisea* cross 2539 × Guy11 (Smith and Leong, 1994) (Table 6.1). This analysis along with additional test crosses (data not shown) indicated that a cultivar specificity gene for CO39 was segregating in this mapping population. *avr-CO39* was mapped to one end of chromosome 1 of *M. grisea* (Fig. 6.1) and is linked to RFLP markers CH5-120H, 1.2H and 5–10F. Valent and coworkers

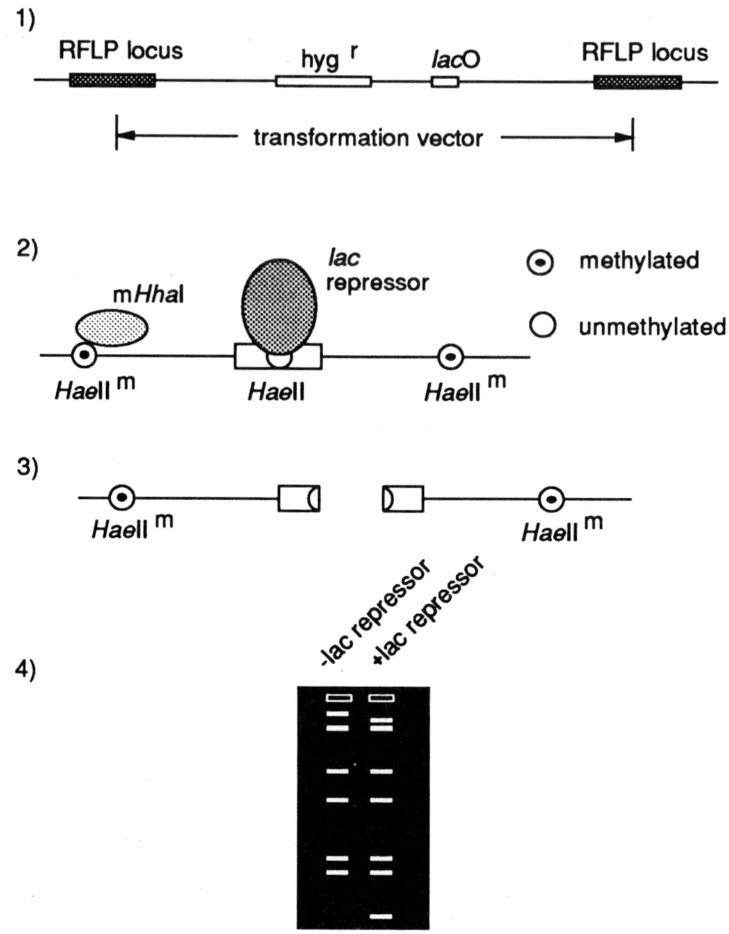

Fig. 6.4. *lac*O-assisted Achilles' cleavage (Koob and Szybalski, 1990): (1) A vector carrying the *lacO* sequence from *Escherichia coli* is targeted to a chosen RFLP locus by transformation. Intact chromosomes are prepared from this strain in agarose microbeads. (2) Chromosomal DNA is methylated with m*Hha*I in the presence of the *lac* repressor. (3) The *lac* repressor is denatured and the DNA is digested with *Hae*II. (4) Chromosomal DNA is analyzed by CHEF electrophoresis.

(Valent *et al.*, 1991) have also identified a gene controlling cultivar specificity toward CO39 in a different cross and a repetitive DNA sequence which maps approximately 7.6 cM away from the locus (Chumley and Valent, 1991). Whether *avr-CO39* is the specificity determinant or a gene modifying the activity of another gene that conditions cultivar specificity on CO39 (see chapter 4, this volume) remains to be confirmed by outcrossing and the eventual molecular characterization of the gene. However, the finding that a single gene mapping to the end of chromosome 1 conditions cultivar specificity toward CO39 in two separate crosses (Valent *et al.*, 1991; Smith and Leong, 1994; M. Orbach, Madison,

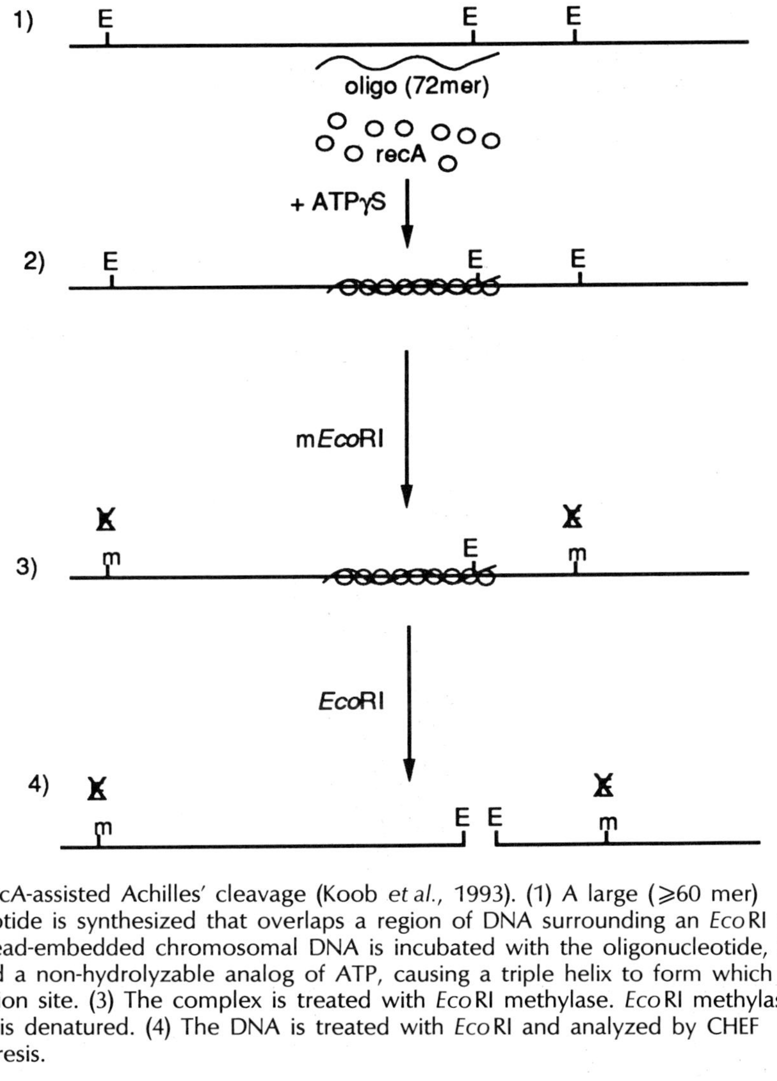

Fig. 6.5. RecA-assisted Achilles' cleavage (Koob *et al.*, 1993). (1) A large (≥60 mer) oligonucleotide is synthesized that overlaps a region of DNA surrounding an *Eco*RI site. (2) Microbead-embedded chromosomal DNA is incubated with the oligonucleotide, RecA protein and a non-hydrolyzable analog of ATP, causing a triple helix to form which covers the restriction site. (3) The complex is treated with *Eco*RI methylase. *Eco*RI methylase and the triplex is denatured. (4) The DNA is treated with *Eco*RI and analyzed by CHEF electrophoresis.

1993, personal communication), suggests that a simple gene-for-gene model (Flor, 1971) may apply in this case.

Unfortunately, no differential interaction phenotype was observed on the five isolines studied (Table 6.1) indicating that additional cultivar specificity genes corresponding to resistance genes in these isolines were not segregating in our mapping population. Progeny that were virulent on CO39 were also virulent on the isolines; likewise, progeny that were avirulent on CO39 were avirulent on the isolines. Because 2539 is already avirulent on CO39, it would not have been possible to make conclusions about the *avr* gene complement of 2539 without studying the interaction phenotypes of the progeny strains on the isolines. For example, if two unlinked *avr* genes had been segregating, a 3:1 ratio of avirulent:virulent progenies would have been observed when inoculated on the isoline carrying resistance genes corresponding to these *avr* genes.

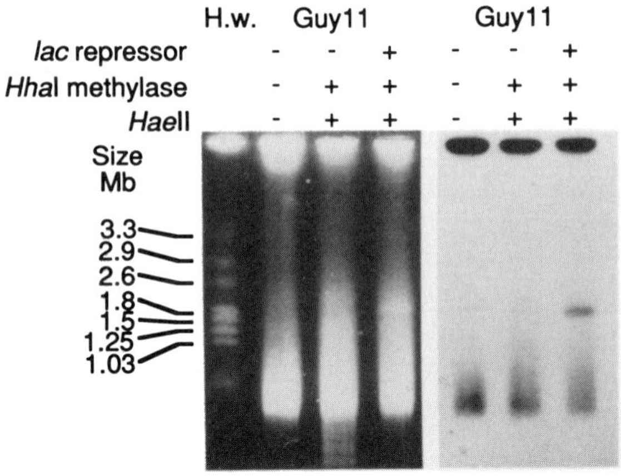

Fig. 6.6. *lac*O-mediated cleavage of the genome of *M. grisea* isolate Guy11 at marker 11. The left panel shows an ethidium bromide-stained gel of the products of Achilles' cleavage of chromosomal DNA of Guy11 containing the *lac*O integrated at marker 11. DNA was treated with *Hha*I methylase in the presence or absence of *lac* repressor as indicated above the lane. Each sample was subsequently digested with *Hae*II and analyzed by CHEF electrophoresis. *Hansenula wingei* chromosomes were employed as size standards. DNA from the resulting gel was transfered to nylon and probed as indicated with pSPlac11 (the *lac*O targeting vector containing marker 11). Panel on the right show the autoradiogram of the probed blot. DNA treated with *Hha*I methylase in the absence of *lac* repressor was completely insensitive to *Hae*II digestion. Inclusion of *lac* repressor in the methylation reaction protected the *Hae*II site in the *lac* operator from methylation which resulted in cleavage at marker 11 and release of a 1.8 mgbp fragment.

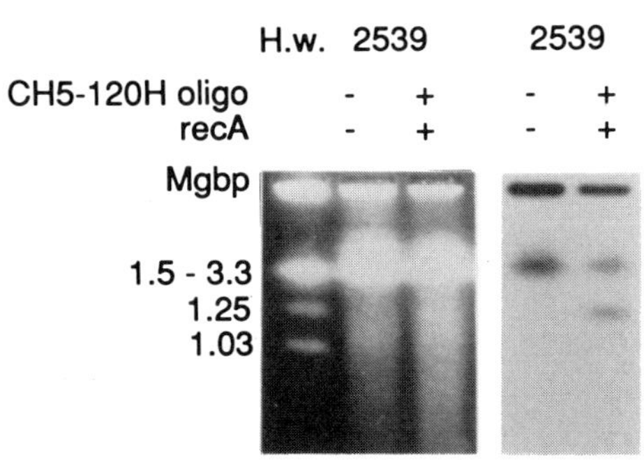

Fig. 6.7. RecA-mediated cleavage of the genome of *M. grisea* strain 2539 at marker CH5-120H. The panel on the left shows the ethidium bromide-stained gel of chromosomes of 2539 treated with *Eco*RI methylase in the presence or absence of nucleoprotein filament as indicated above the lane. Each sample was subsequently digested with *Eco*RI and analyzed by CHEF electrophoresis. *Hansenula wingei* chromosomes were run as size standards. DNA from the resulting gel was transferred to nylon and hybridized with marker 1.2 H. The panel on the right shows the resulting autoradiogram. Complete methylation occurs in the absence of nucleoprotein filament. By contrast, in the presence of the filament a 1.2 mgbp fragment is liberated upon treatment with *Eco*RI.

Table 6.1. Segregation of virulence on seven rice cultivars in *M. grisea* 'HL' progeny obtained from a cross between isolates 2539 × Guy11[a, b].

Cultivar	Gene[c]	Interaction phenotype[d] 2539	Guy11	Progeny segregation (A:V)[d]	X^2 (1:1)
C101LAC	*Pi-1*	1	3	25:18	1.1395
C101A51	*Pi-2*	1	3	25:18	1.1395
C104PKT	*Pi-3*	1	4	25:18	1.1395
C101PKT	*Pi-4a*	1	3	25:18	1.1395
C105TTP-4	*Pi-4b*	1	3	25:18	1.1395
C039		1	3	25:18	1.1395
51583[e]		4	4	0:43[f]	.....

[a] Rice seedling pathogenicity assays were conducted essentially as described in Valent *et al.* (1991).
[b] 'HL' progeny were developed by Hei Leung (Leung *et al.*, 1988) from a cross of *M. grisea* isolates Guy11 × 2539.
[c] Mackill and Bonman (1992).
[d] Interaction phenotypes are: Type 0 – no visible symptoms; Type 1 – small dark brown pinpoint-sized, non-sporulating lesions; Type 2 – dark brown, non-sporulating lesions 2–3mm in length; Type 3 – circular, sporulating lesions with tan centers and dark brown margins; and Type 4 – large diamond-shaped, sporulating lesions with tan centers and dark brown margins. *M. grisea* isolates causing lesion types 3 and 4 were considered virulent (V) while those causing types 0, 1, and 2 were considered avirulent (A).
[e] Indica rice cultivar obtained from Hei Leung (Leung *et al.*, 1988).
[f] Paul Tooley, Al Rossi, and Francis Latterell tested 32 of the 'HL' progeny on 51583 and obtained similar results to those reported here (Paul Tooley, 1990, Maryland, personal communication).

No differential interaction was observed between parental and progeny isolates on the Russian rice cultivar 51583 (Table 6.1). Our data clearly demonstrate that 2539 is a pathogen of rice and contrast with those obtained by Leung *et al.* (1988) who observed a 1:1 segregation for pathogenicity on 51583 in the same population of progeny isolates. This discrepancy may relate to differences in the environmental conditions used to assess pathogenicity as, in our system, generally fewer type 4 lesions were produced on 51583 by strain 2539 (Smith and Leong, 1994). Interestingly, this reduction in number of lesions was not manifested on 51583 by Guy11 or by any of the progeny tested.

To further investigate the pathogenicity of 2539 on rice, 765 additional rice genotypes were evaluated for susceptibility to strain 2539 (Smith and Leong, 1994). Of these, only 29 were found to be moderately susceptible to this strain. These data indicate that 2539 may be a rich source of cultivar specificity genes as the remaining 736 accessions were resistant to infection by strain 2539. Alternatively, 2539, which is a laboratory strain, may be deficient in factors affecting lesion development in rice (Valent *et al.*, 1991; Roumen, 1992). These must be cultivar-specific as 2539 is capable of producing type 4 lesions on 51583 (Table 6.1).

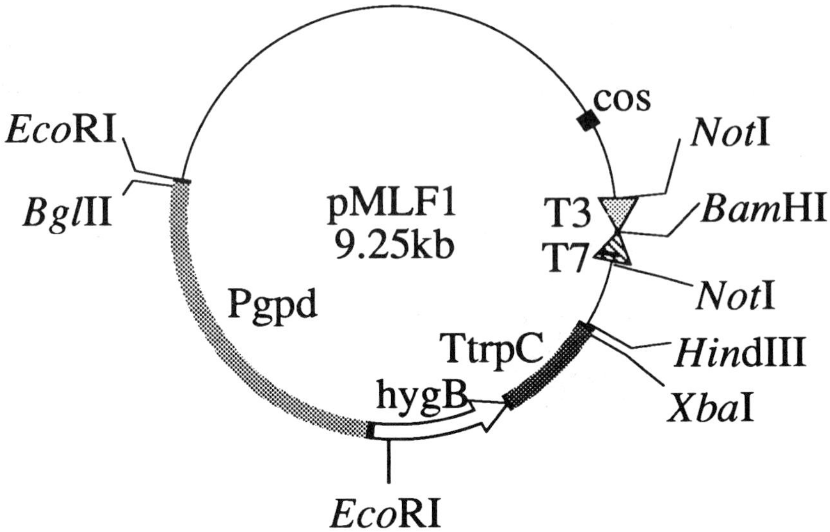

Fig. 6.8. pMLF1, a cosmid vector for *M. grisea*. pMLF1 was constructed by deleting the *Bam*HI site of cosmid pAN7-2 (P. Punt, unpublished vector) and inserting a blunt-ended, *Eco*RI fragment containing the bacteriophage T7 and T3 promoters flanking a *Bam*HI cloning site from pWE15 (Stratagene Inc., La Jolla, California) into the *Eco*RV site of pAN7-2. Thus radioactively labeled RNAs homologous to the *M. grisea* DNA flanking the *Bam*HI cloning site can be synthesized *in vitro* for use as probes in chromosome walking studies. In the walking studies described in the text, we have isolated the *M. grisea* DNA fragments flanking the *Bam*HI site by digestion of the cosmid-containing *M. grisea* insert DNA with *Apa*I, religation of the vector containing the flanking genomic DNA, transformation of *Escherichia coli*, and digestion of the resulting cosmid clones with *Apa*I and *Not*I, which cuts in the vector DNA immediately adjacent to the *Bam*HI cloning site. The resulting fragments are gel purified and radiolabeled by standard methods (Sambrook *et al.*, 1989).

Chromosome Walk to *AVR-CO39*

Chromosome walking (see Fig. 6.9) is a procedure that allows one to physically define a contiguous set of overlapping genomic DNA clones (Sambrook *et al.*, 1989). In practice one initiates the walk from a defined site in the genome, i.e. an RFLP marker which is linked to a gene of interest. Genomic DNA clones that contain this DNA are identified by hybridization of a genomic DNA library with the RFLP marker. Genomic insert DNA that is flanking the DNA cloning site of the vector is identified and used to reprobe the genomic DNA library to identify cross-hybridizing, overlapping segments of genomic DNA. This process is reiterated. The resulting set of genomic DNA clones is also surveyed for their ability to identify new RFLPs which can be used to define both the orientation and the progress of the chomosome walk.

A cosmid vector pMLF1 (Fig. 6.8) useful for chromosome walking experiments was constructed and employed to prepare a *M. grisea* genomic DNA library for strain 2539. Cosmids are plasmid-based DNA cloning vectors that contain the bacteriophage Lambda *cos* site (Sambrook *et al.*, 1989). This sequence is involved

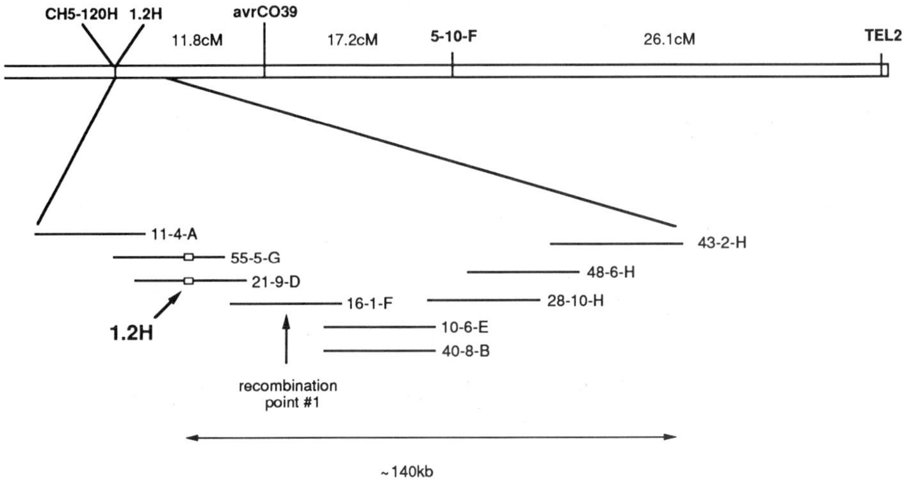

Fig. 6.9. Progress in the chromosome walk to *avr-CO39*. The orientation of the walk was established by the detection of a recombination point in the interval between marker 1.2H and *avrCO39* using an RFLP identified by cosmid 16-1-F. This analysis has shown that cosegregating markers CH5-120H and 1.2H are separated by at least 50 kbp.

in the packaging of bacteriophage DNA into phage particles. The presence of the *cos* site in the cosmid allows the cosmid clones to be packaged like phage DNA thus improving the cloning efficiency of large DNA as the phage particle can accept ~50 kb-sized DNA molecules. The *M. grisea* library described above contains 5184 clones with an average genomic DNA insert size of 34–40 kbp. These clones have been stored individually at −80°C in microtiter plates as well as in 54 pools, each containing DNA from clones in one plate, and will be made available through the Fungal Genetics Stock Center (University of Kansas Medical Center, Kansas City, KS 66160-7420).

As noted above, the ends of chromosome 1 lack genetic markers. In order to orient the chromosome walk and define the boundaries of the walk, it is useful to have linked markers flanking the gene to which one is walking. We have identified additional cosmid clones which map to chromosome 1 by probing the cosmid library described above with a gel-purified preparation of chromosome 1. Cosmid marker 5-10-F was shown to map 17.2 cM from *avr-CO39* and provides a telomere proximal marker that delimits the chromosome walk. We have now walked 140 kb toward *avr-CO39* from marker 1.2 H (Fig. 6.9). We estimate from the AC experiments at CH5-120H described above that the physical distance between the cosegregating markers CH5-120-H or 1.2 H and *avr-CO39* is 260 kb since the relationship of physical to genetic distance is estimated to be 22 kb/cM (1200 kb/55.1 cM). Recent experiments have supported this interpretation as AC of chromosome 1 of isolate 2539 at markers 5-10F and CH5-120H released a fragment of 600 kb which contains markers 1.2 H and *avrCO39* (M. Farman and S.A. Leong, unpublished findings). *avr-CO39* maps 11.8 and 17.2 cM, respectively, from these markers (Fig. 6.9). We are currently isolating large quantities of this fragment for use as a probe to identify homologous cosmid and cDNA clones spanning this region of the genome and for cloning directly into YAC (yeast artificial chromosome) vectors (Burke *et al.*, 1987) which can accommodate DNA fragments of this size.

Transformation of *M. grisea*

Transformation systems based on hygromycin and phleomycin resistance were developed for *M. grisea* (Leung *et al.*, 1990; M. Farman and S.A. Leong, unpublished findings) using vectors such as pAN7-2, a vector constructed for use in *Aspergillus* by P. Punt by ligating the 4.3 kb *Eco*RI-*Hin*dIII fragment of pAN7-1 (Punt *et al.*, 1987), which contains the hygromycin resistance determinant, into the 5.3 kb *Eco*RI-*Hin*dIII fragment of pJB8 (Ish-Horowicz and Burke, 1981). These vectors make use of the *Aspergillus nidulans* glyceraldehyde-3-phosphate-dehydrogenase promoter to drive expression of a bacterial antibiotic resistance gene. The availability of two dominant selectable markers provides flexibility in molecular genetic analyses as it opens up the possibility of transforming a strain twice. Moreover, it is not necessary to construct an auxotrophic recipient strain as complementation is not the basis of the selection strategy (Parsons *et al.*, 1987). Vector DNA is introduced into protoplasts and integrates into the genome in an apparently random fashion and both single copy and multicopy integration events have been documented (Leung *et al.*, 1990). DNA can be targeted to a particular site by incorporating a fragment of *M. grisea* genomic DNA into the vector (Sweigard *et al.*, 1992; M. Farman and S.A. Leong, unpublished findings). Evidence for meiotic and mitotic instability of transforming DNA in *M. grisea* has been obtained (Tooley *et al.*, 1992). Thus precautions must be taken when evaluating the phenotypes of transformants, especially those cultured under conditions of relaxed selection such as in the host plant.

The *E. coli* β-glucuronidase (GUS) gene under transcriptional control of the *A. nidulans* glyceraldhyde-3-P-dehydrogenase gene promoter (Roberts *et al.*, 1989) was stably introduced into *M. grisea* by transformation and β-glucuronidase activity was readily detected in transformants (data not shown). Preliminary studies have demonstrated the presence of blue-staining hyphae and conidiophores in blast lesions when rice leaves were infected with GUS gene-containing transformants of *M. grisea* and infiltrated with X-glucuronide (Fig. 6.10). These results indicate that the GUS reporter gene may be useful in detecting fungal growth in plants as well as for expression of developmentally regulated genes in planta.

Repetitive DNA in *M. grisea*

In a search for single copy genomic DNA probes for RFLP analysis, we have identified nine classes of repetitive DNA in *M. grisea* (Skinner *et al.*, 1993). Two of these have now been shown to cross-hybridize with MGR (*Magnaporthe grisea* repeated) sequences (Hamer *et al.*, 1989). A second class of element was found

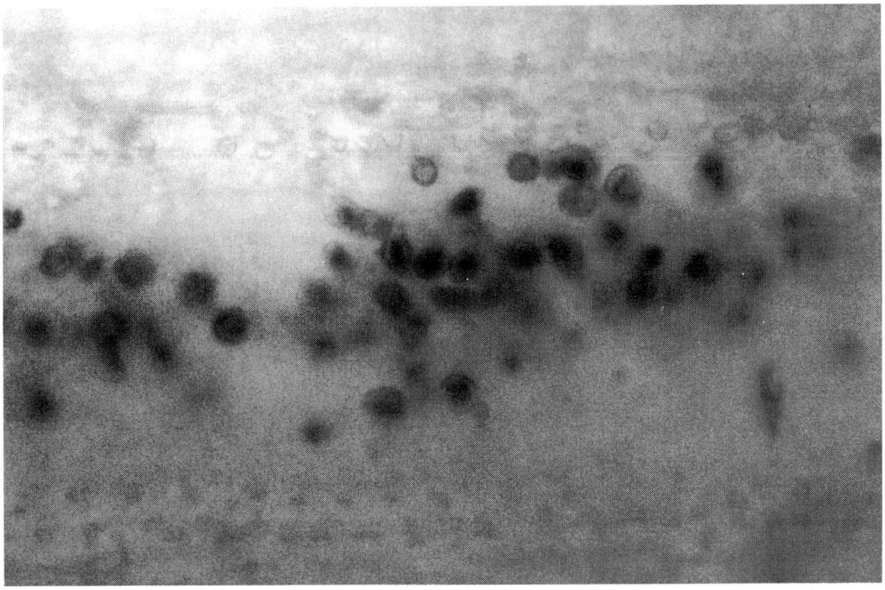

Fig. 6.10. Histochemical staining of a GUS^+ transformant infecting rice leaves. *Magnaporthe grisea* strain Guy11 was transformed with pNOM102 (Roberts *et al.*, 1989) and inoculated onto rice cultivar CO39. After seven days, leaves with lesions were detached, vacuum-infiltrated with X-gluc solution and incubated for four hours at 37°C (Gallagher, 1992). Magnification ×2000.

to have some structural similarity to *Alu* sequences found in primates as well as nucleotide or amino acid sequence similarity to reverse transcriptase of retrotransposon elements from diverse organisms (Skinner *et al.*, 1993). Combined restriction and hybridization analysis of genomic DNA and cosmid clones containing this repeat have indicated that the *Alu*-like repetitive DNA element is part of a larger element containing terminal repeats (M. Farman, Y. Tosa and S.A. Leong, unpublished findings). The complete DNA sequence of one *Alu*-like element has revealed that this element is 5.6 kb in size and is related to the Gypsy class of retrotransposons (Figs 6.11 and 6.12). Sequences of the left and right termini were identical and were directly repeated. This structure is characteristic of Long Terminal Repeat (LTR) sequences found at the termini of retrotransposons. The LTRs are 253 bp long and have 6 bp inverted terminal repeats at their ends. The terminal dinucleotides, TG . . . CA, are typical of the inverted repeats found in most retrotransposons. Imperfect target-site duplications were detected in the one clone examined as shown below.

CGTTAA/GGTTAT
ATTCG/?
GAAAT/?

Four out of the first five flanking bases were repeated at each end. The target-site of two additional cloned elements were different from those of the sequenced

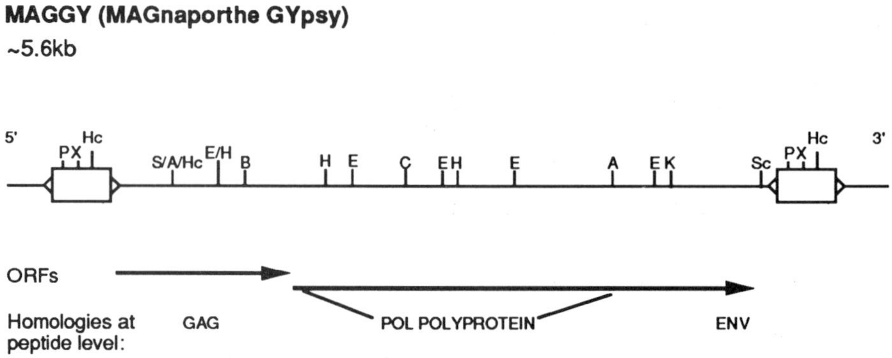

Fig. 6.11. Restriction map and structure of MAGGY (MAGnaporthe GYspy). A = AccI; B = *Bam*HI; C = *Cla*I; E = *Eco*RI; H = *Hind*III; Hc = *Hinc*II; K = *Kpn*I; P = *Pst*I; S = *Sal*I; Sc = *Sac*I; and X = *Xho*I.

element and from each other. The DNA sequences flanking these target sites have suggested that this element may integrate preferentially into AT-rich DNA.

Sequence data from the internal region of the element showed homology, at the peptide level, to reverse transcriptases of the Gypsy class of retrotransposons (Fig. 6.12), the highest homology being to Grasshopper (Dobinson *et al.*, 1993), a retrotransposon found in *M. grisea* grass-infecting isolates. Identities were also found to the retroelement polypeptides GAG, RNase H and Endonuclease (Fig. 6.12). Examination of single copy flanking sequences from three elements revealed that they were located in AT-rich portions of the genome. We have named this element MAGGY (*Mag*naporthe *Gy*psy element). The laboratories of John Hamer (chapter 5, this volume) and Barbara Valent (Lebrun *et al.*, 1993) have also recently identified retrotransposable elements (*Fosbury*) that are structurally related to MAGGY. Their studies have demonstrated that MAGGY is an active transposable element.

The copy number of MAGGY was determined to be ~100 and ~35 in the *M. grisea* mapping strains Guy11 and 2539, respectively (M. Farman and S.A. Leong, unpublished findings). We have mapped two MAGGY elements on the genetic map of *M. grisea* (Fig. 6.1). MAGGYb, which we have sequenced from strain 2539 (see above), was found at the same location on chromosome 1 of both parents (N. Nitta, M. Farman and S.A. Leong, unpublished findings). As this insertion contains imperfect target sequence duplications and is present in the genomes of two unrelated isolates, one from French Guyana and one developed from Asian isolates by intercrossing, this copy of MAGGY may represent an ancient insertion event. A second MAGGY, MAGGYa, has been mapped to chromosome 4 (N. Nitta, M. Farman and S.A. Leong, unpublished findings). Interestingly, a significant portion of the progeny analyzed showed an unusual DNA polymorphism in the DNA flanking this MAGGY (data not shown). As one parent strain appears not to carry a MAGGY at this location of the genome, these results may indicate that the presence of MAGGY elements can cause genome instability in crosses.

```
ORF1   NUCLEIC ACID-BINDING DOMAIN

        codon
MAGGY         382 MLCYRCGSQEHFVAKCPE
CfT-1         258 RTCYGCGKPGHIARDCRG
Copia         230 VKCHHCGREGHIKKDCFH
Ty3-2         265 RLCFTCKKEGHRLNECRA
HIV-1         390 VKCFNCGKEGHTARNCRA


ORF2   REVERSE TRANSCRIPTASE DOMAIN

         codon
MAGGY    350    TAFRTRFGLYESLVMPFGLTGAPATFQRYINDSLREYLDVFCTAYLDDI
grh      177    TAFRTRFGLFEWLVCPFGLAGAPATFQRYVNGVLVDTLGDYASAYLDDI
CfT-1    429    TAFRTRYGLYEFLVMPMGLTNAPASCQDLVNETLRDLLDVCVVAYMDDI
17.6     320    TAFSTKHGHYEYLRMPFGLKNAPATFQRCMNDILRPLLNKHCLVYLDDI
297      319    TAFSTKSGHYEYLRMPFGLRNAPATFQRCMNNILRPLLNKHCLVYLDDI


ORF2   RNase H DOMAIN

         codon
MAGGY    578    NYEIYDKELLAIIRCF
CfT-1    555    NYDIHDKELLAIVAAM
TED      630    NYSTIEKELLAIVWAT
17.6     545    NYSTIEKELLAIVWAT
Gypsy    530    NYATNERELLAIVWAL


ORF2   ENDONUCLEASE DOMAIN

         codon
MAGGY    921    KIDSLLSTAHHPETDGQTERFNASL
CfT-1    958    GIKH-LSTAYHPETDGQTERTNQTI
copia    568    GISYHLTVPHTPQLNGVSERMIRTI
HIV-1    861    GIKQEFGIPYNPQSQGVVESMNKEL
```

Fig. 6.12. Alignment of conserved sequences of protein domains of MAGGY with those of other retroelements (McHale *et al.*, 1992).

A survey of genomic DNA of *M. grisea* for the presence of MAGGY has shown that it was absent in eight different weed isolates, but was present at moderate copy number in two others. By contrast, MAGGY was ubiquitous among rice isolates (25 examined) (Skinner *et al.*, 1993).

Recent collaborative studies with Dr Bharat Chattoo and Mr Pradeep Kachroo of the University of Baroda, India, have resulted in the identification of another transposable element in *M. grisea* (Kachroo *et al.*, 1994). We have compared the partial DNA sequence of this element to those in the available databases and have found 39% similarity at the peptide level to the single ORF of Fot1, a new family of fungal transposons identified in the fungal plant pathogen *Fusarium oxysporum* (Daboussi *et al.*, 1992). This element, based on DNA hybridization studies of genomic DNA, belongs to an abundant class of repeated DNAs in

M. grisea and is present in the genome of both rice and grass-infecting isolates at high copy number.

Our long-term goal is to understand better how transposable elements affect genome evolution. Transposable elements have been shown to inactivate a bacterial avirulence gene in *Xanthomonas campestris* pv. *vesicatoria* (Kearney *et al.*, 1988) thus increasing the virulence of the pathogen. A mutation leading to inactivation of *avr2-YAMO* was recently shown to be caused by insertion of a *Pot2*-like element (S. Kang and B. Valent, personal communication). Mobile elements are also known to affect expression of adjacent genes and can mediate genome rearrangements (Berg and Howe, 1989). Whether pathogenic variability in *M. grisea* can be generally attributed to such transposable element-mediated processes remains to be determined.

Conclusions

Considerable progress has been made in the last decade to develop *M. grisea* into a fungal experimental system that is amenable to molecular genetic studies (Valent and Chumley, 1991). Sexually compatible strains of the fungus have been developed or identified in natural populations, genetic maps have been constructed, molecular karyotypes have been determined, DNA-mediated transformation of the fungus is possible, the population biology of the fungus is being clarified using molecular markers, and progress toward the genetic identification, cloning and characterization of transposable elements and host/cultivar specificity genes has been made. Information gained from these analyses will play a key role in the development of rational breeding strategies (McIntosh, 1992; chapter 16, this volume) to curtail rice blast disease and may offer opportunities to develop transgenic plants with novel kinds of disease resistance as described by de Wit (1992).

Acknowledgments

We thank Hei Leung, Paul Tooley, Francis Latterell, Al Rossi and Tony Marchetti for their cooperative assistance. This work was supported by a grant from the Rockefeller Foundation to S.A.L. and by the United States Department of Agriculture.

References

Barr, M.E. (1977) *Magnaporthe*, *Telimenella* and *Hyponectria* (Physosporellaceae). *Mycologia* 69, 952–956.

Berg, D.E. and Howe, M.M. (eds) (1989) *Mobile Elements.* American Society of Microbiology, Washington DC.

Bonman, M., Khush, G.S. and Nelson, R.J. (1992) Breeding rice for resistance to pests. *Annual Review of Phytopathology* 30, 507–528.

Borromeo, E.S., Nelson, R.J., Bonman, J.M. and Leung, H. (1993) Genetic differentiation among isolates of *Pyricularia* infecting rice and weed hosts. *Phytopathology* 83, 393–399.

Broun, P., Ganal, M.W. and Tanksley, S.D. (1992) Telomeric arrays display high levels of heritable polymorphism among closely related plant varieties. *Proceedings of the National Academy of Sciences of the United States of America* 89, 1354–1357.

Budde, A.D., Smith, J., Farman, M., Skinner, D. and Leong, S.A. (1993) Genetic map and molecular karyotype of the blast fungus *Magnaporthe grisea*. In: O'Brien, S.J. (ed.) *Genetic Maps, Vol. 6.* Cold Spring Harbor Laboratory, New York, pp. 3.110–3.112.

Burke, D.T., Carle, G.F. and Olson, M.V. (1987) Cloning of large segments of exogenous DNA into yeast by means of artificial chromosome vectors. *Science* 260, 806–812.

Chumley, F. and Valent, B. (1991) Strategies for characterizing and cloning host specificity genes in *Magnaporthe grisea*, the rice blast fungus. In: Patil, S.S., Ouchi, S., Mills, D. and Vance, C. (eds). *Molecular Strategies of Pathogens and Host Plants* Springer-Verlag, New York, pp. 131–138.

Daboussi, M.-J., Langin, T. and Brygoo, Y. (1992) Fot1, a new family of fungal transposable elements. *Molecular General Genetics* 232, 12–16.

De Feyter, R.F., Yang, Y. and Gabriel, D.W. (1993) Gene-for-gene interactions between cotton R genes and *Xanthomonas campestris* pv. *malvacearum avr* genes. *Molecular Plant-Microbe Interactions* 6, 225–237.

de Wit, P.J.G.M. (1992) Molecular characterization of gene-for-gene systems in plant-fungus interactions and the application of avirulence genes in control of plant pathogens. *Annual Review of Phytopathology* 30, 391–418.

Dobinson, K., Harris, R. and Hamer, J.E. (1993) Grasshopper, a long terminal repeat (LTR) retroelement in the phytopathogenic fungus *Magnaporthe grisea*. *Molecular Plant-Microbe Interactions* 6, 114–126.

Ellingboe, A.H. (1992) Segregation of avirulence/virulence on three rice cultivars in 16 crosses of *Magnaporthe grisea*. *Phytopathology* 82, 597–601.

Ellingboe, A.H., Wu, B.-C. and Robertson, W. (1990) Inheritance of avirulence/virulence in a cross of two isolate of *Magnaporthe grisea* pathogenic on rice. *Phytopathology* 80, 108–111.

Flor, H.H. (1971) Current status of the gene-for-gene concept. *Annual Review of Phytopathology* 9, 275–296.

Gallagher, S.R. (ed.) (1992) *GUS Protocols.* Academic Press, London.

Hamer, J.E. (1991) Molecular probes for rice blast disease. *Science* 252, 632–633.

Hamer, J.E., Farrall, L., Orbach, M.J., Valent, B. and Chumley, F.G. (1989) Host species-specific conservation of a family of repeated DNA sequences in the genome of a fungal plant pathogen. *Proceedings of the National Academy of Sciences of the United States of America* 86, 9981–9985.

Hayashi, N. and Naito, H. (1993) Genome mapping of *Magnaporthe grisea*, Abstract. In: *Program and Abstracts of the International Symposium on Rice Blast Disease.* Madison, Wisconsin, USA, August 9–13 1993.

Herbers, K., Conrads-Strauch, J. and Bonas, U. (1992) Race-specificity of plant resistance to bacterial spot disease determined by repetitive motifs in a bacterial avirulence protein. *Nature* 356, 172–174.

Hopkins, C.M., White, F.F., Choi, S.-H., Guo, A. and Leach, J.E. (1992) Identification of a family of avirulence genes from *Xanthomonas oryzae* pv. *oryzae*. *Molecular Plant-Microbe Interactions* 5, 451–459.

Horowitz, H., Thorburn, P. and Haber, J.E. (1984) Rearrangements of highly polymorphic regions near telomeres of *Saccharomyces cerevisae. Molecular Cellular Biology* 4, 2509–2517.

Ish-Horowicz, D. and Burke, J.F. (1981) Rapid and efficient cosmid cloning. *Nucleic Acids Research* 9, 2989–2998.

Jones, J.D.G., Hammond-Kosack, K., Jones, D., Thomas, C., Balint-Kurti, P., Dixon, M., Ashfield, T., Dickinson, M. and Hatzixanthis, C. (1993) Analysis of tomato genes for resistance to *Cladosporium fluvum*, Abstract. In: *Abstracts of the Symposium on the Molecular Genetics of Plant-Microbe Interactions*. Rutgers, NJ, April 21–24 1993.

Kachroo, P., Leong, S.A. and Chattoo, B.B. (1994) *Pot2*, an inverted repeat transposon from the rice blast fungus *Magnaporthe grisea. Molecular and General Genetics* (in press).

Kearney, B., Ronald, P.C., Dahlbeck, D. and Staskawicz, B.J. (1988) Molecular basis for evasion of plant host defence in bacterial spot disease of pepper. *Nature* 332, 541–543.

Kobayashi, D.Y., Tamaki, S.J. and Keen, N.J. (1989) Cloned avirulence gene from the tomato pathogen *Pseudomonas syringae* pv. *tomato* confers cultivar specificity on soybeans. *Proceedings of the National Academy of Sciences of the United States of America* 86, 157–161.

Kolmer, J.A. and Ellingboe, A.H. (1988) Genetic relationships between fertility, pathogenicity, and virulence in the rice blast fungus *Magnaporthe grisea. Canadian Journal of Botany* 66, 891–897.

Koob, M. and Szybalski, W. (1990) Cleaving yeast and *Escherichia coli* genomes at a single site. *Science* 250, 271–273.

Koob, M., Burkiewicz, A., Kur, J. and Szybalski, W. (1993) RecA-AC: Single-site cleavage of plasmids and chromosomes at any predetermined restriction site. *Nucleic Acids Research* 20, 5831–5836.

Lander, E.S., Green, P., Abrahamson, J., Barlow, A., Daly, M.J., Lincoln, S.E. and Newburg, L. (1987) MAPMAKER: An interactive computer package for constructing primary genetic linkage maps of experimental and natural populations. *Genomica* 1, 174–181.

Lau, G.W., Chao, C.T. and Ellingboe, A.H. (1993) Interaction of genes controlling avirulence/virulence of *Magnaporthe grisea* on rice cultivar Katy. *Phytopathology* 83, 375–382.

Lebrun, M.-H., Chumley, F. and Valent, B. (1993) Molecular analysis of spontaneous mutations in *Magnaporthe grisea*. Abstract. In: *Program and Abstracts of the International Symposium on Rice Blast Disease*. Madison, Wisconsin, USA, August 9–13 1993.

Leung, H. and Williams, P.H. (1987) Nuclear division and chromosome behavior during meiosis and ascosporogenesis in *Pyricularia oryzae* Carava. *Canadian Journal of Genetics and Cytology* 27, 697–704.

Leung, H., Borromeo, E.S., Bernardo, M.A. and Notteghem, J.L. (1988) Genetic analysis of virulence in the rice blast fungus *Magnaporthe grisea. Phytopathology* 78, 1227–1233.

Leung, H., Lehtinen, U., Karjalainen, R., Skinner, D., Tooley, P., Leong, S.A. and Ellingboe, A.H. (1990) Transformation of the rice blast fungus *Magnaporthe grisea* to hygromycin B resistance. *Current Genetics* 17, 409–411.

Levy, M., Romao, J., Marchetti, M.A. and Hamer, J.E. (1991) DNA fingerprinting with a dispersed repeated sequence resolves pathotype diversity in the rice blast fungus. *Plant Cell* 3, 95–102.

Long, S. and Staskawicz, B. (1993) Prokaryotic plant parasites. *Cell* 73, 921–935.

Mackill, D. and Bonman, J.M. (1986) New hosts of *Pyricularia oryzae*. *Plant Disease* 70, 125–127.

Mackill, D. and Bonman, J.M. (1992) Inheritance of blast resistance in near-isogenic lines of rice. *Phytopathology* 82, 746–749.

McHale, M.T., Roberts, I.N., Noble, S.M., Beaumont, C., Whitehead, M.P., Seth, D, and Oliver, R.P. (1992) CfT-I: an LTR-retrotransposon in *Cladosporium fulvum*, a fungal pathogen of tomato. *Molecular and General Genetics* 233, 337–347.

McIntosh, R.A. (1992) Pre-emptive breeding to control wheat rusts. *Euphytica* 63, 103–113.

Morell, V. (1993) The puzzle of the triple repeats. *Science* 260, 1422–1423.

Parsons, K.A., Chumley, F.G. and Valent, B. (1987) Genetic transformation of the fungal pathogen responsible for rice blast disease. *Proceedings of the National Academy of Sciences of the United States of America* 84, 4161–4165.

Punt, P.J., Oliver, R.P., Dingemanse, M.A., Pouwels, P.H. and van den Hondel, C.A.M.J.J. (1987) Transformation of *Aspergillus* based on the hygromycin B resistance marker from *Escherichia coli*. *Gene* 56, 117–124.

Roberts, I.N., Oliver, R.P., Punt, P.J. and van den Hondel, C.A.M.J.J. (1989) Expression of the *Escherichia coli* β-glucuronidase gene in industrial and phytopathogenic filamentous fungi. *Current Genetics* 15, 177–180.

Romao, J. and Hamer, J.E. (1992) Genetic organization of a repeated DNA sequence family in the rice blast fungus. *Proceedings of the National Academy of Sciences of the United States of America* 89, 5316–5320.

Rossman, A.Y., Howard, R.J. and Valent, B. (1990) *Pyricularia grisea*, the correct name for the rice blast fungus. *Mycologia* 82, 509–512.

Roumen, E.C. (1992) Small differential interactions for partial resistance in rice cultivars to virulent isolates of the blast pathogen. *Euphytica* 64, 143–148.

Sambrook, J., Fritsch, E.F. and Maniatis, T. (1989) *Molecular Cloning, A Laboratory Manual*, 2nd edn. Cold Spring Harbor Laboratory, New York.

Schechtman, M.G. (1989) Segregation patterns of chromosome ends. *Fungal Genetics Newsletter* 36, 71–73.

Shahjahan, A.K.M., Nahar, N.S., Levy, M., Renganathan, N. and Hamer, J.E. (1993) Genetic organization of the rice blast fungus in Bangladesh. Abstract. In: *Abstracts of the Sixth Meeting of the International Program on Rice Biotechnology*, Chiangmai, Thailand, February 1–5 1993.

Silue, D. and Notteghem, J.L. (1992) Identification of a cross between two compatible isolates of *Magnaporthe grisea* (Herbert) Barr and genetic analysis of avirulence/virulence on rice. *Journal of Phytopathology* 135, 77–83.

Silue, D., Tharreau, D. and Notteghem, J.L. (1992a) Evidence for a gene-for-gene relationship in the *Oryza sativa–Magnaporthe grisea* pathosystem. *Phytopathology* 82, 577–580.

Silue, D., Tharreau, D. and Notteghem, J.L. (1992b) Identification of *Magnaporthe grisea* avirulence genes to seven rice cultivars. *Phytopathology* 82, 1462–1467.

Skinner, D.Z., Budde, A.D. and Leong, S.A. (1991) Molecular karyotype analysis of fungi. In: Bennett, J. and Lasure, L. (eds), *More Gene Manipulations in Fungi.* Academic Press, Orlando, pp. 86–103.

Skinner, D.Z., Budde, A.D., Farman, M.F., Smith, J.R., Leung, H. and Leong, S.A. (1993) Genetic map, molecular karyotype and occurrence of repeated DNAs in the rice blast fungus *Magnaporthe grisea*. *Theoretical and Applied Genetics* 87, 545–557.

Smith, J.R. and Leong, S.A. (1994) Mapping of a *Magnaporthe grisea* locus affecting rice (Oryza sativa) cultivar specificity. *Theoretical and Applied Genetics* (in press).

Sweigard, J.A., Chumley, F. and Valent, B. (1992) Disruption of the *Magnaporthe grisea* cutinase gene. *Molecular and General Genetics* 232, 183–190.

Sweigard, J.A., Valent, B., Orbach, M.J., Walter, A.M., Rafalski, A. and Chumley, F. (1993) Genetic map of the rice blast fungus *Magnaporthe grisea* (n = 7). In: O'Brien, S.J. (ed.) *Genetic Maps, Vol. 6*. Cold Spring Harbor Laboratory, New York, pp. 3.113–3.115.

Tharreau, D., Dioh, D., Lebrun, M.-H., Orbach, M. and Notteghem, J.L. (1994) Mapping avirulence genes in the rice blast fungus, *Magnaporthe grisea*. In: *Abstracts of the Seventh International Symposium on Molecular-Plant Microbe Interactions*, Edinburgh, 26 June–1 July 1994.

Tooley, P.W., Leung, H. and Leong, S.A. (1992) Meiotic and mitotic stability of transforming DNA in the phytopathogenic fungus *Magnaporthe grisea*. *Current Genetics* 21, 55–60.

Valent, B. and Chumley, F. (1991) Molecular genetic analysis of the rice blast fungus *Magnaporthe grisea*. *Annual Review of Phytopathology* 29, 443–467.

Valent, B., Farrall, L. and Chumley, F. (1991) *Magnaporthe grisea* genes for pathogenicity and virulence identified through a series of backcrosses. *Genetics* 127, 87–101.

Whalen, M.C., Stall, R.E. and Staskawicz, B.J. (1988) Characterization of a gene from a tomato pathogen determining hypersensitive resistance in a non-host species and genetic analysis of this resistance in bean. *Proceedings of the National Academy of Sciences of the United States of America* 85, 6743–6747.

Whalen, M.C., Innes, R.W., Bent, A.F. and Staskawicz, B.J. (1991) Identification of *Pseudomonas syringae* pathogens of *Arabidopsis* and a bacterial locus determining avirulence on both *Arabidopsis* and soybean. *Plant Cell* 3, 49–59.

Yu, Z.H., Mackill, D.J., Bonman, J.M. and Tanksley, S.D. (1991) Tagging genes for blast resistance in rice via linkage to RFLP markers. *Theoretical and Applied Genetics* 81, 471–476.

7 Avirulence Genes and Mechanisms of Genetic Instability in the Rice Blast Fungus

B. Valent and F.G. Chumley

Central Research and Development, The DuPont Company, PO Box 80402, Experimental Station, Wilmington, DE 19880-0402, USA

Introduction

Host specificity in rice blast disease, which is caused by the fungus *Magnaporthe grisea* (anamorph, *Pyricularia grisea*) (Rossman *et al.*, 1990), has intrigued plant pathologists for generations because of the very large number of cultivar-specific fungal isolates and because of the high potential of the fungus to mutate, gaining the ability to infect previously resistant rice varieties. This chapter will describe how modern genetic analysis is providing answers to key questions about the mechanisms of host specificity and genetic instability in the rice blast fungus.

Cloned genes for host specificity, such as avirulence (AVR) genes, are serving as valuable tools for analyzing molecular genetic mechanisms that determine host-cultivar and host-species specificity. Further studies of AVR gene products will advance the knowledge of biochemical mechanisms that determine race-cultivar specificity. As molecular probes, cloned AVR genes are now being used to study the distribution of these genes in fungal populations. Analyses of this type will permit more reliable determination of the potential for variation in the pathogen population as a whole. All of these studies will provide important insights for controlling rice blast disease in the field.

Genetic variability is a common feature of rice blast cultures in the laboratory. Careful observations have shown that genetic stability or instability is a characteristic of particular loci in particular strains. That is, instability in the rice blast fungus is only associated with certain genetic loci, and genetic loci that are unstable in some strains of the fungus appear to be stable in others. Analysis using cloned genes has revealed the molecular basis for changes at unstable loci. For example, cloning of the melanin biosynthetic gene, *BUF1*, has established that the molecular basis for instability at this genetic locus is frequent deletion. Analysis of mutation events in two unstable host specificity genes, *PWL2* and *AVR2-YAMO*, aided in cloning the two AVR genes and provided insight into the molecular mechanism of instability associated with each gene, as discussed below.

Another valuable approach for studying genetic variation in rice blast fungus has been the use of repetitive DNA sequences as hybridization probes for characterizing strains via 'DNA fingerprinting'. DNA fingerprinting has been used to address questions about the stability of host specificity in the field (Levy *et al.*, 1991, 1993). The rice pathogen population in the field appears to be clonal. That is, field isolates that infect rice can be grouped in distinct lineages (haplotypes), based on the similarity of DNA fingerprints using MGR586 as a probe (Hamer *et al.*, 1989; Levy *et al.*, 1993; Shull and Hamer, this volume). Repetitive DNA sequences are also valuable probes for verification of parent/mutant relationships, as will be discussed in this chapter. In the case of wheat blast in Brazil, the hypothesis that wheat blast pathogens were derived by mutation from rice pathogens was proven false by MGR fingerprinting (F.G. Chumley and B. Valent, unpublished results).

Rice Blast as a Gene-for-Gene System

Individual isolates of *M. grisea* have a limited host range. Races (pathotypes) of the fungus are distinguished among isolates that parasitize rice, depending on the rice cultivars they successfully infect (Atkins *et al.*, 1967; Ling and Ou, 1969; Yamada *et al.*, 1976). The mechanism by which new races of a pathogen arise is an intriguing problem. Although the extent of instability in race determinants has been debated (Latterell, 1975; Ou, 1985), it is clear that new races rapidly appear in the field when new blast resistant rice cultivars are introduced. The working hypothesis for rice blast researchers has been that the rice blast system represents a 'gene-for-gene' system as first defined by Flor (1971). This hypothesis states that for every dominant major resistance gene in the host, there is a corresponding dominant gene for avirulence carried by the pathogen. Thus, there is a one-for-one functional correlation between resistance genes in the host and AVR genes in the pathogen. It was anticipated that a rice pathogen would be capable of infecting a particular rice cultivar only if the pathogen lacks all AVR genes that correspond to the host's resistance genes.

Evidence is accumulating that the rice blast system is, in fact, a gene-for-gene system. Several investigators have now reported Mendelian segregation of single genes that control cultivar specificity in the pathogen (Leung *et al.*, 1988; Ellingboe *et al.*, 1990; Valent *et al.*, 1991; Silué *et al.*, 1992a,b). It also appears that AVR genes identified in the fungus correspond to dominant resistance genes in the host as predicted by the gene-for-gene model (Silué *et al.*, 1992a). Dominance of AVR genes has been harder to establish, because dominance cannot be determined for *M. grisea* genes by classical methods (Crawford *et al.*, 1986). Therefore, establishing dominance first required cloning of AVR genes, and for this purpose it was useful to assume that AVR genes would be dominant. The prediction was that dominant AVR genes could be detected by their ability to transform a virulent strain of the fungus into an avirulent strain (Fig. 7.1a).

A second prediction of the gene-for-gene hypothesis is that mutations should occur that change the phenotype of a strain from avirulence (the dominant phenotype due to the action of an AVR gene) to virulence (the recessive

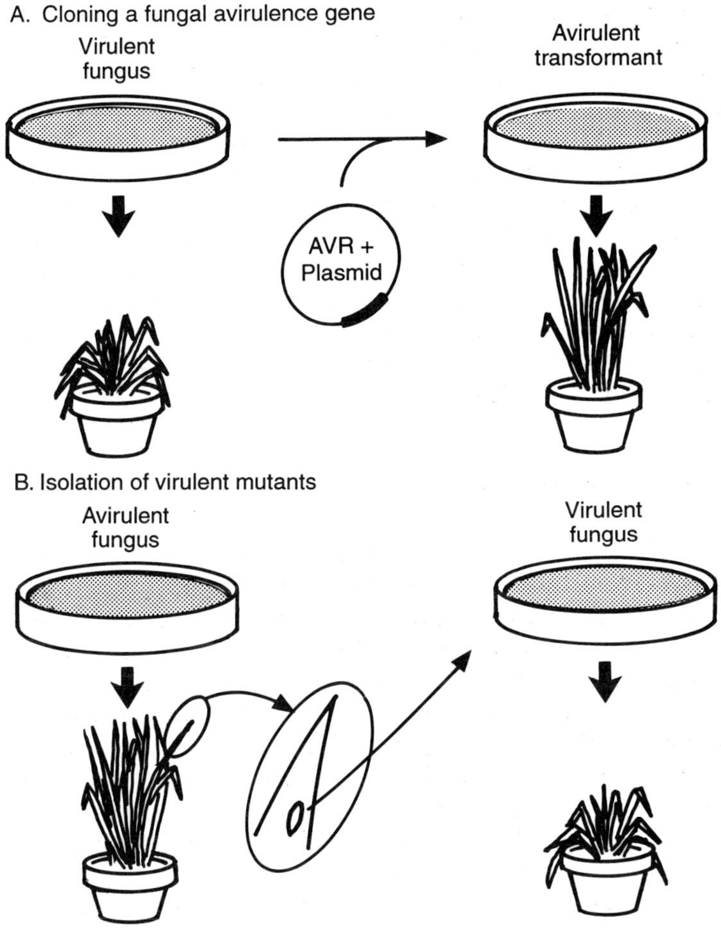

Fig. 7.1. (A) The predicted phenotype for 'avirulence genes' (AVR genes) is that the presence of a functional gene converts a virulent (pathogenic) strain of the fungus into an avirulent (nonpathogenic) strain. Therefore, the only assay for these genes was transformation (Parsons *et al.*, 1987; Leung *et al.*, 1990) of a virulent recipient strain with individual plasmids from a *M. grisea* gene library. Individual transformants were screened for specific loss of the ability to infect the host plant of interest. *AVR2-YAMO*, *PWL2*, and *PWL1* all convert the phenotype of a pathogen from virulence to avirulence. (B) Loss of function of a presumptive AVR gene should result in the gain of ability to infect a specific host. Isolation of fungus from 'rare lesions' obtained after inoculation with an avirulent strain results in recovery of a mutant strain that now devastates that particular host. Some host specificity genes, *PWL2*, *AVR2-YAMO* and *Avr1-TSUY*, appear to be 'unstable'. That is, spontaneous virulent mutants derived from strains carrying the avirulent (or nonpathogenic) alleles of these genes appear frequently in standard assays. No virulent mutants have been recovered from strains carrying *Avr1-CO39*, *Avr1-M201*, *Avr1-YAMO* and *PWL1* in similar infection assays. Thus, these latter four genes are considered to be 'stable' relative to the unstable genes *PWL2*, *AVR2-YAMO* and *Avr1-TSUY* (Table 7.1).

Table 7.1. Genes for host specificity identified in *M. grisea*.

Gene[a]	Specific toward	Source[b]	Stability[c]	Status; reference
PWL1	WLG	GG pathogen, WGG-FA40	Stable	Cloned by homology to *PWL2*; Kang and Valent, unpublished
PWL2	WLG	Rice pathogen, Guy11	Unstable	Cloned from map position; Sweigard, Carroll, Chumley and Valent, unpublished
Avr1-CO39	CO39	WLG pathogen	Stable	Valent *et al.*, 1991
Avr1-M201	M201	WLG pathogen	Stable	Valent *et al.*, 1991
Avr1-YAMO	Yashiro-mochi	WLG pathogen	Stable	Valent *et al.*, 1991
AVR2-YAMO	Yashiro-mochi	Rice pathogen, 0–137	Unstable	Cloned from map position; Orbach, Farrall, Chumley and Valent, unpublished
Avr1-MARA	Maratelli	GG pathogen	Stable	Mapped (see Fig. 7.4)
Avr1-TSUY	Tsuyuake	Rice pathogen, 0–137	Unstable	Mapped (see Fig. 7.4)

[a] Cloned genes are now named with all upper case letters ('AVR'), in keeping with standard genetic nomenclature for dominant genes. Dominance is not proven for uncloned genes; these are still denoted as 'Avr' genes.
[b] The 'source' listed for each gene is based on the infectivity of the parents in the cross in which the gene was identified. The source indicated is the parent strain that carried the allele for nonpathogenicity or avirulence. WLG indicates weeping lovegress, *Eragrostis curvula*. GG indicates goosegrass (*Eleusine indica*) but isolates also infect other *Eleusine* spp. including finger millet. When the source is known to be a field isolate, that strain is named.
[c] A host specificity determinant is listed as 'unstable' if virulent or pathogenic mutants appear at a high frequency, as described in the text and Fig. 7.1b.

phenotype due to loss of AVR gene function; Fig. 7.1b). In fact, such mutants are common from some *M. grisea* strains under laboratory conditions. A high frequency of variation at some genetic loci in the laboratory may or may not reflect the behavior of the fungus in the field. However, mechanisms of genetic variation could be expected to be similar in the laboratory and in the field. Questions of how studies of any plant pathogen system in the laboratory relate to the field situation should be considered carefully.

Understanding the determinants of specificity in a few pathogen strains is the first step toward understanding the properties of a more complex population.

Although this chapter focuses on examples of genetic instability, these examples should be taken as illustrative of the molecular basis for variation, and not as indicative of the behavior of the rice pathogen population as a whole. We describe here the first cloned genes that control host specificity in the rice blast fungus. A functional *PWL2* gene, which encodes a protein containing 145 amino acids, prevents the fungus from infecting weeping lovegrass, *Eragrostis curvula*. We refer to *PWL2* as an AVR gene because it has formal genetic properties similar to those of AVR genes. A functional *AVR2-YAMO* gene, which encodes a protein with 223 amino acids, prevents the fungus from infecting rice variety Yashiro-mochi. DNA sequences with strong homology to both *PWL2* and *AVR2-YAMO* are present in most rice pathogens and in most *Digitaria* pathogens. Functional homologs of both genes, including the previously identified gene *PWL1*, have been cloned from strains of the fungus that infect grasses other than rice.

Stable and Unstable Genetic Loci

M. grisea isolates tend to be unstable with regard to some characteristics relating to morphology, fertility and pathogenicity during subculture in the laboratory (Valent and Chumley, 1991), as is typical of many fungal pathogens. For example, some strains of the fungus, including pathogens of rice as well as pathogens of other grasses, show frequent spontaneous appearance of mutants with altered spore morphology due to mutations in the *SMO1* gene (Hamer *et al.*, 1989; Hamer and Givan, 1990). These mutants show altered pathogenicity phenotypes on rice. During subculture in the laboratory, many rice pathogens collected from fields around the world show genetic instability for expression of the *BUF1* melanin biosynthetic gene, which is essential for infecting unwounded plants (Chumley and Valent, 1990; Howard and Ferrari, 1989; chapter 1, this volume). Because mutations occur as the fungus is subcultured on agar media, it is critical that subculturing of the fungus be minimized, preferably by maintaining permanent stocks in a frozen, non-metabolizing condition (Valent *et al.*, 1991).

High mutability is not, however, a general feature of *M. grisea* genes. Many genes appear to mutate at normal frequencies, including ALB^+ and RSY^+, which encode melanin biosynthetic enzymes (Chumley and Valent, 1990). Auxotrophic colonies were recovered at a frequency of approximately 0.35% of the survivors of UV mutagenesis in which about 95% of the starting population are killed (Crawford *et al.*, 1986). This frequency is consistent with expectations for a typical haploid fungus. Spontaneous mutants resistant to 6-methylpurine, the sulfonylurea chlorimuron ethyl (DuPont Agricultural Products; 'DPX-F6025') or benomyl were isolated in independent experiments at frequencies of 10^{-6}–10^{-7} (F.G. Chumley, unpublished results). These 'normal' frequencies for recovery of mutants in a haploid fungus stand in sharp contrast to the very high frequency of spontaneous buf^- and smo^- mutants, and to the high frequency of some host specificity mutants.

A molecular analysis was undertaken with the goal of understanding why the *BUF1* gene is unstable in some strains of the pathogen. The Chinese rice pathogen O-137 shows the most extreme instability at this locus. About 0.25% of the conidia produced by O-137 during subculture in the laboratory carry *buf*$^-$ mutations. Spontaneous *buf*$^-$ mutants occasionally occur during vegetative growth of other rice pathogens, including O-135 and Guy11, but at a lower frequency than with O-137. Many rice pathogens cross with other normally pigmented strains to produce 5–25% *buf*$^-$ progeny (Leung *et al.*, 1988; Chumley and Valent, 1990; Valent *et al.*, 1991), indicating that instability also occurs during meiosis. In order to understand the nature of the mutations, Leonard Farrall cloned the *BUF1* gene from a gene library by complementation of a *buf*$^-$ defect in a mutant strain. Southern hybridization analysis was conducted using a *BUF1* plasmid as a probe to detect structural changes at the *BUF1* locus (Fig. 7.2). Mutants derived from three field isolates that infect rice and two laboratory strains that infect weeping lovegrass and/or finger millet (*Eleusine coracana*) all contained deletions of the *BUF1* gene and adjacent genomic sequences (L. Farrall, F.G. Chumley, and B. Valent, unpublished results). The deletions can be quite large. Mutants obtained from O-135 and Guy11 lost all homology to the 40 kb of genomic sequences included in the hybridization probe. For a particular fungal strain, deletion events appeared to be identical in independently derived mutants that arose during vegetative growth or in the sexual cycle. We hypothesize that unstable *BUF1* genes are deleted due to homologous recombination of flanking repetitive DNA sequences. It is well documented that repetitive DNA sequences can destabilize regions of the genome in many organisms (Petes and Hill, 1988). We speculate that in some cases the difference between stable and unstable loci is that unstable loci are flanked by repetitive DNA sequences whereas stable loci are not.

Major Genes that Control Host Specificity are Common

Several laboratories have invested considerable effort to identify genes that control host specificity in *M. grisea* (Leung *et al.*, 1988; Ellingboe *et al.*, 1990; Valent *et al.*, 1991; Silué *et al.*, 1992a,b). In the following discussion, we will focus on the 'major genes' (single genes that have an all-or-nothing effect on phenotype) that have been identified in our laboratory (Table 7.1). 'Minor' genes with individually small effect have also been detected (Valent *et al.*, 1991), but these will not be discussed here. In a cross between the finger millet pathogen, WGG-FA40, and the weeping lovegrass pathogen, K76-79, Valent *et al.* (1986) defined a single gene, named *PWL1*, that determined pathogenicity toward weeping lovegrass. Segregation of *PWL1* was confirmed in subsequent crosses. This gene is likely to be the same gene described by Yaegashi (1978). A similar gene, *PWL2*, controlling the ability to infect weeping lovegrass was identified in the RFLP mapping cross described below.

Three avirulence genes, *Avr1-CO39*, *Avr1-M201* and *Avr1-YAMO* (Table 7.1), control specificity toward rice cultivars CO39, M201 and Yashiro-mochi, respectively. These genes were identified in an extensive backcrossing study undertaken to determine the genetic differences between O-135, a Chinese rice pathogen, and strain 4091-5-8, a laboratory strain that infects weeping lovegrass, but produces no visible symptoms on rice (Valent *et al.*, 1991). *Avr1-CO39*, *Avr1-M201* and *Avr1-YAMO* appear to be derived from strain 4091-5-8, so they

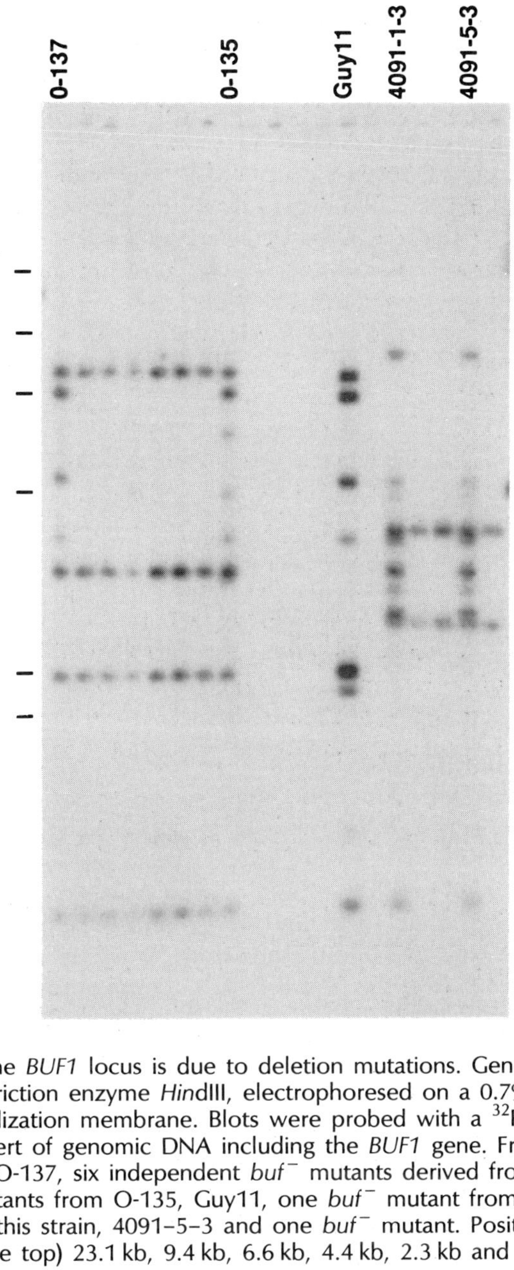

Fig 7.2. Instability at the *BUF1* locus is due to deletion mutations. Genomic DNAs were digested with the restriction enzyme *Hin*dIII, electrophoresed on a 0.7% agarose gel, and transferred to a hybridization membrane. Blots were probed with a ^{32}P-labeled plasmid containing a 40 kb insert of genomic DNA including the *BUF1* gene. From left to right, lanes contained DNA from O-137, six independent *buf*$^{-}$ mutants derived from O-137, O-135, 4 independent *buf*$^{-}$ mutants from O-135, Guy11, one *buf*$^{-}$ mutant from Guy11, 4091–1–3 and 2 *buf*$^{-}$ mutants from this strain, 4091–5–3 and one *buf*$^{-}$ mutant. Positions marked for size standards are (from the top) 23.1 kb, 9.4 kb, 6.6 kb, 4.4 kb, 2.3 kb and 2.0 kb.

either came originally from the *Eragrostis* pathogen K76–79 or the *Eleusine* pathogen WGG-FA40 (Valent *et al.*, 1986). Yaegashi and Asaga (1981) reported strong circumstantial evidence that a finger millet pathogen, WGG-FA40, carries an AVR gene corresponding to the rice blast resistance gene *Pi-a*. These genetic results all suggested that AVR genes specific for rice cultivars are common in *M. grisea* pathogens of grasses other than rice.

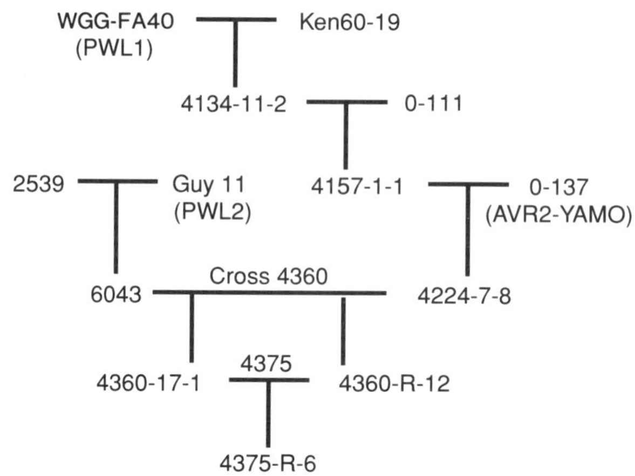

Fig. 7.3. Pedigree of *M. grisea* strains. The parents in cross 4360 (used to generate the RFLP genetic map in Fig. 7.5) were strains 4224–7–8 and 6043. Strain 4224–7–8 was derived from a series of crosses involving four field isolates, the *Eleusine* pathogen WGG-FA40 (from H. Yaegashi, Tsukuba, Japan), and the rice pathogens Ken60-19 (from H. Yaegashi, Tsukuba, Japan), O-111 (CH-104-2 from S.H. Ou), and O-137 (collected by B. Valent and Shen Ying, China National Rice Research Institute, Hangzhou, China). Strain 6043 was generously provided by H. Leung (Washington State University). See Leung *et al.* (1988) for the pedigree of 6043. Source strains for the cloned avirulence genes are listed.

Other AVR genes were identified in a second series of crosses summarized in Fig. 7.3. In cross 4360, parental strain 4224–7–8 was avirulent on rice cultivars Yashiro-mochi, Maratelli and Tsuyuake, but virulent on rice cultivars CO39, M201 and Sariceltik. The second parent, strain 6043 (Leung *et al.*, 1988), was virulent on all of these rice cultivars. Strain 4224–7–8 was a weeping lovegrass pathogen, but strain 6043 was not. Data from five complete tetrads (a representative tetrad is shown in Table 7.2) and from 67 random ascospore cultures suggest that different single genes control each example of host specificity. Simple segregation was observed for all-or-nothing differences in the ability to infect each of the three rice cultivars and weeping lovegrass. Other crosses confirmed segregation of each of these putative AVR genes (B. Valent, L. Farrall, and F.G. Chumley, unpublished results).

As summarized in Table 7.1, two genes that control the ability to infect rice cultivar Yashiro-mochi were identified from different genetic backgrounds. Genetic crosses were performed to determine if *AVR2-YAMO*, which was derived from a rice pathogen, is allelic to *Avr1-YAMO*, which is derived from a grass pathogen. Data from three different crosses indicated that the two genes are not linked to one another. Preliminary data also suggested that *PWL1*, from a finger millet pathogen, and *PWL2*, from a rice pathogen, were unlinked. Therefore, in both cases there are either two distinct genes, or the same gene resides in separate locations in the parental strains.

Table 7.2. Lesion types produced by parents of cross 4360 and progeny from one tetrad.

Strain	Mating type	Lesion types on rice cultivar[a]			Lesion type on weeping lovegrass[a]
		Sariceltik	Yashiro-mochi	Maratelli	
Parents:					
4224–7–8	1	5	0	0	5
6043	1	5	5	5	0
Progeny from tetrad 1:					
4360–1–1	2	5	0	5	0
4360–1–4	2	5	0	5	0
4360–1–2	2	5	5	0	5
4360–1–8	2	5	5	0	5
4360–1–3	1	5	5	5	0
4360–1–7	1	5	5	5	0
4360–1–5	1	5	0	0	5
4360–1–6	1	5	0	0	5

[a] A scale of lesion types has been defined according to the size and sporulation of lesions formed, ranging from type 0, no visible symptoms, to type 5, the largest spreading lesions characteristic of the host variety (Valent *et al.*, 1991).

Some Host Specificity Genes are Unstable

Some rice pathogens isolated from the field in central China frequently change specificity toward certain rice cultivars in laboratory studies (B. Valent, unpublished observations). That is, spontaneous virulent mutants appear frequently in standard infection assays (Fig. 7.1b). For example, strain O-137 mutates to become virulent on rice cultivar Yashiro-mochi with no change in specificity toward several other cultivars. Similarly, we isolated multiple independent mutants that had become virulent on cultivar Tsuyuake, and multiple mutants that have become pathogens of weeping lovegrass. In a typical experiment, a pot of rice seedlings or weeping lovegrass was inoculated with a suspension containing a total of 2×10^6 conidia. Eighty percent of such inoculations yielded one or more fully developed blast lesions. Fungus isolated from such lesions devastated the original host upon reinoculation (Fig. 7.1b), but had not changed in specificity toward numerous other host plants. Instability of these host specificity determinants was retained through several generations of genetic crosses (Fig. 7.3). For example, avirulent progeny from crosses 4360 and 4375 (two and three generations away from their unstable field isolate parent) frequently mutate to virulence on weeping lovegrass, Yashiro-mochi, or Tsuyuake.

In our experiments, MGR586 fingerprints (Fig. 7.4), as well as fertility, colony characteristics, and other host specificities confirmed that presumptive mutants were indeed derived from the nominal parental strains. MGR586 DNA

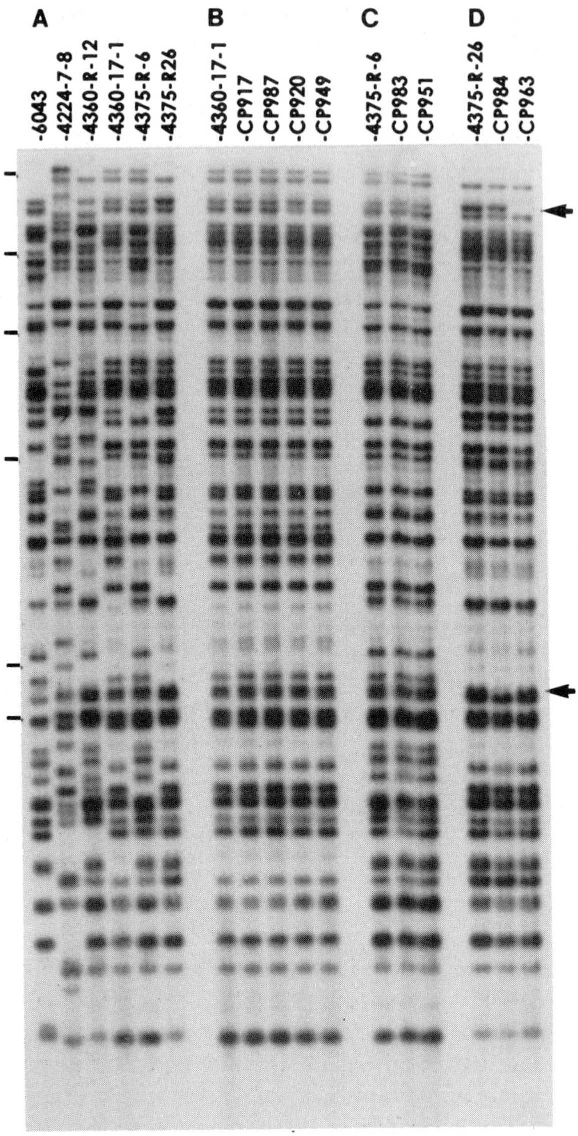

Fig. 7.4. MGR586 DNA fingerprints confirm parent–mutant relationships. Genomic DNAs from the strains indicated were digested with *Eco*RI, subjected to high resolution agarose gel electrophoresis, blotted to a hybridization membrane and probed with radiolabeled MGR586 DNA sequences (Hamer *et al.*, 1989). (A) *Mendelian segregation of MGR sequences*. Parental strains 6043 and 4224-7-8 were compared to their progeny, strains 4360-R-12 and 4360-17-1. The latter two strains were also compared to their progeny (strains 4375-R-6 and 4375-R-26). (B–D) *Parent-mutant families*. Strains 4360-17-1, 4375-R-6 and 4375-R-26 are all avirulent on Yashiro-mochi and weeping lovegrass. Strain CP917 is a spontaneous mutant of 4360-17-1 that gained virulence on Yashiro-mochi, and strains CP920 and CP949 had become pathogenic on weeping lovegrass. CP987 is a mutant of CP917 that infects weeping lovegrass. Strains CP983 and CP951 are mutants of 4375-R-6 that were selected for virulence on Yashiro-mochi and on weeping lovegrass, respectively. Similarly, strains CP984 and CP963

fingerprints sample genome structure at many loci (Hamer *et al.*, 1989; Romao and Hamer, 1992; Sweigard *et al.*, 1993) and represent definitive proof of the clonal identity of two strains.

The instability of genetic loci in the laboratory strains under study allowed us to obtain families of mutants derived from the same pathogen strain. An example is the 4360–17–1 mutant family consisting of a parental strain avirulent on Yashiro-mochi and weeping lovegrass, mutants that have changed specificity toward Yashiro-mochi or toward weeping lovegrass, and mutants that have changed specificity toward both hosts (Fig. 7.4). The small 4375-R-26 family has two mutants. In this case, both mutants also have a single obvious change in the MGR fingerprint (arrows in Fig. 7.4). These laboratory results appear to be analogous to results from studies of the association of MGR586 'lineage specific networks' with pathotypes in Colombia (Levy *et al.*, 1993). In these studies a limited number of pathotypes was associated with each lineage, and the associated pathotypes differed from each other by single-step changes in specificity toward single rice cultivars.

Mapping of Avirulence Genes

Cloning genes of unknown biochemical function by 'chromosome walking' involves isolation of overlapping chromosomal DNA segments spanning the region from a previously cloned linked marker to the gene of interest. A reliable and densely-marked genetic map is a prerequisite to having closely linked genetic markers that flank any gene of interest (chapter 7, this volume). A major *M. grisea* RFLP mapping effort was undertaken using the parents and progeny of cross 4360 to map the host specificity genes, *AVR2-YAMO*, *Avr1-MARA*, *Avr1-TSUY* and *PWL2* (Fig. 7.5) (Sweigard *et al.*, 1993). Seven linkage groups were defined with over 200 physical and genetic markers. Telomeres were mapped using a 24 bp oligonucleotide containing four repeats of the hexamer 5′-(AACCCT)-3′, homologous to the highly conserved telomere repeat sequence (Schechtman, 1989). All host specificity genes mapped as normal Mendelian markers (indicated by arrows in Fig. 7.5). Interestingly, the unstable genes *AVR2-YAMO* and *Avr1-TSUY* cosegregated with telomeric markers. Evidence also suggests that *PWL2* resides near a chromosome tip, whereas the stable gene *Avr1-MARA* resides in an internal region.

Cloning of *PWL2* and Analysis of Virulent Mutants

James Sweigard and Anne Carroll cloned *PWL2* by chromosome walking beginning with a linked physical marker, cos222 (Fig. 7.5; J.A. Sweigard, A.M. Carroll,

---

are mutants of 4375-R-26 that gained virulence on Yashiro-mochi and on weeping lovegrass, respectively. Each parent-mutant family is marked by a distinctive MGR586 fingerprint, which is identical or nearly identical in parent and mutants. The fingerprints of the two mutants isolated from 4375-R-26 are each missing one band present in the parent (arrows). Size standards are the same as in Fig. 7.2.

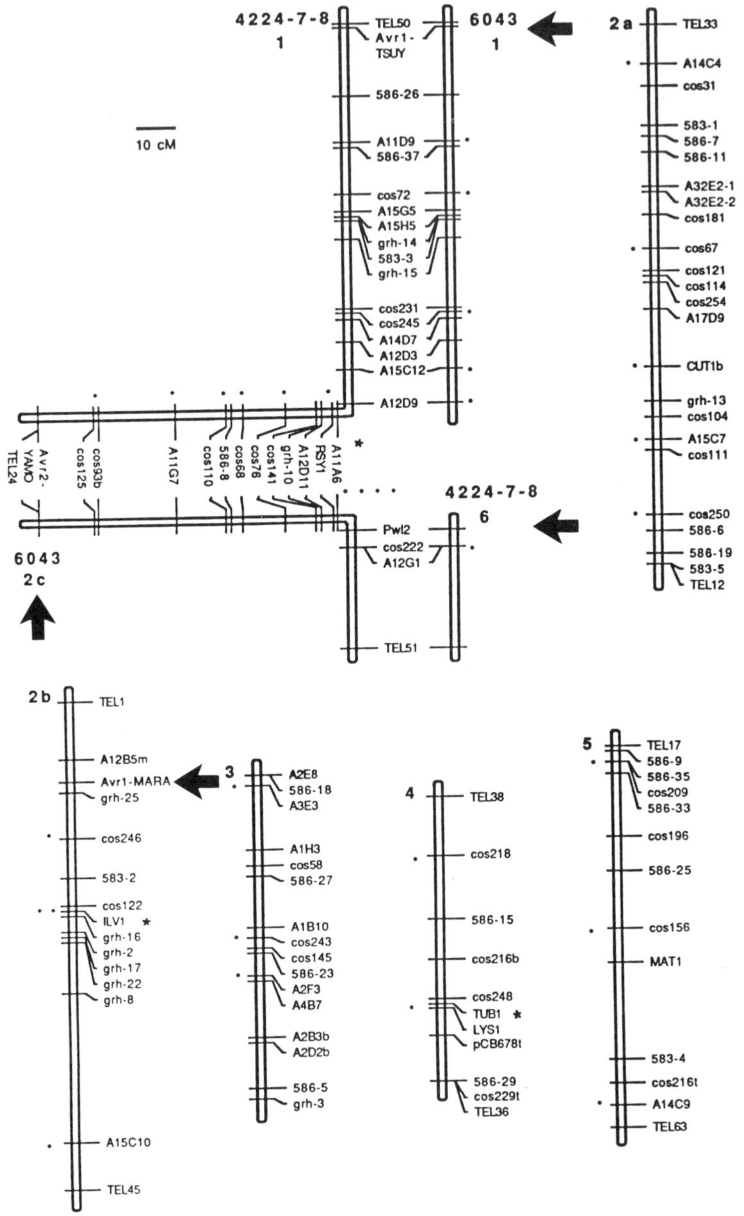

Fig. 7.5. Genetic map of the rice blast fungus. Segregation of over 200 markers including cloned genes, random cosmid clones, repetitive DNA elements, a telomeric repeat sequence and host specificity genes (arrows), was scored for 66 progeny of cross 4360 (Sweigard *et al.*, 1993). Not all markers are shown here. Linkage group assignment of some markers (denoted •) was verified by hybridization to blots of chromosome-size DNAs separated on CHEF gels. Chromosomes have been numbered based on their relative rates of migration in CHEF gels, with the slowest migrating band numbered chromosome 1 and the fastest numbered chromosome 6. Some chromosomes migrate coincidentally during electrophoresis (designated 2a, 2b and 2c). CHEF gel hybridization analysis showed that the parental strains differed by a reciprocal translocation involving chromosomes 1 and 6 in 4224–7–8 and 1 and 2c in 6043. (*) denotes clusters of markers with no recombination.

F.G. Chumley, and B. Valent, unpublished results). Clones with homology to cos222 were identified in a cosmid gene library, which permitted the screening of large fragments (approximately 40 kb) of the *M. grisea* genome. Walking was difficult because all cosmids with homology to cos222 contained one or more of several different middle repetitive DNA sequences including a previously unidentified sequence, MGR768. Attempts to isolate single copy sequences to use as hybridization probes for identifying additional overlapping cosmid clones were unsuccessful. Potential single copy clones were determined to be low copy repeat sequences present at 4–6 unlinked locations in the *M. grisea* genome. An example is shown in Fig. 7.6, in which a low copy fragment from a cosmid linked to *PWL2* was used as a hybridization probe against genomic DNAs from parents and mutants that gained the ability to infect weeping lovegrass (see Fig. 7.4 above). Six homologous bands were detected. The third largest band ('Band 3') was linked to *PWL2* in the RFLP mapping population, and this fragment was missing in all of the mutants. Interestingly, Band 2 (which is faint) mapped near the *Avr1-MARA* gene on linkage group 2b, and Band 5 was linked to the *AVR2-YAMO* gene. Other low copy repeat sequences showed a similar tendency to map near AVR genes and at telomeric regions, although different regions of the genome were identified. As a consequence of using low copy repeat probes, the chromosome walking effort was in reality 'chromosome wandering' in four or five genomic regions at the same time. Nevertheless, Sweigard and Carroll identified cosmids with the properties expected for *PWL2* clones. That is, transformation of a strain that was pathogenic on weeping lovegrass produced transformants that were no longer able to infect weeping lovegrass (Fig. 7.1a). This result clearly demonstrated that the presence of a functional *PWL2* gene prevented infection of weeping lovegrass by a previously virulent pathogen.

Further analysis, including the study of a *pwl2*$^-$ point mutation, has confirmed that the *PWL2* gene encodes a glycine-rich polypeptide containing 145 amino acids. The protein encoded by the *PWL2* gene has no significant homology to proteins in sequence databases. A putative signal peptide suggests that the protein might be secreted. However, the biochemical function of this protein remains to be determined.

Results such as those shown in Fig. 7.6 showed that spontaneous high-frequency mutation of the *PWL2* gene was due to deletion. Other types of mutants, such as mutants with altered specificity toward Yashiro-mochi, did not show deletions in the *PWL2* region. All spontaneous *pwl2*$^-$ mutants tested so far have deleted the *PWL2* gene and at least 30 kb of flanking DNA sequences. Two or three distinct classes of deletions were found. Thus, instability at the *PWL2* locus in the strains under study resembles that of unstable *BUF1* loci. The one virulent *pwl2*$^-$ allele that is not a deletion came from the RFLP mapping parent that infects weeping lovegrass, 4224–7–8. This allele has a single base pair change in the protein coding sequence, converting an aspartic acid residue to asparagine.

Cloning of *AVR2-YAMO* and Analysis of Mutants

The *AVR2-YAMO* gene, which prevents infection of the rice cultivar Yashiro-mochi, cosegregated with a cluster of physical markers at the end of a linkage

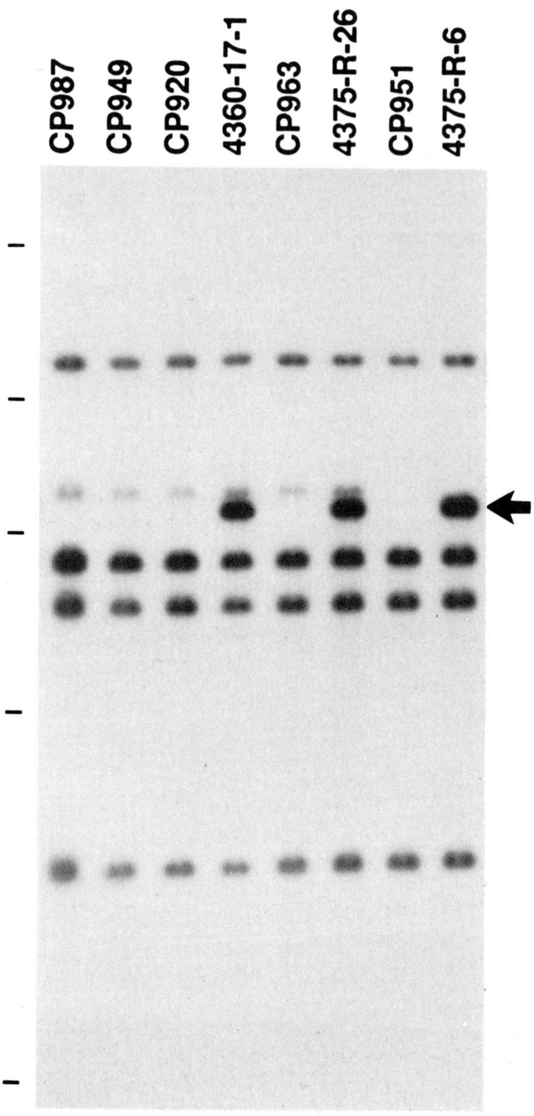

Fig. 7.6. Loss of function of *PWL2* is associated with deletion of genomic sequences. Southern analysis was performed on genomic DNAs from parents and mutants (see Fig. 7.4) that gained the ability to infect weeping lovegrass. Methods were as described in the legend to Fig. 7.2, except that DNAs were digested with *Eco*RI, and the blot was hybridized with a radiolabeled 7 kb *Eco*RI fragment isolated from a cosmid containing DNA sequences that map near *PWL2*. The bands that could be mapped segregated independently. The third largest band (arrow) cosegregated with the *PWL2* gene, and it had been deleted from all spontaneous pathogenic mutants tested. Positions of DNA size standards marked correspond to the largest five in Fig. 7.2.

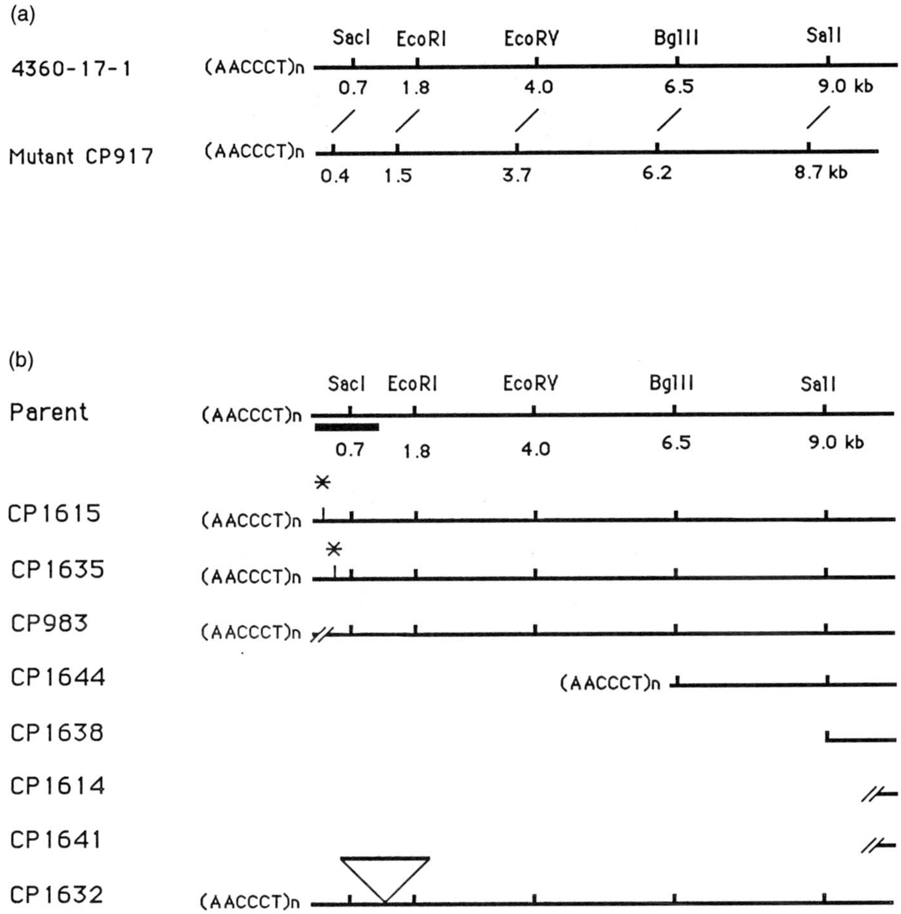

Fig. 7.7. (a) Mutant analysis suggests *AVR2-YAMO* resides at the tip of a chromosome. Using Southern analysis, a synthesized oligonucleotide probe homologous to the conserved 5′(AACCCT)3′ repeat sequence at the ends of chromosomes identifies terminal restriction fragments corresponding to the telomere linked to *AVR2-YAMO*. The sizes of the terminal fragments of the avirulent parent and the virulent mutant CP917 (Fig. 7.4) are diagrammed. For all restriction enzymes used in this analysis, the telomeric fragment for CP917 was approximately 300 bp smaller than the corresponding fragment from the avirulent parent. This suggests that a deletion within 700 bp of the chromosome end correlated with loss of function of the *AVR2-YAMO* gene. For all restriction enzymes tested, mutant CP984 (Fig. 7.4) had lost the telomeric fragment linked to *AVR2-YAMO* and had not gained a new one. (b) Multiple events result in gain of virulence toward rice cultivar Yashiro-mochi. Diagrammatic representation of molecular events associated with loss of function of *AVR2-YAMO* in eight spontaneous mutants derived from strain 4375-R-6. CP1615 and CP1635 contain single base pair changes (*) in the coding sequence of the gene. CP983 (Fig. 7.4) has a small deletion at the distal end of the gene, and CP1632 contains an insertion of approximately 1.5 kb (triangle). Other mutants have deletions ranging in size from 6.5 kb to 9 kb to larger deletions of undetermined size.

group (Fig. 7.5). Marc Orbach discovered from mutant analysis that this gene resided very near the tip of a chromosome (Fig. 7.7; M.J. Orbach, L. Farrall, B. Valent, and F.G. Chumley, unpublished results). The chromosomes of organisms ranging from humans to the fungus *Neurospora crassa* end with multiple copies of a highly conserved oligonucleotide repeat sequence (Blackburn, 1990). Thus, telomeric DNA restriction fragments can be visualized by Southern analysis using a synthesized oligonucleotide complementary to the conserved telomeric repeat sequence as a hybridization probe. Orbach discovered that spontaneous mutants that had become virulent on rice cultivar Yashiro-mochi showed structural changes in telomeric restriction fragments that mapped with *AVR2-YAMO*. In eight of 11 independently isolated spontaneous mutants, loss of expression of *AVR2-YAMO* correlated with the disappearance of the linked telomeric DNA fragment. In some cases new telomeric fragments were detected. Analysis of mutants CP917 and CP984 and their parental strains identified the linked telomeric restriction fragments (Fig. 7.7a) produced by a variety of restriction endonucleases. This analysis led to the hypothesis that *AVR2-YAMO* resided near the end of linkage group 1/2c, apparently within 1–2 kb of the tip. This hypothesis was tested by cloning a telomeric 6.5 kb *Bgl*II DNA fragment. When the cloned fragment was introduced into a pathogen that was able to infect Yashiro-mochi, transformants specifically lost the ability to infect Yashiro-mochi (Fig. 7.1a), proving that this DNA fragment contained the *AVR2-YAMO* gene. Again the prediction of dominance of the avirulence phenotype was borne out.

The *AVR2-YAMO* gene encodes a putative protein containing 223 amino acids. The primary amino acid sequence of this protein lacks significant homology to other known proteins. The gene resides totally within 1.5 kb of the tip of its chromosome in the strains under study here. Unlike the relatively uniform deletion events occurring at the unstable *BUF1* and *PWL2* loci, a wide variety of genetic events result in inactivation of the *AVR2-YAMO* gene. CP917 has a deletion of approximately 300 bp and CP984 has a deletion of approximately 12 kb. Fig. 7.7b summarizes a number of events in one lineage specific family. Analysis of eight independent mutants from strain 4375-R-6 detected two with point mutations, one with an insertion and five with deletions. The deletions ranged in size from approximately 100 base pairs to over 12.5 kb. This type of instability is characteristic of telomeric regions in other organisms (Horowitz *et al.*, 1984).

Homologous Sequences from Pathogens of other Grasses

The cloning of AVR genes allowed Seogchan Kang to study the distribution of homologous sequences in *M. grisea* strains that differ in host specificity (Fig. 7.8; S. Kang and B. Valent, unpublished results). *AVR2-YAMO* and *PWL2* were first identified in field isolates that infect rice, and most rice pathogens contain sequences with strong homology to both genes. Interestingly, *Digitaria* pathogens typically contain genomic sequences with strong homology to *PWL2*, and many also have sequences with strong homology to *AVR2-YAMO*. *Eleusine* pathogens typically lack strong homology to both cloned AVR genes, but many have

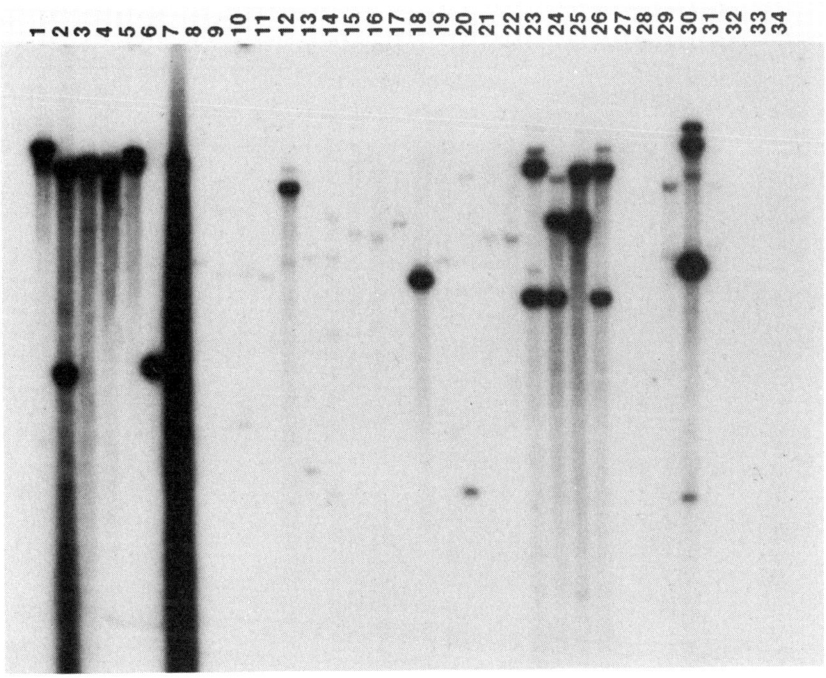

Fig. 7.8. Distribution of sequences with homology to *PWL2*. Genomic DNAs from 34 field isolates were subjected to Southern analysis as described in the legend to Fig. 7.2, except that the DNAs were digested with *Eco*RI, and the *PWL2* gene was used as a probe. The host specificity of the field isolates is listed in Table 7.3. Strains with strong homology include the rice pathogens (Lanes 1–4), a barley pathogen (Lane 5), a *Leersia* pathogen (Lane 6), a wheat pathogen (Lane 7; badly degraded in this analysis), a maize pathogen (Lane 12), a *Paspalum* pathogen (Lane 18), the four *Digitaria* pathogens (Lanes 23–26) and a *Pennisetum* pathogen (Lane 30). Table 7.3 summarizes the strains with strong homology to *AVR2-YAMO* and compares them to the results described here. The *Eleusine* pathogen WGG-FA40 (Lane 13), showed weak homology in two bands, which turned out to be associated with the *PWL1* gene. The *PWL1* gene, cloned from a WGG-FA40 gene library exploiting this limited homology with *PWL2*, has an *Eco*RI site in it, accounting for the two bands in Lane 13.

sequences with weak homology to one or both genes. Table 7.3 summarizes results of hybridization studies of the strains in Fig. 7.8 with both AVR gene probes. It seems surprising that *Digitaria* pathogens contain AVR genes in common with rice pathogens while *Eleusine* pathogens do not. Several studies suggest that the relationship between rice pathogens and *Eleusine* pathogens is much closer than the relationship between rice pathogens and *Digitaria* pathogens (A. Kubelik, J.A. Sweigard, F.G. Chumley, and B. Valent, unpublished results; M.H. Lebrun, University of Paris, personal communication; Chapter 5, this volume).

Kang predicted that genomic sequences in the *Eleusine* pathogen WGG-FA40 that were weakly homologous to *PWL2* (Fig. 7.8, Lane 13) would be the *PWL1* gene. He cloned the homologous sequence from a gene library and demonstrated by RFLP analysis that it cosegregated with the *PWL1* gene in genetic crosses. The cloned sequence from WGG-FA40 functioned in

Table 7.3. AVR gene homologies in field isolates with diverse host specificities.

Strain	Host	*PWL2* homology[a,b]	*AVR2-YAMO* homology[a]
1 0–135	Rice	+	
2 0–137	Rice	+	+
3 Guy11	Rice	+	+
4 0–316	Rice (wild)	+	+
5 G-197	Barley	+	+
6 G-160	*Leersia*	+	+
7 T-5	Wheat	+	
8 G-16	*Eragrostis*		
9 G-77	*Eleusine*		
10 G-199	*Eleusine*		
11 G-17	*Eragrostis*		
12 G-226	Maize	+	
13 WGG-FA40	*Eleusine*		
14 G-217	*Panicum*		
15 G-48	*Setaria*		
16 G-188	*Setaria*		+
17 G-165	*Brachiaria*		
18 G-216	*Paspalum*	+	
19 G-58	*Panicum*		+
20 G-219	*Panicum*		
21 G-221	*Pennisetum*		+
22 G-225	*Setaria*		
23 G-156	*Digitaria*	+	+
24 G-32	*Digitaria*	+	+
25 G-190	*Digitaria*	+	+
26 G-212	*Digitaria*	+	
27 G-123	*Pennisetum*		+/–
28 G-161	*Pennisetum*		+
29 G-209	*Digitaria*		
30 G-223	*Pennisetum*	+	+
31 G-194	*Leersia*		
32 G-227	Ginger		
33 G-229	Cyprus		
34 G-231	Cyprus		

[a] A plus sign (+) indicates strong homology. A plus–minus sign (+/–) indicates intermediate homology. Weak homologies are not indicated.
[b] Summarized from data in Fig. 7.8.

transforming a pathogen of weeping lovegrass into a nonpathogen, confirming the presence of *PWL1*. Subsequent analysis has permitted identification of the *PWL1* coding region. The amino acid sequence of the putative *PWL1* protein shows 85% similarity and 74% identity with the amino acid sequence of the *PWL2* protein. Functional homologs of *PWL2* have also been cloned from *Pennisetum* pathogens.

Functional homologs of *AVR2-YAMO* have been cloned from *Digitaria* pathogens and *Pennisetum* pathogens (S. Kang and B. Valent, unpublished results). Thus, the genetic prediction that pathogens of grasses other than rice contain AVR genes that confer specificity toward particular rice cultivars (Yaegashi and Asaga, 1981; Valent *et al.*, 1991) has been proven.

Wheat Blast in Brazil

One of the first recognized features of the middle repetitive DNA sequence MGR586 was its presence in 45–50 copies per genome in rice pathogens, but in only 1–3 copies per genome in pathogens of other grasses (Hamer *et al.*, 1989). The utility of MGR586 and other DNA probes specific for particular classes of pathogens is illustrated by the clear and rapid answer that the use of MGR586 provided for questions concerning the origin of wheat blast in Brazil. Blast disease caused by *M. grisea* became a serious problem for wheat farmers in Brazil during the mid-1980s. Because wheat blast first appeared in the state of Paraná, a rice farming region where rice blast was prevalent, it seemed possible that the rice blast fungus had mutated to parasitize wheat. MGR586 was used to generate DNA fingerprints of Brazilian rice pathogens and wheat pathogens. The results clearly demonstrated that the Brazilian wheat pathogens were not derived from rice pathogens indigenous to the area (Fig. 7.9; F.G. Chumley, L. Farrall, and B. Valent, unpublished results). Brazilian rice pathogens resembled rice pathogens collected from various parts of the world in having many copies of MGR sequences. In contrast, the wheat pathogens exhibited MGR fingerprints typical of pathogens of grasses other than rice, containing very few bands with MGR sequence homology.

Summary and Conclusions

The cloning of several host specificity genes from *M. grisea* has provided a first look into the molecular nature of an apparently complex recognition system that controls host specificity in rice blast. As in bacterial plant pathogens (Kobayashi *et al.*, 1989), AVR genes corresponding to specific rice cultivars appear to be common in *M. grisea* pathogens of rice or other grasses. The working hypothesis has been that the product of a fungal AVR gene interacts, directly or indirectly, with the product of a plant resistance gene. This interaction triggers defense responses in the plant that block infection by the fungus. The first cloned rice blast AVR genes confirm a key feature of this hypothesis, the dominance of AVR genes. That is, avirulence results from the presence of the gene product of an active AVR gene. Interestingly, similar genes control the ability of the fungus to infect a second host species, weeping lovegrass (*Eragrostis curvula*). We refer to these 'host specificity genes', *PWL1* and *PWL2*, as AVR genes because their mode of action seems to be analogous to that of classical AVR genes.

Clearly, AVR genes in the pathogen did not evolve to prevent the pathogen from infecting various genotypes of its host. Instead, it is likely that plants have

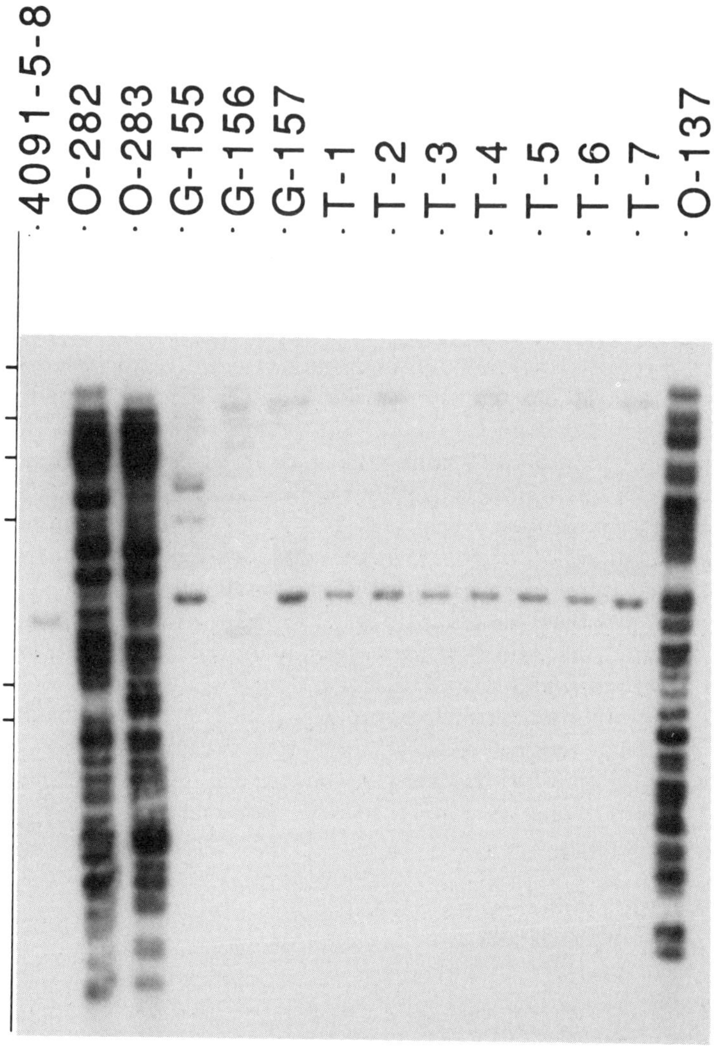

Fig. 7.9. MGR586 fingerprints demonstrate that wheat pathogens in Brazil did not arise from rice pathogens. Southern analysis was performed as described in the legend to Fig. 7.2, except that genomic DNAs were digested with *Eco*RI, and the blot was hybridized with radiolabeled MGR586 sequences. From left to right, strains are 4091–5–8, a weeping lovegrass pathogen; O-282 and O-283, Brazilian rice pathogens; G-155–G-157, Brazilian grass pathogens; T-1–T-7, Brazilian wheat pathogens; and O-137, Chinese rice pathogen. Strains O-282, O-283, G-155 (from *Rhynchelytrum roseum*), G-156 (from *Digitaria horizontalis*) and G-157 (from feral Triticale), and some of the wheat pathogens tested were obtained from Dr A.S. Prabhu (EMBRAPA/CNPAF, Brazil). Other wheat pathogens were obtained from Dr S. Igarashi of the Agronomic Institute of Panará, Brazil). Size standards marked as in Fig. 7.2.

gained the ability to recognize pathogen molecules with various functions. Determining what these functions are could give important clues as to which resistance genes might be durable. For example, a resistance gene that keys on an AVR gene that the fungus requires to be competitive might be difficult for the fungus to evade. There obviously are strains of the pathogen in the field that do not contain the AVR genes we have identified, demonstrating that they are not required for survival in the field. Neither the *PWL2* nor the *AVR2-YAMO* proteins have significant primary amino acid sequence homology to proteins of known function. Thus, these sequences do not immediately provide clues to the function of these genes. Antibodies to the AVR proteins can now be used for cytological localization of the proteins in fungus-infected plant tissues, perhaps providing a clue to their function. Cytological studies are necessary to identify cell responses that are correlated with the function of each AVR gene. Heath *et al.* (1990) reported the first efforts toward identifying cytological feature(s) affected by *PWL1* in parental strains and two full tetrads. Pathogenicity or nonpathogenicity correlated perfectly with the absence or presence, respectively, of brown cells around developing colony margins. Other cytological features clearly did not cosegregate with *PWL1*.

An immediate benefit of cloning AVR genes from the rice blast pathogen has been derived from using the cloned genes as molecular probes to study the distribution of these genes in the fungal population. Both *AVR2-YAMO* and *PWL2* were identified in the genomes of rice pathogens: *AVR2-YAMO* in the Chinese field isolate O-137 and *PWL2* in the female fertile rice pathogen Guy11 (Leung *et al.*, 1988; Silué *et al.*, 1992a). *AVR2-YAMO* is not required for pathogenicity toward rice in the field because the Chinese field isolate O-135 lacks all sequence homology to this AVR gene. It is an interesting question as to why these AVR genes also occur in *Digitaria* pathogens, but not in pathogens more closely related to rice pathogens, such as *Eleusine* pathogens.

An extensive ongoing effort has been devoted to the identification of clonal lineages in the world population of *M. grisea* rice pathogens (Levy *et al.*, 1991, 1993). Studies have now been initiated to determine the relationship between AVR gene restriction fragment length polymorphisms (RFLPs) and MGR586 lineages. Initial results suggest that the sizes of restriction fragments homologous to the *AVR2-YAMO* and *PWL2* genes are often conserved among strains in the same lineage (M. Levy and B. Valent, unpublished results; Levy *et al.*, 1993). Some lineages from China have the same size *AVR2-YAMO* *Eco*RI fragment (approximately 1.8 kb) as the Chinese rice pathogen that was the source of this AVR gene. In one of these lineages, the *AVR2-YAMO* fragment is missing in some strains and present in others. It will be interesting to learn if strains in this MGR lineage have *AVR2-YAMO* in a telomeric location, with consequent genetic instabilities similar to those of *AVR2-YAMO* in strain O-137. Studies such as these will test the hypothesis that the mutant families we are studying serve as relevant models for diversity within lineage specific networks in the field. These model studies may also provide insights into the basis for diversity in MGR fingerprints within a lineage. In the 4375-R-26 mutant family, MGR fingerprints of each of two mutants have single bands missing. In the case of the *avr2-YAMO*$^-$ mutant, strain CP984, the missing MGR586 fragment is consistent with the fact that this mutant contains a deletion of approximately 12 kb, including an identified region of MGR586 homology.

Major resistance genes have had limited utility in controlling rice blast disease because of variability in the pathogen. Understanding the potential for variation in the fungus could help direct the attention of plant breeders to major genes that confer more durable resistance. In the cases studied so far, genetic instability in the fungus appears to be a consequence of the chromosomal location of the gene. The abundance of repetitive DNA sequences in the fungal genome may account for some variability via homologous recombination leading to deletion of genomic segments. We suggest that unstable *PWL2* and *BUF1* loci are flanked by repetitive sequences that recombine to delete large genomic regions including these genes. The telomeric location of *AVR2-YAMO* appears to account for instability in this case. More cases of hyper-mutability will have to be analyzed at the molecular level before one can safely generalize about mechanisms for genetic instability in the rice blast fungus. However, it seems likely that mechanisms that have already been detected in laboratory studies can be expected to be operable in the field, even if absolute frequencies of mutation turn out to be different in the laboratory and in the field.

Acknowledgments

This chapter discusses recent work by members of the Molecular Plant Pathology Group at DuPont. James A. Sweigard has made many significant contributions, especially on studies of *PWL2*. Anne Carroll and Leonard Farrall have provided expert technical assistance for most of the projects discussed. Postdoctoral fellows who have made important contributions include Seogchan Kang and Marc J. Orbach, who is now at the University of Arizona, Department of Plant Pathology.

References

Atkins, J.G., Robert, A.L., Adair, C.R., Goto, K., Kozaka, T., Yanagida, R., Yamada, M. and Matsumoto, S. (1967) An international set of rice varieties for differentiating races of *Piricularia oryzae*. *Phytopathology* 57, 297–301.

Blackburn, E.H. (1990) Telomeres: Structure and synthesis. *Journal of Biological Chemistry* 265, 5919–5921.

Chumley, F.G. and Valent, B. (1990) Genetic analysis of melanin-deficient, nonpathogenic mutants of *Magnaporthe grisea*. *Molecular Plant-Microbe Interactions* 3, 135–143.

Crawford, M.S., Chumley, F.G., Weaver, C.G. and Valent, B. (1986) Characterization of the heterokaryotic and vegetative diploid phases of *Magnaporthe grisea*. *Genetics* 114, 1111–1129.

Ellingboe, A.H., Wu, B.C. and Robertson, W. (1990) Inheritance of avirulence/virulence in a cross of two isolates of *Magnaporthe grisea* pathogenic to rice. *Phytopathology* 80, 108–111.

Flor, H.H. (1971) Current status of the gene-for-gene concept. *Annual Review of Phytopathology* 9, 275–296.

Hamer, J.E. and Givan, S. (1990) Genetic mapping with dispersed repeated sequences in the rice blast fungus: Mapping the SMO locus. *Molecular and General Genetics* 223, 487–495.

Hamer, J.E., Farrall, L., Orbach, M.J., Valent, B. and Chumley, F.G. (1989) Host species-specific conservation of a family of repeated DNA sequences in the genome of a fungal plant pathogen. *Proceedings of the National Academy of Sciences of the United States of America* 86, 9981–9985.

Hamer, J.E., Valent, B. and Chumley, F.G. (1989) Mutations at the SMO genetic locus affect the shape of diverse cell types in the rice blast fungus. *Genetics* 122, 351–361.

Heath, M.C., Valent, B., Howard, R.J. and Chumley, F.G. (1990) Correlations between cytologically detected plant-fungal interactions and pathogenicity of *Magnaporthe grisea* toward weeping lovegrass. *Phytopathology* 80, 1382–1386.

Horowitz, H., Thorburn, P. and Haber, J.E. (1984) Rearrangements of highly polymorphic regions near telomeres of *Saccharomyces cerevisiae. Molecular and Cellular Biology* 4, 2509–2517.

Howard, R.J. and Ferrari, M.A. (1989) Role of melanin in appressorium function. *Experimental Mycology* 13, 403–418.

Kobayashi, D.Y., Tamaki, S.J. and Keen, N.T. (1989) Cloned avirulence genes from the tomato pathogen *Pseudomonas syringae* pv. *tomato* confer cultivar specificity. *Proceedings of the National Academy of Sciences of the United States of America* 86, 157–161.

Latterell, F.M. (1975) Phenotypic stability of pathogenic races of *Pyricularia oryzae*, and its implications for breeding of blast resistant rice varieties. In: *Proceedings, Seminar on Horizontal Resistance to the Blast Disease of Rice*. Centro Internacional de Agricultura Tropical, Cali, Columbia, pp. 199–234.

Leung, H., Borromeo, E.S., Bernardo, M.A. and Notteghem J.L. (1988) Genetic analysis of virulence in the rice blast fungus *Magnaporthe grisea. Phytopathology* 78, 1227–1233.

Leung, H., Lehtinen, U., Karjalainen, R., Skinner, D., Tooley, P., Leong, S. and Ellingboe, A. (1990) Transformation of the rice blast fungus *Magnaporthe grisea* to hygromycin B resistance. *Current Genetics* 17, 409–411.

Levy, M., Correa-Victoria, F.J., Zeigler, R.S., Xu, S. and Hamer, J.E. (1993) Genetic diversity of the rice blast fungus in a disease nursery in Colombia. *Phytopathology* 83, 1427–1433.

Levy, M., Romao, J., Marchetti, M.A. and Hamer, J.E. (1991) DNA fingerprinting with a dispersed repeated sequence resolves pathotype diversity in the rice blast fungus. *Plant Cell* 3, 95–102.

Ling, K.C. and Ou, S.H. (1969) Standardization of the international race numbers of *Pyricularia oryzae. Phytopathology* 59, 339–342.

Ou, S.H. (1985) *Rice Diseases*, 2nd edn. Commonwealth Mycological Institute, Kew, UK, 380 pp.

Parsons, K.A., Chumley, F.G. and Valent, B. (1987) Genetic transformation of the fungal pathogen responsible for rice blast disease. *Proceedings of the National Academy of Sciences of the United States of America* 84, 4161–4165.

Petes, T.D. and Hill, C.W. (1988) Recombination between repeated genes in microorganisms. *Annual Review of Genetics* 22, 147–168.

Romao, J. and Hamer, J.E. (1992) Genetic organization of a repeated DNA sequence family in the rice blast fungus. *Proceedings of the National Academy of Sciences of the United States of America* 89, 5316–5320.

Rossman, A.Y., Howard, R.J. and Valent, B. (1990) *Pyricularia grisea*, the correct name for the rice blast disease fungus. *Mycologia* 82, 509–512.

Schechtman, M.G. (1989) Segregation patterns of *Neurospora* chromosome ends: mapping chromosome tips. *Fungal Genetics Newsletter* 36, 71–73.

Silué, D., Notteghem, J.L. and Tharreau, D. (1992a) Evidence for a gene-for-gene relationship in the *Oryza sativa–Magnaporthe grisea* pathosystem. *Phytopathology* 82, 577–580.

Silué, D., Tharreau, D. and Notteghem, J.L. (1992b) Identification of *Magnaporthe grisea* avirulence genes to seven rice cultivars. *Phytopathology* 82, 1462–1467.

Sweigard, J.A., Valent, B., Orbach, M.J., Walter, A.M., Rafalski, A. and Chumley, F.G. (1993) Genetic map of the rice blast fungus *Magnaporthe grisea*. In: O'Brien, S.J. (ed.), *Genetic Maps*. Cold Spring Harbor Laboratory, New York, pp. 3.112–3.117.

Valent, B. and Chumley, F.G. (1991) Molecular genetic analysis of the rice blast fungus, *Magnaporthe grisea*. *Annual Review of Phytopathology* 29, 443–467.

Valent, B., Crawford, M.S., Weaver, C.G. and Chumley, F.G. (1986) Genetic studies of fertility and pathogenicity in *Magnaporthe grisea* (*Pyricularia oryzae*). *Iowa State Journal of Research* 60, 569–594.

Valent, B., Farrall, L. and Chumley, F.G. (1991) *Magnaporthe grisea* genes for pathogenicity and virulence identified through a series of backcrosses. *Genetics* 127, 87–101.

Yaegashi, H. (1978) Inheritance of pathogenicity in crosses of *Pyricularia* isolates from weeping lovegrass and finger millet. *Annals of the Phytopathological Society of Japan* 44, 626–632.

Yaegashi, H. and Asaga, K. (1981) Further studies on the inheritance of pathogenicity in crosses of *Pyricularia oryzae* with *Pyricularia* sp. from finger millet. *Annals of the Phytopathological Society of Japan* 47, 677–679.

Yamada, M., Kiyosawa, S., Yamaguchi, T., Hirano, T., Kobayashi, T., Kushibuchi, K. and Watanabe, S. (1976) Proposal of a new method for differentiating races of *Pyricularia oryzae* Cavara in Japan. *Annals of the Phytopathological Society of Japan* 42, 216–219.

II Host Plant Resistance

8 International Collaboration on Breeding for Resistance to Rice Blast

S.W. Ahn

International Network for Genetic Evaluation of Rice, Genetic Resources Center, International Rice Research Institute, PO Box 933, 1099 Manila, The Philippines

Introduction

Developing rice varieties with highly effective and durable resistance to blast is the most economically feasible and environmentally sound management approach in most blast-prone rice ecosystems. Dynamic changes in race composition of the blast pathogen *Pyricularia grisea* has often resulted in short-lived efficiency of host resistance in improved varieties (Ahn, 1982). Thus, breeding for resistance to rice blast is directed towards enhancement of durability of host resistance through the use of genetically diverse blast resistance sources.

Because of the global importance and complexity of this disease, international collaboration on breeding for resistance to rice blast has been underway since the 1960s and has played an important role in developing cultivars with blast resistance. Since 1975, the International Network for Genetic Evaluation of Rice (INGER, formerly the International Rice Testing Program), a collaborative effort of world rice scientists in national and international agricultural research systems, facilitates the exchange, evaluation, and utilization of rice germplasm regardless of political and geographical barriers. The International Rice Blast Nursery (IRBN), an integral part of INGER trials, was specifically designed to provide rice scientists working on rice blast easy access to diverse resistance donors and improved breeding lines originating from different national and international programs. It also allows the identification of promising entries useful for rice breeders worldwide.

This chapter summarizes activities and results of multilocation evaluations of the IRBN with emphasis on analysis, interpretation, and utilization of the information.

Needs for the International Exchange and Evaluation of Rice Germplasm

The blast pathogen population responds to a resistant rice cultivar in a highly dynamic manner. The stability of resistance in a given rice cultivar is determined by how the blast pathogen population changes and the way resistance is deployed. Thus, an adequate evaluation scheme and access to the different sources of resistance become fundamental requirements for resistance breeding.

Many rice scientists working in developing countries have limited access to donors and breeding lines from different sources. Also, diverse testing environments are not readily available, and systematic evaluation schemes are often not well established. The INGER scheme has greatly improved accessibility of germplasm and allowed rice breeders to test their own materials across a wide range of environments. Since 1981, more than half of the globally released rice varieties were obtained through INGER or were bred from parents included in INGER trial nurseries.

Many rice cultivars with highly race-specific resistance have often been severely damaged due to the emergence of new compatible races. The breeding efforts and evaluation scheme for blast resistance in the past were reactive rather than anticipatory. Little attention was given to determine the efficiency of resistance in breeding lines against new races which may occur after the release of breeding lines as varieties. A wide variation of blast pathogen populations present in test sites of IRBN allows anticipatory evaluation of breeding lines against a wide variety of blast populations in a short span of time. Susceptibility at a different site of a locally resistant line may predict the occurrence of new compatible races in the target site.

The origin of IRBN can be traced back to the initiation of the Uniform Blast Nursery (UBN) program recommended at the 9th meeting of the International Rice Commission (IRC) in 1961 (Seshu *et al.*, 1986). After several modifications in nursery designation and evaluation procedures, the current form of systematic exchange and evaluation for blast resistance in rice cultivars was started in 1975. The specific objectives of the IRBN are to evaluate promising blast-resistant breeding lines and identify genetic donors for blast resistance.

Composition, Distribution and Utilization of IRBN

Test entries for IRBN are nominated yearly by rice scientists working on breeding for blast resistance. The majority of nominated entries are advanced generations of breeding lines. Small quantities of seeds sent to INGER are multiplied at IRRI. After proper seed processing and seed health testing, all test entries are assembled and dispatched to different sites upon request. Total number of entries per year including differential varieties and check cultivars varies from 150 to 350. Fieldbooks containing standard procedures, varietal information, and data sheets are provided along with the seed.

More than 3000 entries were evaluated at 126 test sites in 31 countries from 1975 to 1992 (Tables 8.1 and 8.2). In recent years, more than 60% of test entries

Table 8.1. Number and origin of test entries evaluated in the International Rice Blast Nursery and the number of test sites.

Period	No. of entries[a]	Proportion by origin (%)		Tested in	
		National program	International center	Sites (no.)	Countries (no.)
1975–80	1280	54	46	77	30
1981–86	1010	64	36	106	30
1987–92	874	64	36	95	29

[a] Generally entries were tested twice. Differential varieties were included in each test.

were from national programs and the rest were from international centers (primarily IRRI). While many sites are located in blast-prone areas, others are located away from endemic sites due to logistic and operational reasons. In those areas, artificial measures are taken to establish blast-conducive environments (Ou, 1985).

Evaluation results of the IRBN are recorded in the fieldbook. A copy of each data set is returned to INGER headquarters at IRRI. Data from different test sites are processed for storage and retrieval in the INGER database. Results are analyzed by location, country, region, and across locations and presented in preliminary and final reports published annually. A total of 522 entries of the IRBN were utilized in hybridization in 18 countries from 1984 to 1992. The most widely utilized entries are listed in Table 8.3.

IRBN Evaluation Procedures

IRBN evaluation currently employs a uniform blast nursery test method (Ou, 1965). This method accommodates a large number of test entries while requiring a small quantity of seed, ensures relatively uniform infection, and demands relatively little space.

The main features of evaluation procedures for leaf blast are:

- The evaluation is generally done under upland conditions. However, wet nursery bed is also used when the dry land preparation is difficult but the climatic conditions are highly favorable for blast.
- Each test entry is planted in one row (30–100 cm length) spaced at 10–30 cm. The resistant and susceptible checks are planted after every 25 or 50 test entries. A mixture of several broadly susceptible local cultivars are planted along the borders as spreaders to ensure presence of inoculum consisting of diverse races of the blast pathogen including those in farmers' fields. Extremely susceptible cultivars are avoided as spreaders since plants are rapidly killed.
- Initial inoculum can be introduced by transplanting infected plants or spreading plant parts such as leaves, nodes, or panicles in the spreader rows. Spore suspension of specific isolates can be applied, if necessary.

Table 8.2. International Rice Blast Nursery test sites, 1984–1991[a].

Region/country	Trial sites
East Asia	
Japan	Omagari (5)[b]
Korea	Iri (8), Suweon (8), Milyang (8), Icheon (8)
Taiwan (China)	Chia-yi (8)
China	Nanning City (6), Nanjing (2), Guangzhou (7), Hangzhou (6), Fuzhou (7), Luzhou (5), Neijiang (5), Changsha (7), Chengdu (4), Fuyang (2)
South-East Asia	
Myanmar	Yezin (2), Heho (3)
Indonesia	Sitiung (3), Maros (2), Kendari (3)
Malaysia	Bumbong Lima (6)
Philippines	Los Baños (8)
Thailand	Phrae (5), San Pa Tong (3), Ubon Rachathani (2), Bangkhen (8), Chainat (7), Pathumthani (3), Nakhon Si Thammarat (2), Kuan Gut (5), Prachinburi (3), Muang (4), Samoeng (6)
Vietnam	Omon (5), Quang Nam (2), Chau Thanh (2)
South Asia	
Bangladesh	Joydebpur (7)
India	Hyderabad (2), Maruteru (4), Nellore (5), Nawagam (2), Ponnampet (4), Lonavala (2), Almora (3), Hawalbagh (3), Malan (3), Shillong (3), Hazaribagh (3), Aduthurai (2)
Nepal	Kathmandu (2), Kankai (2), Birganj (4)
Pakistan	Kala Shah Kaku (3)
Sri Lanka	Batalagoda (4), Getambe (4)
West Asia and North Africa	
Egypt	Sakha (6)
Iran	Rasht (3), Tonekabon (4)
Sub-Saharan Africa	
Nigeria	Ibadan (2)
Zanzibar	Bumbwi Sudi (2)
Zaire	Yangambi (3)
Europe	
Italy	Mortara (8)
Latin America	
Brazil	Goiania (4)
Colombia	Sta. Rosa (6)
Guatemala	Los Angeles (2)
Mexico	Tabasco (2), Huimanguillo (2), Chetumal (2)
Panama	Chiriqui (2)
Honduras	Yoro (2)

[a] A total of 54 trial sites participated only once.
[b] Figures in parentheses refer to the number of trials conducted.

Table 8.3. International Rice Blast Nursery entries utilized in hybridization in four or more countries, 1984–1992.

Origin	Cultivars[a]
Bangladesh	Kataktara DA2
Indonesia	B5592F-5-ST-69
Ivory Coast	IRAT 144, IRAT 146, IRAT 147
China	Chokoto
IITA	ITA 212
IRRI	IR17525-7-19-1-3-3-2, IR25604-99-1-3-2-2, IR28210-96-4-3-3, IR32429-47-3-2-2, IR37865-29-3-1-3, IR42, IR60
Korea	Iri 370, Iri 371, Iri 387, Milyang 77, Milyang 79, Milyang 82, Suweon 351, Suweon 355, Unbong 6
Vietnam	Tetep

[a] A total of 522 IRBN entries were utilized in hybridization in 18 countries from 1984 to 1992.

- Proper management to establish and maintain blast-conducive environments includes dense planting, high rate of nitrogen application, prolonged dew period by covering the plots with plastic film at night and/or supplemental overhead sprinkler irrigation 2–3 times each rainless day.
- Known differential varieties, key commercial varieties, and cultivars with known level of partial resistance are included to monitor the change of virulence and disease pressure at the time of evaluation.

Evaluation is done about 30 days after seeding using the Standard Evaluation System (SES) based on a 0–9 scale (International Rice Testing Program, 1988). Scales 0–4 are based on lesion types (qualitative assessment) and scales 4–9 are based on severity caused by compatible isolates (quantitative assessment). Reaction scores of 0–3 indicate the absence of compatible type lesions, hence high efficiency of qualitative resistance at a given test site. Scores of 4–9 indicate the presence of compatible races and possibly different levels of quantitative resistance as well.

Panicle blast is also evaluated using transplanted or direct-seeded plants where natural conditions are favorable. Several susceptible checks of different maturity should be included to determine the effect of climatic conditions on panicle blast. Proper timing of evaluation (20–25 days after heading) is important, since other diseases such as brown spot may interfere with accurate evaluation.

Since there is no replication, uniform distribution of inoculum and proper management are crucial for a precise evaluation. Also a diverse spectrum of races representing the blast populations present in farmers' fields should be maintained in the test site. For this purpose, a careful selection of susceptible cultivars and a periodic monitoring of races in test sites is required.

Table 8.4. Rice cultivars with high frequency of incompatible reactions to blast in the International Rice Blast Nursery, 1984–1992.

Cultivar[a]	Origin	Score frequency 0–3	4–6	7–9	Location with high scores (7–9)
Hsia-Chioh-Keh-tu (Acc.08240)	Taiwan (China)	66	5	1	Lonavala
IR37704-131-2-1-3-2	IRRI	74	7	0	–
IR39422-163-1-3-2-2	IRRI	72	7	1	Chiriqui
IR47686-9-1-B-1	IRRI	79	9	1	Lonavala
Iri 386	Korea	71	6	1	Hazaribagh
Iri 387	Korea	105	15	0	–
Milyang 82	Korea	118	9	0	–
P5589-1-1-3P-3-6P	CIAT	74	9	0	–
Ta-poo-cho-z	Taiwan (China)	110	9	2	Sitiung, Maruteru
Tetep	Vietnam	615	56	13	Sta, Rosa, Changsha, Lonavala, Fuzhou
TKY 1024	Taiwan (China)	70	7	0	–
Unbong 9	Korea	57	6	2	Tirur, Lonavala

[a] Cultivars with equal or more than 85% of incompatible reactions in the IRBN.

Analysis of IRBN Results

The evaluation procedures of IRBN are not structured to generate a detailed analysis of the results. Nevertheless, repeated tests over a large number of test sites provide an overview of the expression of resistance in different genotypes in different environments. The results of IRBN from 1975 to 1983 were summarized by Seshu *et al.* (1986).

Evaluation parameters

Different strategies for developing blast-resistant cultivars are pursued by different rice improvement programs (Ahn and Mukelar, 1986). Thus INGER only provides summarized information on various aspects of blast reaction of test entries. The following major criteria are used to identify promising lines or donors for further testing and utilization:

- Frequency of incompatible reaction (scores 0–3) across test sites.
- Average disease severity index (DSI) based on scores at test sites where compatible reactions (scores 4–9) are observed.

Promising entries are identified based on the above categories (Tables 8.4 and 8.5). Breeding lines or cultivars exhibiting high frequency of incompatible reaction across many test sites over several years are considered to possess broad

Table 8.5. Rice cultivars that showed low disease severity to blast in the International Rice Blast Nursery, 1984–1992.

Cultivar[a]	Origin	Disease severity index (DSI)	Score frequency 0–3	4–6	7–9
B5592-F-5-ST-69	Indonesia	4.5	55	15	0
BG367-4	Sri Lanka	4.7	63	18	0
Fukunishiki	Japan	4.3	52	14	0
Fukuhikari	Japan	4.5	48	14	0
IR28228-96-3-2-1-3	IRRI	4.4	57	13	0
IR37721-3-3-3-3-1	IRRI	4.5	70	19	0
Suweon 351	Korea	4.7	62	14	0
Tadukan	Philippines	4.6	77	22	0
UPL Ri-5	Philippines	4.6	59	20	0
IR50 (check)	IRRI	6.4	125	69	64
CO 39 (check)	India	7.4	28	123	290

[a] Cultivars with incompatible reactions less than 85% of, and with DSI value lower than 5 were selected.

spectrum resistance. This type of resistance is generally qualitative and broadly effective against many prevailing races in test sites (Ahn and Ou, 1982). Qualitative resistance is governed by major gene(s), and the usefulness will decrease as the proportion of matching virulent race(s) increases.

Some cultivars are consistently less affected than others even though their spectrum of qualitative resistance is not so broad (Table 8.5). Various terminologies such as field resistance, slow blasting, partial resistance, and quantitative resistance were used to describe this type of resistance (Ahn, 1982). The level of quantitative resistance in rice observed under natural field conditions can be partitioned into two components (Ahn and Koch, 1988): (i) the efficiency of qualitative resistance in eliminating the avirulent portion of the available inoculum; and (ii) partial resistance to lower the infection efficiency of virulent isolates. Disease severity index (DSI) is used to assess the quantitative resistance using the following formula:

$$\text{DSI} = \frac{\text{Sum of compatible reaction scores}}{\text{Total number of trials showing compatible reactions}}$$

DSI was found to be a useful parameter in identifying potential cultivars with high partial resistance (Ahn *et al.*, 1989). The DSI is also utilized as an indicator of disease pressure at test sites.

Panicle blast is evaluated at some test sites. Entries frequently showing low panicle blast are listed in Table 8.6. Results of panicle blast evaluation are often quite different from leaf blast evaluation under field conditions. A critical comparison should be made to eliminate the effect of different weather or temporal changes in the composition of races. Staggered planting method to synchronize the evaluation of leaf and panicle blast has been used to demonstrate good

Table 8.6. Rice cultivars with low panicle blast scores in the International Rice Blast Nursery, 1986–1991.

Cultivar	Origin	Year tested	Average panicle blast score	Score frequency			Leaf blast disease severity index
				0–3	4–6	7–9	
C73110	Taiwan (China)	1987–89	2.1	23	6	0	5.4
IR37721-9-2-1-3	IRRI	1986–88	1.5	18	5	0	4.6
Milyang 82	Korea	1986–89	1.7	17	2	1	4.6
Tetep	Vietnam	1989–91	1.3	28	5	0	5.1

correlation between leaf and panicle blast resistance under field conditions (Ahn and Rubiano, 1984).

IRBN data and varietal ranking in farmers' fields

A fundamental assumption for any resistance evaluation is that the assessment must be highly correlated to actual performance of a cultivar in farmers' fields. Nevertheless, few studies have been done to determine how blast evaluation results of promising breeding lines relate to actual field performance after releasing those cultivars to farmers. Twenty-six Tongil-type Korean varieties (Korea indicas) were included in IRBN trials prior to the release to and cultivation by Korean farmers. Varietal ranking of these cultivars in farmers' fields was indirectly obtained since no data on direct comparison of blast severity are available (Ahn *et al.*, 1988a). Among different evaluation parameters, DSI of IRBN has the highest positive rank correlation with varietal ranking of Tongil type varieties in farmers' fields (Ahn *et al.*, 1988b).

Durable resistance in IRBN

Except in a retrospective way, it is not possible to predict the durability of blast resistance. Many rice cultivars, however, are known to possess durable blast resistance under blast-prone environments (Lee *et al.*, 1989). Some of these cultivars were also tested in the IRBN at many locations for several years.

Moroberekan, OS6 and other tropical japonica cultivars have been cultivated without severe and continuous blast damage for many years in upland areas of West Africa. IR36, IR42, Milyang 30, and Milyang 42, which exhibit relatively high levels of partial resistance, are known as durably resistant cultivars under irrigated or rainfed lowland conditions. IR20 and IAC47 have been grown in a large area for long periods; however, they gradually became susceptible. Several cultivars showed a short span of blast resistance.

Most of the durably resistant cultivars in the IRBN test showed low DSI

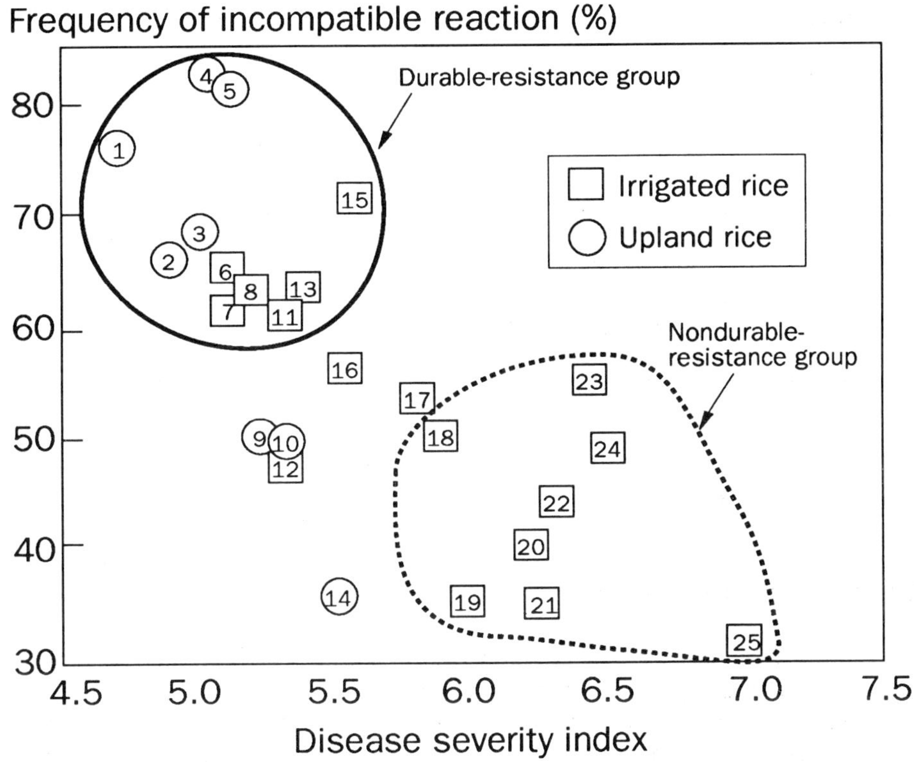

Fig. 8.1. Reaction to leaf blast of rice cultivars with and without durable resistance in the International Rice Blast Nursery, 1975–88. 1 = Moroberekan, 2 = Dourado Precoce, 3 = OS6, 4 = IRAT104, 5 = IRAT 13, 6 = IR36, 7 = Taebaeg, 8 = Baegyang, 9 = IAC47, 10 = Pratao, 11 = IR42, 12 = Milyang 42, 13 = Chupung, 14 = Denorado, 15 = Samgang, 16 = Milyang 30, 17 = IR20, 18 = IR50, 19 = Milyang 23, 20 = Yushin, 21 = IR8, 22 = CICA4, 23 = Yeongpung, 24 = IR24, 25 = Nopung.

values, mostly near 5.0, while the DSIs of all the known susceptible cultivars are higher than 6.0 (Fig. 8.1). Frequency of incompatible reactions of durably resistant cultivars is higher than 60%, while the frequency of extremely high scores, 8–9, is much lower than those of nondurable resistant cultivars (Table 8.7). These results suggest that the combination of broadly effective qualitative resistance controlled by major gene(s) and high level of partial resistance may constitute host resistance in durably resistant cultivars (Ahn and Koch, 1988). Rice cultivars grown under high blast pressure such as upland conditions would require the combination of two different resistance mechanisms. This concept was supported by a study on the genetics of blast resistance in Moroberekan using molecular markers (Wang, 1992; see McCouch *et al.*, this volume).

Known rice cultivars with durable blast resistance are extensively utilized in many breeding programs. IRAT 13, Moroberekan, IRAT 104 (IRAT 13/Moroberekan), and several lines from IR47686 (IRAT 104/Palawan) are frequently used in upland breeding programs at IRRI. In the past, many highly resistant

Table 8.7. Blast reaction of durably resistant rice cultivars in the International Rice Blast Nursery.

Cultivar	Origin	No. of trials	Frequency of reaction scores (%)							Disease severity inded (DSI)	
			0–2	3	4	5	6	7	8–9	Mean	SD
Moreberekan	West Africa	108	63	12	16	4	2	2	1	4.7	1.3
Dourado Precoce	Brazil	35	51	14	17	9	3	6	0	4.9	1.2
OS 6	Zaire	114	53	17	13	10	3	2	2	5.0	1.2
IRAT 104[a]	Ivory Coast	117	73	10	8	4	2	2	1	5.1	1.2
IRAT 13[a]	Ivory Coast	153	70	12	8	6	1	1	2	5.1	1.4
IR36	Philippines	640	50	14	14	11	6	3	2	5.1	1.2
Taebaeg (Suweon 287)[b]	Korea	177	50	11	16	11	5	5	2	5.1	1.3
Baegyang[a] (IRI 347)	Korea	153	53	11	15	11	2	7	1	5.2	1.3
Samgang (Milyang 55)	Korea	118	59	12	10	7	3	6	3	5.6	1.6
CICA 4[c]	Colombia	157	32	12	9	14	10	6	17	6.3	1.7

[a] Not widely cultivated but known resistant cultivar.
[b] Line number.
[c] Known susceptible cultivar.

breeding lines were selected from the crosses that include Tetep and Gam Pai 15 (Khush, 1989). In recent years, IR36 and IR64, known as durably resistant cultivars, are the major donor sources for blast resistance for tropical irrigated environments. Due to their complex geneology, it is rather difficult to identify the original donor sources of these cultivars without the aid of molecular markers.

Estimation of partial resistance

Partial resistance is the relative ability of the host genotype to limit the proportion of diseased leaf area caused by susceptible types of lesions in all growth stages of the plant (Parlevliet, 1975). Once qualitative resistance of a cultivar is overcome, the level of partial resistance then determines the degree of damage. A high level of partial resistance to rice blast is a common feature of durably resistant rice cultivars. Partial resistance alone can usually adequately protect the irrigated rice crops in the tropics; however, it is not efficient under very blast-conducive conditions (Yeh and Bonman, 1986).

One prerequisite to assess the expression of partial resistance to blast is the availability of matching virulent isolate(s). Often partial resistance cannot be assessed due to the absence of virulent isolates in test sites. The IRBN multilocation tests, therefore, provide an indirect estimate of partial resistance. DSI obtained from IRBN results was positively correlated with the partial resistance levels observed under field conditions at IRRI (Ahn *et al.*, 1989).

Since the IRBN evaluation is generally done under upland conditions, partial resistance of lowland-adapted rice cultivars may not be readily recognized due to their poor adaptation to upland conditions and often extremely high disease pressure (Fig. 8.2). Local irrigated rice with high partial resistance must be included as a check to determine acceptable disease severity levels. Subsequent assessment in irrigated conditions should follow for confirmation.

Virulence patterns over IRBN test sites

Qualitative varietal reaction patterns observed at different IRBN test sites are primarily determined by the virulence composition of the field population of *P. grisea* at test sites during evaluation. The presence of distinct pathogenic races of blast pathogen in different countries is well recognized (Matsumoto *et al.*, 1969). The term 'geovirulence' was proposed to refer to the overall virulence spectrum of rice blast populations at a given location based on the absence or presence of compatible reactions of a set of diverse rice cultivars over a range of sites (Ahn *et al.*, 1988a,b).

Four major location groups were observed among 24 test sites in 14 countries based on multivariate analysis of the 1985 IRBN results (Ahn and Seshu, 1987). Similar results were obtained in a cluster analysis of 37 test sites in 19 countries based on combined results of blast reactions of 71 test entries in the IRBN from 1984 to 1992 (Fig. 8.3). At 65% of similarity levels using simple matching coefficient, four major location groups were recognized. Geovirulence patterns of four

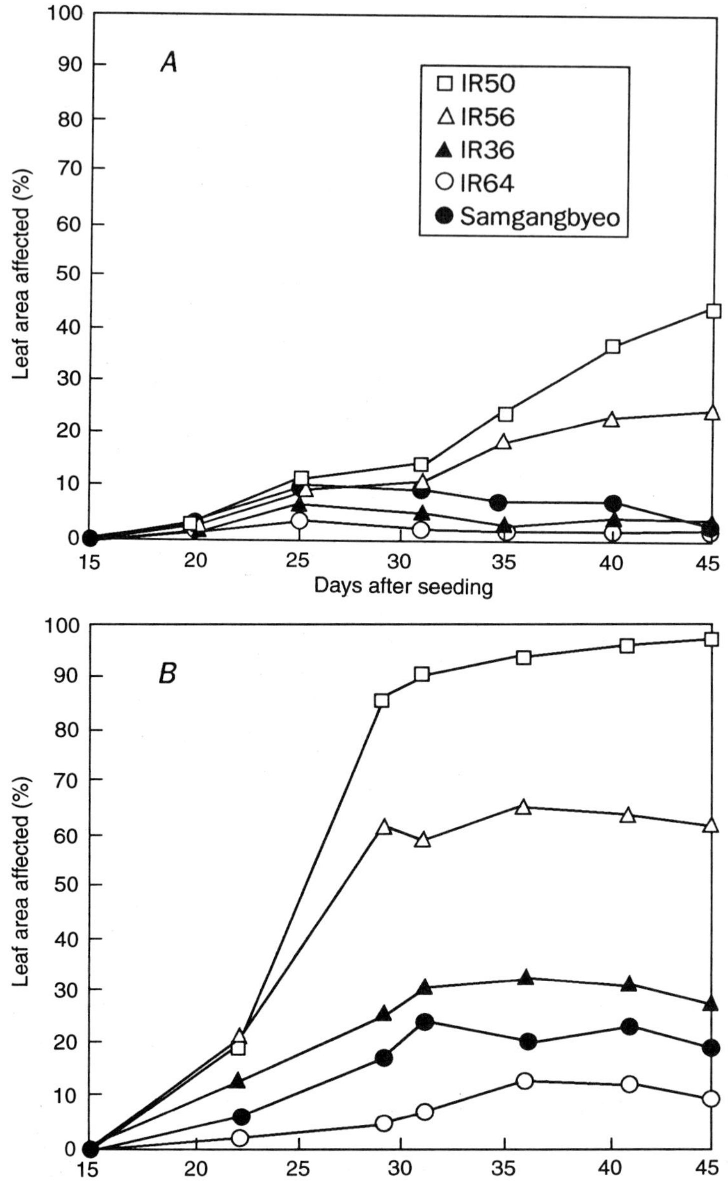

Fig. 8.2. Leaf blast disease progress curves of selected rice cultivars under irrigated (A) and upland conditions (B).

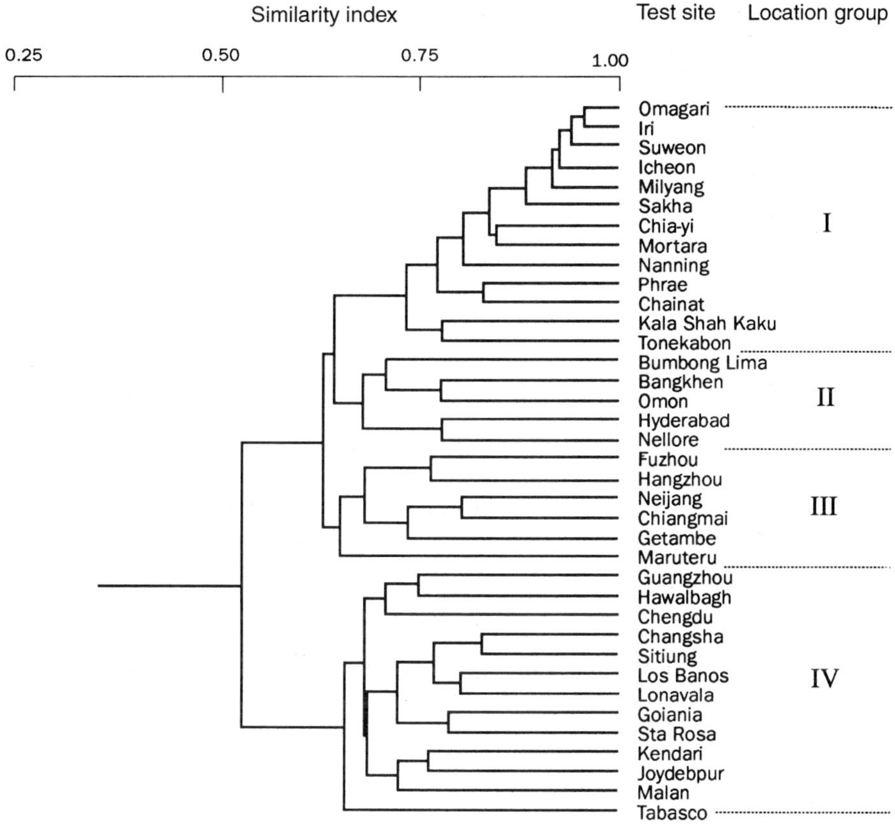

Fig. 8.3. Phenogram constructed from the International Rice Blast Nursery results from 1984 to 1992, indicating the similarity of varietal reaction at different test sites.

major location groups are shown in Fig. 8.4. At 75% level of similarity, 16 groups of test sites were formed indicating a wide variation of geovirulence patterns in the IRBN test sites.

Among test sites in location group I, eight are situated in japonica rice growing areas. Also included are two test sites, Kala Shah Kaku (Pakistan) and Tonekabon (Iran), where Basmati-type rice is cultivated. Indica varieties are cultivated in sites of other patterns. Wide virulence spectra are found in test sites belonging to group IV. Several test sites in this group such as Sitiung (Indonesia), Goiania (Brazil), Sta. Rosa (Colombia), and Changsha (China) are known as 'blast hot spots' as they have been major blast evaluation sites for many years. Test sites in China and India were classified into three different groups, indicating a wide virulence variation within a country.

All the IRBN test sites in Korea were clustered into group I. The same result was obtained in a study where 17 blast test sites in Korea were grouped into five location groups (Park, 1993). To obtain a wider range of pathogenic variation of japonica-infecting isolates, current IRBN test sites in Korea should include other

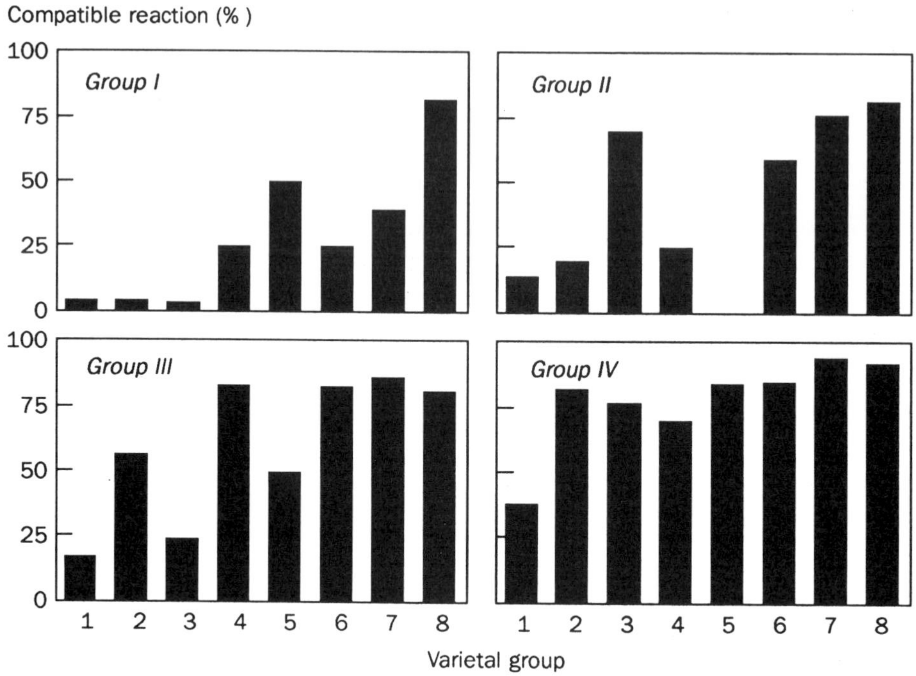

Fig. 8.4. Percentage frequency of compatible reactions (scores 4–9) of eight varietal groups of entries in location groups I, II, III, and IV. The analysis is based on the International Rice Blast Nursery results from 1984 to 1992.

test sites representing distinct location groups while reducing the current sites to only one.

Based on these results, key test sites for an intensive evaluation could be selected from each distinct geovirulence group without reducing the pathogenic variation currently present in the IRBN test sites. However, extrapolation of evaluation results obtained from these selected sites should be done carefully among test sites since the reaction similarity could be changed due to the shift of major rice genotypes cultivated in the region.

Many conventional race differential varieties and near isogenic lines do not appear suitable for the monitoring of field populations of the blast pathogen in the IRBN test. Isogenic lines with various combinations of resistance genes would be more suitable than only single-gene-incorporated ones for a rapid monitoring of the field population of blast pathogen. Virulence monitoring should provide a guide for the breeders in selecting effective combinations of resistance genes.

All 71 test entries used for geovirulence analysis can be grouped into eight major varietal groups including two single-entry groups (Table 8.8). All known source donors and cultivars exhibiting a broad spectrum of resistance were clustered into varietal group 1. Several IR lines and varieties released in the early 1980s were grouped into varietal group 3. They showed high frequency of compatible reactions in the majority of indica rice growing locations of groups 2 and 4 indicating the low effectiveness of qualitative resistance involved in these

Table 8.8. Major varietal groups based on cluster analysis of the blast reactions of 71 rice cultivars included in the International Rice Blast Nursery evaluations from 1984 to 1992.

Varietal group	Rice cultivar
1	Carreon, Raminad Str. 3, Ta-poo-cho-z, Tetep, IRAT 13, IRAT 144, ITA 212, Fukunishiki, BG367-4, Milyang 82, IR9852-93-2-2-2-3, IR28228-96-3-2-1-3, IR31851-63-1-2-3-2, IR31892-46-3-2, IR31899-30-1-1-3, and IR32429-122-3-1-2
2	Dular, Kataktara DA-2, Salumpikit, PTB33, Intan, IR64, UPL Ri-5, Chianung Si-Pi 661020, IR13146-45-2-3, BR203-26-2, B5526F-ST-17, IR10011-16-3, IR14632-2-3, IR22082-41-2, and IR28210-154-3-2-2-3
3	IR36, IR42, IR50, IR54, IR60, C1158-7, IR18349-22-1-2-1-1, IR21820-154-3-2-2-3, IR28118-138-2-3, IR28128-45-3-3-2, IR29658-43-3-2-1, and IR32429-47-3-2-2
4	Toride 1
5	NP 125
6	C662083, PAU50-B-25-1
7	Peta, Usen, ADT30, B2433B-KN-10-1-1-1, BR220-1-1, IR10068-18-2, IR1552, RAU4004-127, and S287B-39-1-3
8	Chokoto, Kanto 51, Pi No. 4, Tsuyuake, Shin 2, Yashiromochi, Taichung T.C.W.C., C039, B40, PY2, TNAU9039-14, HAU16-20-3, and RNR74229

cultivars. Apparently the combination of *Pi-z* and *Pi-sh* genes in Fukunishiki confers seemingly wide resistance. Although this may suggest that wide resistance could be due to one or a few unmatched resistance genes, broad resistance of other cultivars such as Tetep and other breeding lines may be due to several resistance genes of qualitative (Kiyosawa, 1972) and quantitative nature. Since no genetic information on IRBN test materials is available, it is difficult to know what specific resistance gene combinations determine the differences in distinct varietal groupings. Nevertheless, the efficiency of qualitative resistance in different test materials can be easily differentiated by this kind of variety × location pattern analysis (Fig. 8.4).

Future Aspects of International Collaboration

Data generated from IRBN provide only limited information on host resistance to rice blast of test entries. Due to major emphasis on the exchange of diverse genetic materials among rice breeders, and a highly variable and inconsistent data return from participating countries, comprehensive analysis and interpretation of IRBN data is rather difficult. Nevertheless, many rice improvement programs fully utilize IRBN as a major source of improved breeding materials for hybridization.

IRBN results are used by cooperating scientists to monitor and assess the progress in development of blast-resistant cultivars.

Recently INGER has initiated a supplementary IRBN trial on partial resistance in selected test sites. Unlike conventional IRBN, it contained a reduced number of test entries and proper experimental procedures to determine the stability of partial resistance across test environments (Bonman *et al.*, 1989).

To maximize the use of information generated by IRBN, each rice program should integrate IRBN as part of its overall blast evaluation scheme. This scheme also requires integration and coordination of related disciplines such as plant breeding, plant pathology, molecular biology, and epidemiology. Linkage between individual breeding programs and basic research on host–pathogen interaction should be strengthened for rapid progress in breeding for resistance to rice blast.

Genetically diverse germplasm is a valuable resource for rice breeders and other scientists working on blast resistance. A systematic evaluation at appropriate test sites would greatly accelerate progress in enhancing durable blast resistance and generating detailed information on G × E interaction. Research efforts in biotechnology for rice improvement is making steady progress (Khush *et al.*, 1993) and various products are to be evaluated in different areas over time. International collaboration on exchange and evaluation of these materials would facilitate and expand utilization of these novel genotypes in the future. To ensure stabilization of past yield advances and to allow future progress of varietal improvement, breeding efforts will be continued. In breeding for resistance to rice blast, diverse resistant sources, systematic evaluation schemes, and suitable test environments are essential. INGER's IRBN provide these required key factors in global and regional scales.

Research is still needed to improve methods for evaluation of durable resistance and to develop gene deployment schemes to prolong the efficiency of blast resistance. International collaboration on this area of research will permit the evaluation of different strategies under a range of environmental conditions.

References

Ahn, S.W. (1982) The slow blasting resistance to rice blast. In: IRAT/GERDAT (ed.), *Proceedings of the Symposium on Rice Resistance to Blast.* 18–21 March 1981, IRAT/GERDAT, Montpellier, France, pp. 343–370.

Ahn, S.W. and Koch, M.F. (1988) A conceptual model of disease resistance in rice pathosystems, and its implications for evaluating resistance. *International Rice Research Newsletter* 13(6), 12–13.

Ahn, S.W. and Mukelar, A. (1986) Rice blast management under upland conditions. In: *Progress in Upland Research.* International Rice Research Institute, PO Box 933, Manila, The Philippines, pp. 363–374.

Ahn, S.W. and Ou, S.H. (1982) Quantitative resistance of rice to blast disease. *Phytopathology* 72(3), 279–282.

Ahn, S.W. and Rubiano, M. (1984) Relationship between susceptibility to leaf blast and panicle blast severity. *International Rice Research Newsletter* 9(5), 13.

Ahn, S.W. and Seshu, D.V. (1987) Geographical variation of virulence patterns of the rice blast fungus. In: *Abstracts of the 11th International Congress on Plant Protection*, Manila, The Philippines, p. 152.

Ahn, S.W., Kim, C.H., Park, N.K., Choi, H.C. and Cho, S.Y. (1988a) Varietal ranking of blast severity in Korean farmers' fields. *International Rice Research Newsletter* 13(5), 11–12.

Ahn, S.W., Seshu, D.V., Park, N.K. and Kim, C.H. (1988b) Analysis of multilocation evaluation for rice blast resistance of Korean varieties. In: *Abstract of the 5th International Congress of Plant Pathology*, Kyoto, Japan, p. 267.

Ahn, S.W., Seshu, D.V. and Bonman, J.M. (1989) Prediction of partial resistance to rice blast using multilocation trials. *Phytopathology* 79, 1183 (Abstract).

Bonman, J.M., Bandong, J.M., Lee, E.J. and Valent, B. (1989) Race-specific partial resistance to blast in temperate japonica rice cultivars. *Plant Disease* 73, 496–499.

International Rice Testing Program (1988) *Standard Evaluation System for Rice*, 3rd edn. International Rice Research Institute, PO Box 933, Manila, The Philippines. 54 pp.

Khush, G.S. (1989) Multiple disease and insect resistance for increased yield stability in rice. In: *Progress in Irrigated Rice Research*. International Rice Research Institute, PO Box 933, Manila, The Philippines, pp. 79–92.

Khush, G.S., Brar, D.S., Zapata, F.J., Nelson, R., McCouch, S. and Bottrell, D.G. (1993) Biotechnology for rice improvement. In: *Focused Plant Improvement*. Proceedings (vol. 1) of the Tenth Australian Plant Breeding Conference, Gold Coast, 18–23 April, pp. 258–279.

Kiyosawa, S. (1972) Genetics of blast resistance. In: *Rice Breeding*. International Rice Research Institute, PO Box 933, Manila, The Philippines, pp. 203–225.

Lee, E.J., Zhang, Q. and Mew, T.W. (1989) Durable resistance to rice diseases in irrigated environments. In: *Progress in Irrigated Rice Research*. International Rice Research Institute, PO Box 933, Manila, The Philippines, pp. 93–110.

Matsumoto, S., Kozaka, T. and Yamada, M. (1969) Pathogenic races of *Pyricularia oryzae* Cav. in Asian and some other countries. *Bulletin of the National Institute of Agricultural Sciences*. Tokyo C23, 1–36.

Ou, S.H. (1965) A proposal for an international program of research on the rice blast disease. In: *The Rice Blast Disease*. The Johns Hopkins Press, Baltimore, Maryland, pp. 441–446.

Ou, S.H. (1985) *Rice Diseases*, 2nd edn. Commonwealth Mycological Institute, Kew, UK, 380 pp.

Park, N.K. (1993) Classification of rice cultivars by the rice leaf blast reaction and its utilization for breeding. PhD. thesis, Chungnam National University, Taejon, Korea.

Parlevliet, J.E. (1975) Partial resistance of barley to leaf rust, *Puccinia hordei*. 1. Effect of cultivar and development stage on latent period. *Euphytica* 24, 21–27.

Seshu, D.V., Kwak, T.S. and Mackill, D.J. (1986) Global evaluation of rice varietal reactions to blast disease. In: *Progress in Upland Rice Research*. International Rice Research Institute, PO Box 933, Manila, The Philippines, pp. 335–351.

Wang, G. (1992) RFLP mapping of major and minor genes for blast resistance in a durably resistant rice cultivar. PhD Thesis, University of the Philippines at Los Baños, Laguna.

Yeh, W.H. and Bonman, J.M. (1986) Assessment of partial resistance to *Pyricularia oryzae* in six rice cultivars. *Plant Pathology* 35, 319–323.

9 Present Knowledge of Rice Resistance Genetics and Strategies for *Magnaporthe grisea* Pathogenicity and Avirulence Gene Analysis

J.L. Notteghem, D. Tharreau, D. Silué and E. Roumen

CIRAD-CA, Unité de Recherche Phytopathologie-Malherbologie, BP 5035, 34032 Montpellier, Cedex 1, France

Introduction

Knowledge of pathogenicity and avirulence genes for plant pathogenic fungi is not as advanced as for viruses and bacteria. However, important progress in the field of genetics of plant pathogenic fungi has been made over the last years, particularly for *Magnaporthe grisea*, the rice blast pathogen. Differences in virulence between isolates were first observed by Sasaki in 1922 and resistance genes of rice to blast disease were described soon after (e.g. Sasaki, 1923; Nakatomi, 1926; and others cited in: Takahashi, 1965). Since then, a series of resistance genes has been identified and analyzed. An important contribution was made by S. Kiyosawa and his co-workers, who identified a total of 12 resistance genes (Kiyosawa, 1984). Several sets of differential rice cultivars were proposed to distinguish races between isolates of *M. grisea*, and the virulence pattern of many isolates has been evaluated. The low fertility of rice pathogenic isolates prevented genetic analysis of virulence in the fungus for many years. But since the successful culturing of the perfect stage *in vitro* in 1970 (Hebert, 1971), many attempts to carry out genetic analysis of *M. grisea* have been made (Leung *et al.*, Valent *et al.*, 1991b; Ellingboe, 1992; Silué *et al.*, 1992a). Up to the present, these analyses remain much easier for isolates pathogenic to certain grasses, such as *Eragrostis* spp. or *Eleusine* spp. than for isolates pathogenic to rice. However, it is now possible to carry out genetic analysis for a limited number of rice pathogenic isolates. A better knowledge of the plant–parasite interaction is expected from analysis of the host resistance genes and the corresponding avirulence genes, as well as pathogenicity genes, at the molecular level.

Resistance to *M. grisea* in *Oryza sativa*

Similar to the leaf rust diseases of wheat and barley, two types of resistance to the blast pathogen in rice have been described: hypersensitivity resistance that is often complete, characterized by a resistant infection type and usually monogenic, and partial resistance that is incomplete, characterized by a susceptible infection type, and usually polygenic (Ezuka, 1972; Parlevliet, 1979). Various combinations of these two types of resistance occur in almost all cultivars, so when studying rice–*M. grisea* relationships, the choice of cultivar is very important. Depending on the objective, several kinds of resistant or susceptible cultivars are useful.

A cultivar without any gene inducing a hypersensitivity reaction, and with a low level of partial resistance, is very useful as a susceptible check, or as representative of the *O. sativa* species for pathogen host specificity studies. Among the *japonica* group, cultivar Maratelli meets these criteria. This cultivar was tested in our laboratory with about 200 isolates from widely different geographic origin without expressing any hypersensitivity resistance and its partial resistance is low. Sariceltik, used for genetic analysis by Valent *et al.* (this volume) is another *japonica* cultivar that might lack major gene resistance. Its partial resistance appears considerably lower than that of Maratelli. In a recent test done at CIRAD, the number of sporulating lesions that developed on Sariceltik after spray inoculation was consistently about 20-fold more than obtained on Maratelli for three different pathogen races (E.C. Roumen, unpublished findings). Among the *indica* group, CO39 showed a susceptible reaction to many pathogen races in tests done at IRRI where it is used as the susceptible check (Bonman *et al.*, 1986).

Breakdown of the resistance of an improved cultivar frequently occurs some years after its release. Thus studies of virulence of the rice blast population were initiated, as well as genetic analysis of host resistance. This resulted in the identification of differential rice cultivars, of which several sets were developed. The Japanese set (Table 9.1) was set up by Kiyosawa (1984) after identification and genetic analysis of the resistance gene(s) of each of these cultivars. The international set (Table 9.2), based on differential reactions of some cultivars described by Latterel *et al.* (1965), was set up by Atkins *et al.* (1967). It is used in many countries despite the fact that the genetic analysis for this set is incomplete. In many places, local cultivars showed differential reaction when the differential sets did not, and local differential cultivars were added for identification of races (e.g. Bonman *et al.*, 1986).

When two isolates show virulence to a cultivar known to possess a certain host resistance gene, these isolates are considered to lack the corresponding avirulence gene. However, when two isolates are avirulent on a cultivar, they could have the same avirulence gene or different avirulence genes. An avirulent isolate is necessary to identify a host resistance gene and the use of new isolates may reveal host resistance genes that have so far escaped detection. For example, the gene named *Pi-sh* (Imbe and Matsumoto, 1985), present in several of the Japanese differentials, was not discovered until a long time after their intensive use, because no avirulent isolate to the gene *Pi-sh* was observed during that time.

Similarly, identical cultivar reactions to different isolates are an indication, but not proof, that these cultivars contain the same host resistance gene(s). Tests

Table 9.1. Differential set of Japanese cultivars, their resistance genes, and corresponding race code (Kiyosawa, 1984).

Variety	Resistance gene	Code number
Shin 2	*Pi-k*^{s}	1
Aichi-asahi	*Pi-a*	2
Fujisaka 5	*Pi-i, Pi-k*^{s}	4
Kusabue	*Pi-k*	10
Tsuyuake	*Pi-k*^{m}	20
Fukunishiki	*Pi-z*	40
K1	*Pi-ta*	100
Pi n°4	*Pi-ta*^{2}	200
Toride 1	*Pi-z*^{t}	400
K60	*Pi-k*^{p}	.1
BL1	*Pi-b*	.2
K59	*Pi-t*	.4

Table 9.2. International set of differential cultivars (Atkins *et al.*, 1967).

Variety	Known resistance genes
Raminad Str. 3	?
Zenith.	*Pi-a, Pi-z* (Kiyosawa, 1967)
NP 125	?
Usen	*Pi-a* + ? (Kiyosawa, 1972)
Dular	*Pi-k** (Kiyosawa, 1972)
Kanto 51	*Pi-k* (Kiyosawa, 1984)
Sha-Tiao-Tsao	*Pi-k*^{s} (Kiyosawa, 1972)
Caloro	*Pi-k*^{s} (Kiyosawa, 1972)

*Unidentified gene at the locus *Pi-k*.

for allelism by crosses between the cultivars are necessary to prove that they have identical genes. An appropriate method to identify 'new' genes was described by Kiyosawa (1967) and by Toriyama *et al.* (1986). The work required is considerable when a 'new' gene must be compared to many known genes. This explains why very few new genes have been described since the work done by Kiyosawa and co-workers. None the less, inoculation results of many tropical isolates and cultivars indicate that a considerable number of major genes probably exist in addition to those so far described.

Meanwhile, several new methods have been used to identify resistance genes with the objective of mapping or cloning them (McCouch *et al.*, this volume). One of these methods is the creation of near isogenic lines (NILs) which, ideally, are sets of cultivars differing only for genes of interest, in this case genes for resistance. This type of material is very useful because the expression of a host resistance gene can be compared with the reaction of the 'susceptible' allele in the same genetic background. NILs with resistance genes of Japanese differentials have been obtained in Japan by introgressing them into the background of the cultivar Nipponbare (Ise, 1992). Resistance genes from other cultivars were introduced in cultivar CO39 at IRRI (Mackill *et al.*, 1988; Mackill and Bonman, 1992). Tests for allelism between the resistance genes in these series of NILs have been initiated (Inukai *et al.*, 1992a, b), but are yet far from complete. Another method is the use of single seed descent (SSD) or of doubled haploid (DH) lines derived from crosses between resistant and susceptible cultivars, to identify and map host resistance genes by means of molecular markers. Analysing DH lines from the cross IRAT 177 × Apura, a gene named *Pi-6t* was identified in the cultivar Apura (Abadassi *et al.*, 1990). Carefully chosen DH or SSD lines with low partial resistance and well-characterized resistance genes, as well as NILs, are promising host material candidates for establishing an improved set of differential cultivars.

The above research has focused on complete, monogenic resistance inducing hypersensitivity reactions, but so far, none of these hypersensitivity genes have provided a resistance durable enough to be of benefit for resource-poor farmers. Exceptions to this rule are rare, but include the cultivar Tongil that remained resistant in Korea for 6 years while being widely grown. Efficient management strategies for these resistance genes remain to be developed. The newest strategy proposes bringing together into cultivars those genes that are each effective against an entire 'lineage' of the blast population at one location (Levy *et al.*, 1993; see chapters 13 and 16, this volume). Cultivars bred following this strategy are not yet available, but it will be interesting to monitor the durability of their resistance. The most durable resistance found so far is partial polygenic resistance. Unfortunately, little is known about possible interactions between this type of resistance and different fungal races.

Among the cultivars with effective partial resistance, the traditional upland cultivar Moroberekan is often mentioned (Bidaux, 1978; Notteghem, 1985), and this cultivar could serve as a reference. Some others can be used also as check cultivars, such as IR36 and IRAT 216, which also have an adequate level of partial resistance. However, when used as checks for partial resistance, all of these cultivars also have hypersensitivity genes inducing complete resistance against an important proportion of the pathogen races in many places. Some upland *japonica* cultivars such as Tainung Chuen 2 or Lung Sheng 1 have a moderate level of partial resistance and were found to develop a susceptible infection type to the majority of isolates tested in our experiments. Another upland cultivar that often develops a susceptible infection type is Azucena (J.L. Notteghem, unpublished findings). These cultivars are good candidates for checks in partial resistance experiments.

Pathogenicity and Host Specificity Genes of *M. grisea*

Pathogenicity genes have been studied by mutant analysis. The mutations named 'buff', 'rosy' and 'alb' correspond to melanin synthesis deficiency, and a failure to penetrate host cells due to weakness in appressorium cell walls (Chumley and

Valent, 1990). Other mutations include those for spore morphology such as the 'SMO' mutations (Hamer *et al.*, 1989) or for appressorium deficiency (Lee and Dean, 1993). Differences in gene expression *in vitro* and *in vivo* can be evaluated by comparing mRNAs. This method used by Talbot *et al.* (1993) revealed the importance of the gene *Mpg1* coding for an hydrophobin which is necessary for pathogenicity. Many other genes will probably be identified in the near future.

The good fertility of isolates pathogenic to some grasses, in particular weeping lovegrass (*Eleusine indica*) and goose grass (*Eragrostis curvula*), was used to study host specificity by intercrossing them, and also by crossing them with rice pathogenic isolates. In most cases, host specificity appears controlled by complex genetic systems. Crosses between isolates pathogenic to rice, *Eleusine*, wheat, and maize revealed a polygenic control of host specificity, and it was not possible to identify any major gene (Notteghem, 1990). In other experiments, host specificity appeared to be simply inherited (Yaegashi, 1978). Backcrosses allowed Valent *et al.* (1991) to identify the host specificity genes as *Pwl1*, *Pwl2*, as well as the genes *AVR1-Mara*, and *AVR2-Mara*.

Fertility of *M. grisea* Isolates Pathogenic to Rice: Key for Avirulence Analysis

In the absence of the sexual stage, avirulence was studied by comparing field isolates or by studying mutants (Yamasaki and Kiyosawa, 1966; Kiyosawa, 1976; Kwon and Oh, 1977; Notteghem, 1977). After the discovery of the perfect stage of *Pyricularia grisea* by Hebert (1971), evaluation of the fertility of many rice isolates was carried out in Japan (Ueyama *et al.*, 1977; Kato and Yamaguchi, 1982; Yaegashi and Yamada, 1986), and in France (Silué and Notteghem, 1991). Isolates obtained from rice and pathogenic to rice were shown to have a low fertility, much lower than that of *M. grisea* isolates pathogenic to *Eleusine coracana* or to *Eragrostis curvula*.

Recently, several investigations of the genetic relationships of *Magnaporthe* sp. pathogenic to various grasses were conducted. The use of various molecular markers led to the conclusion that there is a group of genetically closely related isolates, pathogenic to either *Eleusine* sp., to wheat, or to rice, and this group is distinct from groups of isolates pathogenic to *Pennisetum* sp. or to *Digitaria* sp. (Capy *et al.*, 1993; chapter 5, this volume). The relatedness of isolates pathogenic to rice and to some grasses facilitates the strategy of improving the fertility of rice pathogenic isolates through crosses with fertile ones from the same group, but pathogenic to other grasses. In general, initial crosses between rice pathogenic and non-rice pathogenic isolates are followed by several intercrosses between progeny, and by backcrosses to the rice pathogenic isolate. At the same time, continued efforts are made to identify more fertile rice pathogenic isolates for use as recurrent rice parents in backcrosses. The improvement of fertility of rice isolates was achieved step by step. At first, attempts to interbreed rice isolates were a complete failure. Then, using fertile ragi (*Eleusine coracana*) isolates as fertility testers, Kato and Yamaguchi (1982) were able to screen hundreds of isolates and identified some rice isolates that were fertile enough to mate together. However, viability of ascospores resulting from these crosses was low. Valent and Chumley (1991) used two rice isolates from China, named O-111 and O-135, that are relatively fertile, but in our laboratory they show very poor sporulation. The isolate GUY11, collected in 1979 in French Guyana, appeared fertile and hermaphroditic when tested for mating type in 1986 (Notteghem and Silué, 1992). GUY11 produces viable ascospores when mated with some other rice isolates (which act as the male parent) from various origins. This finding was the start of several genetic studies with this isolate or its progenies. More recently, another isolate, CHNOS 37-1-1, collected about 6 years ago in China (H. Kato, personal communication), was identified as very fertile (Hayashi and Naito, 1993). Like GUY11, this isolate was crossed with other rice isolates which act as the male parent. Variability for avirulence genes in all available crosses seems limited and it will be very useful to identify other fertile rice isolates. Except for GUY11, which may have been introduced in French Guyana through seed exchanges, the fertile isolates all originate from the south of China. Fertility tests of new isolates are now routinely conducted by several laboratories and population analysis of *M. grisea* is in progress. This should help identify new fertile rice isolates.

The isolates now available made possible identification, mapping, and now cloning of the host specificity and avirulence genes listed in Table 9.3. Genetic analysis of (hypersensitivity) resistance of rice cultivars has demonstrated the presence of a gene-for-gene relationship (Silué *et al.*, 1992a; Hayashi and Naito 1993). We observe that the number of genes identified in each cross is small compared with the number of known resistance genes, or with the number of resistant rice cultivars. This is partially due to improving fertility by backcrosses which reduces the diversity of avirulence genes. In the case of the cross GUY11 × 2/0/3 (Silué *et al.*, 1992b), 2/0/3 was obtained by crossing GUY11 with the field isolate ML25, followed by intercrossing the progeny. The parental isolates of this cross were tested on 250 resistant rice cultivars. Most of the rice cultivars were not suitable for identification of avirulence genes, since in 60% of the cases no differential reaction for the parental isolates was observed. Intermediate (no clear cut) differential reactions were observed on another 25%, and some of the remaining cultivars could not be studied further because of lack of seed. Due to these limitations, we were able to do a genetic analysis for avirulence using only seven cultivars (Silué *et al.*, 1992b). Avirulence to these cultivars is controlled by three genes, two of which are different from those so far described. The use of molecular markers, RFLP, RAPD, and repeated sequences corresponding to probe MGR 586 and MGR 583, has allowed us to map these genes (D. Tharreau *et al.*, unpublished findings).

Analysis of Aggressiveness of *M. grisea* Isolates

Unlike the variation of virulence among isolates of the blast pathogen, variation for aggressiveness has rarely been studied. Nevertheless, differences for aggressiveness among isolates occur, and quantitative interactions between rice and *M. grisea* have been demonstrated (Roumen, 1992). More studies on the frequency

Table 9.3. Avirulence genes identified in *Magnaporthe grisea*.

Cross	AVR gene	Corresponding rice variety	Reference
2359 × GUY11 2 1	*POS1* *POS2*	51583 Sha-tiao-tsao	Leung *et al.*, 1988
GUY11 × CH 104-3 1 2	*POS3* *POS4*	K59 Kinandong Patong	
4091-5-8 × 0-135 4 1	*Avr1-CO39* *Avr1-M201* *Avr1-YAMO*	CO39 M201 Yashiro-mochi	Valent *et al.*, 1991, and this volume
42024-7-8 × 6043 2 2	*Avr2-YAMO* *Avr1-MARA* *Avr1-TSUY* *Avr1-MINE*	Yashiro-mochi Maratelli Tsuyuake Minehikari	
GUY11 × 2/0/3 1 3*	*Avr1-CIC6* *Avr1-KU86* *Avr1-IrT7*	CICA 6 CICA 8 Med Noï Ku 86 IRAT 7 DJ 8-341	Silué *et al.*, 1992b
GUY11 × ML 25 1 2	*Avr1-Pin°4*	Pi n°4	Silué *et al.*, 1992a
GUY 11 × 66.10 1 2	*P11, P12, S11, A12*	Katy	Lau *et al.*, 1993
Tm4 × 70-6 1 2*	2 genes 3 genes	Leah Newbonnet Lemont Lebonnet	Chao *et al.*, 1993
CHNOS 37-1-1 × CHNOS 12-2-1 1 1	*Avr1-KB* *Avr1-TA*	Kusabe Tsuyuake	Hayashi and Naito, 1993
CHNOS 37-11 × CHNOS 32-2-4 1 1	*Avr1-XM* *Avr1-K59*	Yoshiro-mochi K59	
GUY11 × 2359 1 2	1 gene	CO39	Leong *et al.*, this volume

1 rice field isolate.
2 rice isolate improved for fertility by crosses with non-rice isolates.
3 rice isolate derived from crosses with non-rice isolates.
4 non-rice isolate.
* progeny from crosses involving GUY11.

and magnitude of possible host–pathogen genotype interactions for aggressiveness are needed, especially in the view of breeding programs that aim to improve partial resistance. Genetic analysis of aggressiveness in crosses between rice isolates would be very informative regarding the durability of partial resistance. Accurately assessing the aggressiveness of *M. grisea* isolates is not an easy job due to the strong impact of uncontrollable environmental factors. On the other hand, good progress has been made in measuring quantitative resistance (Roumen, 1993, and chapter 15, this volume) and available methods can be adapted for evaluation of aggressiveness among progeny of crosses between rice isolates. Conducting genetic analysis of aggressiveness will probably be feasible within a few years after enough experience in genetic analysis of avirulence factors has been obtained, and appropriate strains are developed for this analysis. Although perhaps less attractive for a geneticist than a monogenic avirulence study, genetic analysis of aggressiveness is important and must be pursued.

Conclusion

Because of low fertility of *M. grisea* isolates pathogenic to rice, the genetic control of virulence was for a long time poorly understood. However, in the last 5 years rapid progress has been made in genetic analysis of pathogenicity, host specificity, and avirulence. The identification of rice resistance genes is important for several resistance breeding strategies. Efficient resistance breeding aiming at 'lineage exclusion' requires a better knowledge of the hypersensitivity genes that are present in many cultivars of agronomic importance. Resistance breeding aimed at improving partial resistance also requires good knowledge of genes for hypersensitivity resistance in cultivars used by breeders, because of the epistatic effects of these genes on the expression of partial resistance. As a by-product, genetic analysis of avirulence genes can produce progeny isolates that could be used for more efficient characterization of host resistance genes, as a complementary tool for genetic analysis of resistance of rice cultivars. This approach is more easily employed when more avirulence genes will be characterized.

The other interest in gaining more knowledge of the avirulence mechanism could be the discovery of new and nonconventional methods for control of the rice blast disease. Regarding this matter, it is important to clone, sequence, and study the function of avirulence genes. Conducting this work for several genes is highly desirable.

For all these objectives, obtaining more fertile rice pathogenic isolates having broader variability for avirulence genes will be extremely useful. The search for other fertile rice isolates should be continued, but this search could be more efficient if results of population analysis are considered. Population analysis could indicate areas where sexual reproduction occurs in a rice pathogenic population or where a sexual reproduction on other grasses generates new rice pathogenic isolates that are (still?) highly fertile. At present, there is no evidence for such places, but the identification of at least some fertile rice isolates indicates that others could be found.

References

Abadassi, J., Glaszmann, J.C., Notteghem, J.L., Guiderdoni, E. and Courtois, B. (1990) Genetic analysis of rice blast resistance using doubled haploid, single seed descent lines, and isozyme marker genes. In: *Rice Genetics II.* International Rice Research Institute, Los Baños, The Philippines, pp. 746–748.

Atkins, J.G., Robert, A.L., Adair, C.R., Goto, K., Kozaka, T., Yamagida, R., Yamada, M. and Matsumoto, S. (1967) An international set of varieties for differentiating races of *Pyricularia oryzae*. *Phytopathology* 57, 297–301.

Bidaux, J.M. (1978) Screening for horizontal resistance to rice blast (*Pyricularia oryzae*) in Africa. In: Buddenhagen, I.W. and Persley, G.J. (eds), *Rice in Africa*. Academic Press, New York, pp. 159–174.

Bonman, J.M., Vergel de Dios, T.I. and Khin, M.M. (1986) Physiologic specialization of *Pyricularia oryzae* in the Philippines. *Plant Disease* 70, 767–769.

Capy, M.P., Lebrun, M.H., Notteghem J.L. and Brygoo, Y. (1993) Phylogenetic relationships among *Magnaporthe grisea* isolates from different hosts plants. In: Program and Abstracts of An International Symposium on Rice Blast Disease. University of Wisconsin, Madison, 9–13 August 1993.

Chao, C.T. and Ellingboe, A.H. (1993) Genetic analysis of avirulence/virulence to several rice cultivars of a field isolate of *Magnaporthe grisea. 6th International Congress of Plant Pathology*. Montreal, Canada.

Chumley, F.G. and Valent, B. (1990) Genetic analysis of melanin-deficient, nonpathogenic mutants of *Magnaporthe grisea. Molecular Plant-Microbe Interactions* 3, 135–143.

Ellingboe, A.H. (1992) Segregation of avirulence/virulence on three rice cultivars in 16 crosses of *Magnaporthe grisea*. *Phytopathology* 82, 597–601.

Ezuka, A. (1972) Field resistance of rice varieties to blast disease. *Review of Plant Protection* 5, 1–2.

Hayashi, N. and Naito, H. (1993) Inheritance of avirulence in a cross of *Magnaporthe grisea* pathogenic to rice. In: *Abstracts of the 6th ICPP*. Montreal, Canada, p. 197.

Hamer, J.E., Valent, B. and Chumley, F.G. (1989) Mutations at the SMO genetic locus affect the shape of diverse cell types in the blast fungus. *Genetics* 122, 351–361.

Hebert, T.T. (1971) The perfect stage of *Pyricularia grisea*. *Phytopathology* 61, 83–87.

Imbe, T. and Matsumoto, S. (1985) Inheritance of resistance of rice varieties to the blast fungus strains virulent to the variety Reiho. *Japanese Journal of Breeding* 35, 332–339.

Inukai, T., Mackill, D.J., Bonman, J.M., Sarkarung, S., Zeigler, R., Nelson, R.J., Takamure, I. and Kinoshita, T. (1992a) Blast resistance genes *Pi-2(t)* and *Pi-z* may be allelic. *Rice Genetics Newsletter* 9, 90–91.

Inukai, T., Mackill, D.J., Bonman, J.M., Sarkarung, S., Zeigler, R., Nelson, R.J., Takamure, I. and Kinoshita, T. (1992b) A blast resistance gene *Pi-3(t)* in near isogenic line C-104PKT. *Rice Genetics Newsletter* 9, 94–95.

Ise, K. (1992) Reexamination of linkage relationships between blast (Bl) resistance genes *Pi-i* and *Pi-z* using near isogenic lines (NILs) of rice. *International Rice Research Newsletter* 17, 8–9.

Kato, H. and Yamaguchi, T. (1982) The perfect-state of *Pyricularia oryzae* Cav. from rice plants in culture. *Annals of the Phytopathological Society of Japan* 48, 607–612.

Kiyosawa, S. (1967) Genetic studies on host-pathogen relationship in the rice blast disease. In: Ogura, T. (ed.), Proceedings, *Rice Diseases and Their Control by Growing Resistant Varieties and Other Measures*. Ministry of Agriculture and Forestry, Tokyo, Japan, pp. 137–153.

Kiyosawa, S. (1972) The inheritance of blast resistance transferred from some indica varieties in rice. *Bulletin of the National Institute of Agricultural Sciences* 23, 69–96.

Kiyosawa, S. (1976) Pathogenic variations of *Pyricularia oryzae* and their use in genetic and breeding studies. *SABRAO Journal* 8, 53–67.

Kiyosawa, S. (1984) Establishment of differential varieties for pathogenicity test of rice blast fungus. *Rice Genetics Newsletter* 1, 95–97.

Kwon, S.H. and Oh, J.H. (1977) Selection for blast resistant mutants in irradiated rice populations. In: *Induced Mutations Against Plant Diseases*. IAEA, Vienna, pp. 130–140.

Latterel, F.M., Marchetti, M.A. and Grove, B.R. (1965) Coordination of effort to establish an international system for race identification in *Pyricularia oryzae*. In: *The Rice Blast Disease*. Proceedings of a symposium at the International Rice Research Institute, Los Baños, Philippines, July 1963. The Johns Hopkins Press, Baltimore, Maryland, pp. 257–274.

Lau, G.W., Chao, C.T. and Ellingboe, A.H. (1993) Interaction of genes controlling avirulence/virulence of *Magnaporthe grisea* on rice cultivar Katy. *Phytopathology* 83(4), 375–382.

Lee, Y.H. and Dean, R.A. (1993) AMP regulates infection structure formation in the plant pathogenic fungus *Magnaporthe grisea. The Plant Cell* 5, 693–700.

Leung, H., Borromeo, E.S., Bernardo, M.A. and Notteghem, J.L. (1988) Genetic analysis of virulence in the rice blast fungus *Magnaporthe grisea. Phytopathology* 78, 1227–1233.

Mackill, D.J. and Bonman, J.M. (1992) Inheritance of blast resistance in near-isogenic lines of rice. *Phytopathology* 82, 746–749.

Mackill, D.J., Bonman, J.M., Tenorio, P.D. and Vergel de Dios, T.I. (1988) Near-isogenic *Indica* rice lines with blast resistance genes. *Rice Genetics Newsletter* 5, 98–101.

Nakatomi, S. (1926) On the variability and inheritance of their resistance of rice plants to rice blast disease. *Japanese Genetics* 4(1), 31–38 (in Japanese).

Notteghem, J.L. (1977) Etude de la résistance d'*Oryza sativa* a *Pyricularia oryzae* par application au parasite des techniques de mutagenese. In: *Induced Mutations Against Plant Diseases*. IAEA, Vienna, pp. 121–128.

Notteghem, J.L. (1985) Définition d'une strategie d'utilisation de la résistance par analyse génétique des relations hôte-parasite. Cas du couple riz-*Pyricularia oryzae*. *Agronomie Tropicale* 40, 129–147.

Notteghem, J.L. (1990) Results and orientation of E.C. projects on rice resistance to blast. *Summa Phytopathologica* 16, Jan. Maco., 1990, 57–72.

Notteghem, J.L. and Silué, D. (1992) Distribution of mating type alleles in *Magnaporthe grisea* populations pathogenic on rice. *Phytopathology* 82, 421–424.

Parlevliet, J.E. (1979) Components of resistance that reduce the rate of epidemic development. *Annual Review in Phytopathology* 17, 203–222.

Roumen, E.C. (1992) Small differential interactions for partial resistance in rice cultivars to virulent isolates of the blast pathogen. *Euphytica* 64, 143–148.

Roumen, E.C. (1993) Partial resistance in rice to blast and how to select for it. PhD Thesis, Wageningen Agricultural University, Wageningen, The Netherlands, 108 pp.

Sasaki, R. (1922) Inheritance of resistance to *Pyricularia oryzae* in different varieties of rice. *Japanese Genetics, Japan* 1, 81–85 (in Japanese).

Sasaki, R. (1923) Existence of strains in rice blast fungus II. *Journal of Plant Protection* 10, 1–10 (in Japanese).

Silué, D. and Notteghem, J.L. (1991) Compatibilité et fertilité de souches de *Magnaporthe grisea*, agent de la pyriculariose du riz. *Cryptogamie-Mycologie* 12, 87–95.

Silué, D., Notteghem, J.L. and Tharreau, D. (1992a) Evidence of a gene-for-gene relationship in the *Oryza sativa–Magnaporthe grisea* pathosystem. *Phytopathology* 82, 577–580.

Silué, D., Tharreau, D. and Notteghem, J.L. (1992b) Identification of *Magnaporthe grisea* avirulence genes to seven rice cultivars. *Phytopathology* 82, 1462–1467.

Takahashi, Y. (1965) Genetics of resistance to the rice blast disease. In: *The Rice Blast Disease*. Proceedings of a symposium at International Rice Research Institute, Los Baños, The Philippines. The Johns Hopkins Press, Baltimore, Maryland, pp. 303–329.

Talbot, N.J., Ebbole, J. and Hamer, J.E. (1993) Identification and characterization of *MPG1*, a gene involved in pathogenicity from the rice blast fungus *Magnaporthe grisea*. *The Plant Cell* 5, 1575–1590.

Toriyama, K., Ezuka, A., Asaga, K. and Yokoo, M. (1986) A method for estimating true resistance genes to blast in rice varieties. In: *Rice Genetics*. International Rice Research Institute, Los Baños, The Philippines, pp. 441–449.

Ueyama, A., Tsuda, M., Taga, M. and Nakagawa, H. (1977) Mating types for perfect state of *Pyricularia oryzae*. *Transactions of the Mycological Society of Japan* 18, 312–317.

Valent, B. and Chumley, F.G. (1991) Molecular genetic analysis of the rice blast fungus, *Magnaporthe grisea*. *Annual Review of Phytopathology* 29, 443–467.

Valent, B., Farral, L. and Chumley, F.G. (1991) *Magnaporthe grisea* genes for pathogenicity and virulence identified through a series of backcrosses. *Genetics* 127, 87–101.

Yaegashi, H. (1978) Inheritance of pathogenicity in crosses of Pyricularia isolates from weeping lovegrass and finger millet. *Annals of the Phytopathological Society of Japan* 44, 626–632.

Yaegashi, H. and Yamada, M. (1986) Pathogenic race and mating type of *Pyricularia oryzae* from Soviet Union, China, Nepal, Thailand, Indonesia and Colombia. *Annals of the Phytopathological Society of Japan* 52, 225–234.

Yamasaki, Y. and Kiyosawa, S. (1966) Studies on inheritance of resistance of rice varieties to blast. I. Inheritance of resistance of Japanese varieties to several strains of the fungus. *Bulletin of the National Institute of Agricultural Science, Japan, Series D* 14, 39–69.

10 Mapping of Blast Resistance Genes in Rice

S.R. McCouch[1,2], R.J. Nelson[1], J. Tohme[3] and R.S. Zeigler[1]

[1]*Plant Breeding, Genetics, and Biochemistry Division, IRRI, PO Box 933, Manila, The Philippines;* [2]*Plant Breeding Department, Cornell University, Ithaca, NY 14853-1901, USA;* [3]*Biotechnology Unit, CIAT, Apartado Aereo 6713, Cali, Colombia*

Introduction

Over the last several decades, much has been learned about the genetics of resistance to the blast fungus. Many major genes for resistance have been identified. Breeding for resistance to blast has, however, often been frustrated by the rapid adaptation of the pathogen population to resistant cultivars. Especially when resistance is based on single major genes, rice varieties may be rapidly overcome by compatible races of the pathogen (Kiyosawa, 1982). Certain rice cultivars are considered to possess 'durable resistance', as defined by Johnson (1981), and this resistance is thought to be conferred by 'minor genes' conferring partial resistance (Parlevliet, 1988). The genetic constitution of these cultivars, however, has not been well understood, and it has been difficult to attain reliable, long-lasting resistance by conventional breeding methods (Johnson, 1984). Although the desirability of combining multiple major and minor genes conferring both complete and partial resistance in rice breeding has long been recognized (Ikehashi and Khush, 1979), this has been difficult to achieve with predictability.

The difficulties associated with breeding for blast resistance can be attributed to two general problems. First, we are not entirely sure what genetic constitution results in durable resistance. Second, it is not necessarily possible to select for the presence of multiple genes, or for specific gene combinations, because of the epistatic interactions among genes. Molecular marker technology offers the opportunity to overcome both problems. First, molecular markers can improve the efficiency and resolution of genetic analysis, particularly when multiple resistance genes are present in a single cultivar (Paterson *et al.*, 1991). For instance, we are using molecular markers to dissect traditional rice genotypes which are associated with durable blast resistance. As described below, marker-assisted genetic analysis has already contributed to the identification and mapping of genes conferring both complete and partial resistance, and has provided insight into the genetic basis of durable resistance (Wang *et al.*, 1994). Second, markers linked to resistance genes provide tools to facilitate the selection of lines carrying resistance genes in desirable combinations (Abenes *et al.*, 1993). By making it possible to identify discrete genetic factors that are the building blocks of complex, quantitatively-inherited forms of resistance, the use of molecular markers allows us to approach the problem of blast resistance from both reductive (taking genotypes apart) and constructive (putting genes together) points of view. Using this approach, new tools enable us to mimic ancient strategies for durable resistance breeding.

The availability of a molecular genetic map of rice (McCouch *et al.*, 1988; Causse *et al.*, 1994) provides the groundwork for identifying the chromosomal location(s) of blast resistance genes via linkage to molecular markers (Fig. 10.1). The identification of a gene's chromosomal location is known as 'gene mapping'. The identification of marker(s) very closely linked to a gene, such that the marker(s) can be used for indirect selection of the gene, is known as 'gene tagging' (Yu *et al.*, 1991; McCouch *et al.*, 1992; Mackill *et al.*, 1993). Information about the chromosomal location of a gene can be useful both for genetic analysis and for resistance breeding. The map location of a gene can indicate when genes are obviously non-allelic, or when rigorous complementation tests are needed to confirm allelism (Inukai *et al.*, 1992a). Information on allelism and linkage relationships can also be used to select donors and target loci for a gene pyramiding program (one cannot combine different alleles at the same locus in an inbred variety), and can provide guidance about the population sizes needed to obtain recombinants carrying multiple genes. Markers tightly linked to a gene can be used for indirect selection in a breeding program.

Our gene mapping and tagging efforts are aimed at providing molecular markers tightly linked to blast resistance genes in rice. The work has the following three main objectives: (i) marker-assisted genetic analysis; (ii) marker-assisted breeding; and (iii) marker-assisted gene cloning.

The Genetics of Blast Resistance

Resistance to blast may be conditioned by major genes or by quantitative trait loci (QTLs). Major genes are those that prevent completion of the life cycle of incompatible races of *P. grisea*; QTLs are those that reduce the sporulation of the pathogen within the context of a compatible interaction. Complete resistance to rice blast may be conferred by dominant (Kiyosawa, 1981; Mackill *et al.*, 1985; Yu *et al.*, 1987; Mackill and Bonman, 1992), incompletely dominant or recessive (Oka and Lin, 1957) genes. Many major genes have been well characterized in Japan (Kiyosawa, 1981; Ou, 1985; Kinoshita, 1986), but linkage information for blast resistance genes identified in the tropical regions has been limited (Mackill and Bonman, 1992; Yu *et al.*, 1991). In-depth analysis of specific QTLs with the aid of molecular markers is only just beginning (Wang *et al.*, 1994).

At least 30 blast resistance loci have been identified in rice (*Oryza sativa* L.) (Kinoshita, 1991; Mackill and Bonman, 1992; L. Zhu, Academia Sinica, Beijing,

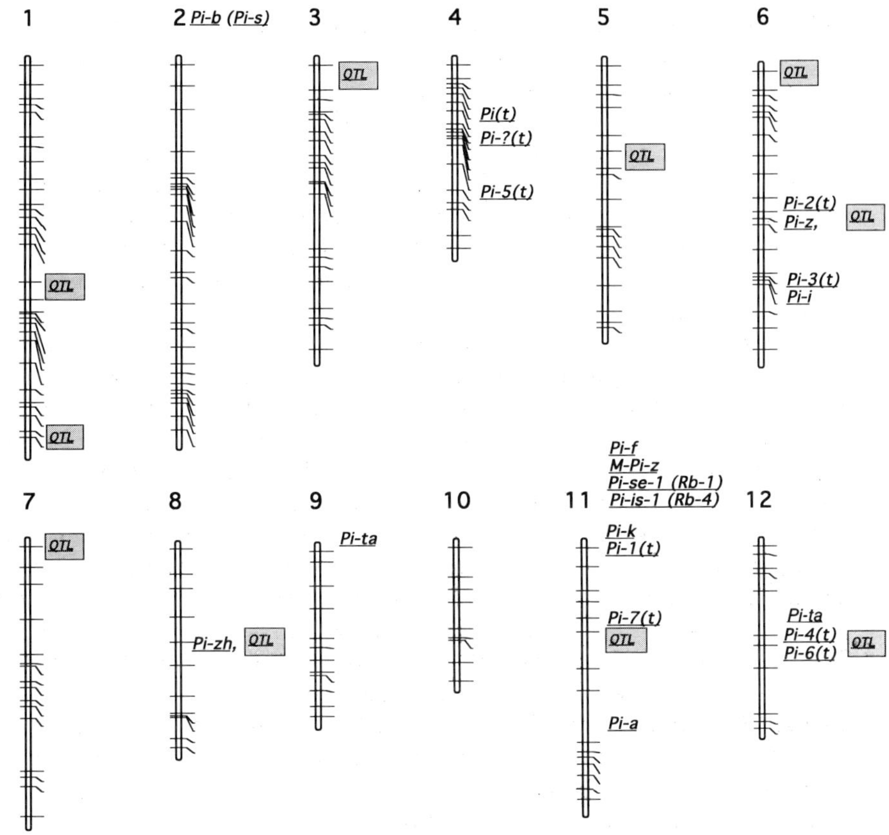

Fig. 10.1. Map showing locations of known blast resistance genes along the 12 chromosomes of rice. Where no linkage information is available, genes are listed above the chromosomes on which they are located.

China, 1993, personal communication; Wang *et al.*, 1994). Of these, 20 are major genes and 10 are putative QTLs. Twelve of the major genes have been confirmed to be non-allelic and are officially registered with the Rice Genetics Cooperative (Kinoshita, 1990, 1991). Within this group, several resistance loci are known to harbor more than one resistance allele. For example, five different alleles have been identified at the *Pi-k* locus, two at the *Pi-z* locus, and two at the *Pi-ta* locus (Kiyosawa, 1989). The different alleles are distinguishable from each other because they mediate resistance (an incompatible host–pathogen interaction) to different spectra of blast isolates. Chromosomal positions via linkage to genetic markers have been reported for 14 of the loci conferring complete resistance and for 10 QTLs conferring partial resistance (Kinoshita, 1991; Mackill and Bonman, 1992; L. Zhu, Academia Sinica, Beijing, China, 1993, personal communication; Wang *et al.*, 1994) (Fig. 10.1).

As the map positions of these resistance genes are determined, it is becoming apparent that they occur in clusters in several regions of the rice genome. Eight loci have been reported on chromosome 11, including the complex *Pi-k* locus and one QTL (Table 10.1 and Fig. 10.1). Although *Pi-k* and *Pi-1(t)* are located near

Table 10.1. Blast resistance genes and their chromosomal locations in rice.

Locus	Phenotype/product	Chromosome	Reference
Pi-a	*Pyricularia oryzae* resistance-a	11	Shinoda *et al.*, 1971; Kiyosawa, 1972
Pi-b (Pi-s)	*Pyricularia oryzae* resistance-b	2	Shinoda *et al.*, 1971; Kiyosawa, 1972
Pi-f	*Pyricularia oryzae* resistance-f	11	Yunoki *et al.*, 1970; Shinoda *et al.*, 1971
Pi-i	*Pyricularia oryzae* resistance-i	6	Shinoda *et al.*, 1971; Kiyosawa, 1972
Pi-k (*Pi-k*, $Pi\text{-}k^{m}$, $Pi\text{-}k^{s}$, $Pi\text{-}k^{k}$, $Pi\text{-}k^{p}$)	*Pyricularia oryzae* resistance-k	11	Shinoda *et al.*, 1971; Kiyosawa, 1972
Pi-ta (=sl)	*Pyricularia oryzae* resistance-ta	9 or 12?	Shinoda *et al.*, 1971; Kiyosawa, 1972; Ise, 1993
M-Pi-z	*Pyricularia oryzae* resistance-z	11	Goto, 1976
Pi-z	*Pyricularia oryzae* resistance-z	6	Kiyosawa, 1967, 1972; Shinoda *et al.*, 1971
Pi-is-l (Rb-4)	*Pyricularia oryzae* resistance-is	11	Goto, 1970
Pi-se-1 (Rb-1)	*Pyricularia oryzae* resistance-se	11	Goto and Baluch, 1983, 1984
*Pi(t)**	*Pyricularia oryzae* resistance	4	Hsieh, 1976
*Pi-?(t)**	*Pyricularia oryzae* resistance	4	Tohme *et al.*, 1993
*Pi-1(t)**	*Pyricularia oryzae* resistance-1	11	Yu, 1991
*Pi-2(t)**	*Pyricularia oryzae* resistance-2	6	Yu *et al.*, 1991
*Pi-3(t)**	*Pyricularia oryzae* resistance-3	6	Inukai *et al.*, 1992b
*Pi-4(t)**	*Pyricularia oryzae* resistance-4	12	Yu *et al.*, 1991
*Pi-5(t)**	*Pyricularia oryzae* resistance-5	4	Wang *et al.*, 1994
*Pi-6(t)**	*Pyricularia oryzae* resistance-6	12	Yu, 1991
*Pi-7(t)**	*Pyricularia oryzae* resistance-7	11	Wang *et al.*, 1994
*Pi-zh(t)**	*Pyricularia oryzae* resistance-zh	8	L. Zhu, personal communication

(t) = Tentative designation, pending alleleism tests with previously identified genes.
*Not registered with the Committee on Gene Symbolization, Nomenclature, and Linkage Groups of the Rice Genetics Cooperative.

each other at the end of the chromosome (Goto *et al.*, 1981; Yu *et al.*, 1994), they appear to be non-allelic (T. Inukai, Hokkaido University, Japan, 1993, personal communication). The relationship between the QTL and *Pi-7(t)* remains to be elucidated. Four blast resistance genes have been reported on chromosome 6. Of these *Pi-z* and *Pi-2(t)* may be allelic (Inukai *et al.*, 1992a), while *Pi-i* is reported to be independent of *Pi-z*. A recent report suggests that *Pi-3* is linked to *Pi-i*, but is nonallelic (Inukai *et al.*, 1992b). Similarly, three to four genes are reported to lie in a common region of chromosome 12 (Shen *et al.*, 1986; Yu *et al.*, 1994), including a recent report that *Pi-ta* may be allelic to *Pi-4(t)* (Ise, 1993).

Genetic analysis of blast resistance has indicted that many cultivars carry multiple genes for resistance. Of 51 cultivar/isolate combinations examined in a study conducted at the International Rice Research Institute (IRRI, 1981; Mackill *et al.*, 1985), resistance was conditioned by two dominant genes in 30 cases, and by one dominant gene in 15 cases. Recessive genes were identified in only a few cases. Linkage analysis indicated that distinct resistance genes were effective against different blast isolates. Similar results were reported from genetic studies on various traditional and improved rice cultivars (Yu *et al.*, 1987; Moldenhauer *et al.*, 1992). Phenotypic evaluation of near-isogenic lines (Mackill and Bonman, 1992) suggested that as many as four resistance genes could be identified in Tetep, three in Pai-Kan-Tao, and two in LAC23, while only one reaction pattern was observed among the five lines developed from the A5173 donor parent. The complexity of the genetic basis of blast resistance in these and other varieties has made it difficult to identify and analyze individual genes using traditional approaches.

The inheritance of partial resistance to blast has also been shown to be conditioned by multiple genes (Goto, 1970; Higashi and Saito, 1985; Notteghem, 1985; Lin, 1986; Roumen, 1993). Although genetic analyses have indicted that partial resistance may be inherited polygenically, the specific loci involved in conditioning resistance were not identified in these studies. Molecular marker technology makes it possible to dissect complex genotypes and to identify approximate map positions of individual genetic factors that contribute to a quantitatively inherited phenotype. Both major genes and QTLs are mapped as they are identified. Once identified via linkage to molecular markers, individual QTLs can be treated similarly to other Mendelian loci.

Mapping of Single Genes Conferring Complete Resistance

Use of near-isogenic lines for mapping major genes

In order to characterize blast resistance genes used in tropical rice improvements, Mackill *et al.* (1985) developed sets of nearly isogenic lines (NILs), each containing a single major dominant gene. These genes were introgressed from different donors into the *indica* cultivar 'CO39', a short growth-duration variety (95 days to maturity) developed in Coimbatore, India, which is highly susceptible in the Philippines. NILs used in the gene tagging experiments reported here were the product of six backcrosses to the recurrent parent, CO39, followed by three selfings ($B6F_4$). In addition, $B6F_3$ segregating populations were scored for blast reaction and used to confirm whether donor-derived regions of DNA identified in the isolines actually contained resistance genes.

Four donor varieties were used to develop four NILs, as reported in Mackill *et al.* (1985) and Mackill and Bonman (1992). The donors were: (i) 'LAC23', a tropical *japonica* cultivar originating in Liberia, from which *Pi-1(t)*, was derived; (ii) '5173', an *indica* cultivar, originating in Colombia, from which *Pi-2(t)* was derived; (iii) Pai-Kan-Tao, a temperate *japonica* cultivar from China, from which *Pi-3(t)* and *Pi-4a(t)* was derived; and (iv) 'Tetep', an *indica* cultivar which originated in Vietnam, from which *Pi-4b(t)* was derived. Allelism tests demonstrated that the four differently numbered genes isolated in this set of NILs were non-allelic.

Five isolates of the pathogen (designated 101 for IK81–3, 102 for IK81–25, 103 for PO6–6, 104 for PO3–82–51, and 105 for 43) were used individually in inoculation experiments to identify resistant or susceptible phenotypes during NIL development (Mackill and Bonman, 1992). The four donor parents were completely resistant to the *P. grisea* isolates used, with the exception of Pai-kan-tao, which was susceptible to isolate 103 (PO6–6). The total number of blast resistance genes contained in the four donor varieties was not known, but previous work suggested that more than one was involved in the resistance of Pai-Kan-Tao and Tetep (Mackill *et al.*, 1985; Yu *et al.*, 1987). Kiyosawa (1989) reported that Pai-Kan-Tao carried the gene *Pi-ta*, among others, and that Tetep carried an unspecified allele of *Pi-k*, among others. Mackill *et al.* (1985) also reported more than one gene in Tetep. Therefore, backcross F_1 plants from each cross in each generation were inoculated with each of the five pathogen isolates in an effort to try to isolate more than one gene from each of the donors. This scheme suggested the development of 20 NILs, but some of the donor/isolate combinations were dropped and additional lines from some combinations were added, resulting in a total of 22 $BC6F_3$ NILs being developed.

The 22 NILs were inoculated with 73 Philippine isolates of *P. grisea* from the Plant Pathology collection at IRRI and classified into seven groups based on the reaction pattern to these isolates. In addition, the reaction of the lines was evaluated in response to non-Philippine isolates by J.L. Notteghem (CIRAD, Montpellier, France, personal communication). Two isolates maintained by CIRAD researchers were capable of further differentiating NILs that gave identical reaction patterns to all 73 Philippine blast isolates, indicating that there were at least 10 different reaction patterns among the 22 NILs (Mackill and Bonman, 1992).

$BC6F_3$ segregating populations were scored for blast reaction based on the method outlined in Bonman *et al.* (1986). Disease reactions were scored seven days after inoculation on a scale from 0 to 5 in which 0 = no evidence of infection, 1 = brown specks smaller than 0.5 mm in diameter, 2 = brown specks about 0.5–1.00 mm in diameter, 3 = roundish to elliptical lesions about 1–3 mm in diameter with gray centers and brown margins, 4 = typical spindle-shaped blast lesions, and 5 = same as 4 but half of one or more leaves was killed by coalescence of lesions. Plants rated 0–3 were considered resistant and those rated 4–5 were considered susceptible. (Resistant, or no sporulating lesions were coded 'R', segregating lines were coded 'Seg', and susceptible, or lines with fully sporulating lesions 'S'.) Thirty-six homozygous individual $BC6F_3$ lines per cross (15 individuals per line) with known R/S phenotypes were subsequently used to analyze segregation of blast resistance with molecular markers and to estimate the recombination value between molecular markers and resistance genes.

A molecular map of the rice genome, constructed at Cornell University (McCouch *et al.* 1988), was used as the basis for RFLP analysis in these gene tagging studies. The most recent version of the rice map consists of approximately 700 DNA markers (Causse *et al.*, 1994), including both genomic and cDNA clones. In studies aimed at identifying linkage with the major blast resistance genes contained in the NILs, 150–200 DNA clones, distributed at approximately 15 centimorgan (cM) intervals, were surveyed on sets of NILs, their respective donor varieties, and their common recurrent parent, CO39 (Yu *et al.*, 1991, 1994).

RFLP analysis by Yu *et al.* (1994) confirmed that *Pi-1(t)* was linked to a single-copy locus hybridizing to the cDNA clone, *RZ536*, at a distance of 14.0 + 4.5 cM. *RZ536* maps to the very end of chromosome 11. A second RFLP locus, *RZ424*, was also found to be linked to *Pi-1(t)*, at 19.6 + 5.3 cM. *RZ536* and *RZ424* were estimated to be 5.5 + 2.8 cM apart, based on data from the $BC6F_3$ population (Fig. 10.2). The orientation of *Pi-1(t)* with respect to these two RFLP markers on the Cornell map remains to be clarified.

The same region of the rice genome that carries *Pi-1(t)* also contains the blast resistance gene, *Pi-k*. Inukai *et al.* (in preparation) analyzed the allelism relationship between *Pi-1(t)* and *Pi-k* in crosses between C101LAC and three of Kiyosawa's differential cultivars carrying alleles at the *Pi-k* locus. One susceptible recombinant was detected among the 557 F_2 progeny of Tsuyuake/C101LAC, suggesting that the loci were not allelic (no recombinants were obtained in a total of 970 progeny obtained from the two other crosses). To confirm this, F_3 families of C101LAC/K60 (K60 is the differential carrying *Pi-k*^{p}) were tested. Four of the 142 F_3 families analyzed showed segregation for resistance, indicating that *Pi-1(t)* and *Pi-k* were linked at approximately 4.7 ± 1.9 cM ± SE.

The *Pi-2(t)* gene was found to be tightly linked to RG64 on chromosome 6, as reported by Yu *et al.* (1991), at a distance of 2.8 ± 1.4 cM. Three crossovers were found in the F_2 generation among 36 F_3 lines from a cross between CO39 and C101A51–2, and only one crossover was found among 36 F_3 lines from a cross between CO39 and C104A51–4 (Yu *et al.*, 1991). Earlier genetic studies suggested that the isolines used in these two crosses had probably obtained the same gene (*Pi-2(t)*) from their donor 5173, and therefore combined data from the 72 plants in both populations was used to estimate the map distance. The location of this gene coincides with the available information about the location of *Pi-z* from the classical linkage map (Kinoshita, 1991). Recently Inukai *et al.* (1992a) provided data from 876 F_2 individuals suggesting that these two genes may reside at the same locus.

Further allelism studies of Inukai *et al.* (1992b) indicated that *Pi-3(t)* is not identical to any of the loci present in Kiyosawa's differential series. *Pi-3(t)* was found, however, to be linked to a resistance gene in Fujisaka 5, which carries *Pi-i*. Linkage analysis between RFLP markers and *Pi-3(t)* is underway (T. Inukai, Hokkaido University, Japan, 1993, personal communication).

Yu *et al.* (1991) reported that *Pi-4(t)* was linked to *RG869* on chromosome

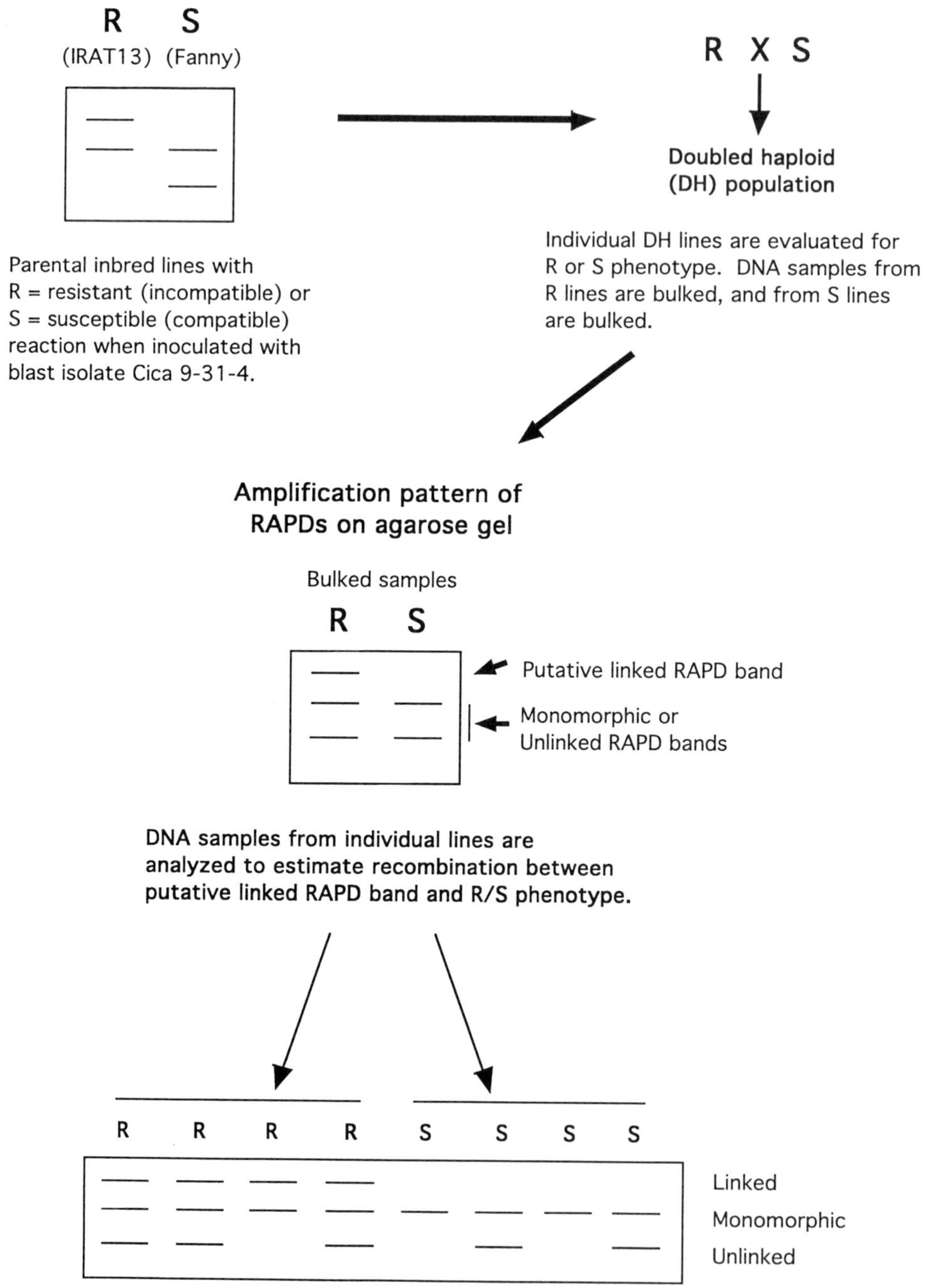

Fig. 10.2. Flow diagram of gene tagging approach using bulked segregant analysis on doubled haploid lines.

12 at a distance of 15.3 + 4.7 cM. *Pi-4(t)* was subsequently found to be linked to another cDNA clone, *RZ397*, at a distance of 18.1 + 5.1 cM. *RG869* and *RZ397* are only 5.6 + 2.8 cM apart, suggesting that *RG869* is located between *Pi-4(t)* and *RZ397* on chromosome 12. In China, Shen *et al.* (1986) and K.L. Zheng (China National Rice Research Institute, Hangzhou, China, 1993, personal communication) previously reported a cluster of blast resistance genes in a similar location on chromosome 12. The relationship between *Pi-4(t)* and those genes remains to be elucidated.

Blast resistance derived from wild species

Many wild relatives of rice have been found to be tolerant, resistant, or immune to rice pests and diseases. One such wild rice is the grassy species *Oryza minuta*, which is resistant to blast, blight, and some insect pests (Sitch *et al.*, 1989; Amante-Bordeos *et al.*, 1992). A diploid line (hereafter referred to as *Om*R) recovered from a cross between the allotetraploid *O. minuta* and an elite *O. sativa* breeding line was found be highly resistant to the blast isolate *PO6–6*, as well as to the field populations of the fungus in test sites in Bangladesh, Colombia, Indonesia, the Philippines, and Thailand (Amante-Bordeos *et al.*, 1992). Resistance in *Om*R was shown to be due to the presence of a single gene (Amante-Bordeos *et al.*, 1992). This gene, which is incompletely dominant, was designated *Pi-9(t)*. Analysis of an *Om*R backcross population with RFLP and RAPD markers demonstrated that the line was remarkably isogenic to its recurrent parent, the elite rice line IR31917. More than 100 RFLP markers distributed an average of 25 cM apart failed to reveal any introgressions. In order to uncover the region carrying the resistance gene, 540 RAPD primers were used to analyze the genome. On average, each primer amplified 13 loci. Assuming that a RAPD band is equivalent to a locus (Williams *et al.*, 1990), approximately 7020 loci were thereby tested. This survey revealed four loci putatively linked to resistance. If the RAPD loci tested during the course of this study are randomly distributed throughout the genome, the introgression carrying *Pi-9(t)* would be approximately 1 cM in length, given a genome size of 1950 cM for rice (Tanksley *et al.*, 1992).

Use of doubled haploid and recombinant inbred populations for mapping major genes

The studies described above made use of NILs, F_2 and F_3 populations to map single major genes for blast resistance. As mentioned above, however, many resistant cultivars carry multiple and/or unknown resistance genes, and/or other traits of interest for mapping and marker-aided selection. It is difficult to assess multiple traits in F_2 populations. Complex resistances can, however, be analyzed effectively using populations of fixed recombinants.

Two types of fixed populations can be developed for this purpose. Anther culture-derived doubled haploid (DH) populations provide the advantages of fixed lines after only one generation of meiosis, but can only be produced efficiently for some rice genotypes. Recombinant inbred (RI) populations can be obtained by single-seed descent from any cross, but the process requires seven or more generations of meiosis. Both DH and RI populations may show distorted segregation for some or many alleles as a result of both genetic and/or environmental factors (Zamir and Tadmor, 1986; Sato *et al.*, 1990; Lin and Ikehashi, 1993; Wang *et al.*, 1994; M. Champoux, IRRI and Cornell University, personal communication). Statistical adjustments are required when skewing is extreme (Wagner *et al.*, 1992; Wang *et al.*, 1994).

Use of doubled haploid (DH) lines for mapping

A population of 75 DH individuals was used to map *Pi-6(t)* (Yu, 1991). The population was derived from a cross between the *indica* cultivar 'Apura' and the *japonica* cultivar 'IRAT177' (Guiderdoni *et al.*, 1989). Abadassi *et al.* (1991) demonstrated that Apura was resistant to the two Brazilian *P. grisea* isolates, BR14 and BR26, while IRAT177 was susceptible to both. Inoculation and disease evaluation with these isolates was done in France at the Centre International de Recherche Agronomique pour le Developpement (CIRAD-CA) by J.L. Notteghem, according to the procedures of Silué and Notteghem (1992). Unlike the inoculation procedure at IRRI which involved spraying blast inocula on to rice seedlings, BR14 and BR26 were injected into the sheaths of greenhouse-grown rice plants (4th or 5th leaf stage) using spore suspensions of 2.5×10^4 conidia ml^{-1}.

A resistance gene conferring complete resistance to both BR14 and BR26 was identified on chromosome 12 based on RFLP analysis (Yu, 1991). The locus was tentatively named *Pi-6(t)* and is located at about 15.3 cM from *RG869* and about 18.4 cM from the isozyme marker, *Sdh1*. *Pi-6(t)* is located in the same chromosomal region as *Pi-4(t)* (reported above), but these two resistance genes are believed to lie on opposite sides of *RG869* (Z. Yu, Cornell University, personal communication). Based on RFLP analysis using the same set of molecular markers, at least two other genes conferring complete resistance (Shen *et al.*, 1986; K.L. Zheng, China National Rice Research Institute, Hangzhou, China, 1993, personal communication) and partial resistance (Wang *et al.*, 1994) have also been reported to lie in this region. These independent studies suggest that several blast resistance genes or alleles may be clustered in this region of chromosome 12. Further evaluation and complementation tests are required to confirm this possibility.

Tohme *et al.* (1993) used a DH population and 'bulked segregant analysis' (Michelmore *et al.*, 1991; Fig. 10.2) to map a blast resistance locus derived from 'IRAT13'. A population of 90 DH lines was produced by anther culture of F_1 hybrids from a cross between IRAT13 and Fanny. The DH population was evaluated for blast reaction using seven isolates of *P. grisea* from two locations in Colombia. Fanny was susceptible to the five isolates from Santa Rosa but resistant to the two isolates from Altillanura, while the reaction of IRAT13 was the opposite. The lines showed either type 1 or type 4 lesions; no intermediate reactions were observed. Reaction of the DH lines to isolates FY 29–1 (from Fanny) and 1–84 (from Oryzica) were similar, but not identical to, that of Cica 9–31–4, suggesting a cluster of closely linked genes in this region. The reactions of the two isolates from Altillanura were identical. Two hundred 10-base primers (Operon Technologies Inc., Vancouver, British Columbia) were used for RAPD analysis (Williams *et al.*, 1990) of pooled samples of resistant and susceptible plants (Fig. 10.2). Two markers were found to cosegregate with resistance to Cica 9–31–4 derived from IRAT13. These markers were mapped relative to RFLP markers of known chromosomal location. A resistance locus, *Pi-?(t)*, was mapped to chromosome 4 (Fig. 10.1). Allelic relationships between this blast resistance locus and others known to reside on chromosome 4 are not known.

Use of recombinant inbred (RI) lines for mapping

A population of recombinant inbred lines derived from a cross between Moroberekan and CO39 was used for mapping genes conferring both complete and partial resistance (Wang *et al.*, 1994). Moroberekan is a *japonica* rice cultivar from Guinea with durable resistance to blast disease in Asia. It was completely resistant to over 300 Philippine isolates of *P. grisea* in greenhouse inoculation tests, while CO39 was susceptible to most of them. Five isolates, collected from diverse regions of the Philippines, were used to inoculate the entire RI population in monocyclic tests in the greenhouse. Plants were scored 6–7 days after inoculation using a modified scoring system based on Bonman *et al.* (1986). Lines were given a score according to the most susceptible lesion type that was abundantly represented. Approximately half of the population showed complete resistance to all five isolates.

Based on linear modeling (Statistical Analysis Systems, SAS) two dominant loci associated with qualitative resistance to all five isolates of the fungus were identified in this study. One locus, tentatively designated *Pi-5(t)*, was mapped to chromosome 4, near *RG788*. Another major gene, bracketed by RG103A and RG16 on chromosome 11, was tentatively designated *Pi-7(t)*. In both cases, the resistant allele derived from Moroberekan. Further inoculation experiments suggest there are three additional genes conferring complete resistance (R.S. Zeigler and R. Nelson, International Rice Research Institute, Philippines, 1993, unpublished data).

In addition to assessing the reaction to *PO6–6*, Wang *et al.* (1994), tested the RI population with four other *P. grisea* isolates. Although isolates were selected from diverse locations in the Philippines, they were later shown to represent only two of the many genetic lineages of the pathogen found in the Philippines (Levy *et al.*, 1991; Chen and Nelson, unpublished data). Subsequently, molecular marker information on the pathogen population was used as a basis for selecting diverse isolates. Inoculation of the RI population with isolates representing different pathogen lineages has recently suggested the presence of at least one additional resistance gene in Moroberekan, which cosegregates with RG448. This locus is near RG788 (*Pi-5(t)*) and in the same vicinity as the resistance gene from IRAT13 on chromosome 4. Thus, it appears that a cluster of blast resistance genes with different specificities resides on rice chromosome 4.

Mapping of quantitative trait loci conferring partial resistance

The aforementioned population of 281 F_7 recombinant inbred (RI) lines was produced via single seed as described in Wang *et al.* (1994). As mentioned above, the RI population was inoculated with five pathogen isolates in monocyclic tests. Approximately half the population showed complete resistance to all five isolates. Because quantitative resistance could only be evaluated in the absence of genes governing qualitative resistance to *PO6–6*, RI lines showing complete resistance were excluded from subsequent evaluation for partial resistance. The remaining 131 plants were evaluated for lesion number, lesion size, and diseased leaf area in polycyclic tests using isolate *PO6–6*. This isolate was selected because it was known to have a broad spectrum of virulence, to be genetically stable, and was routinely used for studies at IRRI. The polycyclic test was developed in order to measure cumulative disease development on a population of test plants after several cycles of pathogen reproduction. It is useful because small differences in the degree of susceptibility are amplified, making phenotypic evaluation more informative and reliable (Wang *et al.*, 1994).

Plants were scored for partial resistance 14 days after inoculation based on the three parameters measured. Diseased leaf area was visually estimated for 10 randomly selected plants from each line, and lesion size and lesion number were both measured for three leaves from each of four plants (12 observations) for each line.

Based on the polycyclic tests, ten chromosomal segments were found to be significantly associated with effects on lesion number ($P < 0.0001$, $F > 30.0$, and log of the odds (LOD) > 6.0) (Fig. 10.1). Both single marker (ANOVA; Statistical Analysis Systems) and Mapmaker/QTL (Lincoln *et al.*, 1992) analysis gave similar results. Linear models suggested that individual QTLs each explained between 19 and 50% of the phenotypic variance. Therefore, interaction among the putative QTLs was tested using two-way ANOVAs and multiple regression analysis. Results of these analyses suggested that there were epistatic interactions among the genes associated with partial resistance in Moroberekan.

We observed that the genes for partial resistance not only demonstrated epistatic interactions among themselves, but their presence could also affect the accuracy of classification of lines for complete resistance to blast. Lines showing resistant scores of 0–2, or susceptible scores of 4–5, showed consistent reactions to all five isolates tested. However, 41 lines that had intermediate scores (3 or 3+) showed inconsistent reactions in different inoculation tests. Subsequent analysis indicated that most of these lines had the CO39 allele (susceptible) at one or both of the loci associated with the gene for complete resistance, and had the Moroberekan allele (resistant) at many loci associated with genes for partial resistance. Because of the epistatic effects of genes for partial and complete resistance, these 41 lines were not used to map complete resistance.

It was previously known that a qualitative resistance gene can obscure the presence of quantitative resistance genes. It was not clear, however, how the presence of genes conferring quantitative resistance would affect the detection of genes for qualitative resistance. The combined effect of genes conferring partial resistance was difficult to distinguish from the effect of a single gene conferring complete resistance. The observation that the presence of partial resistance genes can lead to inconsistent scoring results might explain some of the variable reactions that have frustrated blast researchers for many years. Thus, the level of quantitative resistance should be considered in a cultivar/isolate interaction when studies on the genetics of resistance are undertaken.

The identity and behavior of genes conferring blast resistance in rice is highly dependent on the genotype in which the genes reside (i.e. genetic background), the blast isolate(s) used to evaluate the host reaction, and the particular phenological and environmental conditions under which plants are inoculated and evaluated. Indeed, three of the markers identified with QTLs for partial resistance in this study had been reported to be linked to complete blast resistance in previous studies (Fig. 10.1). This raises the question of whether genes conferring complete and partial resistance are, indeed, distinct sets of loci, or whether the same gene might confer complete resistance in one genetic background, in one set of environmental conditions, or in response to a specific blast isolate(s), and partial resistance in a different genetic and physical environment. Further analysis is needed to resolve these issues and to determine the allelism relations between complete and partial blast resistance genes known to reside in the same chromosomal regions.

In addition to the greenhouse tests, the population of RILs was evaluated at two independent field locations, with two replications in each location using a miniplot technique (Marchetti, 1983). One trial was conducted at the upland screening site at Cavinti, the Philippines, and one in Sitiung, Indonesia. The percentage of diseased leaf area was visually estimated at 3, 4, 5, and 6 weeks after sowing.

The study demonstrated that QTLs identified in polycyclic tests in the greenhouse were good predictors of blast resistance at two field sites. Eight of the markers associated with complete resistance to *PO6-6* and four of the markers associated with partial resistance were also associated with reduced DLA at both sites. A highly significant correlation was found between the number of QTLs and disease reaction in the field ($r = 0.64^{**}$ and 0.67^{**} respectively for the two sites (asterisks denote significance at $P < 0.01$)). In view of the known complexity of populations of the blast fungus at the field sites, it is encouraging that the QTLs identified in greenhouse studies showed effectiveness in the field.

This study of quantitatively inherited blast resistance illustrates the usefulness of RI lines for mapping a complex trait such as blast resistance, and suggests that durable resistance in Moroberekan involves a combination of both complete and partial resistance genes. From this initial QTL analysis of the RI population, however, many new questions arise. The exact locations of the putative QTLs identified here remain to be determined, and the behavior of these resistance factors when moved into a different genetic background and inoculated with different blast isolates is under investigation. Lines are being selected based on phenotypic and RFLP data and these lines are being backcrossed to CO39

(S. Hittalmani and N. Huang, 1993, IRRI, personal communication). Genetic analysis will be conducted in the backcross and derived populations. In addition, crosses designed to move these QTLs into more useful genetic backgrounds will enable us to test the hypothesis that the same genes may confer either 'complete' or 'partial' resistance, depending on the complex set of genetic and environmental conditions determining the evaluation of phenotype.

Marker-assisted selection

Marker-assisted selection can be used for producing plants with genotypes of interest, either for research purposes, or in an applied breeding program. In both cases, the use of markers can increase the efficiency with which a desired genotype is produced. In addition, indirect selection of resistance genes based on linked markers can facilitate the combining of genes in ways that are difficult or impossible to achieve based on phenotypic analysis alone. To date, markers have been used in the production of research materials. With rapid advances in gene tagging efforts and marker technologies, we hope that marker-assisted selection will also soon contribute directly to rice improvement through breeding.

One research application of marker-assisted selection is in the production of near-isogenic lines. Such lines are useful for the detailed characterization of resistance genes, and also as defined differentials for the classification of pathogen strains (Reimers and Leach, 1991; Yu *et al.*, 1991). NILs are conventionally produced by several generations of backcrossing and selection. Two problems have been encountered with this approach. First, while backcrossing and selection is effective for the transfer of single genes, it may not be efficient when multiple genes are present in the donor parent. If multiple genes are not differentiated by the isolates used for selection, some of the loci may be 'lost'. Second, it is difficult to evaluate the extent of return to the recurrent parental genotype by conventional means. Although the amount of donor chromatin is expected to be reduced by half with each generation of backcrossing to the recurrent parent, the observed level of donor DNA may be substantially higher than expected. For instance, based on restriction fragment length polymorphism data, the NILs for blast resistance carried surprisingly high numbers of loci from the donor parents (Yu, 1991). The use of molecular markers can overcome both of these problems, thus contributing to the efficiency of NIL production.

Development of NILs from populations of fixed recombinants (recombinant inbreds or doubled haploids) that have been used for gene mapping may allow more comprehensive capture of genes from resistance sources of interest. For instance, based on phenotypic and RFLP data, particular recombinant inbred lines derived from the Moroberekan/CO39 cross described above were identified as 'pre-isogenic lines' (PILs) for specific major genes and QTLs. Because of the skewed segregation favoring *indica* alleles, lines could be selected that had inherited chromosomal regions carrying putative resistance loci from Moroberekan, but with few other loci from the donor, such that the resistance genes existed in a genetic background that was mostly derived from CO39. The number of introgressed markers in these skewed RILs which were not linked to resistance was similar to or less than that found for the NILs obtained by backcrossing (Yu, 1991). The selected recombinant inbred lines, including two lines carrying the putative QTL linked to RG64, have been added to the set of NILs for blast resistance. As other qualitative and quantitative resistance genes are identified in the RI population through inoculation with more diverse pathogen isolates, lines carrying these genes in a largely CO39 background will be identified. In addition to allowing the identification of PILs from fixed mapping populations, markers can be used to eliminate extraneous regions of introgressed donor DNA detected by RFLPs (McCouch *et al.*, 1992) in PILs or NILs.

It is also of interest for geneticists to be able to use markers tightly linked with blast resistance genes to facilitate the development of lines carrying combinations of resistance genes. Rigorous genetic analysis of lines carrying various combinations of complete and/or partial resistance genes with respect to a wide array of pathogen isolates promises to provide information about which combinations are likely to be most stable in the field. Similarly, if specific 'molecular tags' are demonstrated to be diagnostic of the presence of specific resistance genes in unknown germplasm, this will greatly facilitate analysis and interpretation of years of field data which has been collected from locations around the world as part of the International Rice Testing Program (IRTP) at IRRI. This data provides key insights into the usefulness of specific varieties with respect to the level and longevity blast resistance in specific geographic locations.

Although marker-assisted selection is currently in use for research purposes, it has not yet been widely applied to resistance breeding. Further work is needed to provide more markers closely linked to genes, and to make the detection of polymorphisms easier and more cost-effective. Because molecular marker-assisted selection offers a way of determining the genotype of an individual without inoculation or phenotypic analysis, it is important to identify markers that tightly bracket a gene of interest. By bracketing a target gene with markers on either side of it, the probability that a cross-over event will separate the gene from both markers is greatly reduced. By providing markers linked within 1–2 cM of a gene, the likelihood that the markers will reliably detect the presence of the gene is around 99%. Marker-assisted selection does not replace phenotypic analysis in a breeding program, but can be used for one or two generations to speed up the process of accumulating valuable genes (M. Murray, Agrigenetics, 1993, personal communication).

Markers have been successfully used to identify genes segregating in a population derived from two fixed parents, or to predict the presence of genes in lines derived from known parental genotypes. It would also be extremely useful to rice breeders if molecular markers identified in one population could also be used to 'diagnose' the presence of resistance genes in uncharacterized germplasm. Work is underway to investigate the extent to which this is empirically useful with respect to blast resistance in rice.

Databases

Rapid developments in molecular biology and biotechnology have been paralleled by equally profound developments in the field of information science. Computer databases and networking capability provide powerful tools for assembling and retrieving information in a highly efficient manner. In conjunction with the USDA's Plant Genome Project, we have begun to develop an internationally accessible Rice Genome Database, called RiceGenes. The organizing principle of RiceGenes is the genetic map of rice and it currently consists of information on over 700 DNA and isozyme markers, four genetic maps of rice, relevant bibliographic citations, and digitized images of autoradiograms, disease symptoms, and plant morphologies. Over 80 genes conferring monogenically inherited characters and approximately 30 loci associated with quantitatively inherited characters have been located on the genetic map of rice via linkage to molecular markers. This list includes 15 blast resistance genes (Fig. 10.1). We intend to load information related to the diversity of rice germplasm, genetic stocks, pedigrees, sequence information related to markers or known genes of rice, information related to quantitative trait evaluations and QTLs, and to develop interrelated genome databases for important pests and pathogens of rice, such as blast and bacterial blight.

In collaboration with other groups, we aim to initiate a database for rice blast which would interface directly with RiceGenes. Several genetic maps of *P. grisea* are available to serve as the foundation for this database. We expect to incorporate information about the origin, collection and storage of isolates, the fingerprinting, identification, and geographic occurrence of specific isolates, the availability of sexually compatible strains and mapping populations, inoculation and scoring methodologies, virulence spectra, and bibliographic references, as well as images of lesion types. The development of a RiceBlast database is in its infancy, but it represents an opportunity to help organize and make accessible a growing body of information that is of interest to a wide range of researchers throughout the world.

In conclusion, substantial progress has been made towards understanding the genetics of blast resistance, and identifying and refining the linkages between molecular markers and blast resistance genes. Further work is needed to identify distinctive polymorphisms that reliably identify specific genes in divergent rice germplasm, to characterize the behavior of blast resistance genes in different genetic and physical environments, to isolate the gene(s), and to develop dependable, cost-effective technologies that allow efficient screening for blast resistance in breeding programs. Our ultimate goal is to learn how to most effectively combine, deploy, and care for natural occurring resistance genes in order to provide rice growers with a more durable form of blast resistance. It is expected that this will be achieved as we understand more about the evolution and behavior of these resistance genes in rice, about the evolution and population biology of the blast fungus, and about the molecular basis of the rice host–blast-pathogen interaction.

References

Abenes, M.L.P., Angeles, E.R., Khush, G.S. and Huang, N. (1993) Selection of bacterial blight resistant plants in the F2 generation via their linkage to molecular markers. *Rice Genetics Newsletter* 10, 120–123.

Abadassi, J., Glaszmann, J.C., Notteghem, J.L., Guiderdoni, E. and Courtois, B. (1991) Genetic analysis of rice blast resistance using doubled haploids, single seed descent lines, and isozyme marker genes. In: *Rice Genetics II, Proceedings of the Second International Rice Genetics Symposium*. May 14–18 1990. International Rice Research Institute, Manila, The Philippines, pp. 746–748.

Amante-Bordeos, A.D., Nelson, R.J., Oliva, N.P., Dalmacio, R.D., Leung, H. and Sitch, L.A. (1992) Transfer of blast and bacterial blight resistance from *Oryza minuta* to *Oryza sativa*. In: *Rice Genetics II, Proceedings of the Second International Rice Genetics Symposium*. May 14–18 1990. International Rice Research Institute, Manila, The Philippines, pp. 237–246.

Bonman, J.M., Vergel de Dios, T.I. and Khin, M.M. (1986) Physiologic specialization of *Pyricularia oryzae* in the Philippines. *Plant Disease* 70, 767–769.

Causse, M., Fulton, T.M., Cho, Y.G., Ahn, S.N., Wu, K., Xiao, J., Chunwongse, Z.Yu., Ronald, P.C., Harrington, S.B., Second, G.A., McCouch, S.R. and Tanksley, S.D. (1994) Saturated molecular map of the rice genome based on an interspecific backcross population. *Genetics* (in press).

Goto, I. (1970) Genetic studies on the resistance of rice plant to the blast fungus. I. Inheritance of resistance in crosses Sensho X H-79 and Imochi-shirazu X H-79. *Annals of the Phytopathological Society of Japan* 36(5), 304–312.

Goto, I. (1976) Genetic studies on the resistance of rice plant to the blast fungus II. Difference in resistance to the blast disease between Fukunishiki and its parental cultivar, Zenith. *Annals of the Phytopathological Society of Japan* 42(3), 253–260.

Goto, I. and Baluch, A.A. (1983) Genetic studies on the resistance of rice plant to the blast fungus. V. Introduction of Pi-se of Sensho to Rikuto-norin-mochi-No. 4 and to No. 26. *Bulletin of the Yamagata University Agricultural Science* 9(2), 121–125. (In Japanese/English.)

Goto, I. and Baluch, A.A. (1984) Genetic studies on resistance of rice plant to blast fungus. VI. Additive effect of blast resistance genes of Sensho under natural infection. (Preliminary report). *Bulletin of the Yamagata University Agricultural Science* 9(3), 273–283. (In Japanese with English summary.)

Goto, I., Jaw, Y.L. and Baluch, A.A. (1981) Genetic studies on resistance of rice plant to blast fungus. IV. Linkage analysis of four genes, *Pi-a*, *Pi-k*, *Pi-z*, and *Pi-i*. *Annals of the Phytopathological Society of Japan* 47(2), 252–254.

Guiderdoni, E., Glaszmann, J.C. and Courtois, B. (1989) Segregation of 12 isozyme genes among doubled haploid lines derived from a *japonica* × *indica* cross of rice (*Oryza sativa* L.). *Euphytica* 42, 45–53.

Higashi, T. and Saito, S. (1985) Linkage of field resistance genes of upland rice variety 'Sensho' to leaf blast caused by *Pyricularia oryzae* Cav. *Japanese Journal of Breeding* 35, 438–448. (In Japanese with English summary.)

Hsieh, S.C. (1976) Recent advances in rice breeding and genetical studies in Taiwan. *Scientific Agriculture, Taipei* 4(12), 48–68.

Ikehashi, I. and Khush, G.S. (1979) Breeding for blast resistance at IRRI. In: *Proceedings of the Rice Blast Workshop*. International Rice Research Institute, PO Box 933, Manila, The Philippines, p. 220.

International Rice Research Institute (1981) Annual Report for 1980. IRRI, Los Baños, Laguna, The Philippines, 467 pp.

Inukai, T., Mackill, D.J., Bonman, J.M., Sarkarung, S., Zeigler, R., Nelson, R., Takamure, I. and Kinoshita, T. (1992a) Blast resistance genes *Pi-2(t)* and *Pi-z* may be allelic. *Rice Genetics Newsletter* 9, 90–92.

Inukai, T., Mackill, D.J., Bonman, J.M., Sarkarung, S., Zeigler, R., Nelson, R., Takamure, I. and Kinoshita, T. (1992b) A blast resistance gene *Pi-3(t)* in near-isogenic line C104PKT. *Rice Genetics Newsletter* 9, 94–95.

Ise, K. (1993) A close linkage between the rice blast resistance gene *Pi-ta*^{2} and a marker on chromosome 12 in japonica rice. *International Rice Research Institute* 18(2), 14.

Johnson, R. (1981) Durable resistance: definition of, genetic control, and attainment in plant breeding. *Phytopathology* 71, 567–568.

Johnson, R. (1984) A critical analysis of durable resistance. *Annual Review of Phytopathology* 22, 309–330.

Kinoshita, T. (1986) Standardization of gene symbols and linkage maps in rice. In: *Rice Genetics, Proceedings International Rice Genetics Symposium*. May 1985, International Rice Research Institute, Los Baños, The Philippines, pp. 215–228.

Kinoshita, T. (1990) Report of the Committee on Gene Symbolization, Nomenclature, and Linkage Groups. *Rice Genetics Newsletter* 7, 16–57.

Kinoshita, T. (1991) Report of the Committee on Gene Symbolization, Nomenclature, and Linkage Groups. *Rice Genetics Newsletter* 8, 2–37.

Kiyosawa, S. (1967) The inheritance of resistance of the Zenith type varieties of rice to the blast fungus. *Japanese Journal of Breeding* 17(2), 99–107.

Kiyosawa, S. (1972) Genetics of blast resistance. In: *Rice Breeding*. International Rice Research Institute, Manila, The Philippines, pp. 203–225.

Kiyosawa, S. (1981) Gene analysis for blast resistance. *Oryza* 18, 196–203.

Kiyosawa, S. (1982) Genetic and epidemiological modeling of breakdown of plant disease resistance. *Annual Review of Phytopathology* 20, 93–117.

Kiyosawa, S. (1989) Breakdown of blast resistance in rice in relation to general strategies of resistance gene deployment to prolong effectiveness of disease resistance in plants. In: Leonard, K.J. and Fry, W.E. (eds), *Plant Disease Epidemiology, Genetics, Resistance, and Management*. Macmillan, New York, pp. 251–283.

Levy, M., Romao, J., Marchetti, M.A. and Hamer, J.E. (1991) DNA fingerprinting with a dispersed repeated sequence resolves pathotype diversity in the rice blast fungus. *The Plant Cell* 3, 95–102.

Lin, S.C. (1986) Genetic analysis of minor gene resistance to blast in *japonica* rice. In: *Rice Genetics*. International Rice Research Institute, Los Baños, The Philippines, pp. 451–469.

Lin, S.Y. and Ikehashi, H. (1993) A gamete abortion locus detected by segregation distortion of isozyme locus *Est-9* in wide crosses of rice (*Oryza sativa* L.). *Euphytica* 67, 35–40.

Lincoln, S., Daly, M. and Lander, E. (1992) Mapping genes controlling quantitative traits with MAPMAKER/QTL 1.1. *Whitehead Institute Technical Report*, 2nd edn.

Mackill, D.J. and Bonman, J.M. (1992) Inheritance of blast resistance in near-isogenic lines of rice. *Phytopathology* 82, 746–749.

Mackill, D.J., Bonman, J.M., Suh, H.S. and Srilingam, R. (1985) Genes for resistance to the Philippine isolates of the rice blast pathogen. *Rice Genetics Newsletter* 2, 80–81.

Mackill, D.J., Salam, M.A., Wang, Z.Y. and Tanksley, S.D. (1993) A major photoperiod-sensitivity gene tagged with RFLP and isozyme markers in rice. *Theoretical and Applied Genetics* 85, 536–540.

Marchetti, M.A. (1983) Dilatory resistance to rice blast in USA rice. *Phytopathology* 73(5), 645–649.

McCouch, S.R., Kochert, G., Yu, Z.H., Wang, Z.Y., Coffman, R., Khush, G.S. and Tanksley, S.D. (1988) Molecular mapping of rice chromosomes. *Theoretical and Applied Genetics* 76, 815–829.

McCouch, S.R., Abenes, M.L., Angeles, R., Khush, G.S. and Tanksley, S.D. (1992) Molecular tagging of a recessive gene, *xa-5*, for resistance to bacterial blight of rice. *Rice Genetics Newsletter* 8, 143–145.

Michelmore, R.W., Paran, I. and Kesseli, K.V. (1991) Identification of markers linked to disease-resistance genes by bulked segregant analysis: A rapid method to detect markers in specific genomic regions by using segregating populations. *Proceedings of the National Academy of Sciences of the United States of America*, 88, 9828–9832.

Moldenhauer, K.A.K., Bastawisi, A.O. and Lee, F.N. (1992) Inheritance of resistance in rice to races IB-49 and IC-17 of *Pyricularia grisea* rice blast. *Crop Science* 32(3), 584–588.

Notteghem, J.L. (1985) Definition d'une strategie d'utilisation de la resistance par analyse genetique des relations hoteparasite. Cas du couple riz-*Pyricularia oryzae. Agronomie Tropicale* 40, 129–147.

Oka, H.I. and Lin, K.I. (1957) Genetic analysis of resistance to blast disease in rice (by biometrical genetic method). *Japanese Journal of Genetics* 32, 20–27.

Ou, S.H. (1985) *Rice Diseases*, 2nd edn. Commonwealth Mycological Institute, Kew, UK, 380 pp.

Parlevliet, J.E. (1988) Identification and evaluation of quantitative resistance. In: Leonard, K.J. and Fry, W.G. (eds), *Plant Disease Epidemiology, Genetics, Resistance, and Management*, Vol. 2. McGraw-Hill, New York, 377 pp.

Paterson, A.H., Tanksley, S.D. and Sorrells, M.E. (1991) DNA markers in plant improvement. *Advances in Agronomy* 46, 39–90.

Reimers, P.J. and Leach, J.E. (1991) Race-specific resistance to *Xanthomonas oryzae* pv. *oryzae* conferred by bacterial blight resistance gene Xa-10 in rice (*Oryza sativa*) involves accumulation of a lignin-like substance in host tissues. *Physiological and Molecular Plant Pathology* 38, 39–55.

Roumen, E.C. (1993) Partial resistance in rice to blast and how to select for it. PhD Thesis, Wageningen Agricultural University, Wageningen, The Netherlands, 108 pp.

Sato, Y.I., Ishikawa, R. and Morishima, H. (1990) Nonrandom association of genes and characters found in *indica* × *japonica* hybrids of rice. *Heredity* 65, 75–79.

Shen, Z., Zhang, W., He, Z., Sun, S., Tao, R. and Shi, D. (1986) Genetic analysis for blast resistance in some indica rice varieties (*Oryza sativa* L.). *Chinese Journal of Rice Science* 1, 1–7.

Shinoda, H., Toriyama, K., Yunoki, T., Ezuka, A. and Sakurai, Y. (1971) Studies on the varietal resistance of rice to blast, 6. Linkage relationship of blast resistance genes. *Bulletin Chugoku Agricultural Experiment Station Series* A 20, 1–25. (In Japanese/English.)

Silué, D. and Notteghem, J.L. (1992) Identification of a cross between two compatible isolates of *Magnaporthe grisea* (Hebert) Barr and genetic analysis of avirulence/virulence of rice. *Journal of Phytopathology* 135, 77–83.

Sitch, L.A., Amante, A.D., Dalmacio, R.D. and Leung, H. (1989) *Oryza minuta*, a source of disease resistance for rice improvement. In: Mujeeb-Kazi, A. and Sitch, L.A. (eds), *Review of Advances in Plant Biotechnology 1985–1988*. Proceedings of the 2nd International Symposium on Genetic Manipulation in Crops. International Maize and Wheat Improvement Center, Lisbon 27, 06600 Mexico DF, Mexico and International Rice Research Institute, PO Box 933, Manila, The Philippines, pp. 315–322.

Tanksley, S., Causse, M., Fulton, T., Ahn, N., Wang, Z., Wu, K., Xiao, J., Yu, Z., Second, G. and McCouch, S. (1992) A high density molecular map of the rice genome. *Rice Genetics Newsletter* 9, 111–115.

Tohme, J, Montenegro, M.V., Correa, F., Martinez, C., Zeigler, R. and Roca, W. (1993) Implications of molecular markers in breeding for Colombian lineages of rice blast. Abstract presented at the Sixth Annual Meeting of the International Program on Rice Biotechnology, Feb. 1–5 1991, Chiang Mai, Thailand.

Wagner, H., Weber, W.E. and Wricke, G. (1992) Estimating linkage relationship of isozyme markers and morphological markers in sugar beet (*Beta vulgaris* L.) including families with distorted segregations. *Plant Breeding* 108, 89–96.

Wang, G.L., Mackill, D.J., Bonman, J.M., McCouch, S.R., Champoux, M.C. and Nelson, R.J. (1994) RFLP mapping of genes conferring complete and partial resistance to blast in a durably resistant rice cultivar. *Genetics* 136, 1421–1434.

Williams, J.G.K., Kubelik, A.R., Livak, K.J., Rafalski, J.A. and Tingey, S.V. (1990) DNA polymorphisms amplified by arbitrary primers are useful as genetic markers. *Nucleic Acids Research* 18, 6531–6535.

Yu, Z.H. (1991) Molecular mapping of rice (*Orzya sativa* L.) genes via linkage to restriction fragment length polymorphism (RFLP) markers. PhD Dissertation, Cornell University, 107 pp.

Yu, Z.H., Mackill, D.J. and Bonman, J.M. (1987) Inheritance of resistance to blast in some traditional and improved rice cultivars. *Phytopathology* 77, 323–326.

Yu, Z.H., Mackill, D.J., Bonman J.M. and Tanksley, S.D. (1991) Tagging genes for blast resistance in rice via linkage to RFLP markers. *Theoretical and Applied Genetics* 81, 471–476.

Yunoki, T., Ezuka, A., Morinaka, T., Sakurai, Y., Shinoda, H. and Toriyama, K. (1970) Studies on the varietal resistance to rice blast. 4. Variation of field resistance due to fungus strains. *Bulletin Chugoku Agricultural Experiment Station Series* E 6, 21–41 (In Japanese/English).

Zamir, D. and Tadmor, Y. (1986) Unequal segregation of nuclear genes in plants. *Botanical Gazette* 147(3), 355–358.

11 Molecular Genetic Analysis of the Rice Bacterial Blight Resistance Locus, *Xa21*

P.C. Ronald, T. Holsten, J. Scambray, W. Song
G. Wang and C. Williams

Department of Plant Pathology, University of California, Davis, CA 95616, USA

Rice: A Model for Map-Based Cloning in Monocots

Many agronomically important genes (such as those conferring heterosis, resistance to insects and pathogens and improved crop quality) have been identified solely on the basis of their phenotypes. Successful isolation of these genes will lead to an eventual understanding of their function at the molecular level. The goal of our research, therefore, is to develop a map-based cloning system in rice that will have general applicability for isolation of genes conferring heterosis, improved crop quality and disease resistance. The cloned genes may then be manipulated to improve specific characters and transferred to other crop varieties or species for use in crop improvement programs (McIntosh, 1992).

Map-based cloning is broadly applicable to genes for which other cloning methods are not available. Essential for the success of map-based cloning is the availability of a high density RFLP map, precise mapping of the target gene, establishment of recombinant stocks, availability of genomic libraries containing large insert DNA, and transformation methods for identifying subclones containing the target gene. Recently, several laboratories have demonstrated the feasibility of map-based cloning in plants (Arodel *et al.*, 1992; Giraudet *et al.*, 1992; Ho *et al.*, 1992; Martin *et al.*, 1993).

Among crop plants, rice has several attributes that make it especially amenable to gene isolation by map-based cloning. Rice has a DNA content smaller than any other grain species (C = 450 Mbp) and is among the smallest of any crop plant (Arumanagathan and Earle, 1991; Table 11.1). Based on the genome size and the total map units of rice, 1 cM averages 265 kbp. In comparison, in *Arabidopsis*, a widely used model system for plant molecular genetics, 1 cM equals 290 kbp (Arumuganathan and Earle, 1991). The small genome of rice includes a large percentage (*c.* 75%) of single copy DNA (Deshpande and Ranjekar, 1980; McCouch *et al.*, 1988). It is a diploid species ($2n = 24$), which simplifies genetic studies. A vast reservoir of germplasm (>200,000 accessions)

Table 11.1. Genome size of selected plants.

Species	Mbp*	cM	kbp/cM
Wheat	16000	5000	3000
Maize	2700	1500	1800
Tomato	1000	1400	600
Arabidopsis	145	500	290
Rice	450	1700	265

* Arumanagathan and Earle, 1991.
Mbp = megabasepairs; cM = centimorgan; kbp = kilobasepairs.

of both domestic and wild rices is available for genetic and breeding research (Vaughan, 1989). Rice has proven to be the most readily transformable (by exogenous DNA) of any graminaceous species (Zhang *et al.*, 1988; Shimamoto *et al.*, 1989). Transformation is important for studying the expression of modified or novel genes and for identifying subclones (from a chromosome walk) that contain the gene of interest. Finally, rice is the most widely grown crop plant worldwide, so that applied benefits of genome research are likely to have broad applications and impact (Coffman and Juliano, 1987).

Several high-density RFLP maps have been developed for rice and are being applied to plant breeding and map-based cloning approaches. The Cornell genetic map of rice contains 600 markers (Tanksley *et al.*, 1992). The current Tsukuba RFLP map of rice constructed by Dr Saito and the NIAR Rice Genome Research Program contains more than 1500 DNA markers (Sasaki, 1994). Tanksley *et al.* (1992) estimate the total map units in the rice genome in intra-specific crosses to be 1950 cM. A subset of markers from the Tsukuba map have been located on the Cornell map (Xiao *et al.*, 1992), making the marker density of the integrated RFLP map to be on average one marker per cM (265 kbp), a distance well within the walking range permissible with yeast artificial chromosome (YAC), bacterial artificial chromosome (BAC) or cosmid libraries.

Rice Resistance Against *Xanthomonas oryzae* pv. *oryzae*

Disease resistance loci have been identified in virtually every plant species examined. Genetic analysis of many plant–pathogen interactions has demonstrated that resistant plants often contain single loci that specify resistance against pathogens with corresponding avirulence genes: the 'gene-for-gene' interaction (Flor, 1971; see chapter 7 this volume). Despite extensive genetic studies of many host–pathogen interactions, only two disease resistance genes, the maize *Hm1* locus and the tomato *Pto* locus, have been cloned (Johal and Briggs, 1992; Martin *et al.*, 1993). Recently, several other laboratories are making good progress towards cloning disease resistance genes including isolation of a gene conferring resistance to *Pyricularia oryzae* (H. Hashimoto and K. Shimamoto, 1993, personal communication). The lack of cloned resistance genes has restricted a detailed

Table 11.2. Reaction of Philippine *Xoo* races on rice differential cultivars.*

Xoo races	Rice cultivars					
	Ir24	*Xa3*	*Xa4*	*Xa5*	*Xa10*	*Xa21*
Race 1	S	R	R	R	S	R
Race 2	S	R	S	R	R	R
Race 3	S	R	S	R	S	R
Race 4	S	R	S	S	S	R
Race 5	S	R	R	R	R	R
Race 6	S	S	S	S	S	R

*From Mew, 1987.
R = resistant; S = susceptible.

understanding of the mechanisms by which plant resistance gene products mediate recognition of bacterial avirulence gene products.

Elucidation of the molecular basis of disease resistance and susceptibility ultimately will involve the cloning and characterization of a resistance/avirulence gene pair. In the bacterial blight disease of rice, both the host and pathogen (*Xanthomonas oryzae* pv. *oryzae*, hereafter called *Xoo*) are amenable to molecular genetic analysis, making this an attractive system for studies of disease resistance. Genetic resistance is the most economical and effective control for bacterial blight, which is the most destructive bacterial disease of rice in Asia and Africa and has been isolated recently from fields in Texas and Louisiana (Jones, 1988). Six Philippine races of *Xoo* have been identified that induce resistant or susceptible reactions on a set of rice differential cultivars (Mew, 1987; Table 11.2). Nineteen genes for resistance have been identified and eight of these genes are clustered among three linkage groups (T.W. Mew, personal communication; Yoshimura *et al.*, 1984; Ikeda *et al.*, 1991). The physical and functional relationships between genes within and between groups is unknown.

Recently, a new source of resistance was identified in the wild species *O. longistaminata* (Ikeda *et al.*, 1990). This locus, *Xa21*, has been introduced into the commercial cultivar IR24 that is susceptible to all known Philippine *Xoo* races (Table 11.2) and near isogenic lines are available. *Xa21* is linked to three other bacterial blight resistance genes (*Xa3*, *Xa4*, *Xa10*) on chromosome 11 (Yoshimura *et al.*, 1984; Ikeda *et al.*, 1990, 1991). Unlike the other genes identified, the dominant locus *Xa21* confers resistance to all isolates of *Xoo* tested (Table 11.2). The mechanism by which the *Xa21* locus can confer multi-isolate resistance is unknown. It may be that the *Xa21* locus consists of a complex locus of several tightly linked genes, or alternatively, the *Xa21* locus may encode a single gene product that can recognize a determinant present in all isolates of the pathogen. Isolation of the *Xa21* locus will provide an opportunity to begin to understand the molecular basis for the broad spectrum resistance of *Xa21*. Molecular characterization of this locus will allow us further to address questions regarding the complexity of the resistance locus and to answer fundamental questions regarding the molecular basis of disease resistance in this important crop plant.

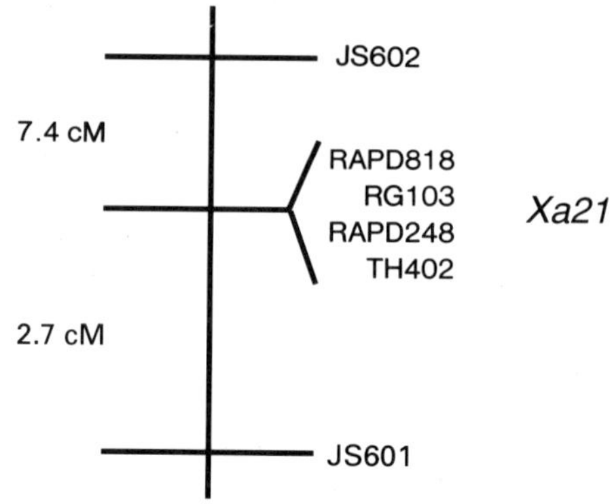

Fig. 11.1. Molecular map of the *Xa21* genomic region on rice chromosome 11. The distance between the markers is shown in centimorgans (cM) on the left.

Chromosomal Location of the *Xa21* Locus

Map-based cloning depends on the tight linkage of a gene of interest to genetically mapped DNA probes. Near isogenic lines (NILs) can be used to find markers that are very close to an introgressed gene from a wild relative because the polymorphism between lines is limited to such a segment (Young *et al.*, 1988). Khush and Ikeda at the International Rice Research Institute (IRRI) constructed NILs for *Xa21* by backcrossing for five generations to the recurrent parent IR24 followed by six generations of selfing (Ikeda *et al.*, 1990, 1991). We surveyed these NILs with 123 rice genomic clones that were evenly distributed on the rice RFLP map (McCouch *et al.*, 1988). One marker (RG103) that maps to chromosome 11 showed polymorphism between the NILs (Fig. 11.1).

In an effort to isolate additional markers linked to *Xa21*, we used the random amplified polymorphic DNA (RAPD) technique (Williams *et al.*, 1990; Martin *et al.*, 1991) to identify PCR products that are polymorphic between the NILs. From a survey of 1700 random primers (corresponding to 5657 amplification products observed after separation by gel electrophoresis), we identified five primers (RAPD248, RAPD818, TH401, JS601, JS602) that are amplified polymorphic bands between the NILs.

The polymorphic DNA markers identified through RFLP and RAPD analysis of the NILs were mapped in populations segregating for these markers and for the bacterial blight resistance locus *Xa21*. In our standard mapping population (Tanksley *et al.*, 1992), RAPD248 cosegregated with marker RG103 and mapped to chromosome 11 (Fig. 11.1). A second F_2 population, derived from a cross between the NILs, was used to determine linkage of RG103, RAPD248, RAPD818, TH401, JS601, JS602 and *Xa21*. The combined data from the two populations led to the consensus RFLP map shown in Fig. 11.1. No recombinants between RAPD818, RAPD248, RG103 and *Xa21* were found out of 386 F_2 progeny scored. These results indicate that the three markers and the *Xa21* locus are located within 1.2 cm, or approximately 300 kb (99% upper confidence limit) from one another on chromosome 11.

Estimated Physical Size of the Introgressed Region of Rice Chromosome 11

Genetic linkage maps (including those based on RFLPs) reflect the frequency of meiotic crossovers between chromosomal loci. However, map distances provide only a very rough estimate of physical distances along a chromosome, since the kbp/map-unit ratio can vary over at least an order of magnitude depending on what part of the chromosome is being examined (Ganal *et al.*, 1989). Studies have shown that the recombination is not constant throughout the genome and that suppression of recombination frequently occurs in regions introgressed from wild species (Ganal *et al.*, 1989; Messeguer *et al.*, 1991). We have therefore used two methods to estimate the physical size of the region introgressed from *O. longistaminata*.

Out of a total of 5657 RAPD and 123 RFLP loci analyzed, we identified six DNA fragments that are polymorphic between the NILs. Based on the rice genome size, the 65% polymorphism between the recurrent and donor parents, and the frequency with which we have recoverd polymorphic *Xa21*-linked loci 6/5780, we have estimated the physical size of the introgressed region to be 700 kbp ($6/5780 \times 1/0.65 \times 450{,}000$). Using the exact binomial method (Mendenhall and Scheaffer, 1973), this value lies within a 95% confidence interval of 263–1563 kbp.

Pulsed field gel electrophoresis provides a second method for determining the physical distance between genetically linked markers. Three of the closest linked markers (RAPD818, RAPD248 and RG103) hybridized to common DNA fragments generated from eight different enzymes (data not shown). None of the enzymes tested generated a DNA fragment that hybridized with all markers linked to *Xa21* indicating that the region of introgression containing the resistance locus *Xa21* is greater than 300 kbp (Ronald *et al.*, 1992).

Chromosome Walking in Rice

The availability of large insert rice genomic libraries facilitates chromosome walking approaches. Procedures for isolation of high molecular weight DNA have been developed for rice (Wu *et al.*, 1992; G. Wang, personal communication).

Due to the relatively small genome of *Arabidopsis*, successful map-based cloning has been carried out using only cosmid clones (Giraudat *et al.*, 1992). Such a strategy is also promising for rice which has approximately three times the genome size of *Arabidopsis*. In addition, cosmid clones are useful for making a finely detailed physical map in the region of interest contained on a larger YAC or BAC clones and for use as RFLP markers (Giraudat *et al.*, 1992). We have therefore constructed a DNA cosmid library from an *Indica* variety containing the resistance gene *Xa21*. The genomic DNA was partially digested with *Sau*3AI and cloned into a derivative of the vector pHC79 (kindly provided by Scott Hulbert). Individual cosmids were subjected to gel electrophoresis to obtain an estimate of the average insert size. Clones from the pHC79 library have an average DNA insert size of 40 kb. We have also constructed a rice BAC library with an average insert of 120 kb.

Cosmid and BAC clones corresponding to the *Xa21*-linked markers have been isolated and construction of a contiguous set of clones spanning the *Xa21* locus is in progress. Transformation of a susceptible cultivar with putative *Xa21* candidate clones and subsequent complementation to resistance will provide confirmation of successful cloning of the *Xa21* disease resistance locus.

Conclusions

Disease resistance is an environmentally sound disease management approach that is available to farmers everywhere. Recently, two plant disease resistance genes have been cloned from maize and tomato (Johal and Briggs, 1992; Martin *et al.*, 1993). It is likely that in the near future, disease resistance genes from rice and many other species will be isolated and their mechanism of resistance characterized. In addition, many researchers are constructing novel resistance genes derived from pathogen or cytotoxic molecules. Interspecific transfer of these 'new' resistance genes will provide access to resistance germplasm that was previously inaccessible for a particular crop plant.

Incorporation of these 'new' resistance genes into crop improvement programs needs to be carefully implemented. Although the durability of these 'new' resistance genes is hard to predict, it is highly probable that many of these genes will not provide durable resistance under high selection pressures. As was presented at this meeting, the use of varietal mixtures can reduce the pathogen selection pressures and the need for pesticides (see chapter 17, this volume). Thus, mixing engineered varieties with 'traditional' resistance cultivars in anticipatory breeding programs may provide a useful method of disease control (McIntosh, 1992). Successful breeding for disease resistance using new and traditional resistance genes will ultimately depend on knowledge of the pathogenicity phenotype and host resistance genes together with constant monitoring of pathogen populations.

References

Arondel, V., Lemieux, B., Hwang, I., Gibson S., Goodman, H.M. and Somerville, C.R. (1992) Map-based cloning of a gene controlling omega-3-fatty acid desaturation in *Arabidopsis*. *Science* 258, 1353–1355.

Arumanagathan, K. and Earle, E.D. (1991) Nuclear DNA content of some important plant species. *Plant Molecular Biology Reporter* 9, 229–241.

Coffman, W.R. and Juliano, O.J. (1987) Rice. In: *Nutritional Quality of Cereal Grains: Genetic and Agronomic Improvement*. Agronomy Monograph No. 28, ASA-CSSA-SSSA, Madison, WI, USA, pp. 101–131.

Deshpande, V.G. and Ranjekar, P.K. (1980) Repetitive DNA in three Gramineae species with low DNA content. *Hoppe Seyler's Zeitschrift für Physiologische Chemie* 361, 1223–1233.

Flor, H.H. (1971) Current status of the gene-for-gene concept. *Annual Review of Phytopathology* 9, 275–296.

Ganal, M.W., Young, N.D. and Tanksley, S.D. (1989) Pulsed field gel electrophoresis and physical mapping of large DNA fragments in the *Tm-2a* region of chromosome 9 in tomato. *Molecular and General Genetics* 215, 395–400.

Giraudat, J., Hauge, B., Valon, C., Smalle, J., Parcy, F. and Goodman, H. (1992) Isolation of *Arabidopsis AB13* gene by positional cloning. *Plant Cell* 4, 1251–1261.

Ho, J.-Y., Weide, R., Ma, H.M., Wordragen, M.F. van, Lambert, K.N., Koornneef, M., Zabel, P. and Williamson, V.M. (1992) The rootknot nematode resistance gene (Mi) in tomato: construction of a molecular linkage map and identification of dominant cDNA markers in resistant genotypes. *Plant Journal* 2(6), 971–982.

Ikeda, R., Khush, G.S., Tabien, R.E. and Ogawa, T. (1990) A new resistance gene to bacterial blight from *Oryza longistaminata*. *Japanese Journal of Breeding* 40 (suppl. 1), 280–281.

Ikeda, R., Tabien, T.E. and Khush, G.S. (1991) Chromosomal location of *Xa21*. *Rice Genetics Newsletter* 8, 102–103.

Johal, G.S. and Briggs, S.P. (1992) Reductase activity encoded by the *Hm1* disease resistance gene in maize. *Science* 258, 985–987.

Jones, R. (1988) New disease hits Texas. *Rice Journal*, February, pp. 4–5.

Kurata, N., Harushim, Y., Yamamot, K., Wu, J., Antonio, B., Sue, N., Shomura, A., Toyama, T., Miyamoto, T., Miyao, A., Sasaki, T., Yano, N. and Minobe, Y. (1992) Our current RFLP maps of rice. *Rice Genetics Newsletter* 9, 130–132.

Martin, G., Williams, J.G.K. and Tanksley, S.D. (1991) Rapid identification of markers linked to a Pseudomonas resistance gene in tomato by using random primers and nearly isogenic lines. *Proceedings of the National Academy of Sciences of the United States of America* 88, 2336–2340.

Martin, G.B., Brommonschenkel, S.H., Chunwongse, J., Frary, A., Ganal, M.W., Spirey, R., Wu, T., Earle, E.D. and Tanksley, S.D. (1993) Map-based cloning of a protein kinase gene conferring disease resistance in tomato. *Science* 262, 1432–1436.

McCouch, S.R., Kochert, G., Yu, Z.H., Khush, G.S., Coffman, W.R. and Tanksley, S.D. (1988) Molecular mapping of rice chromosomes. *Theoretical and Applied Genetics* 76, 815–829.

McIntosh, R.A. (1992) Pre-emptive breeding to control wheat rusts. *Euphytica* 63, 103–113.

Mendenhall, W. and Scheaffer, R.L. (1973) *Mathematical Statistics with Applications*. Duxbury Press.

Messeguer, R., Ganal, M.W., Steffens, J.C. and Tanksley, S.D. (1991) Characterization of the level, target sites and inheritance of cytosine methylationin tomato nuclear DNA. *Plant Molecular Biology* 16, 753–770.

Mew, T.W. (1987) Current status and future prospects of research on bacterial blight of rice. *Annual Review of Phytopathology* 25, 359–382.

Ronald, P., Albano, B., Tabien, R., Abenes, L., Wu, K., McCouch, S. and Tanksley, S. (1992) Genetic and physical analysis of the rice bacterial blight resistance locus, *Xa21*. *Molecular and General Genetics* 236, 113–120.

Tanksley, S., Causse, M., Fulton, T., Ahn, N., Wu, K., Xiao, J., Yu, Z., Second, G. and McCouch, S., (1992) A high density molecular map of the rice genome. *Rice Genetics Newsletter* 9, 111–115.

Sasaki, T. (1994) An overview of rice genome research program: three year progress report. In: *Seventh Meeting of the International Program on Rice Technology*. Bali, Indonesia.

Shimamoto, K., Terada, R., Izawa, J. and Fujimoto, H. (1989) Fertile transgenic rice plants regenerated from transformed protoplasts. *Nature* 338, 274–276.

Vaughan, D.A. (1989) Collection, conservation and potential use of the wild relatives of rice in Asia and Australia. In: Muzech-Kazi, and Sitch (eds), *Recent Advances in Plant*

Biotechnology 1985–1988. 2nd International Symposium on Genetics Manipulation in Crops, pp. 179–190.

Williams, J.G.K., Kubelik, A.R., Livak, K.J., Rafalski, J.A. and Tingey, S.V. (1990) DNA polymorphisms amplified by arbitrary primers are useful genetic markers. *Nucleic Acids Research* 18, 6351–6355.

Wu, K.S., Roder, M.S. and Ganal, M.W. (1992) Isolation of plant DNA for pulsed field gel electrophoresis. In: Burmeister, M. and Ulanovsky, L. (eds), *Methods in Molecular Biology* Vol 12: Pulsed-field gel electrophoresis. Humana Press Inc., Totowa, New Jersey.

Xiao, J., Fulton, T., McCouch, S., Tanksley, S., Kishimoto, N., Oshawa, R., Ukai, Y. and Saito, A. (1992) Progress in integration of the molecular maps of rice. *Rice Genetics Newsletter* 9, 124–128.

Yoshimura, A., Mew, T.W., Khush, G.S. and Omura, T. (1984) Inheritance of resistance to bacterial blight in rice cultivar, Cas 209. *Phytopathology* 73, 1409–1412.

Young, N.D., Zamir, D., Ganal, M.W. and Tanksley, S.D. (1988) Use of isogenic lines and simultaneous probing to identify DNA markers tightly linked to the *Tm-2a* gene in tomato. *Genetics* 120, 579–585.

Zhang, H.M., Yang, H., Rech, E.L., Golds, T.J., Davis, A.S., Mulligan, B.J., Cocking, E.C. and Davey, M.R. (1988) Transgenic rice plants produced by electroporation-mediated plasmid uptake into protoplast. *Plant Cell Reports* 7, 379–384.

12 Current Status for Gene Transfer into Rice Utilizing Variety-Independent Delivery Systems

T.L. Ford, J.T. Cooley and P. Christou

Agracetus Inc., Middleton, WI 53562, USA

Introduction

The cultivated rice plant, *Oryza sativa*, is one of approximately 25 *Oryza* species. It was most probably domesticated in India around 3000 BC and its cultivation spread north and east to reach China and Indochina, as well as south to Indonesia and to the Japanese islands. During the early Middle Ages the plant was introduced to Southern Europe. In the United States it was first planted in 1609 (Taylor, 1990). A number of subspecies of rice exist, the most important being *indica*, *japonica* and *javanica*. The *indica* type accounts for 80% of the cultivated rice and is the staple food for more than two billion people (Swaminathan, 1982; Wu *et al.*, 1990).

Rice is truly a crop of global importance. Almost half the world's population, particularly in East and South-East Asia, depends on rice as the major source of nutritional calories. Only 5% of the rice grown is used in processed foods, industrial products or alcoholic beverages; 95% is consumed by humans as an unprocessed food. Between 1930 and 1980, rice production more than doubled (Witt, 1985) with current world production standing at approximately 420 million tonnes, and the cultivated area reaching 350 million acres (Anonymous, 1986). Production increases over the 50-year period between 1930 and 1980 were primarily due to the utilization of improved varieties, the new semi-dwarf and dwarf varieties and more efficient agronomic practices. These varieties exhibited disease and lodging resistance as well as increased productivity. Like the high yielding wheat varieties developed during the Green Revolution, these modern rice varieties tended to be restricted to areas with adequate irrigation and to farms with sufficient resources to apply high levels of fertilizer and other inputs. They were not adopted under less-favorable rainfed and temperate regions where soil, water and temperature conditions were unsuitable.

Despite the extremely important role that rice plays in mankind's food supply, it has traditionally been thinly traded with a market of less than 5% of the 345 or so million tonnes of milled rice worldwide (Christou and Timmins, 1992). This is largely due to the significant social and cultural role rice plays as a subsistence crop throughout Asia. China and India produce approximately 200 million tonnes, or almost 60% of total world production. Rice is also the world's single largest market for agrochemicals, consuming nearly $3 billion annually. Because of the interdependence of the crop and local water resources, the elimination of potentially contaminative agrochemicals from rice cultivation represents, in itself, a major opportunity for biotechnology. An even more important goal, however, is the reduction of the estimated 50% loss each year due to insects, fungi and weeds (Grayson *et al.*, 1990). Genetic engineering could facilitate the transfer of useful genes conferring resistance and other agronomically important traits.

It has been estimated that annual rice production must increase from 460 million tonnes in 1987, to 560 million tonnes by the year 2000 and 760 million tonnes by the year 2020, in order to keep up with increasing population growth (International Rice Research Institute, 1988). Worldwide experimental yield trends indicate we are very near the limit of genetic improvement possible by conventional methods. Biotechnology represents an attractive approach to attain dramatic increases in rice productivity by expanding the genetic base of rice germplasm (Murashige, 1989). Areas where rice improvement can benefit substantially from biotechnology include: *in vitro* production of haploids, isolation of mutant cell lines, protoplast regeneration and somatic hybridization, cryopreservation and gene transfer technology, development of linkage maps for rice genes, understanding the molecular basis of resistance to diseases, development of varieties resistant to pathogens and insects, study and manipulation of rice storage proteins, regulation of gene expression, and development of practical hybrid rice technology. Altogether, the combination of agrochemical expenditures and crop losses total tens of billions of dollars, an extraordinary opportunity for significant contributions from biotechnology.

Assessing Targets for Rice Improvement Through Biotechnology

The Rockefeller Foundation has provided a major boost to rice biotechnology research by initiating and funding the Rice Biotechnology Network. Toenniessen and Herdt (1988) discussed some of the criteria the Rockefeller Foundation used to analyze the potential impact of, and assign priorities to, rice biotechnology research programs supported by the Foundation:

1. the driving force has been that benefits of new production technologies should go not only to producers through reductions in production costs, but also to consumers through reductions in the price of rice;
2. the targeted improvements should take place within a reasonable time frame, requiring little initial capital investment;
3. the amount of additional rice produced impacts directly on price reductions;

4. desired traits that are difficult to introduce through alternative technologies were to be given higher priority;
5. technology development was prioritized based on whether it addressed specifically problems in upland, deep-water and rain-fed or irrigated conditions, in that order;
6. the likelihood of success utilizing a specific biotechnology approach.

With these criteria the Foundation developed a ranking of alternative priority listings for the problems included in the analysis. Another area of importance was deemed to be the social and economic impact such improvements would bring to rice growing countries, particularly in Asia and South America. Such factors include future supply and demand for rice, differential impact of new rice technology in various countries, labor force migration, etc. Major targets for rice biotechnology ranked in accordance with the above criteria include: resistance to tungro virus, yellow stemborer and gall midge, cytoplasmic male sterility and drought tolerance (Toenniessen and Herdt, 1988).

Existing Rice Transformation Methods and Their Limitations

Successful generation of callus from rice protoplasts (Deka and Sen, 1976; Cai *et al.*, 1978; Toriyama and Hinata, 1985; Baba *et al.*, 1986; Thompson *et al.*, 1986) and subsequent recovery of plants, originally from *japonica* varieties (Fujimura *et al.*, 1985; Abdullah *et al.*, 1986; Coulibali and Demarly, 1986; Toriyama *et al.*, 1986; Yamada *et al.*, 1986; Kyozuka *et al.*, 1987) and later from a number of *indicas* (Kyozuka *et al.*, 1988; Lee *et al.*, 1989; Wang *et al.*, 1989) provided the necessary prerequisite for the development of protoplast-based methodology for gene transfer into rice. Original reports described transient expression of marker genes in rice protoplasts using direct DNA transfer. Milestones in achieving variety-independent gene transfer into rice included demonstration of transient activity in protoplasts, utilizing direct DNA transfer with polyethylene glycol (PEG) (Ou-Lee *et al.*, 1986; Hayashimoto *et al.*, 1990). In the same year, stably transformed callus was recovered by the same methodology (Uchimiya *et al.*, 1986). Reports of the recovery of the first transgenic rice plants utilizing protoplasts and electroporation or polyethylene glycol (PEG) fusion were first published in 1988 by several groups (Toriyama *et al.*, 1988; Zhang and Wu, 1988; Zhang *et al.*, 1988; Shimamoto *et al.*, 1989). Datta *et al.* (1990) published the first report of the recovery of transgenic *indica* plants from the variety Chinsurah Boro II and Christou *et al.* (1991, 1992) developed a variety-independent method based on bombardment of immature embryos and recovery of fertile transgenic plants from many varieties, including *indicas* and *japonicas*, at high frequencies.

Subsequent research favored electroporation and PEG-mediated gene transfer into protoplasts as methods of choice for gene transfer into rice protoplasts and recovery of fertile transgenic plants (Matsuki *et al.*, 1989; Peng *et al.*, 1990; Tada *et al.*, 1990; Davey *et al.*, 1989; Meijer *et al.*, 1990). A severe limitation of these methods, however, is the requirement to establish systems to regenerate plants from protoplasts. Even though such systems do exist in some of the *japonica* varieties, e.g. Taipei 309, most elite *japonicas* as well as the vast majority of *indica* varieties are very difficult to regenerate from protoplasts. In most experiments in which regeneration of *indica* varieties from protoplasts was accomplished, transgenic plant recovery was unsuccessful. A notable exception is the report by Datta *et al.* (1990) which described recovery of fertile transgenic plants from the variety Chinsurah Boro II (*indica* type) and, more recently, primary transformants from the elite variety IR72, which were infertile (Datta *et al.*, 1992). In both cases embryogenic suspension cultures were established from immature pollen grains and these cultures were used to isolate protoplasts for subsequent transformation experiments. The principal constraint to transformation using protoplasts is the strong genotype- and culture-dependent components of rice *in vitro* culture and regeneration. The predominant rice cultivars in the world are *indicas* and, therefore, are largely excluded from transformation via these methods. Another serious problem of protoplast-based system was identified as reduced fertility or complete infertility of primary transformants.

Putative transgenic callus was recovered from immature rice embryos using *Agrobacterium*, but regeneration of plants was not reported (Raineri *et al.*, 1990). Experiments describing a simple method using pollen tubes as a vehicle to introduce exogenous DNA into rice plants, lack essential molecular and genetic data to demonstrate convincingly integration of foreign genes into high molecular weight DNA and remains unconfirmed (Luo and Wu, 1988).

Genotype-Independent Gene Transfer into Rice via Particle Bombardment

It is clear, therefore, that in order to apply conveniently recombinant DNA methods in major *indica* rice improvement programs a system must be developed that is genotype independent. Currently this implies that the system should not depend on protoplasts or embryogenic suspension culture. Such a system would permit introduction of useful agronomic and other traits into elite varieties and consequently accelerate rice improvement programs. The development and commercialization of a variety-independent gene transfer method for soybean using electric discharge particle bombardment (Christou *et al.*, 1990) prompted us to investigate its suitability for inserting foreign genes into regenerable rice tissue. Due to the nature of the DNA delivery process, genotype should not be a variable (Christou, 1992). Transformation of other important species, especially the more recalcitrant cereals, has focused exclusively on gene transfer into embryogenic suspension cultures. A typical example is transformation of elite breeding lines of maize which requires development of such regenerable cultures (Fromm *et al.*, 1990; Gordon-Kamm *et al.*, 1990). So far this approach has under-utilized the potential of this technology by restricting particle bombardment to elite varieties rather than developing transformants into breeding populations.

Efficient systems for the regeneration of rice plants from the scutellum of immature embryos through somatic embryogenesis have been described (Hartke and Lorz, 1989; Datta *et al.*, 1990). These studies demonstrated that essentially

Table 12.1. Rice transformation frequencies.

No. of explants	Genotype	Independently derived transformed plants	Transformation frequency (%)
100	IR72	12	12
200	IR72	25	12
150	IR72	15	10
100	IR72	12	12
100	IR72	11	11
100	IR72	11	11
200	IR36	18	9
150	IR36	8	5
100	IR26	8	8
100	IR54	10	10
100	Gulfmont	15	15
150	Gulfmont	14	9
200	Gulfmont	25	12
100	Gulfmont	11	11
100	CM101	13	13
100	CM101	5	5
250	S201	8	3
150	Koshihikari	9	6
250	Koshihikari	15	6
150	Koshihikari	10	7
100	Koshihikari	12	12
100	Koshihikari	15	15
150	Koshihikari	9	6

all rice cultivars could be regenerated from immature embryos even though regeneration efficiencies vary substantially. Exposure of immature rice embryos to the auxin 2,4D results in the development of embryogenic callus which can be converted to plants upon transfer to media supplemented with indole-3-acetic acid (IAA) and zeatin (Z). Following particle bombardment as described (Christou *et al.*, 1991) we were able to recover transformed embryogenic callus and plants either with or without selection. Transformation frequencies in experiments in which selection was applied were very high, in the range of 2–15% calculated in terms of independently-derived transgenic plants per unit number of bombarded explants (Table 12.1). The antibiotic hygromycin B was used to select embryogenic callus transformed with plasmids carrying the gene for antibiotic resistance: hygromycin phosphotransferase (*hmr*). Transgenic callus and plants were also recovered from experiments in which selection was not used, illustrating the high efficiency of the transformation process.

In a typical experiment, 12–15-day-old immature embryos were excised from fully expanded panicles and subjected to particle bombardment. Use of the β-glucuronidase (*gus*) gene permitted monitoring of transformation events and their stable integration into organized embryogenic callus, either with or without selection by a simple histochemical assay (Jefferson *et al.*, 1987). Development of transgenic tissues and plants was followed utilizing a combination of selectable hygromycin resistance (*hmr*) and screenable (*gus*) markers. A large number of independently derived transgenic plants from many diverse *indica* and *japonica* varieties were obtained and characterized phenotypically, molecularly, and genetically. All recovered plants were clonal in nature, demonstrating that each transformation event was derived from single or a small number of cells. Chimeric plants were never obtained in this system; however, multiple regenerants from a single transformation event were frequently obtained.

Aspects of these procedures were recently reproduced by another group utilizing a Biolistics® instrument. However, transformation frequencies were significantly lower, the range of genotypes transformed limited, and fertility of transgenic plants was very low, with some important varieties exhibiting total sterility (Li *et al.*, 1993).

Molecular and Genetic Characterization of Transgenic Rice Plants

We have shown that it is possible to produce large numbers of independently derived transgenic rice plants from all elite varieties tested by utilizing particle bombardment. This method results in the introduction of DNA into regenerable scutellar tissue capable of differentiating into clonally transformed plants (Christou *et al.*, 1991). We have characterized the integration and expression patterns of marker genes as well as genes conferring agronomically important traits through several generations. The integration pattern of exogenous DNA in primary regenerates and progeny was determined by Southern blot analysis. Expression of the *gus* marker gene in RO plants was scored visually by GUS histochemical assay of leaf, root and floral tissue. Inheritance of an agronomically important trait was demonstrated by using the *bar* gene (De Block *et al.*, 1987) which confers resistance to the herbicide phosphinothricin (Basta®) in greenhouse and field trials, using commercial rates of application.

Southern blot analysis was designed to identify independent transformation events by comparing patterns of bands, the sizes of which represented those of the chimeric gene insert and flanking rice genomic DNA (Fig. 12.1). Plants transformed with one of several plasmids were analyzed. All plasmids had the pUC19 backbone in common. Plasmid pWRG2426 contains the *gus*, *bar* and *hmr* genes all fused to their own CaMV (cauliflower mosaic virus; Gardner *et al.*, 1981) 35s promoter. The *gus* and *hmr* genes contain the AMV (alfalfa mosaic virus) 5′ untranslated leader sequence. The *bar* gene has the maize *adh* (alcohol dehydrogenase; Callis *et al.*, 1987) 5′ untranslated leader. The *bar* and *hmr* genes have the *Agrobacterium nos* (nopaline synthase; Depicker *et al.*, 1982) poly A addition site. Transcription of the *gus* gene is terminated by the poly A addition site from the soy *ssu* (small subunit of Rubisco; Berry-Lowe *et al.*, 1982) gene. The three genes are immediately adjacent to each other with the *hmr* gene positioned between *bar* and *gus*. Both *bar* and *hmr* are transcribed in a clockwise direction while the *gus* gene is transcribed counter-clockwise. Plasmid pWRG4517

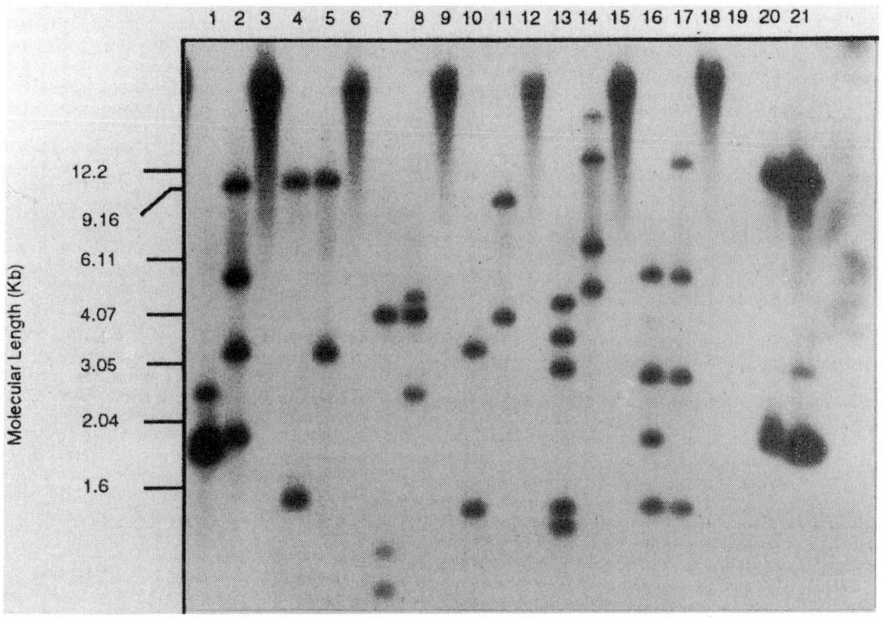

Fig. 12.1. Southern analysis of IR72 transgenic plants transformed with pWRG2426 or pWRG4517. 5 μg of appropriately digested genomic DNA were subjected to electrophoresis, transferred to nylon membranes and probed with ^{32}P-labeled *bar* gene DNA. Lanes 1, 4, 7, 10, 13 and 16 contain DNA doubly digested with *Xba*1 and *Xho*1 from independent transformation events 533–2, 533–3, 534–1, 534–3, 534–11 and 536–1 respectively. Lanes 2, 5, 8, 11, 14 and 17 contain DNA singly digested with Xba1. Lanes 3, 6, 9, 12, 15 and 18 contain undigested high molecular weight DNA. Lane 19 contains non-transformed IR72 genomic DNA digested with *Xba*1. Plants 533–2, 533–3 and 536–1 are transformed with pWRG2426. Plants 534–1, 534–3 and 534–11 are transformed with pWRG4517. 1X and 5X copy standard lanes 20 and 21, respectively contain 5 μg of non-transformed IR72 genomic DNA with appropriate amounts of an equimolar mixture of singly and doubly digested pWRG2426. Singly digested pWRG2426 and pWRG4517 is 9260 and 4520 bps respectively. Double digestion of pWRG2426 and pWRG4517 releases a *bar* fragment of 1900 and 1360 bps respectively.

contains both the *bar* and *hmr* genes fused to 35s promoters and soy *ssu* transcription terminators. The *bar* gene has the AMV leader and the *hmr* gene has the leader from CaMV 35s mRNA. The genes are immediately adjacent and are transcribed toward each other. Additional plasmids are derivatives of pWRG4517 and were constructed by replacement of the 35S-*bar*-soy poly A cassette with *gus* fused to four different promoters. pWRG4515, pWRG4520, pWRG4522 and pWRG4524 have *gus* genes driven by barley *Gt* (Leisy *et al.*, 1989), soy *ssu*, Arabadopsis *ssu* (Timko *et al.*, 1988) and rice *ssu* (de Pater *et al.*, 1990) promoters, respectively. All *gus* constructs end with the *nos* transcription termination site, with the *gus* and *hmr* genes transcribing toward each other.

Multiple Southern blots to probe for all chimeric genes of interest (*gus*, *bar* and *hmr*) were done to confirm the identity of all independent transformants and to determine cointegration of selectable marker gene(s) and the gene(s) of

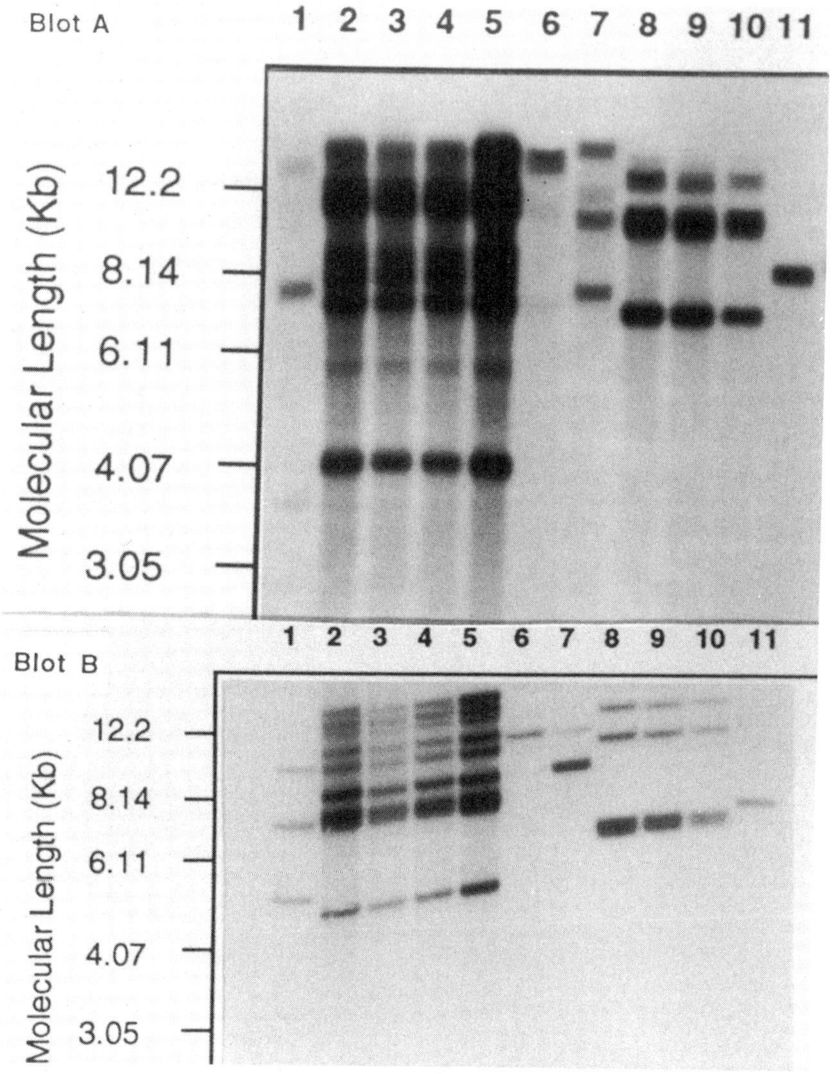

Fig. 12.2. Southern blots of genomic DNA from GUS positive transgenic rice plants transformed with pWRG2426. DNA from R1 progeny was digested with *Sac*I (Blot A) or *Xba*I (Blot B), transferred to nylon membranes, and probed with ^{32}P-labeled *gus* gene DNA (Blot A) or similarly labeled *bar* gene DNA (Blot B). *Sac*I and *Xba*I cut pWRG2426 once. Lanes 2–5 are DNAs from individual progreny from the RO transgenic parent 496–1. Lanes 8–10 are DNAs from individual progeny from the RO transgenic parent 496–4. Lanes 1, 6, and 7 are DNAs from the transgenic RO plants 495–1, 496–2, and 496–3 respectively. Lane 11 is plasmid DNA at one copy/haploid genome digested with *Sac*I (Blot A) or *Xba*I (Blot B).

interest. All plasmids contained the *hmr* gene (selectable marker) and one or both of the other two genes of interest. Multiple Southern blots probing for genes of interest on the specific transforming DNA demonstrated that all genes were integrated at least once in all independent events (100% cointegration frequency; Fig. 12.2). For most individuals, the complexity of the Southern band patterns was comparable from one probe to the other, indicating similar copy-numbers of the co-transferred genes. This suggests that the transforming plasmids generally integrate as a complete unit. Estimation of copy-number was based on the intensity and number of bands in Southerns blots in which the genomic DNA was digested with an enzyme that cuts once in the transforming plasmid. A high proportion (38%) of all plants analyzed had one or two stably integrated copies of the gene. Forty-two percent had between three and nine copies and 19% had ten or more copies.

Molecular and genetic analyses of transgenic progeny indicated that the transgenes were inherited as a single locus. Southern blot analysis of progeny from individual transgenic plants exhibited the same banding pattern as the primary transformant. About 25% of the progeny tested showed no hybridization which is consistent with Mendelian segregation of a single dominant gene. Patterns of GUS expression in successive generations were identical to that observed in the original transformant; likewise, levels of resistance to the herbicide Basta® were also consistent with the RO plant. First generation R_1 plants homozygous for exogenous genes did not differ from heterozygotes in expression patterns.

Thirty four hygromycin-resistant individuals containing *gus* transgenes driven by either 35s or Arabidopsis *ssu* promoters were tested for GUS expression by histochemical staining of leaf, root, and/or floral tissues. Plants with strong blue staining were recorded as positive and those with no staining or a few localized spots were designated as negative. Resistance to the herbicide Basta® was determined by spraying plants at 1000 p.p.m. and was subjectively ranked as high (no damage), low (some damage), or none (death) by the relative degree of necrosis observed after 5 days. For the purposes of this analysis, both high and low resistance were tallied as positive for BAR expression. GUS and HMR were co-expressed in 73% of the *35s-gus* plants, as compared to 81% co-expression in *35s-bar* and *hmr* individuals. These results suggest that co-expression rates of GUS/HMR and BAR/HMR were similar when comparing genes with similar promoters. The larger size (threefold) of the *gus* gene may be responsible for lower co-expression rates, compared to *bar*, due to increased probability of truncation during integration.

The general conclusion from this analysis is that molecular and genetic properties of transgenic rice plants derived through particle bombardment experiments were very similar to those of plants obtained through alternative direct DNA transfer procedures. Variation in expression and co-expression patterns between different transformants exists, but these distinct patterns are inherited by both homo- and heterozygous progeny. The system described here, however, displays a 100% cointegration rate and 73–80% co-expression of the selectable marker gene and gene(s) of interest. The majority of transgenic rice plants recovered through particle bombardment using ACCELL® technology have a relatively low number of input gene(s) per haploid genome.

Concerns with Transgenic Rice Expressing Herbicide Resistance

In applying recombinant DNA methods to rice improvement, it is important to consider potential risks that transgenic rice plants expressing a given trait may pose to the environment, the farmer, or the consumer. One of the major issues that needs to be addressed in the case of rice is its close relative red rice. This weedy variety is found commonly in the southern United States and also in many other rice growing areas around the world. Red rice is the same species (*Oryza sativa*) as the cultivated rice and as such shares many morphological, physiological, biochemical and other traits. In addition to competing with cultivated rice for essential resources thereby reducing crop yields, red rice reduces the quality of the harvested product (Pantone and Baker, 1991). Engineering herbicide-resistant rice cultivars would allow management of red rice in the field by herbicide application. However, in such situations there is the real possibility of pollen transfer of herbicide resistance from cultivated to red rice. Comprehensive weed management practices must be developed for safe cultivation of herbicide-resistant rice. It should be possible to arrive at an optimum window for herbicide application by gaining as much understanding as possible on pollen dispersal and out-crossing rates between cultivated and red rice. Transgenic rice plants with easily scorable characteristics will provide some of the answers that are needed to make a critical assessment on the risks and benefits of targeting herbicide resistance as a goal in rice improvement programs. Results from field trials in which several of these issues are being addressed will be presented in detail elsewhere.

Conclusion

Total world population is projected to increase from 5.3 billion in 1990 to 6 billion by the year 2000 and up to 10 billion by 2025. All of this increase will most likely take place in the developing world, thus putting additional pressure on already limited and rapidly depleting food supplies. It is obvious that in order to prevent massive famine and starvation, incremental improvements in crop yields will not be adequate. Advances in rice biotechnology in combination with conventional breeding, including the introduction of hybrid rice, provide a realistic hope for assuring the availability of adequate food supplies particularly in developing countries as we move into the next century. The development of efficient gene transfer methodology for the creation of large numbers of transgenic rice plants from varieties that are being cultivated today, provides an invaluable tool to plant breeders and pathologists to address problems that are otherwise very difficult to solve using conventional breeding and traditional gene transfer methodology. It is apparent that for any gene transfer method to be useful and practical it must possess the capability to engineer large numbers of plants rapidly, economically and in elite varieties that are in wide cultivation. Among the methods that are currently being used for rice genetic engineering, bombardment-based technologies fulfill these criteria.

REFERENCES

Abdullah, R., Cocking, E.C. and Thompson, J.A. (1986) Efficient plant regeneration from rice protoplasts through somatic embryogenesis. *Bio/Technology* 4, 1087–1090.

Anonymous (1986) *Encyclopedia Britannica*, 15th edn. Chicago University, Chicago, vol. 10, p. 41.

Baba, A., Hasezawa, S. and Syono, K. (1986) Cultivation of rice protoplasts and their transformation mediated by *Agrobacterium* spheroplasts. *Plant Cell Physiology* 27, 463–471.

Berry-Lowe, S.L., McKnight, T.D. and Meager, R.B. (1982) The nucleotide sequence, expression, and evolution of one member of a multigene family encoding the small subunit of ribulose-1,5-biphosphate carboxylase in soybean. *Journal of Molecular Applied Genetics* 1, 483–498.

Cai, Q., Zhou, Y. and Wu, R. (1978) A further study on the isolation and culture of rice (*Oryza sativa* L.) protoplasts. *Acta Botanica Sinica* 20, 97–102.

Callis, J., Fromm, M. and Walbot, V. (1987) Introns increase gene expression in cultured maize cells. *Genes and Development* 1, 1183–1200.

Christou, P. (1992) Genetic transformation of crop plants using microprojectile bombardment. *Plant Journal* 2, 275–281.

Christou, P. and Timmins, J. (1992) A commercial process for the generation of transgenic rice plants utilizing ACCELL™ technology. *AgBiotech News and Information* 4, 159–163.

Christou, P., Ford, T.L. and Kofron, M. (1991) Production of transgenic rice (*Oryza sativa* L.) plants from agronomically important *indica* and *japonica* varieties via electric discharge particle acceleration of exogenous DNA into immature zygotic embryos. *Bio/Technology* 9, 957–962.

Christou, P., Ford, T.L. and Kofron, M. (1992) The development of a variety independent gene transfer method for rice. *Trends in Biotechnology* 10, 239–246.

Christou, P., McCabe, D.E., Martinell, B.J. and Swain, W.F. (1990) Soybean genetic engineering – Commercial production of transgenic plants. *Trends in Biotechnology* 8, 145–151.

Coulibaly, M.Y. and Demarly, Y. (1986) Regeneration of plantlets from protoplasts of rice, *Oryza sativa* L. *Zeitschrift für Pflanzenzüchtung* 96, 79–81.

Datta, S.K., Datta, K. and Potrykus, I. (1990) Embryogenesis and plant regeneration from microspores of both *Indica* and *Japonica* rice (*Oryza sativa*). *Plant Science* 67, 83–88.

Datta, S.K., Datta, K., Soltanifar, N., Donn, G. and Potrykus, I. (1992) Herbicide-resistant *Indica* rice plants from IRRI breeding line IR72 after PEG-mediated transformation of protoplasts. *Plant Molecular Biology* 20, 619–629.

Datta, S.K., Peterhans, A., Data, K. and Potrykus, I. (1990) Genetically engineered fertile *Indica*-rice recovered from protoplasts. *Bio/Technology* 8, 736–740.

Davey, M.R., Kothari, S.L., Zhang, H., Rech, E.L., Cocking, E.C. and Lynch, P.T. (1991) Transgenic rice: Characterization of protoplast-derived plants and their seed progeny. *Journal of Experimental Botany* 42, 1159–1169.

De Block, M., Botterman, J., Vandewiele, M., Dockx, J., Thoen, C., Gossele, V., Rao Movva, N., Van Montagu, M. and Leemans, J. (1987) Engineering herbicide resistance in plants by expression of a detoxifying enzyme. *EMBO Journal* 6, 2513–2518.

de Pater, S., Hensgens, L.A.M. and Schilperoort, R.A. (1990) Structure and expression of a light-inducible shoot-specific rice gene. *Plant Molecular Biology* 15, 399–406.

Deka, P.C. and Sen, S.K. (1976) Differentiation of calli originated from isolated protoplasts of rice (*Oryza sativa* L.) through plating techniques. *Molecular and General Genetics* 145, 239–243.

Depicker, A., Stachel, S., Dhaese, P., Zambryski, P. and Goodman, H.M. (1982) Nopaline synthase: Transcript mapping and DNA sequence. *Journal of Molecular and Applied Genetics* 1, 499–512.

Fromm, M.E., Morrish, F., Armstrong, C., Williams, R., Thomas, J. and Klein, T.M. (1990) Inheritance and expression of chimeric genes in the progeny of transgenic maize plants. *Bio/Technology* 8, 833–844.

Fujimura, T., Sakurai, M., Akagi, H., Negishi, T. and Hirose, A. (1985) Regeneration of rice plants from protoplasts. *Plant Tissue Culture Letters* 2, 74–75.

Gardner, R., Howarth, A., Hahn, P., Brown-Luedi, M., Shepherd, R. and Messing, J. (1981) The complete nucleotide sequence of an infectious clone of cauliflower mosiac virus by M13mp7 shotgun sequencing. *Nucleic Acids Research* 9, 2871–2888.

Grayson, B.T., Green, M.B. and Copping, L.G. (1990) *Pest Management in Rice*. Elsevier, Barking, UK.

Gordon-Kamm, W.J., Spencer, T.M., Mangano, M.L., Adams, T.R., Daines, R.J., Start, W.G., O'Brien, J.V., Chambers, S.A., Adams, W.R., Willetts, N.G., Rice, T.B., Mackey, C.J., Krueger, R.W., Kausch, A.P. and Lemaux, P.G. (1990) Transformation of maize cells and regeneration of fertile transgenic plants. *The Plant Cell* 2, 603–618.

Hartke, S. and Lorz, H. (1989) Somatic embryogenesis and plant regeneration from various *indica* rice (*Oryza sativa* L.) genotypes. *Journal of Genetics and Breeding* 43, 205–214.

Hayashimoto, A., Li, Z. and Murai, N. (1990) A PEG-mediated protoplast transformation system for production of fertile transgenic rice plants. *Plant Physiology* 93, 857–863.

International Rice Research Institute (1988) *Rice Facts*. International Rice Research Institute, Los Baños, The Philippines.

Jefferson, R.A., Kavanagh, T.A. and Bevan, M.W. (1987) GUS fusions: β-glucuronidase as a sensitive and versatile gene fusion marker in higher plants. *EMBO Journal* 6, 3901–3907.

Kyozuka, J., Hayashi, Y. and Shimamoto, K. (1987) High frequency plant regeneration from rice protoplasts by novel nurse culture methods. *Molecular and General Genetics* 206, 408–413.

Kyozuka, J., Otoo, O. and Shimamoto, K. (1988) Plant regeneration from protoplasts of *indica* rice: genotypic differences in culture response. *Theoretical Applied Genetics* 76, 887–890.

Lee, L., Schroll, R.E., Grimes, H.D. and Hodges, T.K. (1989) Plant regeneration from *indica* rice (*Oryza sativa* L.) protoplasts. *Planta* 178, 325–333.

Leisy, D.J., Hnilo, J., Zhao, Y. and Okita, T. (1989) Expression of a rice glutenin promoter in transgenic tobacco. *Plant Molecular Biology* 14, 41–50.

Li, L., Qu, R., de Kochko, A., Fauquet, C. and Beachy, R.N. (1993) An improved rice transformation system using the biolistic method. *Plant Cell Reports* 12, 250–255.

Luo, Z. and Wu, R. (1988) A simple method for the transformation of rice via the pollen-tube pathway. *Plant Molecular Biology Reporter* 6, 165–174.

Matsuki, R., Onodera, H., Yamauchi, T. and Uchimiya, H. (1989) Tissue-specific expression of the rolC promoter of the Ri plasmid in transgenic rice plants. *Molecular and General Genetics* 220, 12–16.

Meijer, E.G.M., Schilperoort, R.A., Rueb, S., van Os-Ruygrok, P.E. and Hensgens, L.A.M. (1991) Transgenic rice cell lines and plants: expression of transgenic chimeric genes. *Plant Molecular Biology* 16, 807–820.

Murashige, T. (1989) Rice improvement through biotechnology: Discussion of the Rockefeller Foundation's International Program. In: *Proceedings of the Midwest Plant*

Biotechnology Consortium, 1989 Symposium. Purdue University, West Lafayette, Indiana, pp. 124–136.

Ou-Lee, T.M., Turgeon, R. and Wu, R. (1986) Expression of a foreign gene linked to either a plant virus or a Drosophila promoter after electroporation of protoplasts of rice, wheat and sorghum. *Proceedings of the National Academy of Sciences of the United States of America* 83, 6815–6819.

Pantone, D.J. and Baker, J.B. (1991) Weed-crop competition models and response-surface analysis of red rice competition in cultivated rice: A review. *Crop Science* 31, 1105–1110.

Peng, J., Lyznik, L.A., Lee, L. and Hodges, T.K. (1990) Cotransformation of *indica* rice protoplasts with gus A and Neo genes. *Plant Cell Reports* 9, 168–172.

Raineri, D.M., Bottino, P., Gordon, M.P. and Nester, E.W. (1990) *Agrobacterium*-mediated transformation of rice (*Oryza sativa* L.). *Bio/Technology* 8, 33–38.

Shimamoto, K., Teda, R., Izawa, T. and Fujimoto, H. (1989) Fertile transgenic rice plants regenerated from transformed protoplasts. *Nature* 338, 274–277.

Swaminathan, M.S. (1982) Biotechnology research and third world agriculture. *Science* 218, 967–972.

Tada, Y., Sakamoto, M. and Fujimura, T. (1990) Efficient gene introduction into rice by electroporation and analysis of transgenic plants; use of electroporation buffer lacking chloride ions. *Theoretical and Applied Genetics* 80, 475–480.

Taylor, H.R. (1990) Rice. In: Halsey, W.D. and Johnston, B. (eds.), *Collier's Encyclopedia*, Vol. 20. MacMillan, New York.

Thompson, J.A., Abdullah, R. and Cocking, E.C. (1986) Protoplast culture of rice (*Oryza sativa* L.) using media solidified with agarose. *Plant Science* 47, 123–134.

Timko, M.P., Herdies, L., de Almeida, E., Cashmore, A.R., Leemans, J. and Krebbers, E. (1988) Genetic engineering of nuclear-encoded components of the photosynthetic apparatus in *Arabidopsis*. In: *The Impact of Chemistry on Biotechnology*. American Chemical Society, pp. 278–295.

Toenniessen, G.H. and Herdt, R.W. (1988) The Rockefeller Foundation's international program on rice biotechnology. In: *Strengthening Collaboration in Biotechnology: International Agricultural Research and The Private Sector*. Proceedings of a Conference, Rosslyn, VA (April 17–21). Bureau for Science and Technology, Agency for International Development, Washington DC, pp. 291–317.

Toriyama, K. and Hinata, K. (1985) Cell suspension and protoplast culture of rice. *Plant Science* 41, 179–183.

Toriyama, K., Hinata, K. and Sasaki, T. (1986) Haploid and diploid plant regeneration from protoplasts of anther callus in rice. *Theoretical and Applied Genetics* 73, 16–19.

Toriyama, K., Arimoto, Y., Uchimiya, H. and Hinata, K. (1988) Transgenic rice plants after direct gene transfer into protoplasts. *Bio/Technology* 6, 1072–1074.

Uchimiya, H., Fushimi, T., Hashimoto, H., Harada, H., Syono, K. and Sugawara, Y. (1986) Expression of a foreign gene in callus derived from DNA-treated protoplasts of rice (*Oryza sativa* L.). *Molecular and General Genetics* 16, 204–207.

Wang, D., Miller, P.D., Sondahl, M.R. (1989) Plant regeneration from protoplasts of *Indica* type rice and CMS rice. *Plant Cell Reports* 8, 329–332.

Witt, S.C. (1985) *BriefBook, Biotechnology and Genetic Diversity*, p. 15, California Agricultural Lands Project.

Wu, R., Kemmerer, E. and McElroy, D. (1990) Transformation and regeneration of important crop plants: Rice as the model system for monocots. *Gene Manipulation in Plant Improvement II*, 251–263.

Yamada, Y., Yang, Z.Q. and Tang, T.D. (1986) Plant regeneration from protoplast-derived callus of rice (*Oryza sativa* L.). *Plant Cell Reports* 5, 85–88.

Zhang, W. and Wu, R. (1988) Efficient regeneration of transgenic plants from rice protoplasts and correctly regulated expression of the foreign gene in the plants. *Theoretical and Applied Genetics* 76, 835–840.

Zhang, H.M., Yang, H., Rech, E.L., Golds, T.J., Davis, A.S., Mulligan, B.J. and Cocking, E.C. (1988) Transgenic rice plants produced by electroporation-mediated plasmid uptake into protoplasts. *Plant Cell Reports* 7, 379–384.

III Pathogen Population Dynamics and Utilization of Host Plant Resistance

13 Virulence Characteristics of Genetic Families of *Pyricularia grisea* in Colombia

F.J. Correa-Victoria[1], R.S. Zeigler[2] and M. Levy[3]

[1]Rice Program, Centro Internacional de Agricultura Tropical (CIAT), Apdo aereo 6713, Cali, Colombia; [2]Rainfed Lowland Rice Program, International Rice Research Institute (IRRI), PO Box 933, Manila, The Philippines; [3]Department of Biological Sciences, University of Purdue, West Lafayette, IN 47907, USA

Introduction

Rice blast caused by *Pyricularia grisea* is one of the most economically important and well-studied diseases of rice (Ou, 1985). Numerous races, or pathotypes, of the pathogen exist and are usually defined by their ability or inability to cause disease on a set of host differentials (Ling and Ou, 1969). As breeding resistant rice cultivars is a principal tool for blast management, major efforts have been made to understand and describe this pathotypic diversity.

There is a lingering controversy over the degree of pathotype instability in the fungus. Great pathogenic variation has been reported from single spore isolates originating from single lesions and monoconidial subcultures (Ou and Ayad, 1968; Ou *et al.*, 1970), while other studies have shown isolates to be pathogenically stable (Latterell, 1971; Bonman *et al.*, 1986). One of the suggested reasons for these discrepancies is that the pathotype assay is very sensitive to variations in inoculation techniques, and environmental conditions, while scoring for the disease is inherently subjective. None the less, pathotype remains the most widely used method of classifying virulence, and until recently has been the only practical means to characterize pathogen populations to which resistant cultivars are exposed.

Restriction fragment length polymorphism (RFLP) analysis offers a new tool for studying pathogen population structure. A family of dispersed repetitive DNA sequences in the genome of *P. grisea* was described by Hamer *et al.* (1989) and was found to be largely limited to rice-infecting members of the species (Dobinson *et al.*, 1993). A DNA probe specific for one of these sequences MGR586 (hereafter referred to as MGR) was used by Levy *et al.* (1991) to construct restriction length profiles from field isolates of this fungus. They were able to group field isolates collected from the southern USA according to the similarity of their RFLP profiles or 'fingerprints'. Those groups showing a high degree of internal similarity were statistically distinct from one another. Considering the clonal propagation habit of the pathogen, they inferred that the groups consisted of genetically related individuals descended from a common ancestor. An analysis of the pathotypes of these groups, hereafter referred to as 'lineages', showed that each had an associated specific pathotype, or small groups of similar pathotypes.

If *P. grisea* populations are generally composed of limited number of lineages and if there is a close relationship between lineage and virulence characteristics of the constituent individuals, population analysis at the lineage level could aid in directing resistance breeding projects that target the pathogen population in question. In this paper we examine the correspondence between lineage structure and virulence diversity in rice blast fungus populations collected at two geographically distinct sites in Eastern Colombia. One site is a breeding facility selected for reliable and severe occurrence of rice blast (a 'hot spot') and the other is a region where rice had never been grown before 1989.

Breeding for Rice Blast Resistance at the Centro Internacional De Agricultura Tropical (CIAT), Colombia

The blast disease is present in all Latin American and Caribbean (LAC) countries where rice is grown. The use of resistant cultivars has been recognized as the most practical and economic method by which the farmer can control the disease. Large breeding populations have been managed at CIAT, together with the Colombia Institute of Agriculture (ICA) from 1969 to 1976 aiming at combining high yield with blast resistance. The first cultivars released were IR 8 and IR 22 obtained directly from the International Rice Research Institute (IRRI) or derived directly from IRRI germplasm, such as the cultivar Cica 4. As these cultivars rapidly became susceptible to blast, additional resistance was incorporated, first using Tadukan, producing Cica 6. However, after one cycle of planting, it became susceptible not only in Colombia but in other LAC countries (Rosero, 1979).

The CIAT Rice Program then incorporated resistance sources such as Colombia 1, Tetep, Dissi Hatif, Mamoriaka, and C 46–15. Several high yielding, blast-resistant lines were selected and then released as the cultivars Cica 7, Cica 8 and Cica 9, yet these cultivars became susceptible to blast within two or three seasons of commercial planting (Rosero, 1979).

The breeding approach was then modified to combine or 'pyramid', multiple sources of resistance into new cultivars. Sources included Tetep, C 46–15, Dissi Hatif, Colombia 1, and Carreon. The success of the pyramiding approach taken by CIAT was limited by the absence of information on the genetics of resistance in the donor lines (Weeraratne *et al.*, 1981). Even though distinct resistant donors were used, whether functionally different or identical genes controlled their resistance was unknown. Screening techniques were not suitable for identification of lines with combinations of different resistance genes (Weeraratne *et al.*, 1981).

A second approach was focused on producing multiline cultivars where high yielding cultivars such as Cica 9 were selected as recurrent parents. Selected isogenic lines from different backcrosses were chosen and evaluated at different locations in the region expecting better possibilities for selecting cultivars with a broad spectrum of resistance to the blast disease (Rosero, 1979). The multiline approach was finally abandoned as a majority of plants were more susceptible than the parents after three backcrosses. Other problems such as development and subsequent management of multilines by national programs also argued against continuing this approach.

Other approaches taken by the CIAT Rice Program during the 1980s included concentration of slow blasting components of diverse geographic background such as Africa, Surinam, Japan and the Philippines into high yielding widely adapted suceptible cultivars. Combination of major gene resistance and slow blasting was also expected to impart better stability since genes controlling them are believed to increase the effectiveness of each other. Detection of the combination of the two types of resistance has been the major difficulty as the expression of lower levels of resistance is presumed to be masked by the presence of the major genes. Furthermore, identification of slow blasting segregants in segregating populations was difficult, and screening methods were not available. Results also indicated that leaf resistance does not necessarily predict resistance in the most damaging panicle stage (Weeraratne *et al.*, 1981).

All the above strategies for the development of stable resistance were seen to be at best only partially successful, as most derived cultivars did not have stable blast resistance. Of those cultivars released from these projects, Metica 1, Metica 2, Oryzica 1, Oryzica 2, and Oryzica 3, only Oryzica 1 and Oryzica 3 are now being grown commercially, although farmers need to control blast with fungicides.

Breeding for Blast Resistance at a 'Hot Spot' Site in Colombia

Although blast resistance screening was conducted in a highly favorable, disease epidemic location, breeding activities and selection were conducted under irrigated conditions where leaf blast was minimal.

Failure of breeding lines to encounter low frequency compatible pathotypes in a population may explain frequent resistance breakdown (Buddenhagen, 1983). Breeding strategies to minimize the probability of such escape include line selection in 'hot spot' sites favorable for the pathogen, where populations remain high throughout the season, and pathogenic diversity is high (Correa-Victoria and Zeigler, 1993a,b). Following a decade of unsatisfactory progress, it was proposed to grow the segregating populations under upland conditions to ensure heavy blast pressure (Weeraratne *et al.*, 1981).

In 1985 the CIAT Rice Program began breeding for rice blast resistance at a hot spot site, the Santa Rosa experimental and breeding farm in eastern Colombia (Correa-Victoria and Zeigler, 1993a,b). The farm is in an important rice-growing area where rice blast is one of the principal production constraints. Rice blast levels are maintained at high levels in breeding plots (F_2–F_6) during the entire season using spreader rows composed of commercial cultivars and resistance sources with complementing susceptibility to different pathotypes of the blast fungus (Correa-Victoria and Zeigler, 1993a,b). Escape is avoided by continuous exposure to the pathogen during the season, at all growth stages, and in several generations.

Two commercial rice cultivars have been developed at Santa Rosa and were released for the Eastern Plains of Colombia in 1989. One of the cultivars, Oryzica Llanos 5, preferred by rice growers, is being planted yearly in more than 50,000 ha, remaining blast resistant since its release. Controlled inoculations under greenhouse conditions with more than 500 isolates representing all the known blast pathogen diversity at Santa Rosa have supported the observed resistance of Oryzica Llanos 5.

Although the approach thus far appears to be successful (Correa-Victoria and Zeigler, 1993b), improvement is hampered by a lack of understanding of pathogen population structure. Critical evaluation of this approach as a practical means to develop durable blast resistance began with the characterization and monitoring of the pathogenic variability at Santa Rosa. More than 45 international races possessing virulences for at least 13 known resistance genes are present in this pathogen population (Correa-Victoria and Zeigler, 1993a). No cultivar was susceptible to all isolates tested and low compatibility frequencies were associated with combinations of resistance genes. There was a marked specialization for compatibility of a blast isolate and the cultivar from which it was recovered. Matching virulence for all resistance genes is already present in the pathogen population. Virulence factors were accumulated in large numbers in the most virulent issolates, but no isolate was virulent to all cultivars tested. Rare compatibilities with particular cultivars may suggest that certain virulence combinations are not possible in the pathogen and may be associated with poor fitness (Correa-Victoria and Zeigler, 1993a). The complexity of the virulence structure prompted the attempt to describe the population in terms of its lineage structure and determine if there was a correspondence between lineage and virulence patterns.

Pyricularia grisea Population Structure at Santa Rosa

An extensive study of Colombia *P. grisea* diversity was initiated by analyzing the population structure at Santa Rosa. MGR-DNA fingerprints of more than 100 isolates collected and pathotyped from Santa Rosa reduced isolate diversity to only six distinct genetic lineages, designed SRL 1–6 (Levy *et al.*, 1993). Isolates within each lineage had a 92% or greater average fingerprint similarity, with no significant subclustering among lineage members. In contrast, the maximum consensus similarity between lineages was 85%, the minimum was 37%, and the average was 49%. All genetic lineages were significantly different from each other for genetic distance (Levy *et al.*, 1993).

Each lineage was associated with a specific subset of cultivars and a specific subset of pathotypes as determined by reaction of eight international differentials; 90% of all pathotypes were lineage-specific and most were cultivar-specific within a lineage. The distribution of pathotypes within lineages was non-random. In each lineage, the pathotypes formed a network or continuum of virulence variants that were related by single differences in pathogenicity on the international differentials. Further, with few exceptions, each network was an aggregate composed of cultivar-specific subnetworks of similarity-related virulence variants (Levy *et al.*, 1993). Sampling elsewhere in Colombia has yielded an additional set of lineages, designated ALL 7–13.

The chronic epidemic environment and available hosts at Santa Rosa apparently has promoted an accelerated and exaggerated version of the pathotype evolution that may occur in rice field monocultures. However, the apparent virulence changes that are observed within lineages followed identifiable routes that may be reproduced in more typical agricultural plantings. MGR polymorphism analysis may be an important and applicable new tool for deciphering the population structure and monitoring the epidemiology of the rice blast fungus (Levy *et al.*, 1993). However, since pathotype-lineage relationships to date are based upon the very limited pathotypes determined by the international set of differentials, it remains to be seen if there is a general relationship between lineage and the full virulence spectrum of the constituent individuals.

Virulence Spectrum in Genetic Lineages of the Blast Fungus at Santa Rosa

The purpose of this analysis was to determine if lineages have distinct virulence spectra, and to determine if a particular isolate of an individual lineage can be used as representative of lineage pathotype diversity for use in a resistance screening program.

We have analyzed extensively the virulence diversity of more than 100 blast isolates collected from different rice cultivars, mainly commercial cultivars, at the Santa Rosa breeding station in Colombia. An additional set of isolates was used from the seven ALL lineages. These isolates were inoculated under greenhouse conditions onto rice cultivars with known resistance genes, rice commercial cultivars released in Colombia over a period of 21 years, sources of durable blast resistance, and the international set of host differentials. (Inoculation and evaluation methods are presented in Correa-Victoria and Zeigler, 1993a, and Levy *et al.*, 1993.)

The virulence spectrum of six MGR-genetic families corresponding to 127 blast isolates on 20 rice cultivars with at least 13 resistance genes (Kiyosawa, 1972; Yamada *et al.*, 1976; Kiyosawa, 1981) are presented in Table 13.1. In most cases there was variability with respect to virulence on any cultivar. That is, rarely were all isolates in a lineage compatible with a given cultivar. Occasionally only one or two were compatible. Compatibility for all known resistance genes was present in the pathogen population. Genetic lineages SRL-4, SRL-5, and SRL-6 exhibited the broadest spectrum of virulence with small differences among them, although lineage SRL-6 had high virulence frequencies on most cultivars. Lineage SRL-6 seems to be the only lineage with virulence factors compatible with most of the resistance genes tested. Since resistance genes *Pi-i*, *Pi-K*^{s} were ineffective against some isolates of SRL-6 when present in different backgrounds either singly or in combination with other genes, incompatibility with Fujisaka 5 may indicate these

Table 13.1. Virulence spectrum of six genetic lineages of Colombian rice blast isolates on international differentials and on rice cultivars with known resistance genes.

		Genetic lineage (no. of isolates)[a]					
Cultivar	Resistance gene	SRL-1(16)	SRL-2(10)	SRL-3(7)	SRL-4(26)	SRL-5(22)	SRL-6(46)
Aichi Asahi	*Pi-a*	+	+		+	+[32][b]	+
BL-1	*Pi-b*		+		+[19]		+
Caloro[c]	*Pi-K*^{s}				+[12]	+[23]	+
Chokoto	*Pi-a, Pi-k*				+[8]	+	+[40]
Dular[c]					+[4]	+	+
Fukunishiki	*Pi-z*				+[38]		+[26]
Fujisaka 5	*Pi-i, Pi-K*^{s}				+[42]	+[9]	
Kanto 51[c]	*Pi-K*			+[14]		+[50]	+[30]
Kusabue	*Pi-K*	+		+[29]	+[50]	+[18]	+
K 1	*Pi-Ta*			+[29]	+[42]	+[18]	+
K 59	*Pi-T*	+			+	+	+
NP-125[c]				+	+[4]	+[5]	+[17]
Pi No. 4	*Pi-Ta*^{2}	+	+[20]		+[19]	+[27]	+
Raminad STR 3[c]		+			+	+[27]	+
Sha-Tiao-Tsao[c]	*Pi-K*^{s}				+	+	+[24]
Shin 2	*Pi-K*^{s}	+[13]		+	+[50]	+[5]	+
Tetep	*Pi-K*^{h}				+[4]	+	+[17]
Tsuyuake	*Pi-K*^{m}, *Pi-m*					+[27]	+[28]
Usen[c]	*Pi-a*				+[21]	+[18]	+
Zenith[c]	*Pi-z, Pi-a*	+[6]			+[8]	+[9]	+[4]

[a] Genetic lineages as described by Levy *et al.* (1993), and number of isolates tested for each lineage.
[b] Figures in square brackets indicate percentage of virulent isolates. Percentages above 50% are not indicated.
+ = virulent; blank = avirulent.
[c] International differential cultivar.

genes combined pose a major, perhaps insurmountable, barrier to the entire lineage. Resistance genes present in cultivars BL-1 and Tsuyuake are effective against lineages SRL-5 and SRL-4 respectively. Although none of the resistance genes were effective against the three most virulent lineages, it can be seen that cultivar Zenith, which combines at least two resistance genes, was susceptible to a low proportion of the isolates within each of the genetic lineages present at Santa Rosa.

That Tetep has been the only resistance source among all cultivars in Table 13.1 used in the breeding program at CIAT suggests that virulence/avirulence genes may be retained, or perhaps generated, in the absence of a compatible host. Controlled inoculations showed that the whole virulence spectrum of a particular lineage may be present in a single isolate, although this type of isolate tends to be less frequent in the pathogen population. It is more common to find this type of isolate on cultivars exhibiting high levels of resistance than on highly susceptible cultivars. Most isolates within a lineage exhibit individual pathotypes that are a subset of virulence corresponding to the full spectrum of that lineage. The constituent pathotypes are arrayed in a network of variants related by single virulence differences on the rice cultivars tested as described by Levy *et al.* (1993). It is noteworthy that the MGR fingerprints of lineages SRL-4, SRL-5, and SRL-6 which have very similar virulence spectra are statistically more similar to each other than to the other three lineages present at Santa Rosa (Levy *et al.*, 1993).

Virulence spectra of SRL-1, SRL-2, and SRL-3 were much narrower compared to SRL-4, SRL-5, and SRL-6 based on the cultivars used (Table 13.1). The former are characterized by greater incompatibility with known resistance genes than compatibility. The number of pathotypes observed in lineages SRL-1, SRL-2, and SRL-3 is also fewer than the other three lineages; however, pathotypes also form networks of complementing virulences. In general, cultivars which exhibited resistance to the three most virulent lineages also exhibit resistance to the other three. It is clear from Table 13.1 that additional genes are involved in some of the cultivars used. In each of the following pair of cultivars, Caloro and Shin-2, Kanto 51 and Kusabue, Aichi Asahi and Usen, one is susceptible and the other resistant to the same isolates within SRL-1 despite the fact that the same resistance gene has been reported for each pair of cultivars.

No cultivar with known resistance genes was either resistant or susceptible to all genetic lineages of the blast pathogen present at Santa Rosa. Cultivars were susceptible to at least two different genetic lineages. Directed crosses based on the avirulence/virulence information generated on these cultivars with known resistance genes suggest that useful combinations of resistance genes derived from these sources could be developed from multiple crosses between cultivars BL-1, Fukunishiki, Fujisaka 5, Kanto 51, and Tsuyuake. Breeding lines from these crosses should theoretically confer resistance to all lineages of the pathogen present at Santa Rosa.

If shifts in virulence are driven by changes in the hosts upon which the pathogen has to survive, then the virulence spectrum of lineages on widely cultivated Colombian commercial rice cultivars should reflect these changes. The virulence spectrum of six genetic lineages present at Santa Rosa is shown in Table 13.2. As observed from virulence on cultivars with known resistance genes, genetic

Table 13.2. Virulence spectrum of six genetic lineages of the rice blast pathogen on Colombian commercial rice cultivars.

Variety	Year of release	Genetic lineage (no. of isolates)[a]					
		SRL-1(16)	SRL-2(10)	SRL-3(7)	SRL-4(26)	SRL-5(22)	SRL-6(46)
Oryzica Llanos 5	1989						
Oryzica Llanos 4	1989						
Linea 2	1989		+				
Oryzica 3	1987	+[13][b]			+[8]	+[27]	+
Oryzica 2	1984					+[5]	+[26]
Oryzica 1	1982		+		+[19]		+
Metica 1	1981			+	+	+[18]	+
Cica 8	1978				+[12]	+	+[9]
Cica 7	1976			+[14]	+[19]		+
Cica 9	1976	+	+				
Cica 6	1974					+	+[41]
Cica 4	1971	+	+		+[4]	+[50]	+
IR 22	1969	+[13]			+[4]	+[14]	+
IR 8	1968	+	+		+[12]	+[14]	+
Bluebonnet 50	1951	+[6]			+[4]	+[9]	+[9]
Fanny[c]	–	+		+	+	+[50]	+

[a] Genetic lineages as described by Levy *et al.* (1993), and number of isolates tested for each lineage.
[b] Figures in square brackets indicate percentage of virulent isolates. Percentages above 50% are not indicated.
[c] Susceptible check, non-commercial cultivar.
+ = virulent; blank = avirulent.

Table 13.3. Virulence spectrum of six genetic lineages of Colombian rice blast fungus on blast resistance sources.

	Genetic lineage (no. of isolates)[a]					
Cultivar	SRL-1(16)	SRL-2(10)	SRL-3(7)	SRL-4(26)	SRL-5(22)	SRL-6(46)
IR-42				+[35]	+[14]	+[4]
K 8				+[4]	+[5]	+[7]
Peta	+[13][b]					
Ceysvoni						+[24]
Colombia 1				+[12]		
IRAT 13						
Moroberekan						

[a] Genetic lineages as described by Levy *et al.* (1993), and number of isolates tested for each lineage.
[b] Figures in square brackets indicate percentage of virulent isolates. Percentages above 50% are not indicated.
+ = virulent; blank = avirulent.

lineages SRL-4, SRL-5, and SRL-6 exhibited the largest spectrum of virulence on the Colombian rice cultivars. In several cases compatibility was detected with only one isolate from the lineage. Oryzica 1, Oryzica 2, Cica 6 and Cica 7 show differential compatibility with SRL 4 and 5, and Cica 9 is incompatible with all but SRL 1 and 2. As in the cultivars shown in Table 13.1, some are incompatible with all isolates tested from some lineages.

A high specialization or specific interaction between genetic lineage and host of origin is observed under field conditions. For example, isolates retrieved from the rice cultivars Oryzica 1, Cica 8, Metica 1, and which reinfect those cultivars under greenhouse conditions, belong always to the genetic lineages SRL-6, SRL-5, and SRL-3 and SRL-4, respectively. This is observed despite the fact that other lineages are virulent on Oryzica 1, Metica 1, and Cica 8 when inoculated under greenhouse conditions. This apparent specificity may simply reflect relative frequencies and small sample sizes. Thus, to determine the range of lineages with the capacity to infect a cultivar, it is not sufficient to consider only those recovered after field sampling from the cultivar in question.

We can deduce from Table 13.2 that a logical source of resistance for a large part of the blast population in Colombia would be the cultivars Linea 2 and Cica 9. These two cultivars could be crossed with other cultivars such as Oryzica 2, Cica 6, Cica 7, Cica 8, and Metica 1 in single or multiple crosses with the possibilities of combining resistance to all blast lineages of the pathogen.

The virulence spectrum of the six lineages found at Santa Rosa on some commonly used resistance donors again shows the similarity in spectrum among SRL-4, SRL-5, and SRL-6, as well as specificity between donors and lineage (Table 13.3). Cultivars Ceysvoni and Colombia 1 are the two most common sources used in the CIAT rice program. Cultivars IRAT 13 and Moroberekan show incompatibility to all the Santa Rosa lineages. The cultivars in Tables 13.1 and 13.2 are from distinct subgrouping within the *Oryza sativa* species (Glaszmann, 1987), and share similar compatibility spectra. However, IRAT 13 and Moroberekan are

Table 13.4. Virulence spectrum of genetic lineages of the rice blast pathogen collected in the Eastern Plains (Altillanura) of Colombia on international differentials and on cultivars with known resistance genes.

Cultivar	Resistance gene	Genetic lineage (no. of isolates)[a]								
		ALL-7(13)	ALL-8(1)	ALL-9(1)	ALL-10(4)	ALL-11(1)	ALL-12(1)	ALL-13(4)	SRL-5(5)	SRL-6(15)
Aichi Asahi	*Pi-a*	+	+	+	+	+	+	+	+	+
BL-1	*Pi-b*									+
Caloro[c]	*Pi-K^s*									
Chokoto	*Pi-a, Pi-k*					+		+	+	
Dular[c]						+		+	+	+
Fukunishiki	*Pi-z*									+[7]
Fujisaka 5	*Pi-i, Pi-K^s*	+[8][b]	+	+	+		+	+	+	+[7]
Kanto 51[c]	*Pi-K*	+[23]	+	+		+	+	+	+	
Kusabue	*Pi-K*								+[20]	
K 1	*Pi-Ta*	+[31]	+	+	+	+		+	+	+
K 59	*Pi-t*		+	+	+	+		+	+	+
NP-125[c]			+		+			+[25]	+[40]	+[7]
Pi No. 4	*Pi-Ta^2*								+[20]	+
Raminad STR 3[c]		+[23]	+		+		+	+	+	+
Sha-Tiao-Tsao[c]	*Pi-K^s*	+[50]	+	+	+[50]			+	+	+
Shin 2	*Pi-K^s*	+[15]								
Tetep	*Pi-K^h*							+	+	
Tsuyuake	*Pi-K^m, Pi-m*					+		+	+	
Usen[c]	*Pi-a*	+	+	+	+		+	+	+[40]	+
Zenith[c]	*Pi-z, Pi-a*	+			+	+		+	+	+[40]

[a] Genetic lineages as described by Levy *et al.* (unpublished), and number of isolates tested for each lineage.
[b] Figures in brackets indicate percentage of virulent isolates. Percentages above 50% are not indicated.
[c] Internal differential cultivar.
+ = virulent; blank = avirulent.

from an entirely different group, the tropical *japonicas*, and are derived from traditional African upland cultivars. The complete lack of compatibility among Santa Rosa lineages and IRAT 13 and Moroberekan suggest that a fundamental incompatibility with entire grouping of lineage and *O. sativa* population may exist.

The relation between virulence spectrum and genetic lineages of the blast pathogen can also be observed in Table 13.1 on the international set of host differentials. SRL-4, SRL-5, and SRL-6 again exhibit the largest and similar spectrum of virulence. No blast differential was resistant to all isolates of lineage SRL-6, while no isolate of lineage SRL-2 was virulent on any differential.

A breeding program directed to acid soil upland of LAC was initiated at CIAT in the mid-1980s. Similar methodology for developing blast resistance was employed but the parental material was largely African and Brazilian upland lines and improved traditional upland cultivars. The advanced breeding lines, often multiplied in Santa Rosa, and the eventual released cultivars underwent extensive testing and evaluation in the Altillanura, a rolling savanna beginning approximately 200 km east of Santa Rosa, where rice has never been grown before. Monitoring the blast population at this new site yielded seven genetic lineages of the pathogen, designated ALL 7–13, distinct from those identified in Santa Rosa (Table 13.4) (see Manry *et al.*, poster abstract, this volume). Lineages SRL-5, and SRL-6 were also found in the Altillanura. Compatibility to most known resistance genes is found in this new blast population; however, the spectrum of virulence found in any lineage is somewhat narrower than lineages at Santa Rosa when tested on the same sets of cultivars. Since the cultivars from which the ALL lineages were recovered are upland cultivars with very different backgrounds from those previously discussed, the set used to determine virulence spectrum in this study may not be the best. Likewise, the low number of isolates tested may not allow full appreciation of the virulence spectra of the ALL lineages.

Virulence on cultivars with known resistance genes show very similar spectrum among several lineages found in the Altillanura (Table 13.4). The origin of these lineages, if as believed rice was never grown at this site, is difficult to explain at the moment. The lack of virulences on the commercial cultivars released in Colombia as shown in Table 13.5 suggests that these lineages have not been surviving, at least as pathogens, on these cultivars. This minimizes the possibilities that new genetic lineages found in the Altillanura are coming from the nearby rice growing areas. Cultivar Fanny, which is highly susceptible to most lineages, is not grown commercially but only as a check cultivar at the research station. New virulences are, however, characteristic of the new blast population. The upland cultivar IRAT 13, resistant to all lineages from Santa Rosa, is defeated by a large proportion of isolates in the lineage ALL-7 present in the Altillanura (Table 13.6). It is also interesting to note that the cultivar Zenith, which is susceptible to a low proportion of isolates in Santa Rosa, is susceptible to blast isolates of several lineages in Altillanura fields and in the greenhouse to the same isolates (Table 13.4). Altillanura genetic lineages had virulence factors on most of the international differentials as lineages found at Santa Rosa (Table 13.4).

Table 13.5. Virulence spectrum of genetic lineages of the rice blast pathogen collected in the Eastern Plains (Altillanura) of Colombia on commercial rice cultivars.

	Genetic lineage (no. of isolates)[a]								
Cultivar	ALL-7(13)	ALL-8(1)	ALL-9(1)	ALL-10(4)	ALL-11(1)	ALL-12(1)	ALL-13(4)	SRL-5(5)	SRL-6(15)
Oryzica Llanos 5									
Oryzica Llanos 4									
Linea 2									
Oryzica 3									+
Oryzica 2									
Oryzica 1									+
Metica 1	+[38][b]			+[50]				+[20]	+
Cica 8									
Cica 7									+[33]
Cica 9									
Cica 6					+		+	+[40]	+
Cica 4					+		+	+	+
IR 22						+			+
IR 8									+
Bluebonnet 50	+					+			
Fanny	+[38]	+	+	+	+	+	+	+	+

[a] Genetic lineages as described by Levy *et al.* (unpublished), and number of isolates tested for each lineage.
[b] Figures in brackets indicate percentage of virulent isolates. Percentages above 50% are not indicated.
+ = virulent; blank = avirulent.

Table 13.6. Virulence spectrum of genetic lineages of the rice blast pathogen collected in the Eastern Plains (Altillanura) of Colombia on blast resistance sources.

	Genetic lineage (no. of isolates)[a]								
Cultivar	ALL-7(13)	ALL-8(1)	ALL-9(1)	ALL-10(4)	ALL-11(1)	ALL-12(1)	ALL-13(4)	SRL-5(5)	SRL-6(15)
IR-42				+				+[40]	+[7]
Peta	+[38][b]								
Ceysvoni						+			
IRAT 13	+[38]								
K 8									
Moroberekan									
Colombia 1									

[a] Genetic lineages as described by Levy *et al.* (unpublished), and number of isolates tested for each lineage.
[b] Figures in brackets indicate percentage of virulent isolates. Percentages above 50% are not indicated.
+ = virulent; blank = avirulent.

Table 13.7. Average linkage cluster analysis of the virulence of 110 *Pyricularia grisea* isolates on 37 rice cultivars.

Similarity[a] (%)	Clusters[b] (N)
73	22
85	49
90	72

[a] Similarity percentage of the compatible–incompatible reactions in the inoculation of 110 blast isolates on 37 rice cultivars.
[b] Virulence clusters determined by average linkage cluster analysis of the distances (dissimilarities) between two isolates and the average of the whole group. Distance between two isolates was calculated as 1 – 2 Nxy/Nx + Ny, where Nxy = number of test cultivars compatible to both isolate x and y; Nx =number of test cultivars compatible with isolate x and Ny = number of test cultivars compatible with isolate y.

Virulence Spectrum and Genetic Lineage Relationship

Virulence clusters were determined by average linkage cluster analysis of the distance (dissimilarity) between two isolates and the average of the whole group. Distance between two isolates was calculated as 1 − 2 Nxy/Nx + Ny where Nxy = number of test cultivars compatible to both isolates 'x' and 'y'; Nx = number of test cultivars compatible with isolate 'x', and Ny = number of test cultivars compatible with isolate 'y' (Johnson and Wichern, 1982; SAS, 1989). The number of clusters based on 73% virulence similarity was 22 and doubled and triplicated for 85% and 90% similarity, respectively (Table 13.7). Clustering of blast isolates with 73% virulence similarity groups isolates from several genetic lineages into one virulence cluster. This is most extreme in the case of virulence cluster 1 which grouped 12 isolates of six different genetic lineages into only one cluster. Most of the other virulence clusters grouped only isolates from one genetic lineage, or from two to three lineages. In the latter case, lineages are genetically related as observed in virulence clusters 3 and 6 (Table 13.8).

Isolates within individual genetic lineages were clustered in several virulence groups. This is to be expected since a large number of cultivars were compatible across several lineages; and within lineages both compatibility and incompatibility with these cultivars was observed. We can infer that a number of virulence factors are shared among lineages. To minimize the effect of similar rice cultivars as well as lineages with only one compatible isolate in determining the virulence clusters, we performed a discriminant analysis using R^2 and F statistics to select cultivars that discriminate isolates, based on their virulences, corresponding to the DNA–genetic lineage classification (Table 13.9). The technique used was a backward and stepwise discriminant analysis (Klecka, 1980; SAS, 1989). More isolates from poorly represented lineages (not included in the table) are needed to identify the most suitable cultivars for determining the diversity of the blast fungus in Colombia.

Table 13.8. Relationship between virulence clusters and genetic lineages of 110 isolates of *Pyricularia grisea*.

Virulence cluster[b]	Genetic lineage[a]											
	SRL-2	SRL-4	SRL-5	SRL-6	ALL-7	ALL-8	ALL-9	ALL-10	ALL-11	ALL-12	ALL-13	Total
1		4[c]	1	1		1	1	4				12
2			4									4
3			9						1		4	14
4					7							7
5	7											7
6		3		37								40
7				3								3
8					3							3
9		2			1							3
10		2										2
11				2								2
12			1	1								2
13				2								2
14		1										1
15					1							1
16		1										1
17				1								1
18					1							1
19										1		1
20			1									1
21		1										1
22			1									1
Total	7	14	17	47	13	1	1	4	1	1	4	110

[a] Genetic lineages as determined by Levy *et al.* (1993 and unpublished).
[b] Virulence clusters determined by average linkage cluster analysis of the distances between two isolates and the average of the whole group. Distance between two isolates was calculated as 1 – 2 Nxy/Nx + Ny, where Nxy = number of test cultivars compatible to both isolate x and y; Nx = number of test cultivars compatible with isolate x and Ny = number of test cultivars compatible with isolate y.
[c] Frequency of isolates per virulence cluster within each genetic lineage.

Table 13.9. Rice cultivars selected by backward and stepwise discriminant analysis giving similar clustering between virulence and genetic lineage[a].

		Genetic lineage[b]						
Cultivar	Resistance gene	SRL-2	SRL-4	SRL-5	SRL-6	ALL-7	ALL-10	ALL-13
Cica 9		+[c]						
Cica 8			+	+	+			
Cica 4		+	+	+	+			+
IR 22			+	+	+			
Raminad STR 3			+	+	+	+	+	+
Zenith	*Pi-a*, *Pi-z*		+	+	+	+	+	+
NP-125			+	+	+		+	+
Usen	*Pi-a*		+	+	+	+	+	+
Dular			+		+			+
Kanto 51	*Pi-K*			+	+	+		+
Sha-Tiao-Tsao	*Pi-K*^{s}		+	+	+	+	+	+
Aichi Asahi	*Pi-a*	+	+	+	+	+	+	+
BL-1	*Pi-b*	+	+		+			
Fujisaka 5	*Pi-i*, *Pi-K*^{s}		+	+		+	+	+
IR 42			+	+	+		+	
K 59	*Pi-T*		+	+	+		+	+
Pi No. 4	*Pi-Ta*^{2}	+	+	+	+			
Peta						+		
Bluebonnet 50			+	+	+	+		
Chokoto	*Pi-a*, *Pi-k*		+	+	+			+
Shin-2	*Pi-K*^{s}		+	+	+	+		
Taichung			+	+	+	+	+	+
Fanny			+	+	+	+	+	+

[a] Discriminant analysis using R^2 and F statistics to select cultivars that discriminate isolates, based on their virulences, corresponding to DNA-genetic lineages.
[b] Genetic lineages as determined by Levy *et al.* (1993 and unpublished).
[c] + = virulent; blank = avirulent. Each column indicates the potential spectrum of virulence of the lineage.

Table 13.10. Classification of 106 blast isolates into MGR-DNA genetic lineages according to their virulence on 23 rice cultivars.

Original genetic lineage[b]	Genetic lineage according to virulence[a] — Number of isolates (% classified into correct lineage)						
	ALL-7	ALL-10	ALL-13	SRL-2	SRL-4	SRL-5	SRL-6
ALL-7	13(100)						
ALL-10		4(100)					
ALL-13			4(100)				
SRL-2				7(100)			
SRL-4					13(93)	1	
SRL-5			2			15(88)	
SRL-6		1			1	1	44(94)

[a] Discriminant analysis using 23 rice cultivars selected by the backward and stepwise analysis giving a discriminant model where the probability of classifying an isolate into its correct lineage is 100% in most cases.
[b] Genetic lineage classification according to Levy *et al.* (1993 and unpublished) using MGR-586.

A discriminant analysis (Cooley and Lohnes, 1971; SAS, 1989) of the virulence of blast isolates on the 23 rice cultivars selected by the backward and stepwise analysis yielded a very tight correspondence between these groups and those determined solely by MGR fingerprint pattern. The selected cultivars give a discriminant model highly similar to the clustering by genetic lineages (Table 13.10), where the probability of classifying an isolate into its correct lineage is 100% in most cases. All isolates of the ALL-7, ALL-10, ALL-13, and SRL-2 lineages were grouped correctly into their lineages. Only six isolates are mistaken in different genetic groups (Table 13.10). Three of them belong to the most common genetic lineage found in Colombia, SRL-6. Thus there appears to be an important relationship between virulence characteristics among isolates within a lineage.

The groupings generated using the 23 cultivars in Table 13.9 are essentially determined by those cultivar-lineage interactions which are fully incompatible. Thus, when the capacity of a cultivar to completely exclude a lineage is considered, the relationship between MGR-lineage and virulence becomes clear. This suggests that perhaps the most important feature of lineage virulence spectrum is its limits, rather than the variation within.

The effective exploitation of virulence spectrum of lineages for breeding purposes then would be to combine those resistance genes in combinations which are effective against (form the limits to the spectic of) all lineages within a target population. This concept is discussed in more detail in Chapter 16 of this volume.

Conclusions

Achieving the stability of resistance in a commercial cultivar such as Oryzica Llanos 5 has required an intensive crossing, evaluation, and selection scheme under field conditions. A strategy combining pathotyping, MGR-fingerprinting and molecular marker-assisted gene tagging is providing a way to define the genetic organization and distribution of rice blast pathogen diversity and to incorporate appropriate resistance genes. Focusing on resistance genes to lineages of the pathogen, rather than on genes resistant to specific pathotypes, may allow the identification of suitable resistance sources to include in a breeding program for developing stable blast resistance.

Selecting high levels of resistance from populations with diverse combinations of resistance in the presence of a diverse and abundant pathogen population is a key to develop stable blast resistance. In this sense, then, resistance stability and durability are not only functions of host genotype, but also depend upon the dynamics of pathogenicity and virulence within the pathogen population. Understanding the components of stable resistance will require then an understanding of the mechanisms of pathogenic variability in *P. grisea* (Robinson, 1969; Crill and Khush, 1982; Correa-Victoria and Zeigler, 1993a,b). Development of a system for differentiating host resistance genes and pathogenic variability will be essential to develop feasible strategies to control blast with genetic resistance. Ultimately, from ecological, epidemiological, and breeding perspectives it must be understood how genetic diversity is maintained in the blast pathogen and how new pathotypes arise.

References

Bonman, J.M., Vergel de Dios, T.I. and Khin, M.M. (1986) Physiologic specialization of *Pyricularia oryzae* in the Philippines. *Plant Disease* 70, 767–769.

Buddenhagen I.W. (1983) Disease resistance in rice. In: Lamberti, F., Waller, J.M. and Van der Graaff, N.A. (eds), *Durable Resistance in Crops*. Plenum Press, New York, pp. 401–428.

Cooley, W.W. and Lohnes P.R. (1971) *Multivariate Data Analysis*. John Wiley and Sons, New York.

Correa-Victoria, F.J. and Zeigler, R.S. (1993a) Pathogenic variability in *Pyricularia grisea* at a rice blast 'hot spot' site. *Plant Disease* 77, 1029–1034.

Correa-Victoria, F.J. and Zeigler, R.S. (1993b) Field breeding for durable rice blast resistance in the presence of diverse pathogen populations. In: Jacobs, Th. and Parlevliet, J.E. (eds), *Durability of Disease Resistance*. Kluwer Academic Publishers, The Netherlands, pp. 215–218.

Crill, J.P. and Khush, G.S. (1982) Effective and stable control of rice blast with monogenic resistance. In: *Evolution of the Gene Rotation Concept for Rice Blast Control*. International Rice Research Institute, PO Box 933, Manila, The Philippines, pp. 87–102.

Dobinson, K.F., Harris, R.E. and Hamer, J.E. (1993) Grasshopper, a long terminal repeat (LTR) retroelement in the phytopathogenic fungus, *Magnaporthe grisea. Molecular Plant Pathogenic Interactions* 6, 114–126.

Glaszmann, J.C. (1987) Isozymes and classification of Asian rice varieties. *Theoretical Applied Genetics* 74, 51–66.

Hamer, J.E., Farral, L.L., Orbach, M.J., Valent, B. and Chumley, F.G. (1989) Host species-specific conservation of a family of repeated DNA sequences in the genome of a fungal plant pathogen. *Proceedings of the National Academy of Sciences of the United States of America* 86, 9981–9985.

Johnson, R.A. and Wichern, D.W. (1982) *Applied Multivariate Statistical Analysis*. Prentice-Hall, Englewood Cliffs, New Jersey.

Kiyosawa, S. (1972) Genetics of blast resistance. In: *Rice Breeding*. International Rice Research Institute, PO Box 933, Manila, The Philippines, pp. 203–225.

Kiyosawa, S. (1981) Gene analysis for blast resistance. *Oryza* 18, 196–203.

Klecka, W.R. (1980) Discriminant analysis. *Paper Series on Quantitative Applications in the Social Sciences*. Saga University, Beverly Hills, California.

Latterell, F.M. (1971) Phenotypic stability of pathogenic race of *Pyricularia oryzae* and its implications for breeding of blast resistant rice varieties. In: *Proceedings of the Seminar on Horizontal Resistance to the Blast Disease of Rice*. Centro Internacional de Agricultura Tropical, A.A. 6713, Cali, Colombia, pp. 199–234.

Levy, M. Romao, J., Marchetti, M.A. and Hamer, J.E. (1991) DNA fingerprinting with a dispersed repeated sequence resolves pathotype diversity in the rice blast fungus. *Plant Cell* 3, 95–102.

Levy, M., Correa-Victoria, F.J., Zeigler, R.S., Xu, S. and Hamer, J.E. (1993) Genetic diversity of the rice blast fungus in a disease nursery in Colombia. *Phytopathology* 83, 1427–1433.

Ling, K.C. and Ou, S.H. (1969) Standardization of the international race numbers of *Pyricularia oryzae*. *Phytopathology* 59, 339–342.

Ou, S.H. (1985) *Rice Diseases*, 2nd edn. Commonwealth Mycological Institute, Kew, UK, 380 pp.

Ou, S.H. and Ayad, M.R. (1968) Pathogenic races of *Pyricularia oryzae* originating from single lesions and monoconidial cultures. *Phytopathology* 58, 179–182.

Ou, S.H., Nuque, F.L., Ebron, T.T. and Awoderu, U. (1970) Pathogenic races of *Pyricularia oryzae* derived from monoconidial cultures. *Plant Disease Reporter* 54, 1045–1049.

Robinson, R.A. (1969) Disease resistance terminology. *Review Applied Mycology* 48, 593–606.

Rosero, M.J. (1979) Breeding for blast resistance at CIAT. In: *Proceedings of the Rice Blast Workshop*. International Rice Research Institute, PO Box 933, Manila, The Philippines, pp. 63–67.

SAS Institute Inc. (1989) *SAS/STAT Users Guide*, Version 6. Volume 1, 4th edn. Cay, North Carolina.

Yamada, M.S., Kiyosawa, S., Yamaguchi, T., Hirano, T., Kobayashi, T., Kushibuchi, K. and Watanabe, S. (1976) Proposal of a new method for differentiating races of *Pyricularia oryzae* Cavara in Japan. *Annals of the Phytopathological Society of Japan* 42, 216–219.

Weeraratne H., Martinez, C. and Jennings, P.R. (1981) Genetic strategies in breeding for resistance to rice blast *Pyricularia oryzae* in Colombia. In: *Proceedings of the Symposium on Rice Resistance to Blast*. IRAT, Montpellier, France, pp. 305–311.

14 Race-Specific and Rate-Reducing Resistance to Rice Blast in US Rice Cultivars

M.A. Marchetti

USDA-ARS, Texas Agricultural Experiment Station, Rt. 7, Box 999, Beaumont, TX 77713-8530, USA

Introduction

The terms rate-reducing resistance and race-specific resistance are just one of many pairs of terms generated by plant pathologists to refine or in some way qualify the meanings of horizontal and vertical resistance *sensu* Vanderplank (1963). Others include general and specific resistance, dilatory and discriminatory resistance, partial and complete resistance, field and true resistance. Although no two pairs have identical definitions they share common features. In general, race-specific resistance (and its approximate synonyms) is characterized as being qualitative, frequently a hypersensitive response to infection, simply inherited, relatively insensitive to environmental conditions, and effective against some but not all pathotypes of a pathogen. Conversely, rate-reducing resistance is characterized as quantitative, reducing the rate of disease development, polygenic, sensitive to environment and at least partially effective against all pathotypes of a pathogen. In the case of rice blast disease, we know that these terms are not mutually exclusive, as will be discussed later. In this chapter I use some of the synonyms interchangeably with rate-reducing and race-specific when I feel one of the synonyms is more descriptive of the trait being discussed.

A Brief History of Blast Resistance Breeding in the US

Rice blast, caused by the fungus *Pyricularia grisea*, has been present in the southern US since the beginning of the US rice industry (Metcalf, 1906). Although severe outbreaks have been reported, usually on 'new' or 'rested' rice land, the disease has been sporadic in importance in the US (Atkins *et al.*, 1965). Pathogenic specialization was first recognized in the US in 1953 by F.M. Latterell (Latterell *et al.*, 1960). Atkins and Johnston (1965) identified simply inherited resistances in 'Zenith' and 'Northrose' (and 'Nato') to US races

6 (ID-13 [Ling and Ou 1969; Latterell, 1975]) and 1(IB-54), respectively, in 1963.

Although improved blast resistance was noted in such varieties as 'Nova' (Johnston *et al.*, 1965) and 'Saturn' (Jodon, 1965) (two US medium grain cultivars released in 1964), the cultivar 'Dawn' was the first US cultivar developed with improved blast resistance as the major goal (Bollich *et al.*, 1968). Its most effective resistance (*Pi-*) gene ($Pi\text{-}k^h$) was derived from an introduction from Punjab, CI 5309, which was also one of the cultivars used by Latterell *et al.* (1965) to differentiate US pathotypes before the establishment of the International Differentials (Atkins *et al.*, 1965). Kiyosawa (1974) identified not only the $Pi\text{-}k^h$ gene, but also the *Pi-i* and *Pi-a* genes in Dawn. In a later study, we detected a recessive gene which confers resistance to race IB-1, identified as *pi-d* (Marchetti *et al.*, 1987). These *Pi*-genes were first utilized broadly with the widespread acceptance of the cultivar 'Labelle' after its release in 1972 (Bollich *et al.*, 1973). It became the most widely grown cultivar in southern US during 1975–1983, until it was replaced by the newly developed US semidwarf cultivars 'Lemont' and 'Gulfmont' which possess these same *Pi*-genes (Bolich *et al.*, 1985, 1990). In Arkansas, the Dawn genes were incorporated into the cultivar 'Newbonnet' (Johnston *et al.*, 1984). It became the most widely grown cultivar in Arkansas and parts of Louisiana and Mississippi in the 1980s.

The concept of horizontal resistance (*sensu* Vanderplank, 1963) was not given serious consideration in the US rice breeding programs until 1977, although years earlier certain cultivars were recognized as being 'challenge' varieties, i.e. difficult to infect under field conditions, and therefore avoided as subjects for early epidemiological studies and eliminated as candidates for inclusion as blast differentials (Latterell *et al.*, 1965). Owing to the generally unfavorable weather for blast and the apparent effectiveness of race-specific resistance, disease research efforts were channeled elsewhere. Our interest in horizontal resistance was stimulated by discussions held at the Rice Blast Workshop at the International Rice Research Institute (IRRI) in 1977 (IRRI, 1979) and by reports of the work of Bidaux (1976) on breeding for horizontal resistance in Ivory Coast, West Africa. We developed our so-called miniplot system for evaluating rate-reducing resistance based on Bidaux's methods (Marchetti, 1983).

Emergence of Pathotypes Driven by Pathodemes

We frequently hear of newly developed blast resistant rice cultivars being overcome by 'new races' of the blast pathogen, inferring that the new pathotype did not exist before the new cultivar was released. The controversy regarding the pathogenic stability and variability in *P. grisea* (Latterell, 1975; Ou, 1975) has been ongoing for decades, but its proper resolution is of little consequence in the realm of blast-resistance breeding. The fact is that there are many pathotypes, and given the numbers of propagules produced by the pathogen in a field, it is likely that thousands of variants arise by spontaneous mutation. Most of them offer no advantage in fitness and consequently we never see them. A new pathodeme (*sensu* Robinson, 1976), however, provides an additional selection

Table 14.1. A chronology of the release of US rice cultivars and the pathotypes of *Pyricularia grisea* that emerged or re-emerged into prominence with each in the southern US.

Cultivar	Year of release	*P. grisea* pathotype(s)
Nato	1956	ID-13, IG-1
Gulfrose	1960	IB-54, IB-49
Saturn	1964	IB-45, ID-13[a]
Nova 66	1966	IH-1
Labelle	1972	IC-17
Mars	1977	IB-1
Newbonnet	1983	IC-17, IB-49
Katy	1989	IB-33[b], IC-17[a]

[a] Rottenenck only.
[b] Known to occur, but not yet collected from Katy in commercial fields.

pressure favoring those pathotypes that are compatible with it. Pathotypes which can infect the new pathodeme increase in prevalence to the point that they are detectable. Thus, we have a 'new' race. No cultivar has been released in the US which was not susceptible to one or more known pathotypes. But pathotype prevalence has changed, and has been driven by pathodeme (Table 14.1). It is self-evident that one will not collect isolates of a pathotype from cultivars with which it is incompatible. But what we have also experienced is that even among the range of compatible pathotypes, only one or perhaps two will prevail on a particular cultivar. For example, although the cultivar 'Brazos' (Bollich *et al.*, 1975) is susceptible to all but one US pathotype, only races IG-1 and IH-1 were isolated from it in field collections (Marchetti *et al.*, 1976).

The discovery of the existence of pathogenic specialization resulted in extensive collections of diseased specimens in order to ascertain the numbers of races present. Almost every race found to date in the US was characterized by the early 1960s (Atkins, 1965), though some only rarely. Races IB-54, ID-13, IG-1 and IH-1 were the most common; all the cultivars grown in the 1950s and 60s were compatible with at least one of those pathotypes. In 1964 the cultivar Saturn was released (Jodon, 1965), and Dawn was released in 1965 (Bollich *et al.*, 1968). Both of these cultivars were resistant to all the above pathotypes, but susceptible to other known pathotypes: Saturn to IB-45 and Dawn to IC-17. Saturn became a popular cultivar, and though Dawn itself was never grown widely, its blast resistance was incorporated into most of the long-grain cultivars developed in Texas as well as some released in recent years from Arkansas and Louisiana. Consequently, the races prevalent in the 1960s disappeared from commercial fields, replaced by races IB-45 and IC-17. In 1976, 'Mars' was released in Arkansas (Johnston *et al.*, 1979), and has replaced Saturn to become the prevalent medium grain grown in the southern US. With Mars came field resistance to race IB-45, but susceptibility to race IB-1.

Race IB-49 has been collected sporadically in the southern US since the late 1950s. It is pathogenic to all US cultivars except one, 'Katy' from Arkansas

(Moldenhauer *et al.*, 1990). Race IB-49 became common in Arkansas and Louisiana during the mid-1980s particularly on the widely grown cultivar Newbonnet. It was found only rarely in Texas, where almost no Newbonnet was grown. Newbonnet is also susceptible to IC-17, and the two races have been collected on Newbonnet with about equal frequency.

Katy, with its *Pi-ta2* gene for blast resistance from the Indochinese cultivar 'Tetep', was released in 1989 as a replacement for Newbonnet in blast-prone areas. It should be pointed out, however, that in spite of the relatively severe blast in 1986 in Arkansas with statewide losses estimated at about 10%, the average per hectare yield was among the highest on record for the State.

Race IE-1 has become more common since 1991. It is similar in virulence to races IB-49 and IC-17 on US long-grain cultivars, but lacks virulence on 'Zenith' and 'NP-125', differentials which separate these three pathotypes from each other. One might speculate that the emergence of IE-1 is an example of improved fitness through the loss of unnecessary virulence (Vanderplank, 1975). DNA-fingerprinting indicates IE-1 was derived from or shares ancestry with IB-49 (M. Levy, Purdue University, 1993, personal communication).

Race-Specific (Complete) Resistance

For the first 25 years following the discovery of pathogenic specialization in the blast fungus, US rice breeders relied mainly on simply-inherited race-specific resistance genes to combat blast. There are several factors which may have contributed to the overall success of this strategy without experiencing the 'boom-and-bust' cycles frequently associated with vertical resistance. Usually our climate is not conducive to epidemic development. In Texas, the 1991–1992 seasons, with their many showers (as opposed to torrential downpours), were the exceptions that proved the rule. During those seasons we saw more leaf blast than ever before; usually we see none. We have a discrete growing season, and usually several hard freezes during the winter, although the past several winters have been warmer than usual with no hard freezes. Rice is grown in rotation with other crops and pasture, and in Texas, for example, less than 10% of the acreage in the rice-growing area is actually planted to rice in any one year. Virtually all US rice is grown under flood. And finally, many US cultivars have considerable levels of partial resistance.

Table 14.2 summarizes some of the *Pi*-genes found in representative US cultivars. Most of the genes were identified by Kiyosawa (1967, 1969, 1971) in cultivars which were used as parents of our more recent cultivars. In studies here, we identified several recessive resistance genes including the *pi-d* gene found linked to the *Pi-k*^{h} gene in Dawn (Marchetti *et al.*, 1987).

Because of rigid quality requirements imposed by the US rice industry, and the adaptability of local cultivars to US cultural conditions, practices and mechanization, there is justifiable reluctance to resort to exotic germplasm for disease resistance. Breeders try to avoid reintroducing undesirable agronomic and quality traits which they have spent decades eliminating from parental material. In the last 30 years only the *Pi-k*^{h} and *Pi-ta2* genes have been introduced into

Table 14.2. Some *Pi*-genes conferring race-specific resistance to pathotypes of *Pyricularia grisea* in representative US rice cultivars.

Gene	Effective against US pathotypes	Cultivars
Pi-i	IH-1	Bluebelle, Rexmont
Pi-k^s	IB-54	Brazos, M-201, A-301
Pi-k^h	all except IB-1, IB-33, IB-49, IC-17	Newbonnet, Skybonnet, Lemont, Lacassine, Gulfmont, Cypress
Pi-ta2	all except IB-33	Katy
Pi-z	all except IB-races	Mars, Bengal, Orion
pi-d	IB-1	Skybonnet, Lemont, Gulfmont

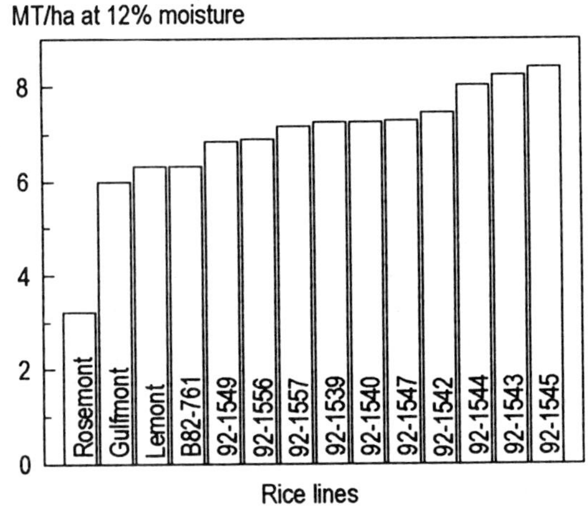

Fig. 14.1. Yields of blast resistant F_7 lines from the cross B82-761/'Rosemont' compared to those of the parents and two major US cultivars, 'Gulfmont' and 'Lemont'. B82-761 is from the cross 'Lebonnet'/'Vista' and combines the *Pi-z* and *Pi-k^h* blast resistance genes. Rosemont is a very high yielding semidwarf in the absence of blast. In this test, Rosemont sustained about 80% panicle blast. Beaumont, Texas, 1992.

US rice cultivars. Already present were the *Pi-z, Pi-i, Pi-a, and Pi-k^s* genes. The most effective of these against US pathotypes of *P. grisea* are *Pi-ta2*, *Pi-k^h* and *Pi-z*. These are being combined in breeding lines in various combinations into high yielding semidwarf long-grain lines with promising results, as illustrated in Fig. 14.1. The line B82-761 carries the *Pi-k^h* and *Pi-z* resistance genes from 'Lebonnet' (Bollich *et al.*, 1975) and 'Vista' (Jodon *et al.*, 1973), respectively. Katy and Lemont have been crossed to put the *Pi-ta2* gene in a semidwarf plant type and combine it with the *Pi-k^h* and other resistance genes in Lemont. From here it should not be difficult to combine all three resistance genes in high-yielding semidwarf lines with typical US long-grain quality.

Table 14.3. Mean percentage of rice blast diseased leaf area (DLA) among US cultivars and parental lines with no major-gene resistance to race IG-1 of *Pyricularia oryzae*, rated 48 days after seedling emergence in blast nursery miniplots at Beaumont, Texas, in 1978.

Variety	Symbol	Cross[a]	DLA[b] (%)
Nova	NOVA	LACR//ZNTH/NIRA	3 a
Bluebonnet 50	BB50	*RXOR*/FRTA[c]	4 ab
Bluebelle	BBLE	9214//*CP231*/9122	6 ab
Belle Patna	BLPT	9122/*RXOR*	7 ab
PI331581	3-81	*BBLE**6/TN-1[c]	12 abc
Nortai	NTAI	*NROS/T487*	13 abc
CI9879	9879	*9545/NROS*	18 bc
Tainan-iku 487	T487	Taiwan Introd.	22 bc
Starbonnet	STBN	*CP231*/BBNT	27 cd
Magnolia	MGNL	IMBR/FRTA[c]	27 cd
Texas Patna 49	TP49	TXPT//*RXOR*/7689	32 cde
Brazos	BRAZ	*9545/NOVA*	32 cde
Rexoro	RXOR	Philippines Introd.	33 def
Northrose	NROS	LACR/AROS	50 def
CI9545	9545	9214/*T487*	53 ef
Nato	NATO	*RXOR*/PRLF//*MGNL*	57 f
Century Patna 231	CP31	TXPT//*RXOR*/SBLR	57 f
PI325893	3-93	PETA/*T487*	73 g
Calrose 76	CR76	Irrad. CLDY*2/CALO	98 h

[a] Italicized ancestors in pedigree were among lines rated for dilatory resistance in this experiment.
[b] Means of three replications. Analysis of variance and Duncan's multiple range test performed after arc sine-square root transformation of the data. Values followed by the same letter are not significantly different ($P = 0.05$).
[c] Cultivars Fortuna (FRTA) and Taichung Native 1 (TN-1) originated in Taiwan.

Rate-Reducing (Partial) Resistance

Partial blast resistance was not given serious consideration in the US until 1977, partly because of personal skepticism and partly because major gene resistance was easier to manipulate and seemed to be filling our needs in the US. This was also about the time that a 0–9 disease rating scale similar to the IRRI scale (IRRI, 1975) was adopted, allowing for further refinement over the previous 0–5 scale (Latterell *et al.*, 1965). The new scale effectively increased fivefold the number of gradations between susceptible (4) and killed (5) since 0–3 on both scales were the same. Because we routinely tested large numbers of cultivars many times across seasons, we were able to recognize that quantitative differences in severity, e.g. the difference between a '5' and a '7' rating, were real and not merely artifacts of environmental influence on suscept-pathogen interactions. We recognized that some cultivars were difficult to infect except under the most favorable conditions

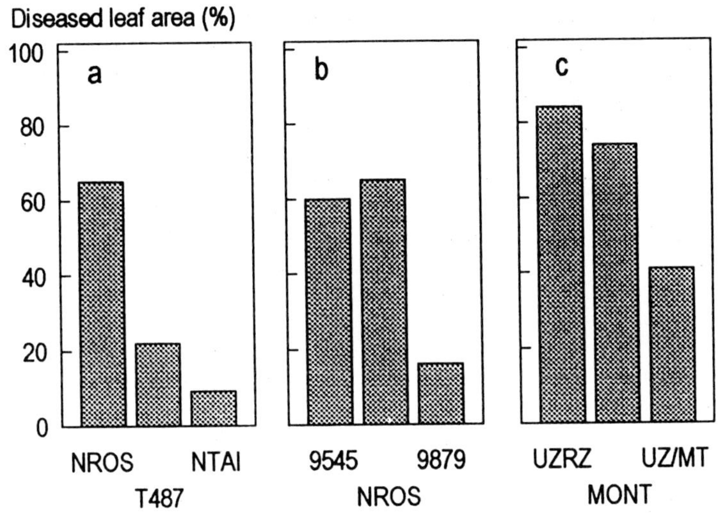

Fig. 14.2. Transgressive segregation for improved partial resistance to rice blast. Diseased leaf areas (DLAs) recorded in an outdoor blast nursery (Marchetti, 1983) when susceptible check M-101 showed 100% DLA. (a) NTAI is a selection from Northrose/Tainan-iku 487; (b) CI 9879 is from CI 9545/Northrose; (c) Uz/Mt is from Uzroz/Monticelli; all three were grown in the former Yugoslavia. NTAI = Nortai, NROS = Northrose, UZRZ = Uzroz, MONT = Monticelli, UZ/MT = Uzroz/Monticelli.

(Latterell *et al.*, 1965; Latterell, 1975). Discussions at the 1977 Rice Blast Workshop at IRRI (1979), knowledge of Bidaux's work (1976) in the Ivory Coast, and the above observations stimulated us to investigate partial resistance in US cultivars.

The first experiments showed that many of the ancestors of current cultivars had considerable levels of partial resistance, as evidenced by slower disease development in polycyclic tests in the blast nursery (Table 14.3, from Marchetti, 1983). Another experiment demonstrated that partial resistance could be selected for even in the F_3 generation, indicating oligogenic (as opposed to polygenic) control, at least in certain crosses (Marchetti, 1983). We also found several cases of transgressive segregation for superior partial resistance, in which cultivars were more resistant than either of their parents, as well as several examples of race-specific partial resistance (Figs 14.2 and 14.3, from Marchetti and Bonman, 1989).

Excellent work has been and is being done on quantifying the various components of partial resistance, most recently by Roumen (this volume). It is important to know the modes of inheritance and relative impact of the various components, e.g. latent period, infection efficiency, spores per lesion, in order to breed for improved partial resistance more intelligently, if not more efficiently. We recognized that the cultivar 'Nortai' exhibited a so-called 'few lesions' trait (Marchetti, 1983). Infections resulted in typical large susceptible lesions, but relatively few per plant even under favorable blast nursery conditions. With this one small experiment, it became apparent that we could not afford the resources required to analyze the components of partial resistance and still fulfill our

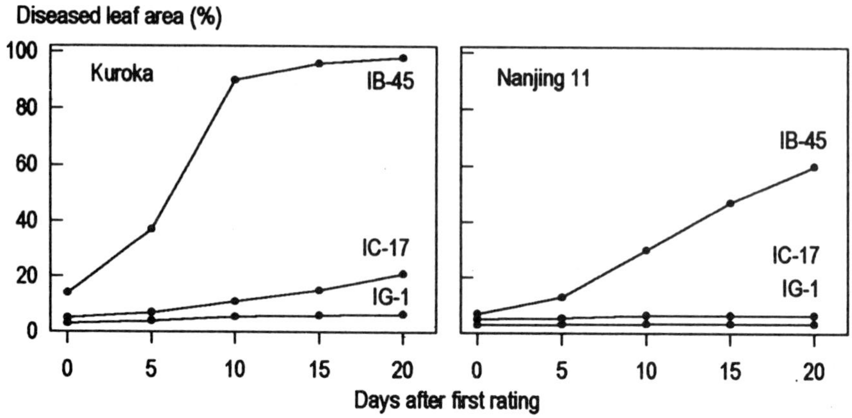

Fig. 14.3. Two blast-resistant rice varieties with race-specific partial resistance, indicated by the differences between disease progress curves of rice infected with race IB-45 and rice infected with races IC-17 or IG-1 of *Pyricularia oryzae*. Kuroka is an upland variety from Japan; Nanjing 11 is a very high-yielding semidwarf variety from China. Tests were conducted in an outdoor blast nursery in Beaumont, Texas, one race tested per year, 1981–84. Intervals from seedling emergence to initial disease rating varied by year from 13 to 19 days. Reactions of race-nonspecific partially resistant and susceptible varieties were consistent across three tests.

mission in varietal improvement. Therefore, we took an integrated 'brown-to-green ratio' approach to evaluating partial resistance, not determining the size and number of lesions, but rather estimating the ratio of diseased to healthy plant tissue (Marchetti and Lai, 1986).

We became convinced that partial resistance was a major factor in the management of blast in the southern US, even though we had not bred for it intentionally prior to 1978. Screening for blast resistance had not become routine in our rice breeding programs until the mid-1960s. Prior to that, breeders and pathologists took a 'holistic' approach to varietal improvement. The main selection criteria were healthy appearance, yield, and grain quality. If any disease, blast included, adversely affected any of these traits in breeding lines, chances were that those lines were eliminated in favor of the less affected. From the mid-1960s until 1978, emphasis was on saving only the most blast-resistant lines. With the recognition of partial resistance came refinements in strategy, one being the elimination of only the most susceptible breeding lines rather than the selection of only the most resistant. Thus, we preserved lines whose partial resistance was not masked by race-specific complete resistance and were less likely to lose partial resistance whose absence was masked by complete resistance, the so-called 'Vertifolia effect' (Vanderplank, 1963).

Early-generation breeding lines are evaluated in single-row plots in the blast nursery, with every third row planted to a highly susceptible spreader. Prior to planting breeding material, several broad-spectrum pathotypes are established in the windward nursery bed by spread from greenhouse-inoculated flats of susceptible seedlings. The breeding lines are rated once, at the time that the spreader rows are killed by leaf blast. These data are considered by the breeders, along

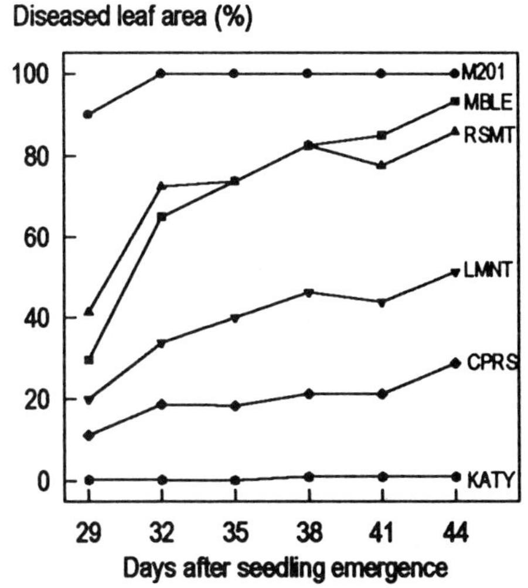

Fig. 14.4. Leaf blast disease development curves of representative southern US long-grain cultivars compared to that of the California cultivar M-201. 'Katy' has complete resistance to the four pathotypes present in the blast nursery (IB-1, IB-45, IB-49 and IC-17). 'Maybelle' and 'Rosemont' have no effective major (*Pi-*) gene resistance. 'Lemont' and 'Cypress' have the *pi-d* and *Pi-k*^{h} genes, which confer resistance to IB-1 and IB-45. Points represent mean values of four replications. Beaumont, Texas, 1992.

with other disease and agronomic data, in line selection for advancement. As promising lines advance they are observed repeatedly in additional and more rigorous disease resistance evaluations. Lines selected for potential release as cultivars are entered in replicated partial resistance trials in which serial ratings are recorded in order to plot disease progress curves (Fig. 14.4) and areas under disease progress curves (AUDPC) (Table 14.4). Sometimes the disease progress curves themselves are more revealing than the AUDPC or the final percent diseased leaf area (% DLA) (Marchetti and Lai, 1986). Figure 14.5 illustrates three disease progress curves and their respective AUDPCs and % DLAs. Although curves *a* and *b* have the same final % DLA and curve *a* has the lower AUDPC, in this instance one should favor the line producing the *b* curve. In all likelihood, the leveling out of curve *b* represents an increase in resistance to infection commonly associated with plant age (Roumen, 1992). Had the test been extended in time, curve *a* probably would have continued to climb, perhaps at a reduced rate, and curve *b* would have flattened or turned downward, indicating plant recovery.

These types of data are contributing to the selection of lines combining improved blast resistance with the yield, quality and plant type traits demanded by the US rice producers and processors.

General Considerations

As we attempt to broaden our genetic base, seek desirable yield and quality traits, and incorporate new sources of disease resistance through hybridization with

Table 14.4. A comparison of leaf blast development in current US rice cultivars under blast nursery conditions (1992) with observations of rotten neck blast in commercial fields (1990–1992). Leaf blast is shown as area under the disease progress curve (AUDPC). The '*', '+', and '–' symbols indicate that blast in the field was as expected, more severe than expected, and less than expected, respectively, based on blast nursery performance.

Variety	AUDPC	LSD (0.5)	Field panicle blast
M-201	99	a	*
RT7015	78	b	*
Rosemont	74	bc	+
Maybelle	74	bc	–
Newbonnet	70	bc	+
Bengal	70	c	–
Lacassine	43	d	*
Lemont	40	d	*
Texmont	40	d	++
Cypress	20	e	*
Orion	12	ef	–
Katy	<1	f	*

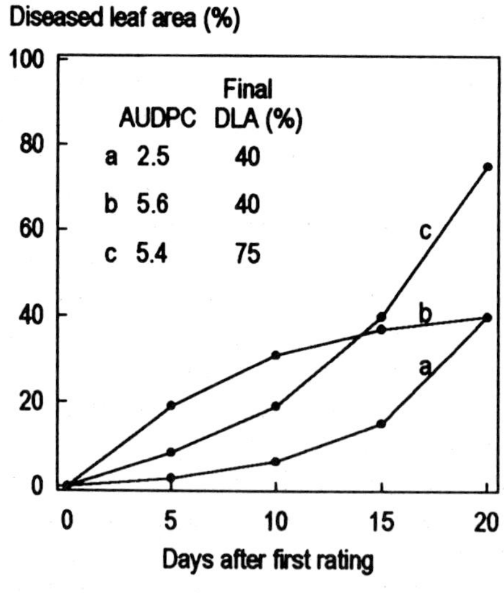

Fig. 14.5. Hypothetical disease progress curves and associated AUDPCs and final % DLAs. In the absence of the curves, the numerical values can be misleading indicators of comparative field resistance.

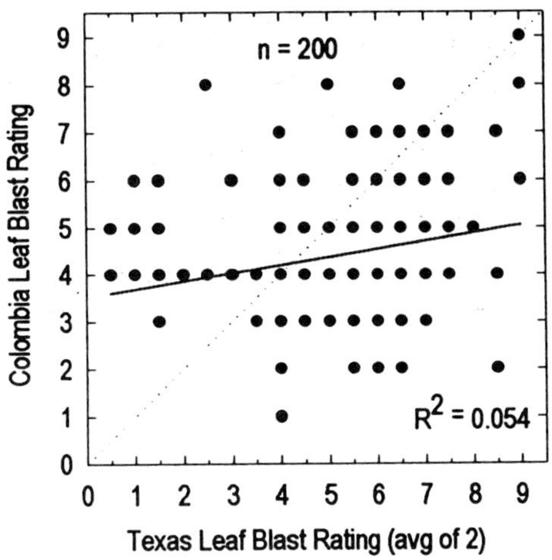

Fig. 14.6. Comparative leaf blast ratings reported from blast nurseries in Santa Rosa, Colombia, and Beaumont, Texas, USA, on the 1992 US Uniform Regional Nursery consisting of advanced breeding lines and commercial cultivars from the Arkansas, Louisiana, Mississippi, and Texas rice breeding programs, 1992.

exotic germplasm (as was the case with Katy), we must be aware of the possibility of losing partial resistance during the selection process. So far we have compatible pathotypes for every *Pi*-gene found in US cultivars, which gives us the ability to assess partial resistance directly. But even in the absence of a compatible pathotype we can get a fair indication of partial resistance by observing the array of susceptible types among progeny lines, particularly when hybridization involves a line with poor partial resistance. If progeny lines are either highly resistant (major gene present) or highly susceptible (major gene absent), then the prospect of finding useful partial resistance is remote. If, on the other hand, one observes many lines with moderate susceptibility (major gene absent), then it will be possible to supplement the major-gene resistance with partial resistance and thereby reduce the probability of a serious blast epidemic, should the major-gene resistance be overcome by 'new' pathotypes.

Partial resistance is known to be environmentally sensitive (Marchetti and Lai, 1986); it is important that evaluation for partial resistance be conducted near the areas where they are to be grown. In addition to plant adaptedness, one must also consider the pathogen population and the chance that what appears to be effective partial resistance is also race-specific (Fig. 14.3). Toriyama reported that traditional Japanese upland cultivars with durable partial resistance in Japan fared poorly when tested in Latin America (in Marchetti and Bonman, 1989). In 1992, 200 US rice lines were evaluated for blast resistance in Santa Rosa, Colombia, a blast 'hot spot' used by Dr Fernando Correa-Victoria and his colleagues at CIAT (1993, personal communication). Figure 14.6 illustrates that selections for partial resistance would vary considerably depending upon whether the selections were made at Santa Rosa or at Beaumont, Texas. The greater the correlation between ratings at Santa Rosa and Beaumont, the more the points would have clustered along the diagonal. In fact, many of the crosses with Katy which showed good partial resistance (2–4 ratings) in Texas were mediocre to poor (4–7) in Colombia. A question to be answered is whether the difference in partial resistance is due to differences in pathotype population or in the adaptedness of southern US rices to the Colombian environment, or both. Regardless of the reasons, the fact remains that partial resistance can be location-specific.

A weakness in the US system for blast evaluation is that we screen seedlings under upland conditions, in order to ensure that conditions favor blast development and to permit screening large numbers of lines. All of our commercial production, however, is grown under flood-irrigation, and the crop usually is in the mid- to late-tillering stage before the weather begins to favor blast development. The effect of plant age on susceptibility is, therefore, not considered. In addition, panicle blast (rottenneck) – the most common manifestation of the disease – is not assessed. As demonstrated by Kim (1986), rice plant nutrition is affected profoundly by flooding, effecting changes in the color and hardness of the plants, and also increasing resistance to blast. Generally, most researchers have reported high correlations between leaf and panicle susceptibility, but exceptions also have been noted (Latterell, 1975). In general, we believe that our upland blast nursery with sprinklers to extend leaf-wetness periods provides a fair estimate of relative resistance among rice lines, but we cannot account for potential differential effects of flood irrigation on partial resistance. Kim (1986) showed the effect of flooding on resistance was greater on 'Brazos' than on 'M-101', even though flooding improved resistance in both cultivars. We have experienced inconsistencies in both directions between blast nursery ratings and actual field experience (Table 14.4).

In summary, we have effective major (*Pi-*) resistance genes available in lines adapted to the southern US, and we have significant levels of partial resistance which were accumulated in ancestral lines before there was a conscious effort to screen for it. We propose to utilize both in our efforts to improve blast resistance in high-yielding superior-quality semidwarf rices for our part in effective integrated pest management, which includes sound cultural practices and the judicious use of fungicides.

References

Atkins, J.G. (1965) Physiologic races of *Pyricularia oryzae* in the Western Hemisphere. In: *The Rice Blast Disease.* The Johns Hopkins Press, Baltimore, Maryland, p. 243–244.

Atkins, J.G., Bollich, C.N., Johnston, T.H., Jodon, N.E., Beachell, H.M. and Templeton, G.E. (1965) Breeding for blast resistance in the United States. In: *The Rice Blast Disease.* Johns Hopkins Press, Baltimore, Maryland, pp. 333–341.

Atkins, J.G. and Johnston, T.H. (1965) Inheritance in rice of reaction to races 1 and 6 of *Pyricularia oryzae. Phytopathology* 55, 993–995.

Bidaux, J.M. (1976) Phytopathologie du riz. Rapport Annuel 1974–1975. IRAT (Institut de Recherches Agronomiques Tropicales), Cote d'Ivoire (in French with English summary), 75 pp.

Bollich, C.N., Atkins, J.G., Scott, J.E. and Webb, B.D. (1968) Registration of 'Dawn' rice. *Crop Science* 8, 400.

Bollich, C.N., Atkins, J.G., Scott, J.E. and Webb, B.D. (1973) Registration of 'Labelle' rice. *Crop Science* 13, 773–774.

Bollich, C.N., Webb, B.D., Scott, J.E. and Atkins, J.G. (1975) Registration of 'Lebonnet' rice. *Crop Science* 15, 886.

Bollich, C.N., Scott, J.E., Webb, B.D. and Atkins, J.G. (1975) Registration of 'Brazos' rice. *Crop Science* 15, 887.

Bollich, C.N., Webb, B.D., Marchetti, M.A. and Scott, J.E. (1985) Registration of 'Lemont' rice. *Crop Science* 25, 883–885.

Bollich, C.N., Webb, B.D., Marchetti, M.A. and Scott, J.E. (1990) Registration of 'Gulfmont' rice. *Crop Science* 30, 1159–1160.

IRRI (1975) *Standard Evaluation System for Rice*. International Rice Research Institute, PO Box 933, Manila, Philippines, 64 pp.

IRRI (1979) *Proceedings, Rice Blast Workshop, 22–24 February 1977*. International Rice Research Institute, PO Box 933, Manila, Philippines, 222 pp.

Jodon, N.E. (1965) Registration of 'Saturn' rice. *Crop Science* 5, 288.

Jodon, N.E., Sonnier, E.A. and McIlrath, W.O. (1973) Registration of 'Vista' rice. *Crop Science* 13, 774–775.

Johnston, T.H., Kuenzel, K.A., Lee, F.N., Wells, B.R., Henry, S.E. and Dilday, R.H. (1984) Registration of 'Newbonnet' rice. *Crop Science* 24, 209–210.

Johnston, T.H., Templeton, G.E. and Atkins, J.G. (1965) Registration of 'Nova' rice. *Crop Science* 5, 285–286.

Johnston, T.H., Wells, B.R., Marchetti, M.A., Lee, F.N. and Henry, S.E. (1979) Registration of 'Mars' rice. *Crop Science* 19, 743–744.

Kim, C.H. (1986) Effect of water management on the etiology and epidemiology of rice blast caused by *Pyricularia oryzae* Cav. PhD Thesis, Louisiana State University, Louisiana, USA, 170 pp.

Kiyosawa, S. (1967) Inheritance of blast resistance in Zenith type rice varieties. *Ikushugaku Zasshi* 17, 25–35.

Kiyosawa, S. (1969) Inheritance of resistance of rice varieties to a Philippines strain of *Pyricularia oryzae*. *Ikushugaku Zasshi* 19, 61–73.

Kiyosawa, S. (1971) Gene analysis of blast resistance in exotic varieties of rice. *Japanese Agriculture Research Quarterly* 6, 1–8.

Kiyosawa, S. (1974) Inheritance of blast resistance of a US rice variety, Dawn. *Ikushugaku Zasshi* 24, 117–124.

Latterell, F.M. (1975) Phenotypic stability of pathogenic races of *Pyricularia oryzae* and its implications for breeding of blast resistant varieties. In: *Proceedings of a Seminar on Horizontal Resistance to the Blast Disease of Rice*, 8–12 October 1971. Centro Internacional de Agricultura Tropical (CIAT), Cali, Colombia. CE-9, pp. 199–234.

Latterell, F.M., Tullis, E.C. and Collier, J.W. (1960) Physiologic races of *Pyricularia oryzae* Cav. *US Department Agricultural Plant Disease Reporter* 44, 679–683.

Latterell, F.M., Marchetti, M.A. and Grove, B.R. Jr. (1965) Coordination of effort to establish an international system for race identification in *Pyricularia oryzae*. In: *The Rice Blast Disease*. The Johns Hopkins Press, Baltimore, Maryland, pp. 257–274.

Ling, K.C. and Ou, S.H. (1969) Standardization of the international race numbers of *Pyricularia oryzae*. *Phytopathology* 59, 339–342.

Marchetti, M.A. (1983) Dilatory resistance to rice blast in USA rice. *Phytopathology* 73, 645–649.

Marchetti, M.A. and Bonman, J.M. (1989) Rice blast disease management. In: *Rice Farming System: New Directions*. International Rice Research Institute, PO Box 933, Manila, The Philippines, pp. 175–183.

Marchetti, M.A. and Lai, X.H. (1986) Screening techniques to identify slow-blasting rice lines. In: *Progress in Upland Rice Research*. International Rice Research Institute, PO Box 933, Manila, The Philippines, pp. 317–326.

Marchetti, M.A., Rush, M.C. and Hunter, W.E. (1976) Current status of rice blast in the southern United States. *Plant Disease Reporter* 60, 721–725.

Marchetti, M.A., Lai, X.H. and Bollich, C.N. (1987) Inheritance of resistance to *Pyricularia oryzae* in rice cultivars grown in the United States. *Phytopathology* 77, 799–804.

Metcalf, H. (1906) A preliminary report on the blast of rice, with notes on other rice diseases. *South Carolina Agricultural Experimental Station Bulletin*, 121, 43 pp.

Moldenhauer, K.A.K., Lee, F.N., Norman, R.J., Helms, R.S., Wells, B.R., Dilday, R.H., Rohman, P.L. and Marchetti, M.A. (1990) Registration of 'Katy' rice. *Crop Science* 30, 747–748.

Ou, S.H. (1975) Variability of *Pyricularia oryzae* Cav. and its relation to varietal resistance. *Proceedings of a Seminar on Horizontal Resistance to the Blast Disease of Rice*, 8–12 October 1971. Centro Internacional de Agricultura Tropical (CIAT), Cali, Colombia, CE-9, pp. 49–64.

Robinson, R.A. (1976) *Plant Pathosystem. Advanced Series in Agricultural Sciences 3*. Springer-Verlag, Berlin 184 pp.

Roumen, E.C. (1992) Effect of age on components of partial resistance in rice to leaf blast. *Euphytica* 63, 271–279.

Vanderplank, J.E. (1963) *Plant Disease: Epidemics and Control*. Academic Press, New York, 349 pp.

Vanderplank, J.E. (1975) *Principle of Plant Infection*. Academic Press, New York, 216 pp.

15 A Strategy for Accumulating Genes for Partial Resistance to Blast Disease in Rice within a Conventional Breeding Program

E.C. Roumen

Unité de phytopathologie et malherbologie, Centre International en Recherche Agriculture pour le Développement (CIRAD-CA), B.P. 5035, 34032 Montpellier Cedex 1, France

Durable Resistance to Blast is Associated with Partial Resistance

Unlike hypersensitivity resistance (HSR) based on major genes, which often becomes ineffective soon after a cultivar is released, available information indicates that partial resistance (PR) to rice blast disease can be durable. PR has been defined by Parlevliet and van Ommeren (1975) as a special type of incomplete resistance (susceptible infection type) that reduces the epidemics. Several examples exist of rice cultivars with PR that have been grown on a large scale over many years that were not severely attacked by blast, thus fitting the definition of durable resistance by Johnson (1984). Examples include IR36, an improved tropical irrigated cultivar, Moroberekan, a traditional African upland cultivar, and IAC-25, an improved cultivar from Brazil (Bonman and Mackill, 1988).

Although PR is of considerable interest to breeders, efficient selection of PR in breeding programs remains unresolved. Genetic studies indicate that PR to rice blast is usually under oligo- or polygenic control (Higashi and Kushibuchi, 1978; Higashi and Saito, 1985; Notteghem, 1985; Wang *et al.*, 1994). On theoretical grounds, for efficient accumulation of favorable PR alleles, selection should be carried out starting in the early generations of a breeding cycle. However, a routine screening method that can be applied in early generations has yet to be developed. The following sections review some of the problems that occur and provide some potential solutions, based on experimental results obtained at the International Rice Research Institute (IRRI) working with tropical rice cultivars.

Constraints for Improving Partial Resistance to Blast Disease by Field Assessment

Large experimental errors

Assessments of PR to blast in the field are characterized by a large error. The susceptibility of rice plants to blast is greatly influenced by uncontrollable environmental fluctuations, such as small variations in soil moisture, nitrogen levels, etc. (Suzuki, 1935; Ou, 1985). Repeated assessments in time and in place are therefore required to evaluate PR properly. Several experimental layouts have been developed for optimizing field evaluation of PR. In Africa, Notteghem developed the 'Decreasing Inoculum Test for Evaluation of Resistance', called 'DITER' in short (Notteghem and Andriatompo, 1977). Plants were grown under upland conditions. Each test plot (three 2-m-long rows) was surrounded by three rows of a cultivar showing complete HSR. Ten spreader rows, grown at one end of the test plots, consisting of a mixture of susceptible cultivars, were planted 8 days prior to the test plots, and inoculated with the pathogen at the fourth leaf stage. Evaluation of PR was obtained by assessing the rate of disease spread into the test plots from the spreader rows.

In the USA, Marchetti (1983) developed a similar approach intended for screening breeding lines. Test plots consisted of three 60 cm long rows, surrounded by two rows of a completely resistant cultivar. This upland miniplot technique was later applied under tropical conditions by Bonman *et al.* (1989a), who tested a range of improved rice genotypes that were found to vary from highly resistant to highly susceptible. The degree of agreement for genotype ranking among trials with 2–4 four replications each was high (mean rank correlation coefficient of 0.91), but the discriminating power of the trials remained too low to reliably detect small differences in PR among cultivars. Additional plots could improve the discriminating power of the test, but the extra seed required would prevent applying this type of test in early breeding generations. Further increasing the number of trials would require more space and time, limiting the number of lines, and therefore the attractiveness to breeders.

Unreliable PR field assessment due to epistatic effects of HSR

Probably even more important than the problem of large experimental errors is the problem that PR to blast cannot be accurately evaluated if the line or cultivar also expresses HSR (Ezuka, 1972). In the field, especially in breeders plots, the pathogen population is usually made up of a mixture of races. Genes for HSR in cultivars may be effective to all, some, or none of the races, causing the effective inoculum dose for the cultivars to depend on which hypersensitivity gene(s) they contain. Consequently, although PR expresses as a reduced epidemic in the field in spite of a susceptible infection type, there is no guarantee that a cultivar showing a low disease severity and a susceptible infection type in the field thus possesses high PR. Just as likely, such cultivars may possess effective HSR to the great majority of the pathogen races in the field population (Parlevliet, 1983).

In work by Ahn and Ou (1982), the amount of disease on cultivars in small field plots was highly correlated ($r = 0.89$) to the proportion of virulent races in the pathogen population.

In Japan, to reduce the effect of HSR, comparison among cultivars in the field has been restricted to those cultivars that are likely to have the same major gene(s) (Narita and Iwata, 1964, in Ezuka, 1972; Ezuka *et al.*, 1969a, b). Information on major genes present in Japanese cultivars is available from extensive and systematic inoculation studies combined with genetic analysis of the HSR in these cultivars (e.g. Goto and Yamanaka, 1968; Kiyosawa, 1972). But even in Japan, characterization of major resistance genes has not been exhaustive, as is demonstrated by the gene *Pi-sh*, found much later than the other genes in some of the Japanese differential cultivars (Imbe and Matsumoto, 1985).

Systematic inoculation studies combined with genetic analysis of HSR are relatively rare in tropical rice cultivars, but the work that has been done indicates that nearly all tropical cultivars contain major genes (Notteghem, 1989). In many, if not in most cases, these are different from those identified in Japan (Kozaka *et al.*, 1970; Kiyosawa *et al.*, 1986). Moreover, cultivars with more than one gene for HSR appear to be the rule rather than the exception (Tanaka, 1986; Yu *et al.*, 1987; Mackill and Bonman, 1992). Notteghem (1986, 1993) concluded that the situation in tropical rice cultivars is much more complicated than in temperate rice cultivars, and considers effects of HSR in tropical rice to be a major constraint for improving PR.

As long as knowledge of the genes controlling HSR in tropical rice cultivars remains limited, field comparison of PR among groups of tropical rice cultivars with the same major gene(s) can not be made. Obviously, proper assessment of PR in segregating populations that may also segregate for HSR will be all the more difficult.

Because of all the complications that occur, some programs fall back on trying to improve PR by avoiding choosing parents with low PR in their crossing program. The impact of this strategy may be limited, since even progeny resulting from a cross involving a PR resistant parent may turn out to be susceptible (IRAT, 1980).

Constraints for Improving PR by Greenhouse Assessment

To avoid the complications of HSR and reduce the experimental error due to environmental fluctuations, Sakurai and Toriyama (1967) developed a method for assessing the PR of Japanese (*Japonica* type) rice cultivars in a greenhouse. A large number of Japanese cultivars was assessed by Yunoki *et al.* (1970a). Plants were grown until leaf stage six or seven, after which they were spray-inoculated using an isolate virulent to all test cultivars. PR was assessed counting the number of lesions with a differentiated centre, since this trait was found to be reasonably correlated with PR as expressed under field conditions. In the Philippines, Yeh and Bonman (1986) used the same approach for assessing the PR of six *Japonica* × *Indica*, and purely *Indica* cultivars. Again, the PR assessed by counting the number of lesions with differentiated centre that developed,

Table 15.1. Integration of infection type scales for blast into a single numeric scale.

Symptoms	Bonman *et al.* (1986)	Cho and Kiyosawa, (1973)	Jeanguyot (1983)	Integrated scale
No visible symptoms	0	–	1	0
Tiny brown pin-head size lesions	1	–	1	1
Small brown lesions up to a few mm diameter	2	b	2	2
Small roundish lesions with gray center and brown margin, less than 2 mm	3	bg	3	3
Small slightly elongated lesions with gray center and brown margin, less than 2 mm	3+		4	4
Elongated lesions with gray center and brown margin, more than 2 mm	4	bG	5	5
Gray elongated lesions without brown margin, more than 2 mm	5	pG	6	6

corresponded well to the level of PR expressed under field conditions. In most studies done in Japan, the ranking order of the cultivars did not change when different virulent isolates were used for inoculation (Asaga and Yoshimura, 1969, cited in Ezuka, 1972; Yunoki *et al.*, 1970b; Hirano and Matsumoto, 1971; Kozaka, 1975). But in some exceptional cases PR was found to be highly race-specific. Toriyama *et al.* (1968) identified a major gene *Pi-f* that resembles polygenic PR when effective. Highly race-specific PR was also reported by Bonman *et al.* (1989b).

It seems then that inoculation in the greenhouse using a single virulent isolate may be an interesting approach to select for higher levels of PR in tropical rice genotypes. Greenhouse assessment has the advantage that it requires less seed than field assessment, making it more suitable for assessing PR in early generations of a breeding cycle. The monocyclic greenhouse test also eliminates any interplot interference, which can be a problem, especially if the amount of seed is limited and only small field plots can be grown (Notteghem, 1985).

Two practical problems arise. First, there is no clear definition of what constitutes a virulent isolate. The classification of isolates as virulent and avirulent depends on what is considered to be a susceptible infection type (IT). Several scales have been developed (e.g. Yamasaki and Kiyosawa, 1966; Cho and Kiyosawa, 1973; Jeanguyot, 1983; Bonman *et al.*, 1986). All scales consider differences in size, shape and color of lesions, and can be integrated into a single numerical scale (Table 15.1). Lesions in rice leaves can vary even in one leaf inoculated with a single isolate (Goto and Yamanaka, 1968; Kozaka, 1979). In older leaves, lesions tend to be of a more resistant type than in the youngest leaf (Goto *et al.*, 1961; Roumen, 1992a). The size and shape of lesions may vary with environmental conditions. Drought stress and high nitrogen fertilization tend to cause a more susceptible IT, and the expression of the major gene *Pi-sh* appears to be temperature sensitive (T. Imbe, 1993, personal communication). The IT is derived from the total picture of the individual lesions, with symptoms in the youngest leaf being the most important. Obviously, with IT being affected by many factors, declaring virulence or avirulence of an isolate based on a single experiment should be avoided.

Isolates that produce sporulating lesions may not necessarily be virulent. According to Jeanguyot (1983), the presence of small sporulating lesions can be caused by a single resistance gene with a large effect, behaving in a similar fashion to genes for complete HSR. In other host–pathogen systems, such as that between barley and *Puccinia hordei*, similar observations have been made (Parlevliet, 1979). Thus, studies on PR to barley leaf rust by Parlevliet and co-workers were restricted to isolate-barley line combinations that show the most susceptible IT possible. Regarding leaf blast, Yamasaki and Kiyosawa (1966) consider the presence of large spindle-shaped lesions with a gray center and brown margin as the minimum requirement for virulence. At the IRRI, an isolate is considered virulent if the IT is 3+ (Bonman *et al.*, 1986) or higher after spray inoculation. The presence of small roundish sporulating lesions is being considered a typical avirulent reaction. Prabhu (1989), however, appears to consider isolates inducing an intermediate IT as virulent since he links intermediate reactions with PR, and regards PR as 'incomplete or intermediate resistance, without any genetic connotation'. Isolates inducing an intermediate IT may also have been considered as virulent by Villareal *et al.* (1981) and Castaño-Zapata (1981). These authors classified the resistance of their host materials according to the SES scale (IRRI, 1988), which is in fact not intended for assessing IT, but for assessing disease severity.

The second practical problem for greenhouse assessment is that it is difficult to find a single isolate showing virulence to a large number of rice entries. In rice, a considerable number of major genes that can confer HSR seem to exist (Kiyosawa *et al.*, 1986; Kozaka *et al.*, 1970). However, this is more of a problem for screening PR of cultivars than for improving PR in segregating generations of a cross. Most known major genes show dominant inheritance, without epistatic interaction. For assessing and improving PR in segregating generations, using an isolate that possesses virulence to both parents will usually do, since such an isolate is likely to be virulent to all the progeny. For routinely identifying suitable isolates for any specific cross, a stock of isolates with a diverse virulence pattern should be generated, requiring a laboratory for the isolation, storage, and multiplication of pathogen isolates. In addition, inoculation results (IT data) of the parental cultivars of rice-crosses exposed to these isolates should be easily accessible.

A Program Aimed at Improving PR in Segregating Generations of Rice Crosses by Greenhouse Assessment

In 1984, a collaborative project involving IRRI, the Netherlands Ministry of International Cooperation (DGIS), and Wageningen Agricultural University (WAU), was initiated at IRRI to develop a methodology to breed for PR to bacterial leaf blight and rice blast in segregating generations in the greenhouse. In principle, the same approach was followed that was applied successfully earlier at the plant breeding department of WAU for barley and wheat, selecting for higher PR to barley (*Puccinia hordei*) and wheat leaf rust (*Puccinia recondita* pv. *tritici*) (Parlevliet *et al.*, 1980; Parlevliet, 1992; Broers and Jacobs, 1989). Within the collaborative project, experiments on PR to bacterial leaf blight were conducted from 1985 to 1989 (Koch, 1990), and on PR to blast from 1987 to 1991. Pure rice lines were studied first to characterize PR to blast and for finding suitable selection criteria that could be applied to segregating populations.

Infection type

Inoculation tests using a large number of rice cultivars and various blast isolates confirmed that resistance expressed as a reduced IT was likely to be race-specific, as previously reported by Jeanguyot (1983). For example, the cultivar 'Iguape Cateto' only developed small, roundish sporulating lesions with isolate P06-6, while it developed large spindle-shaped lesions, sometimes without margin, with isolate V86025. At the same time, there was no difference in infection type between these two isolates on the cultivars 'Malos' or 'Yamada Bake', and several others. It is thus unlikely that the development of small lesions on 'Iguape Cateto' with isolate P06-6 was caused by a difference in environmental conditions. Reduced ITs are probably the result of an incomplete HSR. Because HSR to the blast pathogen is not durable, rice genotypes developing an intermediate IT were excluded for further studies of PR.

Component analysis of partial resistance

Among cultivars that showed a susceptible IT, a component analysis was carried out for latent period (LP), relative infection efficiency (RIE), and the size of the sporulating area in lesions (LS). Plants were grown and inoculated after the recommendations of Sakurai and Toriyama (1967). Unpuddled soil was kept well drained, and when plants developed the second leaf on the main culm, nitrogen was applied at an equivalent of $5\ g\ m^{-2}$ as an ammonium sulfate solution. The same amount of nitrogen was applied again when the plants reached the fourth leaf stage and again one day before inoculation. Inoculation was carried out inside a plastic cage when most plants had developed six leaves on the main culm. Conidia were obtained from sporulating cultures on agar plates as described by Bonman *et al.* (1986). The conidia were harvested by scraping the plates with a rubber spatula in the presence of a little water, and inoculum was prepared by mixing 200 ml suspension adjusted to 1×10^5 conidia ml^{-1} with 200 ml water containing 1% gelatine. Shortly before sunset, a conidia suspension containing 5×10^4 conidia ml^{-1} was sprayed over the plants as a fine mist. About 13 million conidia were dispersed per m^2 cage floor area. The plants were kept in the cage overnight. Leaf wetness duration was about 16 hours and plants were kept dry afterwards.

Latent period

The latent period (LP) is the time between inoculation and the time that the newly emerging lesions begin sporulation. Under these experimental conditions, the LP was about 4 days. Leaf blast thus behaves as a typical compound interest disease (Zadoks and Schein, 1979), and even small differences in LP might result in differences in PR. A large number of rice cultivars of diverse geographic origin, including traditional and improved, upland and lowland, and *Indica* as well as *Japonica* genotypes were inoculated with several blast isolates to identify differences in LP. Depending on rice genotype and isolate, resistant and/or susceptible lesion types developed. Emerging spots that developed into resistant type lesions were observed to appear a few hours to one day ahead of flecks that developed into sporulating lesions. In Japan, the same phenomenon was noticed by Goto and Yamanaka (1968). Resistant-type lesions emerged as minute black or brownish flecks, while sporulating-type lesions emerged as small, light-green depressed areas in the leaves or as tiny whitish flecks. In some cultivars, both resistant and sporulating lesion types developed. The resistant-type lesions in these cultivars appeared at the same time as those in cultivars that developed resistant-type lesions only. In compatible host–pathogen combinations, many sporulating-type lesions became visible within a few hours without any distinct difference in time between the genotypes. For example, the Brazilian tropical upland cultivar IAC 47 with reasonable PR, the highly susceptible Indian lowland cultivar CO39 and even an AUS-type cultivar like Surjamkuhi, all showed the same LP. The observations suggested that LP is not important as a component of PR to leaf blast.

Sporulation on blast lesions cannot be assessed visually, so the time between appearance of sporulating-type lesions and actual production of spores was investigated in more detail. The presence of spores on lesions was assessed by gently pressing 1 mm thick, 3 mm diameter 2% water agar disks on lesions and examining these with a microscope at 100 × magnification. The agar disks were found to be very efficient in removing spores from lesions without causing any damage. Flecks that develop into sporulating lesions were found to be capable of sporulation only a few hours after appearance (Roumen and De Boef, 1993), indicating that the LP of a cultivar is essentially equivalent to the incubation period.

A detailed assessment of the incubation period of sporulating type lesions in three cultivars (CO39, IR50 and IR64) differing in PR in the field was conducted by growing them at two levels of nitrogen (150 and 250 kg ha^{-1} N equivalent). Since nitrogen is known to markedly influence PR to blast, it should also markedly affect any important component of PR. The measurements confirmed the visual

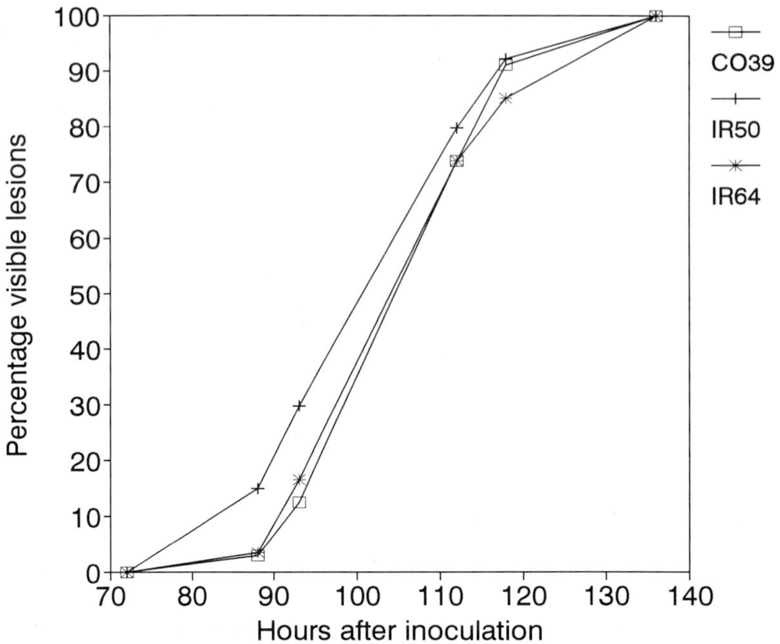

Fig. 15.1. Increase in proportion of visible lesions from the time of inoculation for three rice cultivars inoculated with isolate Po6-6.

impression that most lesions appear within a very short time (Fig. 15.1). The amount of nitrogen did not affect the LP at all. Small differences between cultivars were detected, but these were too small to be of practical use in a breeding program. The LP was longest in IR64 and shortest in IR50. However, if LP is associated to PR, the LP would be expected to be shortest in CO39, by far the most susceptible cultivar. The difference in LP of just 8 hours between the highly susceptible CO39 and partially resistant IR64 suggests LP is not important as a component of PR (Roumen and De Boef, 1993).

In contrast with these results, considerable cultivar differences for LP, up to 5 days between cultivars, have been reported by others (Rodríguez and Gálvez, 1975; Brodny *et al.*, 1988; Castaño *et al.*, 1989). However, the cultivars studied by these authors showed a range of IT, and not all isolates may have been virulent. In the study of Rodríguez and Gálvez (1975), the LP increased when cultivars developed a more resistant IT, but cultivar differences were not very clear when the isolate was virulent. In the study of Castaño *et al.* (1989), pooled across cultivars, the LP also significantly increased with a more resistant classification, but whether significant differences between cultivars within each class were also present was not determined.

Other studies that were restricted to cultivars with a susceptible IT also failed to detect clear cultivar differences for LP. Yeh and Bonman (1986) concluded that the LP is insignificant as a component of PR. Silué *et al.* (1992) did not notice clear genetic variation for incubation period among *Oryza sativa* while finding a 6 hours shorter period in *Oryza glaberrima*.

Table 15.2. Average number of sporulating lesions per plant in leaves of the main culm of four rice cultivars inoculated with isolate Po6-6 of *Magnaporthe grisea* in nine series.

	Cultivar			
Series	CO39	IR50	IR66	IR36
1	17.9 a[a]	4.2 b	–	1.3 c
2	14.9 a	1.6 b	–	1.0 b
3	32.2 a	6.3 b	–	5.0 b
4	22.0 a	8.0 b	6.7 bc	3.8 c
5	46.7 a	10.3 b	10.6 b	6.8 b
7	113.2 a	19.5 b	20.5 b	13.0 b
8	–	20.4 b	17.6 b	9.1 c
9	–	15.8 b	8.4 bc	6.1 c
Weighted means[b]	43.2 k	9.6 l	8.5 l	5.1 m
Relative to CO39	100%	22%	20%	12%

[a] Different letters indicate significantly different cultivar means within series according to the Kruskal–Wallis rank test; multiple comparison at $\alpha = 0.05$.
[b] Different letters indicate significantly different cultivar means across series according to the Wilcoxon signed rank test; pairwise comparison of cultivars at $\alpha = 0.05$.

Relative infection efficiency

The relative infection efficiency (RIE) is assessed as the number of sporulating lesions that develop per plant after exposure to a certain amount of inoculum. In agreement with previous results obtained by others (Yunoki *et al.*, 1970a; Yeh and Bonman, 1986) large differences among cultivars were found for the RIE. The variation in lesion number between the individual plants of the same genotype is usually considerable, the coefficient of variation (CV) among plants typically varying from 80 to 150%, depending on the experiment and the genotype. Standard deviations of populations were found to increase with the mean value. Often, especially when a low number of lesions develop on a cultivar, the number of lesions on the plants does not follow a normal distribution. Consequently, if statistical analysis is performed using the plant values as the experimental unit, applying nonparametric statistical tests is more appropriate to compare rice genotypes than analysis of variance.

Despite the large variation for lesion number from plant to plant, the relative differences of cultivar means were found to be highly repeatable, and the relative differences remained more or less constant at various levels of infection (Table 15.2; Roumen *et al.*, 1992). For easy comparison, a susceptible check was included in each experiment and the value for this check was set at 100%. Inoculation with three different virulent isolates (Po6-6, W6-1, and JMB8401-1) gave similar results, although small race-specific host–pathogen interactions were detected (Roumen, 1992b; Table 15.3). The isolates Po6-6 and W6-1 belong to the same

Table 15.3. Average number of sporulating lesions in all leaves of the main culm of six rice genotypes, relative to the value for CO39 for three isolates of *Magnaporthe grisea* (mean of three experiments).

	Isolate			
Genotype	Po6-6	W6-1	JMB 8401-1	Mean
CO39	100	100	100	100 a[c]
IR50	38	48[a]	33	40 b
IR37704	38	37	21[b]	32 b
IR66	20	24	18	21 c
IR36	12	12	14	14 d
IR64	7	9	6	7 e

[a] Significantly higher ($P \leq 0.01$) than the expected value of 37 (estimated by iteration) without genetic interaction.
[b] Significantly lower ($P \leq 0.01$) than the expected value of 37 (estimated by iteration) without genetic interaction.
[c] Values followed by different letters are significantly different (Bonferroni's test for inequalities; $\alpha = 0.05$).

'lineage', but JMB8401-1 was clearly different (R. Nelson, 1992, personal communication). As found earlier in Japan, expression of PR seemed not to depend much on isolate genotype, suggesting that using a single virulent isolate may usually be sufficient for selecting PR. Comparison of the greenhouse results with those obtained for the same rice cultivars in field experiments by Bonman *et al.* (1989a) showed that differences in RIE reflected differences in PR expressed under field conditions. In general, the results obtained for the tropical *Indica* cultivars were similar to those obtained for temperate *Japonica* cultivars in Japan (Yunoki *et al.*, 1970a). It seems that there are no categorical differences in behavior between the different rice groups. Considering the ample differences among cultivars and the repeatability of results, RIE appears to be a promising, selectable characteristic for improving PR.

Rice leaves are known to become increasingly resistant to blast infection with time (Kahn and Libby, 1958; Goto *et al.*, 1961; Notteghem and Andriatompo, 1979). The increase of resistance with leaf age was shown to differ among cultivars (Fig. 15.2), and the number of lesions that develop was found to be closely associated to the time after which the resistance in newly emerged leaves becomes complete (Roumen, 1992a). The sooner this occurs after leaf emergence, the fewer lesions will develop. Cultivar differences in decline of susceptibility with leaf age not only affect the number of sporulating lesions, but also their position in the different leaves of the main culm. In PR genotypes, relatively more of the lesions were located on the youngest, top-most leaf and, averaged over a number of plants, the number of leaves on the main culm with at least one sporulating lesion was reduced. Counting leaves with lesions is much easier and faster than counting lesion numbers, and the number of leaves with lesions might serve as a fast and easy selection criterion for identifying PR in rice lines (Fig. 15.3).

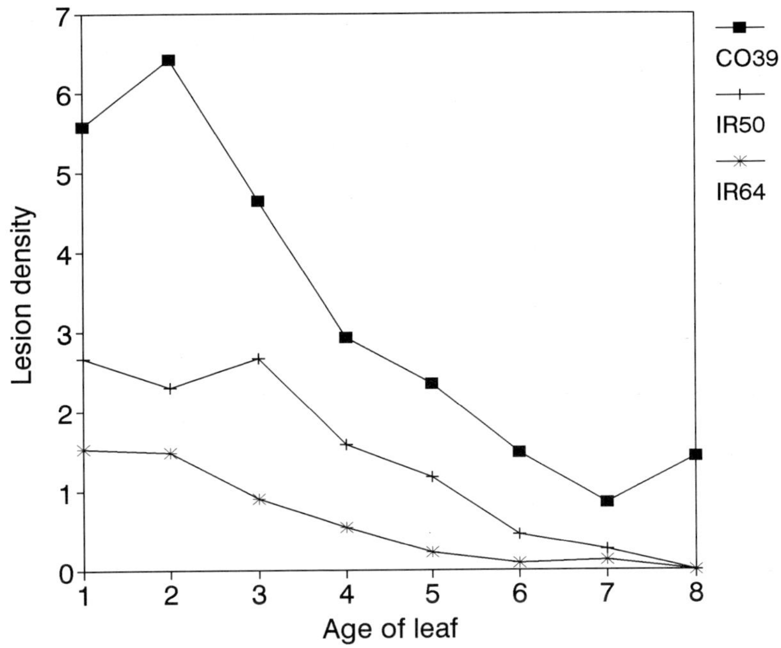

Fig. 15.2. The effect of leaf age on lesion density (number of sporulating lesions per cm^2 leaf) for three rice cultivars after inoculation with a virulent isolate.

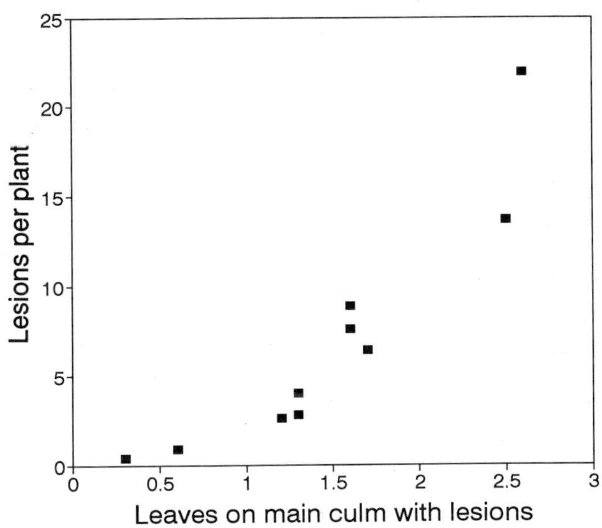

Fig. 15.3. The average number of sporulating lesions per plant of pure lines plotted against the average number of leaves with such lesions.

Lesion size

The lesion size (LS) was assessed as the area of the gray center in lesions. The average size of sporulating lesions showed consistent differences among cultivars, but the magnitude of these differences was less than for the RIE. The average size of the lesions' sporulating area also varied with leaf age (Roumen, 1992a). In most genotypes, the largest mean size was found for 1- and 2-day-old leaves. Subsequent decrease in the size of the lesions' sporulating area with increase of leaf age depended strongly on the genotype. On the whole, the effect of leaf aging was far less pronounced on size of the sporulating area of lesions than on lesion density. The lesion size seems less important as a component of PR and is harder to assess than the RIE. Selection for a smaller lesion size can easily result in selection of rice cultivars with a more resistant IT, which is best avoided since lower IT resistance to blast is not durable. The risks of selecting for smaller lesion size might be minimized by choosing genotypes showing a rapidly declining lesion size with increase of leaf age, but with a relatively large sporulating area of lesions in the youngest leaf parts (such as in IR36, for example).

Selection in Early Generations of a Breeding Cycle

The effectiveness of using RIE as a trait for improving PR was tested in F_2 and F_3 generations.

Selection in F_2

Divergent selection for higher and lower RIE was applied in F_2 populations of crosses between the cultivars IR36, IR64 (both PR) and CO39 (highly susceptible) after exposure to a virulent isolate. The low RIE in IR36 and IR64 is probably under control of several genes (Roumen, 1993). Following the procedure outlined above, at least 350 F_2 plants of each cross were grown and inoculated together with 60 plants of each parent. The F_3 was grown and inoculated in the same manner as the F_2. Fifteen F_3 plants were grown from each selected F_2 plant in a completely randomized design, also including 30 plants of each parent. The average number of lesions per plant of the selected F_2 fractions and their F_3 progenies relative to the mean value of the more susceptible parent are shown in Table 15.4. The number of lesions was expressed as a percentage of the mean value of the more susceptible parent to standardize for the different number of lesions that developed in the F_2 and F_3 trials. Heritability, h^2_n, varied between 0.14 and 0.20 depending on the cross. Similar heritability values were found by Wang *et al.* (1989), who also applied divergent selection for lesion number in the F_2. Despite the low heritabilities, the response to selection was highly significant for each of the three crosses (Table 15.4). The highly significant response to selection suggests that the low heritability was due to a very large environmental error and not to a lack of genetic variation in the crosses. Gains in PR can be made by selection as early as in the F_2, but the efficiency of selection is probably much improved if repeated measurements are made. Better results are therefore

Table 15.4. Number of sporulating lesions per plant in selected F_2 plants and their F_3 progenies relative to the mean value of the more susceptible parent (100%), and realized heritabilities h_n^2 for divergent selection for lesion number applied in three crosses between rice cultivars. CO39 was the more susceptible parent in the crosses IR36/CO39 and IR64/CO39 and IR36 was the more susceptible parent in the cross IR36/IR64.

	F_2			F_3					
	Selection group			Selection group				Number of lines	
Cross	High	Low	Difference	High	Low	Difference	h_n^2	High	Low
IR36/CO39	63.1	3.8	59.3	48.2	36.5	11.7[a]	0.20	26	24
IR64/CO39	96.3	19.0	77.3	33.1	16.0	17.1[a]	0.22	21	20
IR36/IR64	314.7	8.1	306.6	110.0	68.1	41.9[a]	0.14	15	18

[a] Significant at the 0.005 level of probability, based on an *F*-test of the line means.

expected for selection among F_3 lines. No transgression was observed in the F_3 of the crosses IR36/CO39 or IR64/CO39. On the other hand, transgression in the direction of both higher and lower RIE was detected in the cross IR36/IR64.

Selection in F_3

The inheritance of the RIE was investigated in the crosses IR36/CO39 and IR36/IR64, comparing the number of lesions in 50 random F_3 lines and their F_5 progeny. Of a randomly selected F_3 plant per line, five F_4 plants were grown. The seed of four of these F_4 plants was harvested to produce 50 groups of four F_5 lines (hereafter called F_5 groups). For the evaluation of the inheritance, reserve seed of the 50 F_3 lines and seed of the four corresponding F_5 lines per F_3 line was sown together with seed of the parents and of the F_1 in a completely randomized design. Per cross, two trials of 25 F_3 lines and corresponding F_5 lines were planted. The estimated heritabilities for individual plants in the F_2 were 0.08 and 0.12 for the trials 1 and 2 of the cross IR36/CO39, and 0.16 and 0.06 for those of the cross IR36/IR64. The corresponding heritabilities for F_3 lines were 0.47 and 0.58 in the cross IR36/CO39, and 0.68 and 0.41 in the cross IR36/IR64 (Roumen, 1993). In both crosses, the F_3 line means and the F_5 group means that were derived from the same F_2 plant were reasonably well correlated (Fig. 15.4). The results confirmed that the efficiency of selecting individual plants (such as in the F_2) is low, and that efficiency is much improved for line selection. Combining plant and line selection in the F_3, such as selection of the best plants within the better lines, should improve PR, since the heritability of F_3 lines was reasonably high, and because a reasonably high correlation was observed between the F_3 line means and the F_5 group means derived from the same F_2 plant for both crosses.

A disadvantage of greenhouse assessment of PR as performed in Japan and in the analysis of PR performed at IRRI is that counting the number of lesions

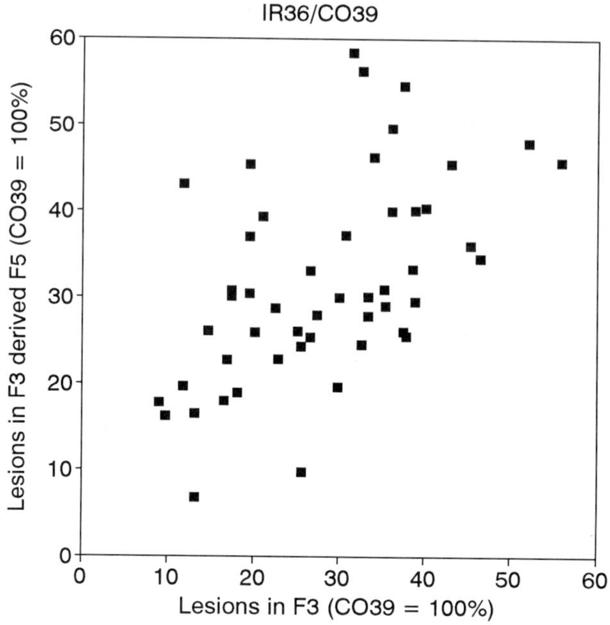

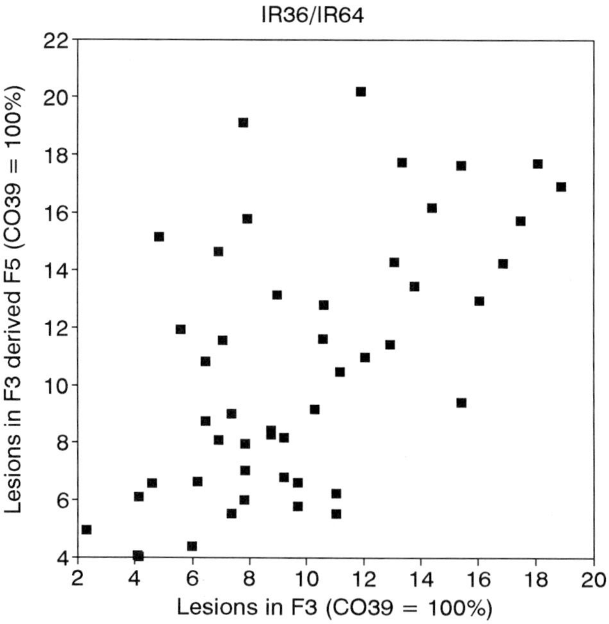

Fig. 15.4. The relative number of sporulating lesions per plant (CO39 = 100%) in F_3 lines and F_5 lines derived from the same F_2 plant for the crosses IR36/CO39 (upper part) and IR36/IR64 (lower part).

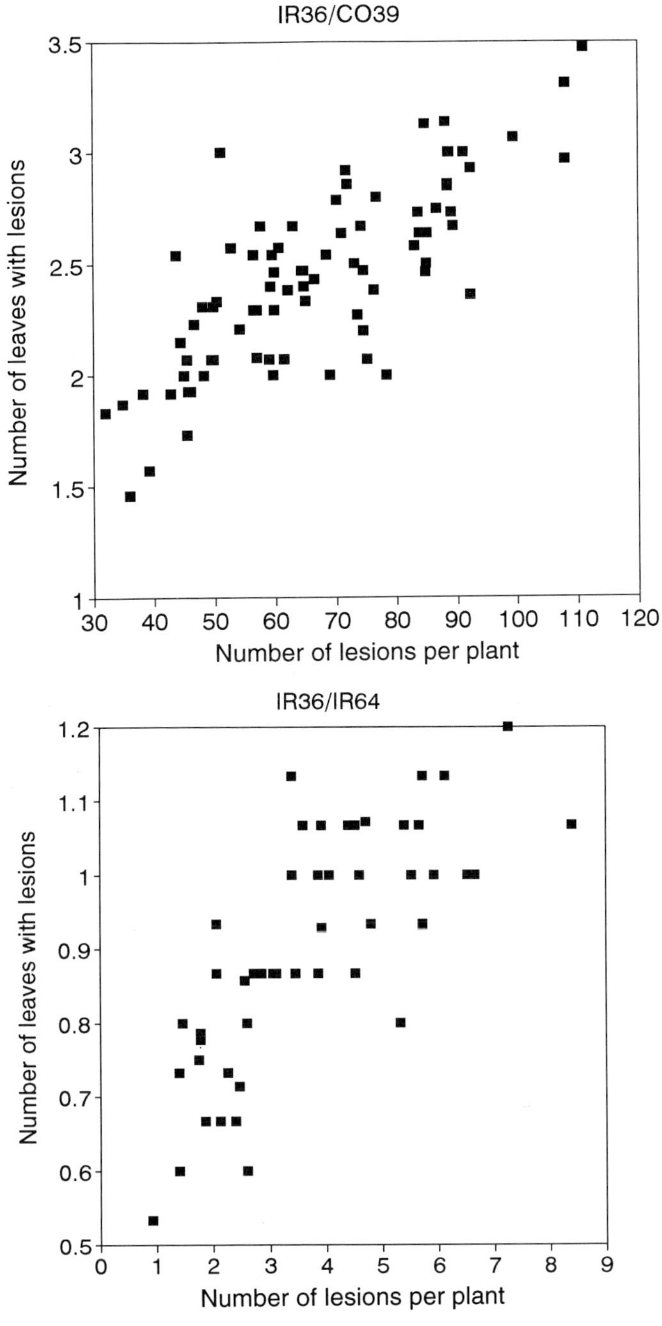

Fig. 15.5. The average number of sporulating lesions per plant of F_3 lines of the crosses IR36/CO39 (upper part) and IR36/IR64 (lower part) plotted against the average number of leaves with such lesions.

Table 15.5. Mean number of sporulating lesions per plant in the best 12.5% and the best 25% of the F_3 lines relative to the mean of all F_3 lines (unselected control) of three crosses using the number of leaves with sporulating lesions (leaves) or the number of sporulating lesions itself (lesions) as criteria.

	Cross					
	IR36/CO39		IR64/CO39		IR36/IR64	
Selection criterion	12.5%	25%	12.5%	25%	12.5%	25%
Lesions	60	66	23	35	39	45
Leaves	64	75	27	38	50	50
Mean of all F_3 lines	100 = 66.9 lesions/plant		100 = 2.6 lesions/plant		100 = 3.8 lesions/plant	

is too time-consuming to use in a breeding program. Because the number of lesions within a given genotype usually varies markedly from plant to plant, only large differences between genotypes can be assessed reliably by visual observation. However, when observed in detail, even small cultivar differences in number of sporulating lesions appear reproducible.

Such small differences might be more easily assessed by scoring the number of leaves with sporulating lesions. As in pure lines, the average number of leaves with sporulating lesions in segregating lines was highly correlated with the number of sporulating lesions that developed (Fig. 15.5). Using the number of leaves with lesions as the selection criterion instead of the number of lesions itself, the average number of sporulating lesions was reduced by nearly as much (Table 15.5), suggesting that this may be a good criterion for F_3 selection.

Due to the limited time-span of the project, a question that could not be addressed was how gains for a reduced RIE in the greenhouse correspond to gains for PR in the field. Some of the lines resulting from the cross IR36/IR64 showed a significantly lower RIE than either of the parental cultivars (up to 60% less than the most resistant parent IR64). It would be interesting to take a sample of sister-lines from this cross representing a range of RIE in the greenhouse and evaluate the PR of these lines in the field, not only in the Philippines, but at other locations as well.

Accumulating PR in a Breeding Program

Based on the results of the experiments done at IRRI, the following procedure is recommended for selecting PR in breeding programs.

Choice of isolate(s)

If possible, only isolates that produce a susceptible IT (type 4 or higher using the integrated scale of Table 15.1) on both parents of a cross should be used for selecting PR. If such isolates are not available, the next best choice is using an isolate that shows a highly susceptible IT on one parent and a highly resistant IT on the other. In this case, the progeny will usually show a clear Mendelian segregation for IT. Plants or lines that develop a resistant or intermediate IT are discarded and PR is assessed only in that part of the progeny showing an IT of 4 or higher. Isolates that produce an intermediate IT are best avoided at all times. Whenever possible, selection in different generations should be done with a different virulent isolate (belonging to a different race or 'lineage') to avoid the risk of selecting a highly race-specific resistance resembling polygenic PR such as that conferred by the *Pi-f* gene.

Plant cultivation and inoculation

Plants can be grown in the greenhouse, screenhouse, or field, as long as there is no risk of natural infection. Growing the plants at high levels of nitrogen fertilizer promotes a clear symptom development after inoculation. Applying a last split of a rapidly acting nitrogen fertilizer 1 or 2 days before inoculation gave very good results. Moisture stress and flooding should both be avoided at all times. A minimal plant distance of 4 cm is recommended. Inoculation is best carried out in evening hours when plants have developed five or six leaves on the main culm. Plants that were scored and selected at this plant stage usually showed a vigorous regrowth and good seed production after transplanting them in nurseries or in the field, thus allowing selection for other characters than blast resistance. Recovery of selected plants can be improved by treating with a blast-specific systemic fungicide and removal of all diseased leaves. It is recommended to include both parents as well as a highly susceptible check cultivar in each test for comparison. Plant F_3 lines and later generations in single rows of 20–30 plants each and add rows of the parental genotypes and the check at regular intervals.

Selection

Although heritability of individual plant selection is very low, selection in the F_2 was shown to be highly effective. A mild negative mass selection in the F_2 is recommended using the RIE as criterion, eliminating the most susceptible 25–33% of the plants. Although selection in the F_2 is optional, selection should definitely be carried out in the F_3.

First, a large portion of the more susceptible F_3 lines can be easily identified and discarded by visual observation of the lines using the RIE as criterion. Further narrowing down the number of lines can be done by assessing the average number of leaves from the top that show at least one sporulating lesion. As a practical guideline, any F_3 line that is more susceptible than the most resistant parent of the cross could be discarded. As a last step, the best plants from within the best lines can be rapidly selected by visual observation using the RIE as criterion. The procedure for the F_3 can also be applied to subsequent selection in the F_4 or later generations.

REFERENCES

Ahn, S.W. and Ou, S.H. (1982) Quantitative resistance of rice blast disease. *Phytopathology* 72, 279–282.

Asaga, K. and Yoshimura, S. (1969) Relation between field resistance of rice varieties to blast and the races or strains of the causal fungus. *Annals of the Phytopathological Society of Japan* 35, 385–386 (in Japanese).

Bonman, J.M. and Mackill, D.J. (1988) Durable resistance to rice blast disease. *Oryza* 25, 103–110.

Bonman, J.M., Vergel de Dios, T.J. and Khin, M.M. (1986) Physiologic specialization of *Pyricularia oryzae* in the Philippines. *Plant Disease* 70, 767–769.

Bonman, J.M., Estrada, B.A. and Bandong, J.M. (1989a) Leaf and neck blast resistance in tropical lowland rice cultivars. *Plant Disease* 73, 388–390.

Bonman, J.M, Bandong, J.M., Lee, Y.H., Lee, E.J. and Valent, B. (1989b) Race specific partial resistance to blast in temperate japonica cultivars. *Plant Disease* 73, 496–499.

Brodny, U., Castaño-Zapata, J. and Nelson, R.R. (1988) Evaluation of rate limiting resistance to blast disease in rice cultivars using sub-lethal concentrations of a systemic fungicide. *Journal of Phytopathology* 122, 174–180.

Broers, L.H.M. and Jacobs, Th. (1989) Histological, genetical and epidemiological studies on partial resistance in wheat to wheat leaf rust. PhD Thesis, Wageningen Agricultural University, Wageningen, The Netherlands, 201 pp.

Castaño-Zapata, J. (1981) Identification and testing of rate-limiting resistance of rice (*Oryza sativa* L.) to blast disease caused by *Pyricularia oryzae* Cav. PhD Thesis, Pennsylvania State University, Pennsylvania, USA.

Castaño, J., Mackenzie, D.R. and Nelson, R.R. (1989) Components analysis of race non-specific resistance to blast disease of rice caused by *Pyricularia oryzae*. *Journal of Phytopathology* 127, 89–99.

Cho, C.I. and Kiyosawa, S. (1973) Blast resistance of Korean rice varieties. *Research Reports of the Office of Rural Development, Korea* 15, 83–88.

Ezuka, A. (1972) Field resistance of rice varieties to blast disease. *Review of Plant Protection Research, Tokyo* 5, 1–21.

Ezuka, A., Yunoki, T., Sakurai, Y., Shinoda, H. and Toriyama, K. (1969a) Studies on the varietal resistance to rice blast. 1. Tests for genotype of 'true resistance'. *Bulletin Chugoku Agricultural Experimental Station, Japan*, Series E 4, 1–32.

Ezuka, A., Yunoki, T., Sakurai, Y., Shinoda, H. and Toriyama, K. (1969b) Studies on the varietal resistance to rice blast. 2. Tests for field resistance in paddy field and upland nursery beds. *Bulletin Chugoku Agricultural Experimental Station, Japan*, Series E 4, 31–53.

Goto, K. and Yamanaka, S. (1968) Studies on the race of rice blast fungus. *Bulletin of the College of Agriculture Utsunomiya University* 7, 27–82.

Goto, K., Hirano, K. and Ohata, K. (1961) Susceptibility of leaf of rice to blast in reference to leaf age and position. I. Variation of susceptibility among different leaf positions and grades of emergence of top leaf. *Special Bulletin Okayama Prefecture Agricultural Experiment Station, Japan* 58, 77–88.

Higashi, T. and Kushibuchi, K. (1978) Genetic analysis of field resistance to leaf blast (*Pyricularia oryzae*) in rice. *Japanese Journal of Breeding* 28, 277–286.

Higashi, T. and Saito, S. (1985) Linkage groups of field resistance genes of upland rice variety Sensho to leaf blast caused by *Pyricularia oryzae*. *Japanese Journal of Breeding* 35, 438–448.

Hirano, T. and Matsumoto, K (1971) Resistance of the rice varieties to several C races of blast fungus. I. Studies on variation of field resistance by races and fungus strains. *Japanese Journal of Breeding* 21, 110–111 (in Japanese).

Imbe, T. and Matsumoto, S. (1985) Inheritance of resistance of rice varieties to the blast fungus strains virulent to the variety Reiho. *Japanese Journal of Breeding* 35, 332–339.

IRAT (1980) Lutte contre la pyriculariose. Bilan des travaux de recherche sur le riz pluvial 1960 à 1980. IRAT, Montpellier, France, 52 pp.

IRRI (1988). *Standard Evaluation System for Rice*. International Rice Testing Program, 3rd edn. International Rice Research Institute, Los Baños, The Philippines, 54 pp.

Jeanguyot, M. (1983) Lutte contre la pyriculariose du riz. IRAT, Nogent-sur-Marne, France, 55 pp.

Johnson, R. (1984) A critical analysis of durable resistance. *Annual Review of Phytopathology* 22, 309–330.

Kahn, R.P. and Libby, J.L. (1958) The effect of environmental factors and plant age on the infection of rice by the blast fungus, *Piricularia oryzae*. *Phytopathology* 48, 25–30.

Kiyosawa, S. (1972) Genetics of blast resistance. In: *Rice Breeding*. International Rice Research Institute, Los Baños, The Philippines, pp. 203–225.

Kiyosawa, S., Mackill, D.J., Bonman, J.M., Tanaka, Y. and Ling, Z.Z. (1986) An attempt of classification of world's rice varieties based on reaction pattern to blast fungus strains. *Bulletin National Institute of Agrobiological Resources, Japan* 2, 13–37.

Koch, M.F. (1990) Aspects of quantitative resistance to *Xanthomonas campestris* pv. *oryzae* in rice. PhD Thesis, Wageningen Agricultural University, Wageningen, The Netherlands, 174 pp.

Kozaka, T. (1975) Recent advances in studies on horizontal resistance to blast disease of rice in Japan. In: Gálvez, G.E. (ed.), *Proceedings, Seminar on Horizontal Resistance. Blast Disease of Rice*. Centro Internacional Agricultura Tropical, Cali, Colombia, pp. 101–116.

Kozaka, T. (1979) The nature of the blast fungus and varietal resistance in Japan. In: *Proceedings of the Rice Blast Workshop*. International Rice Research Institute, Los Baños, The Philippines, pp. 3–26.

Kozaka, T., Matsumoto, S. and Yamada, M. (1970) Reactions of foreign rice varieties to major races of *Pyricularia oryzae* in Asia. *Bulletin National Institute Agricultural Science, Japan*, Series C 24, 113–152.

Mackill, D.J. and Bonman, J.M. (1992) Inheritance of blast resistance in near-isogenic lines of rice. *Phytopathology* 82, 746–749.

Marchetti, M.A. (1983) Dilatory resistance to rice blast in U.S.A. rice. *Phytopathology* 73, 645–649.

Narita, T. and Iwata, T. (1964) Varietal difference in severity of blast occurrence in late sown upland nursery beds. Special report on diseases and insect pests outbreak forecast work. *Ministry of Agriculture and Forestry, Tokyo, Japan*, 59–60.

Notteghem, J.L. (1985) Définition d'une stratégie d'utilisation de la résistance par analyse génétique des relations hôte-parasite. Cas du couple riz-*Pyricularia oryzae*. *Agronomie Tropicale* 40, 129–147.

Notteghem, J.L. (1986) Blast resistance methodology in the Ivory Coast. In: Shideler, F.J. (ed.), *Progress in upland rice research. Proceedings of the 1985 Jakarta Conference*. International Rice Research Institute, Los Baños, The Philippines, pp. 305–316.

Notteghem, J.L. (1989) La création de variétés résistantes. Cas du riz et de la Pyriculariose. Colloque les apportes récents de la biologie végétale en régions tropicales. UPSA Orsay, 16–18 Nov. 1989. *Bulletin de la Société Botanique de France. Actualités Botaniques* 136, 227–237.

Notteghem, J.L. (1993) Durable resistance to rice blast disease. In: Jacobs, Th. and Parlevliet, J.E. (eds), *Durability of Disease Resistance*. Kluwer Academic Publishers, Dordrecht, The Netherlands, pp. 125–134.

Notteghem, J.L. and Andriatompo, G.M. (1977) Mesure au champ de la résistance horizontale du riz a *Pyricularia oryzae*. *Agronomie Tropicale* 32, 400–412.

Notteghem, J.L. and Andriatompo, G.M. (1979) Etude de quelques facteurs de la résistance horizontale du riz a la Pyriculariose. *Agronomie Tropicale*, 34, 180–195.

Ou, S.H. (1985) *Rice Diseases*, 2nd edn. Commonwealth Mycological Institute, Kew, UK, 380 pp.

Parlevliet, J.E. (1979) Components of resistance that reduce the rate of epidemic development. *Annual Review of Phytopathology* 17, 203–222.

Parlevliet, J.E. (1983) Can horizontal resistance be recognized in the presence of vertical resistance in plants exposed to a mixture of pathogen races? *Phytopathology* 73, 379.

Parlevliet, J.E. (1992) Selecting components of partial resistance. In: Stalker, H.T. and Murphy, J.P. (eds), *Plant Breeding in the 1990s, Proceedings of the Symposium on Plant Breeding in the 1990s*. CAB INTERNATIONAL, Wallingford, Oxon, UK, pp. 281–302.

Parlevliet, J.E. and van Ommeren, A. (1975) Partial resistance of barley to leaf rust *Puccinia hordei*. II Relationship between field trials micro plot tests and latent period. *Euphytica* 24, 293–303.

Parlevliet, J.E., Lindhhout, W.H., van Ommeren, A. and Kuiper, H.J. (1980) Level of partial resistance of leaf rust *Puccinia hordei* in West-European barley and how to select for it. *Euphytica* 29, 1–8.

Prabhu, A.S. (1989) Methods for evaluating resistance to *Pyricularia oryzae* in rice. *International Rice Research Newsletter* 14, 18–19.

Rodríguez, M. and Gâlvez, G.E. (1975) Indications of partial resistance of rice to the fungus *Pyricularia oryzae*. In: Gâlvez, G.E. (ed), *Proceedings, Seminar on Horizontal resistance. Blast Disease of Rice*. Series CE-No. 9. Centro Internacional Agricultura Tropical, Cali, Colombia, pp. 137–154.

Roumen, E.C. (1992a) Effect of leaf age on components of partial resistance in rice to leaf blast. *Euphytica* 63, 271–279.

Roumen, E.C. (1992b) Small differential interactions for partial resistance in rice cultivars to virulent isolates of the blast pathogen. *Euphytica* 64, 143–148.

Roumen, E.C. (1993) Partial resistance in rice to blast and how to select for it. PhD Thesis, Wageningen Agricultural University, Wageningen, The Netherlands, 108 pp.

Roumen, E.C., Bonman, J.M. and Parlevliet, J.E. (1992) Leaf age related partial resistance to *Pyricularia oryzae* in tropical lowland cultivars as measured by the number of sporulating lesions. *Phytopathology* 82, 1414–1417.

Sakurai, Y. and Toriyama, K. (1967) Field resistance of the rice plant to *Piricularia oryzae* and its testing method. In: Ogura, T. (ed.), *Rice Diseases and Their Control by Growing Resistance Varieties and Other Measures. Proceedings, Symposium Tropical Agriculture Researches*. Ministry of Agriculture and Forestry, Tokyo, Japan, pp. 123–135.

Silué, D., Notteghem, J.L. and Chaume, J. (1992) Etude de la résistance spécifique d'*Oryza glaberrima* Steud a la pyriculariose du riz. *Agronomie Tropicale* 46, 121–129.

Suzuki, H. (1935) Studies on the influence of some environmental factors on the susceptibility of the rice to blast and Helminthosporium diseases on the anatomical characters of the plant. II. Influence of differences in soil moisture and in the amount of nitrogenous fertilizer given. *Journal College Agriculture Tokyo Imperial University* 13, 235–276.

Tanaka, Y. (1986) Gene analysis for major resistance genes to rice blast in some Brazilian upland varieties. In: Shideler, F.J. (ed), *Progress in Upland Rice Research, Proceedings of the 1985 Jakarta Conference*. International Rice Research Institute, Los Baños, The Philippines, pp. 295–304.

Toriyama, K., Yunoki, T. and Shinoda, H. (1968) Breeding rice varieties for resistance to blast. II. Inheritance of high field resistance of Chugoku No. 31. *Japanese Journal of Breeding* 18, 145–146.

Villareal, R.L., Nelson, R.R., MacKenzie, D.R. and Coffman, W.R. (1981) Some components of slow blasting resistance in rice. *Phytopathology* 71, 608–611.

Wang, Z., Mackill, D.J. and Bonman, J.M. (1989) Inheritance of partial resistance to blast in indica rice cultivars. *Crop Science* 29, 848–853.

Wang, G., Mackill, D.J., Bonman, J.M., McCouch, S.R., Champoux, M.C. and Nelson, R.J. (1994) RFLP mapping of genes conferring complete and partial resistance to blast in a durably resistant rice cultivar. *Genetics* 136, 1421–1434.

Yamasaki, Y. and Kiyosawa, S. (1966). Studies on inheritance of resistance of rice varieties to blast. I. Inheritance of resistance of Japanese varieties to several strains of the fungus. *Bulletin National Institute Agricultural Science, Japan,* Series D 14, 39–69.

Yeh, W.H. and Bonman, J.M. (1989) Assessment of partial resistance to *Pyricularia oryzae* in six rice cultivars. *Plant Pathology* 35, 319–323.

Yu, Z.H., Mackill, D.J. and Bonman, J.M. (1987) Inheritance of resistance to blast in some traditional and improved rice cultivars. *Phytopathology* 77, 323–326.

Yunoki, T., Ezuka, A., Sakurai, Y., Shinoda, H. and Toriyama, K. (1970a) Studies on the varietal resistance to rice blast. 3. Testing methods for field resistance on young seedlings grown in greenhouse. *Bulletin Chugoku Agricultural Experiment Station, Japan,* Series, E 6, 1–19.

Yunoki, T., Ezuka, A., Morinaka, T., Sakurai, Y., Shinoda, H. and Toriyama, K. (1970b) Studies on the varietal resistance to rice blast. 4. Variation of field resistance due to fungus strain. *Bulletin Chugoku Agricultural Experiment Station, Japan,* Series E 6, 21–45.

Zadoks, J.C. and Schein, R.D. (1979) *Epidemiology and Plant Disease Management.* Oxford University Press, New York, 427 pp.

16 Lineage Exclusion: A Proposal for Linking Blast Population Analysis to Resistance Breeding

R.S. Zeigler[1], J. Tohme[2], R. Nelson[1], M. Levy[3] and F.J. Correa-Victoria[2]

[1]*International Rice Research Institute, PO Box 933, Manila, The Philippines;* [2]*Centro Internacional de Agricultura Tropical, Apdo aereo 6713, Cali, Colombia;* [3]*Department of Biological Sciences, Purdue University, West Lafayette, Indiana 47907, USA*

Introduction

The rice blast disease, caused by the filamentous fungus *Pyricularia grisea* (perfect state *Magnaporthe grisea*) has plagued farmers since the beginning of recorded rice cultivation (as reported in Ou, 1985) and challenged and intrigued rice breeders and pathologists since the beginning of modern rice improvement. Chemical control of the disease remains economically impractical for small, resource-poor farmers, and environmentally undesirable for most farmers. Devastating epidemics of the disease have been exacerbated by high input management, particularly high nitrogenous fertilizer applications, associated with modern rice cultivation. Considerable effort, therefore, has been directed towards developing blast-resistant rice cultivars as a means to achieve low-cost and environmentally safe blast management.

Blast resistance, however, has tended to be unreliable, with resistance often failing, or 'breaking down', under field conditions (Kiyosawa, 1972). Blast-resistance breakdown has been attributed to extreme heterogeneity and/or variability of virulence in the pathogen, inadequate resistance screening and evaluation methods, inappropriate use of rice resistance genes, or a combination of these. Numerous proposals have been made for developing durable resistance, but these have met with only limited success (Ou, 1979; Buddenhagen, 1981a,b, 1983a,b; Marchetti, 1983; Ezuka, 1979; Ikehashi and Khush, 1979; Notteghem, 1985; Correa-Victoria and Zeigler, 1991; Bonman *et al.*, 1992).

In this chapter we will briefly examine the diversity and variability of the pathogen as it relates to resistance breeding, and summarize our current understanding of resistance to rice blast. We will then introduce the breeding approaches that have been proposed for developing durable resistance. Each topic will incorporate a short discussion of the most recent relevant findings in molecular and population biology as they relate to resistance breeding. Finally, we propose to apply these latest findings to the breeding approaches that have been proposed by earlier researchers.

Pathogen Diversity and Variability

Virulence diversity

The differential susceptibility of some cultivars to strains of *P. grisea* was reported in the early 1920s in Japan (Kiyosawa, 1972). Strains with different virulence (compatibility or incompatibility) on standard sets of cultivars were considered to represent different pathotypes (or races, pathogenic races, physiological races, etc.) (Kiyosawa, 1972; Ezuka, 1979). Heterogeneity within the pathogen population was further indicated by the range of lesion types observed on susceptible cultivars in the field. Plants typically show a range of symptoms, from small, non-sporulating lesions, associated with a resistance response, to large and abundantly sporulating lesions, associated with a fully compatible, or susceptible, response. Kiyosawa (1972) proposed that incompatibility with rice is mediated by avirulence genes, and that a classical gene-for-gene (Flor, 1971) relationship governs rice–*P. grisea* interactions.

Serious attention was directed towards pathotype analysis when the first cultivars bred to be blast resistant rapidly showed susceptibility in the field (Ou, 1979; Kiyosaka, 1972). Several researchers (Atkins *et al.*, 1967; Ling and Ou, 1969; Ezuka, 1979) developed standard sets of differential cultivars in order to classify local isolates into pathotypes according to their respective pathogenicity on the differentials. In turn, controlled inoculations with particular races would specify the resistance of a cultivar and help direct further resistance breeding. However, in the absence of a standard set of differential cultivars it was impossible to compare pathotypic structure of populations across countries. It appeared that by simply adding differential cultivars, previously uniform groups of pathotypes could be further partitioned (Ou, 1979).

In the 1960s Japanese and US scientists formed a working group to generate a set of international differentials in an attempt to standardize pathotype nomenclature and classification. From this work emerged a set of eight differentials (Atkins *et al.*, 1967) from which a simple nomenclature was developed to describe 256 possible pathotypes (Ling and Ou, 1969). This differential set failed to adequately describe pathotype diversity in many areas. Furthermore, in the tropics, use of this largely *japonica* set of differential varieties was limited by difficulties in seed production and adaptation. By the early 1970s, researchers in countries with active blast resistance breeding programs were again using very different sets of differentials (Ou, 1972; Bonman *et al.*, 1986).

A new set of differentials was proposed for Japan in 1976 (Yamada *et al.*, 1976) based on lines believed to have single major resistance genes. Work continued into the 1980s to develop 'ideal' differential lines by developing near-isogenic lines (NILs), each containing a single major blast resistance gene (Mackill and Bonman, 1992). This offered the possibility of developing a standard on which the absolute virulence spectrum of pathotypes could be established. Although it is agreed that *P. grisea* populations may be very diverse and are composed of individuals with distinct virulence characteristics, there is as yet no practical means to compare different pathogen populations based on the published results of pathotype analysis. Proposals to establish centralized facilities for pathotyping and population characterization (Latterell, 1971) were never implemented. Despite some recent attempts to classify sites relative to blast severity (Cuevas-Perez and Gaona, 1988), it is still not possible to predict how a cultivar will perform in one environment, based upon its performance in another.

Virulence instability

The diversity of pathotypes across rice growing regions, and difficulty of comparing them, was complicated by numerous reports of the instability of pathotypes (Ou and Ayad, 1968; Giatgong and Frederiksen, 1969; Ou, 1971; Ou *et al.*, 1970). That is, single spore isolates from a single lesion (caused by an isolate of a known pathotype) were reported to give rise to pathotypes different from the parental strain. In some reports a bewildering array of new pathotypes was obtained from a series of single spore isolates recovered from a lesion derived from a single spore isolate. Typically the parental type was the predominant form recovered, with other types recovered in lower frequency. New pathotypes showed both gain and loss of virulence on differential cultivars, but the trend was towards a gain in virulence over succeeding generations (Giatgong and Frederiksen, 1969; Kiyosawa, 1972).

This pathotype instability was viewed by some as the principal explanation for the instability of resistance in the field (Ou, 1971; Kiyosawa, 1972). That is, cultivars released were indeed resistant to the pathotypes against which they were tested, and to the pathogen populations present at the time of their field evaluation; however, as new pathotypes emerged the resistance was quickly rendered ineffective, and the resistance broke down. Pathotypic variation was also used to explain the large diversity of pathotypes observed, and the range of lesion types found on any susceptible plant in the field. Any *P. grisea* population was constantly generating new pathotypes, some of which were compatible (large lesions) and some of which were incompatible (small lesions) with a given cultivar or breeding line. It was suggested that this variability was the mechanism by which some cultivars with putative complex major gene resistance showed a more durable resistance. This model proposed that the rapid rate of pathotypic change resulted in only transient and low frequency complex virulence configurations required for compatibility with these hosts (Ou *et al.*, 1971, 1975).

The variability in pathotypes arising from a single source was consistent with the observed variation in colony morphology. However, the frequency and degree of change exceeded the rate that would be expected from simple genetic mutations. Various mechanisms were advanced to explain the observed variability in virulence. Large numbers of nuclei were reported in some strains (Ou, 1979), but this was later disproven when advanced microscopy techniques were applied. Lagging chromosomes were also reported during meiosis, and it was proposed that aneuploidy could explain differences in virulence (Row *et al.*, 1985). Giatgong and Frederiksen (1969) proposed that chromosome anomalies were driving pathotypic variation. Hyphal anastomosis was observed and was reported as a possible means of exchange or addition of genetic material that could form the basis for variability. Parasexual recombination was also demonstrated to occur between laboratory strains (Crawford *et al.*, 1986), providing another plausible mechanism for generating variability. However, it has not been a trivial matter to generate this process in the laboratory, and it is difficult to see how it could explain the observed variation between single spore progeny from the same parent. The perfect state has been achieved *in vitro* between both grass-infecting and rice-infecting strains (Kato and Yamaguchi, 1982; Notteghem and Silué, 1992), and is routinely used by several laboratories for genetic studies (Leung and Williams, 1986; Valent *et al.*, 1986; Valent, 1990; Valent *et al.*, 1991). However, no evidence for its occurrence in nature has been found. Indeed, most field isolates are infertile (Notteghem, 1985; Valent and Chumley, 1991; Valent *et al.*, 1991; Notteghem and Silué, 1992).

The degree of variability described by Ou and others was by no means accepted by all blast researchers. Latterell and colleagues (Latterell, 1971, 1972; Latterell and Rossi, 1986) indicated that while changes occurred, shifts in virulence were rare, and only occurred in some strains. In their studies, virulence stability was the norm. They attributed much of the observed diversity to the ambiguity in the inoculation assay upon which the virulence assessments were based. Valent (1990; Valent *et al.*, 1991) reported that some variation in pathotype does occur, but concurred that the rate is much lower than the extremely high rates discussed earlier. Bonman *et al.* (1986) revisited the controversy, testing strains from the Philippines and Korea on differential cultivars from Korea and the Philippines. They found that with replication and repeated inoculation of questionable reactions, most of the strains were quite stable. They indicated that some differential cultivars typically gave intermediate reactions, and that different readings of these by different researchers could give the impression of variation in the pathogen. Furthermore, they stressed that the assay is very sensitive to subtle changes in the environment (temperature, relative humidity, etc.), and that escape from inoculation is fairly common when the preferred spray inoculation method is used. The controversy over virulence instability concerned the degree or frequency of pathotypic change, rather than over whether or not it occurs. Most researchers today accept that pathotypic change occurs in nature.

Molecular perspectives on virulence and variability

Several groups have begun describing the genetics of pathogenicity and virulence in *M. grisea* (Leung *et al.*, 1988; Heath *et al.*, 1990; Valent, 1990; Valent and Chumley, 1991; Valent *et al.*, 1991). Further research (Ellingboe *et al.*, 1990; Valent, 1990; Valent and Chumley, 1991; Valent *et al.*, 1991; Ellingboe, 1992;

Silué and Notteghem, 1992; Silué *et al.*, 1992; Valent and Chumley, this volume) has shown that, in strains pathogenic to rice, compatibility is mediated by classical avirulence (*avr*) loci. That is, virulence is determined by either the absence of, or a dysfunctional *avr* (*avr-*) gene, when a matching resistance gene is present. This is consistent with a 'gene-for-gene' relationship between the pathogen and the host, wherein for each *avr* gene in the pathogen there is a corresponding resistance gene in the host (Flor, 1971). Compatibility is unaffected by the state of the *avr* locus when no matching resistance gene is present. Kiyosawa (1972) predicted this relationship and noted that observed virulence shifts tend to be from avirulent to virulent. Presence of the *avr* gene product apparently initiates a resistance response in the host with the corresponding resistance gene, resulting in incompatibility. Hosts lacking the resistance gene corresponding to an *avr* gene are susceptible to the pathogen regardless of its *avr/avr-* configuration. The product of an *avr* gene probably plays an important role in fungal metabolism (otherwise, the pathogen would have sets of genes whose only purpose would be to prevent it from surviving). Multiple *avr* genes may reflect alternative means for meeting metabolic needs, with only a subset of the *avr* genes required to be functional in order for the pathogen to be viable.

The few *avr* genes that have been identified so far tend to be located near the telomeres and are therefore prone to deletion (Valent and Chumley, this volume). Deletion of the *avr* gene, of course, results in a virulent genotype, and compatibility with the host. In the absence of recombination in *P. grisea*, most *avr* to *avr-* changes would be irreversible. An *avr* to *avr-* shift may not entail complete loss of function but may result in decreased efficiency and/or require a complementary shift at another locus. If the general mode of accumulation of virulence is through loss or disruption of genes coding for, presumably, important products, additional virulence might come at a cost. Strains with a narrow virulence spectrum could gradually broaden their spectrum through loss of *avr* genes. This process would appear to be degenerative, leading to a minimum viable configuration, which would be stable. That is, since the only further changes from *avr* to *avr-* would be lethal, all non-parental *avr* types would be non-viable. As a consequence, strains could evolve into an *avr* configuration where there is an absolute limit to their virulence spectra. Such limits would be characteristic of the parent strain and all its descendants. This simple model would be complicated by duplication of *avr* alleles, that could mutate freely into functional but *avr-* alleles. Indeed, the first probing with putative *avr* clones (Valent and Chumley, this volume) of *Eco*RI digests of *P. grisea* DNA has shown multiple *avr* copies in some isolates (Levy *et al.*, 1993b).

At least one dispersed, repeated element (called MGR for *Magnaporthe grisea* repeats; Hamer *et al.*, 1989a), cloned in pCB583 (Valent and Chumley, 1991) seems to be part of a retro transposon, the mobility of which could generate variability as parts of the genome are disrupted. A gene mediating spore morphology (SMO) and one related to pigment biosynthesis have been shown to have particularly high mutation rates (Hamer *et al.*, 1989b). Ou (1985) suggested that similar rates in genes mediating virulence could explain the high rates of variation in virulence that he observed.

Karyotypic analysis of the pathogen conducted using pulsed-field gel electrophoresis (PFGE) (Talbot *et al.*, 1993) confirmed the reports of mini-chromosomes (Row *et al.*, 1985) that could conceivably be lost during mitosis (Talbot *et al.*, 1993) and supported the suggestion by Giatgong and Frederiksen (1969) that chromosome anomalies are responsible for pathotypic variation. Furthermore, although chromosome number appeared to be stable, length was quite variable. Karyotype changes resulted from deletions and translocations of varying complexity. Analysis of isolates following repeated sub-culturing indicated that karyotype changes could occur within a given strain, but this was not related to any change in pathotype based on compatibility with the international set of differentials. Talbot *et al.* (1993) concluded that detectable, persistent chromosome changes are retained during asexual reproduction only if they are neutral in terms of fitness (we note that changes with beneficial effects would be conserved, as well). As such, they speculate that chromosome rearrangement and maintenance indicated by karyotypic diversity is simply a result of the clonal propagation of the pathogen, and are unrelated to observed virulence variability. Such a situation, if generalized throughout the species, might preclude sexual reproduction and greatly reduce the amount of genetic material that could be exchanged in parasexual event.

The isolates were judged to be pathotypically stable based only on the limited virulence description provided by the international differentials. Furthermore, if karyotypic variation derives from random events, then only a portion should be expected to affect loci associated with pathogenesis and virulence, and the small set of strains utilized and the small number of new karyotypic changes observed might not have detected these events. Certainly karyotypic variation could result in the deletions or other disrupting rearrangements that would transform the *avr* loci described by Valent and Chumley (this volume) into an *avr-* form.

Pathogen Population Structure

Rice breeders and pathologists recognize that rice cultivars are exposed to pathogen populations in the field, and that resistance must be effective against the entire population if it is to be durable. Development of techniques to monitor populations has been a major constraint to progress (Leung and Williams, 1985, 1986), although Chin (1985) suggested that 'mobile nurseries' could be used to characterize populations. Until very recently the only means of describing populations was to sample and determine pathotype based on differential hosts. No technique was adopted to assure that those isolates sampled reflected the complete array of pathotypes within the population. Buddenhagen (1981a) suggested that most cases of resistance breakdown could be attributed to difficulties in screening techniques related to a multitude of infrequent pathotypes within populations. Those that are compatible with putatively resistant lines, but do not encounter the lines during screening, are responsible for breakdown as their populations increase with commercial cultivation of the selected lines. He termed this phenomenon 'cryptic error', while most others simply call it 'escape'.

The application of DNA fingerprinting and phylogenetic analyses using dispersed repetitive DNA has opened a new avenue for *P. grisea* population analysis (Levy *et al.*, 1991a,b). The element pCB586, now called MGR586, and hereafter referred to simply as MGR, has been particularly useful for this (Hamer *et al.*, 1989a). The MGR element is dispersed rather uniformly across the genome of the pathogen (Romao and Hamer, 1992). Strains from rice typically show between 50 and 80 bands when digested with a restriction enzyme (usually *Eco*RI) and probed with MGR. It was noted that the high MGR copy number appeared to be restricted to rice-infecting strains (Hamer *et al.*, 1989a; Borromeo *et al.*, 1993). Those *P. grisea* strains recovered from non-rice hosts typically had very few MGR bands, suggesting that strains from non-rice hosts are genetically isolated from those infecting rice (Hamer *et al.*, 1989a; Borromeo *et al.*, 1993). It has been postulated that amplification of MGR in rice-infecting *P. grisea* occurred after specialization on the host (Dobinson *et al.*, 1993). Recently, however, some strains with few MGR copies have been recovered from rice (Levy *et al.*, 1993a), and rice-type profiles have been recovered from finger millet and weeds in the foothills of the Himalayas (R.S. Zeigler and M. Levy, unpublished) and from non-rice hosts in Korea (Han and R.R. Nelson, unpublished).

There is no evidence to suggest that MGR is other than a neutral marker. Thus it is ideal for generating 'fingerprints' that may be used to judge genetic relatedness among strains. Estimates of strain relatedness are generated from pair-wise comparisons between fingerprints, and grouped accordingly using cluster analysis (Levy *et al.*, 1991b). From these analyses can be derived phenograms which graphically illustrate putative phylogenetic relationships among clusters. Considering that only asexual (clonal) reproduction has been observed in nature, it is inferred that these discrete groups represent genetically related lineages. That is, the pathogen population is composed of distinct groups of individuals, with each group (lineage) derived through clonal propagation from a common ancestor. Distinct, but relatively closely related lineages may well share a common ancestor. If sexual recombination were occurring frequently in nature, a continuous variation in MGR profiles would be expected, but this is not the case. *Phytophthora infestans*, the late blight pathogen of potato, shows discrete lineage-like groupings where it reproduces in an asexual mode (Fry *et al.*, 1991), but continuous fingerprint variation in a sexually reproducing population (Matuszak *et al.*, 1990, 1991). It is conceivable that parasexual recombination in which whole chromosomes are exchanged could give rise to discrete sets of MGR profiles similar to those observed. However, in light of the apparently major chromosome rearrangements that have taken place, it is questionable how viable large-scale parasexual recombinants would be. None the less, small-scale parasexual exchanges between individuals could occur and be virtually indistinguishable from clonal lineages. Likewise, if the blast pathogen has 'epidemic' population structure (Maynard-Smith *et al.*, 1993) in which recombination occurs followed by explosive reproduction of the most fit recombinant, truly clonal descendants would be difficult to distinguish from recombinants.

A small set of strains from the south-eastern United States collected over 30 years from commercial fields and experimental plots, and previously characterized for the pathotypes on the international differential cultivar set, were analyzed using MGR (Levy *et al.*, 1991b). In a second study, 154 strains from Colombia (Levy *et al.*, 1991a,b), collected over two years from a blast-resistance breeding research station, were analyzed (Correa-Victoria and Zeigler, 1991). These studies grouped the isolates into eight lineages in the US and six lineages in Colombia, based on cluster analysis of MGR fingerprint similarities (Levy *et al.*, 1991a,b). Further analyses of collections from many different rice growing areas support the hypothesis that blast populations are organized as sets of discrete lineages (Levy *et al.*, 1993b).

Lineage structure of a population would be primarily an intellectual curiosity were it not somehow related to virulence potential. The US lineage analysis (Levy *et al.*, 1991b) showed a clear relationship between pathotype and lineage; in most cases there was a one-to-one correspondence. From Colombia (Levy *et al.*, 1993b) and the Philippines (Borromeo, 1990; E.S. Borromeo *et al.*, unpublished; Chen, 1993), a much more complex relationship between lineage and pathotype emerged. First, a host was often infected by more than one lineage. However, when Colombian isolates were tested for their compatibility with their host of origin, using a greenhouse assay, it was found that many were incompatible. When only those host-strain combinations that were demonstrably compatible were considered, there was a much closer relationship between host and lineage. Second, there were many more pathotypes identified in the site than lineages (Levy *et al.*, 1993a), yielding multiple pathotypes per lineage. However, over 90% of the pathotypes were found to be lineage-specific. Those pathotypes found in more than one lineage were typically those that were compatible on none, or only one of the international differentials. These might best be considered as 'undefined' pathotypes, rather than as a single pathotype.

Analysis of the multiple, and essentially unique, pathotypes found within each lineage in Colombia (Correa-Victoria and Zeigler, 1991) revealed that they were rather similar. Pathotypes within a lineage tended to differ by only one or a few compatible/incompatible (+/−) reactions on the international differential set. The full range of pathotypes in each lineage could be expressed as tight arrays of pathotypes, each separated from another member of the lineage by +/− change in virulence on a single host (Levy *et al.*, 1993a). This suggests that members of a lineage share related pathotypic traits, and suggests that individual lineages may have their particular virulence spectra. The observed networks of pathotypes within lineages could arise from limited variability, or instability, of pathotypes at specific loci.

The international set of differentials and corresponding pathotypes do not allow for a complete appreciation of the relationship between lineage and virulence. Correa-Victoria *et al.* (chapter 13, this volume) present evidence leading to the hypothesis that members of a lineage share a common limit to virulence, and that one may consider a lineage as having a specific virulence spectrum, or potential. That is, the data suggest that while pathotypic change may occur within a lineage, not all changes are possible for any given lineage. It appears that resistance in certain cultivars cannot be overcome by individuals of a particular lineage, while there are other cultivars for which compatibility is frequent in the lineage. Cultivars which fully exclude individuals from one lineage may be susceptible to most strains from another lineage, and vice versa. While the relationship between lineage and the pathotypes of individual isolates is quite complex in some locations, it is possible to construct a 'composite' pathotype of each lineage,

Table 16.1. Genetic diversity of blast fungus population at two sites in the Philippines, during the dry season and wet season, 1992.

	BN/DS[a]		BN/WS		Cavinti/WS	
Lineage	No. of isolates	No. of haplo.	No. of isolates	No. of haplo.	No. of isolates	No. of haplo.
1					44	9
4	18	3	39	2	178	12
7	55	16	537	58	227	10
14	162	6	94	19		
17	8	1	7	4	76	11
21					1	1
25					1	1
30					1	1
43					8	1
44					12	3
45					1	1
46					47	8
Total	243	26	677	83	596	52

[a] BN = Blast nursery at Los Baños, The Philippines; DS = dry season, WS = wet season. haplo. = haplotypes within MGR-defined lineage.

such that only hosts that are resistant to all tested members of a lineage are considered resistant to that lineage. If appropriate cultivars are selected for assessment of the virulence spectra of pathogen lineages, this assessment simultaneously provides useful information on the resistance spectra of host germplasm. Based on the composite pathotypes, resistance genes conditioning resistance to the full spectrum of pathogen lineages can be selected for a gene 'pyramiding' program.

Population characterization – Field composition and virulence spectra

Application of population analysis and the putative pathotype-lineage relationship to breeding will require characterization of target *P. grisea* populations in terms of lineage structure and lineage virulence spectra. To assess the feasibility of this, Chen (1993) began characterizing two breeding nurseries in the Philippines. One lowland nursery (BN) is located at IRRI headquarters in Los Baños, The Philippines, and the other upland nursery is located in Cavinti (Ca), approximately 40 km east of Los Baños. Approximately 900 isolates were collected from BN over two consecutive seasons and almost 600 isolates were collected from Ca during one season (Table 16.1). Isolates were collected from a set of cultivars known to have quite different resistance characteristics. Lineage analysis was conducted following Levy *et al.* (1991b).

Four lineages were recovered from BN, each present in both seasons; all of

Table 16.2. Relationship among cultivars of origin and common lineages sampled from two breeding nurseries in the Philippines.

	1[a]		4		7		14		17		44		
	BN	Ca[b]	BN	Ca	BN	Ca	BN	Ca	BN	Ca	BN	Ca	Total
A. asahi			1	1	28	15	6		1	1			53
Akashi		15		7		1	8						31
Azucena				1	11	5	7		2				26
C101A51											12	12	
C101PKT				43	12	6						61	
C104PKT			12	12		30						54	
C22			8	27		2	11		3	4			55
Carreon			19	10			7						36
Co39			1	2	46	10	10		4	1			74
Fujisaka5	26		14	1		9		3	3			53	
Tetep										37			37

[a] Lineages.
[b] Cultivars: BN = Blast nursery at IRRI, Los Baños, The Philippines; Ca = Cavinti, Laguna, The Philippines.

which were found in Ca (Table 16.1). Seven additional lineages were recovered from Ca, four of which were represented by only one isolate. With the exception of lineage 44, the most common lineages were recovered from more than one cultivar (Table 16.2). In order to characterize population structure accurately, sampling methodology will be a critical problem. Since it can never be proved that all lineages have been detected, the question of the importance of very rare lineages must be addressed.

There was striking apparent specificity between lineage 44 and C101A51 (a near-isogenic line carrying *Pi-2(t)*, a major resistance gene from line 5173 – see discussion on blast resistance below). Only lineage 17 was recovered from Tetep, a well-known resistance donor. Thus, although some lineages have broad spectra, there appear to be cases of specificity or at least restricted spectra – similar to the situation reported from Colombia (Levy *et al.*, 1993b). There was no discernible relationship between the pathotypes (as determined by inoculation on a set of 11 differentials, nine of which are near-isogenic for different major blast resistance genes), haplotypes and host of origins (Table 16.3), when a subset of isolates from the two sites were compared. This is consistent with the results of earlier studies (E.S. Borromeo, unpublished).

Several experiments were conducted to determine the virulence spectra of the lineages detected in the characterization of the BN and Ca sites. Some cultivars can fully exclude all the tested isolates from a particular lineage (Table 16.4). Also, there appears to be a specific adaptation of lineage 15 to Carreon. Such specificity is similar to that reported in the Colombian study where lineage SRL-1 was limited to the cultivar Cica 9. There are several combinations of resistance sources that, if combined in their entireties, would offer resistance against all these lineages. As seen in Table 16.4, there is complementarity among some of the

Table 16.3. Relationship among lineage, haplotype and pathotype.

Lineage	Site	Isolate no. (*n*)	No. host origin	No. haplo.	No. patho.[a]
1	Cavinti	15	3	5	1
4	Cavinti	9	4	3	7
4	BN	6	3	2	2
7	BN	24	16	6	6
14	BN	42	17	4	12
17	BN	7	5	1	5

[a] Based on 11 differentials: 9 near-isogenic lines, CO39 and C22.
Haplo. = haplotypes; patho. = pathotypes.

Table 16.4. Virulence spectrum of six Philippine lineages tested against 19 cultivars/lines.

	Philippine lineage[a]					
Test line	1 (47)	4 (62)	7 (51)	14 (46)	15 (45)	44 (12)
C101LAC	R[b]	R	24% S	17% S	R	R
C22	57% S	68% S	R	R	R	R
IAC47	60% S	R	R	R	R	R
O. Llanos5	R	R	R	R	R	R
HD14	R	R	R	R	R	R
Carreon	R	2% S	14% S	46% S	100% S	25% S
IRAT 13	R	R	R	R	R	R
IAC165	37% S	2% S	R	R	R	–
BR2	28% S	52% S	R	R	R	R
CO39	R	94% S	84% S	28% S	49% S	R
C101A51	R	R	R	R	R	58% S
IR8	R	76% S	R	20% S	R	R
OS6	64% S	R	2% S	R	R	R
C105TTP-4L-23	R	24% S	76% S	22% S	R	R
C102PKT	R	32% S	91% S	63% S	82% S	75% S
C104PKT	27% S	63% S	26% S	54% S	100% S	83% S
IR50	R	18% S	78% S	17% S	R	R
C101PKT	R	23%	78% S	22% S	4% S	R
IR36	R	11% S	59% S	16% S	R	R

[a] Lineages determined following Levy *et al.* (1991). Numbers in parentheses indicate number of isolates tested.
[b] R = All isolates tested were incompatible; % S = percentage of isolates listed showing compatibility.

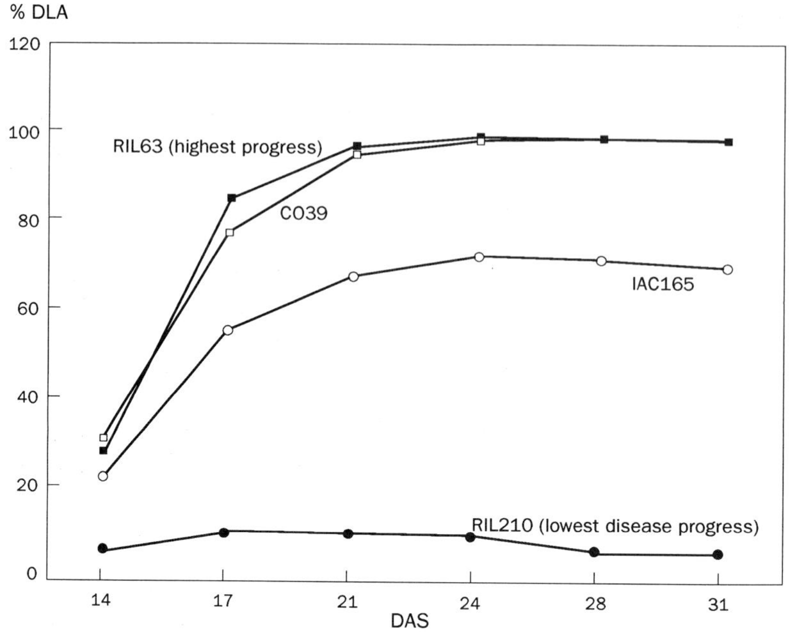

Fig. 16.1. Performance of two recombinant inbred lines (RIL) compared to their two parents when evaluated during the 1993 wet season at the upland rice blast breeding hot spot site in Cavinti, Laguna, The Philippines. DAS = days after seeding; DLA = diseased leaf area.

lineages that attack the *indica* cultivar CO39 and those that attack the *japonica* cultivar IAC165. As part of an RFLP mapping study, a population of recombinant inbred (RI) lines was derived from a cross between these two cultivars. The parental cultivars and the RI population were tested for blast reaction at the upland screening site at Cavinti, the Philippines. While both of the parents were badly blasted in the 1993 wet season (greater than 50% DLA), one of the RI lines showed a high degree of resistance (Fig. 16.1). We speculate that this line carries resistance factors with complementary lineage resistances. Further genetic analysis will allow the genetic factors involved to be identified and characterized relative to pathogen lineages.

A troubling aspect of the data in Table 16.4 is the common occurrence of low frequency compatibility, as in lineages 7 and 14 with C22, and lineage 4 with Carreon and IAC 165. If this reflects true compatibility within the lineage occurring at such low frequencies, then practical assessment of virulence spectrum becomes very difficult. That is, how is it possible to distinguish true lineage exclusion (assuming it is a legitimate phenomenon) from very infrequent, or not-yet-existent, potential virulence in a lineage? Insight can be gained by comparing the stability of the incompatibilities in lineage cultivar interactions characterized by different frequencies of compatibility. For example, are incompatible isolates equally likely to yield compatible progeny in the interactions between Carreon and lineage 1, 4, and 14; or, OS6 and lineages 1, 4 and 7? Such experiments are being undertaken at IRRI as a test of the lineage exclusion hypothesis.

Rice Blast Resistance

Japanese researchers have dedicated substantial effort to identifying and mapping major resistance genes in rice (Yamasaki and Kiyosawa, 1966; Kiyosawa, 1972; Kozaka, 1979). Genes were identified based upon a qualitative reaction of the rice plant to inoculation with a particular isolate. Although a scale (R, MR, M, MS, and S) of reaction was developed giving consideration to the number and relative frequency of lesions of particular types (resistant or susceptible) observed (Cho and Kiyosawa, 1973), there was considerable possibility for rating differences among observers. Data were generally grouped for calculation of R:S ratios. The Japanese researchers considered that 'true' blast resistance was characterized by almost no symptom development, or a hypersensitive reaction (Takase, as reported in Ezuka, 1972).

There have been at least 13 dominant genes at eight loci for blast resistance identified using this approach (Kiyosawa, 1972), some of which have been located in specific regions of rice chromosomes by classical linkage analysis (Yu *et al.*, 1987, 1991; Mackill and Bonman, 1992; Inukai *et al.*, 1993a,c; Wang *et al.*, 1994), and more recently, by RFLP linkage analysis (Tohme *et al.*, 1991; Yu *et al.*, 1991; Inukai, 1993). Allelism tests continue to show the genes have multiple alleles, or are very tightly linked (Inukai *et al.*, 1993a,c). Most genetic analyses of blast resistance have used only a few isolates. The analysis of major blast resistance genes using few isolates is analogous to pathotype analyses using only a few differential cultivars. When different isolates are used to test lines with previously identified genes, evidence for additional genes is typically found (Kiyosawa, 1972; Ou, 1979).

Those cultivars described as highly resistant, or with high levels of resistance observed over wide geographical ranges, are believed to have several major resistance genes (Nelson, 1978; Ou, 1979; Yu *et al.*, 1987). Furthermore, cultivars believed to have the same major gene composition have shown 'qualitative differences' when compared under field conditions (Takeda and Takahashi, 1970, as reported by Sudapak *et al.*, 1993). It is clear that the major gene component of blast resistance is quite complex.

In addition to 'true resistance', the Japanese also described the existence of 'field resistance', a characteristic possessed by cultivars showing less disease development in the field as compared with other cultivars believed to have the same major genes (Ezuka, 1972; Ikehashi and Kiyosawa, 1981). There has been no satisfactory quantification of this type of resistance, and it has been difficult to relate this with the type of 'field resistance' reported for potato (Thurston, 1979), or 'horizontal resistance' as proposed by Van der Plank (1975). Tests of 'field resistant' Japanese varieties in Philippine blast nurseries showed some to be highly susceptible, suggesting that the resistance expressed in the field in Japan was of a pathotype-specific nature, easily overcome by pathotypes present in the Philippines (Ou, 1979). Ou (1979) further speculated that the 'field resistant' cultivars actually contained 'broad spectrum' major genes.

The concept of partial resistance (Kozaka, 1971; Ahn and Ou, 1982; Marchetti, 1983; Johnson, 1984; Notteghem, 1985; Parlevliet, 1988; Bonman and Ahn, 1991) has been introduced into rice breeding for rice blast resistance in recent years. Partial resistance is characterized by compatibility between the pathogen and the plant, with reduced development of disease compared to plants with no partial resistance. This form of resistance is suitable for low to moderately blast conducive environments, but is insufficient in highly blast conducive environments (Bonman *et al.*, 1989, 1991; Bonman and Ahn, 1991). The genetics of partial resistance are very difficult to deduce, since there apparently is a substantial environmental effect on the phenomenon (Wang *et al.*, 1989), and it may be controlled by numerous genes. Parlevliet (1988) currently describes partial resistance as an incomplete, quantitative resistance based on minor genes. Several other researchers have suggested that there are 'minor genes' that play an important role in maintaining an acceptable level of disease under field conditions (Buddenhagen, 1981b). Such genes would be difficult to identify and characterize in the presence of 'major genes'. There has been some question as to whether there really is any difference between 'major' and 'minor' genes (Nelson, 1978; Ou, 1979). None the less, a strategy for identifying and increasing partial resistance in breeding programs is presented by Roumen (this volume).

With the development of molecular maps of the rice genome (McCouch *et al.*, 1988), remarkably rapid progress has been made in mapping of rice blast resistance genes. Near-isogenic lines (NILs) have been developed using some of the most important donors of blast resistance for breeding programs (Yu *et al.*, 1987, 1991; Mackill and Bonman). These are repeatedly back crossed with a susceptible parent, retaining only resistant progeny each generation. This approach has been used to extract single genes from the donor, and some have been mapped (McCouch *et al.*, this volume).

An alternative approach for developing mapping populations has been to develop a large number of recombinant inbred lines (RILs) from a cross between a donor and a susceptible parent. Progeny are advanced to approximately the F_7 generation with no selection (Wang *et al.*, 1994). It is expected that all resistance genes in the donor will be conserved, with some RILs containing only one gene, while others will carry genes in various combinations. A modified approach is to use rice anther culture on F_1 progeny of a resistant × susceptible cross, and generate a sufficient number of completely homozygous progeny to capture the full range of resistance gene combinations (Tohme *et al.*, 1991).

RILs developed from a cross between the 'durably resistant' upland rice cultivar Moroberekan and the susceptible CO39 have been used to dissect the resistance in Moroberekan into that derived from major genes and that derived from what appear to be quantitative trait loci (QTLs) (Wang *et al.*, 1994). Challenging RIL lines with isolates selected from different lineages has facilitated identification of distinct resistance genes (R.S. Zeigler, IRRI, unpublished). This cultivar has been shown now to have at least two major genes (Wang *et al.*, 1994). At least some genes appear to be located close together in what may be a gene cluster. Resistance gene clusters in other crops, e.g. maize (Sudapak *et al.*, 1993) have been shown to provide a means for generating redundancy through unequal meiotic crossing over. This redundancy may permit the development of new resistance genes, without loss of earlier resistance, in those progeny receiving the additional alleles. Major gene evolution in rice may be similarly dynamic.

The strategy used for the analysis of Moroberekan was designed to allow the

Table 16.5. Resistance spectrum of six NILs carrying single major genes for blast resistance.

		Cultivar: major genes					
Lineage[a]	*n*	*Pi-1(t)*	*Pi-2(t)*	*Pi-4b*	*Pi-4a*	*Pi-3*	*Pi-4a*
1	47	47/0[b]	47/0	46/1	46/1	45/2	47/0
4	62	62/0	62/0	47/15	42/20	23/39	48/14
7	46	35/11	46/0	11/35	4/39	34/12	10/36
14	46	38/8	46/0	36/10	16/27	21/25	36/10
15	45	44/1	45/0	45/0	8/37	0/45	43/2
44	12	12/0	5/7	12/0	3/9	2/10	12/0

[a] As determined by MGR fingerprint (Levy *et al.*, 1991b).
[b] Number of isolates showing incompatible reactions/number of isolates showing compatible reactions.

identification and mapping of QTLs (Wang *et al.*, 1994). A number of regions in the genome were identified as being associated with quantitative resistance to the pathogen. That is, RILs inheriting these regions from Moroberekan show reduced lesion number and lesion size when inoculated with a compatible isolate, and reduced disease development in polycyclic tests, when compared to lines with the region inherited from CO39. When tested in the field under blast conducive conditions in Indonesia and in the Philippines, lines with the most powerful QTLs, and with several QTLs, consistently showed reduced disease development when compared to lines with no QTLs. The apparent stability of the putative QTL effect across environment is encouraging, given the environmental sensitivity of other QTL-mediated characters (Paterson *et al.*, 1991). The QTLs as originally described (Wang *et al.*, 1994) were tested against only one isolate. Further tests, using two RILs that approach near isogenicity for the QTL at RG64, and 17 isolates from their different lineages, confirmed the quantitative nature of the resistance (Chen, 1993).

Characterization of resistance genes

Some of the major genes identified in the NIL and RIL studies have been characterized in terms of their virulence spectra against Philippine lineages identified with studies reported above. Rather than simply using different pathotypes, these lines were inoculated with large numbers of isolates representing lineages to determine if single major resistance genes could be effective in excluding some lineages. Some single genes were effective against all tested members of a lineage (Table 16.5). Note that isolates from lineage 44 were the only ones that could infect *Pi-2(t)*. As implied by the analysis of virulence spectrum of lineages presented above, genes which fully exclude members of one lineage may be relatively ineffective against members of another lineage, with the latter lineage being fully excluded by a gene that is ineffective against the first lineage. There is some ambiguity in the data in Table 16.2. Two NILs carrying *Pi-4a(t)* show some

Table 16.6. Proportion of isolates compatible to minor genes and average susceptible lesions relative to CO39.

Lineage	QTL@RG64[a] Comp./Total	QTL@RG64[a] Relative lesion no.	QTL@RG64 Comp./Total	QTL@RG64 Relative lesion no.
4	4/4	0.36[b]	4/4	0.4[b]
7	5/5	0.46[b]	3/3	0.41[b]
14	8/8	0.46[b]	4/4	0.39[b]

[a] QTLs in RIL23 and RIL276 derived from CO39/Moroberekan and linked to marker RG64 on the Cornell rice map (Wang *et al.*, 1994).
[b] Significant difference from the recurrent susceptible parent (CO39) at 0.05 level.

differences in reaction, suggesting that other factors are involved. RFLP analysis of these lines (Yu *et al.*, 1991) showed that frequency of non-recurrent parent RFLP markers ranged from 8% to over 20%. Inukai showed that one of the lines (C105-TTP-L23) has an additional R gene (Inukai *et al.*, 1993d).

It is possible that modifying factors, perhaps QTLs, remaining in the NILs from the non-recurrent parent result in some intermediate reactions that, in an 'R' versus 'S' judgement, are difficult to assess. The infrequent compatible reactions observed could be a result of this ambiguity. This would be far more pronounced in normal cultivars, explaining the variability in reactions (Bonman *et al.*, 1986), and, possibly, those reactions which appear to violate a 'lineage exclusion' relationship. Thus, the confounding effects of background, and possibly QTLs, may explain the 'off types' in the data presented in Table 16.1. While the overwhelming majority of interactions presented in Tables 16.4 and 16.5 are uniformly repeatable, it is these unexpected compatible reactions that may or may not be repeated. This phenomenon must be clarified before we can be reasonably confident that lineage exclusion is a promising avenue for resistance breeding.

Under controlled inoculations with strains from different lineages and with different pathotypes, RILs with QTLs consistently showed reduced disease development, compared to the recombinant susceptible parent, CO39. Table 16.6 shows the general reduction in lesion number on RILs with QTLs, as compared to the recurrent susceptible parent. It is interesting to note that seven isolates from lineage 44 were incompatible with both RILs (data not shown). This suggests that these RILs may carry a previously unidentified major gene, possibly linked to the QTL, with a narrow resistance spectrum. Alternatively, the putative QTL itself may act as a narrow-spectrum major gene.

It was noted that some QTLs were found to be linked with RFLP markers (e.g. RG869B, RG333, RG64) (Wang *et al.*, 1994) previously shown to be linked to major resistance genes (McCouch *et al.*, 1988). This is consistent with previous speculation that QTLs simply reflect the residual effect of defeated major genes (cf. Ellingboe, 1975). If this were the case, then NILs with major genes should consistently show reduced disease development when inoculated with compatible

Table 16.7. Proportion of isolates compatible to respective major genes and average susceptible lesions relative to CO39.

	Pi-1(t)[a]		*Pi-3(t)*		*Pi-4(t)*	
Lineage	Comp./ Total	Relative lesion no.	Comp./ Total	Relative lesion no.	Comp./ Total	Relative lesion no.
4	0/4	0	1/4	1.93[b]	3/4	0.54[b]
7	3/6	1.67[b]	1/5	1.72[b]	5/5	1.24[b]
14	5/7	1.11	3/8	1.84[b]	8/8	0.78[b]

[a] Sources of resistance are: *Pi-1(t)* = LAC23; *Pi-2(t)* = 5173; *Pi-3(t)* = Pai Kan Tao; $Pi\text{-}4_b(t)$ = Tetep; in CO39 background.
[b] Significant difference at 0.05 level.

isolates, as compared to the susceptible recurrent parent. This hypothesis was tested by inoculating NILs containing *Pi-1(t)*, *Pi-3(t)*, and *Pi-4(t)* with compatible isolates and comparing disease development (number of susceptible type lesions) with that of CO39, the recombinant susceptible parent, inoculated with the same isolates. Again, those RILs that were nearly isogenic at RG64 were tested.

It was observed that usually there were significantly more compatible (type 4) lesions on NILS then on CO39 (Table 16.7). Even in those sets where the mean relative lesion number is less then 1 (e.g. lineage 14 on *Pi-4(t)*), with some isolates relative lesion number on the NIL was equal to or greater than on CO39. In no instance was a consistent reduction in disease observed. Overall, there were as many lines showing increased susceptibility as there were lines showing reduced susceptibility. Therefore, for these blast resistance genes there is no evidence to support the hypothesis that they will confer residual protection to the plant when challenged with compatible isolates.

Apparently, once major genes are overcome, they may pose no further obstacle to the pathogen's colonization of the host. It seems likely that the observed differences among cultivars with the same major resistance genes in the rates and severity of disease development are functions of other sets of resistance genes, as has been suggested by a number of authors (e.g. Kiyosawa, 1972; Ou, 1979). The QTLs identified in the Moroberekan analysis fit the description of these genes, and these same kinds of genes may well be those responsible for partial resistance (Wang *et al.*, 1989). Resistance in rice to blast may be conditioned by major genes and/or QTLs, and its overall effectiveness depends upon the combined resistance spectrum of the major genes, their durability, and the strength of the underlying QTLs. An effective breeding strategy for durable resistance should be directed to combining both these and maximizing their contributions.

Breeding for Resistance

With such uncertainty surrounding the variability of the pathogen and the history of resistance breakdown it is not surprising that a number of different plant breeding approaches have been proposed to achieve durable blast resistance. Most approaches have common features, and we believe that all the elements for a successful approach have been included when the full range of concepts and approaches is examined. Until very recently, however, the tools to apply the concepts for breeding for durable blast resistance have been lacking. Molecular techniques for characterizing pathogen virulence, pathogen populations, and host plant resistance, should offer the means to test hypotheses on the nature of durable resistance.

Proposals for the use of major genes have moved beyond developing lines with single genes effective against a few pathotypes. Robinson's (1973) and Nelson's (1978) general proposal that major genes may be combined, or 'pyramided', to confer 'horizontal' resistance effective against all pathotypes of a pathogen, was supported by Ou (1979) as a means to achieve reasonably durable resistance to blast. Crill *et al.* (1981) proposed that varieties with different major genes could be rotated in and out of production in a particular region to take advantage of virulence shifts in the pathogen population driven by the changes in varietal resistance. MacKenzie (1979) proposed the use of multilines, or a mixture of lines, each carrying one or a few major genes. He emphasized the care with which the resistance genes should be manipulated, and that their usefulness and durability will depend on how they are deployed. Each of these approaches demands that the effectiveness of the resistance gene be known, that there be a means of manipulating the genes, and that the composition and the full virulence characteristics of the pathogen in the presumed target region be understood.

Other rice scientists have argued that major genes obscure the underlying resistance of the plant and that use of these genes may result in lines that are extremely susceptible when the major genes are overcome (Roumen, this volume) (the so-called 'vertifolia effect') (Van der Plank, 1968). That is, there will be no means to select for the minor genes, and in crosses where the parents differ markedly in their composition there is a high probability that most minor genes of interest will be lost. We can presume that this underlying resistance is functionally equivalent to that mediated by QTLs. Various proposals have been put forth to avoid erosion of resistance, escape, and the 'vertifolia effect'. These include: breeding in environments that are extremely conducive to blast disease development ('hot spots' in which narrow spectrum, 'vertical' resistance will break down quickly); rejection of breeding lines that are symptomless, or show signs of the hypersensitive reaction typical of extreme vertical resistance; use of exotic, preferably traditional, and 'unimproved', sources of resistance obtained from blast conducive environments; and good characterization of the target environment (Kiyosawa, 1972; Ezuka, 1979; Ou, 1979; Buddenhagen, 1981a,b, 1983a,b; Bonman *et al.*, 1992). An approach integrating most of these suggestions has been applied with some success at the Centro Internacional de Agricultura Tropical (CIAT) rice breeding research farm in eastern Colombia (Correa-Victoria *et al.*, 1989, 1992; Correa-Victoria and Zeigler, 1991) where materials are developed for distribution to rice programs in Latin America. Latin America tends to have serious blast problems because nitrogen application rates are high, water control in irrigated fields is usually poor, and vast areas are grown under upland conditions.

A word of caution should be raised regarding the use of 'hot spot' breeding sites. By developing complex resistance under conditions of high pathogen diversity and unknown variability, breeders may be allowing the pathogen to adapt to novel resistance gene combinations. Conditions in these nurseries promote coexistence of diverse lineages at relatively high frequencies, as compared to normal field conditions. Furthermore, 'neutral' cultivars that may be susceptible to a relatively large number of lineages are commonly used to increase the inoculum load. Thus, while in nature it may be rare for different lineages to come into contact needed for a recombination event to occur, we may be creating just the conditions that will favor what was formerly very rare.

In contrast to CIAT, IRRI has opted to focus on developing partial resistance for the less blast conducive irrigated rice environments. Breeding lines are evaluated in blast nurseries (Bonman *et al.*, 1992) and those with relatively reduced disease development are selected as partially resistant. Cultivars developed in this manner have generally performed well in the field, but when conditions are particularly favorable for disease development, economic losses may occur.

Lineage Exclusion Hypothesis

'Lineage exclusion' is a term coined to summarize the following hypotheses, which were formulated to explain the observations described in the previous text:

- Rice blast populations are composed of sets of discrete lineages.
- Each lineage has a specific virulence spectrum, characterized by uniform incompatibility with one or a combination of host resistance.
- In every lineage certain incompatibilities are very difficult to overcome.
- Rice blast resistance genes may be effective against all isolates in a lineage.
- Those resistance genes effective against members of one lineage may be relatively ineffective against members of another lineage.
- Combinations of resistance genes in a cultivar that, in sum, confer resistance to every lineage, will confer resistance to the entire population.
- The durability of resistance developed by combining resistances effective against all members of all lineages will be substantially longer than those combined based on field performance by donors or incompatibility with selected races.

Application of Lineage Exclusion to Breeding

Although the success of breeding for blast resistance has improved over the past decades, there is substantial room for improvement. Current approaches are working virtually blind, in that there is little understanding of the virulence characteristics of the target pathogen population, and it is impossible in a practical sense to know which major resistance genes and/or QTLs are being carried by a particular breeding line. We propose that a breeding program with blast resistance as one of its principal objectives should be structured such that:

1. major genes are combined to exclude the known lineages in a target region; resistance gene combinations that cannot confer a lineage exclusion effect are discarded;
2. major genes are combined in such a way that overlapping resistance spectra form at least two barriers to the existing lineages; and
3. since lineage exclusion can never be established without doubt, the major gene resistance is supported by a high level of general resistance conferred by QTLs.

We further suggest that this approach be incorporated into mainstream breeding programs when blast resistance is a priority. If resistance of this complexity is generated via a separate project, it will surely be lost when attempts are made to incorporate it into advanced lines. The development of such an approach will require the following interrelated and interdependent steps:

1. *Characterization of the pathogen population*. The pathogen population from a specifically delineated target region must be sampled, its lineage composition determined, and the virulence spectra of the different lineages determined. This will require the development of an effective sampling technique that will assure that all lineages present in a region are likely to be included in the sample. Ultimately, we might work towards developing trap nurseries composed of lines with combinations of resistance genes known to select for the full range of viable pathotype configurations (assuming that there is a finite and 'knowable' set).
2. *Mapping of major and QTL resistance genes*. Resistance genes must be fine mapped and identified with very tightly linked flanking markers. Eventually probes developed from cloned segment of the genes will permit certain identification of a gene in a breeding line.
3. *Characterization of resistance genes*. The resistance spectra of the major virulence genes must be established in NILs. Keep testing diverse isolates from each lineage until compatibility is encountered, or a large number of incompatible reactions have been scored. Their functioning in combination must be assessed. This will be required to develop an effective lineage sampling approach for pathogen population analysis. The most effective QTLs must be identified and a manageable number selected for inclusion in a breeding program.
4. *Development of a marker-aided breeding program*. To effectively combine genes in a breeding program there must be an inexpensive and convenient means to determine if breeding lines carry the genes of interest. Presently available methods are still too cumbersome for incorporation into routine breeding programs.
5. *Combination of major genes and QTLs*, such that they exclude the lineages present in a target population. The combination should be such that the resistance spectra overlap for each lineage (Table 16.5).

The approach described is a complex and long-term endeavor. Fortunately much of the arduous task of resistance gene mapping and tagging upon which much of this strategy depends will have to be done regardless of which breeding strategy eventually predominates. The concept of lineage exclusion as a means to achieve durable resistance remains to be proved, and depends upon the truly clonal nature of reproduction in the pathogen. If significant amounts of recombination occur such that limits to virulence are readily overcome, the concept will have much less applicability. Fortunately, the evidence to date indicates that such recombination in the field is extremely uncommon. Since each geographical region seems to have its own set of lineages, the description of population structure will have to be undertaken locally. With careful planning important questions regarding the relationship between underlying pathotypic configurations and lineage structure can be answered in the process.

References

Ahn, S.W. and Ou, S.H. (1982) Quantitative resistance of rice to blast disease. *Phytopathology* 72, 279–282.

Atkins, J.G., Robert, A.L., Adair, C.R., Goto, K., Kozaka, T., Yanagida, R., Yamada, M. and Matsumoto, S. (1967) An international set of rice varieties for differentiating races of *Pyricularia oryzae*. *Phytopathology* 57, 297–301.

Bonman, J.M. and Ahn, S.W. (1991) Managing rice blast disease–a continuing challenge. In: *Focus on Irrigated Rice. Proceedings, 1990 International Rice Research Conference*. International Rice Research Institute, PO Box 933, Manila, The Philippines.

Bonman, J.M., Vergel de Dios, T.I. and Khin, M.M. (1986) Physiologic specialization of *Pyricularia oryzae* in the Philippines. *Plant Disease* 70, 767–769.

Bonman, J.M., Bandong, J.M., Lee, Y.H., Lee, E.J. and Valent, B. (1989) Race-specific partial resistance to blast in temperate japonica rice cultivars. *Plant Disease* 73, 496–499.

Bonman, J.M., Estrada, B.A., Kim, C.K., Ra, D.S. and Lee, E.J. (1991) Assessment of blast disease and yield loss in susceptible and partially resistant rice cultivars in two irrigated lowland environments. *Plant Disease* 75, 462–466.

Bonman, J.M., Khush, G.S. and Nelson, R.J. (1992) Breeding rice for resistance to pests. *Annual Review of Phytopathology* 30, 507–528.

Borromeo, E.S. (1990) Molecular characterization of *Pyricularia oryzae* Cav. population from rice and other hosts. Unpublished PhD Thesis, University of the Philippines at Los Baños, The Philippines, 124 pp.

Borromeo, E.S., Nelson, R.J., Bonman, J.M. and Leung, H. (1993) Genetic differentiation among isolates of *Pyricularia* infecting rice and weed hosts. *Phytopathology* 83, 393–399.

Buddenhagen, I.W. (1981a) Conceptual and practical considerations when breeding for tolerance or resistance. In: *Plant Disease Control: Resistance and Susceptibility*. Wiley, New York, pp. 221–234.

Buddenhagen, I.W. (1981b) Practical breeding for yield stability and durable resistance to rice blast. Paper presented at the IRAT Blast Symposium, Paris, 23 pp.

Buddenhagen, I.W. (1983a) Breeding strategies for stress and disease resistance in developing countries. *Annual Review of Phytopathology* 21, 385–409.

Buddenhagen, I.W. (1983b) Agroecosystems, disease resistance, and crop improvement. In: Kommedahl, T. and Williams, P.H. (eds), *Challenging Problems in Plant Health*. American Phytopathological Society, St. Paul, Minnesota, pp. 450–460.

Chen, D.-H. (1993) Population structure of *Pyricularia grisea* at two screening sites and quantitative characterization of major and minor resistance genes. PhD Thesis, University of the Philippines at Los Baños, Laguna, The Philippines.

Chin, K.M. (1985) Virulence analysis as a tool in disease management. In: *Proceedings*,

1985 Jakarta Conference. International Rice Research Institute, PO Box 933, Manila, The Philippines, pp. 393–401.

Cho, C.I. and Kiyosawa, S. (1973) Blast resistance in Korean rice varieties. *Research Reports of the Office of Rural Development (Korea)* 15, 83–88.

Correa-Victoria, F.J. and Zeigler, R.S. (1991) Stable resistance and pathogenic variability in the rice–*Pyricularia oryzae* complex. In: *Rice in Latin America: Improvement, Management, and Marketing*. Centro Internacional de Agricultura Tropical (CIAT), Cali, Colombia, p. 240.

Correa-Victoria, F.J., Zeigler, R.S., Weber, G. and Sarkarung, S. (1989) Metodologias desarrolladas por el Programa de Arroz del CIAT para caracterizacion de germoplasma. In: *Evaluacion Cooperativa del Germoplasma de Arroz en America Latina*. Centro Internacional de Agricultura Tropical (CIAT), Cali, Colombia, 286 pp.

Correa-Victoria, F.J., Zeigler, R.S., Tohme, J. and Levy, M. (1992) Field breeding for durable blast resistance in the presence of diverse and high pathogen populations. Paper presented at the *Symposium on Durability of Disease Resistance*, 24–28 February 1992, IAC, Wageningen.

Crawford, M.S., Chumley, F.G., Weaver, C.G. and Valent, B. (1986) Characterization of the heterokaryotic and vegetative diploid phases of *Magnaporthe grisea*. *Genetics* 114, 1111–1129.

Crill, P., Ham, Y.S. and Beachell, H.M. (1981) The rice blast disease in Korea and its control with race prediction and gene rotation. *Korean Journal of Breeding* 13(2), 106–114.

Cuevas-Perez, F. and Gaona, J.S. (1988) Disease selection in rice in Colombia and Central America. *International Rice Research Newsletter* 13(1), 14–15.

Dobinson, K.F., Harris, R.E. and Hamer, J.E. (1993) Grasshopper, a long terminal repeat (LTR) retroelement in the phytopathogenic fungus *Magnaporthe grisea*. *Molecular Plant-Microbe Interactions* 6(1), 114–126.

Ellingboe, A.H. (1975) Horizontal resistance: An artifact of experimental procedure? *Australian Plant Pathology Society Newsletter* 4, 44–46.

Ellingboe, A.H. (1992) Segregation of avirulence/virulence on three rice cultivars in 16 crosses of *Magnaporthe grisea*. *Phytopathology* 82, 597–601.

Ellingboe, A.H., Wu, B.C. and Robertson, W. (1990) Inheritance avirulence/virulence in a cross of two isolates of *Magnaporthe grisea* pathogenic to rice. *Phytopathology* 80, 108–111.

Ezuka, A. (1972) Field resistance of rice varieties to blast disease. *Review of Plant Protection* 5, 1–12.

Ezuka, A. (1979) Breeding for and genetics of blast resistance in Japan. In: *Proceedings, Rice Blast Workshop*. International Rice Research Institute, PO Box 933, Manila, The Philippines, pp. 27–48.

Flor, H.H. (1971) Current status of the gene-for-gene concept. *Annual Review of Phytopathology* 9, 275–296.

Fry, W.E., Drenth, A., Spielman, L.V., Mantel, B.C., Davidse, L.C. and Goodwin, S.B. (1991) Population genetic structure of *Phytophthora infestans* in the Netherlands. *Phytopathology* 81, 1330–1336.

Giatgong, P. and Frederiksen, R.A. (1969) Pathogenic variability and cytology of monoconidial subcultures of *Pyricularia oryzae*. *Phytopathology* 59, 1152–1157.

Hamer, J.E., Farrall, L., Orbach, M.J., Valent, B. and Chumley, F.G. (1989a) Host species-specific conservation of a family of repeated DNA sequences in the genome of a fungal plant pathogen. *Proceedings of the National Academy of Sciences of the United States of America* 86, 9981–9985.

Hamer, J.E., Valent, B. and Chumley, F.G. (1989b) Mutations at the SMO genetic locus affect the shape of diverse cell types in the rice blast fungus. *Genetics* 122, 351–361.

Heath, M.C., Valent, B., Howard, R.J. and Chumley, F.G. (1990) Interactions of two strains of *Magnaporthe grisea* with rice, goosegrass, and weeping lovegrass. *Canadian Journal of Botany* 68, 1627–1637.

Ikehashi, I. and Khush, G.S. (1979) Breeding for blast resistance at IRRI. In: *Proceedings, Rice Blast Workshop*. International Rice Research Institute, PO Box 933, Manila, The Philippines, pp. 69–80.

Ikehashi, H. and Kiyosawa, S. (1981) Strain-specific reaction of field resistance of Japanese rice varieties revealed with Philippine strains of rice blast fungus, *Pyricularia oryzae* Cav. *Japan Journal of Breeding* 31(3), 293–301.

Inukai, T., Mackill, D.J., Bonman, J.M., Sarkarung, S., Zeigler, R.S., Nelson, R.J., Takamure, I. and Kinoshita, T. (1993a) Blast resistance genes (pi-2(t) and Pi-z) may be allelic. *Rice Genetics Newsletter* 9, 90–92.

Inukai, T., Mackill, D.J., Bonman, J.M., Sarkarung, S., Zeigler, R.S., Nelson, R.J., Takamure, I., Kinoshita, T. (1993b) A true blast resistance gene Pi-3(t) in near-isogenic line C104PKT. *Rice Genetics Newsletter* 9, 94–95.

Inukai, T., Nelson, R., Sarkarung, S. and Zeigler, R.S. (1993c) Allelism between blast (bl) resistance genes Pi-4a(t) and Pi-ta. *International Rice Research Newsletter* 18, 15.

Inukai, T., Nelson, R.J., Sarkarung, S., Mackill, D.J., Bonman, J.M., Takamure, I. and Kinoshita, T. (1993d) Allelism of blast resistance genes in near-isogenic lines of rice. *Phytopathology* (in press).

Johnson, R. (1984) A critical analysis of durable resistance. *Annual Review of Phytopathology* 22, 309–330.

Kato, H. and Yamaguchi, T. (1982) The perfect state of *Pyricularia oryzae* Cav. from rice plants in culture. *Annals of the Phytopathology Society of Japan* 48, 607–612.

Kiyosawa, S. (1972) Genetics of blast resistance. In: *Rice Breeding*. International Rice Research Institute, PO Box 933, Manila, The Philippines, pp. 203–225.

Kozaka, T. (1971) Recent advances in studies on horizontal resistance to blast disease of rice in Japan. *Proceedings, Seminar on Horizontal Resistance to the Blast Disease of Rice*. Centro Internacional de Agricultura Tropical (CIAT), Colombia.

Kozaka, T. (1979) The nature of the blast fungus and varietal resistance in Japan. In: *Proceedings, Rice Blast Workshop*. International Rice Research Institute, PO Box 933, Manila, The Philippines, pp. 3–25.

Latterell, F.M. (1971) Phenotypic stability of pathogenic races of *Pyricularia oryzae*, and its implications for breeding of blast resistant rice varieties. In: *Proceedings, Seminar on Horizontal Resistance to Blast Disease of Rice*. Centro Internacional de Agricultura Tropical (CIAT), Cali, Colombia, pp. 199–234.

Latterell, F.M. (1972) Two views of pathogenic stability in *Pyricularia oryzae*. *Phytopathology* 62, 771 (Abstract).

Latterell, F.M. and Rossi, A.E. (1986) Longevity and pathogenic stability of *Pyricularia oryzae*. *Phytopathology* 76, 231–235.

Levy, M., Correa, F.J., Zeigler, R.S., Xu, S. and Hamer, J.E. (1991a) Organization of genetic and pathotype variation in the rice blast fungus at a Colombian 'hot spot'. *Phytopathology* 81, 1236–1237.

Levy, M., Romao, J., Marchetti, M.A. and Hamer, J.E. (1991b) DNA fingerprinting with dispersed repeated sequence resolves pathotype diversity in the rice blast fungus. *Plant Cell* 3, 95–102.

Levy, M., Correa-Victoria, F.J., Zeigler, R.S., Shen, Y., Shahjahan, A.K.M., Gnanamanickam, S.S., Nelson, R.J., Manry, J. and Hamer, J.E. (1993a) International atlas of genetic diversity in the rice blast fungus: 1992–1993. Paper presented at the Annual Rockefeller Foundation Rice Biotechnology Network, Chiang Mai, Thailand, February 2–5 1993.

Levy, M., Correa, F.J., Zeigler, R.S., Xu, S. and Hamer, J.E. (1993b) Genetic diversity of the rice blast fungus in a disease nursery in Colombia. *Phytopathology* 83, 1427–1433.

Leung, H. and Williams, P.H. (1985) Genetic analyses of electrophoretic enzyme variants, mating type, and hermaphroditism in *Pyricularia oryzae* Cavara. *Canadian Journal of Genetics and Cytology* 27, 697–704.

Leung, H. and Williams, P.H. (1986) Enzyme polymorphism and genetic differentiation among geographic isolates of the rice blast fungus. *Phytopathology* 76, 778–783.

Leung, H., Borromeo, E.S., Bernardo, M.A. and Notteghem, J.L. (1988) Genetic analysis of virulence in the rice blast fungus *Magnaporthe grisea. Phytopathology* 78, 1227–1233.

Ling, K.C. and Ou, S.H. (1969) Standardization of the international race numbers of *Pyricularia oryzae. Phytopathology* 59, 339–342.

MacKenzie, D.R. (1979) The multiline approach to the control of some cereal disease. In: *Proceedings, Rice Blast Workshop*. International Rice Research Institute, PO Box 933, Manila, The Philippines, pp. 199–216.

Mackill, D.J. and Bonman, J.M. (1992) Inheritance of blast resistance in near-isogenic lines of rice. *Phytopathology* 82, 746–749.

Marchetti, M.A. (1983) Dilatory resistance to rice blast in USA rice. *Phytopathology* 73, 645–649.

Matuszak, J.M., Goodwin, S.B., Fry, W.E. and Villareal-Gomez, M.J. (1990) Changes in the genetic diversity of *Phytophthora infestans* during an epidemic in Central Mexico as determined by DNA fingerprints. 1990 Annual Meeting of the American Phytopathological Society, 4–8 August 1990, Grand Rapids, Michigan (Abstract).

Matuszak, J.M., Goodwin, S.B., Fry, W.E., Villareal-Gomez, M.J. and Fernandez-Elguezabal, V. (1991) The genetic structure of the *Phytophthora infestans* population in the Toluca Valley as determined by molecular markers. American Phytopathology Society Annual Meeting, 17–21 August 1991, St. Louis, Missouri.

Maynard-Smith, J., Smith, N., O'Rourke, M. and Spratt, B. (1993) How clonal are bacteria? *Proceedings of the National Academy of Sciences of the United States of America* 90, 4384–4388.

McCouch, S.R., Kochert, G., Yu, Z.H., Wang, Z.Y., Khush, G.S., Coffman, W.R. and Tanksley, S.D. (1988) Molecular mapping of rice chromosomes. *Theoretical and Applied Genetics* 76, 815–829.

Nelson, R.R. (1978) Genetics of horizontal resistance to plant diseases. *Annual Review of Phytopathology* 16, 359–378.

Notteghem, J.L. (1985) Definition d'une strategie d'utilisation de la resistance par analyse genetique des relations hote-parasite cas due couple riz-*Pyricularia oryzae. L'Agronomie Tropicale* 40(2), 129–147.

Notteghem, J.L. and Silué, D. (1992) Distribution of the mating type alleles in *Magnaporthe grisea* populations pathogenic on rice. *Phytopathology* 82, 421–424.

Ou, S.H. (1971) Variability of *Pyricularia oryzae* Cav. and its relation to varietal resistance. *Proceedings, Seminar on Horizontal Resistance to the Blast Disease of Rice*. Centro Internacional de Agricultura Tropical (CIAT), Colombia.

Ou, S.H. (1972) Studies on stable resistance to rice blast disease. In: *Rice Breeding*. International Rice Research Institute, PO Box 933, Manila, The Philippines, pp. 227–237.

Ou, S.H. (1979) Breeding rice for resistance to blast – a critical review. In: *Proceedings, Rice Blast Workshop*. International Rice Research Institute, PO Box 933, Manila, The Philippines, pp. 81–137.

Ou, S.H. (1985) *Rice Diseases*, 2nd edn. Commonwealth Mycological Institute, Kew, UK, 380 pp.

Ou, S.H. and Ayad, M.R. (1968) Pathogenic races of *Pyricularia oryzae* originating from single lesions and monoconidal cultures. *Phytopathology* 58, 179–182.

Ou, S.H., Nuque, F.L., Ebron, T.T. and Awoderu, V. (1970) Pathogenic races of *Pyricularia oryzae* derived from monoconidial cultures. *Plant Disease Reporter* 54(12), 1045–1049.

Ou, S.H., Nuque, F.L., Ebron, T.T. and Awoderu, V.A. (1971) A type of stable resistance to blast disease of rice. *Phytopathology* 61, 703–706.

Ou, S.H., Nuque, F.L. and Bandong, J.M. (1975) Relation between qualitative and quantitative resistance to rice blast. *Phytopathology* 65, 1315–1316.

Parlevliet, J.E. (1988) Identification and evaluation of quantitative resistance. In: Leonard, K.J. and Fry, W.E. (eds), *Plant Disease Epidemiology, Genetics, Resistance, and Management*, Vol. 2. McGraw-Hill, New York, 377 pp.

Paterson, A.H., Damon, S., Hewitt, J.D., Zamir, D., Rabinowitch, H.D., Lincoln, S.E., Lander, E.S. and Tanksley, S.D. (1991) Mendelian factors underlying quantitative traits in tomato: comparison across species, generations, and environments. *Genetics* 127, 181–197.

Robinson, R.A. (1973) Horizontal resistance. *Review of Plant Pathology* 52, 483–501.

Romao, J. and Hamer, J.E. (1992) Genetic organization of a repeated DNA sequence family in the rice blast fungus. *Proceedings of the National Academy of Sciences of the United States of America* 89(12), 5316–5320.

Row, K.V.S.R.K., Aist, J.R. and Crill, J.P. (1985) Mitosis in the rice blast fungus and its possible implications for pathogenic variability. *Canadian Journal of Botany* 63, 1129–1134.

Silué, D. and Notteghem, J.L. (1992) Identification of a cross between two compatible isolates of *Magnaporthe grisea* (Hebert) Barr and genetic analysis of avirulence/virulence of rice. *Journal of Phytopathology* 135, 77–83.

Silué, D., Notteghem, J.L. and Tharreau, D. (1992) Evidence of a gene-for-gene relationship in the *Oryza sativa–Magnaporthe grisea* pathosystem. *Phytopathology* 82, 577–580.

Sudapak, M.A., Bennetzen, J.L. and Hulbert, S.H. (1993) Unequal exchange and meiotic instability of disease resistance genes in the Rp1 region of maize. *Genetics* 133, 119–125.

Talbot, N.J., Salch, Y.P., Ma, M. and Hamer, J.E. (1993) Karyotypic variation within clonal lineages of the rice blast fungus, *Magnaporthe grisea. Applied and Environmental Microbiology* 59, 585–593.

Thurston, H.D. (1979) Breeding potatoes for late blight resistance. In: *Proceedings, Rice Blast Workshop*. International Rice Research Institute, PO Box 933, Manila, The Philippines, pp. 141–152.

Tohme, J., Montenegro, M., Correa, F., Martinez, C., Zeigler, R.S. and Roca, W. (1991) Tagging resistance genes to Rice Hoja Blanca Virus and Colombian isolates of rice blast with RFLP and RAPD markers. Fifth Annual Meeting of the Rockefeller Foundation Biotechnology Network, Tucson, Arizona 5 (Abstract).

Valent, B. (1990) Rice blast as a model system for plant pathology. *Phytopathology* 80(1), 33–36.

Valent, B. and Chumley, F.G. (1991) Molecular genetic analysis of the rice blast fungus, *Magnaporthe grisea. Annual Review of Phytopathology* 29, 443–467.

Valent, B., Crawford, M.S., Weaver, C.G. and Chumley, F.G. (1986) Genetic studies of fertility and pathogenicity in *Magnaporthe grisea* (*Pyricularia oryzae*). *Iowa State Journal of Research* 60, 569–594.

Valent, B., Farrall, L. and Chumley, F.G. (1991) *Magnaporthe grisea* genes for pathogenicity and virulence identified through a series of backcrosses. *Genetics* 127, 87–101.

Van der Plank, J.E. (1968) *Disease Resistance in Plants*. Academic Press, New York, 206 pp.

Van der Plank, J.E. (1975) Horizontal resistance: six projects in relation to blast disease of rice. In: *Horizontal Resistance to the Blast Disease of Rice*. CIAT Ser. CE-9. Centro Internacional de Agricultura Tropical (CIAT), Cali, Colombia, pp. 21–26.

Wang, Z., Mackill, D.J. and Bonman, J.M. (1989) Inheritance of partial resistance to blast in indica rice cultivars. *Crop Science* 29, 848–853.

Wang, G-L., Mackill, D.J., Bonman, J.M., McCouch, S. and Nelson, R. (1994) RFLP mapping of genes conferring complete and partial resistance to blast in a durably resistant rice cultivar. *Genetics* 136, 1421–1434.

Yamada, M.S., Kiyosawa, S., Yamaguchi, T., Hirano, T., Kobayashi, T., Kushibuchi, K., and Watanabe, S. (1976) Proposal of a new method for differentiating races of *Pyricularia oryzae* Cavara in Japan. *Annals of the Phytopathology Society of Japan* 42, 216–219.

Yamasaki, Y. and Kiyosawa, S. (1966) Studies on inheritance of resistance of rice varieties to blast. I. Inheritance of resistance of Japanese varieties to several strains of the fungus. *Bulletin of the National Institute of Agricultural Sciences Ser.* D 14, 39–69.

Yu, Z.H., Mackill, D.J. and Bonman, J.M. (1987) Inheritance of resistance to blast in some traditional and improved rice cultivars. *Phytopathology* 77, 323–326.

Yu, Z.H., Mackill, D.J., Bonman, J.M. and Tanksley, S.D. (1991) Tagging genes for blast resistance in rice via linkage to RFLP markers. *Theoretical and Applied Genetics* 81, 471–476.

17 Use of Host Genetic Diversity to Control Cereal Diseases: Implications for Rice Blast

C.C. MUNDT

Department of Botany and Plant Pathology, Oregon State University, Cordley Hall 2082, Corvallis, OR 97331-2902, USA, and Division of Plant Pathology, International Rice Research Institute, PO Box 933, Manila, The Philippines

INTRODUCTION

Selection within a population results in a decrease of genetic diversity. This is merely the definition of selection, and is true of selection directed both by nature and by humans. What has distinguished agricultural genetics in the 20th century, however, has been the rapidity and degree of selection imposed on crop populations by humans. This process of intensive selection has resulted in both benefits and costs. For example, modern plant breeders have achieved yield increases that are nothing short of remarkable. In field crops, plant breeding has generally accounted for about half of the total yield advances that have occurred over the past century (e.g. Jensen, 1978).

The reduction of crop genetic diversity through plant breeding has sometimes also resulted in problems, however. Genetic uniformity can make crops vulnerable to biotic and abiotic stresses. The most striking examples are the 'boom and bust' cycles brought about by the release and subsequent extensive culture of crop cultivars with race-specific disease or pest resistance (e.g. Browning *et al.*, 1969; Mew, 1987; Bonman *et al.*, 1992). These boom-and-bust cycles are all too familiar to those working with rice (*Oryza sativa*) blast (caused by *Magnaporthe grisea*), and are a major justification for this symposium.

Though the view has been questioned (Kranz, 1990), many agriculturalists agree that genetic diversity contributes to significant protection from disease in natural ecosystems and in traditional agriculture (Browning, 1974; Schmidt, 1978; Wolfe, 1985). Thus, many have proposed the purposeful increase of genetic diversity in agriculture to manage plant diseases and stabilize crop yields (Browning and Frey, 1969; Mundt and Browning, 1985; Wolfe, 1985).

There are two polar attitudes that have restricted the adoption of genetic diversity in modern agricultural systems. The first is the 'pure-line mentality' that genetic diversity is somehow 'dirty' or 'immoral' (Harlan, 1972). Fortunately, this bias against diversity by agricultural scientists has decreased considerably over the past several years, and I am hopeful that this trend will continue. At the other extreme is the view that natural ecosystems are genetically diverse and tend to be buffered against disease. Therefore, plant pathogens can easily be brought into balance if we can merely increase the genetic diversity of agricultural systems. Though this view is helpful for the purpose of initiating interest in agricultural diversity, it can actually hinder the application of diversity in agriculture. A more useful view was expressed by May (1975), who stated that 'The instability of so many man-made [sic] agricultural monocultures is likely to stem not from their simplicity, as such, but rather from their lack of any significant history of coevolution with pests and pathogens'. Thus, the diversity of natural ecosystems may be considered to be 'functional' (*sensu* Schmidt, 1978), in that protracted evolution has resulted in diversity that contributes specific 'functions', such as protection against disease. 'Haphazard or unplanned diversity will not necessarily provide stability' (Mundt and Browning, 1985).

Evolution, of course, operates over a very long time scale and is driven by natural selection. Thus, our challenge is to use knowledge of genetics and population biology, as a substitute for natural selection, so as to introduce functional genetic diversity into modern agriculture within a time frame to meet the urgent need for development of a sustainable society. The purpose of this chapter is to briefly review successes in use of functional genetic diversity in small grains, particularly wheat (*Triticum aestivum*) and barley (*Hordeum vulgare*), and to explore the potential for managing rice blast through use of host genetic diversity.

Mechanisms of Disease Control in Diverse Crops

Most research regarding the effects of crop genetic diversity on disease has been with intraspecific mixtures of genotypes possessing different race-specific resistance genes (see reviews in Browning and Frey, 1969; Mundt and Browning, 1985; Wolfe, 1985). The basis for this approach is the assumption that no single pathogen race will be highly virulent on all host genotypes in the mixture. Thus, the rate of epidemic development will be slowed, hopefully to below the economic threshold. Mechanisms by which such disease reductions occur have been reviewed (Burdon, 1978; Mundt and Browning, 1985) and include: (i) an increased distance between plants of the same genotype, which dilutes inoculum of compatible pathogen races; (ii) interactions between pathogen races, such as induced resistance and competition; and (iii) a barrier effect of resistant plants in blocking spore dispersal among compatible hosts. With barley powdery mildew (caused by *Erysiphe graminis*) the increased distance between plants of the same genotype in cultivar mixtures was found to be the most important mechanism, especially early in the season (Chin and Wolfe, 1984).

In addition to the above interactions expected in classic gene-for-gene systems, there are other mechanisms of disease modification operable in mixtures as well. Interactions between components of 'partial' or 'race non-specific' resistancc can also result in benefits due to mixing (Jeger *et al.*, 1981a,b). Further, competition between different plant genotypes in mixtures has been found to influence disease levels via differential survival of plant genotypes with different levels of disease resistance (Finckh and Mundt, 1992b) and through plant competitive interactions that may alter the physiological susceptibility of plant genotypes (Finckh and Mundt, 1992a).

Historical Perspectives

One of the first to formally promote the use of genetic diversity in agriculture was Jensen (1952), who proposed the development of multiline oat (*Avena sativa*) cultivars. Such multilines would be mixtures of genotypes that are selected for phenotypic uniformity, but genetically diverse as possible for other traits, such as disease resistance. Later, Borlaug (1959) and Frey *et al.* (1973) utilized a backcrossing approach to produce multilines of wheat and oats, respectively, that contained different genes for race-specific resistance to rust (*Puccinia* spp.) pathogens.

Multiline cultivars have had significant commercial success (Browning and Frey, 1981; Mundt and Browning, 1985). However, recent attention has turned to the use of cultivar mixtures, without further selection for uniformity among mixture components. The advantages of cultivar mixtures over multilines have been elegantly described by Martin Wolfe and colleagues (Wolfe and Barrett, 1980; Wolfe, 1985). Most importantly, cultivar mixtures do not require additional breeding for agronomic uniformity. Thus, diversity for disease resistance can be attained immediately, utilizing the most agronomically superior genotypes, provided that these cultivars display diversity for reaction to disease(s) and are sufficiently similar to be cultured within the same field. Components can be easily added to a mixture as new, agronomically improved cultivars become available. In addition, the additional genetic diversity present in a cultivar mixture, as compared to a mixture of backcrossed lines, may provide protection against the primary and secondary diseases, limit the evolution of pathogen races with multiple virulence, and provide increased possibilities for yield synergisms due to differential resource utilization by the component cultivars.

Experiments with Wheat Cultivar Mixtures

Wheat cultivar mixtures are becoming increasingly popular in Oregon, USA, following work with barley in Europe (Wolfe, 1985). Though wheat cultivar mixtures are being used for different market classes in Oregon, this chapter will focus on the 'club wheats', which received the first emphasis in Oregon.

Club wheat is a subclass of the soft white winter wheat market class. The compact or club head derives from *Triticum compactum* and results from a shortened peduncle. Associated with this head type is a flour quality that is strongly favored in Asia, where greater than 80% of Oregon wheat is sold. In fact, Western White Wheat, which has a very high market demand, must contain at least 10–15% club wheat. Though Oregon wheat farmers often obtain substantial premiums for producing club wheat, the area planted to this wheat has declined from the vast

Table 17.1. Reaction of five winter wheat cultivars to two *Puccinia striiformis* races used in mixture trials.

	Reaction[a] to	
Cultivar	Race 5	Race 27
Faro	S/R[b]	R
Jacmar	S	R
Moro	S	R
Tres	MR	R
Tyee	R	S

[a] S = susceptible; R = resistant; MR = moderately resistant.
[b] Approximately one-half of Faro plants were susceptible and one-half resistant to this race.

majority to a small minority over the past 30 years. The primary reason for this decline has been problems with stripe rust, caused by *Puccinia striiformis*. Most resistant club wheat cultivars have contained a single, race-specific gene for resistance to stripe rust. Thus, resistance has been short-lived, with recent cultivars maintaining effective resistance for only 2–3 years in commercial production.

In 1986, my research group began to evaluate the potential for club wheat cultivar mixtures to control stripe rust and stabilize crop yield. Our approach was to work with the five most commonly grown club wheat cultivars, as we felt that a successful mixture system would need to utilize cultivars that are agronomically accepted as pure-lines. These five cultivars were grown as pure-lines and in all of the possible 26 equiproportional mixtures at three locations. The locations were chosen to represent a diversity of winter wheat environments in Oregon, and average annual rainfall ranged from 280 to 1000 mm among the sites.

Two experiments with four replications per treatment were conducted at each of the three sites. In the first experiment, each plot was inoculated in the fall with susceptible, transplanted seedlings that were infected with two stripe rust races. These races were similar to those found in commercial production, and reacted differently on the five cultivars (Table 17.1). The other experiment at each location was not inoculated and was sprayed with a systemic fungicide as necessary to maintain the plots rust-free for evaluating mixture yield in the absence of rust. Each plot was 1.5 × 4.3 m and no attempt was made to separate plots to avoid interplot interference, a topic that will be discussed in more detail later in this chapter.

In this preliminary screening trial, cultivar mixtures showed positive benefits for both disease control (Table 17.2) and yield (Table 17.3). On average, the mixtures reduced the severity of stripe rust by 39% relative to the mean of the components grown in a pure stand. However, there was considerable variation among mixtures, with disease reduction ranging from 13 to 64%. In addition, disease reduction due to mixing improved with increasing number of cultivars in the mixture. Interestingly, a naturally occurring leaf rust epidemic (caused by *Puccinia recondita*) at one location suggested that the mixtures controlled leaf rust to the same or greater degree as stripe rust.

Table 17.2. Percent reduction of rust severity (relative to mean of component cultivars in pure stand) for wheat cultivar mixtures tested in field plots at three locations.

Number of components in mixture	Number of mixtures	Stripe rust			Leaf rust
		Mean[a]	High[b]	Low[c]	Mean[d]
2	10	31	59	14	37
3	10	42	64	19	42
4	5	49	59	28	63
5	1	48	48	48	70
Mean of mixtures	26	39	64	14	45

[a] Mean over three locations and 3–4 replications per location for all mixtures with a given number of components.
[b] Highest disease reduction for the mean of an individual mixture over locations and replications.
[c] Lowest disease reduction for the mean of an individual mixture over locations and replications.
[d] Mean over four replications at the one location where a leaf rust epidemic occurred.

Table 17.3. Yield (kg plot^{-1}) of pure-line wheat cultivars and mixtures in inoculated plots and in non-inoculated plots that received multiple fungicide applications.

Number of cultivars in population	Number of populations	Inoculated			Fungicide-treated		
		Mean[a]	High[b]	Low[c]	Mean	High	Low
1	5	1.88	2.28	1.52	2.66	2.94	1.91
2	10	2.01	2.43	1.56	2.68	3.01	2.18
3	10	2.07	2.31	1.79	2.72	2.92	2.54
4	5	2.13	2.26	2.03	2.74	2.90	2.70
5	1	2.15	2.15	2.15	2.79	2.79	2.79

[a] Mean over three locations for all populations with a given number of components.
[b] Highest yield for the mean of an individual population over locations and replications.
[c] Lowest yield for the mean of an individual population over locations and replications.

It should be noted that the disease reductions shown in Table 17.2 are probably only minimum estimates, as spores from the more heavily diseased pure stands influenced the mixture plots. Field studies done in Oregon (Table 17.4) show that plot size, distance between plots, and wind conditions can all have an important effect on disease control measured in mixture plots. Probably the most important variable is the ratio of plot size to distance between plots (Table 17.4; Wolfe, 1985). In subsequent studies where we have planted one buffer plot of

Table 17.4. Effect of plot size and spacing on percent reduction of stripe rust severity by wheat cultivar mixtures relative to component pure stands in three locations with different wind intensities. Mixtures were either grown adjacent to or separated from pure-line susceptible plots.

	Plot size								
	1.2 × 1.2 m			2.4 × 2.4 m			4.9 × 4.9 m		
	Wind intensity			Wind intensity			Wind intensity		
Plot arrangement	Low	Med	High	Low	Med	High	Low	Med	High
Adjacent	23	10	39	−62	50	3	13	5	14
6-m separation	–	50	64	74	53	5	77	20	−7

resistant plants between each pair of test plots, stripe rust reductions due to mixing have been larger than that shown in Table 17.2.

Field trials with wheat cultivar mixtures also showed positive yield benefits in both the presence and absence of stripe rust (Table 17.3). On average, the mixtures increased yield, relative to the mean of their component pure stands, by 10.1% in the presence of stripe rust and 2.5% in the absence of stripe rust. The observed yield increases in the absence of stripe rust could be due to improved resource utilization by the mixtures (e.g. improved light interception as compared to mean of the pure stands) or, perhaps, due to unobserved interactions with soil-borne pathogens. As with disease control, there was considerable variation among mixtures for yield increase and yield increases due to mixing improved with increasing number of components in the mixtures.

An important benefit from the use of cultivar mixtures is risk reduction. By growing mixtures, farmers may reduce the probability that a low yield will be obtained due to an unpredictable biotic or abiotic stress. This can be demonstrated with the data of Table 17.3, where the yield per plot of pure stands and mixtures averaged over inoculated and non-inoculated plots and three locations is used to represent the yield performance of the different wheat populations over a range of environments. Absolute yield increased with the number of cultivars grown, and the range of yields measured decreased with increasing number of components.

A recurring observation in our work with wheat cultivar mixtures is that of specificity of interactions between wheat genotypes. We have found that certain cultivars and certain combinations of cultivars provide greater benefits due to mixing than do others. This observation has been formalized in the concept of general and specific mixing ability (Knott and Mundt, 1990). A cultivar is said to have positive general mixing ability if it provides greater than average disease control and/or yield benefits averaged over all cultivars it is mixed with. Specific mixing ability refers to certain combinations of cultivars that provide disease control and/or yield benefits which are greater than other combinations.

We have used the concept of general and specific mixing ability to help wheat farmers choose appropriate cultivar mixtures in Oregon. For example, we have

Table 17.5. Yield (bushels acre^{-1}) of two pure-line wheat cultivars and a 1:1 mixture of those same two cultivars in a regional yield trial. Data are from P.K. Zwer, Crop and Soil Sciences Department, Oregon State University.

	Treatment		
Location	Hyak	Tres	Hyak/Tres
Arlington	26	26	26
Athena	81	69	88
Helix	80	70	73
Heppner	61	58	65
La Grande	104	111	112
Lexington	48	52	53
Moro	27	32	37
Pendleton	75	77	78
Mean	63	62	67

found that the cultivars Tres and Tyee have positive general mixing ability for both stripe rust control and yield. Further, the combination of Tres and Tyee shows good specific mixing ability. As a result, the treatment that showed the highest yield of all populations in Table 17.3 was the mixture of Tres and Tyee. Several years ago, Oregon growers were in the situation of having no stripe rust resistant wheat cultivars and needed advice regarding cultivars to grow in mixture. At that time, seed of Tyee was no longer commercially available. However, there was available a newer cultivar, Hyak, which resulted from a backcrossing program to incorporate strawbreaker foot rot (caused by *Pseudocercosporella herpotrichoides*) resistance into the Tyee background. The cultivars Tres and Hyak also have different genes for stripe rust resistance. A blend of these two cultivars showed good performance in a regional yield trial (Table 17.5) and has been grown successfully on a commercial basis, despite the fact that the individual component cultivars have 'broken down' to stripe rust.

Mixtures at Commercial Scale

Current evidence suggests that mixtures look far better at commercial scale than in experimental field plots. There are several possible reasons for this. The first is the interplot interference phenomenon in small-scale experiments that was mentioned earlier. Wolfe (1985) has claimed that mixture trials are more sensitive to the effects of interplot interference than are other types of trials. From a theoretical standpoint, this should be the case. The sole epidemiological effect of mixtures is to decrease the infection efficiency of the pathogen, while the effect of interplot interference is the opposite. In contrast, other methods of control, such as partial resistance, also influence factors such as latent period and sporulation, which are not directly affected by interplot interference.

Artificial inoculation in experiments may also reduce the effectiveness of cultivar mixtures. For example, to ensure that all plots developed epidemics in our wheat stripe rust studies, we have had to utilize initial inoculum levels much greater than those which occur naturally. These increased inoculum levels reduce the number of pathogen generations that are impacted by the mixtures and, hence, the effectiveness of mixtures for disease control (Mundt, 1990). Finally, small field plots may not allow identification of the ability of mixtures to buffer against the diversity of biotic and abiotic environments present at larger scales.

Our experience with wheat cultivar mixtures in Oregon suggests the potential for considerable success at the commercial scale, and the use of cultivar mixtures is in the early adoption phase in Oregon. For example, one wheat farmer compared 25 ha of a susceptible cultivar that he had traditionally grown with 25 ha of a three-way mixture that contained that same susceptible cultivar. The pure stand showed stripe rust infections beginning on nearly every flag leaf during grain filling and was sprayed with fungicide. In contrast, the mixture was not sprayed, and it was difficult to find rust in this field. This farmer is now planting his entire farm (150 ha) to cultivar mixtures. The practice has now been implemented on an area of between 10,000 and 100,000 ha. Though interest in mixtures is clearly increasing, it is too early to predict the final level of acceptance.

The most impressive and best documented example of the commercial use of cultivar mixtures for disease control is that of barley powdery mildew in eastern Germany (Wolfe, 1992). Genetic uniformity of barley had resulted in severe powdery mildew problems, and a move towards cultivar mixtures beginning in 1984. From 1983 to 1990, the proportion of the barley area sown to mixtures increased from 0 to 92% on an area of more than 300,000 ha. Powdery mildew incidence on a national scale declined rapidly during this time, from an average of greater than 50% before the adoption of mixtures to less than 10% from 1987 to 1990 when mixtures were widely grown. Fungicide use also declined severalfold during this period, but barley yields were maintained and malt quality was sufficient for domestic beer production and for export.

Effects on Pathogen Populations

Despite the generally successful field performance of mixtures, there remain two major areas of concern. The first is whether the cultivation of mixed crop populations will extend the useful life of new resistance genes as compared with deploying the genes sequentially in monocultures. Though the question is extremely difficult to address experimentally, there are some observations which suggest that mixtures do extend the useful life of resistance genes. For example, the club wheat cultivar Tres was first deployed commercially as one of ten components of the multiline cultivar crew (Allan *et al.*, 1983). This wheat genotype maintained its resistance until released as a pure-line cultivar, after which its resistance 'broke down' in 2–3 years (C.C. Mundt, unpublished). Similarly, the oat genotype X421 has maintained its resistance to crown rust (caused by *Puccinia recondita*) as part of oat multilines despite the fact that race-specific resistance to oat crown rust is very ephemeral in monoculture (personal observation).

The second issue regarding the effects of host mixtures on pathogen populations is whether such mixtures will select for the dominance of complex races that can overcome many or all resistance genes and erode the effectiveness of the mixtures for disease control. Many mathematical models have addressed this issue (Leonard and Czochor, 1980; Kiyosawa, 1989; Marshall, 1989) with inconclusive results. In contrast, most field evidence, from both natural and agricultural ecosystems, suggests that mixtures will not cause rapid selection for complex races (Mundt and Browning, 1985; Mundt, 1990). Most of the models rest on Vanderplank's (1968) hypothesis that there is a fitness cost associated with pathogen virulence, a hypothesis which lacks strong empirical support (Parlevliet, 1981). However, there are several other mechanisms besides a cost of virulence that could result in selection against complex races (Mundt and Browning, 1985; Mundt, 1990, 1994). In field plots, we have found that selection for or against complex races of stripe rust in wheat cultivar mixtures depends on factors such as the genetic background of the race, interactions with the physical environment, and interactions with other races in the population (DiLeone and Mundt, in review). Chin and Wolfe (1984) failed to detect rapid selection for complex powdery mildew races in barley cultivar mixtures. They also presented data to support the hypothesis that differential adaptation to the genetic background of the different cultivars caused disruptive selection and reduced the fitness of complex races in mixtures. Recent computer simulation studies suggest that even relatively small degrees of such differential adaptation can significantly slow evolution towards the dominance of complex races (Lannou and C.C. Mundt, unpublished).

Practical Concerns

Modern agricultural practices are based on the paradigm of a genetically and phenotypically uniform crop. Thus, certain practical concerns arise in considering the reintroduction of diversity into modern agricultural systems. Most of the concerns, however, should not limit the use of diversity in practice or can be addressed in a fairly straightforward manner.

A common practical concern regarding mixtures involves agronomic uniformity, especially for height and crop maturity. The solution to this problem is to choose mixture components with compatible heights and maturities for the farming system and environment in question. It is important to keep in mind that, for small grains, maturity differences are often based on heading dates. Very often, large differences in heading date translate into much smaller differences in maturity times required for harvest. In fact, the determining factor for harvest can often be the availability of favorable drying weather, rather than physiologic maturity. It is often quite feasible to harvest, by hand or mechanically, genotypes of differing heights. However, extremes of height can sometimes cause difficulties, for example, by resulting in an excess of straw being run through a mechanical harvester in an attempt to harvest heads of differing heights. In Oregon, we have rarely had difficulty in using mixtures due to lack of agronomic uniformity. In fact, some of our wheat farmers actually *prefer* diversity in height to provide a more uniform feeding of heads through harvesters and diversity in flowering dates to avoid risk due to abiotic stresses.

Marketing restrictions are often cited as the major limitation to the use of genetic diversity in agriculture. In reality, however, grains are often stored and sold by market class (e.g. soft white wheat vs. hard red wheat), and not by cultivar. For example, in Oregon, wheat cultivars of the same market class are *not* segregated in storage. Thus, there is no reason why different cultivars cannot be mixed and harvested within the same field. We have tested this hypothesis in Oregon and found that the quality characteristics of wheat grown as mixtures are equal to or better than the average of the pure-line cultivars that constitute the mixture. It is also important to note that many quality characteristics are influenced more by environment than by genotype. For example, variation in grain protein was found to vary much more among fields of the same cultivar than among different cultivars grown in the same fields (R. Karow, Oregon State University, 1993, personal communication). Thus, it seems unlikely that mixtures would generally be detrimental to marketing. In fact, adequate quality of mixtures can be maintained even for a demanding commodity such as barley malt used for beer making (Wolfe, 1992).

The maintenance and distribution of seed often requires more effort and thought for mixtures compared to pure stands. This is especially true for self-pollinated crops, for which farmers may save part of their crop as seed for the next crop. Seed of multilines has been maintained by universities in Iowa (Browning and Frey, 1981) and Washington (Allan *et al.*, 1983), USA. Seed of cultivar mixtures has been sold by a major seed company in the UK. In eastern Germany, it has been suggested that cultivar mixture seed be produced by mechanically harvesting perpendicular to alternating strips of the component pure stands (Wolfe, 1985).

Seed of wheat cultivar mixtures has been provided in various ways in Oregon. Some farmers have the facilities to grow pure-lines and mix their own seed, sometimes providing seed to neighbors who lack such facilities. Many growers, perhaps the majority using mixtures, are saving seed of the harvested mixture to plant next year's crop. Standard dogma has been that such a practice is unacceptable because the composition of the mixture will change rapidly over the years. Some of the evidence for this view is based on the work of Harlan and Martini (1938), who found that one or two components will dominate a barley mixture over time. However, their mixture components were not all agronomically adapted at the test sites, differed greatly in yield, and were apparently chosen for morphological characters which allowed identification in mixture. In mixing wheat cultivars of more equal yield and which were bred to be grown in the same environment, we have found evidence for development of equilibrium frequencies of cultivars (Finckh and Mundt, 1993). Further, there may be some adaptation of mixtures for better performance on a specific farm when mixture seed is regrown (Mundt, unpublished data). Currently, we inform farmers that reusing their own mixture seed could entail some risk, but can probably be done safely for three seasons.

Mixtures and Rice Blast

One might expect rice mixtures to provide disease control similar to that found for the rusts and powdery mildews of small grains, as the pathosystems are very similar. All involve a grass host with similar growth habits, and a pathogen that is aerially dispersed and which interacts in a clear gene-for-gene manner with its host. Further, all of the pathosystems involve classic polycyclic epidemics that can cause severe disease from a small amount of initial infection. However, with neck blast, a single infection can cause severe crop loss. With rusts and powdery mildews, a single infection is of very small consequence to the crop.

In Table 17.6, I have summarized data on blast in rice cultivar mixtures, accounting for all papers on the topic of which I was aware. In doing so, only data that were in tabular form were used so that results could be accurately evaluated. For simplicity, I considered only data for mixtures of equal proportions of cultivars, even though a few papers reported results of differing proportions. Some of the studies were mixtures of rice cultivars resistant to all races mixed with cultivars that were susceptible. Other studies involved mixtures where each cultivar was susceptible to one or more races. When information was provided, I recorded data on the mixing method, i.e. whether the cultivars were mixed within hills or whether the hills were mixed, with each hill being genetically uniform.

The small-field plot studies with rice blast summarized in Table 17.6 show results extremely similar to those obtained with rusts and mildews of other cereals. As with other cereals, I would expect the performance of rice cultivar mixtures to improve considerably when scaled up to commercial levels. Interestingly, there seems to be no consistent difference in blast control for intra-hill vs. extra-hill mixtures. Thus, rice farmers may have the flexibility of using either approach to mixing. On average, the mixtures were somewhat less effective in controlling neck blast than in controlling leaf blast; in some specific studies, this difference was quite large. Bonman *et al.* (1986) suggested that interplot interference may have greater effects on neck blast than on leaf blast. I expect that this is true. In addition, however, it may require higher levels of inoculum reduction to adequately control neck blast as compared to leaf blast. Thus, it may be highly important to consider neck blast resistance when choosing cultivars to include in mixtures used for blast control. The mean yield increase due to mixing (8.2%) is probably lower than one might expect. However, it should be noted that, in some of the experiments, blast severity was not sufficient to cause significant yield reductions even in the pure-line plots.

Chuke and Bonman (1988) followed the change in frequency of blast races in pure stands and a mixture of three rice cultivars when races with virulence against one, two, or three of the cultivars were present. There was no evidence of selection for the 'superrace' in field experiments or in isolation cages. However, in nurseries sequentially planted to the same treatments and sampled for 195 days, there was some evidence of selection for the race capable of attacking all three components of the mixture. In that case, however, there was also strong selection for the same race in one of the pure stands.

Chin and Husin (1982) conducted a detailed analysis of the effects of a

Table 17.6. Rice blast reduction and crop yield increase relative to mean of component pure stands reported for rice cultivar mixtures.

Author(s)	Number of cultivars in mixture	Mixing[a] method	Reduction of blast (%) Leaf	Reduction of blast (%) Neck or panicle	Yield increase (%)
Bonman *et al.*, 1986	3	?	–	68	28.6
	3	?	33	26	19.4
	3	?	25	60	−3.0
Chin and Husin, 1982	3	Intra-hill	65	69	5.7
	3	Intra-hill	79	63	6.7
	3	Extra-hill (unreplicated)	80	40	−6.1
	3	Extra-hill (unreplicated)	100	65	3.2
Chung *et al.*, 1987	2	Intra-hill	40	11	15.4
	2	Extra-hill	65	40	22.9
	2	Intra-hill (blast-free)	–	–	21.8
	2	Extra-hill (blast-free)	–	–	12.2
CIAT, 1983					
Cali location	5	Intra-hill	48	−39	5
	5	Intra-hill	98	44	−17.9
	5	Intra-hill	60	39	−0.1
La Libertad location	5	Intra-hill	77	21	–
	5	Intra-hill	51	23	–
	5	Intra-hill	70	7	–
	8	Intra-hill	62	94	–
	7	Intra-hill	20	73	–
	7	Intra-hill	−64	83	–
	6	Intra-hill	30	−29	–
Koizuma and Kato, 1987	2	Intra-hill	49	–	–
Unpublished data of Shindo cited in Ezuka, 1979	3	?	85	86	16.5
	3	?	72	65	1.7
Mean over-all studies	–	–	55	43	8.2

[a] Intra-hill = cultivars mixed within hills; extra-hill = mixture of hills that were genetically uniform within.

three-way cultivar mixture on rice quality. A test panel found no significant difference between the mixture and two of the component pure stands in consumer acceptability based on taste, aroma, tenderness, and stickiness. The mixture was significantly superior to the third pure stand in terms of consumer acceptability. Further, the panel was unable to recognize the mixed nature of grain from the cultivar mixture. The amylose content and gelatinization temperature of grain from the mixture were between that of the highest and lowest mixture component. As in many other parts of Asia, Chin and Husin (1982) noted that rice cultivars in Malaysia are currently blended by wholesalers and retailers after milling before being marketed. Thus, it is unlikely that consumers would notice a difference in the quality of grain harvested from cultivars that were mixed in the field provided there was not excessive breakage during milling. Nevertheless, more work needs to be done, as maintenance of adequate crop quality may be more difficult for rice mixtures than for other crops that are utilized for feed or ground into flour.

In implementing cultivar mixtures for rice blast control, seed technology is a critical issue, as the target farmers may lack income for yearly seed purchase and sophisticated seed production facilities may not be readily available. There are many ways that mixed seed can be provided to rice farmers, including encouraging farmers to grow small plots of pure source seeds, developing local seed cooperatives, providing mixtures with components that shift little over time, providing morphological markers to detect component shifts, planting mixtures as alternating hills of pure stands that can be harvested by hand separately, etc. Mixtures of a small number of cultivars are commonly utilized in traditional rice culture (Bonman *et al.*, 1986). Thus, the technology must be feasible. However, it may take some effort and thought on the part of researchers and extension personnel to develop technology that is appropriate to the local situation and that reaches the farmers most in need.

Conclusions

Experimental evidence indicates that functional genetic diversity contributes greatly to the control of rusts and powdery mildews of wheat, barley, and oats, and to higher yields and increased yield stability of those crops. The performance of mixtures tends to improve when scaled-up from experimental plots to commercial production. Despite the prevailing viewpoint, practice suggests that there are few agronomic and marketing problems associated with the use of crop diversity in modern agricultural systems. Presently, available data suggest that the same successes may be attainable with rice blast as well.

Though genetic diversity can contribute greatly to the management of rice blast, the approach will not be an effortless one. The foundation of the diversity approach is an active, successful breeding program. Without an influx of agronomically superior germplasm with a diversity of resistance to diseases, the mixture approach will have little to offer. Similarly, implementation of practices such as cultivar mixtures requires an investment in thoroughly understanding the system to be managed, including all of the agronomic, marketing, and pest problems involved. In this regard, there is no substitute for a thorough familiarity with the germplasm to be utilized, gained by considerable field observations and direct interactions with farmers who will adopt the technology.

Finally, it is important to emphasize a diversity of approaches as well as a diversity of germplasm. For example, a mixture of cultivars each containing a different race-specific resistance gene will provide considerable protection against rice blast. However, a mixture of cultivars containing resistance gene combinations or a background level of partial resistance in addition to major genes will provide even more obstacles to the pathogen. Other chapters in this volume demonstrate the potential of gene combinations for blast control, and that partial resistance is easier to breed for and is much more effective than thought 20 years ago. Further, modern techniques such as marker-facilitated selection will allow all of these approaches to be used more quickly. Molecular analyses have also allowed us to understand better the evolution of the blast pathogen. Clearly, the knowledge and tools exist to make rice blast a disease of minor importance. What remains to be done is to invest in a concerted effort to bring this technology to the farmers who need it.

References

Allan, R.E., Line, R.F., Peterson, C.J., Jr., Rubenthaler, G.L., Morrison, K.J. and Rohde, C.R. (1983) Crew, a multiline wheat cultivar. *Crop Science* 23, 1015–1016.

Bonman, J.M., Estrada, B.A. and Denton, R.I. (1986) Blast management with upland rice cultivars. *Proceedings of the Symposium on Progress in Upland Rice Research*. International Rice Research Institute, Los Baños, Laguna, The Philippines, pp. 375–382.

Bonman, J.M., Khush, G.S. and Nelson, R.J. (1992) Breeding for resistance to rice pests. *Annual Review of Phytopathology* 30, 507–528.

Borlaug, N.E. (1959) The use of multilineal or composite varieties to control airborne epidemic diseases of self-pollinated crop plants. *Proceedings of the International Wheat Genetics Symposium* 1, 12–27.

Browning, J.A. (1974) Relevance of knowledge about natural ecosystems to development of pest management programs for agroecosystems. *Proceedings of the American Phytopathological Society* 1, 191–199.

Browning, J.A. and Frey, K.J. (1969) Multiline cultivars as a means of disease control. *Annual Review of Phytopathology* 7, 355–382.

Browning, J.A. and Frey, K.J. (1981) The multiline concept in theory and practice. In: Jenkyn, J.F. and Plumb, R.T. (eds), *Strategies for the Control of Cereal Disease*. Blackwell Scientific Publications, Oxford, pp. 37–46.

Browning, J.A. Simons, M.D., Frey, K.J. and Murphy, H.C. (1969) Regional deployment for conservation of oat crown rust resistance genes. *Iowa Agricultural and Home Economics Experiment Station Special Report* 64, 49–56.

Burdon, J.J. (1978) Mechanisms of disease control in heterogeneous plant populations – an ecologist's view. In: Scott, P.R. and Bainbridge, A. (eds), *Plant Disease Epidemiology*. Blackwell Scientific Publications, Oxford, pp. 193–200.

Centro Internacional de Agricultura Tropical (1983). *Annual Report for 1983*. CIAT, Cali, Columbia.

Chin, K.M. and Husin, A.N. (1982) Rice variety mixtures. *Proceedings of the Padi Workshop*. Bumbong Lima, Province Wellesley, Malaysia, pp. 203–216.

Chin, K.M. and Wolfe, M.S. (1984) The spread of *Erysiphe graminis* f.sp. *hordei* in mixtures of barley varieties. *Plant Pathology* 33, 161–171.

Chuke, K.C. and Bonman, J.M. (1988) Changes in virulence frequencies of *Pyricularia oryzae* in pure and mixed stands of rice. *Journal of Plant Protection in the Tropics* 5, 23–29.

Chung, H.S., Lee, C.Y., Cho, J.Y. and Lee, E.W. (1987) Effects of cultivar mixtures on rice blast epidemics and yields in the paddy field. *Journal of the National Academy of Sciences of Korea* 26, 341–361.

Ezuka, A. (1979) Breeding for and genetics of blast resistance in Japan. *Proceedings of Rice Blast Workshop*. International Rice Research Institute, Los Baños, Laguna, The Philippines, pp. 27–49.

Finckh, M.R. and Mundt, C.C. (1992a) Plant competition and disease in genetically diverse wheat populations. *Oecologia* 91, 82–92.

Finckh, M.R. and Mundt, C.C. (1992b) Stripe rust, yield, and plant competition in wheat cultivar mixtures. *Phytopathology* 82, 905–913.

Finckh, M.R. and Mundt, C.C. (1993) Effects of stripe rust on the evolution of genetically diverse wheat populations. *Theoretical and Applied Genetics* 85, 809–821.

Frey, K.J., Browning, J.A. and Simons, M.D. (1973) Management of host resistance genes to control diseases. *Zeitschrift für Pflanzenkheiten und Pflanzenschutz* 80, 160–180.

Harlan, H.V. and Martini, M.L. (1938) The effect of natural selection in a mixture of barley varieties. *Journal of Agricultural Research* 57, 189–199.

Harlan, J.R. (1972) Genetics of disaster. *Journal of Environmental Quality* 1, 212–215.

Jeger, M.J., Griffiths, E. and Jones, D.G. (1981a) Disease progress of non-specialised fungal pathogens in intraspecific mixed stands of cereal cultivars. I. Models. *Annals of Applied Biology* 98, 187–198.

Jeger, M.J., Jones, D.G. and Griffiths, E. (1981b) Disease progress of non-specialised pathogens in intraspecific mixed stands of cereal cultivars. II. Field experiments. *Annals of Applied Biology* 98, 199–210.

Jensen, N.F. (1952) Intra-varietal diversification in oat breeding. *Agronomy Journal* 44, 30–44.

Jensen, N.F. (1978) Limits to growth in world food production. *Science* 201, 317–320.

Kiyosawa, S. (1989) Breakdown of blast resistance in relation to general strategies of resistance gene deployment to prolong effectiveness of disease resistance in plants. In: Leonard, K.J. and Fry, W.E. (eds), *Plant Disease Epidemiology*, Vol. 2. McGraw-Hill, New York, pp. 251–283.

Knott, E.A. and Mundt, C.C. (1990) Mixing ability analysis of wheat cultivar mixtures under diseased and nondiseased conditions. *Theoretical and Applied Genetics* 80, 313–320.

Koizuma, S. and Kato, H. (1987) Effect of mixed plantings of susceptible and resistant rice cultivars on leaf blast development. *Annals of the Phytopathological Society of Japan* 53, 28–38.

Kranz, J. (1990) Fungal diseases in multispecies plant communities. *New Phytologist* 116, 383–405.

Leonard, K.J. and Czochor, R.J. (1980) Theory of genetic interactions among populations of plants and their parasites. *Annual Review of Phytopathology* 18, 237–258.

Marshall, D.R. (1989) Modeling the effects of multiline varieties on the population genetics of plant pathogens. In: Leonard, K.J. and Fry, W.E. (eds), *Plant Disease Epidemiology*, Vol. 2. McGraw-Hill, New York, pp. 284–317.

May, R.M. (1975) Stability in ecosystems: Some comments. In: van Dobben, W.H. and Lowe-McConnel, R.H. (eds), *Unifying Concepts in Ecology*. PUDOC, Wageningen, pp. 161–168.

Mew, T.W. (1987) Current status and future prospects of research on bacterial blight of rice. *Annual Review of Phytopathology* 25, 359–382.

Mundt, C.C. (1990) Disease dynamics in agroecosystems. In: Carroll, C.R., Vandermeer, J.H. and Rossett, P.M. (eds), *Agroecology*. McGraw-Hill, New York, pp. 263–299.

Mundt, C.C. (1994) Techniques for managing pathogen co-evolution with host plants to prolong resistance. *Proceedings of the 21st International Rice Research Conference*. International Rice Research Institute, Los Baños, Laguna, The Philippines (in press).

Mundt, C.C. and Browning, J.A. (1985) Genetic diversity and cereal rust management. In: Roelfs, A.P. and Bushnell, W.R. (eds), *The Cereal Rusts*, Vol. 2. Academic Press, Orlando, pp. 527–560.

Parlevliet, J.E. (1981) Stabilizing selection in crop patho-systems: An empty concept or a reality? *Euphytica* 30, 259–269.

Schmidt, R.A. (1978) Diseases in forest ecosystems: The importance of functional diversity. In: Horsfall, J.G. and Cowling, E.B. (eds), *Plant Disease: An Advanced Treatise*, Vol. 2. Academic Press, New York, pp. 287–315.

Vanderplank, J.E. (1968) *Disease Resistance in Plants*. Academic Press, New York.

Wolfe, M.S. (1985) The current status and prospects of multiline cultivars and variety mixtures for disease resistance. *Annual Review of Phytopathology* 23, 251–273.

Wolfe, M.S. (1992) Barley diseases: Maintaining the value of our varieties. In: Munck, L. (ed.), *Barley Genetics VI*. Munksgaard International Publishers, Copenhagen, pp. 1055–1067.

Wolfe, M.S. and Barrett, J.A. (1980) Can we lead the pathogen astray? *Plant Disease* 64, 148–155.

18 Figs, Wasps, Nematodes and Sitting Ducks: Rice Blast, from the Outside Looking in

E.A. Herre

Smithsonian Tropical Research Institute, Apartado 2072, Balboa, Republic of Panama, and Smithsonian Tropical Research Institute, Unit 0948, APO AA 34002-0948, USA

Introduction

If the human population is to continue to grow at its present rate, that growth will require an increase in the capacity to produce food of roughly 2% per year. Given present patterns of consumption, a large portion of that increase will have to come from an increase in the production of rice. During the last decades, the introduction of intensive agricultural practices, the application of increasingly refined breeding programs, and the increased conversion of land and diversion of water to growing rice have all contributed to impressive increases in production. However, despite tremendous effort devoted to the breeding of resistant cultivars and other control programs, rice blast disease (*Pyricularia grisea*) continues to present an enigmatic threat to that productivity. Cultivars that appear to be resistant and perform remarkably well in one setting, often fail miserably in others. Even within sites, resistance can break down from one year to the next, often with devastating results (Bonman *et al.*, 1992). Effective and safe fungicides take many years to develop, and even then they can be very expensive to apply and are generally useful only for the few years it takes for the blast to develop resistance (May and Dobson, 1985). It appears as though blast poses problems that require running as fast as possible, just to stay in place.

I accepted the invitation to write this chapter, acutely aware of my ignorance of the specifics of applied rice cultivation and the blast fungus. For most of the last ten years, I have been studying natural, non-agricultural host–parasite systems in the tropical forest of Panama; for example, the species-specific nematodes that parasitize fig-pollinating wasps. My work has focused on 15 species of figs, each of which is pollinated by unique pollinator wasp species, each of which, in turn, is parasitized by its own species-specific nematode (Herre, 1985, 1987, 1989, 1993; Poinar and Herre, 1991). The figs depend completely on the wasps for pollination and the resulting production of their fruit. That fruit is a major component of the diet of many species of birds, bats, monkeys, and other frugivores. Therefore, any negative effect that the nematodes have on their direct host, the wasp, makes them indirect parasites of the figs and all of the animals that eat fig fruit, as well. Roughly speaking, the nematodes that I have studied are to the bats and monkeys that eat figs as rice blast is to people who eat rice.

My studies have been largely motivated from theoretical and evolutionary, rather than applied, perspectives. It has not been my objective to find ways to maximize yields of wasps or of figs or to minimize the loss due to the nematodes. However, my research program shares many themes in common with those of the researchers working on the disparate aspects of the interactions of rice and blast. For example, by combining existing biogeographical and paleontological work with results from ongoing molecular work on the figs, wasps, and nematodes that I and my collaborators have conducted, I have attempted to determine the identities and phylogenetic affinities of the different hosts and their parasites. This parallels the efforts to employ a variety of molecular and more traditional techniques to determine, first, which strains of *P. grisea* attack which rice cultivars and, second, whether there is any potentially useful information within that pattern. I have also conducted fairly detailed epidemiological studies of the fig/wasp/nematode systems, following the proportion of individual hosts infected over scores of host and parasite generations attempting to identify correlates of infection rate. This mirrors the tremendous effort devoted to blast monitoring programs. Further, and most relevant to the following discussion, by studying a wide range of species that differ in a few key parameters, I have been able to identify conditions that seem to predispose the nematodes to greater or lesser degrees of virulence. Specifically, my studies have demonstrated that increased opportunities for parasite transmission are linked with more virulent strains of parasites (Herre, 1993). With respect to rice and its blast disease, my hope is that comparing and contrasting these widely divergent systems will help to identify conditions that will contribute to the stable reduction of blast damage to rice. With this background in mind, I will present what I see as the important contrasts between the fig and rice systems, and suggest lines of research that might profitably be pursued.

The Fig–Wasp–Nematode System

The over 700 recognized fig species (*Ficus*, Moraceae) comprise one of the most diverse genera of flowering plants, and, with their species-specific wasps (Agaonidae, Chalcidoidae), and nematodes (Parasitodiplogasteridae, Rhabidae) are native to nearly all tropical habitats on earth (Berg, 1989; Poinar and Herre, 1991). This global distribution suggests an ancient origin of the group. That antiquity has been confirmed by the identification of fossil figs from both New and Old World strata that have been determined to be at least 40 million years old and by fig pollinating wasps (as well as their nematodes) preserved in amber dating back at least 20 million years. Moreover, molecular studies provide an additional dimension of resolution and allow us to glimpse the degree of coevolution among these ancient organisms. Phylogenies of the wasps and their associated nematodes that are based on sequences of homologous mitochondrial genes show overall congruence. This suggests that the particular pairs of hosts and parasites have had the opportunity to coevolve over exceedingly long periods of time, possibly millions of years (Herre, 1993; Herre *et al.*, unpublished data). It is worth emphasizing the fact that, even after so much time, neither figs nor wasps nor nematodes have been eliminated, but neither have the nematode parasites become entirely benign.

In studies relating to a more recent time frame, I have sampled the proportion of wasps that are infected over a period that roughly corresponds to 80–100 fig, wasp and nematode generations. The host fig wasps are dispersed among the thousands of fruits that are produced on each tree. Each fruit is pollinated by some number of female wasps ('foundresses') that additionally lay their eggs inside the fruits. Their offspring develop, hatch and mate within the fruit. If the wasp foundress(es) are infected by nematodes, then the wasp offspring are liable to be infected by the offspring of the nematodes. Thus, the overall population of the host wasps, as well as the parasitic nematodes, is subdivided into thousands of semi-isolated smaller ones. Increasingly, during the course of my studies, the proportion of the fig wasps that are infected by nematodes have remained fairly constant. That is, there do not seem to be any sweeping epidemics. More importantly, I have determined that the most virulent nematode species are associated with the host species that provide their parasites with the highest opportunities for transmission.

If the number of foundresses per fruit is small, then the system is extremely subdivided. If, on the other hand, the number of foundresses is relatively large, then the system is less subdivided, and approaches a more panmictic, open system. The degree of subdivision is important because it is causally linked to the opportunities for nematode transmission. In the more subdivided systems, the opportunities for parasite transmission are comparatively restricted. In a less subdivided, more open system, the nematodes have greater oppportunities for transmission. It is in those species that are characterized by relatively higher numbers of foundressses and, therefore, increased opportunities for parasite transmission that the nematodes are the most virulent. This finding is consistent with a broad body of theory that predicts that increased opportunities for transmission should promote the evolution of increased virulence in parasitic and disease-producing organisms (Herre, 1993, and references therein). The essential elements of the system, therefore, are long coevolutionary history of given nematode–wasp pairs, relative constancy of the dynamics of infection and low incidence of epidemiological outbreaks, and a high degree of population subdivision in the hosts (Table 18.1). The most important point is that systems that are characterized by high opportunities for parasite transmission possess virulent parasites.

The Rice Blast System

The roughly 20 wild species of rice (*Oryza*) have a wide geographic distribution, spanning Asia, Africa, and South America (Morishima *et al.*, 1991). The wild

Table 18.1. A comparison of major features of the fig–wasp and rice–host parasite systems.

Fig–wasp–nematode system	Rice blast system
Ancient, tightly coevolved	Ancient, loosely coevolved
Subdivided population of hosts	Open population of hosts
Low but variable opportunities for nematode transmission	Extremely high opportunities for blast transmission
Relatively constant infection rate	Highly variable infection rate
Relatively low virulence of nematodes	Highly virulent blast
No external input into system	High human input into system

species (*O. rufipogon* and *O. barthii*) that are the ancestors of the currently cultivated varieties (*O. sativa* and *O. glaberrima*) are found in Asia and Africa, respectively. As with the figs and wasps, it is clear that some form of rice has also been around for quite a long time. Less can be inferred about the long-term coevolutionary history of rice with the blast fungus (*Pyricularia grisea*). The fact that nearly all known forms of blast that attack rice are asexual complicates traditional genetic analyses.

What is clear, however, is that molecular techniques are absolutely essential for the rapid and reliable identification of strains of blast. On the one hand, blast has been reported to infect not only wild relatives of cultivated rice, but also entirely different grain species (wheat) and grassy weeds from a variety of genera (e.g. *Digiteria*). Considerable effort has been expended to determine whether the weeds surrounding rice fields can act as reservoirs of *P. grisea* capable of infecting rice (Borromeo *et al.*, 1993). On the other hand, different strains of the fungus are known to show a great deal of species- or even cultivar-specificity (Valent and Chumley, 1991). Although the results from molecular studies generally support the idea of specificity (as will be further discussed below), when compared to the very tight coevolution of nematodes and wasps, blast appears to be much more evolutionarily mobile.

At a fine scale, specificity can probably be understood in terms of the mechanisms that underlie the infection of rice tissues. This involves a series of environmental factors that may predispose plants to fungal attack as well as a series of steps during which either major or minor genes in both the plants and the fungus may play a role in determining the presence and extent of fungal damage to the plant. After a blast spore lands and germinates, it begins the process of infecting the plant by producing an apressorium that literally glues itself to the rice plant's surface (chapter 1, this volume). The capacity of this structure to penetrate the exterior cell walls depends largely on the environment being sufficiently humid for a sufficiently long time to allow the build-up of tremendous osmotic pressures due to the accumulation of melanin within the apressorium. Under these pressures, the fungal hyphae are able to rupture the epidermal cell walls (possibly weakened by enzymatic acivity), enter, and begin propagating within the interior cells. Successful infection results in significant colonization of host tissue and, ultimately, the production of spores. This dependence on humidity accounts for many of the observations of cultivars that are successful in one environment yet often fail miserably in others and the general observation that higher altitudes and higher latitudes with their longer periods of dew are notoriously difficult areas to grow blast-free rice. That is, cultivars that are susceptible may none the less be grown in environments that are not conducive to infection and subsequent epidemic development (such as the comparatively arid central valley of California). Another environmental factor that predisposes otherwise resistant cultivars to blast is the application of nitrogen-based fertilizers, although the basis of nitrogen enhancement of infection is less well understood. The genetic basis of resistance to penetration is likely to be largely polygenic (see below), and probably acts by affecting outer cell wall toughness, silica content, and surface characteristics which may deter apressorium adhesion such as epicuticular waxes, degree of hairiness, etc.

Once inside the cell, the fungus either does or does not evoke a hypersensitivity response in which the invaded cells, along with the invaded and adjacent cells, along with the invading hyphae are destroyed by the plant. These responses are, fairly clearly, triggered by the presence of molecules on the surface of the fungus (e.g. proteins or lectins on the cell wall of membrane) or fungal metabolic products. A plant-mediated response/nonresponse based on presence or absence of these fungal products almost certainly forms the molecular basis for the classical 'gene-for-gene' forms of resistance. The proposed elements of this paradigm have been strongly supported in at least one case by the demonstration that the deletion of an 'avirulence gene site' is associated with the transformation to virulence of a previously nonvirulent blast strain (chapter 7, this volume). Therefore, although the response/no response 'trigger' appears to be under major gene control, the magnitude of the response itself is probably under polygenic control.

This major gene resistance based on a hypersensitive response appears to be distinct from that mediated by quantitative trait loci (QTLs) in which a compatible infection reaction occurs, but is characterized by reduced disease development. This resistance, although not complete, is apparently effective across all blast strains and may reduce the epidemic potential in cultivars with adequate QTLs. The most durably blast resistant of the traditional rice cultivars have been shown to possess complex configurations of multiple major gene and multiple QTLs (chapter 10, this volume). The observation that these traditional cultivars are frequently more resistant to blast than the products of breeding programs that have produced high yielding varieties has an interesting parallel in natural systems. A broad survey of nonagricultural tropical plant species indicates that increasing levels of host defense against pathogens and herbivores is correlated with decreased growth rates (Coley, 1988), suggesting that genetically based characteristics that promote disease resistance often retard growth rate (see below).

None the less, despite the fact that the phenomenology and genetics of resistance are increasingly better understood (chapters 4, 6, 7, 9, 11, 13, 15 and 16, this volume; Levy *et al.*, 1993), it is still the case that the mechanistic basis of that resistance is largely unknown. For example, it is known that the tissue destruction that characterizes the hypersensitivity response is mediated through the release of a combination of enzymes and phenolics. What is less clear is the relative extent to which these responses are based on induced, *de novo* synthesis or on the release of already present, but sequestered, products and how that balance might vary across cultivars. Further, despite remarkable advances (chapter 7, this volume), the identities and functions of the fungal molecules that induce the hypersensitivity responses in the rice are still unknown (see below). QTL-based resistance is less clearly understood still. Perhaps some of this partial resistance is explainable by standing levels of toxins or rice products that induce the fungus to sporulate rather than propagate, thus limiting blast damage. However, characterizing the mechanisms that underlie the different forms of resistance that have been to a greater or lesser degree characterized genetically should be a high research priority. For example, depending on the specific antifungal mechanism associated with a given set of resistance genes, it may be intrinsically difficult to use those genes to breed cultivars of rice that are not only disease resistant but are also fast growing or possess the wide variety of other desirable properties that growers seek as well (chapters 14 and 28, this volume).

At a larger scale, in most agricultural settings, the opportunities for blast transmission are enormous, even when compared to the most outbred fig systems. At the extreme, in those cases of extensive, near-monoculture planting, the rice cultivar is essentially a genetic 'sitting duck', waiting for a killer pathotype to evolve or to spread. Once a fungal pathotype that is capable of bypassing the defenses of a particular cultivar is present, the results can be devastating. The field's genetic uniformity facilitates the transmission and propagation of those strain types that are increasingly able to infect, reproduce, and spread most rapidly. These are almost invariably the strains that produce the most damage to the rice.

The repeated breakdown of resistance in existing cultivars followed by the repeated introduction of new host genotypes by growers and breeders every few seasons contrasts sharply with the relative constancy found in the wasp and nematode systems which are characterized by short-term equilibrium and long-term coevolution. Among the wasp–nematode systems, those with the highest transmission are those in which the nematodes are most virulent. The normal agricultural setting of rice permits even greater transmission opportunities for the blast pathogen and, without human intervention, it is capable of producing orders of magnitude more damage to its host. In short, the fig system is a highly subdivided, largely equilibrated system as opposed to the open and very nonequilibrated agricultural systems of the rice (Table 18.1). Perhaps the most striking contrast lies in the simple observation that, without considerable human effort, it is almost certain that the existing rice cultivars would vanish. Is it possible to apply that effort more efficiently, perhaps even reducing it, and yet to decrease loss in rice yields due to blast?

It is obviously not practical to grow rice with individual plants isolated from others in a close mimic to the conditions that seem to promote lower virulence in the nematode parasites of the wasps. However, the principle that greater opportunities of transmission open greater possibilities for pathogen virulence seems to apply and it is worthwhile to try to identify steps that could be taken that would both lower blast transmission (a desirable objective in itself) and (thereby) promote lower levels of virulence. In order to do this, it is useful to consider the potential of combining the products of the 'lineage exclusion' (chapters 13 and 16, this volume; Levy *et al.*, 1993) approach to breeding resistant cultivars with multiple-cultivar planting (chapter 17, this volume).

The key insight underlying the lineage exclusion approach has involved the integration of molecular information with an evolutionary perspective. The above-mentioned inference that the host resistance depends on the presence of specific genetic products in the fungus is combined with the recognition that all strains of blast are not totipotent in their capacity to adapt to a range of hosts. Moreover, as the molecularly determined phylogenetic relationships of different blast pathotypes have become known, it has become obvious that closely related pathotypes generally possess similar suites of hosts that they can infect. The clear implication is that evolutionarily similar entities are biochemically similar and it is the biochemical similarity that either permits or inhibits them from infecting particular hosts. This, in turn, suggests that for a pathotype to bypass the defenses of a given cultivar would require it to lose the surface molecules or shift the metabolic pathways that were the signals for the host to initiate a response to it. Within lineages, some such shifts or losses are almost certainly easier than others, as they require the alteration of less essential pathways. Minor, relatively common and easily accomplished shifts might differentiate closely related strains within a lineage. However, major shifts involving fundamental changes in important pathways may be the factors that differentiate lineages, or caused the original divergence that gave rise to discrete lineages. The implications of this line of thought suggest a fairly clear program for improving the efficiency of breeding for long-term resistance.

The key elements of this approach are to: (i) use molecular techniques to identify as broad a sample as possible of different strains of blast; (ii) sort them into phylogenetically related 'lineages'; (iii) identify cultivars that possess resistance to all strains in given 'lineages'; (iv) use these cultivars to breed new cultivars possessing combinations of genes that provide resistance to all identified lineages; (v) test these new cultivars against combinations of known pathotype lineages; and (vi) when successful, introduce these for general cultivation. The net effect of this directed breeding approach is to increase the diversity of resistance genes within individual cultivars in a deliberate fashion. Previous to the adoption of this approach which emphasizes utilizing phylogenetic knowledge of the potentially infective blast pathotypes, the average resistance of newly introduced rice cultivars was 2 years or less in Colombia. New cultivars that reflect the lineage exclusion approach have maintained high levels of resistance in the field for eight growing seasons. These results seem promising, to say the least.

However, even with these genetically diverse cultivars, if they are planted in vast monocultures, the individual cultivar is still a 'genetic sitting duck', albeit now a much more formidable duck. Even a complex and diverse set of major resistance genes and QTLs may prove to be a soluble problem for the blast. If this turns out to be the case, it may be useful to take the process a few steps further. The increased within-cultivar diversity of resistance genes produced by the 'lineage exclusion' breeding programs could be augmented by simultaneously planting several such cultivars that differ in their resistance genes (chapter 17, this volume). The introduction of multiple cultivars of rice in single fields is expected to have a twofold benefit. First, it will reduce the rate of direct transmission of blast within fields because individual plants are no longer certain to be susceptible to the spores coming from their infected neighbors. This effect is observed in multiple-cultivar planted wheat and acts to reduce the impact of rust outbreaks as well as the subsequent need for fungicide application. Second, a more complex suite of hosts imposes a much more complex selective regime on the fungi which will prevent strains from becoming as specialized on any one cultivar as effectively as it could in a monoculture. Therefore, due to the restrictions that it places on transmission, multiple cultivar planting is expected to retard the spread of extremely virulent blast genotypes on evolutionary as well as ecological timescales. A third level of diversity of resistance should also be considered. In addition to maintaining higher diversity of resistance genes within individual cultivars and growing mixes of cultivars, it may be useful to rotate cultivars before the emergence of blast pathotypes capable of infecting them. It is not clear whether virulent pathotypes are continually being produced *de novo* by evolutionary change or whether most pathotypes are already in the environment and simply take some time to become sufficiently abundant to cause problems (May and Dobson, 1985). In either case, but particularly in the latter, changing cultivars before virulent pathotypes become established will prevent the permanent loss of any particular cultivar for productive use in cultivation.

However, there are several difficulties with implementing such a multi-tiered system. Whether a one, two, or three level system is adopted depends on the answers to a number of questions. First, can all of the important lineages and their constituent pathotypes be identified within a target region? At what cost? The diversity of blast lineages is apparently much higher in Asia (not incidently the origin of rice cultivation) than in South America where lineage exclusion has thus far been successful. Second, how discrete and stable is the spectrum of cultivars (with their characteristic resistance alleles) that are resistant to the range of different pathotypes within a lineage? Does the fact that the blast does not reproduce sexually lower the capacity of lineages to develop the capacity to infect new, initially resistant cultivars? Third, how rapidly and at what cost can new cultivars be produced by lineage-exclusion-based breeding? Techniques described during the course of the conference for marker-facilitated plant breeding has demonstrated the capacity to speed up the production of new cultivars several fold (see McCouch *et al.*, this volume and personal communication). None the less, the level of effort that is expended in breeding will depend on what planting strategies (multicultivar planting and/or cultivar cycling) are employed and how effective they prove to be. Fourth, how long do the products of this breeding program stay resistant? Once a given cultivar's defenses have been breeched, can it ever be successfully reintroduced? How quickly can a virulent pathotype that exists in low levels spread to a reintroduced sensitive host (May and Dobson, 1985)? If the cost to produce resistant cultivars is high and if once resistance is lost the cultivar is lost, then cycling cultivars before resistance is lost becomes more worth the extra organizational effort needed to maintain and keep track of the cycling of the cultivars. Finally, the unpleasant question, can blast pathotypes evolve that bypass even the best laid plans for within-cultivar, within-field, and temporal diversity of resistance?

An approach requiring a large investment in laboratories and centralized organization, such as any of those outlined above, will by its very nature be aimed at large-scale production areas. However, effective, low-investment, low-technology approaches are also needed for small national research programs with more limited resources working at much smaller scales. One example is a program that was outlined for by accumulating genes which give partial resistance (see chapter 15, this volume). Although it is a less directed approach than lineage exclusion and perhaps less effective, it is simple, can be easily implemented, and requires very little 'high technology' or the capital-intensive investment that comes with it.

None the less, even inexpensive techniques such as these require an organizational structure that can effectively introduce them to a frequently very conservative and wary populace. This point was driven home to me by a conversation with a rice breeder about the reluctance of rural African farmers to incorporate new technologies or new strains of rice, preferring to plant their own varieties. Local varieties of rice that have been grown in traditional fashions frequently suffer less loss to blast than the high yield varieties that have been developed under different settings. After stating how difficult this made the agents' work, he stated, 'Well, this is because many times we have given them varieties that failed completely and so they are justifiably suspicious. If they will plant what we give them at all, they will only do it on a small part of their land. They prefer to grow several types of rice at once so that they do not suffer from a complete disaster'.

Conclusions

A growing body of theoretical and empirical work suggests that transmission breeds virulence. A well-established body of observation indicates that nature abhors a monoculture, almost as much as a vacuum. This means substantial effort must be expended to maintain monocultures in the face of highly transmissible pathogens. How best to expend that effort effectively? In the relative absence of such basic information as the identities of strains of blast, the range of cultivars that they can attack, and their spatial and temporal distribution, it is difficult to clearly understand the nature of the blast, much less form a coherent counter-strategy. With the introduction of molecular identification techniques and the incorporation of the resulting identifications into an evolutionary perspective, the fungus is no longer such a mysterious opponent. This understanding allows the formulation of an informed strategy which has the potential of approximating intelligently guided coevolution. If the experiences of the Colombian lineage exclusion breeding program prove to be robust and if the results of the multiple cultivar planting from the wheat indicative, there is every reason to think that this combination of strategies that, on the one hand, intelligently build the diversity of resistance genes within cultivars and, on the other, act to retard within-field transmission of pathogens will dramatically stabilize the pattern of blast outbreak and lower the overall loss due to this fungus. However, at a much more fundamental level, if attempts at human population control are not ultimately successful, it is very likely that our fates will be determined in ways that will probably be neither under our control nor to our liking. Toward avoiding this unpleasant possibility we should try to increase crop productivity through understanding and controlling their diseases; however, we should not lose track of the larger and arguably much more urgent problem. Increasing agricultural output will buy time, but the effort will be for naught if that time is not used wisely.

Acknowledgments

I would like to thank Sally Leong for giving me the opportunity to attend the conference and for copious background information. I also thank John Mumford, Barbara Valent, Monte Jones, Morris Levy, Fernando Correa-Victoria, Chris Mundt, Edward Roumen, Sally Leong, and Robert Zeigler for stimulating discussions. I thank Elisabeth Kalko for reading earlier versions of this manuscript.

References

Berg, C.C. (1989) Classification and distribution of Ficus. *Experientia* 45, 605–611.

Borromeo, E.S., Nelson, R.J., Bonman, J.M. and Leung, H. (1993) Genetic differentiation among isolates of *Pyricularia* infecting rice and weed hosts. *Phytopathology* 83, 393–399.

Bonman, J.M., Khush, G.S. and Nelson, R.J. (1992) Breeding rice for resistance to pests. *Annual Review of Phytopathology* 30, 507–528.

Coley, P.D. (1988) Effects of plant growth rate and leaf lifetime on the amount and type of anti-herbivore defense. *Oecologia* 74, 531–536.

Herre, E.A. (1985) Sex ratio adjustment in fig wasps. *Science* 228, 896–898.

Herre, E.A. (1987) Optimality, plasticity, and selective regime in fig wasp sex ratios. *Nature* 329, 627–629.

Herre, E.A. (1989) Coevolution of reproductive characteristics in 12 species of New World figs and their pollinator wasps. *Experientia* 45, 637–647.

Herre, E.A. (1993) Population structure and the evolution of virulence in nematode parasites of fig wasps. *Science* 259, 1442–1445.

May, R.M. and Dobson, A.P. (1985) Population dynamics and the rate of evolution of pesticide resistance. In: *Pesticide Resistance: Strategies and Tactics for Management*. National Academy Press, Washington DC, pp. 170–193.

Morishima, H., Sano, Y. and Oka, H. (1991) Evolutionary studies in cultivated rice and its wild relatives. In: Futuyma, D. and Antonovics, J. (eds), *Oxford Surveys in Evolutionary Biology*. Oxford University Press, Oxford, pp. 135–184.

Levy, M., Correa-Victoria, F.J., Zeigler, R.S., Xu, S. and Hamer, J. (1993) Genetic diversity of the rice blast fungus in a disease nursery in Colombia. *Phytopathology* 83, 1427–1433.

Poinar, G.O., Jr and Herre, E.A. (1991) Speciation and adaptive radiation in the fig wasp nematode parasitodiplogaster. *Revue de Nematologie* 14, 361–374.

Valent, B. and Chumley, F.G. (1991) Molecular genetic analysis of the rice blast fungus, *Magnaporthe grisea*. *Annual Review of Phytopathology* 29, 443–467.

IV Epidemiology, Loss Assessment, and Management

19 The Economic Impact of Rice Blast Disease in China

MINGGAO SHEN AND JUSTIN YIFU LIN

Development Research Center of the State Council, China, No. 9 Xihuangchenggen Nanjie, Beijing 100032, China

INTRODUCTION

For the past 40 years, Chinese grain producers have successfully provided food for over 20% of the world's population with only 7% of the world's arable land. China's population increased from 552 million in 1950 to 1143 million in 1990. However, from 1950 to 1990, the nation's total grain output increased from 132 million tonnes to 446 million tonnes. During this period, the annual growth rate of grain production was 3.09%, significantly higher than China's 1.84% population growth rate. As a result, the per capita food production had grown from 239 kg in 1950 to 393 kg in 1990.

Rice is the most important grain crop in China. It has the highest yield of all grain crops. Due to the high population pressure, almost all of China's irrigated land is sown with rice. Rice area makes up 30% of the area sown to grain, but it contributes over 40% of China's grain output. The annual growth rate of rice output from 1950 to 1990 was 3.13%. Of the total increase in grain output in the period, 42.7% can be attributed to increases in rice production.

In the past 40 years, the area of sown rice in China only increased slightly from 26.1 million hectares to 33.1 million hectares. The main source of increase in rice output has been yield increase, which can be attributed to the increased uses of modern varieties and modern inputs. As China's agriculture changed from a traditional to a modern one, a number of yield constraints were ameliorated, including unfavorable climatic factors (such as temperature and light), water, soil conditions, insects, pests and diseases. The main purpose of China's rice production and research system is to increase rice output further. The above problems remain the focus of China's rice research program. Of the diseases affecting rice production in China today, blast is one of the most serious. Its persistence continues to be a potential threat to the nation's ability to further increase and stabilize rice output.

This chapter discusses the occurrence of rice blast disease in China, including its change over time and its prevention and control, and examines the extent of its economic impact.

The Occurrence of Rice Blast Disease in China

Prior to the 1960s, China's agriculture was based on traditional methods. The utilization of modern inputs was nearly non-existent. Since then, the modernization of Chinese agriculture has been rapid. In addition to extending a range of new varieties and the uses of modern inputs, a number of other technological improvements in the system of cultivation and plant management have been introduced. These changes in part have served to create an environment in which rice blast disease has evolved from a regional disease to one capable of creating a nationwide disaster.

The spread of modern varieties

China's agricultural research was fairly weak until the end of the 1950s. The main varieties of rice were bred by farmers themselves and typically used for several generations. Although these varieties were well adapted to their region of origin, their yields were low. Under the conditions in the 1950s and 1960s, rice blast remained a regional disease occurring mainly in mountainous areas with low temperature and high humidity (GSPP, 1988). Since the 1960s, with the development of agricultural research in China, modern varieties with high-yield potential began to replace traditional farm-bred varieties. The release of modern varieties was much faster than was possible with traditional methods. During this period, Chinese rice economy experienced two major breakthroughs: one was the substitution of semidwarf fertilizer-responsive varieties for long-stalked varieties in the 1960s; the other was the diffusion of hybrid varieties in the 1970s. In the 1980s, the proportion of total area sown to improved varieties was more than 80% of total sown area (Lin and Shen, 1990).

The expansion of area sown to improved varieties significantly increased rice yield, but simultaneously decreased the number of varieties. This increase in relative genetic homogeneity has resulted in some incidence of widespread disease occurrence. Also, since blast physiological races change in composition with varieties in different rice regions, the substitution of modern varieties has caused a differentiation of race groups. This differentiation process has included a loss of varietal resistance, facilitating the spread of rice blast disease.

Changes in the cropping system

Chinese producers have switched from single-season rice to two-season rice to increase grain output since the 1950s. In the southern Yangzi River areas, farmers have moved from rotations that included other crops to continuous cropping systems. North of the Yangzi River and south of the Huai River, winter wheat and rapeseed areas were expanded, and rice–wheat systems was popularized. In the 1970s, in the middle and lower reaches of the Yangzi River and in the southern rice regions, a three-harvest system (including two rice crops) was developed. Production rotations such as rice–rice–green manure, rice–rice–wheat and rice–rice–rapeseed spread rapidly (Lin and Shen, 1990). With the rise of the cropping index, single-season rice and double-harvest rice coexisted. Early-season rice overlapped with single-season rice and single-season rice overlapped with late-season rice in a location in many parts of China. This crop pattern led to a serious incidence of cross-infection. Atavism of rice blast disease was common (GSPP, 1988). The change of cropping systems also altered the growing season for rice, which increased the possibility of weather-induced blast disease. Early-season rice and single-season rice varieties have been especially vulnerable since a large part of the crop's growth would then occur during the rainy season which is characterized by long spells of fog and less sunlight.

Changes in cultural practices

Research has shown that the occurrence of rice blast disease is closely related to environmental factors, among which temperature and relative humidity are most important. Also, the increasing application of nitrogen fertilizer makes the rice stems more succulent and the plant produces more leafy material making infection more probable (Huang and Miao, 1993). With the spread of modern varieties, rice cultivation practice in China has radically changed. The shortening of the growing seasons, the compact planting, the increased availability of fertilizers, and expansion of irrigated area has made rice blast disease more likely to occur.

Due to these changes, rice blast disease changed from a regional disease in the 1950s and 1960s to one which threatened China with nationwide disaster in the 1970s. By the late 1970s and the early 1980s, rice blast disease epidemics began occurring nationwide. Figure 19.1 shows the proportion of rice area in China affected by leaf blast and neck blast in 1980–1990, based on the data from an intensive survey of all of the 176 rice-producing prefectures in China carried out by the authors in 1991. In 1985, leaf blast affected 12% of national rice acreage and neck blast affected another 12%. From the figures we find that the incidence of leaf blast disease and neck blast disease is highly correlated. The incidence of blast disease fluctuated from year to year. Table 19.1 shows that in 1984 and 1985 blast disease was a widespread disease in the prefectures which consisted of about 40% of China's rice area. In 1988, however, blast disease was a serious problem only in the prefectures that covered 7% of China's rice area and in almost half of the rice-producing prefectures in China, blast disease did not occur at all. Table 19.2 lists the occurrence separately for early-, late- and single-season rice. The information shows that early-season rice and single-season rice, which are grown during periods of low temperature and continuous rain, are more likely to be affected by blast disease than late-season rice.

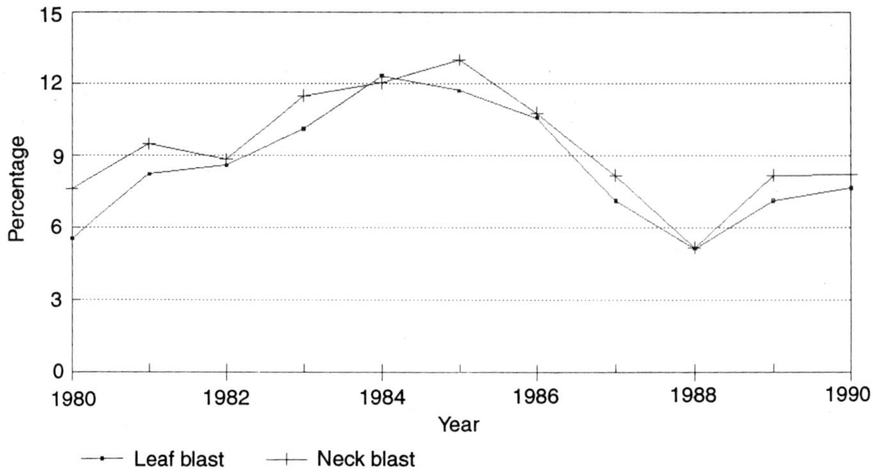

Fig. 19.1. Percentage of rice acreage in China infected by blast disease.

Table 19.1. The occurrence of rice blast disease in China (%).

Year	Rank 1	Rank 2	Rank 3	Rank 4	Rank 5
1980	54	19	12	6	9
1981	41	23	13	7	16
1982	30	26	14	9	21
1983	29	24	12	10	25
1984	23	14	12	14	37
1985	22	19	13	8	38
1986	28	19	10	10	32
1987	36	23	17	8	16
1988	50	25	11	8	7
1989	32	21	14	14	18
1990	31	26	18	10	16

Rank 1: Blast disease occurred in 0–5% of the rice area in a prefecture's total rice area.
Rank 2: Blast disease occurred in 5.1–10% of the rice area in a prefecture's total rice area.
Rank 3: Blast disease occurred in 10.1–15% of the rice area in a prefecture's total rice area.
Rank 4: Blast disease occurred in 15.1–20% of the rice area in a prefecture's total rice area.
Rank 5: Blast disease occurred in 20.1% and more of the rice area in a prefecture's total rice area.
Numbers in the rank columns indicate the percentage of the respective rank's rice acreage in the nation's total rice acreage.
Source: The data reported in the table are taken from an intensive survey of all 176 rice-producing prefectures in China by the authors in 1991 for the project 'Agricultural Research Priorities in China'.

Table 19.2. The occurrence of blast disease for early-, late-, and single-season rice in China (%).

	Early-season		Late-season		Single-season	
Year	Rank 1	Rank 2	Rank 1	Rank 2	Rank 1	Rank 2
1980	98	02	85	15	89	11
1981	79	21	80	20	94	06
1982	64	36	84	16	90	10
1983	67	33	69	31	86	14
1984	44	56	82	18	62	38
1985	63	37	71	29	54	46
1986	51	49	83	17	69	31
1987	82	18	87	13	83	17
1988	96	04	93	07	91	09
1989	80	20	88	12	79	21
1990	75	25	94	06	84	16

Rank 1: Blast disease occurred in 0–20% of the rice area in a prefecture's total rice area.
Rank 2: Blast disease occurred in more than 20% of the rice area in a prefecture's total rice area.
Numbers in the rank columns indicate that the percentage of the respective rank's rice acreage in the nation's total rice acreage.
Source: As for Table 19.1.

The Prevention and Control of Rice Blast Disease

The prevention and control of rice blast disease became an important concern of the Chinese government in its attempt to increase and stabilize rice output. Control of rice diseases was one of the primary functions of the Chinese plant protection system as the frequency and incidence of various diseases increased after the extension of modern varieties. The prevention and control of rice diseases switched from being primarily dependent on chemical application in the 1960s to an integrated management which concentrated on varietal improvement with disease resistance and adjustments to cultivation techniques. Comprehensive prevention and control of rice blast disease have been achieved in three ways:

1. The agriculture research system was organized to breed disease-resistant varieties for cool, moist climates.
2. Better and more effective cultivation methods were extended to farmers.
3. The plant protection sector was organized to use chemical pesticides more effectively, basing application on disease forecasts and encouraging appropriate application levels.

Breeding varieties with disease resistance

Early in the 1950s and 1960s, China began to investigate and appraise the extent of blast disease resistance in the national rice germplasm collection. But success in this kind of research was limited until the knowledge became available on blast race composition. There was a lack of understanding of the concept of resistance and therefore utilization of resistant material was inefficient. Moreover, in those early years, experiments with multiple blast race resistance in different rice regions had not been started. Consequently, selection of resistant material was somewhat restricted and varietal resistance tended to be unstable. As rice blast disease incidence increased, the Ministry of Agriculture organized a nationwide, multi-disciplinary coordinating research group for rice blast disease to appraise and select varieties. This group was charged with the task of breeding higher disease-resistant varieties, and studying the ideas of 'resistance heritage' and 'biological race'.

Under this intense research on the biological races of rice blast disease, eight groups and 66 types were identified (Sun *et al.*, 1988) using domestic and international differentials. Agricultural scientists implemented breeding programs that included traditional breeding, radiation-induced mutation breeding and cross breeding. Consequently, a series of semidwarf, high-yield blast-resistant varieties was extended in the 1970s and 1980s. The program met with remarkable results as rice blast disease was effectively controlled (Sun *et al.*, 1988). The decline in the occurrence of areas affected by rice blast disease by the late 1980s can be largely attributed to the increase in disease-resistant varieties.

Improving agricultural cultivation technology

After a long period of experimentation, researchers discovered that the occurrence of rice blast disease was also closely connected with compact cropping, fertilization levels, and irrigation conditions. Combining the experience of researchers and farmers, the agricultural production sector developed a set of recommendations concerning the cultivation of rice to help prevent outbreaks of rice blast disease. Practices emphasized the attainment of high yields while attempting to create an environment unfavorable to blast disease occurrence. In this way targets of raising disease resistance and yields were simultaneously pursued. These measures included: (i) adjusting cropping patterns and rotations to prevent disease spread from one crop to another; (ii) appropriate irrigation – encouraging gutter irrigation and reducing chain irrigation that involved shallow and repeat irrigation in the early part of the rice growing season; (iii) rational fertilizing that involved moderate basal fertilizing, higher fertilizing in the tillering stage, but no nitrogen fertilizing after spike formation, appropriate phosphorus and potassium fertilizing to increase the disease resistance capability of stalks; and (iv) eliminating survival forms of the pathogen – destroying infected straw and using seed treated with chemicals.

Prevention with chemical pesticides

Specific data regarding the use of pesticides for rice blast disease do not exist at the prefectures that we surveyed in 1991. Only general observations and experimental data are available. After 1979, the yearly chemical usage on rice was 65% of the total chemical usage in the whole country. In 1989, the average national plant protection cost per ha for rice pests and diseases was 6.4 yuan, about 7% of total rice cost (Ministry of Agriculture, 1991). Rice blast disease usually occurs suddenly, so a rapid response is critical if prevention is to be successful. Once the disease center of leaf blast forms, under favorable conditions, it will spread in 7–9 days, and break out in 10–14 days. If there is any sign of disease on the leaves during the late seedling or tillering stage, the disease will typically break out in 3–5 days. Effective control requires that the 'disease center' be completely eliminated. Neck blast control needs considerable attention during the heading stage. If leaf blast continues during the spike forming stage, or if 10% or more of the leaves are infected or affected, chemical pesticides are recommended to be used once each during the heading and post-heading stages. If the weather is favorable for developing blast, Chinese farmers typically make one more application during the grain filling stage (Huang and Miao, 1993). Appropriate utilization of chemicals can prevent rice blast disease from spreading and also reduce losses significantly. Leaf blast is relatively easy to control with chemicals, while neck blast requires more attention to the timing of application. Proper timing can be facilitated by accurate disease forecasting. In order to prevent and control loss due to diseases and pests, the Chinese government has established a complete network of forecasting and control. The Ministry of Agriculture runs the National Plant Protection Agency. Every province, prefecture, county, and district has a station set up within their jurisdiction. Theoretically every village has a plant protection team and one person who is the liaison officer with the township. Such a system has made considerable contribution to the increasing effectiveness of pest and disease forecasting. Through this system, extension help is also provided to farmers (GSPP, 1988). Table 19.3 summarizes the findings in our national survey. Column 5 of this table shows that area in which control measures are practiced was 1.19–2.28 times the area where blast disease occurred. There is a close relation between the two. The evidence proves that the pest-disease forecasting system is efficient and the plant protection system can meet the demand of pest-disease prevention and control. Column 6 shows that the loss avoided by chemical control was about 0.4–1.6% of the average rice yield.

Yield Loss Caused by Rice Blast Disease and Prospects for its Control

Based on our prefecture-level survey in 1991, we find that the actual output loss caused by rice blast disease has been greatly reduced by the prevention and control measures. The average loss of early-season rice caused by blast disease in 1981–1990 was only 27.2 kg ha^{-1}, about 0.53% of average yields in those

Table 19.3. Rice blast infected area and area treated with fungicides.

Year	Surveyed area (1000 ha)	Infected area (1000 ha)	Treated area (1000 ha)	Treated area/ infected area	Loss retrieved/ yield
1980	20635.51	2134.95	2722.10	128	0.57
1981	21402.70	2946.20	4119.31	140	0.81
1982	22425.05	3069.27	4206.16	137	0.86
1983	22991.15	3954.15	5287.89	134	0.94
1984	22818.43	4550.60	5428.65	119	1.05
1985	22829.29	4653.09	8951.59	192	1.55
1986	22712.95	3974.11	5335.17	134	1.13
1987	22984.93	2783.19	3399.78	122	0.76
1988	22293.51	1905.47	2399.90	126	0.42
1989	22838.67	3104.93	7068.28	228	0.82
1990	22670.27	3235.68	4750.94	147	0.79

Source: As for Table 19.1.

Table 19.4. Actual losses caused by rice diseases (kg ha^{-1}).

Year	Total	Sclerotia blight	Blast	Bacterial leaf blight
1980	55.78	22.09	23.57	8.72
1981	58.66	22.31	28.65	6.97
1982	65.37	25.19	21.30	12.88
1983	66.83	31.55	26.94	5.87
1984	85.53	36.57	40.59	7.01
1985	112.13	64.66	40.93	5.93
1986	56.17	28.41	19.04	3.65
1987	68.23	46.09	11.53	5.93
1988	50.39	35.64	6.95	2.78
1989	72.43	40.62	18.34	8.48
1990	71.36	37.96	15.25	13.60

Source: As for Table 19.1.

10 years. Late-season rice had an even smaller loss, only about 0.23% of the average yields in 1981–1990. Single-season loss was about 0.45% of the average yield. Table 19.4 summarizes the findings in our national survey about the yield losses caused by rice diseases in the 1980s. The table shows that the early half of the 1980s experienced heavier blast disease losses, especially during the two large outbreaks in 1984 and 1985. In those 2 years the output loss was around 0.7–0.8% of the yield. In 1985 the blast disease caused the single-season rice a loss of 130 kg ha^{-1} (about 2.2% of the average yield in China). In 1988, the loss was lightest, less than 0.1% of the yield. The correlation analysis, based on the yearly observations of the prefecture-level data in 1980–1990 in our national survey, shows that the percentage of yield loss in a prefecture is negatively correlated with the yield level in that prefecture but the correlation is not

significant statistically. It suggests that rice blast disease influences the yields, but its impact is controlled.

Overall, rice blast is the number-two rice disease in China, second only to 'banded sclerotia blight', and more important than bacterial leaf blight (Table 19.4). For early-season rice in the 1980s, the loss due to blast was 39.4% of the total losses from diseases (also second to banded sclerotia blight). For late-season rice, the loss was only 19.8%, behind both banded sclerotia blight and bacterial leaf blight. For single-season rice, the loss of rice blast disease was 30.5%, much heavier than bacterial leaf blight. Among all the rice diseases, the loss caused by banded sclerotia blight, blast disease, and bacterial leaf blight totalled between 84 and 97% of total losses from diseases, while the losses caused by all the other diseases were negligible.

The success of controlling rice blast disease in China has been achieved primarily by setting up a comprehensive plant protection system as well as the ability to mobilize farmers. With economic development, however, some of these successes may be weakened or even disappear. The proportion of agricultural output to national economy is decreasing, and the structure of vested interests and investment is changing. The regional governments at different levels, previously called 'agricultural governments', have in many places changed their focus to relatively more profitable sectors, such as industry, business, and the provision of services. The proportion of government investment in agricultural infrastructure to total investment declined from 17.6% in 1963–1965 to 3.3% in 1986–1990 (State Statistic Bureau of China, 1991). The infrastructure accumulated in the 1960s and 1970s may have declined. While the downturn of government investment in agriculture and the dissolution of the collective economy have weakened the agricultural research and extension system, which has public goods characteristics (Lin, 1991), the rural areas, on the other hand, have diversified after rural reforms which caused the proportion of income from grain production in the farmer's overall income to fall. Moreover, the government often suppresses grain markets to lower the grain prices. Grain production is less profitable than other activities. Farmers' incentives to invest in grain production are inadequate. With the shift of labor from the agricultural sector to industries, the quality of the remaining farming population in terms of education decreases. Therefore, it is hard to effectively implement prevention and protection measures (such as blast disease management) over a large area as had been done previously. Before the full development of rural markets and proper adjustment of China's crop structure, the relative profitability of grain production will probably decline further. Along with the weakening of the rice blast disease control system and the declining profitability of rice production, the potential for outbreak is certainly present.

From the chemical control viewpoint, the impact of institutional change on rice blast disease control is even larger than that for other rice diseases. The prevention and control ratio of rice blast disease is much higher than that of banded sclerotia blight. We use the prefecture-level data to undertake a simple correlation analysis between the prevention and control ratio (the 'prevented and controlled area' divided by the occurrence area) and the loss ratio (realized loss divided by avoided and realized loss) for banded sclerotia blight and blast disease.

For banded sclerotia, the correlation coefficients are −0.1152 for early-season rice, −0.1646 for late-season rice, and −0.1033 for single-season rice. All coefficients are significantly different from zero at the 0.1% level. The correlation coefficients for blast disease are, respectively, −0.0213, −0.028, and −0.0167. However, none of them is significantly different from zero. This provides some evidence that the prevention and control of rice blast disease are more difficult, and require more resources and a higher level of organization.

Resistance tests made by the Jiangxi Forecasting Co-ordination Group (1989) in the late 1980s found that the proportions of strong, moderate, weak, and non-resistant hybrid rice varieties among total tested varieties were 35.7%, 57.1%, 7.2%, and 0.0%, and for conventional varieties, they were respectively 19.20%, 41.4%, 19.2%, and 19.2%. One reason for the hybrids' high resistance may arise from the fact that hybrid seeds are replaced freshly each year. The proportion of China's rice area sown to hybrid rice increased from 14% in 1980 to 26% in 1985 and then to 50% in 1990. The large-scale spread of hybrid rice after the mid-1980s may in part be responsible for stopping the outbreak of rice blast disease. Because the quality of hybrid rice currently is inferior to conventional rice and is therefore less profitable, the sown area of hybrid rice has been decreasing since 1990. This trend will probably not be reversed until hybrid varieties with higher qualities are bred and extended. The expansion of area sown to conventional rice varieties threatens to lower the resistance of China's rice crops to blast; meanwhile, seeds of the conventional varieties with blast resistance are in short supply. These may lead to greater outbreaks of rice blast disease in the future.

The burden of controlling rice blast disease in the future falls on rice breeders. Since the blast races are continually changing, one kind of resistant variety may lose its resistance in a period as short as several years. Formerly popular conventional varieties – Nongken 58, Jingying 59, and Zhenshan Series – are now not grown since they have lost their resistance to rice blast disease. The loss of resistance by Xianyou 2 was the main cause of the blast outbreaks in 1984 and 1985. Since breeding a new variety requires at least 5–7 years, a strong research effort is needed now more than ever.

Conclusion

Intensive cultivation of rice is a characteristic of Chinese agriculture. Given the constant population pressure, the Chinese government and the farming community have put much effort into variety-improvement research and controls of diseases and insect pests. As a result, China has one of the world's highest rice yields. Nevertheless, for the time being, the incentive to increase rice output has been weakened by a number of factors. The ability to increase output by traditional cultivation methods has been exhausted, while the rice sector is losing ground in its competition with other farm and non-farm sectors. Rice production needs supports and subsidies from the government. As a result, it must continue to rely on the public sector's research and support system for new advances in science and technology.

With economic development, the requirements of agricultural technology are also changing. The prevention and control of rice blast disease in China has been a great success. But new, appropriate technology may be needed to preserve China's achievement. Other work by the authors show that the technological choices of Chinese farmers are more and more determined by economic considerations (Lin and Shen, 1990). In order to develop and strengthen the prevention and control technology of rice blast disease, it is very important to consider the cost of agricultural technology innovation, the cost of learning and acquiring these technologies, and the economic benefits that can be derived after the application of these new technologies.

Acknowledgment

We are indebted to Paul Teng for helpful comments on an early draft.

References

General Station of Plant Protection (GSPP), the Ministry of Agriculture (1988) Thirty-six years in prevention and control of rice diseases and pests in China. In: Zhen, Z. (ed.), *The Development of Prevention and Control of Rice Diseases and Pests in China.* Zhejiang Science and Technology Press, Zhejiang, pp. 1–8.

Huang, Q. and Miao, F. (1993) *The Disaster Reducing Agriculture*. Chinese Science and Technology Press, Beijing.

Jiangxi Forecasting Co-ordination Group of Rice Blast Disease Race Group and Resistance of Varieties (1989) The appraising results of rice blast disease resistance and its implication. In: *Disease and Pest Forecast.* No. 5 (supplement), National Plant Protection Station, Ministry of Agriculture, Beijing, pp. 32–37.

Lin, J.Y. (1991) The reforming of family responsibility system and the application of hybrid rice in China. *Journal of Development Economics* 36, No. 1.

Lin, J.Y. and Shen, M. (1990) The evidence and policy implication of technology change in Chinese agriculture. *Journal of Comparative Economic and Social System*, No. 2 (in Chinese).

Ministry of Agriculture (1991) *Agricultural Yearbook of China 1991*. Agricultural Press, Beijing.

State Statistic Bureau of China (1991) *Statistical Yearbook of China 1991*. The Statistical Press of China, Beijing.

Sun, S., Tao, R. and Shi, D. (1988) The breeding of blast resistant rice varieties in China. In: Zeng, Z. (ed.), *The Development of Prevention and Control of Rice Diseases and Pests in China*. Zhejiang Science and Technology Press, Zhejiang, pp. 33–45.

20 Current Rice Blast Epidemics and their Management in Thailand

S. Disthaporn

Rice Pathology Research Group, Division of Plant Pathology and Microbiology, Department of Agriculture, Chatuchak, Bangkok 10900, Thailand

Introduction

Rice blast, caused by the fungus *Pyricularia oryzae*, is common in all parts of Thailand. At present, it is considered to be the most important rice disease in the country (Disthaporn, 1989). The status of blast has changed in response to cultivation practices. Thirty years ago, transplanting was the only cultivation method and blast disease was recognized among farmers in relation to the rice growth stage as follows:

1. Seedling blast, which occurs in seedbeds. Leaves of seedlings have numerous small, brown oval lesions. In advanced stages of seedbed infection, there are many dead leaves and leaf sheaths. In very severe infections, large patches of seedlings may be killed.
2. Leaf blast was normally recognized after transplanting. The typical lesions appear on the leaves especially near the upper end of the leaves. Collar blast was commonly found and the leaves may break from the sheath due to rotting at the junction of leaf and sheath.
3. Neck blast and node blast occur in nearly mature plants. If node or neck blast lesions develop early, no grain is formed. If the lesions develop later, poorly developed grain is produced. Neck blast and node blast are the most destructive forms of the disease.

Because of labor shortage, rice cultivation is now commonly done with direct seeding rather than transplanting, particularly in the Central Plains, a major lowland rice growing area where rice can be grown twice a year or five crops in 2 years. In this area, high yielding varieties receiving high inputs have been continuously cultivated, and leaf blast and panicle blast are commonly found, starting 7 days after sowing (DAS) until harvest. To protect their crops, farmers commonly use fungicides, without considering the economic cost. This is in contrast with rice growing in the rainfed areas such as those in the north and

Table 20.1. Rice area damaged by blast disease epidemics in 12 provinces of Thailand in 1992.

Province	Blast damaged area (ha)
Chiangmai	22,762
Chiangrai	121,876
Payao	35,531
Lumpang	5,356
Lumpoon	4,860
Tak	424
Kampeangphet	137
Pitsanukloke	1,205
Pichit	1,579
Petchaboon	1,966
Maehongsom	966
Srisaket	3,422
Total	200,084

Source: Department of Agricultural Extension (DOAE).

north-east, where over 70% of the area is covered with traditional varieties of glutinous rice. Here, rice growing is mainly for family consumption; low inputs are applied and pest problems are uncommon, although average yield is also low, less than 2 tonnes ha^{-1} (Chamras, 1991).

Current Blast Epidemic in Northern Thailand

In 1992, because of a long drought in the northern part of the country, rice planting which normally starts in June had to be postponed to late July. Harvesting time usually in October was also delayed to November and December. The first report of panicle blast outbreaks was from Chiangmai province in November 1992 followed by reports from adjacent provinces until the end of 1992. The blast epidemic was reported officially from Chiangmai, Chiangrai, Payao, Lumpang, Lumpoon, Tak, Kampeangphet, Pitsanuloke, Pichit, Petchaboon, Maehongsorn and Srisaket with the total area damaged of 200,084 ha (Table 20.1) out of the total area in Thailand in 1992 of 9.6 million ha.

A team of plant pathologists investigated the damaged area, especially in Chiangrai province, the area most affected. They found that the six glutinous rice varieties, namely RD6, RD8, RD10, Niew Ubol, Niew Sanpatong and Meir-nong, were prevalent and had been grown in this northern region continuously over the past 20 years. These had been classified as moderately resistant to blast with no severe infections since the introduction year, 1970. The result of disease diagnosis confirmed that about 60% of diseased plants had panicle blast caused by *P. oryzae*. The rest showed very typical symptoms of dirty panicle and sheath-rot diseases. Isolation of diseased seeds collected from the infected fields confirmed that fungi, namely *Helminthosporium oryzae*, *Cercospora oryzae*, *Curvularia lunata*, *Sarocladium oryzae*, *Fusarium semitectum* and *Trichoconis padwickii*, were also the major causal agents. Sixty per cent of total yield was subsequently confirmed to be lost due to the disease, approximately equivalent to 650,000 tonnes of paddy rice.

Meteorological data collected in the rice experiment stations in Prae and Chiangrai provinces indicated that in mid-October there was a depression with rain showers continuously for fours days and three nights with the lowest temperature at 12°C at Chiangrai (Figs 20.1 and 20.2). These conditions were favorable for sudden disease outbreaks, especially for panicle blast when the plants were between flowering and dough stages. Previous observations by Ou (1961) showed that in the field, neck blast and node blast develop in November when there are heavy dews. Previous observations by the Plant Protection Service Unit (PPSU) further showed that in the northern provinces, blast infections occur during the summer and winter crops while in the Central Plains, blast is only observed and may be severe during the winter (dry) season between November to March (Disthaporn, 1987).

Assistance for farmers

The Cabinet of the Royal Thailand Government allocated US$10 million for paddy seeds and fertilizers to be given to farmers whose crops were damaged by blast disease.

Pathogen and Disease Management

Previous research by the Department of Agriculture (DOA) had shown that *P. oryzae* is a highly variable fungus; isolates differ from each other in pathogenicity to rice varieties, and also in physiological characteristics. It was previously believed that single spore isolates of *P. oryzae* were genetically homogenous, and these have been used in race identification. But research by Giatgong and Frederiksen (1969) has shown that eight to eleven races can be identified in sub-monoconidial lines from a single spore. Cytological studies indicated that this variation may be the result of mutation, the parasexual cycle, or heterokaryosis. Variability and diversity in the pathogen have been cited as the means by which the fungus is able to overcome previously effective disease resistant varieties and some fungicides. However, testing rice varieties and breeding lines for blast resistance has been an important part of the rice breeding program and has been part of the activities of the Rice Pathology Research Group (RPRG), DOA, since its establishment in 1959.

Resistant cultivars

Fortunately, during the severe 1992 outbreak of blast disease in the rice experiment station at Sanpatong, Chiangmai, a breeding line KKNLR 175052-PRE-40-1-1-1 showed a moderately resistant reaction to the disease while other

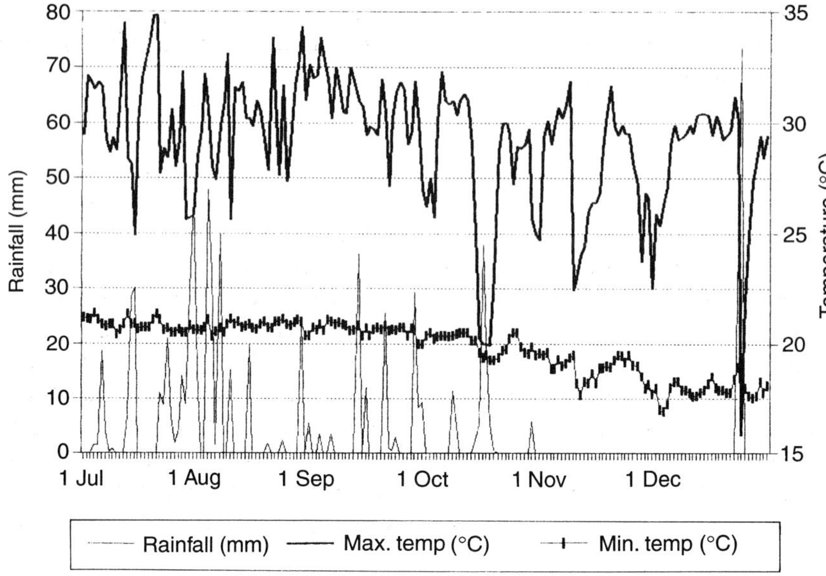

Fig. 20.1. Maximum–minimum temperature and rainfall during July–December 1992, Prae province.

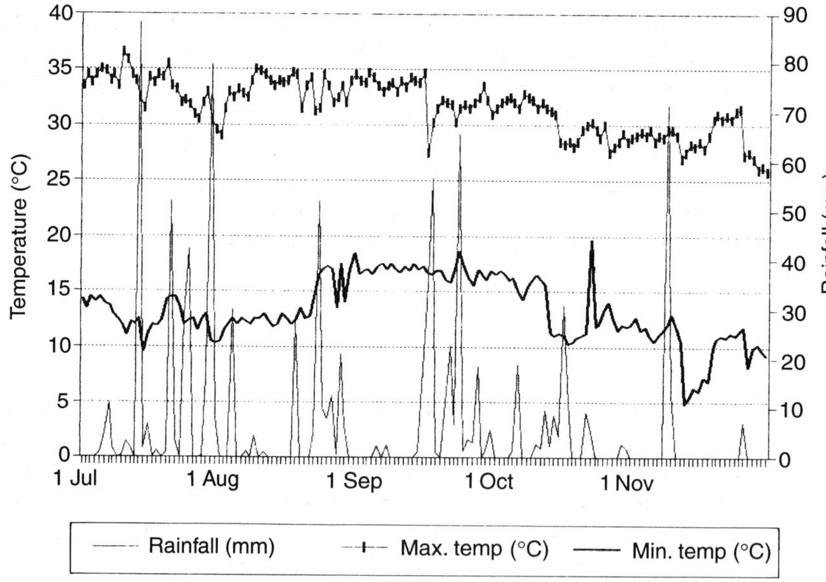

Fig. 20.2. Maximum–minimum temperature and rainfall during July–December 1992, Chiangrai province.

varieties grown at the same time in the same field were entirely destroyed. This line will be multiplied to serve as a resistant source for distribution to farmers in the coming crop season.

Seed treatment

After severe blast damage in 1992 there was insufficient rice seed for the next crop. It became necessary to use seed which escaped from the disease and remained in the infected fields even though they were contaminated with several fungi. In other countries, blast is reported to be a seedborne disease. In Thailand, attempts to culture *P. oryzae* from seed indicated that less than 1% were infested (Giatgong *et al.*, 1971). However, the government allocated US$1.2 million to purchase effective fungicides for rice seed treatment and foliar sprays, against blast and other fungi. Approximately 2000 tonnes of seed of varieties RD6, RD8, RD10 and Niew Sanpatong will be treated with Carbendazim or Mancozeb to minimize the source of inoculum as much as possible, before distribution to farmers.

Alternate hosts

A few isolates of *P. oryzae* from *Panicum repens* were found to be pathogenic to rice; other isolates from this grass were not. Lesions caused by *Pyricularia* spp. are common on various grasses in Thailand, including *Brachiaria mutica*, *B. indica*, *Digitaria ascendens*, *Leersia hexandra*, *Eleusine indica*, *Echinochloa colonum*, *Dactyloctenium aegyptiacum* and *Polytrias amaura*; but their relation to rice blast is not clear. A wild rice, *Oryza officinalis*, sometimes has blast lesions; but no race identification has been made (Sriwasin and Giatgong, 1970). Although the relation between rice and nonrice hosts is not clear, elimination of grasses around the field is recommended for farmers to reduce initial inoculum.

Silica

The physiological function of silica in disease resistance has been investigated by many workers. Ou (1985) indicated that blast spore penetration rates are lower in plants with a high level of silica, possibly as a result of the mechanical barrier attributed to accumulation of silica in the epidermis. In Thailand, our experiments show that silica plays some role in protecting against blast disease infection. Rice plants that received silicon in the form of sodium silicate appeared to have decreased severity of blast. The proportion of dead plants treated with silicon was lower than that of the control. Silicon also increased growth of the rice by increasing plant height and tiller numbers (Table 20.2) (Ratanapol *et al.*, 1990).

Table 20.2. Effect of silica on blast disease severity and growth of rice plants at Prachin Buri Rice Research Center, Thailand, 1989 (Ratanapol *et al.*, 1990).

Level of silica (p.p.m.)	% Disease severity	% Dead plant	% Crude silica	Height (cm)	%Tillering		Dry weight (g)
					pre-inoc.	post-inoc.	
0	10.00 b	97.63 a	0.80 c	90.23 b	115.60 c	39.7 b	33.20 ab
5	15.00 a	93.43 a	1.62 bc	97.93 ab	120.83 bc	44.6 b	30.73 b
10	9.00 b	87.73 ab	1.70 bc	103.37 a	126.13 ab	43.4 b	33.97 ab
25	7.67 b	58.17 b	3.31 b	101.70 a	128.13 ab	60.7 ab	33.55 ab
50	1.67 bc	21.47 c	5.66 a	105.10 a	133.87 a	89.43 a	45.70 ab
100	1.03 c	18.20 c	6.39 a	101.03 a	131.47 a	85.67 a	50.00 a
F. Test	26.35 **	25.23 **	29.62 **	9.10 **	14.63 **	8.52 **	4.70 *
% C.V.	24.19	19.79	22.66	3.50	2.44	21.67	16.75

Letters a–c in columns compare means at $P = 0.05$ level by Duncan's multiple range test.
* Significant difference at 5%.
** Significant at 1%.

Table 20.3. Common fungicides for blast control available in Thailand.

Common name	Trade name
Ediphenfos	Hinosan
Blaticidin	Bla-S
IBP	Kitasin-P
Isoprothiolane	Fuji-one
Kasukamysin	Kasumin
Thiophanate-methyl	Topsin-M
Carbendazim	Bentop, Derosal, Bavisan
Thiabendazole	Pronto
Tricyclazole	Beam
Pyroquilon	Coratop
Kasukamysin+Phosdifen	Kusumiron
Benomyl	Benlate

Source: Disthaporn, 1989.

Farmers practices for blast management

The Department of Agricultural Extension reported that the greatest area affected by leaf and panicle blast was nearly 490,000 ha in 1988 in the Central Plains, where there was widespread use of such susceptible varieties as RD21, RD23 and RD25. To protect their crops, farmers used fungicides heavily, without considering the economic cost. Twelve fungicides are recommended by the DOA for controlling blast (Table 20.3) and are available in the market. Most farmers routinely apply fungicides to prevent any appearance of disease symptoms on the crop.

The following procedures were suggested to convince farmers to use fungicide only when necessary.

1. Rice blast incidence should be assessed frequently at seedling stage.
2. Assessment should start with a systematic diagonal sample of 10 tillers per 0.16 ha.
3. Blast incidence should be measured on the top four leaves of each tiller or on leaf 3 by counting infected leaves.
4. Effective fungicides should be sprayed when more than 70% of the leaves are infected with blast (Surin *et al.*, 1991).

A year after these procedures were recommended, 700 farmers were interviewed to evaluate the percentage who followed the recommendations. No farmers were found to do so. Fortunately, more than 60% of the farmers assessed disease incidence before taking action using their own methods based on experience and after consulting with their neighbours (Sudto and Disthaporn, 1992).

Attempt at blast forecasting

Epidemiology is the study of populations of pathogens in populations of hosts and the resulting disease under the influence of the environment and human interference (Kranz, 1974). The final task of epidemiology is to provide the plant pathologists with refined and reliable knowledge of the behaviour of diseases in the field for optimization of all control measures and more efficient control strategy (Kranz, 1973).

Experiments in Thailand showed that simple spore traps set at 80 cm above the ground can trap spores of *P. oryzae* at any rice growth stage. The fungus conidia were observed throughout the year. If the number of blast conidia is over five per slide, blast incidence will occur in the field within 7–15 days (Arunyanart *et al.*, 1982). The optimum humidity and temperature for disease development was over 90% and 25–28°C, respectively. High rainfall, high humidity and low temperature supported more disease development (Ou, 1985; Suzuki, 1970; Giatgong *et al.*, 1970).

To investigate pathogen movement, approximately 800 simple spore traps were randomly located over disease hot-spot areas in the 12 provinces. District level meteorological data were recorded daily, and stored in the computer for analysis. Disease monitoring was carried out intensively by farmers and extension officers through the whole season. Data on number of spores, microclimatic conditions (relative humidity, max.–min. temperature, rainfall) and disease intensities in the field will be used as a tool for 7-day, short-term early warning as well as for long-term forecasting. The scheme of the disease forecasting network is shown in Fig. 20.3. This activity is only in its early stages and is still being tested for blast modeling and forecasting (Teng *et al.*, 1991).

Conclusion

A panicle blast epidemic was officially reported in January 1993 and affected over 200,000 ha in 12 provinces of the northern part of Thailand. The records of meteorological data indicated that favorable conditions such as widely grown susceptible varieties, namely RD6, RD8 and Niew Ubol, and high humidity from rain showers continuously over 3 days when the plant was at milk stage, are the important factors causing the sudden disease outbreak. To manage the disease and pathogen, about 2000 tonnes of glutinous rice seed varieties, RD6, RD8, RD10 and Niew Sanpatong, were treated with fungicide subsidized by the government before growing. The methodology for proper rice cultivation to avoid blast infestation was low nitrogen use as well as introduction of the new, moderately resistant cultivar, KKNLR175052-PRE-40-1-1-1. Previous research by the DOA has shown that simple spore traps located in farmers' fields gave sufficient early warning about the occurrence of favorable conditions for sporulation. The threshold for leaf blast control is suggested by doing a systematic diagonal sampling of 10 tillers/0.16 ha, recording blast incidence on the top four leaves or on leaf 3 and spraying fungicides when more than 70% of the leaves are infected with blast. Fungicides commonly used are Kasugamycin, Edifenphos,

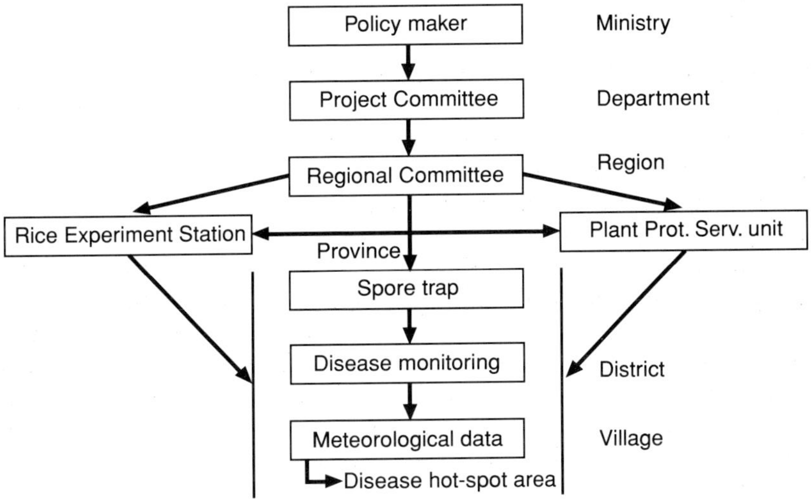

Fig. 20.3. Diagram of blast management network in Thailand.

Tricyclazole, Isoprothiolane, Carbendazim and Benomyl. Previous observations by the PPSU show that, in the northern provinces, blast infections occur during the summer and winter crops while in the Central Plains, blast is only observed during the winter (dry) season between November and March. A collaborative project is in progress jointly to develop an all-Thailand Integrated Blast Management Program, together with IRRI, that uses data collected on the disease, environment and farmers' practices.

References

Arunyanart, P., Surin, A., Dhitikiattipong, R., Rojanahusdin, W. and Disthaporn, S. (1982) *Study on Forecasting of the Rice Blast Disease*. Plant Pathology Division Annual Report, 1982. Department of Agriculture, Bangkok, Thailand (in Thai).

Chamras, P. (1991) *Knowledge of Rice*. Rice Research Institute, Department of Agriculture, Bangkok, Thailand, 267 pp. (in Thai).

Disthaporn, S. (1987) Studies on sampling methods for rice diseases in Thailand. PhD Thesis, University of Giessen, Germany.

Disthaporn, S. (1989) *Rice Disease Control by Farmers*. Thai Phytopathology Society, Thailand, 116 pp. (in Thai).

Giatgong, P. and Frederiksen, R.A. (1969) Pathogenic variability of cytology of monoconidial subcultures of *Pyricularia oryzae*. *Phytopathology* 59(8), 1152–1157.

Giatgong, P., Attabhunyo, A. and Manthachitra, P. (1971) Rice blast, Pyricularia oryzae Cavara. In: *Rice Diseases and Pests of Thailand 1971*. Rice Protection Research Center, Ministry of Agriculture, Thailand, pp. 22–24.

Kranz, J. (1973) Epidemiology, concepts and scope. In: Raychaudhuri, S.P. and Verma, J.P. (eds), *Current Trends in Plant Pathology*. Department of Botany, University of Lucknow, Lucknow, pp. 26–32.

Kranz, J. (1974) Comparison of epidemics. *Annual Review of Phytopathology* 12, 356–374.

Ou, S.H. (1961) Some appects of rice blast and varietal resistance in Thailand. *International Rice Commission Newsletter* 10(3), 11–17.

Ou, S.H. (1985) *Rice Diseases*, 2nd edn. Commonwealth Mycological Institute, Kew, UK, 380 pp.

Ratanapol, S., Sirikajornjaru, V., Dachnuan, S. and Charoendham, P. (1990) Effect of silica on fungus diseases. In: *Proceedings of the 1990 Deepwater Rice Planning Meeting, Thai/IRRI Deepwater Rice Collaborative Project.* pp. 117–123.

Sriwasin, C. and Giatgong, P. (1970) Cytological studies and inoculations of *Pyricularia* spp. PhD Thesis, Kasetsart University, Thailand.

Sudto, P. and Disthaporn, S. (1992) Farmers' pest management practices in the central plain Thailand. A paper presented at IPM Network Meeting on 27–28 April 1992, International Rice Research Institute, Los Baños, The Philippines, pp. 25.

Surin, A., Arunyanart, P., Rojanahusdin, W., Munkong, S., Dhitikiattipong, R. and Disthaporn, S. (1991) Using empirical blast models to establish disease management recommendations in Thailand. In: *Rice Blast Modeling and Forecasting.* International Rice Research Institute, Los Baños, The Philippines, pp. 69–74.

Suzuki, B. (1970) Interrelationship between the occurrence of rice blast disease and meteorological conditions. *Review of Plant Protection Research* 3, 1–11.

Teng, P.S., Klein-Gebbinck, H. and Pinnschmidt, H. (1991) *An Analysis of the Blast Forecasting*. International Rice Research Institute, Los Baños, The Philippines, pp. 1–30.

21 Rice Blast in West Africa: Its Nature and Control

S.N. Fomba and D.R. Taylor

Rice Research Station, PMB 736, Freetown, Sierra Leone, West Africa

Introduction

The West African subregion is very diverse in climate, vegetation and agricultural potential (Leakey and Wills, 1977; Okigbo, 1991). Rice is grown in all but the desert zone, either as rainfed culture in the humid and moist savannah zones or under irrigation in the drier areas including the Sahel. The factors that determine the suitability of these environments for rice growing include surface hydrology, soils, biological stresses and the indigenous rice culture (Buddenhagen, 1978).

Blast Incidence and Severity in the Different Agro-Ecological Situations in West Africa

Rice blast is widespread in tropical West Africa. Incidences of the disease were first reported in the 1930s (Feakin, 1974) when the disease was known to affect seedlings in Sierra Leone but was of negligible importance then. However, with the expansion and intensification of rice cultivation it is now the most serious and destructive fungal disease of rice in the West African subregion (Notteghem and Baudin, 1981). A brief description of blast incidence and damage in the different rice environments in West Africa follows.

Upland / hydromorphic environment

Upland rice constitutes more than 60% of the area cultivated to rice in West Africa (Abifarin *et al.*, 1972; WARDA, 1989). Among all the rice environments blast is most common and destructive in the upland, particularly so in newly cleared uplands (Raymundo and Fomba, 1979; Notteghem and Baudin, 1981).

Upland rice fields are usually contiguous parcels interspersed with bushes which serve as windbreaks thus promoting longer retention of water droplets from rain, drizzle and dew.

Seeding of upland rice is usually during mid-May to mid-July, depending on the location and rainfall. From June onwards there is progressive increase in rainfall. The relative humidity is also usually high (>90%) and the minimum ambient temperatures moderate (21–23°C). These microclimatic conditions promote successful germination of fungal spores and subsequent invasion depending upon host resistance/susceptibility (Ou, 1985). Nutritional imbalance and other stress conditions of rice in this environment invariably contribute to the high incidence of the disease. Susceptible cultivars are usually destroyed (Awoderu, 1974, 1983; Bidaux, 1978; Raymundo and Fomba, 1979). Panicle blast including neck rot becomes serious late in the growth cycle of rice (Notteghem and Baudin, 1981).

Blast disease is relatively less damaging in *boliland* and similar hydromorphic environments, e.g. *Fadamas*, than in the upland, especially as the fields become progressively flooded during the growing season. However, severe outbreaks of blast have been observed from time to time in Sierra Leone (Raymundo, 1980). Rice in *bolilands* is direct seeded and the early stages of growth is essentially dependent on direct rainfall as with upland rice. This subjects the rice to seedling blast damage, especially when dry spells occur due to erratic rainfall.

Similarly high incidences of panicle, neck and node blast have been recorded during unusually short rainy seasons with accompanying dry spells when plants were still in the reproductive stage. The disease situation is aggravated by the cultivation of highly susceptible rices (Raymundo, 1980).

Inland valley swamp (IVS)/paddy

Blast is far less severe in IVS/paddy rice than in upland rice. This is partly due to toughened cell walls resulting from increased silicification when rice is grown under swamp/paddy conditions (Okuda and Takahashi, 1965; Ou, 1985; Kim *et al.*, 1986). However, the disease can be very severe in dry season IVS/paddy rice compared to the wet season crop when panicle blast is of little importance.

Seedling blast, on the other hand, can cause major havoc in either season virtually wiping out nurseries (normally established on the upland) of susceptible varieties (Raymundo and Fomba, 1979; Raymundo, 1980). CCA, a popular introduced cultivar for IVS and irrigated paddy in Sierra Leone, was eventually rejected by farmers because of its high susceptibility to blast (Raymundo, 1978).

Mangrove swamp

As with IVS, complete desiccation and total loss due to seedling blast have been recorded in drybed nurseries established on uplands for transplanting in mangrove swamps, while wetbed nurseries established in hydromorphic areas remain virtually free of the disease (Raymundo and Fomba, 1979; Fomba, 1984). Neck blast is not common except in some short-season swamps, i.e. saline, empoldered areas with less than four months of low salt period to grow rice (Fomba, 1984). In the mangrove associated swamps, neck blast incidence ranged from 10 to 15% in Sierra Leone; the disease incidence was also relatively high (>10%) in the Casamance region of Senegal and at Warri, Nigeria (Fomba, 1984).

Riverain grassland

These rice lands, like the tidal mangrove swamps, also consist of wide expanses of uninterrupted open fields which allow free air movement and therefore reduce chances for blast to infect and spread (Raymundo, 1980). Blast incidence is therefore low. The environment also receives silt deposits brought by flooding from adjacent rivers.

Irrigated Sahel

Blast is rarely a problem in the Sahel because of the low relative humidity, lack of prolonged wetness of aerial surfaces of the rice plant and temperature extremes (Awoderu, 1983; Reckhaus and Adamou, 1986). However, the disease has recently been reported in a southernmost rice scheme in Niger, occurring mainly in the rainy season (Reckhaus and Adamou, 1986). Neck blast also occurs in fields where transplanting has been postponed such that flowering coincides with the cold season. Low temperatures and the cold irrigation water favoured the development of the disease (Reckhaus and Adamou, 1986). The phenomenon may also occur in other Sahelian countries. The incidence and severity rating of rice blast and grain yield in different rice environments in West Africa are summarized in Table 21.1.

Crop loss estimates

Estimates of grain yield loss due to blast provide legitimate justification for disease control. Crop loss data on the rice blast disease in particular areas in the West African subregion have been established, especially on-station, and researcher-managed trials in farmers' fields (Raymundo, 1975; Carpenter, 1977; Raymundo and Fomba, 1979; Akator, 1981; Sere, 1981, Vodouhe *et al.*, 1981; Fomba, 1986; Awoderu, 1990). These losses ranged from 3.2% to 14.5% in several popular upland and lowland rice cultivars in Sierra Leone (losses in excess of 80% were not uncommon in less adapted cultivars or accessions in experimental plots) to over 77% in Liberia (Raymundo, 1975; Carpenter, 1977; Raymundo and Fomba, 1979). Fomba (1986) reported losses in grain yield of mangrove rice due to neck blast at 16.0–30.9%, with the less-adapted cultivar ROK 5 incurring more losses than the local cultivar Pa Nylon in a farmer's field in north-western Sierra Leone. In the Ivory Coast, Awoderu (1990) estimated grain yield

Table 21.1. Mean grain yield of rice and blast incidence/severity in West African rice environments.

Environment	Grain yield (Mt/ha)	Blast incidence[a]
Upland/hydromorphic	1.0	Widespread/severe
Boliland/fadama	1.0	Moderate
Inland valley swamp (IVS)	1.8	Variable
Irrigated Sahel	2.8	Negligible
Mangrove swamp	2.0	Low–moderate
Deep water/floating rice	2.0	Low–moderate

[a] Low = <10% incidence or damage; moderate = 10–25% incidence or damage; severe = >25% incidence or damage.
Sources: Awoderu, 1974; Raymundo, 1980; Fomba, 1984; WARDA, 1989.

losses of 0.5–58.5% on IRAT 112 and IAC 164 in on-station and researcher-managed on-farm trials. Blast could thus become more important in farmers' fields than it is now if less adapted and susceptible rice cultivars are grown.

Blast Control in West Africa

Several methods for control of blast in rice have been developed (Ou, 1985). However, the adoption and use of any one method or combination of methods depend on their reliability, status of rice culture in the environment, the level of technical know-how of farmers and the economics of production. Although fungicides are usually very efficient in controlling the disease, their use hardly fits into the low-input subsistence food production system of West African rice farmers. Fungicides are not easily obtainable, and can be prohibitively expensive for the average farmer. Under this situation the use of resistant cultivars is the most important and easily adoptable method for blast control, in conjunction with appropriate cultural practices (Ou, 1985).

Resistant cultivars

Several traditional African land races including ROK 16 (Ngovie), 63–83, Moroberekan, LAC 23, IRAT 13 (63–83 Mutant), OS6 and some Brazilian upland rices possess high levels of stable resistance to blast (Bidaux, 1978; Ahn, 1981; Alluri *et al.*, 1987). Often these cultivars are crossed with the Asian semidwarfs in attempts to combine high-yield potential of the latter with the stress and disease resistance of the traditional African cultivars in varietal development programs in the subregion.

However, Asian parents are poorly adapted to the African uplands because of extreme susceptibility to blast and other diseases (Winslow, 1989). Recovery of disease resistance in the progenies is usually not satisfactory, mainly due to poor combining ability between *japonica* and *indica* types, unlike crosses between

African and Brazilian parents (Ahn, 1981) which combine well because both are *japonicas* and are genetically close.

At West Africa Rice Development Association (WARDA) and other International Centres (ICs) in the tropics, rice varieties/lines are bred for total adaptation to specific environments. Similar approaches are being taken in National Agricultural Research Systems (NARS) in Ivory Coast, Nigeria, Senegal and Sierra Leone. A sound breeding program for blast resistance necessitates understanding the nature of the disease incidence and severity in the environment for which the cultivars are targeted. In the Casamance region of southern Senegal, for example, the nature of the disease, structure of the pathogen population and virulence spectrum has been well defined (Mbodj *et al.*, 1988a). Only one out of eight race groups (Mbodj *et al.*, 1988a) was absent. Similarly, 13 out of the 14 known genes (Kiyosawa, 1976) for vertical resistance were overcome in this part of the subregion indicating a wide spectrum of virulence in the blast population.

In a study on upland rice cultivars in the same region, Mbodj *et al.* (1988b) observed high levels of quantitative resistance to blast in IRAT 10, IRAT 112, and IRAT 133; Dj 8-14 and Dj 11-509 were moderate in their reaction, while that of Barafita was low. Se 301G, I Kong Pao and Dj 12-539 were highly susceptible to the disease.

In another study on upland rice, Mbodj *et al.* (1989a) showed that Dj 12-223, IRAT 109, IRAT 110, IRAT 146, IRAT 147 and TOX 728-81 had good quantitative resistance which was stable at 17 different locations in the Casamance in 1982–1983. The other cultivars in the study, Ablaye Mano, M 133-162 and TOX 402-44, were highly susceptible to leaf blast. None of the cultivars were, however, completely free of panicle blast; Dj 12-233, IRAT 110 and TOX 728-81 showed more resistance/tolerance than the other cultivars. It was suspected that other important factors contributing to the high level of instability in upland rice yields were the occurrence of other diseases, inappropriate growth cycle and lack of drought tolerance (Mbodj *et al.*, 1989a). Among the lowland rice cultivars, BW 248-1, TOX 103, ITA 123, BKN, 6986-38-1 and BR51-46-5 showed a high degree of partial or incomplete resistance which was stable in time and with rice cultures (Mbodj *et al.*, 1989b).

The morphological and physiological specialization of the blast fungus in Sierra Leone has been established (Onofeghara *et al.*, 1973). Raymundo (1975) also observed variation in severity of leaf blast in multilocational disease screening tests covering major climatic zones and rice ecologies in Sierra Leone. Differences in race composition and/or wide virulence spectrum were evident in the contrasting reactions of some varieties in two or more locations, e.g. Jaya which was susceptible at Bumpeh was found resistant at the Rokupr upland site. A study of the virulence spectrum of the fungus at Rokupr, Sierra Leone, using Kiyosawa differential cultivars with known vertical genes for resistance to blast (Kiyosawa, 1976) showed 10 out of 13 cultivars with susceptible reactions to the disease (Fomba, 1984). Similar results have been reported from Burkina Faso (Sere, 1981), Ivory Coast (Bidaux, 1978) and Nigeria (Awoderu and Ojomo, 1981).

Out of 493 accessions screened for blast resistance in the major agroclimatic zones of Sierra Leone in 1991–1992, 64 showed good resistance in all locations.

Table 21.2. Some blast resistant cultivars in West Africa.

Cultivar/accession	Source	Cross
IRAT 13	IRAT, Ivory Coast	63-83 Mutant
63-83	ISRA, Casamance, Senegal	Traditional cultivar
ROK 16	RRS, Sierra Leone	Selection (Ngovie)
LAC 23	CARI, Liberia	Selection (local variety)
Moroberekan	Ivory Coast	Traditional cultivar
FARO 11	NCRI, Nigeria	Selection (OS 6)
Iguape Cateto	Brazil	
Dourado Precoce	Brazil	
IAC 25	Brazil	

Sources: Bidaux, 1975; Raymundo and Fomba, 1979; Alluri *et al.*, 1987.

Over 70% of these are of African origin, including Ngeyema Yakei, ITA 120, IRAT 147, ROK 16 (Ngovie) and ITA 304. Earlier results of blast screening nurseries showed similar trends (RRS, 1986–1989). Some blast-tolerant rice cultivars in West Africa are listed in Table 21.2.

In the INGER-Africa 1989 African Rice Blast Nursery (ARBN), 96 out of 150 accessions evaluated had average leaf blast scores of 3 (Standard Evaluation System, SES, for rice) or better, across seven locations in Africa. Twelve entries, ITA 116, ITA 212, ITA 257, IRAT 124, IDSA 6 (IRAT 216), ITA 130, Milyang 82, TOX 3093-4-5-4-2-2, TOX 3108-1712-1-2-6, TOX 3118-4-E_4-3-3-3 and TOX 3118-73-2-1-1-1 showed good resistance to blast in all five test sites in West Africa (INGER-Africa, 1992).

Over 600 mangrove rice accessions were screened for seedling blast resistance at Rokupr and Rotifunk, Sierra Leone, from 1982 to 1988. Varietal reactions to seedling blast varied from year to year (John and Fomba, 1989). Part of these variations may have been due to either changes in the racial composition of the blast fungus, or to changes in environmental conditions. Varieties/lines that showed moderate levels of resistance (scores 4–6) during blast conducive years probably possessed high levels of partial resistance. Accessions with low scores (1–3) may be possessing vertical resistance, and/or high level of partial or quantitative resistance to blast (Bonman and Mackill, 1988). Promising varieties or lines obtained from the screening trials included several advanced breeding lines bred by WARDA, Rokupr, for example WAR 1, WAR 77, WAR 81, WAR 115, etc.

Against this background of diverse rice environments and a highly variable and virulent pathogen, the success of any breeding program for resistance to blast will largely depend on the care with which parents are selected. The emphasis should be on selection of progenies/lines which can coexist in a homeostatic relationship with the pathogen rather than vertical resistance *per se* which may break down with time (Vanderplank, 1968). This will lead to highly stable rice cultivars and will minimize the risk of frequent breakdown in resistance as was observed with CCA, a popular irrigated paddy in Sierra Leone. This culture had to be replaced by more blast-tolerant ones like ROK 11 and ROK 12

Table 21.3. Some breeding lines/released varieties involving parents with stable resistance to blast in West Africa.

Accession/Cross	Parents	Country	Environment
Rohyb 104	ROK 1 × ROK 16	Sierra Leone	Upland
Rohyb 113	ROK 3 × ROK 16	Sierra Leone	Upland
Rohyb 116	ROK 5[a] × IRAT 113	Sierra Leone	Lowland
IRAT 109	IRAT 13 × IRAT 10	Ivory Coast	Upland
IRAT 112	IRAT 13 × Dourado Precoce	Ivory Coast	Upland
IRAT 136	IRAT 13 × Moroberekan	Ivory Coast	Upland
IRAT 259	IRAT 112 × Iguape Cateto	Ivory Coast	Upland
IRAT 291	IAC 25 × Chianan 8	Ivory Coast	Upland
ITA 116	63-83 × IR 773-A1-36-2-1	Nigeria	Upland
ITA 122	IR 1529 × Iguape Cateto	Nigeria	Upland
ITA 150	63-83 × Dourado Precoce	Nigeria	Upland
TOX 936-81-6-3-101	IR 1529 × Iguape Cateto	Nigeria	Upland
TOX 1010-5-3-1	IRAT 13 × Dourado Precoce × TOX 494	Nigeria	Upland

[a] ROK 5 = SR 26 × Pa Wellington, a popular mangrove rice cultivar in West Africa.
Sources: IRAT 1977, 1983; Alluri *et al.*, 1987; RRS 1987; INGER-Africa 1992.

(Raymundo, 1978; Jones *et al.*, 1979). A similar story has been reported for IR 8 and IR 442 in Benin Republic (Vodouhe *et al.*, 1981).

Screening techniques

Several screening techniques have been devised to ascertain the level of resistance of rice to blast. In West Africa, two main types of field plot techniques have been used, namely the Uniform Blast Nursery (UBN) (Ou, 1985; Imolehin and John, 1987; John and Fomba, 1989) and the 'Horizontal Resistance' Blast Nursery (HRBN) or 'Un test d'evaluation de la resistance sous inoculum decroissant (DITER)' (Notteghem, 1976; Bidaux, 1978; Fomba, 1989). Several African land races and Brazilian upland rice cultivars have been identified which possess high levels of partial resistance sometimes referred to as 'slow blasting' (Villareal *et al.*, 1980; Weerarante *et al.*, 1981). This type of resistance is being actively incorporated into the new germplasm that is being developed in West Africa (Table 21.3).

Cultural control

Cropping system

Rice is grown under diverse ecological and cultural conditions in West Africa. In the uplands, rice is usually grown in mixtures with other crops including sorghum, maize, cassava, pigeon pea, vegetables, beniseed (*Sesamum*), and spices.

Traditionally, farmers in West Africa, especially in the humid forest and moist savanna zones where the precipitation is favorable, grow several rice cultivars contiguous with one another or along a toposequence. Early maturing cultivars are seeded first to tide over the hungry season (Richards, 1992). This could aggravate blast incidence if the early seeded rices do not have sufficient resistance to blast. They would provide inoculum for the late maturing crops. The cropping systems practiced by farmers in lowlands also involve the growing of vegetables, maize, cassava, sweet potato, groundnut and spices in the dry season following the main season rice crop in June–December.

This crop diversity affects the epidemiology of rice blast disease by creating discontinuity of host tissue. Although concrete data are lacking, however, field observations by the authors suggest that this polyculture in subsistence farming does influence the incidence and severity of rice diseases/insects in general. Evidence of reduction in pest incidence in mixed cropping in Kenya supports this contention (Oloo *et al.*, 1986).

Plant density

This has a significant effect on blast incidence in monocropped rice. High plant density increases the number of host plants and creates a more humid micro-environment conducive for blast proliferation. Our upland experiments at the Rice Research Station (RRS), Rokupr, Sierra Leone, in 1984 showed that twice as many blast lesions were obtained on rice seedlings when seed rate was increased from 80 kg ha^{-1} to 120 kg ha^{-1} (RRS, 1984).

Fertilizers

Application of high levels of nitrogenous fertilizers such as ammonium sulfate leads to increased incidence of blast infection (Ou, 1985). In Sierra Leone, infection of leaf blast could be high on rice in newly cleared uplands (Raymundo *et al.*, 1976). Similarly, blast tends to be more severe on fields planted to newer varieties which invariably are released with recommendation of high nitrogen fertilization.

In a study conducted at Suakoko, Liberia, with 16 rice cultivars, the incidence of blast increased when nitrogen was increased from 60 kg N to 120 kg N ha^{-1} (Awoderu, 1983). This indicates the risk of excessive use of nitrogen fertilizers. This is due probably to the injurious effects of ammonium accumulation in the cells of plants treated with high nitrogen (Ou, 1985). The soluble nitrogen in plants may serve as suitable nutrient for fungus growth. Therefore, to minimize its impact on the rice blast disease, moderate doses of nitrogen are usually recommended.

Soil amendment by the application of silicon ameliorates blast incidence. A study at RRS, Rokupr, Sierra Leone, showed 50% reduction in neck blast incidence when sodium silicate was applied at 100 kg Si ha^{-1}. Incorporation of rice husk in the soil at 5–10 t ha^{-1} similarly reduced neck blast incidence by up to 50% (RRS, 1988).

Time of planting

Changes in weather and seasons invariably have an effect on the incidence of diseases and insects. In double cropping of rice the incidence of blast is generally higher in the second crop which coincides with the dry season (Raymundo, 1980). In 1992 UBN screening sets at Rokupr, Sierra Leone, the mean scores for blast were 2.3 and 6.2 (SES) when trials were established in July and September, respectively. A clearer understanding of the effect of weather parameters affecting blast incidence in the subregion derived from predictive models would prove useful in developing effective control strategies for the blast disease.

Chemical control

In a study conducted on rainfed rice in the Casamance, Senegal, Delassus (1977) failed to observe any significant effect on leaf blast following seed treatment with organomercurial formulations and Benlate (benomyl). The efficacy of NF-48 (thiophanate) on both leaf and panicle blast depended on the disease pressure and precipitation pattern in a given year. Excessive rains wash off blast spores before they germinate and infect the rice plant (John and Fomba, 1989). In 1970, a year characterized by low rainfall in Casamance, Senegal, the fungicide Benlate and Hinosan (edifenphos) applied at 250 g a.i. ha^{-1} and 400 ml a.i. ha^{-1}, respectively reduced the number of diseased necks and produced higher grain yields (Delassus, 1977).

In Sierra Leone, Fongorene-50 (pyroquilon) seed treatment at the rate 4 g kg^{-1} seeds reduced leaf blast incidence by as much as 21% and increased grain yields (RRS, 1989). Seed treatment application in slurry form was more effective in leaf blast control than powder form. Similarly, Kitazin P (IBP) reduced seedling blast damage in an upland nursery at Rokupr, Sierra Leone, by 36.2% followed by Beam (tricyclazole) with a reduction rate of 27.8% in the number of blast lesions formed per seedling (Fomba, 1987). However, systemic fungicides may prove more useful in blast control (Sere, 1981). Ouedraogo (1989) reported significant control of neck blast with Kitazin when applied at 2 l ha^{-1} at heading/flowering.

The indications are that under high blast pressure, application of fungicides could result in marked yield increases (Delassus, 1977) although the results may not be easily transferable to farmers' fields because of various socioeconomic and technical problems.

Summary and Conclusions

Rice blast is the major fungal disease of economic importance across all rice environments in West Africa. It is most severe in the upland environment where drought and soil stresses predispose the rice crop to severe attacks by the pathogen. It is also severe in rainfed lowland rice environments with inadequate water control, especially in the dry season.

Studies on population dynamics at several locations in West Africa showed that the fungus is highly variable with a wide virulence spectrum. None the less, some rice cultivars selected from among local germplasm or bred for West African rice environments have a high degree of adaptation and tolerance to the disease and other stresses. This makes them suitable parents for varietal improvement purposes.

The current emphasis on blast control in West Africa is developing rice cultivars with adequate levels of resistance/tolerance to the disease as well as other diseases/pests and soil stresses. Sound crop management practices would also go a long way to minimize the losses caused by the pathogen to West African rice growers especially under high nitrogen fertilization.

Future directions for rice blast research and control in a hitherto low to moderate input production system in West Africa should focus on:

1. characterizing the relationships between wild grass hosts including wild rices, population dynamics of the fungus, edaphic and weather factors on blast disease development in cultivated rices;
2. conducting more studies on the genetics of resistance;
3. collation of farmers' indigenous knowledge and skills in the management of rice blast and other diseases/pests;
4. studies on the effect of changing cropping practices on disease incidence/intensity; and
5. development of sound and practical integrated management programs for the disease.

The recent collaborative studies on site characterization and population dynamics of the fungus coordinated at WARDA, Bouake, Ivory Coast, is a move in the right direction. This should go a long way towards achieving a better understanding and management of the disease in West Africa, especially as we contemplate the change from shifting cultivation to permanent cropping on the uplands in a sustainable manner.

References

Abifarin, A.O., Chabrolin, R., Jacquot, M., Marie, R. and Moomaw, J.C. (1972) Upland rice improvement in West Africa. In: *Rice Breeding*. International Rice Research Institute, Los Baños, The Philippines, pp. 625–635.

Ahn, S.W. (1981) The slow blasting resistance. In: GERDAT–IRAT (ed.), *Proceedings of the Symposium on Rice Resistance to Blast*. Service de Pathologie Vegetale, Montpellier, France, pp. 343–369.

Akator, S.K. (1981) Methods to combat blast in Togo. In: GERDAT–IRAT (ed.), *Proceedings of the Symposium on Rice Resistance to Blast*. Service de Pathologie Vegetale, Montpellier, France, pp. 35–42.

Alluri, K., Masajo, T.M. and John, V.T. (1987) Breeding for resistance to rice blast. *IITA Research Briefs* 8, 1.

Awoderu, V.A. (1974) Rice diseases in Nigeria. *Pest Articles and News Summaries* 20, 416–424.

Awoderu, V.A. (1983) Technical Service Support: Pathology. *Annual Rice Review Meeting*, 23–27 May 1983. West Africa Rice Development Association, Monrovia, Liberia.

Awoderu, V.A. (1990) Yield loss attributable to neck-rot of rice caused by *Pyricularia oryzae* Cav. in Cote d'Ivoire. *Tropical Pest Management* 36, 394–396.

Awoderu, V.A. and Ojomo, O.A. (1981) Identification of sources of resistance to blast and variability of *Pyricularia oryzae*. In GERDAT–IRAT (ed.), *Proceedings of the Symposium on Rice Resistance to Blast*. Service de Pathologie Vegetale, Montpellier, France, pp. 269–281.

Bidaux, J.M. (1978) Screening for horizontal resistance to rice blast (*Pyricularia oryzae*). In: Buddenhagen, I.W. and Persley, G.J. (eds), *Rice in Africa*. Academic Press, London, pp. 159–174.

Bonman, J.M. and Mackill, D.J. (1988) Durable resistance to rice blast disease. *Oryza* 25, 103–110.

Buddenhagen, I.W. (1978) Rice ecosystems in Africa. In: Buddenhagen, I.W. and Persley, G.J. (eds), *Rice in Africa*. Academic Press, London, pp. 11–27.

Carpenter, A.J. (1977) Crop losses affecting rice in Liberia. In: *Plant Protection for the Rice Crop, Seminar Proceedings No. 4*. West Africa Rice Development Association, Monrovia, Liberia, pp. 160–173.

Delassus, M. (1977) Chemical control of blast of rainfed rice in Casamance, Senegal. In: *Plant Protection for the Rice Crop, Seminar Proceedings No. 4*. West Africa Rice Development Association, Monrovia, Liberia, pp. 207–211.

Feakin, S.D. (1974) *Pest Control in rice, PANS Manual No. 3*. Centre for Overseas Pest Research, Overseas Development Administration, London.

Fomba, S.N. (1984) Rice disease situation in mangrove and associated swamps in Sierra Leone. *Tropical Pest Management* 30, 73–81.

Fomba, S.N. (1986) Grain yield losses caused by neck blast disease, *Pyricularia oryzae* Cav. in unprotected tidal mangrove swamp rice in Sierra Leone. *WARDA Technical Newsletter* 6, 24–25.

Fomba, S.N. (1987) Promising fungicides for the control of seedling blast, *Pyricularia oryzae* Cav. in upland rice nursery in Sierra Leone. *WARDA Technical Newsletter* 7, 14–15.

Fomba, S.N. (1989) Screening rice varieties for blast resistance in mangrove swamp ecologies. In: *Annual Report*. West Africa Rice Development Association, Bouake, Ivory Coast, pp. 81–85.

Imolehin, E.D. and John, V.T. (1987) Screening techniques for disease resistance. Paper presented at the International Rice Testing Programme for Africa (IRTP–Africa) Workshop, St Louis, Senegal 14–16 October 1987.

Institut de recherches agronomiques tropicales et de cultures vivriere (1977) *Rapport annuel*. IRAT, Paris, France.

Institut de recherches agronomiques tropicales et de cultures vivriere (1983) *Rapport annuel*. IRAT, Paris, France.

International Network for Genetic Evaluation for Rice–Africa (1992) *Final Report for 1990*. International Institute of Tropical Agriculture, Ibadan, Nigeria.

John, V.T. and Fomba, S.N. (1989) Screening for blast resistance at IITA and WARDA. Paper presented at the International Rice Testing Programme – Africa Workshop, Nairobi, Kenya, 20–23 March 1989.

Jones, R.A.D., Raymundo, S.A., Mansaray, M.S., Banya, G.S. and Williams, C.D. (1979) Sierra Leone's eight new rice varieties. *International Rice Commission Newsletter* 28, 38.

Kim, C.H., Rush, M.C. and Mackenzie, D.R. (1986) Flood-mediated resistance to the rice blast disease. In: Juo, A.S.R and Lowe, J.A. (eds), *The Wetlands and Rice in Subsaharan Africa*. International Institute of Tropical Agriculture, Ibadan, Nigeria, pp. 151–160.

Kiyosawa, S. (1976) Methods for tests and gene analyses of blast resistance of rice varieties. *Oryza* 13, 1–32.

Leakey, C.L.A. and Wills, J.B. (eds) (1977) *Food Crops of the Lowland Tropics*. Oxford University Press, Oxford.

Mbodj, Y., Gaye, S., Diaw, S. and Faye, A. (1988a) Rice blast in Casamance, Senegal. I. Variability of the pathogen. *L'Agronomie Tropicale* 43, 106–110 (in French, English summary).

Mbodj, Y., Gaye, S. and Diaw, S. (1988b) Rice blast in Casamance, Senegal. II. Characterization of resistance to blast of upland rice varieties already cultivated or to be released for common use in Casamance, Senegal. *L'Agronomie Tropicale* 43, 222–227 (in French, English summary).

Mbodj, Y. Gaye, A. and Demay, G. (1989a) Rice blast in Casamance, Senegal. III. Identification of upland rice varieties with stable yield and durable resistance. *L'Agronomie Tropicale* 44, 43–49 (in French, English summary).

Mbodj, Y., Gaye A., Demay, G., Coly, J.P. and Malou, J.P. (1989b) Rice blast in Casamance, Senegal. IV. Pluriannual screening of lowland rice varieties for their resistance to diseases. *L'Agronomie Tropicale* 44, 51–57 (in French, English summary).

Notteghem, J.L. (1976) Mesure au champ de la resistance horizontale du riz a *Pyricularia oryzae. L'Agronomie Tropicale* 32, 400–412.

Notteghem, J.L. and Baudin, P. (1981) *Main Rice Diseases in West Africa*. West Africa Rice Development Association, Monrovia, Liberia.

Okigbo, B.N. (1991) *Development of Sustainable Agricultural Systems in Africa: Roles of International Agricultural Research Systems*. International Institute of Tropical Agriculture, Ibadan, Nigeria.

Okuda, A. and Takahashi, E. (1965) The role of silicon. In: *The Mineral Nutrition of the Rice Plant*. Proceedings of a Symposium at The International Rice Research Institute, February 1964. The Johns Hopkins Press, Baltimore, Maryland.

Oloo, G., Omolo, E.O., Otieno, W.A. and Oddindo, M.O. (1986) *The Effect of Mixed Cropping System on Pest Population Densities, Incidence of Natural Enemies, Crop Damage and Yield Loss under Subsistence Farming in Kenya. Final Report*. The International Centre of Insect Physiology and Ecology, Nairobi, Kenya.

Onofeghara, F.A., Kapooria, R.G. and Ademokula, D.O. (1973) Studies on *Pyricularia oryzae* Cav. in Sierra Leone: Morphological and physiological variability of some isolates. *Annals of Botany* 37, 193–202.

Ou, S.H. (1985) *Rice Diseases*, 2nd edn. Commonwealth Mycological Institute, Kew, UK, 380 pp.

Ouedraogo, I. (1989) Chemical control of rice blast (*Pyricularia oryzae* Cav.) in western region of Burkina Faso. Paper presented at the International Workshop on Fungicide Resistance, Abidjan, Ivory Coast, 19–24 February 1989.

Raymundo, S.A. (1975) Importance of rice diseases in Sierra Leone. Paper presented at the Seminar on Rice Improvement in West Africa, Rice Research Station, Rokupr, Sierra Leone, 28 September–4 October 1975.

Raymundo, S.A. (1978) Recent advances in knowledge of rice fungous diseases in Sierra Leone. In: *Abstract of Papers*, 3rd International Congress of Plant Pathology, Munich, Germany, 16–23 August 1978, p. 109.

Raymundo, S.A. (1980) *Diseases of Rice in Sierra Leone, Technical Bulletin No. 4*. Rice Research Station, Freetown, Sierra Leone.

Raymundo, S.A. and Fomba, S.N. (1979) Rice blast in Sierra Leone – Identification of horizontal resistance and its utilisation in disease management and varietal improvement programme. Paper presented at the WARDA Seminar on Integrated Disease and Pest Management, Bobo-Dioulasso, Burkina Faso, 17–22 September 1979.

Raymundo, S.A., Mahapatra, I.C. and Jones, R.A.D. (1976) Effects of fertilizers on diseases and disease development on rice. Paper presented at FAO/NORAD/MANR National Seminar on Fertilizer Use and Development in Sierra Leone. Njala University College, Njala, Sierra Leone, 20–29 September 1976.

Reckhaus, P.M. and Adamou, I. (1986) Rice diseases and their economic importance in the Niger. *FAO Plant Protection Bulletin* 34, 77–82.

Rice Research Station (1984) *Annual Report*. Rice Research Station, Freetown, Sierra Leone.

Rice Research Station (1985) *Annual Report*. Rice Research Station, Freetown, Sierra Leone.

Rice Research Station (1986) *Annual Report*. Rice Research Station, Freetown, Sierra Leone.

Rice Research Station (1987) *Annual Report*. Rice Research Station, Freetown, Sierra Leone.

Rice Research Station (1988) *Annual Report*. Rice Research Station, Freetown, Sierra Leone.

Rice Research Station (1989) *Annual Report*. Rice Research Station, Freetown, Sierra Leone.

Richards, P. (1992) Mende names for rice – cultural analysis of an agricultural knowledge-system. Paper presented at the Workshop on Agricultural Knowledge Systems and the Role of Extension, Institut fuer Agrarsoziologie, Universitaet Hohenheim, Bad Boll, Germany, 21–24 May 1991.

Sere, Y. (1981) Rice blast prevention in Upper Volta. In: GERDAT–IRAT (ed.), *Proceedings of the Symposium on Rice Resistance to Blast*. Service de Pathologie Vegetale, Montpellier, France, pp. 51–65.

Vanderplank, J.E. (1968) *Plant Disease Resistance*. Academic Press, London.

Villareal, R.L., Mackenzie, D.R., Nelson, R.R. and Coffman, W. R. (1980) Apparent infection rates of *Pyricularia oryzae* on different rice cultivars. *Phytopathology* 70, 1224–1226.

Vodouhe, S.R., Djegui, N. and Amadji, F. (1981) Impact of blast on rice cultivation in the People's Republic of Benin. In: GERDAT–IRAT (ed.), *Proceedings of the Symposium on Rice Resistance to Blast*. Service de Pathologie Vegetale, Montpellier, France, pp. 27–32.

Weerarante, H., Martinez, C. and Jennings, P.R. (1981) Genetic strategies in breeding for resistance to rice blast. In: GERDAT–IRAT (ed.), *Proceedings of the Symposium on Rice Resistance to Blast*. Service de Pathologie Vegetale, Montpellier, France, pp. 305–311.

West Africa Rice Development Association (1989) *Medium Term Implementation* – 1990–1994. West Africa Rice Development Association, Bouake, Ivory Coast.

Winslow, M.D. (1989) Upland Rice Improvement at IITA. Paper presented at the International Rice Testing Programme – Africa Workshop, Nairobi, Kenya, 20–23 March 1989.

22 Understanding and Modeling Leaf Blast Effects on Crop Physiology and Yield

L. BASTIAANS[1,2,4], R. RABBINGE[1] AND J.C. ZADOKS[3]

[1]Department of Theoretical Production Ecology, Wageningen Agricultural University, PO Box 430, 6700 AK Wageningen, The Netherlands (correspondence address); [2]International Rice Research Institute, Los Baños, The Philippines; [3]Department of Phytopathology, Wageningen Agricultural University, PO Box 8025, 6700 EE Wageningen, The Netherlands; [4]Present address: DLO Research Institute for Plant Protection, PO Box 9060, 6700 GW Wageningen, The Netherlands

INTRODUCTION

Blast disease, caused by the Ascomycete *Magnaporthe grisea* (anamorph *Pyricularia grisea*) (Rossman *et al.*, 1990), is generally considered the principal fungal disease of rice (*Oryza sativa*), because of its wide distribution and its destructiveness under favorable conditions (Ou, 1985). Rice is most susceptible to *P. grisea* in the seedbed, and during the tillering and heading stages of the crop (Anderson *et al.*, 1947). During early growth stages lesions are mainly formed on leaves, whereas after heading the pathogen infects the panicle or the neck node. Accordingly, the rice blast pathosystem is divided into two major subsystems: the leaf and the panicle blast pathosystems (Teng *et al.*, 1991). Lesion formation on leaves is followed by premature leaf senescence of infected leaf tissue, especially in case of heavy infections. The highest fraction of leaf area covered by lesions is usually reached around maximum tillering, followed by a gradual decline in disease severity. This gradual decline has been attributed to adult plant resistance (Yeh and Bonman, 1986; Koh *et al.*, 1987; Roumen, 1993). The resistance of newly formed leaf tissue increases with time, and leaves appearing on physiologically older plants obtain this property faster. Consequently, leaf blast is mainly present before flowering.

After heading, *P. grisea* may infect the panicle or the neck node. Panicle blast causes direct yield losses, since filling of the grains on infected panicles is poor at best. For this reason, and because panicle blast occurs late in the season when the farmer has invested all of his production inputs for the crop, panicle blast is the more serious phase of the blast disease. Previous studies to estimate yield loss due to panicle blast have shown that panicle blast incidence may be linearly related to yield loss, using simple empirical damage functions (Kuribayashi and Ichikawa, 1952; Goto, 1965; Padmanabhan, 1965; Katsube and Koshimizu, 1970; Tsai, 1988; Torres and Teng, 1993). Comparison of the various studies show that the estimated yield loss ranged from 0.4 to 1.0% per percent infected panicles. Ou (1985) ascribed this variation to differences in the time of infection: the earlier the infection, the larger the loss.

Estimation of yield loss due to leaf blast by means of correlation studies has proven more difficult (Ou, 1985). Leaf blast is mainly present before flowering. This period consists of the vegetative and the reproductive phase and is characterized by the formation of source and sink capacity for yield formation. It is only after flowering that the source is used to fill the grains. Leaf blast thus mainly affects grain growth indirectly, through a long-term effect of the disease on crop production. It was the aim of the present study to quantitatively explain the effect of leaf blast on growth and production of a rice crop based on insight into the physiological processes underlying damage (Bastiaans, 1993a).

Methodology

In the early 1970s Zadoks (1973) reviewed attempts to relate yield loss to disease and categorized them as critical point, multiple point and area under the disease progress curve (AUDPC) models. The models differ in the epidemic characteristics that are used to predict crop loss. Critical point models use one measurement of disease severity, multiple point models use several measurements, and AUDPC models use the integrated level of disease for the entire epidemic. All models are used to derive empirical damage functions by applying regression analysis to field data. The models directly relate a measure of disease intensity to yield or yield loss, without considering the causes of yield reduction (Fig. 22.1, method a). For this reason, the question remains whether empirical damage functions are specific for the conditions (time, location, and cultivar) in which they were derived, or whether they have a more general applicability.

In the 1980s awareness grew that a complete understanding of crop loss would not be possible without considering the physiological response of crops to disease (Gaunt, 1978; Madden, 1983). Therefore the crop, rather than the disease, was used as the basis for the determination of crop loss (Fig. 22.1, method b). Several papers appeared in which yield of infected crops was directly related to the green leaf area at a single growth stage (Lim and Gaunt, 1986), to green leaf area duration (Rotem *et al.*, 1983) or to the cumulative amount of radiation intercepted by the green leaf area of the crop (Haverkort and Bicamumpaka, 1986; Waggoner and Berger, 1987). Implicitly this type of relationship assumes that the effect of a disease on crop production is limited to a reduction in photosynthetic leaf area. For some pathosystems, like *Phytophthora infestans* in potato, this assumption has proven valid (van Oijen, 1990). However, apart from a reduction in the amount of intercepted radiation, diseases may also alter the utilization efficiency of intercepted radiation for crop growth (Johnson, 1987). In these situations the relation between yield and intercepted radiation becomes more complex.

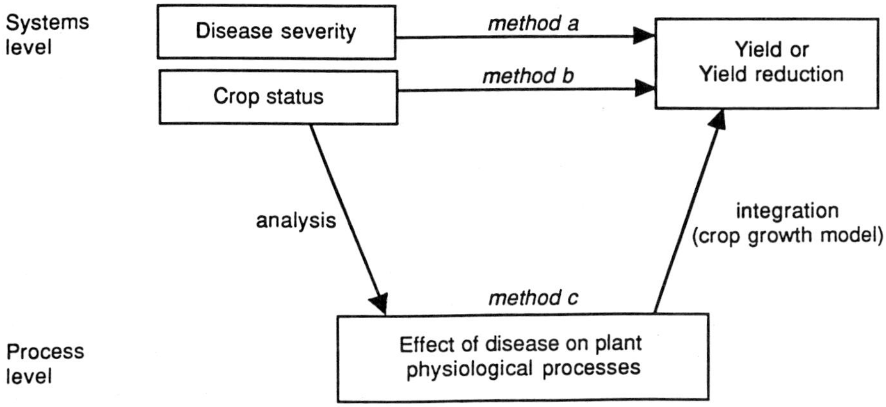

Fig. 22.1. Schematic representation of three methods for determining yield loss due to a foliar pathogen. With the third method research is conducted at two levels of integration: the process and the systems level. Mechanistic crop growth models are required to bridge the gap between both integration levels.

Another methodology in line with a more crop-oriented approach was developed by Rabbinge and Rijsdijk (1981). In their approach, attention is first concentrated on the whole plant level, being the next lower level of integration (Fig. 22.1, method c). At this level the various ways through which a disease interferes with basic plant growth processes (photosynthesis, respiration) or existing biomass (leaf senescence) are identified (Boote *et al.*, 1983). Research is conducted to determine dose-response relationships between the amount of disease and the extent to which a process is affected. Subsequently, the relationships are introduced in a mechanistic crop growth model to determine the consequences of the disease effects for crop growth and production. The model enables the determination of yield reduction due to various epidemics under variable environmental conditions, therewith meeting the limitations of empirical damage functions. It is this approach that was used in the present study on yield reduction in rice due to leaf blast.

Effect of Leaf Blast on the Physiology of Individual Leaves

The main function of leaves is the absorption of light and the conversion of light energy into carbohydrates through photosynthesis. Foliar diseases reduce the photosynthetic rate of leaves, since lesion formation reduces the amount of photosynthetically active leaf tissue. For some pathosystems this is the only mechanism through which the pathogen reduces the leaf photosynthetic rate of the host (*Puccinia recondita* in wheat, Spitters *et al.*, 1990; and *Phytophthora infestans* on potato, van Oijen, 1990). For other pathosystems the leaf photosynthetic rate is further reduced through an effect of the pathogen on photosynthesis of the green leaf tissue surrounding the lesions (*Erisyphe graminis* on wheat, Rabbinge *et al.*, 1985; and *Alternaria alternata* on cotton, Ephrath *et al.*,

1989). As far as the effect of *P. grisea* on leaf photosynthetic rate of rice is concerned, no direct observations have been reported. However, an indirect observation of Goto (1965), who reported that yield loss due to leaf blast exceeded yield loss caused by cutting off of a percentage of leaf area equal to the percentage leaf covered by blast, indicated that leaf blast influences the host plant more than just through a loss in leaf area.

A second effect on the physiology of leaves after infection by a pathogen is an enhanced rate of respiration. This increase can be attributed to the combination of fungal respiration and an increased respiration of the host. Toyoda and Suzuki (1957) reported an increase in the respiration of rice leaves after infection with *P. grisea*. The increase was connected with a shift from the glycolytic to the pentose phosphate pathway, a phenomenon generally observed in plants infected by biotrophic pathogens (Shaw and Samborski, 1957; Smedegaard-Petersen, 1984). In contrast, Burrell and Rees (1974) did not detect an increased respiration of rice leaves after infection by *P. grisea*.

A first step in the present analysis was the quantification of the effect of *P. grisea* on photosynthesis and respiration, by measuring the CO_2 exchange rate of individual rice leaves. The experimental data were used to derive relations between leaf blast severity and photosynthesis and respiration.

Leaf photosynthetic rate

Photosynthesis is the use of energy from light for the production of glucose from water and carbon dioxide, and is generally measured as the rate of CO_2 uptake. In Fig. 22.2 the dependence of leaf CO_2 assimilation on absorbed photosynthetically active radiation (PAR) is given. This photosynthesis light response curve is characterized by the assimilation rate at light saturation (P_{max}; $\mu g\ CO_2\ m^{-2} s^{-1}$) and the initial slope or efficiency at low light intensity (ϵ; $\mu g\ CO_2\ J^{-1}$).

The effect of leaf blast on leaf photosynthetic rate at high light intensities (P_{max}) was measured and related to disease severity (the fraction of leaf area covered by lesions). Measurements were conducted in a field and a greenhouse experiment. The results indicated that the working hypothesis (leaf blast reduces leaf photosynthetic rate through a reduction in green leaf area only) was not valid (Fig. 22.3). The reduction in leaf photosynthetic rate of infected leaves, expressed as a fraction of leaf photosynthetic rate of healthy leaves, surpassed disease severity. The disease thus not only reduced the amount of green leaf area, it also reduced the photosynthetic activity of green leaf tissue surrounding the lesions. This observation resulted in the virtual lesion concept (Bastiaans, 1991). A virtual lesion consists of a visual lesion and a surrounding halo, in which the photosynthetic activity is zero (Fig. 22.3). The size of the halo is chosen such that the reduction in leaf photosynthetic rate is fully accounted for. Parameter β expresses the ratio between virtual and visual lesion area and characterizes the effect of the disease on leaf photosynthetic rate. The parameter can be used to relate the photosynthetic rate of diseased leaf tissue (P_x) to disease severity (x):

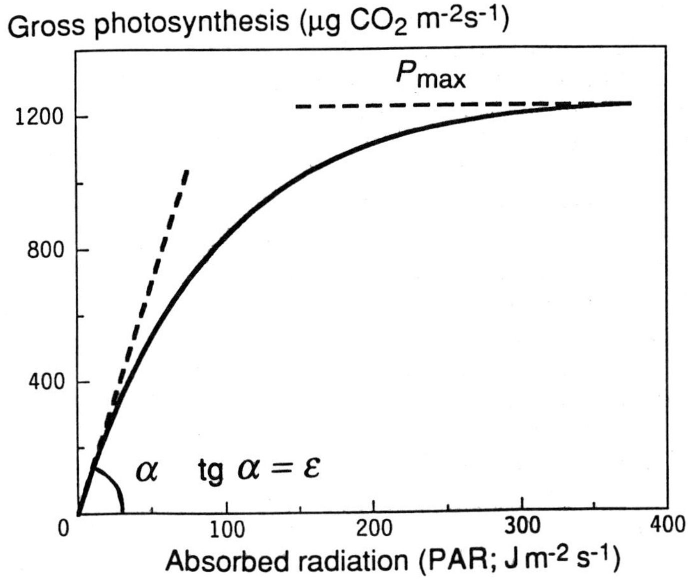

Fig. 22.2. A typical CO_2 assimilation–light response curve for an individual leaf. The response curve is characterized by the initial light use efficiency (ϵ; $\mu g\ CO_2\ J^{-1}$) and the assimilation rate at light saturation (P_{max}; $\mu g\ CO_2\ m^{-2}\ s^{-1}$).

$$P_x = P_0(1 - x)^{\beta} \qquad (1)$$

in which P_0 is the photosynthetic rate of comparable healthy leaf tissue. The reduction in P_{max} and ϵ were characterized by β-values of 3.7 ± 0.6 and 3.0 ± 0.7, respectively (Bastiaans, 1991, 1993b). This demonstrates that the effect of leaf blast on photosynthetic rate of rice leaves is three to four times as high as expected on the basis of visible lesion size.

Differences in β among rice cultivars would imply genotypic differences in tolerance, which can be defined as the ability of the host to endure the presence of the pathogen with reduced disease symptoms and/or damage (Parlevliet, 1979). To investigate the presence of genetic variation in β, three cultivars with a susceptible infection type were selected. The cultivars differed in relative infection efficiency (RIE), a major component of partial resistance to leaf blast in rice (Toriyama, 1975; Ahn and Ou, 1982; Yeh and Bonman, 1986; Roumen, 1993). Photosynthesis measurements did not indicate the presence of genetic variation in β (Bastiaans and Roumen, 1993). Therefore it was concluded that β is not a suitable selection criterion in breeding rice for tolerance to leaf blast. This conclusion is emphasized in Fig. 22.4, where the impact of the observed genetic variation in β and the impact of the genetic variation in RIE on leaf photosynthetic rate are compared.

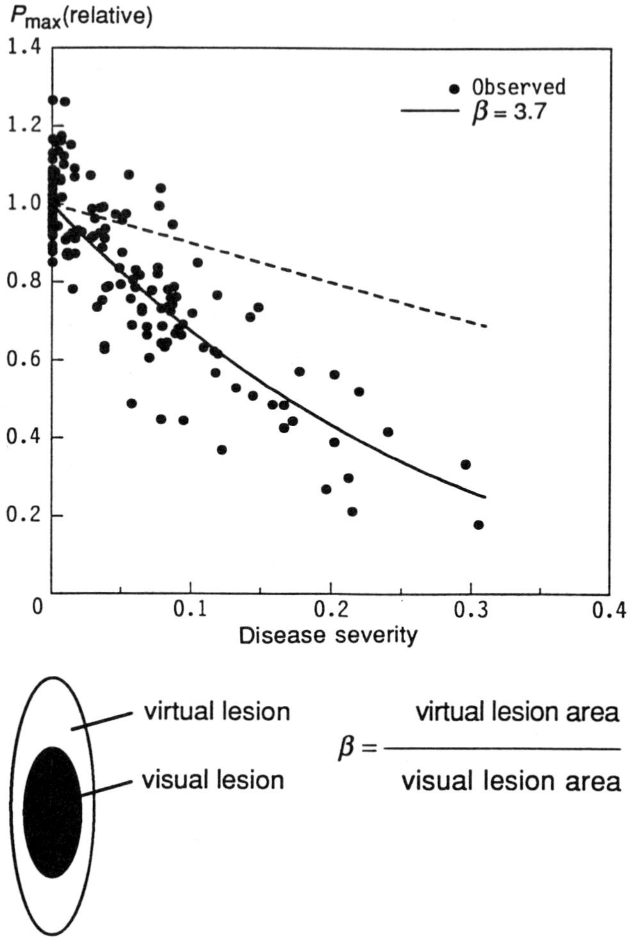

Fig. 22.3. The assimilation rate at light saturation (P_{max}; on a relative scale) of leaves infected by *Pyricularia grisea* in relation to disease severity as measured in a greenhouse experiment. The dotted line represents the working hypothesis: the effect of leaf blast is limited to a reduction in green leaf tissue. The solid line gives the best fit using equation 1. This equation is derived from the virtual lesion concept (see text). Parameter β expresses the ratio between virtual and visual lesion areas.

Leaf dark respiration

The effect of the disease on leaf dark respiration (R_d; $\mu g\ CO_2\ m^{-2}s^{-1}$) was measured in two greenhouse experiments. Dark respiration was defined as the CO_2 production measured after a dark period of 25 min. Both experiments indicated clearly that the rate of dark respiration increased with increasing disease severity (Fig. 22.5). Dark respiration was related to disease severity by:

$$R_x = R_0(1 - x) + \rho R_0 x \qquad (2)$$

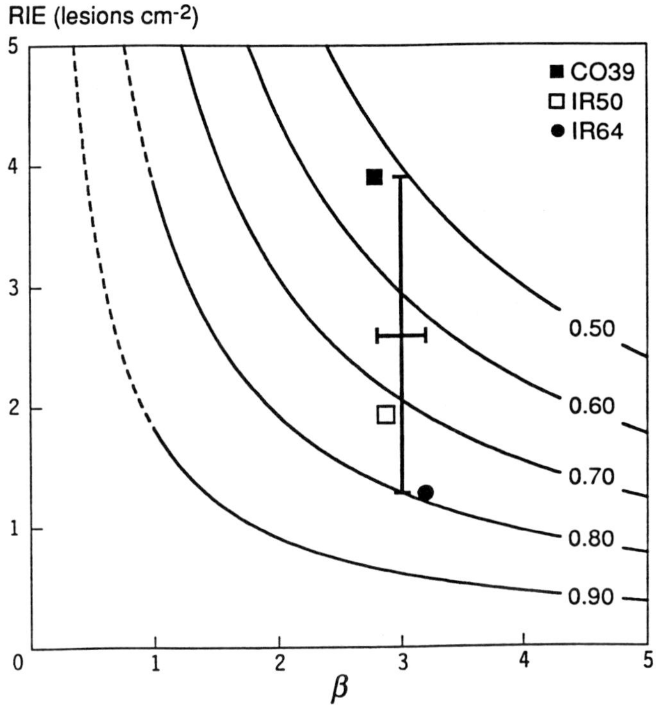

Fig. 22.4. Relative infection efficiency (RIE) and parameter β as measured for three rice cultivars after inoculation with *Pyricularia grisea*, and the implications for relative leaf photosynthetic rate (P_x/P_0). Curves are isolines referring to combinations of RIE and β with an equal relative net leaf photosynthetic rate (Bastiaans and Roumen, 1993; reprinted by permission of Kluwer Academic Publishers).

in which R_0 and R_x are the rate of dark respiration of healthy and diseased leaf tissue, respectively. This function is based on the observation that in general a respiratory increase is restricted to the area of the lesion and its near surrounding area (Samborski and Shaw, 1956; Bushnell and Allen, 1962). Here it is assumed that an increase is restricted to the visible lesion area. The parameter ρ expresses the ratio between the respiration of a lesion and that of an identical area of healthy leaf tissue. The increase in dark respiration was characterized by a ρ-value of 4.5 ± 1.3 (Bastiaans, 1993b).

Integration Step 1: From Individual Leaves to a Rice Canopy

The importance of the measured effects of leaf blast on leaf physiology, as determined by their consequences for crop production and yield, is difficult to assess. Effects on leaf physiology and effects on grain yield represent two different levels of aggregation, and are therefore difficult to relate. Mechanistic crop growth models can help to bridge the gap between these two aggregation levels. The effects of the disease on basic plant growth processes are introduced in a crop

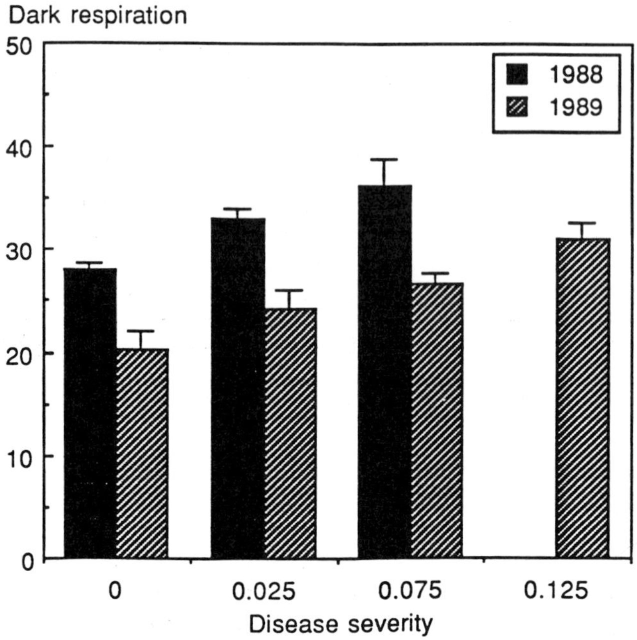

Fig. 22.5. Dark respiration (R_d; μg CO_2 m^{-2} s^{-1}) of rice leaves infected by *Pyricularia grisea* in relation to disease severity, as measured in two greenhouse experiments. Vertical bars represent standard errors of the mean.

growth model, in order to estimate the effect on yield (Rabbinge and Rijsdijk, 1981; Boote *et al.*, 1983; Rabbinge and Bastiaans, 1989). Kropff (1989), in a study on the effects of SO_2 on production of a broad bean crop (*Vicia faba*), refined the described approach. In between the quantification of effects of SO_2 on leaf physiology and the use of the crop growth model for the estimation of effects on yield, he introduced a validation of the assimilation and respiration routines of the model. Canopy photosynthesis of control and fumigated crops was measured in the field and used to evaluate the performance of the specified routines. The validated routines were introduced later into a crop growth model to analyze and predict effects on crop yield.

A similar procedure was used in the present study. The experimentally obtained relations between leaf blast severity and the assimilation rate at light saturation (P_{max}), the initial light use efficiency (ϵ) and the rate of dark respiration (R_d) of rice leaves were introduced into a model to simulate the effects of leaf blast on canopy photosynthesis. Photosynthetic rate of rice canopies infected with *P. grisea* was determined in a field experiment, to evaluate the performance of this model. In this way the first step in the integration from effects of leaf blast on individual leaves to effects of the disease on grain yield was made.

Table 22.1. Characteristics of rice canopies at the time of photosynthesis measurement and measured canopy CO_2 exchange rate (CER) in a control plot (plot A) and in two plots inoculated with *P. grisea* at mid and early tillering (plots B and C, respectively).

	Plot A	Plot B	Plot C
Leaf area index	4.49a[a]	4.80a	4.66a
Disease severity[b]	0.003a	0.046b	0.044b
Fraction dead leaf area	0.01a	0.03b	0.15c
CER ($\mu g\ CO_2\ m^{-2}\ s^{-1}$)	1432.00a	1219.00b	953.00c

[a] Means in the same row followed by the same letter are not significantly different according to *t*-test ($P < 0.05$).
[b] Fraction of green (healthy + diseased) leaf area covered by lesions.

Experimental determination of canopy photosynthesis

Canopy CO_2 exchange rate (CER; $\mu g\ CO_2\ m^{-2}\ s^{-1}$) was measured for a rice crop at maximum tillering in three different plots. The control crop (plot A) was slightly infected with *P. grisea*, due to influx of spores from neighboring plots. A small amount of dead leaf tissue, found in the lower part of the canopy, consisted of naturally senescing leaves (Table 22.1). The other two plots were inoculated with *P. grisea* at mid (plot B) and early tillering (plot C), resulting in a higher disease severity and a higher fraction of dead leaf area at the time of measurement. The amount of dead leaf area was moderate in plot B and high in plot C, reflecting the differences in time of inoculation. Significant differences between the CER of the three plots were observed. Compared to the control plot the CERs of plot B and C were reduced by 15% and 33%, respectively.

Modeling the effect of leaf blast on canopy photosynthesis

More interesting than the observed differences in CER was whether the reduction in canopy photosynthesis could be explained by the adverse effects of leaf blast on photosynthesis and respiration of individual leaves and by the presence of dead leaf tissue. For that purpose a model for the calculation of instantaneous rates of canopy photosynthesis was constructed. The model comprised routines for the calculation of photosynthesis and respiration of an elementary model for crop growth (MACROS-L1D; Penning de Vries *et al.*, 1989), and routines to generate diurnal trends of radiation and temperature on the basis of daily weather data. The model was run with a time interval of 0.25 h. A schematic representation of the model is given in Fig. 22.6. A detailed description of the model is given by Bastiaans and Kropff (1993).

To account for the effect of leaf blast on photosynthesis and respiration, disease severity and the fraction of dead leaf area were introduced into the model. Photosynthesis–light response characteristics of green leaf tissue (healthy + diseased leaves) were related to observed disease severity, using parameter β.

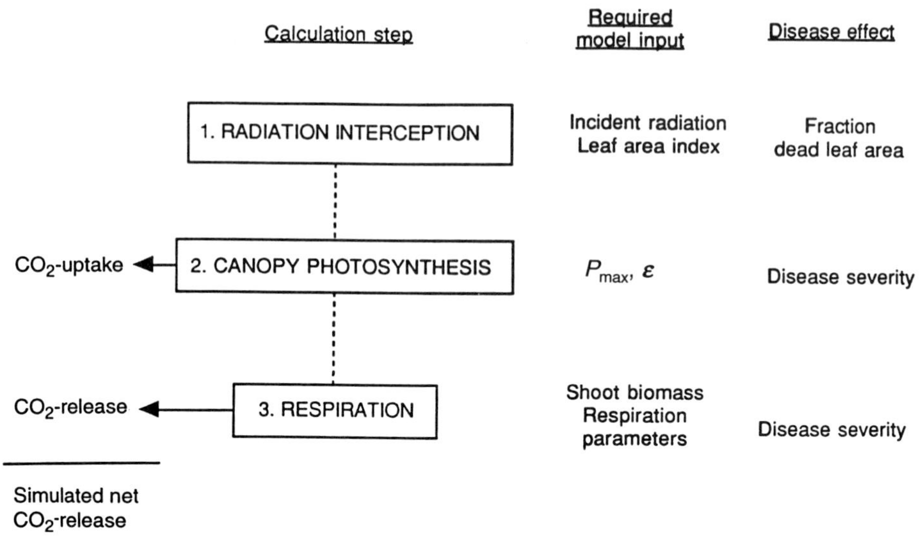

Fig. 22.6. Simplified representation of the most important calculation steps and typical input data for a model on canopy CO_2 exchange rate (CER). Special attention is given to the effect of leaf blast on CER.

Dead leaves do not contribute to photosynthesis, but light interception by this part of the canopy was accounted for. Optical characteristics of diseased and dead leaf tissue were assumed to be identical to the characteristics of healthy leaf tissue.

The effect of leaf blast on respiration was introduced as an increased rate of leaf maintenance respiration. Implicitly, this assumes that assimilate requirement of processes related to the presence of the disease have priority over growth. The increase in respiration was related to disease severity and characterized by parameter ρ. Maintenance requirement of dead leaf tissue is nil and was therefore not considered.

Canopy CO_2 exchange rate for control and inoculated canopies was accurately simulated with the model (Fig. 22.7). The observed reduction in CER can thus be explained by the adverse effect of blast lesions on leaf photosynthesis and respiration and by the existence of dead leaf tissue, resulting from disease-induced senescence. In Table 22.2 a more detailed analysis of the reduction in CER is given. The effects of blast lesions and dead leaf tissue were separated, and further dissected. Lesions cause a decrease in green leaf tissue, a reduction in photosynthetic rate of the remaining green leaf tissue, and an increased respiration. In successive runs these damage mechanisms were removed one after the other, and the subsequent increase in CER was defined as the reduction due to the newly removed damage mechanism. Clearly, the reduction in photosynthetic rate of the remaining green leaf tissue was the most important effect of blast lesions, whereas the increased respiration contributed only marginally to the overall effect.

Dead leaf tissue causes a reduction in photosynthetically active leaf tissue and reduces light intensity in lower leaf layers by light interception (shading). The

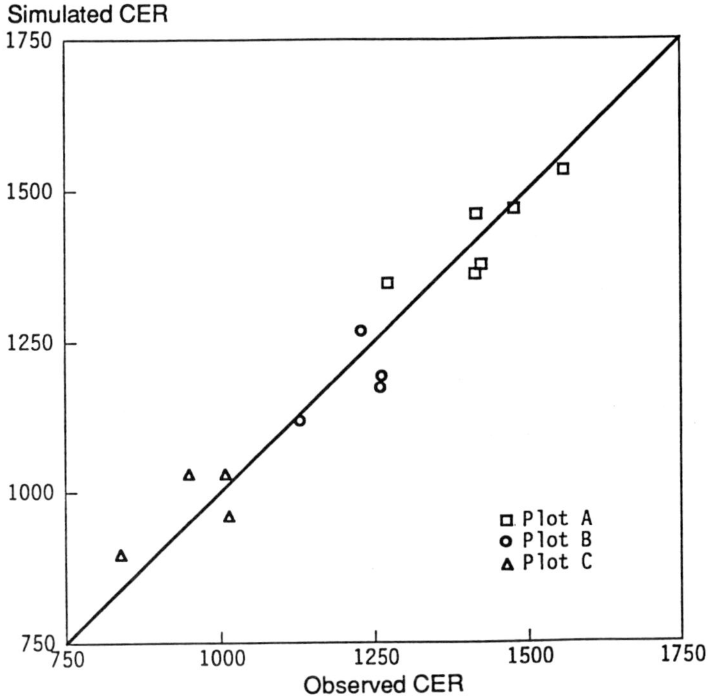

Fig. 22.7. Comparison of observed and simulated canopy CO_2 exchange rate (CER; $\mu g\ CO_2\ m^{-2}\ s^{-1}$) (Bastiaans and Kropff, 1993).

Table 22.2. Contribution of blast lesions and dead leaf tissue to the overall reduction in canopy CO_2 exchange rate (CER; $\mu g\ CO_2\ m^{-2}\ s^{-1}$) of rice crops inoculated with *Pyricularia grisea*. The contribution of the identified damage mechanisms related to blast lesions and dead leaf tissue are specified and expressed as a fraction of the total reduction due to blast lesions and dead leaf tissue, respectively.

	Plot B[a]	Plot C[a]
Total reduction in CER	213 (100%)	479 (100%)
Reduction related to blast lesions	173 (81%)	163 (34%)
Contribution of identified damage mechanisms		
Decrease in green leaf tissue	54 (0.31)	49 (0.30)
Reduction in leaf photosynthetic rate	109 (0.63)	104 (0.64)
Increase in respiration	10 (0.06)	10 (0.06)
Reduction related to dead leaf tissue	40 (19%)	316 (66%)
Contribution of identified damage mechanisms		
Decrease in green leaf tissue	13 (0.32)	95 (0.30)
Shading	27 (0.68)	221 (0.70)

[a] Plot B inoculated at mid tillering and plot C inoculated at early tillering stage.

increase in simulated CER after removal of dead leaf tissue corresponds to the reduction due to shading. The further increase in simulated CER after replacing dead leaf tissue by an identical area of healthy leaf tissue, equals the reduction due to the reduction in photosynthetically active leaf tissue. The analysis showed that shading accounted for two-thirds of the overall reduction due to dead leaf tissue. The importance of this mechanism is determined by the position of dead leaf tissue in the canopy. In the case of natural leaf senescence dead leaf tissue is mainly found in lower leaf layers, and then shading is not an important mechanism.

The three plots in the experiment represent rice crops at various stages of infection, offering the opportunity to theorize about the way a reduction in CER evolves after establishment of lesions of *P. grisea*. The most prominent effect of the lesions is an inhibitory effect on photosynthesis of the surrounding green leaf tissue, which causes a reduced utilization efficiency of intercepted radiation. Newly emerging leaves become infected, whereas earlier infected leaves are subject to an accelerated senescence. As a result, the fraction of diseased leaf area remains more or less constant, whereas the fraction of dead leaf area increases. Consequently, the reduction in CER will increase as a result of an increased reduction in the amount of radiation intercepted by green foliage. The difference in CER of plots B and C is thus mainly due to a difference in intercepted radiation, and reflects the difference in fraction of dead leaf area that existed due to a difference in onset of the epidemic. This analysis suggests that in the long run crop production suffers more from the accelerated senescence of infected leaf tissue than from the initial disturbances in leaf physiology.

Integration Step 2: From a Single Development Stage to a Growing Season

Determining the effect of leaf blast on canopy photosynthesis represents just an intermediate step in the extrapolation of the effects of leaf blast on leaf physiology to effects of the disease on crop production. Since the effect of the disease on canopy photosynthesis is well understood, a second integration step can be made. The effect of the disease on the performance of a canopy is integrated over an entire growing season, to determine the effect on yield. Similar to the previous integration step from individual leaves to a rice canopy, this second integration step was realized by combining experimental research and modeling.

A special feature of leaf blast is that the disease is mainly present before flowering. In the development of a rice crop, flowering marks an important transition. The period prior to flowering is characterized by the construction of a canopy, and the formation of storage organs (grains). It is in this development phase that leaf blast severity is highest, and will have a direct effect on crop growth rate. After flowering, when leaf blast severity has generally dropped to a low level, the canopy is used to fill the grains and grain yield is determined. Leaf blast thus mainly affects grain yield indirectly, through a long-term effect of the disease on crop production. Experimental data were used to determine whether the long-term effect of the disease on crop production is passed on through an effect on the source (carbon or nitrogen economy) or through an effect on the sink.

Identifying the mechanism of yield reduction

The effect of leaf blast on crop production was studied in an experimental field at the lowland research site of the International Rice Research Institute (IRRI), Los Baños, The Philippines. A randomized complete block design with three treatments in five replicates was used (plot size: 12 × 3.6 m). One treatment was used as the control (treatment A), while the other two treatments were inoculated with *P. grisea* during mid (treatment B) and early tillering (treatment C), respectively. From maximum tillering on, fungicides were used to stop the development of the disease, since an undisturbed further development would have destroyed the crop. Neck blast infections were prevented in all treatments by applying fungicides at booting and heading. Disease development and the effect of the disease on crop growth was studied through periodic harvesting (Fig. 22.8). Further experimental details are given by Bastiaans (1993c).

The first aim of this experiment was to identify the mechanism through which the initial reduction in crop growth rate during the presence of the disease affects the formation of grain yield later on. The observations on leaf and canopy photosynthesis demonstrated that leaf blast primarily reduces carbohydrate production. Associated with this reduction is a reduction in crop growth and leaf area formation. The positive feedback between crop growth, leaf area formation and radiation interception further enhances the negative effect of the disease on crop growth (Blackman, 1919). In the case of heavy infestations of leaf blast, the feedback mechanism is further enhanced by an accelerated senescence of infested leaf tissue. Ultimately, all processes result in a reduced amount of green leaf area during ripening, and accordingly carbohydrate supply during grain filling is reduced. This is the most obvious mechanism through which leaf blast, present before flowering, reduces grain yield formation. Lim and Gaunt (1986) reported such a mechanism of yield reduction in the pathosystem barley–mildew in case of early epidemics. The observations on leaf area index (LAI) confirm that during ripening the LAI of the inoculated crops was clearly smaller than the LAI of the control crop (Fig. 22.8A).

The utilization efficiency of intercepted radiation during grain filling is another important determinant of carbohydrate supply in this period. In rice, as in other crop species, this efficiency is largely determined by the N content of the leaves, since the maximum rate of leaf photosynthesis is almost proportional to the fraction of N in leaves (Cook and Evans, 1983). Uptake of N after flowering is generally low and N content of leaf tissue during grain filling is therefore mainly determined by N uptake before flowering. Moreover, N redistribution from leaf blades contributes considerably to N supply during grain filling. A reduced amount of accumulated N at flowering may thus eventually limit yield. This is the second mechanism through which leaf blast may have a long-term effect on yield.

The results of the experiment demonstrate that leaf blast reduces the N uptake of the crop (Table 22.3). This reduction resulted from differences in N uptake before flowering, since total shoot N during ripening was fairly constant in both healthy and inoculated crops. Despite the reduction in N uptake in the inoculated crops, the shoot N content at flowering and maturity was identical or

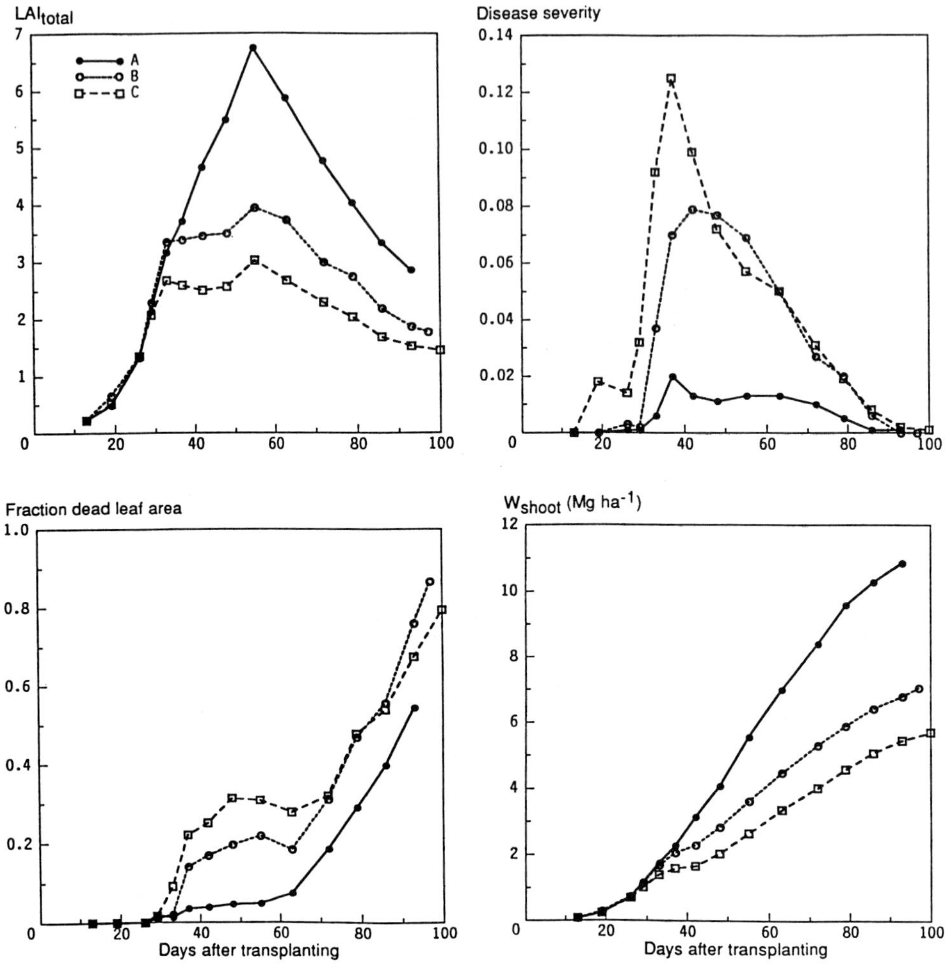

Fig. 22.8. Time course of total leaf area (LAI_{total}; ha ha^{-1}), disease severity, fraction of dead leaf area and shoot dry weight (W_{shoot}; Mg ha^{-1}) of rice in a control crop (treatment A) and in crops inoculated with *Pyricularia grisea* (treatments B and C) (Bastiaans, 1993c).

higher and the same was found for the N content of individual shoot organs. This indicates that the reduced N uptake in the inoculated crops before flowering was not responsible for the yield reduction observed.

A reduced sink formation is the third mechanism through which leaf blast may eventually reduce grain yield. Tiller number, floret number and spikelet number are all determined before flowering. Their numbers are regulated through the carbohydrate supply of the canopy in specific initiation periods (Penning de Vries *et al.*, 1989). Moreover, 1000 grain weight of rice is also determined before flowering, since this characteristic is primarily controlled by the size of the hulls. The product of spikelet mumber and 1000 grain weight of rice is referred to as maximum eventual yield (Matsushima, 1970). If leaf blast reduces

Table 22.3. Nitrogen (N) uptake and nitrogen content of the shoot and relevant shoot organs of rice at flowering and maturity, as observed in a control crop (treatment A) and in crops inoculated with *Pyricularia grisea* at mid and early tillering (treatments B and C, respectively).

	Treatment A	Treatment B	Treatment C
Total N uptake (kg N ha^{-1})	122.0a	88.0b	70.0b
N content (mg g^{-1}) at flowering			
Total shoot	17.5a[a]	18.7a	18.9a
Green leaf tissue	33.4a	36.4b	38.5c
Stem tissue	11.2a	13.7b	14.4b
N content (mg g^{-1}) at maturity			
Total shoot	11.2a	12.1b	12.2b
Panicle	12.9a	13.7b	13.6b

[a] Means in the same row followed by the same letter are not significantly different according to *t*-test ($P < 0.05$).

Table 22.4. Grain yield and yield components of rice cultivar IR50 as observed in a control crop (treatment A) and in crops inoculated with *Pyricularia grisea* at mid and early tillering (treatments B and C, respectively).

	Treatment A	Treatment B	Treatment C
Panicle number (m^{-2})	581.0a[a]	605.0b	623.0c
Spikelets per panicle	65.0a	45.0b	37.0c
1000 grain weight (g)[b]	21.1a	20.1b	19.7b
Maximum eventual yield (Mg ha^{-1})[b]	7.9a	5.4b	4.5c
Fraction filled grains	0.85a	0.80b	0.78b
Actual grain yield (Mg ha^{-1})[b]	6.7a	4.3b	3.5c

[a] Means in the same row followed by the same letter are not significantly different according to *t*-test ($P < 0.05$).
[b] Rough rice at 140 g moisture kg^{-1}.

maximum eventual yield more than proportionally the reduction in grain yield may be sink determined.

Yield component analysis demonstrated that leaf blast reduced maximum eventual yield through a reduction in spikelet number and a reduction in 1000 grain weight (Table 22.4). More important, however, was the observation that in the inoculated treatments the relative reduction in actual yield surpassed the relative reduction in maximum eventual yield. This is reflected in a reduced fraction of filled grains of the inoculated crops and demonstrates that the reduction in grain yield is exclusively source determined. The observations are in line with data on yield reduction reported by Goto (1965), who also observed reductions in spikelet number, 1000 grain weight and fraction filled grains due to leaf blast. Based on this analysis it is concluded that the long-term effect of leaf blast on crop production is solely passed on through an effect on carbon economy. Therefore the model analysis will be restricted to this aspect of yield formation.

Table 22.5. Cumulative radiation interception (PARCUM), radiation use efficiency (RUE) and harvest index (HI) of rice in a control crop (treatment A) and crops inoculated with *Pyricularia grisea* at mid and early tillering (treatments B and C, respectively).

	Treatment A	Treatment B	Treatment C
PARCUM ((MJ PAR)m^{-2})	403.0a	315.0b	260.0c
RUE (g dry matter (MJ PAR)$^{-1}$)	2.69a	2.16b	2.10b
Total shoot dry weight (g m^{-2})	1090.0a	710.0b	570.0c
HI	0.53a	0.52a	0.53a
Grain yield (g m^{-2})[b]	670.0a	430.0b	350.0c

[a] Means in the same row followed by the same letter are not significantly different according to *t*-test ($P < 0.05$).
[b] Rough rice at 140 g moisture kg^{-1}.

Modeling the effect of leaf blast on crop production

A simple model

Crop productivity can be analyzed in terms of solar radiation interception by green leaf area (PARCUM; MJ m^{-2}) and radiation use efficiency (RUE; g dry matter MJ^{-1}) (Monteith, 1977). Accordingly, Johnson (1987) distinguished two broad categories of effects of pests and diseases on crop growth: effects on radiation interception and effects on radiation use efficiency. Multiplication of total shoot dry matter and harvest index (HI) yields grain yield. This simple model was used to make a first analysis of the effects of leaf blast on crop production (Bastiaans, 1993d):

$$\begin{array}{rcccccc} \text{Yield} & = & \text{Radiation Interception} & \times & \text{Radiation Use Efficiency} & \times & \text{Harvest Index} \\ & = & \text{PARCUM} & \times & \text{RUE} & \times & \text{HI} \\ [\text{g m}^{-2}] & = & [(\text{MJ PAR})\ \text{m}^{-2}] & \times & [\text{g (MJ PAR)}^{-1}] & \times & [-] \end{array}$$

Radiation interception of photosynthetically active radiation by green leaf foliage was calculated on a daily basis for every plot (RI; MJ m^{-2} d^{-1}), using a model that consisted of the radiation routines of an elementary model of crop growth. Leaf area index, disease severity and the fraction of dead leaf area were input to this model. Summation of intercepted radiation over the growing season resulted in the cumulative radiation interception (PARCUM; MJ m^{-2}). Radiation use efficiency was calculated by dividing the observed total shoot dry matter by PARCUM. Harvest index was determined at maturity.

The analysis with the RI × RUE model demonstrates that leaf blast in rice is a pathosystem in which the pathogen reduces the dry matter production of the crop through a reduction in cumulative radiation interception and a reduction in radiation use efficiency (Table 22.5). No significant difference was present between the RUE of the inoculated crops (treatments B and C). Therefore, the difference in dry matter production between B and C resulted mainly from a difference in cumulative radiation interception. After initiation of the disease, lesion coverage and disease induced senescence caused a direct reduction of RI. An indirect reduction of RI as a result of leaf blast was caused by a decline in new leaf area formation. This decline resulted from a reduced crop growth rate, which was caused by previous reductions of leaf blast in RI and RUE. The importance of this mechanism depends strongly on the development stage of the crop, since the underlying positive feedback between crop growth, leaf area formation and radiation interception is strongest during early growth stages (Blackman, 1919). Previously, Lim and Gaunt (1986) and van Oijen (1991) pointed at the importance of the available leaf area at initiation of a disease for leaf area duration and cumulative radiation interception. It can be concluded that the difference in disease severity together with the difference in LAI at the initiation of leaf blast, caused by the difference in inoculation date, are the most prominent factors responsible for the observed differences in total dry matter production of the inoculated treatments in this experiment.

Harvest index was not affected in this experiment. This is probably related to the timing of the inoculations. The crop was only inoculated during early development stages. A simulation study demonstrated that leaf blast epidemics late in the reproductive phase are expected to cause a clear reduction in harvest index (Bastiaans and Teng, 1994).

A mechanistic crop growth model

A mechanistic model of crop growth (MACROS-L1D; Penning de Vries *et al.*, 1989) was used to analyze the causes of the reduction in the utilization efficiency of intercepted radiation (Fig. 22.9). The effects of leaf blast on photosynthesis and respiration were introduced into this model in an identical way as in the model on canopy photosynthesis. To account for the effect of leaf blast on photosynthesis and respiration, disease severity and the fraction of dead leaf area were introduced in the model. Similarly, the measured LAI was input to the model. The model was initialized with dry weights of leaves and stems of the first harvest and run until final harvest. Model performance was evaluated by comparing the simulated and observed shoot dry weight over time (Bastiaans, 1993d).

Simulation of dry matter production of the control crop resulted in a good agreement with the observed dry matter production (Fig. 22.10). This demonstrates the thoroughness of the basic crop growth model and indicates that any dissimilarity between simulated and observed shoot dry weight of the inoculated crops is likely to be due to an error in the simulation of leaf blast effects.

Total dry matter production of the inoculated crops was overestimated (Fig. 22.10). The overestimation appeared during the first weeks after the onset of the disease. After that time the difference beteen simulated and observed dry weight stabilized at approximately 550 kg ha^{-1}, in both treatments. The overestimation indicates that during the establishment of the disease the effects of leaf blast on leaf photosynthetic rate and respiration were inadequate to explain fully the reduction in crop growth rate. The stabilization of the overestimation in the consecutive period till maturity indicates that later on the two effects gave a sufficient explanation of the reduction in RUE.

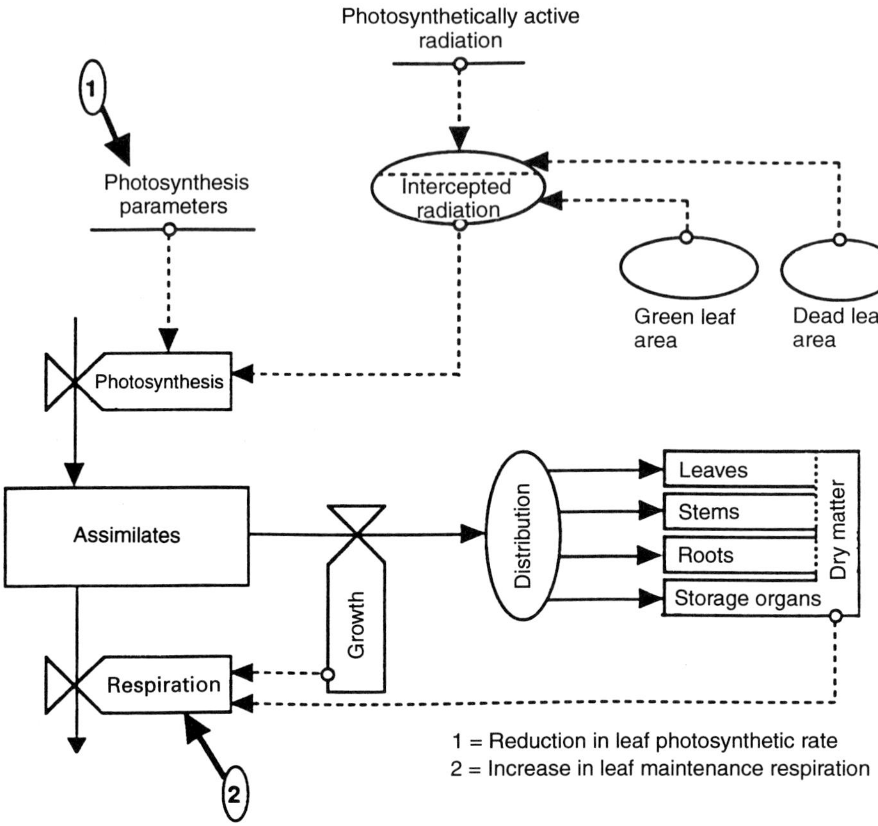

Fig. 22.9. Diagram illustrating schematically the calculation procedure of daily crop growth rate in model MACROS-L1D, and the various ways through which leaf blast affects crop growth rate. Leaf area (green + dead) is introduced as a forcing function. Quantities are shown in boxes, processes in valve symbols, flow of material by lines and flow of information by dotted lines.

The utilization of intercepted radiation for dry matter production can be subdivided into three processes: CO_2 assimilation, allocation of assimilates and utilization of assimilates for dry matter production. The canopy photosynthesis measurements demonstrated that the effects of leaf blast on leaf photosynthetic rate and respiration satisfactorily explained the effect of leaf blast on CO_2 assimilation. The overestimation of shoot dry weight after the onset of the disease observed in this study therefore most likely results from an effect of leaf blast on one of the last two processes.

An increased allocation of assimilates to the roots would be an appropriate explanation for the observed dissimilarity between observed and simulated shoot dry weight. Since root biomass was not determined in the field experiment, this hypothesis could not be validated. However, according to the functional balance theory of Brouwer (1962), an attack by a foliar disease and the succeeding reduction in crop photosynthesis is more likely to be followed by an increased

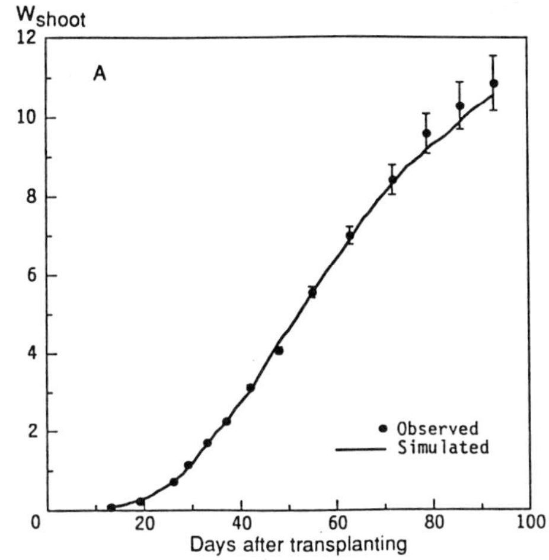

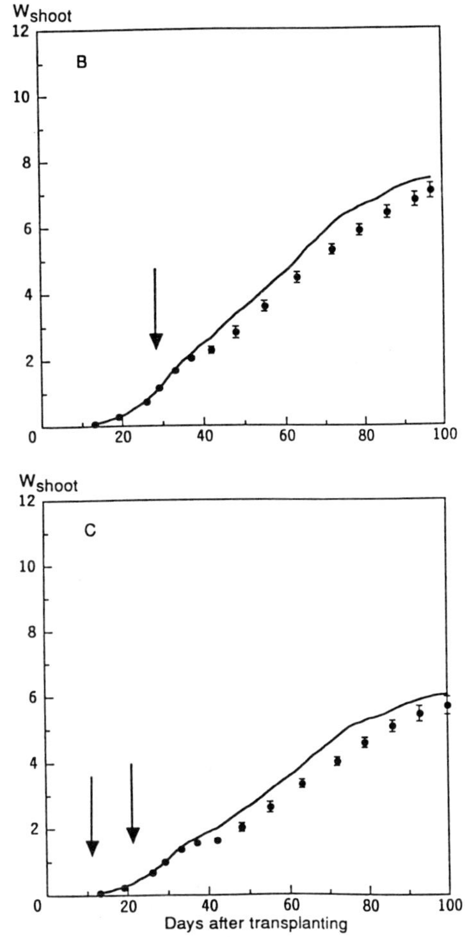

Fig. 22.10. Observed and simulated shoot dry weight (W_{shoot}; Mg ha^{-1}) of rice in a control crop (treatment A) and in a crops inoculated with *Pyricularia grisea* (treatments B and C). Vertical bars represent standard errors of the mean. Arrows indicate inoculation data (Bastiaans, 1993d).

allocation of assimilates to the shoot. Such a shift in the allocation pattern of assimilates was for instance observed in wheat after infection with *Puccinia striiformis* (Siddiqui and Manners, 1971). An increased allocation of assimilates to the roots after infection of rice with leaf blast is therefore considered to be unlikely.

Consequently, the overestimation of shoot dry weight is probably caused by a less efficient utilization of produced carbohydrates for crop growth. In a study on the effect of leaf rust on crop production of winter wheat, van Roermund and Spitters (1990) introduced the uptake of assimilates by the fungus for spore production as a potential damage mechanism. They estimated that the pathogen used 230–570 kg CH_2O ha^{-1} for spore production and related processes. Data on the rate of sporulation of blast lesions are diverse, but data on the production pattern of lesions are rather uniform (Kato *et al.*, 1970; Kato and Kozaka, 1974; Ou, 1985; Kim *et al.*, 1987). Lesions start to produce spores immediately after their appearance, and spore production may last for 20–30 days. Spore production

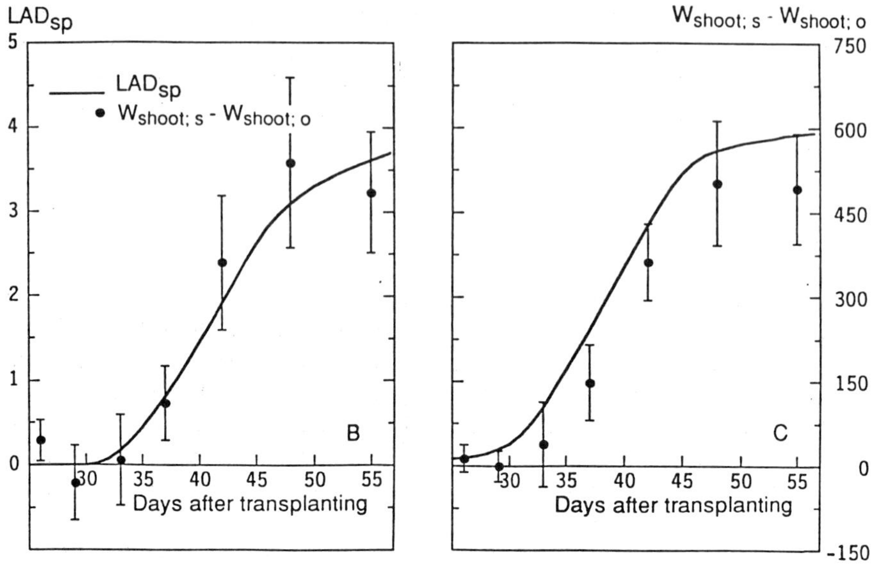

Fig. 22.11. Simulated leaf area duration of sporulating blast lesions (LAD_{sp}; days) and the difference between simulated and observed total shoot dry weight ($W_{shoot;s} - W_{shoot;o}$; kg ha^{-1}) for rice crops inoculated with *Pyricularia grisea* (treatments B and C). Vertical bars represent standard errors of the mean (Bastiaans, 1993d).

reaches its peak between 3 and 8 days after lesion appearance, and 90% or more of the spores are produced during the first 14 days. These last observations were used to simulate the amount of actively sporulating leaf area in the various plots of the experiment. For each day the leaf area occupied with lesions (Diseased Area Index, DAI; Zadoks, 1961) was calculated through multiplication of the green leaf area index and disease severity. The newly colonized leaf area was calculated daily and derived from the increase in DAI. Next, the amount of actively sporulating leaf area for any particular day was estimated through summation of the amount of leaf tissue that was colonized during the last 14 days. Integration of the daily amount of actively sporulating leaf area over time yields the leaf area duration of sporulating leaf tissue (LAD_{sp} in days). The time course of this integral shows a striking similarity with the time course of the difference between simulated and observed shoot dry weight (Fig. 22.11). If the over-estimation of shoot dry weight was completely caused by the use of carbohydrates by the pathogen for spore production, the growth reduction per ha of sporulating leaf tissue would amount to approximately 150 kg dry matter d^{-1}. This corresponds to a carbohydrate withdrawal of around 200 kg CH_2O (ha sporulating lesions)$^{-1}$ d^{-1}.

The present result illustrates the necessity of model validation. Comparison of observed and simulated shoot dry weight in the course of time indicated the absence of an important damage mechanism. Combined with the results of the independent validation of the subroutines on canopy photosynthesis, the missing damage mechanism could be identified. A direct measurement of the costs involved in the production of spores would further strengthen the model. However, detailed information would then be required on the rate of spore production, the weight of individual spores and, most important, the energy efficiency of the spore production process. For the moment, the indirect estimation of the carbohydrate requirement for spore production and related processes seems a usable alternative.

Conclusion

The interaction between experimental research and model development resulted in a better understanding of yield reduction in rice due to leaf blast. This demonstrates the usefulness of the ecophysiological approach to yield loss assessment. Leaf blast was shown to reduce the yield of a rice crop through a reduction in cumulative radiation interception and a reduction in radiation use efficiency. The reduced radiation interception was caused by lesion coverage, premature leaf senescence and a reduced leaf area formation. The reduction in photosynthetic efficiency of the remaining green leaf tissue and the uptake of assimilates by the pathogen for spore production were the main factors responsible for the reduced utilization efficiency of intercepted radiation.

The study resulted in a model that can be used to estimate yield reduction in rice due to leaf blast for various epidemics under variable environmental conditions. Therefore, the model is an appropriate tool in constructing damage functions, which are of primary importance for applying the concept of integrated pest management in blast control.

References

Ahn, S.W. and Ou, S.H. (1982) Quantitative resistance of rice to blast disease. *Phytopathology* 72, 279–282.

Anderson, A.L., Henry, B.W. and Tullis, E.C. (1947) Factors affecting infectivity, spread and persistence of *Pyricularia oryzae* Cav. *Phytopathology* 37, 94–110.

Bastiaans, L. (1991) Ratio between virtual and visual lesion size as a measure to describe reduction in leaf photosynthesis of rice due to leaf blast. *Phytopathology* 81, 611–615.

Bastiaans, L. (1993a) Understanding yield reduction in rice due to leaf blast. PhD Thesis, Wageningen Agricultural University, The Netherlands, 127 pp.

Bastiaans, L. (1993b) Effects of leaf blast on photosynthesis of rice. 1. Leaf photosynthesis. *Netherlands Journal of Plant Pathology* 99, 197–203.

Bastiaans, L. (1993c) Effects of leaf blast on growth and production of a rice crop. 1. Determining the mechanism of yield reduction. *Netherlands Journal of Plant Pathology* 99, 323–334.

Bastiaans, L. (1993d) Effects of leaf blast on growth and production of a rice crop. 2. Analysis of the reduction in dry matter production of rice due to leaf blast, using two models with different complexity. *Netherlands Journal of Plant Pathology* 99, Supplement 3, 19–28.

Bastiaans, L. and Kropff, M.J. (1993) Effects of leaf blast on photosynthesis of rice. 2. Canopy photosynthesis. *Netherlands Journal of Plant Pathology* 99, 205–217.

Bastiaans, L. and Roumen, E.C. (1993) Effect of leaf photosynthetic rate by leaf blast for rice cultivars with different types and levels of resistance. *Euphytica* 66, 81–87.

Bastiaans, L. and Teng, P.S. (1994) Simulation of yield reduction in rice due to leaf blast. *Agricultural Systems*: submitted

Blackman, V.H. (1919) The compound interest law and plant growth. *Annals of Botany* 33, 353–360.

Boote, K.J., Jones, J.W., Mishoe, J.W. and Berger, R.D. (1983) Coupling pests to crop growth simulators to predict yield reductions. *Phytopathology* 73, 1581–1587.

Brouwer, R. (1962) Nutritive influences on the distribution of dry matter in the plant. *Netherlands Journal of Agricultural Science* 10, 399–408.

Burrell, M.M. and Rees, T. (1974) Carbohydrate metabolism of rice leaves infected by *Pyricularia oryzae*. *Physiological Plant Pathology* 4, 489–496.

Bushnell, W.R. and Allen, P.J. (1962) Respiratory changes in barley leaves produced by single colonies of powdery mildew. *Plant Physiology* 37, 751–758.

Cook, M.G. and Evans, L.T. (1983) Nutrient responses of seedlings of wild and cultivated *Oryza* species. *Field Crops Research* 6, 205–218.

Ephrath, J.E., Shteinberg, D., Drieshpoun, J., Dinoor, A. and Marani, A. (1989) *Alternaria alternata* in cotton (*Gossypium hirsutum*) cv. Acala: effects on gas exchange, yield components and yield accumulation. *Netherlands Journal of Plant Pathology* 95, 157–166.

Gaunt, R.E. (1978) Crop physiology: disease effects and yield loss. In: *Proceedings of the APPS Workshop on Epidemiology and Crop Loss Assessment*. Lincoln College, New Zealand, pp. 9/1–12.

Goto, K. (1965) Estimating losses from rice blast in Japan. In: *The Rice Blast Disease*. The Johns Hopkins Press, Baltimore, Maryland, pp. 195–202.

Haverkort, A.J. and Bicamumpaka, M. (1986) Correlation between intercepted radiation and yield of potato crops infested by *Phytophthora infestans* in central Africa. *Netherlands Journal of Plant Pathology* 92, 239–247.

Johnson, K.B. (1987) Defoliation, disease and growth: a reply. *Phytopathology* 77, 1495–1497.

Kato, H. and Kozaka, T. (1974) Effect of temperature on lesion enlargement and sporulation of *Pyricularia oryzae* in rice leaves. *Phytopathology* 64, 828–830.

Kato, H., Sasaki, T. and Koshimizy, Y. (1970) Potential for conidium formation of *Pyricularia oryzae* in lesions of leaves and panicles of rice. *Phytopathology* 60, 608–612.

Katsube, T. and Koshimizu, Y. (1970) Influence of blast disease on harvests in rice plant. 1. Effects of panicle infection on yield components and quality. *Bulletin of the Tohoku National Agricultural Experiment Station* 39, 55–96.

Kim, C.K., Yoshino, R. and Lee, E.J. (1987) Epidemiological studies of rice blast disease caused by *Pyricularia oryzae* Cavara. II. Sporulation and conidia release from naturally infected lesions. *Korean Journal of Plant Protection* 26, 37–41.

Koh, Y.J., Hwang, B.K. and Chung, H.S. (1987) Adult-plant resistance of rice to leaf blast. *Phytopathology* 77, 232–236.

Kropff, M.J. (1989) Quantification of SO_2 effects on physiological processes, plant growth and crop production. PhD Thesis, Wageningen Agricultural University, The Netherlands, 201 pp.

Kuribayashi, K. and Ichikawa, H. (1952) Studies on forecasting of the rice blast disease. Special report. *Nagano Agricultural Experiment Station* 13, 1–229.

Lim, L.G. and Gaunt, R.E. (1986) The effect of powdery mildew (*Erisyphe graminis* f. sp. hordei) and leaf rust (*Puccinia hordei*) on spring barley in New Zealand. I. Epidemic development, green leaf area and yield. *Plant Pathology* 35, 44–53.

Madden, L.V. (1983) Measuring and modeling crop losses at the field level. *Phytopathology* 73, 1591–1596.

Matsushima, S. (1970) *Crop Science in Rice*. Fuji Publishing Co., Tokyo, Japan, 379 pp.

Monteith, J.L. (1977) Climate and the efficiency of crop production in Britain. *Philosophical Transactions of the Royal Society of London, Series B* 281, 277–294.

Oijen, M. van (1990) Photosynthesis is not impaired in healthy tissue of blighted potato plants. *Netherlands Journal of Plant Pathology* 96, 55–63.

Oijen, M. van (1991) Leaf area dynamics of potato cultivars infected by *Phytophthora infestans*. *Netherlands Journal of Plant Pathology* 97, 345–354.

Ou, S.H. (1985) *Rice Diseases*, 2nd edn. Commonwealth Mycological Institute, Kew, UK, 380 pp.

Padmanabhan, S.Y. (1965) Estimating losses from rice blast in India. In: *The Rice Blast Disease*. Johns Hopkins Press, Baltimore, Maryland, pp. 203–221.

Parlevliet, J.E. (1979) Components of resistance that reduce the rate of epidemic development. *Annual Review of Phytopathology* 17, 203–222.

Penning de Vries, F.W.T., Jansen, D.M., ten Berge, H.F.M. and Bakema, A. (1989) *Simulation of Ecophysiological Processes of Growth in Several Annual Crops*. PUDOC, Wageningen, The Netherlands, 271 pp.

Rabbinge, R. and Bastiaans, L. (1989) Combination models, crop growth and pests and diseases. In: Rabbinge, R., Ward, S.A. and van Laar, H.H. (eds), *Simulation and Systems Management in Crop Protection*. PUDOC, Wageningen, pp. 217–239.

Rabbinge, R. and Rijsdijk, F.H. (1981) Disease and crop physiology: a modeller's point of view. In: Ayres, P.G. (ed.), *Effects of Disease on the Physiology of the Growing Plant*. Cambridge University Press, Cambridge, pp. 201–220.

Rabbinge, R., Jorritsma, I.T.M. and Schans, J. (1985) Damage components of powdery mildew in winter wheat. *Netherlands Journal of Plant Pathology* 91, 235–247.

Roermund, H.J.W. van and Spitters, C.J.T. (1990) Simulation of yield reduction by leaf rust in winter wheat, applied to the analysis of genetic variation in partial resistance. *Netherlands Journal of Plant Pathology* 96, 17–28.

Rossman, A.Y., Howard, R.J. and Valent, B. (1990) *Pyricularia grisea*, the correct name for the rice blast fungus. *Mycologia* 82, 509–512.

Rotem, J., Bashi, E. and Kranz, J. (1983) Studies of crop loss in potato late blight caused by *Phytophthora infestans*. *Plant Pathology* 32, 117–122.

Roumen, E.C. (1993) Partial resistance in rice to blast and how to select for it. PhD Thesis, Wageningen Agricultural University, The Netherlands, 108 pp.

Samborski, D.J. and Shaw, M. (1956) The physiology of host–parasite relations. II. The effect of *Puccinia graminis* tritici Eriks. & Henn. on the respiration of the first leaf of resistant and susceptible species of wheat. *Canadian Journal of Botany* 34, 601–619.

Shaw, M. and Samborski, D.J. (1957) The physiology of host–parasite relations. III. The pattern of respiration in rusted and mildewed cereal leaves. *Canadian Journal of Botany* 35, 389–407.

Siddiqui, M.Q. and Manners, J.G. (1971) Some effects of general yellow rust (*Puccinia striiformis*) infection on ^{14}carbon assimilation and growth in spring wheat. *Journal of Experimental Botany* 22, 792–799.

Smedegaard-Petersen, V. (1984) The role of respiration and energy generation in diseased and disease resistant plants. In: Wood, R.K.S. and Jellis, G.J. (eds), *Plant Diseases: Infection, Damage and Loss*. Blackwell Scientific Publications, Oxford, pp. 73–85.

Spitters, C.J.T., Roermund, van H.J.W., Nassau, van H.G.M.G., Schepers, J. and Mesdag, J. (1990) Genetic variation in partial resistance to leaf rust in winter wheat: disease progress, foliage senescence and yield reduction. *Netherlands Journal of Plant Pathology* 96, 3–15.

Teng, P.S., Klein-Gebbinck, H.W. and Pinnschmidt, H. (1991) An analysis of the blast pathosystem to guide modeling and forecasting. In: *Rice Blast Modeling and Forecasting*. International Rice Research Institute, Los Baños, The Philippines, pp. 1–30.

Toriyama, K. (1975) Recent progress of studies on horizontal resistance in rice breeding for blast resistance in Japan. In: Gàlvez, G.E. (ed.), *Proceedings of a Seminar on Horizontal Resistance to the Blast Disease of Rice, CIAT series CE9*. Cali, Colombia. pp. 65–100.

Torres, C.Q. and Teng, P.S. (1993) Path coefficient and regression analysis of the effects of leaf and panicle blast on tropical rice yield. *Crop Protection* 12, 296–302.

Toyoda, S. and Suzuki, N. (1957) Histochemical studies on the lesions of rice blast caused by *Pyricularia oryzae* Cav. III. Changes in the respiration of infected tissues. *Annals of the Phytopathological Society of Japan* 22, 173–177.

Tsai, W.H. (1988) Estimation of rice yield losses caused by panicle blast disease. *Journal of Agricultural Research of China* 37, 86–90.

Yeh, W.H. and Bonman, J.M. (1986) Assessment of partial resitance to *Pyricularia oryzae* in six rice cultivars. *Plant Pathology* 35, 319–323.

Waggoner, P.E. and Berger, R.D. (1987) Defoliation, disease and growth. *Phytopathology* 77, 393–398.

Zadoks, J.C. (1961) Yellow rust on wheat, studies on epidemiology and physiologic specialization. *Tijdschrift over Plantenziekten* 67, 69–256.

Zadoks, J.C. (1973) Modelling the models. *Introduction to Symposium: Developing Models for Measuring Crop Losses (handout)*. 2nd International Congress of Plant Pathology, Minneapolis.

23 Methodology for Quantifying Rice Yield Effects of Blast

H.O. PINNSCHMIDT,[1] P.S. TENG[1] AND LUO YONG[2]

[1]*Entomology and Plant Pathology Division, International Rice Research Institute, PO Box 933, 1099 Manila, The Philippines;* [2]*Present address: Department of Plant Pathology, University of Arkansas, 217 Plant Science, Fayetteville, AR 72701, USA*

INTRODUCTION

Rice blast disease caused by *Pyricularia grisea* has long been considered as a serious constraint to higher rice yields in both tropical and temperate regions (e.g. Barr *et al.*, 1975; Ou, 1985). Some recent surveys (Widawsky and O'Toole, 1990; Geddes and Iles, 1991) confirm that blast remains among the most serious biotic constraints to yield in South Asia. Most yield loss estimates are at the level of region or country, while little information can be found on actual losses in farmers' fields (Teng *et al.*, 1990).

The efficient application of any disease management program hinges on having accurate and precise information on disease intensity and on the relationship between disease intensity and yield loss (Nutter, 1990). Whether for regional planning purposes at the policy level or for on-farm decision-making in practical disease management, appropriate methodology for quantifying yield effects of blast must be developed. In this chapter we will review approaches to the quantification of yield effects from blast. We will also present our own results obtained from field studies conducted from 1990–1991 at the International Rice Research Institute (IRRI) in the Philippines.

BLAST EFFECTS: HOW YIELD IS LOST

Quantification of yield effects of any disease requires information on specific mechanisms of disease damage effects and pathways of disease–host interaction. The most spectacular yield reducing effects are caused by panicle blast which reduces the 1000 grain weight, the percentage of ripe spikelets, and the percentage of fully mature grains (Goto, 1965; Prabhu and de Faria, 1982). It also increases the fraction of chalky kernels, thus lowering the grain quality (Katsube and Koshimizu, 1970). Moreover, panicles infected near the base ('neck') may break causing complete yield loss (Ou, 1985). The amount of loss from panicle blast increases as the position of invasion extends toward larger branches and toward the panicle base, with 'neck blast' being the most severe (Goto, 1965). The earlier the time of panicle infection, the greater the losses (Goto, 1965; Prabhu and de Faria, 1982). Infection of the stem nodes may cause the death of all plant parts above the point of infection. It may also cause a complete yield loss of the affected tiller when the stem breaks at the infected node (Ou, 1985).

Panicle and node blast is usually preceded by leaf blast which causes a reduction in plant height and tiller number, and ultimately, a reduction in the number of grains per panicle and in the 1000 grain weight (Goto, 1965; Prabhu and De Faria, 1982). Goto (1965) also noted that the yield-reducing effect of leaf blast was greater than would have been expected from the fraction of lost leaf area alone. Bastiaans (1991a) was recently able to explain this phenomenon by the fact that leaf blast lesions reduce the net photosynthetic rate of individual leaves to an extent far beyond the visibly diseased leaf fraction. Other recent studies indicated that leaf blast increases plant respiration (Sun *et al.*, 1986; Bastiaans, 1991b) and reduces the maximum photosynthetic rate at light saturation and the initial light use efficiency (Bastiaans, 1991b).

Collar blast, the infection of the junction between leaf blade and leaf sheath, is a fourth and often overlooked type of blast which may have overall effects on yield similar to those caused by severe leaf blast. In particular, collar-blasted leaves seem to have a shorter life span than unaffected leaves (H.O. Pinnschmidt, 1986, unpublished).

METHODOLOGY FOR QUANTIFYING YIELD EFFECTS OF BLAST: A REVIEW

The basic question behind any effort to quantify yield effects of pests and diseases is: how do given levels of disease affect final yield and/or its components? To answer this question requires: (i) disease quantification; (ii) collection of data to measure the relationship between disease level and yield loss; and (iii) modeling of the disease-loss relationship (*sensu* Teng, 1990).

Disease assessment and the anatomy of blast epidemics

Since the effect that a lesion will have on host plant physiology will depend on where on the plant that lesion is located, it seems sensible to subdivide the 'graph of the epidemic' (Kranz, 1974) into the components that describe the epidemic progress on different plant organs. Typical blast epidemics can thus be divided into subepidemics on leaf blades, leaf collars, and panicles including nodes. Leaf blast usually starts at seedling or early tillering, reaching a maximum at around maximum tillering, and declines thereafter. Typically, its disease progress curve (DPC) is unimodal. Collar blast usually sets in later than leaf blast and displays a unimodal- to sigmoidal DPC. Panicle blast becomes visible not before *c.* 10 days

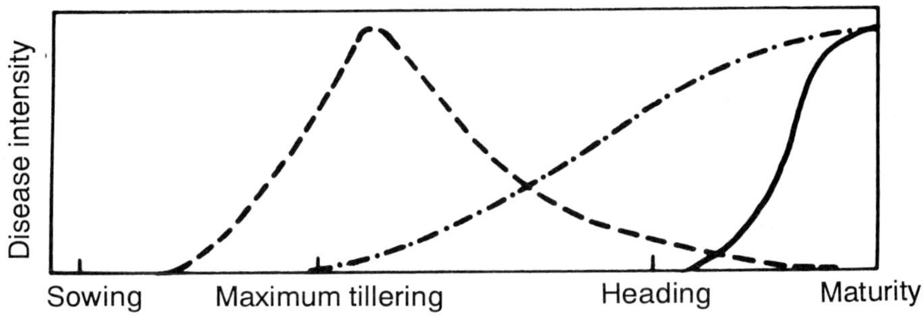

Fig. 23.1. Anatomy of blast epidemics. – –, leaf blast; – · – ·, collar blast; —, panicle and node blast.

after heading. Its DPC is sigmoidal (Fig. 23.1, after Pinnschmidt *et al.*, 1993b). In fact, panicle blast can be further subdivided into blast on different panicle parts, such as uppermost internode, panicle nodes, main axis (=rachis), branches of first and higher order, and spikelets including their pedicles and glumes (Nakamura, 1972; Pinnschmidt, 1989).

Leaf blast

Leaf blast severity can be assessed most conveniently by using either an ordinal scale like that published by IRRI (1988) or by estimating the percentage of diseased leaf area (% DLA) directly. The IRRI scale uses ten classes (0–9) that are assigned to plant samples having corresponding degrees of leaf blast infection. In this scale, a value of 0 represents a blast-free foliage, while a value of 9 corresponds to 75% DLA or more and intermediate values represent intermediate degrees of DLA. A disadvantage of using ordinal scales is that a given class represents a relatively wide range of DLA. Directly assessing % DLA should yield higher precision estimates. Diagrams of different disease severity levels can help standardize direct assessments (Fig. 23.2). Surin *et al.* (1991) found that the DLA of leaf number three (counted from the top) represents a good estimate for the total DLA of a sample. If a higher level of precision is desired, one might determine the blasted leaf area of a sample by measuring total leaf area, then counting the number of leaf blast lesions belonging to different size categories using a key proposed by Pinnschmidt *et al.* (1993a; Fig. 23.3). A very precise estimate of % DLA can then be computed as follows:

$$Y = 100((a_1X_1 + \ldots a_nX_n)/L)$$

where Y = mean % DLA of the sample, $a_1 \ldots a_n$ = mean lesion size of different size classes, X_n = number of lesions falling into the respective size classes, and L = total leaf area of the sample. The method is objective and gives reproducible results, but is very time and labor intensive. One should be cautious in using indirect estimates of % DLA, such as disease–incidence relationships, because the variance increases tremendously with increasing disease incidence (Disthaporn, 1987; H.O. Pinnschmidt, 1988, unpublished).

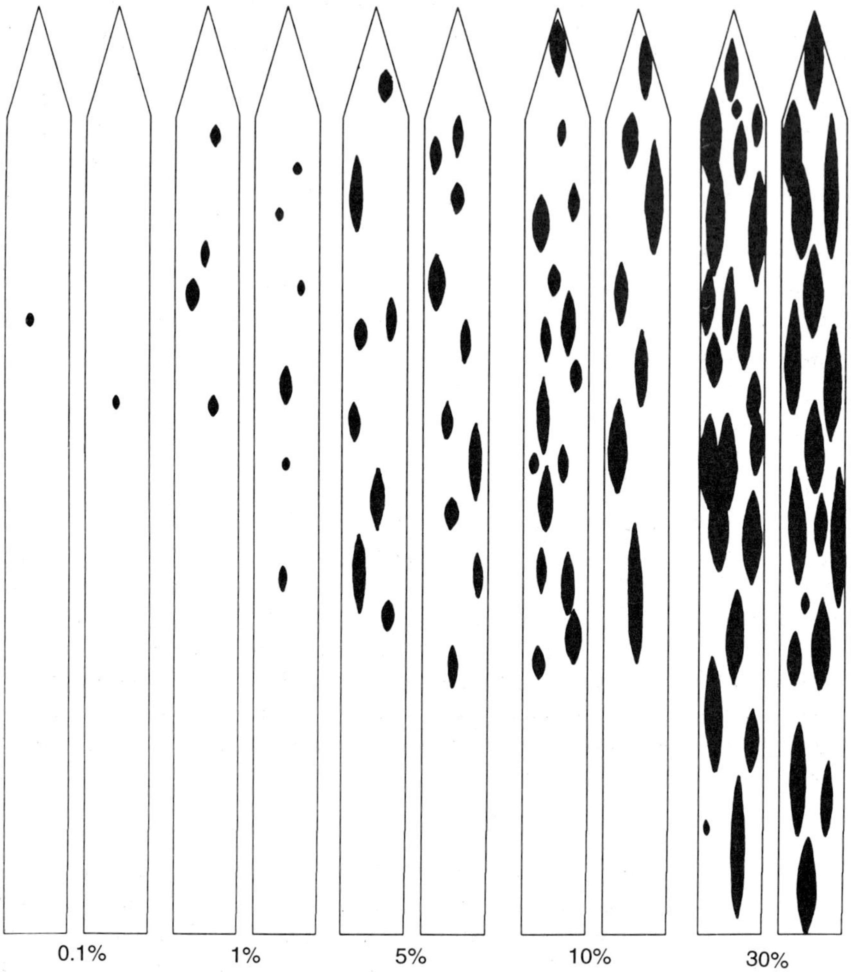

Fig. 23.2. Standard diagram for leaf blast assessment.

Panicle blast

Even though the degree of panicle blast severity is the most meaningful parameter for quantifying yield losses, the usual way to assess panicle blast in yield loss studies has been percentage blasted panicles, i.e. panicle blast incidence (e.g. Kuribayashi and Ichikawa, 1952; Padmanabhan, 1965). The method is fast and objective. Others (Izumi, 1955; Goto, 1965; Ahn and Mukelar, 1986; IRRI, 1988; Tsai, 1988b) have tried to provide better descriptors for the average degree of panicle blast severity and differentiated between partly and completely dead panicles or between panicles showing different degrees of grain filling inhibition or having infections on different portions. For example, Tsai (1988b) computed the mean panicle blast severity scale (x) of a sample as follows:

$$x = [5n_1 + 3n_2 + 2n_3 + 1n_4 + 0(n_5 + n_6)]/N$$

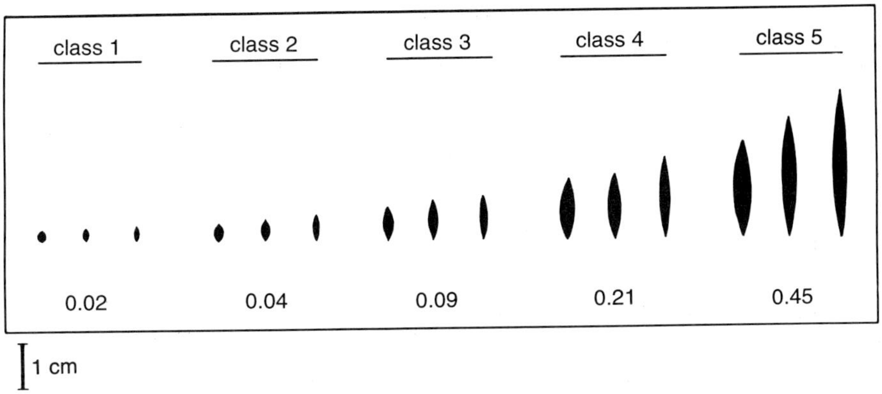

Fig. 23.3. Assessment key for leaf blast lesions. Lower numbers represent cm^2.

where n_1 = number of panicles damaged by 100%, n_2 = number of panicles damaged by $\geq 66.7\% < 100\%$, n_3 = number of panicles damaged by $\geq 50\% < 66.7\%$, n_4 = number of panicles damaged by $\geq 33.3\% < 50\%$, n_5 = number of panicles with necrotic lesions but undamaged grains, n_6 = number of healthy panicles, and N = total number of panicles. A fairly precise and meaningful, though costly, measure for panicle blast severity can be obtained when panicles belonging to different severity classes are counted. The mean panicle blast severity of the sample can then be determined by computing a weighted mean of these classes:

$$Y = (a_1X_1 + \ldots a_nX_n)/\Sigma X_n$$

where Y = mean panicle blast severity of the sample, $a_1 \ldots a_n$ = mean severity of different classes, and X_n = number of panicles of the sample falling into the respective severity class (Ahn and Mukelar, 1986; IRRI, 1988). Even though parameters for panicle blast incidence might be highly correlated with panicle blast severity (Goto, 1965; IRRI, 1986), the coefficient describing the incidence–severity relation might change across data sets (H.O. Pinnschmidt, 1986, unpublished). This will affect the reliability of using panicle blast incidence assessments for purposes of yield loss prediction.

Collar blast

We propose that collar blast be assessed by counting the number of collars in a sample that fall into certain severity classes and relating it to the total number of leaves of the sample:

$$Y = (a_1X_1 + \ldots a_nX_n)/L.$$

Here, Y = mean collar blast severity of the sample, $a_1 \ldots a_n$ = mean severity of different classes, X_n = number of leaves of the sample falling into the respective severity class, and L = total number of leaves of the sample. However, it might be desirable to adjust for the fact that lesions on collars of small leaves will probably have a smaller relative impact on final yield than lesions on big leaves. This can be accounted for by:

$$Y = [\Sigma(a_n L_n)]/L$$

where Y = mean collar blast severity of the sample, a_n = collar blast severity of a given leaf n of the sample, L_n = leaf area of leaf n, and L = total leaf area of the sample. A severity value of 100 would mean that the collars of all leaves of a sample are completely covered by blast lesions while lower severity values have to be interpreted in relation to this standard.

Other, more sophisticated, methods for blast assessment such as video image analysis (Lindow and Webb, 1983) and measurements of canopy reflectance (Pedersen, 1985) have not been tested. Sample size and sampling pattern are serious issues that have been addressed only recently. The sample size required to obtain a resonably good estimate of disease severity was inversely related to the disease incidence; however, the sample mean became fairly stable at a sample size of 40 and above (Disthaporn, 1987). These data suggest a minimum sample size of 10–20 observation units, but the sampling pattern, whether V-pattern, zig-zag, or diagonal, is not really important. However, the random sampling pattern usually yielded mean severity estimates that were free of bias.

Creating and collecting data on host and disease

Data on a wide array of disease scenarios (e.g. disease levels and temporal patterns) plus on corresponding parameters of crop response, such as final yield and/or yield loss, are essential for establishing a sound database in crop loss studies. We can generally subdivide available methods into those that rely on natural blast infestations and into field plot techniques that rely on manipulating blast infestation levels. We can furthermore distinguish, with respect to sampling techniques, whether single tillers, plants, hills, or whole plots are sampled and whether sampling is done at random or systematically. Disease levels should be assessed at experimentally determined frequencies specific for each crop stage.

The single-tiller/single-plant/single-hill method

In this method, many individual tillers, plants, or hills within a given naturally infested field are tagged as observation units covering a wide range of disease levels and patterns (Teng, 1990). Data on disease progress and final yield are collected for each undividual observation unit. The method has been employed by Prabhu and de Faria (1982), Disthaporn (1987), and Surin *et al.* (1991). This method has some advantages because it reflects natural conditions and is economical, but is limited by possible interplant differences in yield introducing high variation (Teng, 1990). This is especially problematic in upland rice where soil variation is very high and infection levels of individual plants might be influenced by their nutritional status. Under such conditions, well-nourished tillers might show higher blast levels than poorly nourished tillers, yet produce higher yields than those. The result can be a positive 'pseudo'-correlation between blast severity and yield.

Fields with natural epidemics

Goto (1965), Padmanabhan (1965), Prabhu and de Faria (1982), Disthaporn (1987), and Surin *et al.* (1991) obtained yield loss estimates by comparing yields of fields or plots having various degrees of natural infection. Padmanabhan (1965) also compared the yields in epidemic and non-epidemic areas. Some of these workers used tagged single tillers, plants, or hills, while others used other methods. Goto (1965) recognized that it is difficult to estimate a blast-free 'standard yield' with this method. The problem of pseudo-correlations might apply here, too.

Field plot techniques

These techniques aim to generate epidemics with a wide array of characteristics (Teng, 1990). A general advantage of field plot techniques over techniques that rely on natural infection is the fact that the researcher has some control over the range and characteristics of the epidemic patterns and that pseudo-correlations can be eliminated. The use of non-replicated 'response surface' designs that emphasize treatment number rather than replication number has been suggested by Teng and Oshima (1983), because replicated standard experimental designs cannot explain the full range of interactions between disease intensity, crop yield, and crop development stage. Also, non-replicated designs match better the statistical methods (e.g. regression techniques) commonly used to quantify disease–loss relationships (Teng, 1990).

Knowledge of some epidemiological principles as outlined by Nutter (1990) can help to maximize the efficiency of field plot techniques. Treatments manipulate the epidemiological parameters I (amount of inoculum), R (efficacy of the inoculum), r (apparent rate of disease increase), y (current level of disease), and t (time). Methods to manipulate these parameters include inoculation techniques which affect I, y, and t, use of spreader rows in connection with pesticide use to affect R and r, manipulation of the environment which usually affects r, use of fungicides to affect I, R, y, and/or r, and host resistance to manipulate R and r (Nutter, 1990). Padmanabhan (1965) used varieties with different degrees of resistance and applied fungicides to obtain different epidemics and study yield losses. Others used fungicides in combination with inoculation techniques and/or environmental control to achieve control over the epidemics to be generated. For example, Prabhu and de Faria (1982) inoculated plots with conidia suspensions at different times while protecting others with spray applications of benomyl at 10-day intervals. At IRRI (IRRI, 1989) blast epidemics were manipulated by applying different amounts of inoculum in the form of chopped blast-infected leaves, timing of inoculation, covering some of the plots with plastic sheets at night, and spraying tricyclazole. Kuerschner *et al.* (1992) exposed infected seedlings in some plots while protecting the control plots by seed treatments with pyroquilon plus spray applications with tricyclazole four to five times during the growing season. Infected seedlings were also exposed by Pinnschmidt (1989) to initiate epidemics. In other studies, Pinnschmidt and Teng (1994) covered plots with plastic sheets at night after applying conidia suspensions obtained either from blast cultures or by washing off conidia from infected leaves after incubating the leaves for 12 hours in a moist chamber.

Specific considerations arise for the use of most of these techniques. The genetic backgrounds including yield potentials and agronomic characteristics of varieties used in loss studies should ideally be identical except for blast resistance genes (Padmanabhan, 1965). Nutter (1990) remarked that varying the number and proportions of component lines in a multiline approach may provide a useful tool for yield loss research. It is important that the fungicide used does not have significant side-effects on crop performance (Padmanabhan, 1965). Inoculation by means of exposure of infected seedlings is probably the most 'natural' and economical inoculation technique, but it is not as efficient as applying inoculum suspensions. On the other hand, the more efficient an inoculation technique and the more uniform an infection resulting from an inoculation, the more the natural compensation mechanisms of a crop will be altered in comparison to a crop suffering from a natural epidemic. Therefore, Nutter (1990) argues that yield losses estimated on the basis of uniform infections resulting from artificial inoculations may overestimate actual losses. Working with an airborne pathogen such as blast requires a rather large plot size (Teng, 1990). While an increase in plot size is associated with a reduced inherent yield variation, the variation due to soil factors might increase. At IRRI (IRRI, 1989; Pinnschmidt and Teng, 1993), we use plots of 4 × 8 m or 5 × 5 m that are surrounded by 'buffer rows' or alleys of *c.* 2 m.

Modeling Yield Effects of Blast: Empirical versus Mechanistic Approaches

Yield losses may be quantified empirically or mechanistically. Teng *et al.* (1991b) provide a detailed review on work done on both approaches in blast yield loss research. In the empirical approach, one single equation or computational rule, usually obtained through regression techniques, relates parameters characterizing the intensity and/or temporal patterns of an epidemic, sometimes in combination with crop variables, to the corresponding expected yield or yield loss. In the simplest case, this translates into:

$$Y' = Y(1 - f(x)) = Y - f(x)Y$$

where Y = potential disease-free yield, x = some measure of disease intensity respectively diseased host fraction, and $Y' = Y$ adjusted for disease effects. Consequently,

$$y = Y - Y' = f(x)Y$$

where y = yield loss. In regression approaches, the corresponding regression models will have the form:

$$Y' = a - bx$$

where a = a constant and b = loss equivalent. Consequently,

$$y = bx.$$

Such 'empirical damage functions' or 'pest-loss equations' quantify the overall effects of pests or diseases on crops without considering specific features of host–pathogen interactions (Pinnschmidt *et al.*, 1990a). Single predictor models use only one independent variable, and multiple predictor models use several independent variables as predictors for yield or yield loss. Within these two categories, we may distinguish between single critical point models that make predictions based on disease intensity at a certain critical crop stage, and multiple critical point models that make predictions based on disease intensities at various crop stages. Multiple predictor models might include variables such as multiple pest intensities, crop stage, etc. in a predictive equation. Area-under-disease-progress-curve (AUDPC) models integrate the total damage that a crop suffered during its lifetime, while healthy-area-duration (HAD), and healthy-area-absorption (HAA) models integrate the total healthiness and/or light interception. They thus certainly add biological realism and flexibility to the empirical approach (see reviews by Madden, 1983; Teng, 1985, 1987; Johnson, 1987; Waggoner and Berger, 1987; and Teng and Johnson, 1988; Waggoner, 1990).

However, most empirical blast-loss models are very simple single-predictor regression equations (single critical point models) where the regression coefficient represents the 'loss equivalent' (= units of yield loss per unit of disease) and the independent variable represents the disease level at a critical crop stage. Examples of such equations were presented by Kingsolver *et al.* (1984), Disthaporn (1987), Tsai (1988a), and Surin *et al.* (1991) for leaf blast and by Kuribayashi and Ichikawa (1952), Goto (1965), Padmanabhan (1965), Katsube and Koshimizu (1970), Klochko *et al.* (1985), and Tsai (1988b) for panicle blast. Other approaches consider weather conditions (Kingsolver *et al.*, 1984) and time of infection (Prabhu and de Faria, 1982) for predictions of yield or yield loss. Some researchers established multiple predictor models that use either disease severity plus crop variables (Disthaporn, 1987), leaf plus panicle blast severity (IRRI, 1989), or multiple pest severities including blast severity (Padmanbhan, 1965) as predictors. The predictive potential of the AUDPC of leaf blast was investigated only recently by IRRI (1989), but it was lower than that of maximum % DLA. Path coefficient analysis was used to study interrelations among leaf blast, panicle blast, yield loss, yield components, and parameters of rice growth (IRRI, 1989). Duve (1989) derived a path coefficient model which included leaf blast and sheath blight severity, as well as parameters of crop growth and plant nutritional status in explaining the variability of the yield components. This type of analysis can help elucidate the complex interrelationships of yield-affecting factors under field conditions, but it does not seem to have resulted in practicable methods for yield loss predictions so far.

The development of dynamic ecophysiological crop simulation models raises the possibility of predicting yield effects of diseases by mimicking their damage effects at the physiological level. This is the mechanistic approach. Crop simulation models compute yield or other parameters of crop performance based on linked equations that calculate the rate and state of relevant crop physiological processes over time as affected by environmental variables and other inputs. Damage effects of pests and diseases can be introduced into crop models via physiological coupling points (Boote *et al.*, 1983) that define which crop model variables will be affected by a given pest. Known pest levels over time, e.g. as depicted by DPCs, are translated into actual daily damage imposed on the respective coupling points by interpolating pest levels between observation dates and by means of damage equations that quantify the relationship between daily pest level and physiological damage effect on a respective day. Thus pest-coupled crop models explicitly quantify specific physiological pest damage effects on a daily basis and integrate them over time into yield losses. Obviously, this approach has great potential for a wide applicability but it also has to be accompanied by research that improves the qualitative and quantitative understanding of the physiological mechanisms of yield loss.

The first authors who quantitatively measured the physiological causes of *leaf blast*-induced yield effects were Sun *et al.* (1986) who reported increased plant respiration due to leaf blast and Bastiaans (1991a, b) who found a reduced net photosynthesis. Chen *et al.* (1991) integrated the findings of Sun *et al.* (1986) and Bastiaans (1991a, b) with the source code of the crop simulation model MACROS (Penning de Vries *et al.*, 1989) and found that the modified model could satisfactorily simulate the impact of leaf blast epidemics on growth and yield of rice. Similarily, Bastiaans (1991a) was able to demonstrate that yield predictions by the MACROS model improved considerably when the model accounted for the effects of leaf blast on photosynthesis. The stochastic *panicle blast* simulator developed by Ishiguro and Hashimoto (1988, 1989) simulates panicle blast progress as driven by host and weather factors. It computes the resulting daily yield loss rates, assuming blockage of grain sinks due to new infections of the various panicle parts and due to the growth of existing panicle lesions as the basic cause for final yield reduction. The simulation model CERES-rice (Alocilja and Ritchie, 1988) was used by Teng *et al.* (1989), Batchelor *et al.* (1992), and Boote *et al.* (1993) to explore possible consequences of *leaf blast* scenarios on final yield with respect to biomass. In addition to leaf blast damage, damage effects of *panicle blast* and/or *collar blast* were simulated with a multiple pest-coupled version of CERES-rice (Pinnschmidt *et al.*, 1990b, 1974; Teng *et al.*, 1990a; Pinnschmidt and Teng, 1994). These authors suggested a generic approach to simulating physiological damage effects of blast and other pests, using the following damage equation:

$$m_t' = m_t(1 - a_1x_{1t})^{b1} \ldots (1 - a_nx_{nt})^{bn}$$

where m_t = rate or state of a given crop physiological process or variable on a given day t, x_{nt} = host fraction diseased or damaged by pest n on day t, a_n = coefficient of damage equivalency corresponding to x_n, b_n = coefficient of

Table 23.1. Physiological damage effects assumed for different blast types in simulation runs of CERES-rice and corresponding estimates of parameters *a* and *b* for damage equations[a].

Blast type	Physiological damage effect assumed respectively simulated	Parameter estimates derived from: Good guess *a*	*b*	Optimization[b] *a*	*b*
Leaf blast	Photosynthesis reduction	1.00	3.04[c]	1.07	4.98
	Leaf senescence acceleration	0.00	0.00	0.57	5.82
Collar blast	Photosynthesis reduction (including blockage of assimilate translocation)	1.00	1.00	1.10	5.10
	Leaf senescence acceleration	0.00	0.00	0.52	1.49
	Xylem blockage (reduced root water uptake)	1.00	1.00	1.15	1.14
Panicle blast	Grain sink blockage (removal of grain sinks)	1.00	1.00	1.23	0.85

[a] Damage equations: $m_t' = m_t(1 - a_1x_{1t})^{b1} \ldots (1 - a_nx_{nt})^{bn}$; m_t = rate or state of crop physiological process or variable on day *t*, x_{nt} = host fraction diseased or damaged by blast type *n* on day *t*, a_n = coefficient of damage equivalency corresponding to x_n, b_n = coefficient of damage proportionality corresponding to x_n, $m_t' = m_t$ after accounting for blast effects. For leaf senescence (s_t): $s_t' = m_t(1 - (1 - s_t/y_t)(1 - a_1x_{1t})^{b1} \ldots (1 - a_nx_{nt})^{bn})$; m_t = healthy leaf area, s_t = 'normal' leaf senescence without blast, s_t' = total m_t senescing on day *t*.
[b] Pinnschmidt and Luo, 1993, unpublished.
[c] Bastiaans, 1991b.

damage proportionality corresponding to x_n, and $m_t' = m_t$ after accounting for pest damage effects. This model is slightly modified if the crop physiological process to be affected is leaf senescence (s_t):

$$s_t' = m_t(1 - (1 - s_t/y_t)(1 - a_1x_{1t})^{b1} \ldots (1 - a_nx_{nt})^{bn}).$$

In this case, m_t = the healthy leaf area, s_t = 'normal' leaf senescence without pest stress, and s_t' = total m_t that will senesce on a given day. Damage effects assumed for the three blast types were: net photosynthesis reduction and leaf senescence acceleration by leaf blast, net photosynthesis reduction including reduction of translocated assimilates, leaf senescence acceleration, and reduced root water uptake due to blockage of water translocation into leaves by collar blast, and reduced number of filling grain sinks due to panicle blast. While Pinnschmidt *et al.* (1994) obtained the parameter estimates for modeling these blast effects from 'good guesses', Pinnschmidt and Teng (1994) additionally used parameter values obtained by an optimization technique (H.O. Pinnschmidt and Y. Luo, 1993, unpublished; Table 23.1). The approach that utilized optimized parameter estimates gave the better yield predictions. In both approaches, the model reacted very sensitively to panicle blast. For a more comprehensive view on the use of pest-coupled crop models, see Loomis and Adams (1983), Rouse (1988), Teng (1988), and Teng and Johnson (1988).

Discussion

The number of publications dealing with quantifying blast yield effects by means of empirical damage functions (EDF) is huge. The advantage of such methods is that they are quite simple and easy to use. However, while useful for purposes such as quick regional or on-farm loss assessments under well-studied conditions, they do not seem to be 'transportable' across years, locations, varieties, and cultivation practices, because the regression coefficients of comparable equations showed large variations across datasets. Thus, for any possible combination of the major components of the agroecosystem (such as variety, cultivation practices, location, etc.), a separate EDF would have to be developed, with some possible interactions of the system components still not yet accounted for. Not much empirical work has been done on blast in combination with other pests and on leaf and panicle blast in combination. Nothing at all has been done on collar blast so far. Also, no multiple point or other models have been developed that possess higher degrees of physiological realism and provide more flexibility. These might include response surface models that use disease variables in connection with crop variables as predictors, AUDPC, HAD, and HAA models. Nevertheless, the empirical work gives important hints about possible yield effect scenarios triggered by blast and their interaction with other factors.

Even though final validation of the mechanistic approaches under diverse conditions has yet to come, the potential uses of these approaches are for adjusting yield expectations and crop simulation model outputs for blast damage effects and for developing decision aids such as iso-loss curves and damage thresholds for disease management. An advantage of mechanistic approaches in comparison to empirical approaches is their apparent applicability across years, locations, cultivation practices, etc. Coupling blast effects to crop simulation models via physiological coupling points probably provides a better way to account for the true nature of the damage effects of the disease on the rice plant than using empirical damage functions, since the physiological reality of host–pathogen interactions is being mimicked. However, the approach requires that exact quantitative information on all important pathways of disease-host interactions be available. The mechanistic approach also requires that the crop model be sophisticated enough to mimick how relevant physiological processes and variables interact with blast. The crop model therefore must have state and/or rate variables that are compatible with the corresponding blast effects to be simulated. Some specific types of crop response, such as active compensation, can not be handled by the current versions of MACROS and CERES. If these phenomena prove to be important, the crop model structures will have to be modified accordingly.

The empirical approach fully relies on field experimental techniques and the above discussion applies for such techniques. It particularly requires a wide array of blast epidemics with many treatments that should be well-distributed within the array. It requires careful field data collection and no sophisticated equipment beyond a computer and a balance. The mechanistic approach as exemplified by the work of Bastiaans (1991a,b) in contrast requires 'micro-level' studies of the mechanisms of blast damage under controlled conditions. These have a high explanatory value and aim to identify the relevant crop physiological processes and variables affected by blast and to quantify the stimulus-response relationships under a wide range of infection levels. This requires sophisticated laboratory equipment and/or delicate analytical procedures in most cases. Parameters derived from this type of research serve as the computational basis for simulating blast damage effects by means of blast-coupled crop models which have to be calibrated and validated at the 'macro-level' by running these models with field data. The field datasets needed for this purpose do not have to represent such a wide, well-distributed array of many different blast epidemics as is desired for empirical approaches, but a few 'cornerstone' treatments might suffice. In contrast to this, the approach of Pinnschmidt and Teng (1994) and H.O. Pinnschmidt and Y. Luo (1993, unpublished) is purely 'macro-level' based. It requires field data material and experimental techniques similar to those of the empirical approach, since it estimates the damage parameters directly from the field data based on comparing observed and simulated systems output. It clearly lacks the evidence and explanatory value of micro-level studies. Nevertheless, it might eventually arrive at similar projections of total blast yield effects.

A Comparison of Empirical and Mechanistic Approaches to Quantify Yield Effects of Blast

In view of the earlier discussed pros and cons for the empirical and the mechanistic approach, a comparison is in order. The experimental part of this work was originally designed to study multiple pest yield effects on rice and to calibrate the pest-coupled version of the CERES-rice crop simulation model (Pinnschmidt and Teng, 1994; Pinnschmidt *et al.*, 1994; H.O. Pinnschmidt and Y. Luo, 1993, unpublished).

Field experiments and data collection

An irrigated lowland field at the IRRI farm in Los Baños, Laguna, The Philippines, was planted to variety IR 72 in the wet season (WS) 1990, dry season (DS) 1991, and (WS) 1991. The crop received 150 kg N ha^{-1} in both wet seasons and 180 kg N ha^{-1} in the 1991 dry season. To keep the pest susceptibility of the crop high throughout the cropping season, N-application was split into five times at 2-week intervals, starting at transplanting. 25 kg P_2O_5, 25 kg K_2O, and 10 kg Zn ha^{-1} were applied in all experiments at transplanting. The field was subdivided into three parts, each representing one replication.

Three leaf blast treatments (zero, low, and high inoculum level) were randomized versus three insect treatments (no, low, and full insecticide protection) across nine plots for each replication, using a strip-plot design. One half of each plot was inoculated for panicle blast by randomizing three panicle blast treatments (inoculation at early, medium, and late heading stage) across these plots. One additional plot per replication remained untreated. In DS 1991, leaf blast and insect treatments were the same as in WS 1990. Additionally, three sheath blight treatments (zero, low, and high inoculum level) were randomized across the plots and one-half of each plot was inoculated with bacterial leaf blight. One-half of each of these sub-plots was inoculated for panicle blast using the same panicle blast treatments as in WS 1990. In WS 1991, four treatments for timing of sheath blight inoculation (maximum tillering, late tillering, early booting, and late booting stage) were randomized with two inoculum levels (low and high) across nine plots per replication. One plot per replication received no sheath blight inoculation. One-half of each plot was inoculated for panicle blast by randomizing three inoculation times (early, medium, and late) in combination with three inoculum levels across the plots of each replication.

To create leaf blast epidemics, plots were inoculated three to five times with conidia suspensions at early tillering stage. To obtain various panicle blast levels, sub-plots were inoculated with conidia suspensions of different densities at different times during or after heading. Inoculation was done at 1700–1900 h and blast-inoculated plots were covered with plastic sheets during the night following the inoculation. Plots not inoculated for leaf blast were sprayed with tricyclazole or benlate at the height of the leaf blast epidemics of the other plots. All plots of experiments dealing with leaf blast alone were sprayed with these fungicides before heading. Sheath blight epidemics were initiated by dispersing sheath blight-infested rice hulls in designated plots. Inoculated plots were sprayed with distilled water at 800 and 1800 h for seven consecutive days following the inoculation. Sheath blight-free plots were twice sprayed with validamycin after maximum tillering stage. Infestation with bacterial leaf blight was initiated in sub-plots of DS 1991 by spraying bacteria suspensions at early tillering stage. Phenazine was applied in control sub-plots three weeks later. Different infestation levels of insect pests, mainly stem borers, leaf folders, and whorl maggots, were created by applying thiodan, hostathion, or brodan at different intensities. Insect-free plots were treated weekly with insecticides.

Eight hills per sub-plot, subdivided into two sub-samples of four hills each, were destructively sampled in intervals of 10–14 days based on a stratified random sampling procedure. Dry matter, leaf area and leaf weight per 12 tillers, yield and yield components, and other crop data of these samples were determined in the laboratory. Severity levels of leaf, collar, and panicle blast were assessed as suggested above. Severity levels of other pests and damage types were also assessed.

Data analysis and modeling

For statistical/empirical analysis, using correlation and regression techniques, variables that served as indicators of crop performance were computed. These were: final yield of filled grains per unit area, yield components (number of panicles per unit area, number of grains per panicle, fraction of filled grains, and weight per filled grain), harvesting index (filled grain yield per total above-ground biomass), total growth duration in days, and healthy-area-duration (HAD) as well as healthy-area-absorption (HAA) (*sensu* Waggoner and Berger, 1987). Whenever analyses were pooled across experiments, all above-mentioned indicators of crop performance were standardized i.e. divided by the maximum value of the respective variable that occurred in a given experiment. For characterizing the intensity of leaf and collar blast infestation and other diseases and pest damage on the vegetative plant body, area-under-disease-progress-curves (AUDPC) were computed:

$$AUDPC_x = \Sigma\{[(d_{xt-1} + d_{xt})/2]\,dt\}$$

where d_{xt} = fraction of leaf area diseased or damaged by pest x on day t. For pooling of data across experiments, $AUDPC_x$ values were divided by the total growth duration in days. The result represents a standardized measure of apparent vegetative pest pressure due to pest x. A value of 1 would mean that the crop was 100% diseased by pest x throughout its entire lifetime. For characterizing the intensity of panicle blast infestation and other pest damage on reproductive plant organs, the maximum relative severity level of the respective pest was taken as a standardized measure of apparent pest pressure. For a given pest x, these standardized measures of apparent pest pressure were named 'fraction of diseased host' (FDH_x).

The pest-coupled version of the ecophysiological crop simulation model CERES-rice (Pinnschmidt and Teng, 1994; Pinnschmidt *et al.*, 1994) was used to simulate plot yields as affected by blast and other pests. Pest progress curves observed in individual plots were fed into the model where specific crop physiological damage effects of the respective pests were mimicked. The computation of pest-specific damage effects followed the procedure briefly described above and outlined in more detail by Pinnschmidt and Teng (1994) and Pinnschmidt *et al.* (1994). The assumed damage effects of leaf, collar, and panicle blast are listed in Table 23.1. Damage coefficients were derived through an optimization procedure (Klepper and Rouse, 1991) by H.O. Pinnschmidt and Y. Luo (1993, unpublished), using the three datasets presented here.

Estimation of the pest-free yield potentials (Y_m) of the experiments depended on the approach. For the empirical approach, Y_m = maximum observed plot yield of the respective experiment, while, for the mechanistic approach, Y_m = yield predicted by a pest-free simulation run of CERES-rice for a given experiment. For comparing the empirical with the mechanistic approach, a multivariate regression model was fitted to the apparent fraction of total yield losses (y_a) estimated per plot. These were computed as:

$$y_a = 1 - Y_o/Y_m$$

where Y_o = observed yield. The regression was forced through the origin and used pooled data of all experiments. Standardized measures of severity levels of all pests such as those described above (FDH_x) served as independent variables. Yield loss fractions (y_x) due to blast and/or other pests were computed by solving the multivariate regression equation only for the pest terms of interest and keeping all other terms zero. The results were re-converted into estimated actual yields (Y_x') as affected by different blast and multiple pest scenarios x:

$$Y'_x = Y_m(1 - y_x).$$

Parallel to this, blast and multiple pest affected yields were estimated by including the damage effects of the respective pests selectively in simulation runs of CERES-rice. The overall quality of prediction of the two approaches was compared by computing the coefficient of variation of the predictions ($\hat{e}_x$) of a given scenario x:

$$\hat{e}_x = [\Sigma(|Y'_x - Y_o|/Y_o)]/n$$

where n = number of plots. For further comparison, the mean estimated fraction of yield losses due to different blast types and other pests ($\hat{y}_x$) was computed for each of the two approaches:

$$\hat{y}_x = \Sigma(1 - Y'_x/Y_m)/n$$

Results and discussion

The three field experiments can be ranked regarding yield, most of the yield components, harvesting index, and HAA as follows: WS 1990 < WS 1991 < DS 1991 (Table 23.2). The relative variation of yield and yield components was comparatively high in WS 1990 which might be mainly due to the very high rat damage and the comparatively high levels of panicle blast, sheath blight, stem borer dead hearts and white heads, and defoliation in that experiment (Fig. 23.4). The blast epidemics observed can clearly be separated into epidemics of leaf, collar, and panicle blast. Leaf blast levels in WS 1990 and DS 1991 were very low, despite inoculation. Because of the lacking leaf blast inoculation in WS 1991, collar blast was low in that experiment and occurred only late in the season as a result of panicle blast inoculation (Fig. 23.4). Standardized measures of the epidemic level of the three blast types (expressed as 'fraction of diseased host' = FDH) that were used as independent variables in correlation and regression analyses are shown in Table 23.2. FDH probably allows a better comparison and judging of the relative importance of a given pest or damage type than do simple progress curves. For example, the maximum FDH observed in the three experiments ranged from 0.43 to 0.69 for panicle blast, while only 0.0 to 0.004 and 0.017 to 0.033 for leaf and collar blast, respectively.

As indicated by simple correlations between crop and blast variables, major yield effects were exerted by panicle blast via reducing the fraction of filled grains (FFG) and the filled grain weight (FGW; Table 23.3). Consequently, panicle blast also correlated highly negatively with harvesting index (HI). The correlations did not indicate high direct effects of leaf and collar blast on yield and yield components, but on healthy-area-duration (HAD) and healthy-area-absorption (HAA), even though FDHs of leaf and collar blast were low. Interaction terms among blast types did not increase the correlations considerably. While AUDPCs as described above refer to the whole canopy, they were also computed for the upper and lower canopy layers separately, but this increase in complexity did not improve the results of the analyses considerably. Also, results did not considerably improve when computations of AUDPCs and parameters of pest intensity on

Table 23.2. Crop performance and standardized measures of apparent blast pressure[a] in three experiments.

Variable	Experiment	Mean	Standard deviation	Minimum	Maximum
Crop performance					
Filled grain yield (kg ha^{-1})	WS 1990	2027	974	531	5028
	DS 1991	5326	1179	2815	7817
	WS 1991	4478	630	2992	5461
Panicles per m^2	WS 1990	329.3	122.0	157.5	899.0
	DS 1991	570.3	92.2	350.0	862.5
	WS 1991	497.4	64.1	358.3	645.8
Grains per panicle	WS 1990	61.4	9.6	41.3	85.3
	DS 1991	62.6	9.0	44.5	87.4
	WS 1991	51.6	5.9	43.2	72.7
Fraction of filled grains	WS 1990	0.511	0.117	0.266	0.763
	DS 1991	0.759	0.095	0.543	0.937
	WS 1991	0.753	0.057	0.593	0.863
Weight per filled grain (g)	WS 1990	0.019	0.002	0.015	0.024
	DS 1991	0.020	0.002	0.015	0.025
	WS 1991	0.023	0.001	0.021	0.028
Harvesting index[b]	WS 1990	0.309	0.098	0.121	0.521
	DS 1991	0.439	0.056	0.268	0.569
	WS 1991	0.544	0.100	0.276	0.692
Healthy area duration[c]	WS 1990	154.9	29.7	91.3	99.9
	DS 1991	222.2	33.0	151.1	296.5
	WS 1991	297.3	30.8	247.7	375.2
Healthy area absorption[d]	WS 1990	593.6	59.7	446.5	681.0
	DS 1991	1134.8	76.8	927.3	1294.8
	WS 1991	921.8	38.0	866.4	1003.7
Total growth duration (days)	WS 1990	115.0	0.82	114.0	116.0
	DS 1991	111.0	0.82	110.0	112.0
	WS 1991	115.7	0.48	115.0	116.0
Blast pressure					
Leaf blast	WS 1990	0.0006	0.0007	0.0000	0.0027
	DS 1991	0.0011	0.0010	0.0000	0.0042
	WS 1991	0.0000	0.0000	0.0000	0.0000
Collar blast	WS 1990	0.0057	0.0043	0.0100	0.0168
	DS 1991	0.0076	0.0069	0.0000	0.0330
	WS 1991	0.0016	0.0021	0.0000	0.0098
Panicle blast	WS 1990	0.0404	0.0795	0.0000	0.4299
	DS 1991	0.2119	0.2393	0.0000	0.6884
	WS 1991	0.1172	0.1690	0.0000	0.5939

[a] i.e. maximum possible value = 1; standardized measures of blast pressure are expressed as 'fraction of diseased host' (FDH); see previous section of text for description.
[b] Harvesting index = grain yield/total above-ground biomass.
[c] See previous section of text for computational procedure.
[d] See previous section of text for computational procedure.

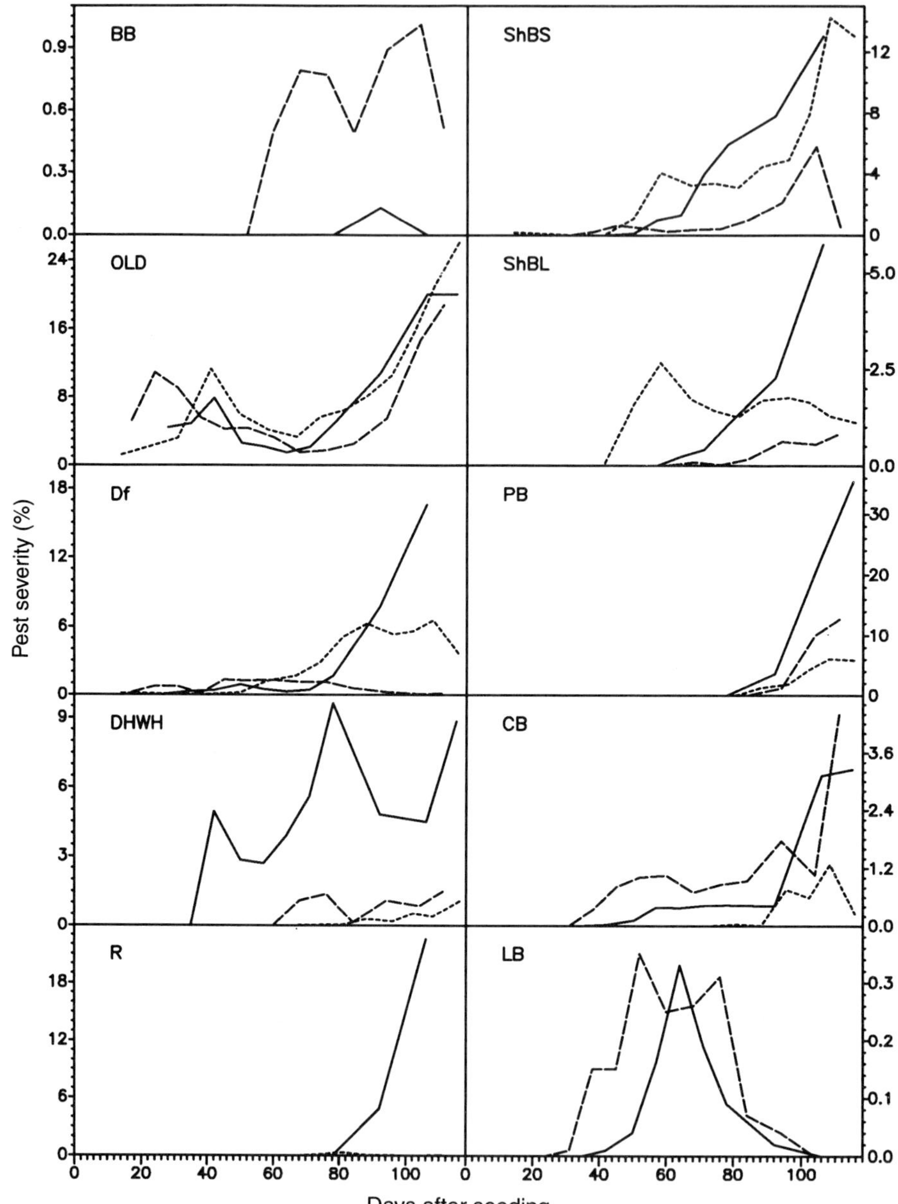

Fig. 23.4. Mean disease or damage progress curves of multiple pest species in three field experiments - - -, wet season 1991; — —, dry season 1991; ——, wet season 1990. LB = leaf blast, CB = collar blast, PB = panicle blast and other panicle diseases, ShBL = sheath blight on leaf blade, ShBS = sheath blight on leaf sheath, DHWH = dead hearts and white heads due to stem borers, R = rat damage, Df = defoliation, BB = bacterial leaf blight, OLD = other diseases and disorders of the leaf blade.

Table 23.3. Correlations[a] between severity levels[b] of different blast types and parameters of crop performance for pooled data of three experiments.

	Yield	PAN	GPP	FFG	FGW	HI	HAD	HAA
Leaf blast (LB)	−0.15	−0.02	−0.13	−0.08	−0.19*	0.08	−0.44**	−0.45**
Collar blast (CB)	−0.22**	−0.06	−0.22**	−0.15	−0.22**	0.01	−0.33**	−0.36**
Panicle blast (PB)	−0.57**	−0.23**	−0.06	−0.63**	−0.65**	−0.45**	−0.06	−0.12
LB · CB	−0.08	0.09	−0.19*	−0.03	−0.16*	0.15	−0.38**	−0.36**
LB · PB	−0.35**	−0.07	−0.11	−0.37**	−0.49**	−0.21*	−0.30**	−0.33**
CB · PB	−0.48**	−0.15	−0.15	−0.53**	−0.61**	−0.34**	−0.16*	−0.21**
LB · CB · PB	−0.24**	0.06	−0.17*	−0.27**	−0.40**	−0.09	−0.24**	−0.25**
HAD	0.21**	0.13	0.08	0.04	0.16*	−0.04	1.00**	0.92**
HAA	0.38**	0.28**	0.10	0.18*	0.21**	0.12	0.92**	1.00**

Yield = grain yield (kg ha^{-1}), PAN = panicles per m^2, GPP = grains per panicle, FFG = fraction of filled grains, FGW = weight per filled grain (g), HI = harvesting index, HAD = healthy area duration (days m^2/m^2), HAA = healthy area absorption (days (MJ m^{-2}/d^{-1})).

[a] Expressed by Pearson's correlation coefficient; * = significant at $P = 0.01$; ** = significant at $P = 0.001$.

[b] Measures of severity for different blast types in terms of FDH (= fraction diseased host), see previous section of text for description.

Table 23.4. Multiple regression equation and test statistics for relating fraction of yield loss[a] to severity levels of multiple pests. Pooled data from three experiments with all variables standardized and regression forced through origin.

Variable respectively term	Regression coefficient B	Standard error of B	Significance of T value
CB · DF · OLD	7901.962	1706.410	0.000
CB · RG	−42.910	13.571	0.002
CB · WH	−39.576	16.222	0.016
$DF^{0.5}$	−0.588	0.213	0.007
LB	−329.406	82.616	0.000
$LB^{0.5}$	12.639	2.539	0.000
LB^2	45956.233	13284.610	0.001
OLD^2	27.135	4.834	0.000
$PB^{0.5}$	0.421	0.031	0.000
RG^2	−1.129	0.335	0.001
RG · OLD	10.007	2.760	0.000
$SHB^{0.5}$	−1.520	0.322	0.000
SHB^2	532.518	88.633	0.000
SHB · SHD · OLD	−4013.809	855.747	0.000
SHD	6.328	0.921	0.000
WH	0.786	0.127	0.000

Adjusted $R^2 = 0.939$; $N = 219$; significance of F value = 0.000.
CB = collar blast, DF = defoliation including leaf folding, LB = leaf blast, OLD = other leaf diseases, PB = panicle blast, RG = rat damage at generative phase, SHB = sheath blight on leaf blade, SHD = sheath blight on leaf sheath, WH = white heads; all variables in terms of FDH (= fraction diseased host); see previous section of text for description. For PB, RG, and WH, FDH is expressed as maximum relative severity level.

[a] See previous section of text for description.

reproductive plant organs included interaction terms with crop age. Such interaction terms take into consideration that crop response to damage might change with crop age as suggested by Shaw and Royle (1987). The results presented here therefore represent only those with the highest explanatory value at the lowest justifiable level of analytic and computational complexity.

The multivariate regression model that was then fitted to the pooled standardized yield loss data became very complex and consisted of terms that considered all pests in the equation. Nevertheless, all variables in the equation were highly significant (Table 23.4). Negative terms in the equation cannot be explained biologically, but have to be considered as correction terms specific for these data.

Yields as affected by different blast types and other pests were estimated with the CERES-rice model and with the multivariate regression model. As Fig. 23.5 shows, the simulation model overestimated observed yields by far more than the regression model, when the estimation referred to a pest-free crop. However, as damage effects of the different blast types and the remaining other pests were

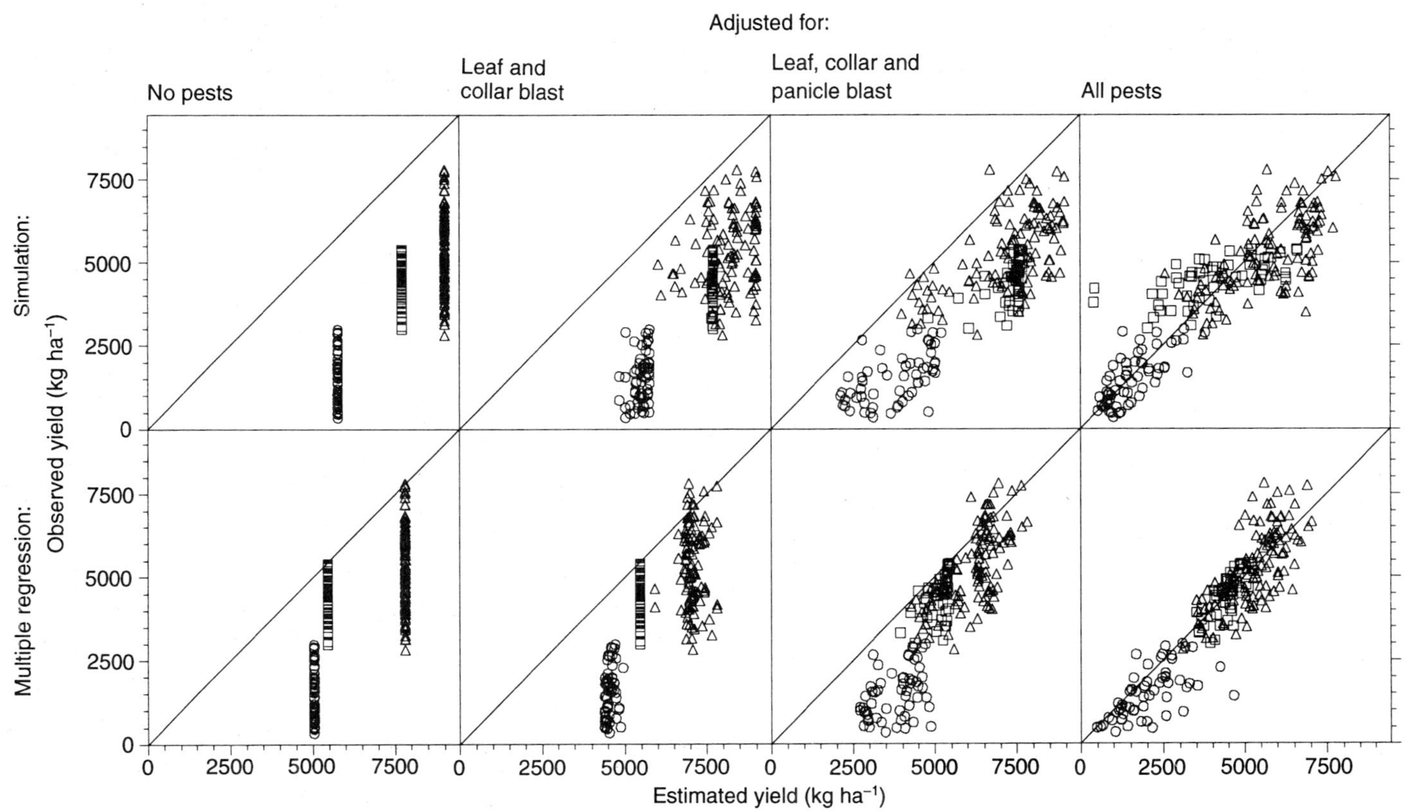

Fig. 23.5. Observed yields versus yields predicted either by the multiple pest-coupled version of CERES-rice (upper portion of figure) or by a multivariate regression model (lower portion of figure). Yield predictions were adjusted for damage effects of different blast types and/or other pests. ○, wet season 1990; △, dry season 1991; □, wet season 1991.

subsequently considered in the simulation, the yield estimates moved from overprediction to a fairly good prediction in a parallel-shift manner. In contrast to this, adjustments of yields predicted by the regression model did not proceed in a strictly parallel-shift manner. Also, the regression model was not able to adjust for collar blast alone or in combination with other blast types, because collar blast entered the regression equation as an interaction term with other pests only. Nevertheless, when adjusting for all pests, it gave a better fit for yield estimates than the simulation model. Obviously, leaf and collar blast, because of their relatively low severity levels, were of minor importance for adjusting yield predictions with either approach, compared to the importance of panicle blast and other pests.

The coefficient of variation of the predictions served as a measure of the performance of the two approaches since it indicates the error of the prediction. As Table 23.5 shows, the coefficient of variation of pooled data could be reduced by almost 50% from 1.67 to 0.92 for the simulation approach and from 1.25 to 0.75 for the regression approach when all blast types were considered in the prediction. This reduction was mainly due to panicle blast. However, when pests other than blast were included in the predictions, the coefficient of variation was 0.56 and 0.42 for pooled data of the two approaches. It dropped down to 0.25 and 0.24 respectively when all pests were considered in the predictions. This indicates that, for the given dataset, pests other than blast had a higher impact on yield than all blast types together. It also indicates that both approaches perform comparably as regards the mean error of the estimate. The data material of WS 1990 was the 'noisiest' one among the three datasets and contributed most to the coefficient of variation of the pooled data. However, the above-stated holds true even for the individual experiments.

The yield loss estimates of the two approaches differ quite a bit. For example, while the simulation estimates a mean fraction of yield loss due to all pests of 0.48 for pooled data, the corresponding figure for the regression is 0.36 (Table 23.5). This can be due either to the fact that the simulation overpredicts pest-free yields or that the regression underpredicts them. Both will result in a biased yield-loss estimate. Since none of the plots was 100% pest or damage free, the regression approach can, strictly speaking, not be applied for estimating pest-free yields and actual yield losses. However, both methods indicate higher yield losses due to pests other than blast. In both cases, panicle blast dominates blast-induced yield losses and the ranking of loss estimates for the different blast types and other pests is similar. Moreover, antagonistic interactions among pests and blast types are indicated by both methods since loss estimates for combined pest infestations are always less than the sum of individual loss estimates.

The simulation model was used to project yield consequences of panicle blast scenarios. Specified disease severities were assumed to set in at specified onset times and to last until maturity. Optimized parameter estimates of H.O. Pinnschmidt and Y. Luo (1993, unpublished; Table 23.1) were used for damage-effect computations in the model. From the results, iso-loss lines were created that indicate equal yield losses resulting from different combinations of onset time and severity level (Fig. 23.6). Such iso-loss lines can be of direct help

Table 23.5. Coefficients of variation of the relationship between estimated and observed yields and mean estimated fraction of yield losses. Estimates are based on either a pest-coupled version of the CERES-rice simulation model or a multivariate regression model and are adjusted for damage effects of different blast types and other pests[a].

	Coefficient of variation as affected by pests whose damage effects were adjusted:					
Experiment	None	Panicle blast	Leaf, collar, and panicle blast	All pests	All pests except blast	
Simulation model:						
All	1.67	1.03	0.92	0.25	0.56	
WS 1990	4.18	2.32	2.15	0.38	1.05	
DS 1991	0.78	0.53	0.39	0.18	0.47	
WS 1991	0.77	0.67	0.67	0.23	0.22	
Multivariate regression model:						
All	1.25	0.94	0.75	0.24	0.42	
WS 1990	3.52	2.75	2.30	0.60	1.00	
DS 1991	0.55	0.35	0.22	0.13	0.28	
WS 1991	0.25	0.17	0.17	0.08	0.10	
		Fraction of yield loss due to:				
		Panicle blast	Leaf, collar, and panicle blast	All pests	All pests except blast	Apparent total yield loss
Simulation model:						
All		0.15	0.20	0.48	0.34	0.50
WS 1990		0.32	0.34	0.72	0.57	0.74
DS 1991		0.12	0.20	0.36	0.17	0.41
WS 1991		0.05	0.05	0.47	0.44	0.42
Multivariate regression model:						
All		0.11	0.18	0.36	0.31	0.36
WS 1990		0.15	0.25	0.59	0.52	0.60
DS 1991		0.11	0.21	0.32	0.26	0.32
WS 1991		0.06	0.06	0.19	0.20	0.18

[a] For computational procedures, see previous section of text.

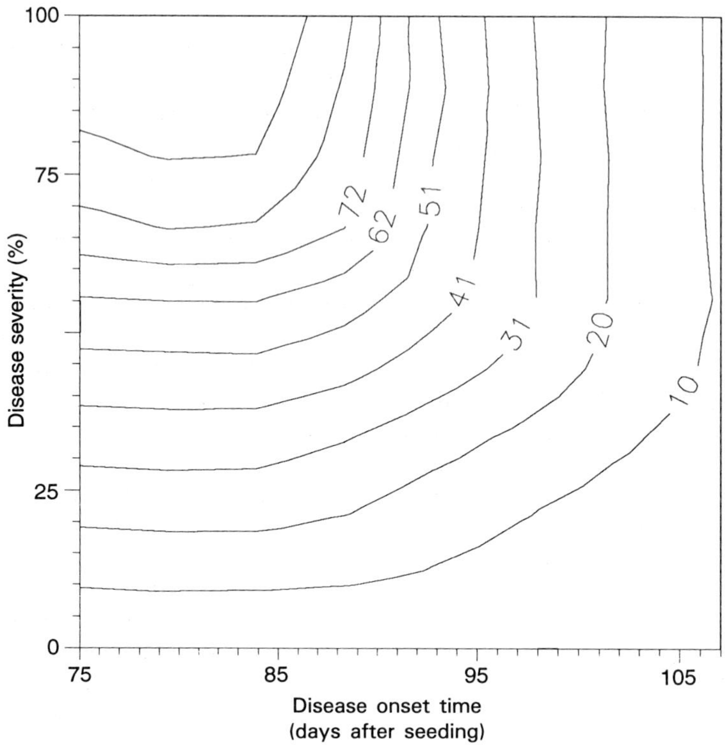

Fig. 23.6. Iso-loss curves depicting simulated yield losses due to panicle blast, depending on onset time and severity level of the disease. Yield losses were estimated with the pest-coupled CERES-rice simulation model. Specified severity levels were assumed to last from specified onset times until harvest. (Iso-loss curves were not computed for leaf and collar blast because of their low and narrow-ranged severity levels.)

for deriving control thresholds. As this example shows, the relationship between severity of panicle blast infestation and yield loss is not static, as suggested by published yield loss equivalents, but greatly depends on the time of infection. These scenario projections would not have been possible with the regression model, because the empirical approach reacted rather insensitively to the timing of blast infestations. For field applications, the approach has to be tested further, validated, and calibrated, using realistic site-specific input data for environment, crop management, cultivar, and epidemic patterns. For leaf and collar blast, datasets with a wider array of severity levels and epidemic patterns have to be obtained.

The empirical and the mechanistic approach gave similar overall results in this study: they both had similar errors of prediction, indicated similar rankings of yield constraints, and indicated less-than-additive yield effects of yield constraints. Obtaining a realistic estimate about the potential yield proved to be crucial for loss appraisal. Because there was no 100% pest-free control, as will usually be the case in field situations, our empirical approach probably underestimated potential yields considerably. Consequently, the magnitude of yield losses was probably underestimated, too. In contrast to this, a 'good' crop simulation model 'automatically' takes care of the potential yield estimate when running a pest-free simulation. The potential yield estimates of CERES-rice were probably realistic for conditions on the IRRI farm at given management practices. While the mechanistic approach is very complex, requiring considerable data input and model development, the empirical approach requires more data manipulation which represents a source of error. For example, deriving a general regression model for pooled data from different experiments requires data to be standardized in order to account for differences in yield potential, growth duration, etc. Because of the nature of regression techniques, the empirical approach will always be very sensitive to the data that were used to build an equation. The mechanistic approach can avoid that trap, provided the crop model used is physiologically detailed enough to mimick all relevant pathways of pest–crop–environment interactions realistically and that it was designed for and validated under the respective cropping conditions. A 'good' pest-coupled simulation model can handle damage effects of intercorrelated pests such as leaf and collar blast quasi-independently by implicitly taking care of interactions, while the empirical approach can not. This requires continued research efforts, especially on physiological processes affected by blast as well as on field validation and calibration. As new findings and/or discrepancies arise, the pest-crop-coupling approach might be subject to modifications. The empirical approach will certainly remain a useful tool for applications that cannot readily rely on simulation models, such as quick loss estimates 'on the spot' and for cross-checking the mechanistic approach, as in the examples presented here.

References

Ahn, S.W. and Mukelar, A. (1986) Rice blast management under upland conditions. In: *Progress in Upland Rice Research*. International Rice Research Institute, Los Baños, The Philippines, pp. 363–374.

Alocilja, E.C. and Ritchie, J.T. (1988) Upland rice simulation and its use in multicriteria optimization. *Research Report Series No. 1*. IBSNAT project, University of Hawaii, Honolulu, USA, 95 pp.

Barr, B.A., Koehler, C.S. and Smith, R.F. (1975) Crop losses of rice: Field losses to insects, diseases, weeds, and other pests. Report of the UC/AID Pest Management and Related Environmental Protection Project at the University of California, Berkeley, under Contract No. AID/csd 3296, to the USAID, 64 pp.

Bastiaans, L. (1991a) Quantifying the effect of rice leaf blast on leaf photosynthesis, crop growth, and final yield using an ecophysiological model of crop growth. In: *Rice Blast Modeling and Forecasting*. International Rice Research Institute, Los Baños, The Philippines, pp. 89–99.

Bastiaans, L. (1991b) Ratio between virtual and visual lesion size as a measure to describe reduction in leaf photosynthesis of rice due to leaf blast. *Phytopathology* 81(6), 611–615.

Batchelor, W.D., Jones, J.W., Boote, K.J. and Pinnschmidt, H.O. (1992) Assessing pest and disease damage with DSSAT. *Agrotechnology Transfer* 16, 1–8.

Boote, K.J., Jones, J.W. and Mishoe, J.W. (1983) Coupling pests to crop growth simulators to predict yield reductions. *Phytopathology* 73, 1581–1587.

Boote, K.J., Batchelor, W.D., Jones, J.W., Pinnschmidt, H. and Bourgeois, G. (1993) Pest damage relations at the field level. In: Penning de Vries, F., Teng, P. and Metselaar, K. (eds), *Systems Approaches for Agricultural Development*. Kluwer Academic Publishers, Dordrecht, pp. 277–296.

Chen Zhongxiao, Hu Guowen and Ma Jufa (1991) Modelling rice growth dynamics and the damage caused by leaf blast (*Pyricularia oryzae* Cavara). In: Penning de Vries, F.W.T., van Laar, H.H. and Kropff, M.J. (eds), *Simulation and Systems Analysis for Rice Production (SARP)*. PUDOC, Wageningen, pp. 305–309.

Disthaporn, S. (1987) Studies on sampling methods for rice disease in Thailand. PhD Thesis, University of Giessen, Germany, 118 pp.

Duve, T. (1989) Der Reisbrenner in Suedkorea (The rice blast disease in South Korea). PhD Thesis, University of Giessen, Germany, 250 pp.

Geddes, A.M.W. and Iles, M. (1991) The relative importance of crop pests in South Asia. *NRI bulletin* No. 39, 102 pp.

Goto, K. (1965) Estimating losses from rice blast in Japan. In: *The Rice Blast Disease*. Johns Hopkins Press, Baltimore, Maryland, pp. 195–202.

IRRI (1986) *Annual Report for 1985*. Los Baños, The Philippines.

IRRI (1988) *Standard Evaluation System for Rice*, 3rd edn. International Rice Testing Program, International Rice Research Institute, Los Baños, The Philippines, 54 pp.

IRRI (1989) *Annual Report for 1988*. Los Baños, The Philippines.

Ishiguro, K. and Hashimoto, A. (1988) Development of a simulation model for rice panicle blast epidemics. 1. Composition of the basic model. *Bulletin Fukushima Prefecture Agricultural Research Station* 27, 1–19.

Ishiguro, K. and Hashimoto, A. (1989) Recent advances in forecasting of rice blast epidemics using computers in Japan. *Tropical Agricultural Research Service* 22, 153–162.

Izumi (1955), cited in Goto, K. (1965) Estimating losses from rice blast in Japan. In: *The Rice Blast Disease*. Johns Hopkins Press, Baltimore, Maryland, pp. 195–202.

Johnson, K.B. (1987) Defoliation, disease, and growth: A reply. *Phytopathology* 77, 1495–1497.

Katsube, T. and Koshimizu, Y. (1970) Influence of blast disease on harvests in rice plant. 1. Effects of panicle infection on yield components and quality. *Bulletin of the Tohoku National Agricultural Experiment Station* 39, 55–96.

Kingsolver, C.H., Barksdale, T.H. and Marchetti, M.A. (1984) Rice blast epidemiology. *Bulletin of the Pennsylvania Agricultural Experiment Station* 853, 1–33.

Klepper, O. and Rouse, D.I. (1991) A procedure to reduce parameter uncertainty for complex models by comparison with real system output illustrated on a potato growth model. *Agricultural Systems* 39, 375–395.

Klochko, A.I., Tikhonova, N.A., Zhuk, O.M., Khalvashi, K.M. and Aleksandridi, D.S. (1985) Methods for determining yield loss from blast in rice. *Selektsiya i semenovodstvo, USSR* 4, 23–24.

Kranz, J. (1974) Comparison of epidemics. *Annual Review of Phytopathology* 12, 355–374.

Kuerschner, E., Bonman, J.M., Garrity, D.P., Tamisin, M.M., Pabale, D. and Estrada, B.A. (1992) Effects of nitrogen timing and split application on blast disease in upland rice. *Plant Disease* 76(4), 384–389.

Kuribayashi, K. and Ichikawa, H. (1952) Studies on forecasting of the rice blast disease. *Special Research, Nagano Agricultural Experimental Station (Japan)* 13, 1–229.

Lindow, S.E. and Webb, R.R. (1983) Use of digital video image analysis in plant disease assessment. *Phytopathology* 73, 520–524.

Loomis, R.S. and Adams, S.S. (1983) Integrative analysis of host-pathogen relations. *Annual Review Phytopathology* 21, 341–362.

Madden, L.V. (1983) Measuring and modeling crop losses at the field level. *Phytopathology* 73, 1591–1596.

Nakamura, K. (1972) On the disease progress curve of the ear blast of rice plant caused by *Pyricularia oryzae* Cav. and the suitable period for various chemical application for the control of rice blast. *Bulletin of the Hiroshima Prefecture Agricultural Experiment Station* 31, 11–29.

Nutter, F.W. Jr. (1990) Generating plant disease epidemics in yield loss experiments. In: *Crop Loss Assessment in Rice*. International Rice Research Institute, Los Baños, The Philippines, pp. 139–150.

Ou, S.H. (1985) *Rice Diseases*, 2nd edn. Commonwealth Mycological Institute, Kew, UK, 380 pp.

Padmanabhan, S.Y. (1965) Estimating losses from rice blast in India. In: *The Rice Blast Disease*. Johns Hopkins Press, Baltimore, Maryland, pp. 203–221.

Pedersen, V.R. (1985) Design and operation of a multispectral radiometer and digital data acquisition system. *Research Reporter 104*. North Dakota State University, Agricultural Experiment Station, Fargo, North Dakota.

Penning de Vries, F.W.T., Jansen, D.M., ten Berge, H.F.M. and Bakema, A. (1989) *Simulation of Ecophysiological Processes of Growth in Several Annual Crops*. PUDOC, Wageningen, 271 pp.

Pinnschmidt, H. (1989) Freilanduntersuchungen zur Epidemiologie von *Pyricularia oryzae* Cav. an Trockenreis (Field studies on the epidemiology of *Pyricularia oryzae* Cav. on upland rice). PhD Thesis, University of Giessen, Germany, 268 pp.

Pinnschmidt, H.O. and Teng, P.S. (1994) Advances in modeling multiple insect-disease-weed effects on rice and implications for research. *Proceedings, International Rice Research Conference*, Los Baños, The Philippines, April 20–25 1992 (in press).

Pinnschmidt, H., Teng, P.S. and Yuen, J.E. (1990a) Pest effects on crop growth and yield. In: Teng, P. and Yuen, J. (eds), *Proceedings of the Workshop on Modeling Pest-Crop Interactions*. University of Hawaii at Manoa, Honolulu, Hawaii, January 7–10 1990. IBSNAT Project, University of Hawaii, Honolulu, Hawaii, USA, pp. 26–29.

Pinnschmidt, H., Teng, P.S., Yuen, J.E. and Djurle, A. (1990b) Coupling pest effects to the IBSNAT CERES crop model for rice. *Phytopathology* 80(10), 997, Abstract no. A311.

Pinnschmidt, H.O., Teng, P.S., Bonman, J.M. and Kranz, J. (1993a) A new assessment key for leaf blast. *International Rice Research Newsletter* 18(1), 45–46.

Pinnschmidt, H.O., Kranz, J. and Bonman, J.M. (1993b) Influence of blast disease in upland rice on the concentration of airborne conidia of *Pyricularia grisea*. *Zeitschrift für Pflanzenkrankheiten und Pfanzenschutz* 100(3), 299–307.

Pinnschmidt, H.O., Batchelor, W.D. and Teng, P.S. (1994) Simulation of multiple species pest damage on rice. *Agricultural Systems* (in press).

Prabhu, A.S. and de Faria, J.C. (1982) Relacionamentos quantitativos entre Brusone nas folhas e paniculas e seus efeitos sobre enchimento e peso dos graos em arroz de sequeiro. *Pesqu. Agropec. Bras., Brazilia* 17(2), 219–223.

Rouse, D.J. (1988) Use of crop growth-models to predict the effects of disease. *Annual Review of Phytopathology* 26, 183–201.

Shaw, M.W. and Royle, D.J. (1987) A new method for disease-loss studies used to estimate effects of *Pyrenophora teres* and *Rhynchosporium secalis* on winter barley yields. *Annals of Applied Biology* 110, 247–262.

Sun, S.Y., Jin, M.Z., Zhang, Z.M., Tao, X.L., Tao, R.X. and Fang, D.F. (1986) *Rice Blast Disease and Its Control*. Shanghai Scientific and Technology Press, Shanghai, 182 pp.

Surin, A., Arunyanart, P., Rojanahusdin, W., Munkong, S., Dhitikiattipong, R. and

Disthaporn, S. (1991) Using empirical blast models to establish disease management recommendations in Thailand. In: *Rice Blast Modeling and Forecasting*. International Rice Research Institute, Los Baños, The Philippines, pp. 69–74.

Teng, P.S. (1985) Construction of predictive models: 2. forecasting crop losses. In: Gilligan, C.A. (ed.), *Mathematic Modeling of Crop Disease*. Academic Press London, pp. 179–206.

Teng, P.S. (1987) Quantifying the relationship between disease intensity and yield loss. In: Teng, P.S. (ed.), *Crop Loss Assessment and Pest Management*. APS Press, St Paul, Minnesota, USA, pp. 105–113.

Teng, P.S. (1988) Pests and pest-loss models. *Agrotechnology Transfer* 8:1, 5–10.

Teng, P.S. (1990) Crop loss assessment: a review of representative approaches and current technology. In: *Crop Loss Assessment in Rice*. International Rice Research Institute, Los Baños, The Philippines, pp. 19–38.

Teng, P.S. and Johnson, K.B. (1988) Analysis of epidemiological components in yield loss assessment. In: Kranz, J. and Rotem, J. (eds), *Experimental Techniques in Plant Disease Epidemiology*. Springer, New York, pp. 179–190.

Teng, P.S. and Oshima, R.J. (1983) Identification and assessment of losses. In: Kommedahl, T.H. and Williams, P.H. (eds), *Challenging Problems in Plant Health*. American Phytopathological Society, St Paul, Minnesota, USA, pp. 69–81.

Teng, P.S., Calvero, S. and Torres, C.Q. (1989) The CERES-rice-blast simulation model. In: *IBSNAT Symposium: The Decision Support System for Agrotechnology Transfer*. Part 2. Poster presentation. American Society of Agronomy Meeting, Oct. 1989, Las Vegas, Nevada.

Teng, P.S., Torres, C.Q., Nuque, F.L. and Calvero, S.B. (1990) Current knowledge on crop losses in tropical rice. In: *Crop Loss Assessment in Rice*. International Rice Research Institute, Los Baños, The Philippines, pp. 39–54.

Teng, P.S., Calvero, S. and Pinnschmidt, H. (1991a) Simulation of rice pathosystems and disease losses. In: Penning de Vries, F.W.T., Kropff, M.J., Teng, P.S. and Kirk, G.J.D. (eds), *Systems Simulation at IRRI. IRRI Research Paper Series No. 151*. International Rice Research Institute, Los Baños, The Philippines, pp. 27–34.

Teng, P.S., Klein-Gebbinck, H. and Pinnschmidt, H.O. (1991b) An analysis of the blast pathosystem to guide modeling and forecasting. In: *Selected Papers from the International Rice Research Conference, 27–31 August 1990 Seoul, Korea*. International Rice Research Institute, Los Baños, The Philippines, pp. 1–31.

Tsai, W.H. (1988a) Estimation of rice yield losses caused by leaf blast disease. *Journal of Agricultural Research of China* 37(2), 207–210.

Tsai, W.H. (1988b) Estimation of rice yield losses caused by panicle blast disease. *Journal of Agricultural Research of China* 37(1), 86–90.

Waggoner, P.E. (1990) Defoliation, disease, and growth. In: Rabbinge, R., Goudriaans, J., van Keulen, H., Penning de Vries, F.W.T. and van Laar, H.H. (eds), *Theoretical Production Ecology, Reflections and Prospects*. Simulation monographs 34, PUDOC, Wageningen, pp. 149–180.

Waggoner, P.E. and Berger, R.D. (1987) Defoliation, disease, and growth. *Phytopathology* 77, 393–398.

Widawsky, D.A. and O'Toole, J.C. (1990) Prioritizing the rice biotechnology research agenda for Eastern India. *Research Report of the Rockefeller Foundation*, 86 pp.

24 The Epidemiological Basis for Blast Management

P.S. Teng

Entomology and Plant Pathology Division, International Rice Research Institute, PO Box 933, Manila 1099, The Philippines

Introduction

In spite of considerable investment in research and extension, rice blast disease remains a problem in many temperate and subtropical rice growing areas because of its apparently unpredictable outbreaks and the resulting economic losses. The variability of the pathogen is well known, the ephemeral nature of host plant resistance in most modern cultivars is recognized, losses caused by blast have been quantified, its epidemiology has been much studied, and yet blast is still commonly considered the most damaging of rice diseases in farmers' fields. Which critical aspects of the disease have not been addressed and what hypotheses have not been tested? It would appear that there has been a less than systematic approach to researching or managing blast in the context of the rice ecosystem. In this chapter, blast management is approached in a systems mode, using epidemiologic principles to emphasize that the most important output from research must be a blast management toolkit derived from and applied based on epidemiology.

Rice production is estimated to provide the world's largest market for pesticides, approximately US$2.4 billion in 1988 (Woodburn, 1990), of which US$570 million were fungicides. Most of the fungicides appear to be used for blast control in temperate irrigated ecosystems (approximately 92% in 1988 of total fungicide value) such as those of Japan, Korea, China and Taiwan. In tropical rice systems, host plant resistance is still the main tool for blast control in lowland (paddy) rice while in upland (dry) rice, blast is not commonly controlled and is a major cause of yield loss. While the perception of blast as an upland rice problem is correct, in reality it causes more tonnage loss in lowland rice under temperate and cool season tropical conditions. This is because upland rice constitutes only 13% of total world rice area and only 4% of total rice production, while the corresponding figures for lowland rice are 80% and 93% respectively, with the remainder made up of flood-prone, deepwater or tidal water ecosystems

(IRRI, 1993). This ecosystem perspective is important for understanding the contribution of epidemiologic studies to blast management, and consequently, to stable rice production, because environmental conditions vary greatly between ecosystems. To date research effort may have been inappropriately proportioned into the pathosystem occurring in an ecosystem that is of minor importance in world food security.

A common evolutionary sequence of cropping systems, *vis-à-vis* plant protection, is from subsistence agricultural systems (typified by utilization of wild species, and indigenous knowledge) to traditional systems (wild and domesticated species and indigenous knowledge) to conventional agriculture (improved high-yielding varieties of domesticated species and modern technology) (Teng *et al.*, 1993). This intensification sequence is closely associated with disease problems and use of plant protection inputs. Rice blast is a disease that appears to be relatively independent of the crop intensification process, occurring throughout most of the rice growing world. It is a recognized problem in upland rice ecosystems with low input use, low yield potential and poor soils, as well as in irrigated rice ecosystems with modern infrastructure, high fertilizer inputs and high yield potential. However, some attributes of crop intensification such as fertilizer and planting density are known to exacerbate the severity of infection (Kingsolver *et al.*, 1984). The development of sustainable blast management strategies requires that the effects of intensification be taken into account even though blast is found in such diverse ecosystems.

Blast epidemics are known to be influenced by crop management, especially nitrogen status, water stress and silicon content (Teng *et al.*, 1991). Therefore, disease management must account for the potential of cropping practices to influence the development of epidemics. Yet epidemics conceptually should be viewed as the result of many matching pathogen–host interactions, i.e. concurrent alloinfections (Robinson, 1976). Epidemics in turn result in numerous blast generations per crop cycle, each of which may confer an evolutionary advantage to a particular subpopulation of the pathogen at the crop level. In the past, epidemiologists and population geneticists have tended to study blast systems separately, with the result that today, little is known on how to integrate population genetics with epidemiology in managing blast.

The International Rice Research Institute (IRRI, 1993) has estimated an increase of 350 million tonnes of rice over 1992 production will be required to feed the world in 2020. In its new Medium Term Plan for 1994–98, IRRI has proposed developing new plant types for different rice ecosystems; that for irrigated tropical rice will be lower tillering, larger panicles and thicker and taller stems, and be grown under high-density direct-seeding conditions with high nitrogen fertilization. It is likely that for these plant types to express their yield potential, input use may have to resemble that of current levels in temperate ecosystems. These future conditions will render much of the current knowledge on blast management obsolete and will require that we extrapolate current knowledge on pathosystem–environment interactions to design new management systems.

Rice pathology, like its sister disciplines, has in recent years been strongly influenced by the concept of Integrated Pest Management (IPM). At IRRI, our interpretation of this concept is that it is, broadly, ecology-based pest management (Teng, 1993). There is a set of circumstances under which the IPM concept will be applied, each of which requires a different response (Teng, 1993):

1. Rice ecosystems currently showing acute problems of fungicide misuse and abuse, and where resistance to fungicides is preventing effective blast management. This situation is still uncommon but may occur in localized areas in temperate zones.
2. Rice ecosystems where blast is a major constraint on yield and is causing severe losses in the absence of control, but where fungicide use is not excessive. This is by far the most common situation in subtropical Asia, as in large parts of Vietnam.
3. Systems where the environment is conducive to blast, but because of low inputs the disease is not problematic. However, significant intensification is anticipated in such ecosystems and with that, blast is a constraint. This is the case with rainfed lowland areas of South and South-East Asia. With rainfed upland rice areas, no fungicides are used, but yields are highly constrained from reaching their potential because of blast. In these situations, there is an urgent need to apply a systems analytic approach for identifying appropriate intervention points to attain sustainability.

IPM may also be considered a pest management system that utilizes a combination of pest control techniques, while enhancing the role of natural regulatory mechanisms in the ecosystem, to produce an economically and socially acceptable yield with no adverse effects on the physical, human and biological environment. The 'best mix' of techniques suggests that a toolkit is needed, consisting of physical tools, decision (knowledge-based) rules, communication tools and policy instruments. As noted previously, research by rice blast pathologists ultimately should contribute to this toolkit, in the form of a tool that is used by beneficiaries such as farmers, extension workers or crop protection specialists.

In this chapter, the concept of a blast management toolkit, to which all research efforts should ultimately contribute, will be discussed with accompanying examples of actual and potential tools. The epidemiologic implication of each tool is identified. Current epidemiological knowledge on blast will then be reviewed in the context of their contribution to tool derivation, and where possible, missing knowledge noted. An integrative concept for blast management will then be discussed and immediate research and development (R&D) needs to implement such a concept determined.

The Blast Management Toolkit

Elements of a blast management toolkit

The broad framework for assembling a blast management toolkit is outlined in Table 24.1. Four types of tools may be discerned: knowledge tools, physical tools, communication tools and policy tools. There is admittedly some overlap between them in this categorization but their definition in this form allows a clear indication of outputs from blast research. Furthermore, the areas of overlap form the

Table 24.1. Elements of a blast management toolkit.

Types of tools	Intervention points: Strategic/tactical (S/T)	Community/ farmer (C/F)	Epidemiologic effect[a] x_0	r	t
Knowledge tools					
Systems analysis and problem-definition techniques	S,T	C,F	+	+	+
Diagnostic aids	T	F	+		
Assessment and sampling guidelines	T	F		+	+
Forecasting/predictive systems	T	C,F		+	+
Thresholds (damage functions, empirical and sliding)	T	F		+	+
Decision-aids (boards, charts)	T	F		+	
Decision-support systems (expert systems, etc.)	S	C,F	+	+	+
Rules for enhancing natural control or suppression	S	F	+	+	
Surveillance and early warning systems	S	C	+	+	
Cultural knowledge	S	C,F	+	+	+
Varietal deployment guidelines	S	C	+	+	+
Environmental risk rules	S	C	+		
Integrated control guidelines	S,T	C,F	+	+	+
Physical tools					
Seeds with known resistance	S	C,F	+	+	+
Biological control agents	T	F	+	+	
Confirmatory serological tools	T	F	+		
Non-chemical seed health treatment	S	F	+		
Pesticides/pesticide application technology	T	C,F		+	
Fertilizer application	T	F		+	+
Silicon (rice hull) amendments	T	F	+	+	
Mechanical control tools	T	F	+	+	
Tools to modify the rice ecosystem	S	C	+	+	+
Communication tools					
Community bulletin boards	S,T	C	+	+	+
Computer bulletin boards	T	C,F	+	+	+
Radio/TV	S,T	C	+	+	+
Farmer group meetings	S,T	C	+	+	+
Scouts	T	F	+	+	+
Farm demonstrations	S	C	+	+	+
Multimedia campaigns	S	C	+	+	+
Management information systems	T	C	+	+	+

Table 24.1. *(cont).*

	Intervention points				
			Epidemiologic effect[a]		
Types of tools	Strategic/tactical (S/T)	Community/ farmer (C/F)	x_0	r	t
Policy tools					
Pesticide regulations	S	C		+	
Pesticide subsidies	S	C		+	
Varietal deployment acts	S	C	+	+	+
Seed health laws	S	C	+		
Plant quarantine laws	S	C	+		

[a] Epidemiologic effects on: x_0 = initial inocolum or disease; r = the epidemic infection rate; t = the duration of the epidemic or period of crop growth when the epidemic is expected to have a negative effect on growth, development or yield.

linkages needed to develop a truly integrated approach. The application of each tool is further rationalized in epidemiologic terms as having an effect either on the initial inoculum or disease (x_0), the epidemic infection rate (r), or the duration of the epidemic or period of crop growth when the blast epidemic is expected to have a negative effect on growth, development or yield (t) (Table 24.1). To be sustainable, every blast management tool should have a discernible epidemiologic effect. *All research should ultimately contribute to improved blast management through development of a new tool or enhancement of an existing tool.* Tools are used in blast management to intervene with the epidemic, either strategically (macro-level and/or longer timeframe than one season) or tactically (within or affecting one season and of smaller area). Some blast management tools are more appropriately applied, in an epidemiologic sense, at the community level, others at the farmer or single farm/field level, while some could apply equally well to both levels.

Knowledge tools

As concerns about sustainability increase and the IPM concept is translated for use in plant pathology, knowledge tools can be expected to gain in importance with farmers and extension workers. The blast pathosystem does not function in isolation from other components in the rice ecosystem; hence a clear definition of the problem that blast poses, in the context of the total ecology of the rice environment, should precede the design of any management strategy. While knowledge tools range from problem-definition techniques to integrated control guidelines (Table 24.1), their potential for blast management only will be discussed here in the context of epidemiology.

Techniques for conceptualization, systems analysis and selection of intervention points are well documented by Norton *et al.* (1991). Because epidemics are the result of the classic man–pathogen–host–environment interaction, management of epidemics must be conceptualized to show linkages to other key elements in the rice ecosystem.

Blast management is predicated on the ability of farmers and extension workers to correctly diagnose the problem. Simple field guidebooks are available in most national programs; IRRI's *Field Problems of Tropical Rice* and *IPM – Rice: Pocket Reference Manual* are examples. Disease assessment to determine the severity required for action, following a specified sampling procedure, is well documented for blast (Surin *et al.*, 1991; Pinnschmidt *et al.*, 1993).

Assessment methods for diseases are considered essential by researchers for developing thresholds, but their use by farmers is still limited. Knowledge on sampling procedures is well developed for insects and weeds, but there is little knowledge on the dispersion of diseases in farmers' fields or on sampling procedures for them; no data is available to show if trained farmers actually use any prescribed sampling procedure before they make a decision. Furthermore, the explosive nature of blast epidemics may limit the available period for intervention.

Surveillance and early warning systems (SEWS) existed in several South-East Asian countries which were precursors of IPM programs (Teng and Heong, 1988). Although they were conceived to provide extensive data on migratory pests for early warning of impending outbreaks, they were not very successful because the concept was ahead of its time. Lack of an adequate infrastructure in many countries led to an inability to utilize the data in a timely manner for decision making. In temperate areas, such as Korea and Japan, wide-area monitoring appears successful for providing information for forecasting in blast management (Kim, 1987; Ishiguro and Hashimoto, 1988). In India and Thailand, simple rules of thumb have been tested but have not been widely adopted. Research appears to be ahead of application in blast management.

Blast forecasting techniques were recently reviewed (Teng *et al.*, 1991) and are based on host-plant criteria, disease features, spore catch and weather, or combinations of these. The systems based on weather appear to be most successful in temperate rice. A method to forecast leaf blast outbreaks (BLASTAM) developed by Koshimizu (1982) relies on daily weather data on rainfall, temperature, sunshine hours, and windspeed. Data from 840 sites all over Japan are collected by Automatic Meteorological Data Acquisition System (AMeDAS). Length of time for leaf wetness, mean temperature during leaf wetness, mean temperature of the five preceding days, and other secondary weather variables that meet certain model criteria are estimated. BLASTAM classifies the weather conditions from favorable to unfavorable for infection. It then predicts if a leaf blast epidemic will start 1 week after the onset of favorable conditions. Uehara *et al.* (1988) tested BLASTAM for leaf and panicle blast. Leaf blast occurrence was predicted well, but not panicle blast occurrence. This led these workers to suggest using hourly weather records for disease forecasting.

Even though BLASTAM does not give quantitative information on disease progress, but only predicts disease outbreaks (starting time of general epidemics and rapid spread of leaf blast), and although it occasionally tends to underestimate wetness duration, it was found to be satisfactory in several prefectures in Japan.

Kim *et al.* (1987, 1988) developed an algorithm to index weather with respect to blast suitability. Criteria for temperature, RH, and leaf wetness duration were used to compute blast units of severity (BUS). The higher the BUS, the more favorable the weather for blast development. Logit-transformed leaf blast progress curve data (% current disease severity) from 2 years and two varieties, under flooded and upland conditions, were regressed on cumulative BUS to develop predictive equations. High R^2 values (about 0.7) that varied across varieties were obtained for the upland treatment; under flooded conditions, the method did not work well.

In India, Padmanabhan (1965) determined the following forecasting rules: (i) seedbed infection occurs if minimum temperature is less than or equal to 24–26°C for 4–7 days; (ii) leaf blast occurs during tillering if minimum temperature is below 24°C for 4–5 days during post-transplanting and tillering, and if RH is greater than or equal to 90%; and (iii) neck infection occurs if conditions in September–October favor leaf infection and temperatures are 20–24°C for a number of days coinciding with RH greater than or equal to 90%. Severe leaf blast is conditional for neck blast to occur. It is not known if these rules have been used in any practical blast management scheme.

Location-specific empirical damage functions and the equivalent disease-loss relationships are available for blast but are not widely used in any country by farmers. For example, at IRRI, Torres and Teng (1993) derived an equation linking leaf and panicle blast.

$$y = 0.21 + 1.012x_1 + 0.51x_2 \; (R^2 = 0.8)$$

where y = percent yield loss, x_1 = percent diseased leaf area at tillering, and x_2 = percent panicle blast incidence at harvest. This equation suggests that the effect of leaf blast is approximately twice that of panicle blast.

Expert systems are still relatively recent innovations in rice pathology; at IRRI, a computer program called DISMAN has, in its menu, modules for managing leaf and panicle blast. Several other knowledge tools are undergoing development at IRRI as part of our new Medium Term Plan, 1994–98: guidelines to enhance natural, microbial control of blast; rules for varietal deployment based on knowledge of pathogen population structures and rice resistance genes; and rules for characterizing environmental risk to blast. For example, using blast epidemic models, historical weather data and mathematical algorithms to produce representative weather sequences, we are able to compare sites using indicators like area-under-curve (Fig. 24.1; Luo Yong and P.S. Teng, unpublished data) or number of severe blast days per defined period (Fig. 24.2; Teng and Yuen, 1990).

Cultural knowledge such as on rotations, varietal mixtures, crop mosaics, and planting times have been part of the local practice in many rice growing areas (Oka, 1988). Studies on the epidemiological basis for the effectiveness of cultural methods are, however, sadly lacking, but these methods appear to have potential under experimental conditions.

Integrated control is probably the weakest aspect of applying the IPM concept; there is relatively little integration of control methods for one or more pests

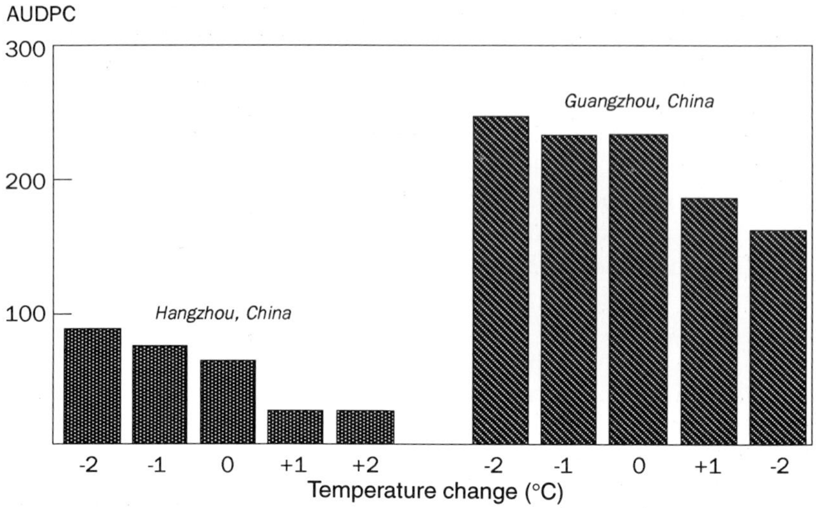

Fig. 24.1. Estimating environmental riskiness using a blast model to generate expected average area-under-blast-progress curves (IRRI, unpublished data).

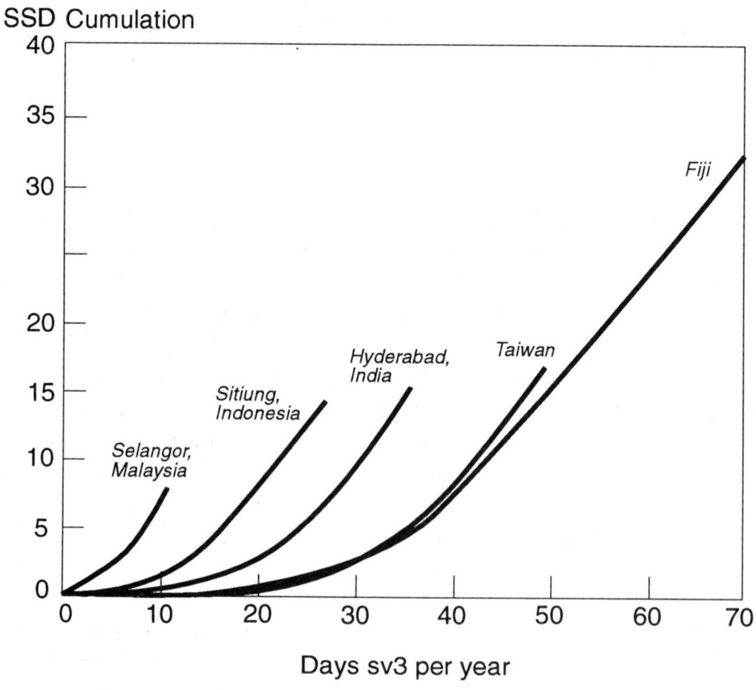

Fig. 24.2 Assessing environmental riskiness using stochastic dominance and number of high severity days (from Teng and Yuen, 1990).

Table 24.2. Matrix of cultural management practices for some pests.

	Pest					
Cultural practice	BL	ShB	RTD	BPH	SB	Rodents
Earlier planting	+	+	+	−	−	−
Synchronized planting (barrio or community)	−	−	+	+	+	+
Crop rotation	−	+	+	−	−	−
Intercropping	−	−	−	+	+	−
Sanitation						
Remove affected plants	−	−	+	−	−	−
Burn/plough stubble	−	+	+	+	+	−
Remove host weeds	−	−	+	−	−	+
No ratoons	−	−	+	+	+	−
Proper water management	+	−	−	+	−	−
Proper fertilizer management	+	+	−	+	+	+
Proper plant spacing	+	+	+	+	−	−

'+' means practice has known effectiveness. BL = blast, ShB = sheath blight, RTD = rice tungro disease, BPH = brown planthopper, SB = stemborer.

in rice farms (Teng, 1993). There is, however, much use of two or more control methods, independently, on the same farm, such as planting a moderately resistant variety which may require less fungicide use. Integrated control would probably be more used by farmers if the research knowledge was available; work is underway at IRRI and national programs to generate guidelines for integration.

Physical tools

These include seed, biological control agent, fungicides, amendments, mechanical tools and cultural practices (Table 24.2). Seed possessing resistance genes to blast have been the basis for plant protection for centuries (Bray, 1986) and this phenomenon was greatly accelerated by the establishment of IRRI in 1960. IRRI genotypes are probably the most widely grown of any crop (IR36 is the most widely grown crop variety in history) but this success has led to the increased selection pressure to overcome the resistance. Much of the HPR in rice is of the 'vertical', single gene type, and while this has been useful, the many epidemics which have occurred attest to its weakness when used as the sole method of plant protection. For example, the blast epidemics in South Korea (1976–1979) resulted from pathogen evolution to overcome HPR in the Tongil-type rice varieties (Kim, 1979; Dalrymple, 1986), and had serious political and social consequences in that country.

Biological control agents and confirmatory serological tools are still very much in an experimental stage for blast control (T.W. Mew, IRRI, personal communication). Blast conidia are commonly found on rice seeds. Their importance in initiating epidemics is not clear and hence the value of seed treatment is uncertain in direct-seeded rice, although seed treatment using hot water

Table 24.3. Fertilizer (nutrient)–fungicide (disease) interactions, Florida, USA (from Datnoff, 1994).

Treatment	% Blast incidence	Yield (kg ha^{-1})
Control	63.3	3268
Si fertilizer	45.3	3700
Benomyl	20.6	3801
Benomyl + Si	16.0	4195

Table 24.4. Fungicide use pattern in Red River Delta, Vietnam, 1992 (unpublished data).

		Spring crop 1992		Summer crop 1992	
Province	No. farms surveyed	% using fungicides	No. of sprays	% using fungicides	No. of sprays
Thai Binh	15	96.7	1.40	100.0	1.15
Ha Tay	15	90.0	1.45	96.7	1.35
Vinh Phu	15	64.4	0.76	46.7	0.52

may be of value to reduce seedbed/seedbox infection in transplanted rice (Ou, 1985).

The effect of nutrient (nitrogen, phosphorus, potassium) and non-nutrient (silica) amendments on blast epidemic development are known in temperate and tropical systems (Datnoff, 1994), and attempts have been made to utilize these in management, especially in relation to cropping practices and fungicide use (Table 24.3). Epidemiologically, these amendments affect several monocyclic processes, as well as x_0 and r.

Fungicides and fungicide application technology are the most popular tactical tools for blast management in temperate rice. Unlike insecticides, resistance to fungicides in the blast pathogen has not been an issue (Sato, 1988), and given the lack of available methods when an epidemic results, chemical control, assisted by predictive systems, will probably continue in such countries as Japan and Korea. In tropical ecosystems, such as those of Vietnam, high disease pressure necessitates the application of one or more (sometimes 3–4) fungicide sprays per season (Table 24.4).

Other results obtained through the IRRI–IPM Network in the Mekong Delta, Vietnam, showed that of 685 farmers surveyed in May 1992, about 38% used one or more fungicides to control blast. The situation in Vietnam is detailed here as it illustrates a situation of intensification in which fungicides appear to be needed to maintain or increase productivity. Fungicide use varied according to variety; in Hue Province, Vietnam, IR 38 averaged 5–6 applications while IR17494 averaged four applications. In the summer crop of Hue province, only 10,000 ha out of 23,000 ha required fungicide application in 1992.

A cultural practice is any farm operation that will make the environment less favorable for pests to develop or multiply but which still favors rice production.

Tools for modifying the rice ecosystem include synchronized planting, crop rotation with intercropping or fallow, sanitation, water management, and changing planting density. Sanitation aims to remove all survival sites for the pathogen, especially in the ratoons or volunteers, by ploughing under or burning. Water management to reduce stress on the plants at blast-susceptible stages is known to many rice farmers (Teng *et al.*, 1991). Close spacing between plants results in a more shaded, cooler, and more humid microenvironment, favorable for blast development even in the lowland, humid tropics. The distance between rice plants depends on the variety. Modern rice varieties with high tillering capacity may be planted further apart than those with less or moderate tillering capacity. Common spacing between rice plants is 20 × 20 cm or 25 × 20 cm. Plant spacing is likely to gain in importance because of the move towards direct seeding and newer plant types. Cultural practices as a group need to be considered in blast management relative to their effects on other key pests in the rice ecosystem because farmers are unlikely to directly adopt a practice for blast alone management (Table 24.2).

Rice has been planted 2–3 times a year in many parts of Asia since the massive introduction of short-duration, modern varieties in the 1960s. However, labor shortages, market prices, water availability and input availability have resulted in these plantings being staggered (i.e. non-synchronized), such that in any region, there is always rice in a range of growth stages. In staggered rice patterns, there is no clear-cut fallow period (i.e. for soil preparation) between any two rice seasons while in the synchronized patterns, there *is* a fallow period between any two rice seasons. Over a region, the blast pathosystem becomes a continuous pathosystem (*sensu* Robinson, 1976) and given the dispersal pattern of conidia, initial inoculum will always be available for matching alloinfections.

Communication tools

These include all the methods used to extend physical or knowledge tools to target audiences – community bulletin boards, computer bulletin boards, radio/television, farmer group meetings, field scouts, farm demonstrations, multimedia campaigns, and management information systems. These tools are included for discussion because they can be used effectively to change farmer behavior towards blast disease.

Community bulletin boards have been used successfully in Malaysia to communicate IPM decisions to groups of villagers. A color coding system for messages is used to denote degrees of urgency (A. Booty, personal communication) by extension and crop protection personnel. This system appears to have potential in many countries but research is lacking to show its suitability to different sociological attitudes. IPM radio 'jingles' and IPM radio drama have been used in the Philippines since 1986 as part of a communication campaign by the Philippine–German Crop Protection Program (Pfuhl, 1988). The usefulness of these does not appear to have been evaluated for blast. Farmer group meetings have been the main method of farmer training in many tropical countries. Skill retention using this technique is relatively high even after 3–4 years.

Demonstration plots to show that IPM (including blast management) works are used in several countries on selected farms using both researcher-managed and farmer-managed plots in irrigated rice (Teng and Heong, 1988). The FAO regional program, IRRI, and National program–German Plant Protection Programs in several South-East Asian countries have produced posters carrying the various pest management messages as well as posters on disease diagnosis. Scouting (private or public) has not been practiced on any significant scale in rice ecosystems. Computer bulletin boards and fact sheets have been limited to extension personnel in industrialized countries and it is unlikely that rice farmers in tropical ecosystems will adopt this technology in the near future. Multimedia strategic campaigns have been held in several South-East Asian countries and appears to have been successful in getting the IPM message across to farmers (Teng, 1993).

Policy tools

Policy tools are particularly important in developing countries because their governments have a more direct effect on the rural population than in more *laissez-faire* developed countries, especially in terms of providing the credit and support services for agricultural production (Teng, 1993). The experience in South/South-East Asia strongly suggests that a conducive policy environment, including an official endorsement, is required in developing countries before effective pest management, with or without external assistance, can start. Policy tools include pesticide regulations and subsidies, varietal deployment acts, seed laws, and plant quarantine laws (Table 24.1).

Epidemiological Knowledge for Blast Management

Epidemiology is the study of population dynamics in pathosystems and how these change in space and time in response to short- or long-term changes in the structure of the pathosystem or its environment. For blast management, knowledge on several aspects of epidemic development are key to its management – the type, source and magnitude of initial inoculum, factors driving the epidemic infection rate, dispersal and spread of fungus propagules, and interactions between the pathogen and the rice plant. In general, it is known that the particular pathogen virulence type and host plant resistance dictates symptomatology and subsequent disease increase (Bonman, 1992). Meteorological factors like rainfall, dew, relative humidity, temperature, mist, wind and solar radiation are also known to affect specific processes in the blast epidemic (Teng *et al.*, 1991). The blast pathosystem is also known to consist of two major subsystems: the leaf pathosystem and the panicle pathosystem. While they have common features, they usually have been treated separately because of time discontinuity and because the relationship between the two is not yet clearly defined. Leaf blast can result in significant yield losses in severe epidemics but it is panicle blast that commonly is the economic pathosystem (Torres and Teng, 1993). In this section, only the salient points of blast epidemic development that have management implications will be discussed; the reader is referred to the papers by Kato (1974), Suzuki (1975) and Teng *et al.* (1991) for more complete reviews of the topic. Missing knowledge and research needs will also be noted where relevant.

Initiation of blast epidemics

Initial inoculum (x_0)

The origin of the initial inoculum still perplexes scientists in those ecosystems where crop rotation is practiced. Alloinfection in each subsystem is thought to occur with inoculum from non-rice hosts of the pathogen or from rice plants in the immediate vicinity which have been successfully infected. The quantitative nature of matching alloinfection is little understood, especially in rice-growing areas where there is not an early season source of infected rice. The work by Borromeo *et al.* (1993) using molecular markers has cast doubts on how common matching alloinfection is from *Pyricularia grisea* spores originating from non-rice hosts. However, once alloinfection has occurred, then disease severity increases via autoinfection. Epidemics resulting from autoinfection are the phenomena that have been most studied.

The blast fungus survives as mycelia in plant residues, conidia or in living plant tissue (Ou, 1985); in tropical and subtropical areas, all three modes are considered important as sources of initial inoculum. Conidia may also be transported by air, water or seed. Mycelia surviving in rice straw have been found to remain viable for up to 3 years at 18–32°C and to produce conidia when moistened, while conidia are reportedly viable for 1 year at 8°C and 20% RH. Although *P. grisea* conidia are associated with seed, there is no evidence that seed infection plays a role in initiating epidemics. In paddy rice ecosystems, puddling greatly reduces survival of conidia in rice refuse or seeds. Long distance dispersal as a source of initial inoculum is not well documented. In irrigated rice areas such those of Tamil Nadu, India, it is common to find germinated rice seedlings around threshing flats in villages which grow two rice crops a year. The storage of rice straw for cattle feed and the use of the straw as thatching in many South Asian villages also provides other sources. In intensively cropped irrigated rice areas, such as the triple-cropped Mekong Delta in Vietnam, the turnaround period between successive rice crops is as short as 15 days in some provinces. Not all fields are fully synchronized in each province and it is common to find live rice tissue throughout the year.

Dispersal and spread

Blast conidia are commonly observed at high numbers over rice crops between 0000 h and 0600 h, and liberation is favored by darkness, high RH and wind speeds $> 3.5 \text{ m s}^{-1}$ (Kingsolver, 1984). Dispersal gradients for blast conidia are strongly related to dominant wind directions and follow the pattern found for other passively-liberated spores. Spores may be released at temperatures above

19 °C (Kato, 1974) while rainfall on rice leaves is known to cause an increase in the conidial concentration over canopies.

The maximum distance that a blast conidium can travel remains controversial (Teng *et al.*, 1991), but is key to determining management strategies. The work of Kato and Sasaki (1974) remains probably the most authoritative. Under average tropical conditions, blast conidia can be dispersed commonly within about 230 m of the source (Kingsolver, 1984). It is very possible that long-distance dispersal of conidia is closely related to weather patterns and low-level upper atmospheric disturbances. This aspect of epidemiology deserves more study in view of modern techniques for spatial analysis of weather and landscape patterns, and for monitoring of fungus and host.

Factors influencing the epidemic infection rate (r)

Describing the component processes of the blast monocycle, and quantifying the effect of specific factors on their rates of change is essential for understanding why epidemics occur and how they may be predicted. Although various levels of detail may be used to characterize blast monocycles, knowledge on four processes are generally sufficient for management: infection, latency, lesion growth and sporulation (Teng *et al.*, 1991).

Infection

This is a composite term to describe subprocesses which include conidia germination, appresorium formation and penetration. Leaf wetness is required for all of the infection subprocesses; the relationship between infection and leaf wetness is well known. If leaf wetness ends before completion of infection, the process is terminated (Yamanaka and Ikeda, 1964). Increasing the leaf wetness period from 12 to 15 h is known to result in a 30% increase in infection (Kato, 1974). The length of the leaf wetness period required for infection is influenced by temperature (Kato, 1974; Kingsolver *et al.*, 1984; Ou, 1985).

Germination commonly begins within 3 h of deposition if leaves are wet (Kato, 1974). Germination is delayed if conidia are exposed to a dry period (Kingsolver *et al.*, 1984). According to Kim (1987), preinfection processes are not affected by plant water stress.

Appressoria formation occurred at an average of 11 h at 24 °C (Kato, 1974) but is rare at temperatures greater than 28 °C. Various optimum temperatures for appressorium formation have been reported to be between 15 and 25 °C (Suzuki, 1969; Kato, 1974).

Temperature and dew period interact to affect infection. A minimum dew period of 6–8 h will initiate infections at the optimum temperature of 25 °C (Kato, 1974). In the tropics, night temperature is not a significant factor as that temperature is almost always in the optimal range (El Refaei, 1977; Ou, 1985). In upland rice, longer dew periods were suggested as a factor in the higher incidence of blast compared with lowland conditions (Ou, 1985). Plant density and fertilizer levels may also increase the dew period (El Refaei, 1977). For

Table 24.5. Effect of soil-applied nitrogen on four plant parameters of rice cultivar, IR72 at IRRI, Los Baños, Philippines (Dilla, 1993).

Nitrogen treatment (kg N ha^{-1})	Plant height (cm)	Tiller no. (per plant)	Leaf area (per plant)	Total biomass (g plant^{-1})
0	96.9 (SD)*	4.6 (SD)	19.8 (M)	52.0 (SD)
45	101.7 (SD)	5.7 (SD)	23.4 (M)	62.2 (SD)
90	107.9 (SD)	6.5 (SD)	29.7 (M)	71.1 (SD)

SD = soft dough, M = milk stage.

example, average dew periods were 12.83, 10.97, and 8.91 h in stands with spacings of 10 × 10 cm, 20 × 20 cm, and 40 × 40 cm, respectively. The amount of dew formed appears to have variable effects on the rate of infection (Kingsolver *et al.*, 1984).

Spore germination and appressorium formation are stimulated in plants fertilized with high levels of nitrogen or potassium (Kawamura and Ono, 1948). Penetration rates are lower in plants with high levels of silica, possibly as a result of the mechanical barrier attributed to accumulation of silica in the epidermis (Ou, 1985). Application of N results in increased infection (Ou, 1985). What is unclear from the literature is whether N confers a direct physiological effect on the pathogen–host interaction or only has an indirect effect by changing plant morphology and consequently, microclimate (Table 24.5).

A critical epidemiologic parameter is the penetration ratio, defined as the number of lesions divided by the corresponding number of successful penetrations. The level of this ratio depends on the particular combination of pathogen and host genotypes, and is a component of partial resistance. Modeling studies confirm that this ratio has a significant effect in modifying the epidemic infection rate; its value commonly ranges from 0.05 to 0.25, and may even change according to plant age (Calvero *et al.*, 1992). Plants subjected to severe water deficit have relative infection efficiencies more than four times that of unstressed controls (Gill and Bonman, 1988).

Latent period

Successful establishment after infection has occurred depends on the age and genotype characteristic of the host, crop nutrition, and environmental factors such as temperature and soil moisture. Latent periods varied from 13–18 d at 9–11°C to 4–6 d at 26–28°C (Teng *et al.*, 1991). Computer simulation studies show that small changes in latent period can significantly change the epidemic infection rate in many pathosystems (Teng and Yuen, 1990). In tropical and subtropical areas where blast is known to be a problem, temperature changes during the different crop seasons (Table 24.6) are sufficient to cause some parts of the year to be more

Table 24.6. Representative temperature and relative humidity data from two sites in Vietnam known to have blast-conducive environments.

	Jan	Feb	Mar	Apr	May
Hue Province, central Vietnam					
1989/90 Daily minimum temperature (°C)	11.0	15.6	15.7	19.0	19.8
Average RH (%)	90	90	88	84	79
Ha Tay Province, northern Vietnam					
1991/92 Average daily temperature (°C)	15.8	16.5	19.8	24.1	27.3
Average RH (%)	79	82	87	86	83

blast conducive. The seasonal profiles for weather in the tropics and subtropics are not as static as previously assumed by blast workers. Small changes such as those shown are important enough epidemiologically to allow more targeted blast management.

Lesion growth

Cultivars with different degrees of partial resistance to blast show differences in infection efficiency and lesion size (Yeh and Villareal *et al.*, 1981; Bonman 1986). Plant age and leaf age, however, influence the expression of partial resistance effects; young leaves are commonly more susceptible (Kahn and Libby 1958; Goto *et al.*, 1961; Kim *et al.*, 1987). Lesion size is known to be affected by temperature (Kato and Kozaka, 1974), with high alternating temperatures of 32/25°C and 32/20°C resulting in initial rapid lesion growth, and water stress increasing the size of lesions (Bonman *et al.*, 1988; Gill and Bonman, 1988). Blast management through cultural practices will therefore be enhanced by minimizing water stress, especially at periods of rapid leaf growth.

Sporulation

Under optimum conditions, conidiophores are formed in 4–6 h, with one conidium formed every 40 min (Toyoda and Suzuki, 1952). The number of spores produced by any one lesion has been found highest on the sixth day after lesion appearance (Villareal *et al.*, 1981). Mature lesions are capable of sporulating when relative humidity (RH) is higher than 89% (with optimum RH > 93%) (Hemmi and Imura, 1939). Maximum sporulation occurs 7–12 d after inoculation, although sporulation may continue at low levels for 60 d (Suzuki, 1975). The cardinal temperatures for sporulation are about 9–12, 25–28, and 34–35°C (Henry and Andersen, 1948).

Sporulation is greater in lesions on leaves that were expanding when inoculated than on lesions from leaves that were fully expanded 3–4 d before inoculation (Kato *et al.*, 1970). Lesions produced on plants inoculated at the tillering stage had a greater sporulation potential than on plants inoculated after the formation of the panicle primordium (Kato *et al.*, 1970). Sporulation potential is related to the level of partial resistance in the host (Yeh and Bonman, 1986). In one study, number of conidia per lesion varied from 4800 on IRAT13 to 33,000 on IR442–2–58 (Villareal *et al.*, 1981). Maximum sporulation is associated with lesions having gray centers and margins changing from dark purple to brown (Kato *et al.*, 1970).

Lesions on plants showing water stress symptoms may produce up to 3.5 times more conidia than those on unstressed plants (Gill and Bonman, 1988). The infectious period is also prolonged as a result of preinfection water stress.

The blast polycycle: Integration of monocyclic knowledge

The net effect of environment, host genotype characteristics and cropping practices on blast pathosystems can be studied by measuring disease severity or disease incidence over a specified period. Epidemics are commonly characterized by population growth curves, or disease progress curves (DPCs), and much work over the last two decades has gone into fitting various mathematical models to DPCs (Teng, 1985). With blast epidemics, the logistic function popularized by Van der Plank (1963) is commonly used to quantify severity-based DPCs for leaf blast and incidence-based DPCs for panicle blast. Field epidemics of blast are polycyclic, with the epidemic rate dependent on the favorability of prevailing conditions. In temperate ecosystems, the blast epidemic may only have 7–8 monocycles per season or year; in the tropics, 10–15 cycles per season with 2–3 seasons is not uncommon. When disease management is considered in a holistic context of gene deployment and tactical measures, this aspect of epidemiology becomes important. Long-term sustainable blast management will have to be concerned with the selection pressure put on the pathogen population in very blast-conducive tropical environments.

To describe the progression dynamics of disease epidemics in space and time requires a systems approach to model building, which commonly results in a computer simulation model. The first leaf blast simulation model published (BLASTL) is probably that of Hashimoto *et al.* (1982, 1984). It was developed using data from the literature and the authors' own experiments. The pathogen's life cycle (sporulation; spore discharge, dispersal, and landing; infection; and lesion growth) were simulated in relation to weather conditions, plant growth, leaf position, and host susceptibility as affected by weather, fertilizer application, plant/leaf age, and leaf position. The dynamics of leaf blast development were calculated as temporal changes in the number of lesions. It was anticipated that the model would predict leaf blast outbreaks as 7-day short-term forecasts. The model was tested in several prefectures in Japan for several years and found to be useful and practical for forecasting disease progress. Since it contains a fungicide submodel, it may become a practical tool for determining the timing and efficiency of fungicide applications.

Two other groups in Japan (Ohta *et al.*, 1982, 1987 – BLASTCAST; Takai *et al.*, 1985) also developed blast simulation models independently, although their approaches appear similar to that of Hashimoto *et al.*

A polycyclic leaf blast simulation model, PYRICULARIA, was developed by

Gunther (1986) using information from the literature. PYRICULARIA was adapted and modified by Tastra *et al.* (1987) for upland rice farming systems in Indonesia (the model was subsequently call PYRNEW). These workers tried to include the effect of nitrogen fertilizer and varietal resistance established by their field experiments. Preliminary results of model validation suggest the need for further work on model structure and stimulus-response relationships, as would be expected. Gunther (1986) had derived most of his structural data from experiments done for temperate climate ecosystems; Tastra *et al.* were adapting it for application in tropical ecosystems.

In Korea, Choi (1987) and Choi *et al.* (1988) used data from their growth chamber experiments and the literature to develop a leaf blast simulation model (LEAFBLST). There appears to be good agreement between the model and the real world for rice cultivar Jinheung. The model does not appear to consider ontogenetic and environmentally caused changes in host susceptibility and does not include collar blast.

At IRRI, a leaf blast simulation model was developed by adding increasing complexity to a logistic growth function. The components of the pathogen's life cycle considered are sporulation, spore dispersal and landing, and infection. Factors that affect epidemic development are x_0, plant age (thus, changing host susceptibility and spore deposition), temperature, dew period, row spacing, and nitrogen applied. The model considers latent period and host area to be constant; the model has been further refined and validated at IRRI (Calvero and Teng, 1992).

Panicle blast appears to have been first simulated by Takasaki (1982), who treated spore deposition and penetration as stochastic processes and subdivided each panicle into small infection site units. Infection is computed according to a probability function and affected panicles are classified into several types. The model does not account for secondary infections.

Probably the most comprehensive panicle blast model developed to date is that of Ishiguro and Hashimoto (1988, 1989). Their stochastic panicle blast simulation model (PBLAST) used the Monte Carlo method; spore deposition and penetration are treated as stochastic processes, and each panicle is subdivided into small infection site units. By using a probability function for spore deposition, as well as considering wetness duration and wetness temperature, and applying the Monte Carlo method, the probability of penetration of each deposited spore into an infection site unit is computed. Heading, fertilization, grain growth, susceptibility of each infection site, appearance and growth of lesions, panicle blast severity, and yield loss were calculated daily; spore formation, discharge, dispersal, deposition, penetration, and colonization were calculated every 3 h. AMeDAS weather data, additional wetness duration data, and data on host development, variety, cultivation practices, and number of spores formed on leaf lesions were used as model inputs. Validation results seem inconsistent, and the model requires a large amount of computer time, even on mainframes. As a preliminary model, it should be considered a tool for epidemiologic research rather than for practical forecasting. Moreover, the model uses some preliminary parameters and functions that have not yet been verified experimentally.

Effect of blast on rice growth, development and yield

This aspect of blast epidemiology is critical for guiding the timing and need to initiate control measures, and is discussed elsewhere in this book. The rice plant is morphologically and physiologically variable in its response to blast infection at different development stages. In general, susceptibility to leaf blast declines from seedling to about heading. Although the rice plant is able to compensate for leaf defoliation early in the season, yield losses have been recorded during the periods in the seasonal profile (Teng and Savary, 1992) when the epidemic is expected. More work is needed to define the lower and upper limits of yield losses due to blast at different growth stages and in this regard, computer simulation offers much potential.

An integrative concept for blast management: Features and R&D needs

A systems view is needed to provide the framework in which often-times, conflicting ideas may be reconciled to meet the same goals for blast management (Teng and Savary, 1992). The blast problem has to be defined in the context of the socioeconomic, biological–physical–technological subsystems in which management is practiced. An *integrative concept for blast management* needs to be developed that recognizes concerns for sustainability and plant health, and combines epidemiologic and genetic aspects of plant pathosystems. Such an integrative concept contains features which address each of the system components of pathogen, host, disease, physical environment, and grower and community, as follows:

1. Management of the pathogen. Key features are:
   (a) reducing opportunities for fitness and reproduction, and
   (b) characterizing population structures.
2. Management of the host. Key features are:
   (a) growing healthy plants from clean seed, with balanced nutrition;
   (b) facilitating associated organisms on plants;
   (c) growing plants with potentially durable resistance; and
   (d) deploying crops for potentially durable resistance.
3. Management of the disease. Key features are:
   (a) Interventions using chemicals (synthetic and natural); and
   (b) Interventions only during loss-susceptible stages.
4. Management of the physical environment. Key features are:
   (a) cultural practices; and
   (b) physical/mechanical practices.
5. Management of the grower and community. Key features are:
   (a) changing growers' misperceptions;
   (b) promoting conditions that favor grower participation in disease management decision-making; and
   (c) promoting conditions that favor community action for disease management.

Translation of an integrative blast management concept into action elements points the way for research leading to a blast management toolkit, and to methods for changing the current situation. The action elements are essentially the different control practices, such as biological practices (host plant resistance, use of biological agents or products, enhancement of natural control, *sensu* Cook, 1988), cultural/mechanical/physical practices, chemical practices (synthetics, natural, botanicals), and regulatory/exclusion practices. Each of these sets of practices have associated with them decision rules or tools, either strategic or tactical in nature, which collectively constitute the management toolkit (Table 24.1; Teng and Savary, 1992). Conceptualization of blast management into a toolkit, developed to address management (points 1–5, above) either strategically or tactically, helps to focus epidemiologic research. A conceptual framework is essential for ensuring that research to improve blast management *per se* has clear indicators for progress towards that goal.

The brief survey of potential and actual blast management tools (previous section) indicates that considerable research is still needed to confirm the wider applicability of these tools and to quantify their epidemiologic basis. Using the concept developed above, topics that require immediate research are as follows:

Managing the pathogen.

- Characterizing and using host genes that reduce pathogen reproduction;
- Development and application of molecular markers to characterize pathogen population structures.

Managing the host.

- Quantifying the relationship between nutritional quality and yield, and disease development;
- Characterizing the role of associated organisms on rice plants;
- Developing rice genotypes with durable resistance;
- Deployment strategies for durable resistance.

Managing the disease.

- Development of simplified disease predictive systems;
- Development of decision tools using sliding disease/yield thresholds.

Managing the physical environment.

- Quantifying the epidemiologic effect of cultural practices;
- Quantifying the epidemiologic effect of physical/mechanical practices.

Managing the grower and community.

- Characterizing growers' perceptions on diseases and disease management;
- Participatory research to change policy and attitudes regarding disease management; and
- Socioeconomic research on IPM (disease) implementation.

This is by no means an exhaustive list of research needs; rather it is an indication of research that has to be done to fill in some serious gaps in the toolkit.

Concluding Remarks

The shifts away from unilateral approaches of pest control, in particular approaches which depend on pesticides or host plant resistance alone, towards an integrated approach of pest management, constitutes an important paradigm shift for plant protection for the 1990s. The IPM concept was a timely one when first propounded in the 1950s and its evolution has paralleled the increasing awareness in the scientific community of sustainability issues. It is in this context that blast management must be viewed if we are to conduct research that has social relevance and will be used to enhance the demand for sustainable agricultural practices. In addition to IPM, rice disease management in general has been strongly influenced by the concept of plant health management as proposed by Cook (1988). While the former has focused attention on the need to account for farmer attitudes and the farming community when designing sustainable practices, the latter has highlighted the role that soils and plant nutrition play in disease management.

Much has been written about the attributes of sustainable agriculture. Sustainable agriculture must make optimal use of the resources available to it to produce an adequate supply of goods at reasonable cost: it must meet social expectations, and it must not overly expend irreplaceable production resources. Sustainable agriculture is founded on the application of an ecological paradigm to agricultural systems. Sustainability in food production is only a small part of the overall picture in attaining sustainable development; the latter must recognize the link between the rural and urban sectors, and the rationalization of modern technology with indigenous knowledge systems. However, the conflicts in thinking that are evident in the broader concern for sustainability are reflected in blast management. For example, Ruttan (1988) has stated that traditional agricultural systems that have met the test of sustainability have not been able to respond to modern rates of growth in demand for agricultural commodities; sustainability must include enhancement of productivity. As we have seen with rice ecosystems, the need to grow more rice under increasingly intensive situations leads to conditions that favor diseases such as blast. Epidemiological research that reconciles the interactions of man, pathogen and the rice plant in a changing environment will have to be the basis for designing sustainable blast management strategies.

References

Bonman, J.M. (1992) Durable resistance to rice blast disease – environmental influences. *Euphytica* 63, 115–123.

Bonman, J.M., Sanchez, L.M. and Mackill, A.O. (1988) Effects of water deficit on rice blast. II. Disease development in the field. *Journal of Plant Protection in the Tropics* 5, 67–73.

Borromeo, E.S., Nelson, R.J., Bonman, J.M. and Leung, H. (1993) Genetic differentiation among isolates of *Pyricularia* infecting rice and weed hosts. *Phytopathology* 83, 393–399.

Bray, F. (1986) *The Rice Economies, Technology and Development in Asian Societies.* Basil Blackwell, Oxford, UK, 254 pp.

Calvero, S.B. Jr, Guico, E.M. and Teng, P.S. (1992) Quantifying rice-leaf blast (Bl) genetic relationship via the receptivity factor. *International Rice Research Newsletter* 17(2), 9.

Calvero, S.B., Jr and Teng, P.S. (1992) Validation of BLASTSIM.2 model in IRRI blast nursery and Cavinti, Laguna, The Philippines. *International Rice Research Newsletter* 17(5), 20.

Choi, W.J. (1987) A computer simulation model for rice leaf blast. MS Thesis, Seoul National University, Suweon, Korea.

Choi, W.J., Park, E.W. and Lee, E.J. (1988) LEAFBLST: a computer simulation model for leaf blast development on rice. *Korean Journal of Plant Pathology* 4(1), 25–32.

Cook, R.J. (1988) Biological control and holistic plant-health care in agriculture. *American Journal of Alternative Agriculture* 3, 51–62.

Dalrymple, D.G. (1986) *Development and Spread of High-Yielding Rice Varieties in Developing Countries*. US AID, Bureau for Science and Technology, Washington DC.

Datnoff, L. (1994) Influence of plant mineral nutrition on rice disease development. In: Teng, P.S., Heong, K.L. and Moody, K. (eds), *Advances in Rice Pest Management*. IRRI, Los Baños, The Philippines (in press).

Dilla, E.M. (1993) Yield loss due to sheath blight in direct-seeded rice as affected by plant density, nitrogen level and inoculum amount. PhD Thesis, University of Philippines, Los Baños.

El Refaei, M.J. (1977) Epidemiology of rice blast disease in the tropics with special reference to leaf wetness in relation to disease development. PhD Thesis, Faculty of the Post-graduate School, Indian Agricultural Research Institute, New Delhi.

Gill, M. and Bonman, J.M. (1988) Effects of water deficit on rice blast. I. Influence of water deficit on components of resistance. *Journal of Plant Protection in the Tropics* 5, 61–66.

Goto, K., Horano, K. and Ohata, K. (1961) Susceptibility of leaf of rice to blast in reference to leaf age and position. *Special Bulletin of the Okayama Prefecture Agricultural Experiment Station* 58, 77–88.

Gunther, H. (1986) *Simulation of the Epidemiology of Pyricularia oryzae in Rice. A First Attempt*. Report of a 3-month MSc project. Department of Theoretical Production Ecology, Agricultural University, Wageningen, The Netherlands.

Hashimoto, A., Matsumoto, K. and Adachi, T. (1982) Some trials for forecasting rice leaf blast epidemics by application of the computer simulation. *Annual Report of the Society of Plant Protection, North Japan* 33, 12–14.

Hashimoto, A., Hirano, K. and Matsumoto, K. (1984) Studies on the forecasting of rice leaf blast development by application of the computer simulation. *Special Bulletin of the Fukushima Prefecture Agricultural Experiment Station* 2, 75–78.

Hemmi, T. and Imura, J. (1939) On the relation of relative humidity to conidial formation in the rice blast fungus, *Pyricularia oryzae*, and the characteristics in the germination of conidia produced by the strains showing different pathogenicity. *Annals of the Phytopathological Society of Japan* 9, 147–156.

Henry, B.W. and Andersen, A.L. (1948) Sporulation by *Pyricularia oryzae*. *Phytopathology* 38, 265–278.

IRRI (1993) *Rice Research in a Time of Change: IRRI's Medium Term Plan for 1994–98*. Los Baños, The Philippines, 80 pp.

Ishiguro, K. and Hashimoto, A. (1988) Development of a simulation model for rice panicle blast epidemics. 1. Composition of the basic model. *Bulletin of the Fukushima Prefecture Agricultural Research Station* 27, 1–19.

Ishiguro, K. and Hashimoto, A. (1989) Recent advances in forecasting of rice blast epidemics using computers in Japan. *Tropical Agriculture Research Series* 22, 153–162.

Kahn, R.P. and Libby, J.L. (1958) The effect of environmental factors and plant age on the infection of rice by the blast fungus, *Pyricularia oryzae*. *Phytopathology* 48, 25–30.

Kato, H. (1974) Epidemiology of rice blast disease. *Review of Plant Protection Research* 7, 1–20.

Kato, H. and Kozaka, T. (1974) Effect of temperature on lesion enlargement and sporulation of *Pyricularia oryzae* in rice leaves. *Phytopathology* 64, 828–840.

Kato, H. and Sasaki, T. (1974) Epidemiological studies of rice blast disease, with special reference to reproductive processes in lesions on rice plants and disease forecast. *Bulletin of the National Institute of Agricultural Science Series C* 28, 1–61.

Kato, H., Sasaki, T. and Koshimizu, Y. (1970) Potential for conidium formation of *Pyricularia oryzae* in lesions on leaves and panicles of rice. *Phytopathology* 60, 608–612.

Kawamura, E. and Ono, K. (1948) Studies on the relation between the preinfection behavior of rice blast fungus, *Pyricularia oryzae*, and water droplets on rice plant leaves. *Bulletin of the National Agriculture Experiment Station* 4, 1–12.

Kim, I.H. (1979) *The Green Revolution in Korea – Development and Dissemination of New Rice Varieties*, 2nd edn. Office of Rural Development, Seweon, Republic of Korea, 184 pp.

Kim, C.H., MacKenzie, D.R. and Rush, M.C. (1987) A model to forecast rice blast disease based on weather indexing. *Korean Journal of Plant Pathology* 3(3), 210–216.

Kim, C.H., MacKenzie, D.R. and Rush, M.C. (1988) Field testing a computerized forecasting system for rice blast disease. *Phytopathology* 78, 931–934.

Kim, C.K. (1987) Disease forecasting: past and future in special reference to rice leaf blast. *Korean Journal of Plant Protection* 3, 304–312.

Kingsolver, C.H., Barksdale, T.H. and Marchetti, M.A. (1984) Rice blast epidemiology. *Bulletin of the Pennsylvania Agricultural Experiment Station* 853, 1–33.

Koshimizu, Y. (1982) A forecasting method for leaf blast outbreak by the use of AMeDAS data. *Kongetsu No Noyaku* 26(1–4), 46–53, 64–67, 74–83, 68–78.

Norton, G.A., Holt, J., Heong, K.L., Cheng, J. and Wareing, D.R. (1991) Systems analysis and rice pest management. In *Rice Insects: Management Strategies*: Heinrichs, E.A. and Miller, T.A. (ed.), CRC Press, Boca Raton, Florida, pp. 287–321.

Ohta, K., Chiba, S. and Shimada, K. (1982) Simulation of rice leaf blast using BLASTCAST, a plant disease simulator. *Annual Report of Plant Protection, North Japan* 33, 9–11.

Ohta, K., Shimada, K. and Chiba, J.I. (1987) A few improvements of the simulation model of rice leaf blast epidemics. *Bulletin of the Aomori Agricultural Experiment Station, Japan* 30, 31–58.

Oka, I. (1988) Role of cultural techniques in rice IPM systems. In: Teng, P.S. and Heong, K.L. (eds), *Pesticide Management and Integrated Pest Management*. Consortium for International Crop Protection, College Park, Maryland, pp. 83–94.

Ou, S.H. (1985) *Rice Diseases*, 2nd edn. Commonwealth Mycological Institute, Kew, UK, 380 pp.

Padmanabhan, S.Y. (1965) Estimating losses from rice blast in India. *The Rice Blast Disease*. Johns Hopkins Press, Baltimore, Maryland, pp. 203–221.

Pfuhl, E.H. (1988) Radio-based communication campaigns: a strategy for training farmers in IPM in the Philippines. In: Teng, P.S. and Heong, K.L. (eds), *Pesticide Management and Integrated Pest Management*. Consortium for International Crop Protection, College Park, MD 20740, pp. 251–256.

Pinnschmidt, H.O., Teng, P.S. and Bonman, J.M. (1993) A new assessment key for leaf blast. *International Rice Research Newsletter* 18(1), 45–46.

Robinson, R.A. (1976) *Plant Pathosystems*. Springer-Verlag, Berlin.

Ruttan, V. (1988) Sustainability is not enough. *American Journal of Alternative Agriculture* 3, 128–130.

Sato, T. (1988) Overview of pesticide resistance problems in southeast Asia. In: Teng, P.S. and Heong, K.L. (eds), *Pesticide Management and Integrated Pest Management in Southeast Asia*. Consortium for International Crop Protection, College Park, MD 20740, pp. 257–267.

Surin, A., Arunyanart, P., Rojanahusdin, W., Munkong, S., Dhitikiattipong R. and Disthaporn, S. (1991) Using empirical blast models to establish disease management recommendations in Thailand. In: *Rice Blast Modeling and Forecasting*. International Rice Research Institute, PO Box 933, Manila, The Philippines, pp. 69–74.

Suzuki, H. (1969) Temperature related to spore germination and appressorium formation of rice blast fungus. *Proceedings, Association of Plant Protection Hokuriku* 17, 6–9.

Suzuki, H. (1975) Meteorological factors in the epidemiology of rice blast. *Annual Review of Phytopathology* 13, 239–256.

Takai, A., Kanai, K., Kobayashi, M., Murata K. and Komori, R. (1985) Studies on the forecasting of rice blast by system dynamic model. Forecast for number of lesions and infection area of rice leaf blast by application of computer simulation. *Bulletin of the Ibararaki Agricultural Experiment Station* 25, 97–131.

Takasaki, T. (1982) Forecasting method of rice blast disease using computer in Fukuoka Prefecture. *Kongetu no Noyaku* 26(8), 33–37.

Tastra, I.K., Irmansyah, R. and Yunizar, Z.H. (1987) Epidemiology of leaf blast (*Pyricularia oryzae*) under various abiotic and biotic conditions. Preliminary report of the SARIF team case study.

Teng, P.S. (1985) A comparison of simulation approaches to epidemic modeling. *Annual Review of Phytopathology* 23, 351–379.

Teng, P.S. (1993) Integrated pest management in rice. *Experimental Agriculture* (in press).

Teng, P.S. and Heong, K.L. (1988) *Pesticide Management and Integrated Pest Management in Southeast Asia*. Consortium for International Crop Protection, College Park, Maryland.

Teng, P.S. and Savary, S. (1992) Implementing the systems approach in pest management. *Agricultural Systems* 40, 237–264.

Teng, P.S. and Yuen, J.E. (1990) Epidemics models: lessons from plant pathology. In: Levin, M. and Strauss, H. (eds), *Risk Assessment in Genetic Engineering: Environmental Release of Organisms*. McGraw-Hill, New York, pp. 211–220.

Teng, P.S., Klein-Gebbinck, H. and Pinnschmidt, H. (1991) An analysis of the blast pathosystem to guide modeling and forecasting. In: *Blast Modeling and Forecasting*. IRRI, Los Baños, The Philippines, pp. 1–30.

Teng, P.S., Savary, S. and Revilla, I. (1993) Systems of plant protection. In: *Crop Protection and Sustainable Agriculture*. CIBA Foundation Symposium 177, Chichester, pp. 116–139.

Torres, C.Q. and Teng, P.S. (1993) Path coefficient and regression analysis of the effects of blast on rice yield. *Crop Protection* 12, 296–302.

Toyoda, S. and Suzuki, N. (1952) Histochemical studies on the lesions of rice blast caused by *Pyricularia oryzae* Cav. I. Some observations on the sporulation on lesions of different types occurring on leaves of the same variety. *Annals of Phytopathological Society of Japan* 17, 1–4.

Uehara, Y., Imoto, M. and Sakai, Y. (1988) Studies on the forecasting of the rice blast development using weather data from AMeDAS. *Bulletin of the Hiroshima Prefecture Agricultural Experiment Station* 51, 1–15.

Van der Plank, J.E. (1963) *Plant Diseases, Epidemics and Control*. Academic Press, New York, 349 pp.

Villareal, R.L., Nelson, R.R., MacKenzie, D.R. and Coffman, W.R. (1981) Some components of slow-blasting resistance in rice. *Phytopathology* 71, 608–611.

Woodburn, A.T. (1990) The current rice agrochemicals market. In: Grayson, B.T., Green, M.B. and Copping, L.G. (eds), *Pest Management in Rice*. Elsevier Applied Sciences, London.

Yamanaka, S. and Ikeda, M. (1964) Viability of water-soaked conidia of the rice blast fungus. *Annual Report of Plant Protection North Japan* 15, 21.

Yeh, W.H. and Bonman, J.M. (1986) Assessment of partial resistance to *Pyricularia oryzae* in six rice cultivars. *Plant Pathology* 35, 319–323.

25 Using Simulation Models to Explore Better Strategies for the Management of Blast Disease in Temperate Rice Pathosystems

KIYOSHI ISHIGURO

National Agriculture Research Center, 3-1-1 Kannondai Tsukuba, Japan, 305

INTRODUCTION

Within the last decade, serious epidemics of rice blast have occurred in 1988 and 1991 in the temperate regions of Japan considerably damaging the local farm economy there. Variation in epidemic severity depends mainly on weather conditions during the epidemics. Breeding for the control of rice blast has reached an impasse, and it is unlikely that new, agronomically acceptable rice cultivars with resistance will be available for growers within the next decade. This means that growers must use fungicides to manage the disease. Fungicides may need to be applied up to five times during an epidemic (Ishiguro and Hashimoto, 1991). More than ten fungicides (not including seed treatments) have been registered for rice blast in Japan (Table 25.1), and most of them are effective even when applied after onset of the epidemic.

Among the many decisions made by growers early in the season, some may affect the severity of subsequent epidemics. Before the onset of the epidemics, they must decide which cultivar to grow, how much nitrogen to apply as a basal fertilizer, and whether or not to make an early application of a systemic fungicide (such as those which would be applied to the nursery boxes or to the water surface of the rice field). After onset of the disease, but before the epidemic becomes serious, growers should determine when, how often, and what kind of fungicide to apply, and when to topdress with nitrogen. Therefore, growers need to make multiple decisions and optimize them based on their own priority. Priorities may differ among decision-makers. For example, one grower may set a priority for high yield, and another for stabilizing income, for low labor intensity, for low fungicide use, or some combination of these factors.

Because of the complex relationship among the components of the rice blast pathosystem, the effects of control measures on the disease epidemic may be nonlinear. Indeed control decisions should be based on knowing how the

Table 25.1. Fungicides for blast disease in Japan.

Common name	Formulation/Application[a]	Notes
Blastcidin-S	Spraying	Antibiotic
EDDP	Spraying	Organic phosphoric
Ferimzone[b]	Spraying	
Fthalide	Spraying	MBI[c]
IBP	Spraying/granular	Organic phosphoric
Isoprothiolane	Spraying/granular	
Kasugamycin	Spraying	Antibiotic
Probenazole	Granular	
Pyroquilon	Granular	MBI
Tricyclazole	Spraying/granular	MBI

[a] Spraying means any formulations for spraying. Granular formulation fungicides are applied to the water surface of paddy fields or on nursery boxes.
[b] Combined with MBIs in all commercial use.
[c] Melanin biosynthesis inhibitor.

epidemic will progress, an uncertain future event; and forecasts should improve the information about the uncertain future event. A systems-analysis approach would therefore be a suitable means of addressing these issues. Fortunately, there is sufficient knowledge about most of the components of the blast pathosystem to enable us to develop a simulation model of blast epidemics.

Nearly a quarter of a century has passed since Waggoner and Horsfall (1969) demonstrated the first simulation model of plant disease epidemics. Campbell and Madden (1990) have pointed out that most simulation models have not yet reached their potential as epidemiology tools. Nevertheless, we still believe that the most realistic way to improve strategies for rice blast disease management is through the application of simulation modeling. Our ultimate goal is to develop better strategies for the management of blast disease, including the design of new disease forecasting systems, and we have used simulation models to obtain this goal.

Leaf blast model, BLASTL (Hashimoto *et al.*, 1984) and panicle blast model, PBLAST, (Ishiguro and Hashimoto, 1988, 1990) have been developed in Japan and, although either may be used eventually in the development of alternative management strategies, the reliability and validity of these models have to be determined. The models should be validated objectively using other datasets not used in the development of the models. The simulation models should be improved to increase the reality of their performance. The original program codes of the models should be rewritten into a more sophisticated programming language to facilitate subsequent modification of the programs.

Finally, the submodel for the counter-measures to control the epidemics should be developed. We have recently organized our research around those priorities.

Leaf Blast Model, BLASTL

The model was originally written by Hashimoto *et al.* (1984) in FORTRAN. This model assumes an arbitrary phenotype of rice plants and the presence of the initial inoculum. It calculates plant growth once a day and the behavior of the fungus (sporulation, liberation, dispersion, deposition, survival, germination, penetration, colonization, and development of lesions) eight times a day based on several weather factors (temperature, wind velocity, duration of sunshine, precipitation, and leaf wetness). It then estimates disease severity, and the number of lesions or lesion area of each leaf position.

Validation of the model

Initial validation of BLASTL was done by Hashimoto *et al.* (1993a) who showed a close correlation between the numbers of lesions per plant predicted by BLASTL using weather records for 19 years, and actual diseased area based on surveys in nearby rice fields (33,000 ha in total) in corresponding years. This assumes a quantitative relationship between disease severity and diseased area which is still unknown. Ishiguro *et al.* (1989) suggested that the number of fields in which disease severity is more than a specific value increased as the mean disease severity increased, based on the data over 5 years from about 150 rice fields. They showed that the frequency distribution of the number of fields versus severity of rice blast disease generally obeyed an exponential distribution with the following probability density function:

$$g(x) = \frac{1}{\lambda} e^{-\frac{x}{\lambda}}$$

where x is disease severity, the definite integral of $g(x)$ is the relative frequency of the fields in a given range, and $1/\lambda$ is the mean and the standard deviation of x. This equation means that the area where the disease occurs more severely than an arbitrary threshold value increases as the mean disease severity increases. Therefore, the general behavior of the model has enough biological reality for prediction in the concerned field and the surrounding area.

Ishiguro *et al.* (1991) have also validated the model objectively, and the next section describes that validation. We define validation after Teng (1981): 'the quantitative comparison of output from the simulation model with empirical results not used in the development of the model'.

BLASTL has several parameters that can be adjusted arbitrarily by the user. The user can adjust how many leaves of a rice plant emerge during a simulation, and how much area each leaf has. These parameters can also be determined using real data. In the model, we can also adjust the susceptibility of each leaf position, a characteristic which varies with cultivars and which is critical for the simulation (evaluating leaf and cultivar susceptibility under field conditions is still difficult). If a model is tested only on the data from which it was constructed, the predictive value of the model is unknown. To determine the predictive utility of BLASTL, we employed the following validation procedure.

First, to estimate the susceptibility parameters, we considered the model as a function:

$$y = f(t; \theta_1, \theta_2, \theta_3)$$

in which t, an independent variable, is the time from the start of simulation (day), y, a dependent variable, is the number of lesions per plant, and θ_i, unknown parameters, are susceptibility of the upper three leaf positions (the subscript, i, refers to each leaf position). The parameters of the nonlinear function were estimated using a nonlinear least squares method (we employed the quasi-Newton method) with a set of temporal changes in the number of lesions per plant in a field. In this case, the number of observed data points has to be more than that of the number of parameters, in order to obtain enough information from the resulting function. Using the parameters determined by this method, we predicted the epidemics in other years using weather data that were not used to obtain the parameters.

Validation was finally performed by a visual comparison of graphs of predicted and observed disease progress curves (Fig. 25.1). Although the result was not adequately satisfactory, there was a similarity between the estimated and observed curves. Therefore we feel it is appropriate to use this model.

Modification of the BLASTL program

After the introduction of our models, several problems arose. The computer language we used was unfamiliar to the other researchers, and comments were written in Japanese instead of English. In addition, the program was too complex for other scientists to understand easily. Therefore, we decided to rewrite the program in Pascal, because most students have learned this language in introductory computer programming courses; it is suitable for structured programming (Kernighan and Plauger, 1978; i.e. it is easily read and modified); the software is cheap; it can be used in most PCs; and it is fairly easy to translate to C language. All the studies reported here were done with the Pascal version of BLASTL.

Initial inoculum in the model

The BLASTL model assumes that initial inoculum is present. There is a controversy about this assumption. Hashimoto *et al.* (1984) found that the timing of initial inoculum is a critical factor in the behavior of the model. Inoculum density was also found to influence disease progress in the model, although the impact is small. Such results are expected for polycyclic diseases (Fry, 1982). Initial inoculum parameters are provided as input by users and are not determined by the model.

Fortunately, Kobayashi (1984) reported detailed information on the initial phase of blast epidemics. He observed two phases of leaf blast epidemics in a temperate rice area in Japan: a focal epidemic followed by a general epidemic.

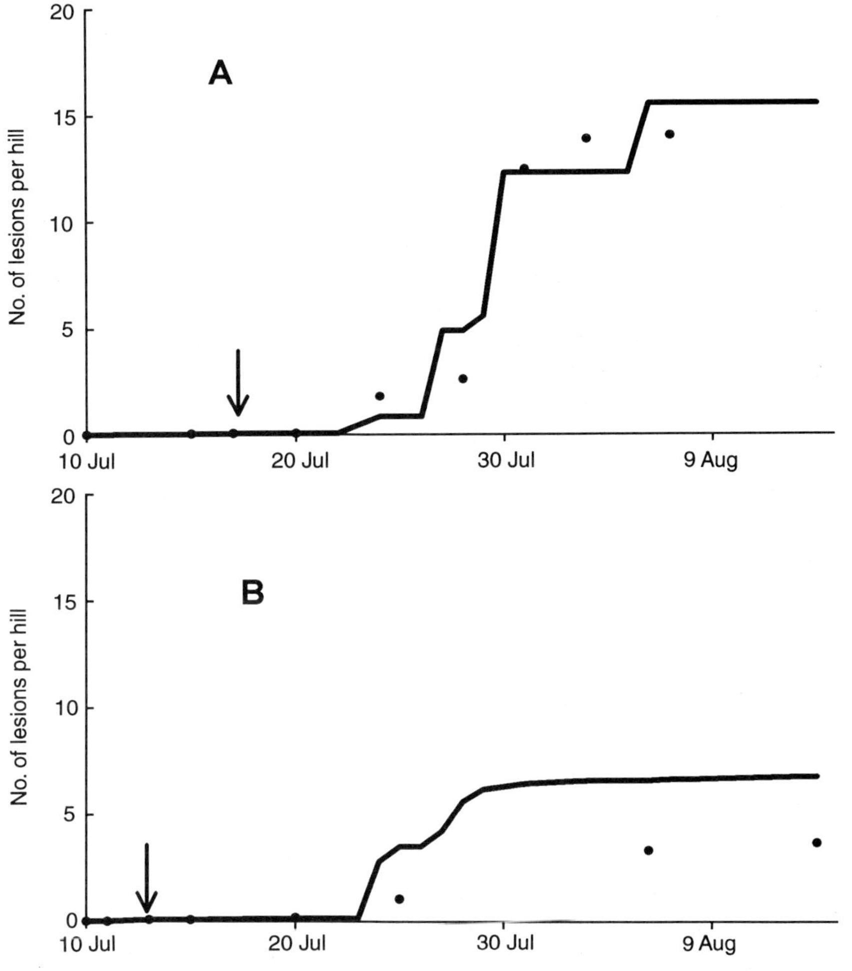

Fig. 25.1. Objective validation of BLASTL. (A) Observed (dots) and predicted (line) disease progression in 1989 season. The leaf susceptibility parameters in BLASTL are adjusted using nonlinear least squares method. (B) Observed (dots) and predicted (line) disease progression of another season (1987). The values of the parameters are the same as above. Arrows indicate the beginning of general epidemics of leaf blast (Ishiguro and Hashimoto, 1991).

He suggested that it is convenient to consider the onset of general epidemics as the onset of leaf blast epidemics in a region. It is believed that most (if not all) inoculum source comes from seedling mats left on paddy fields to fill vacant space in rows. It is possible that the fungus in the mats was introduced on seed, overwintered in infected rice straws piled near the nursery beds, or somewhere else. The initial foci of epidemics are around these seedling mats (focal epidemic). The density of the foci in a region is low (1.0 focus per 10 ha, according to Ishiguro and Hashimoto, 1991), and they expand slowly.

A few weeks after the onset of focal epidemics, general epidemics may develop. Zadoks and Schein (1979) defined this as follows: 'a general epidemic appears all at once, over large areas, regularly distributed within and between fields'. The development of general epidemics of rice leaf blast is marked by the simultaneous appearance of solitary lesions in a random spatial pattern in most fields within a region. These epidemics may be a meso-scale phenomenon. Kobayashi proposed that criteria based on weather conditions should be used to predict the initiation of general epidemics. The criteria are that the duration of wet period is ten or more hours; the wet period occurs due to rain; the period includes a night; and the minimum atmospheric temperature during the period is higher than 16°C (Kobayashi, 1984; Ishiguro and Hashimoto, 1991). Kobayashi (1985) suggested that 10 days after the favorable weather conditions for infection (after the incubation period of the infection), the first generation of rice leaf blast can be detected in most rice fields in a region. Rice fields should be surveyed based on his protocol (to walk in a rice field and observe slowly along a row about 100 m once or twice). If a few solitary lesions are found (as is common), it would suggest that a general epidemic has started. However, there are other possibilities. If aggregations of a parent and daughter lesions (foci) are found, then the general epidemic is more than a week old. Also, if no lesion is found, the general epidemic has not yet occurred. Using these criteria and protocol, BLASTL users can estimate the parameters of initial inoculum.

An important question is source of inoculum for the general epidemics. Kobayashi (Japan, 1991, personal communication) was able to identify a possible inoculum source(s) in each small area whenever he detected a general epidemic. Typically a focus around a diseased seedling mat(s) in a field is a possible inoculum source(s) for a certain area. Takeda (1992) recently reported that in a region (e.g. a village) where there was no identifiable inoculum source, general epidemics started later than predicted using Kobayashi's criteria. These observations suggest that weather conditions that induce general epidemics occur on a meso-scale, but the contribution of an inoculum source to an epidemic is on a micro-scale (within 1 km^2). Therefore, it is inaccurate to predict the initiation of general epidemics based only on weather conditions without including field observations.

Fungicide submodels

Our initial goal was to model the pathosystem without considering the influence of any control measures, and the validation and verification were conducted as such. However, even a highly explanatory or predictive model that fails to improve disease management strategies is of limited practical utility. Therefore, we are constructing submodels for control measures and incorporating them into BLASTL.

The primary means of controlling blast disease after planting is through fungicide application. There are several way to classify fungicides, and we have chosen to group the ten active ingredients (Table 25.1) based on formulation/application as this is convenient from an epidemiological point of view. The fungicides are divided into granular and spray formulations.

The granular fungicides are systemic and applied to the water surface of

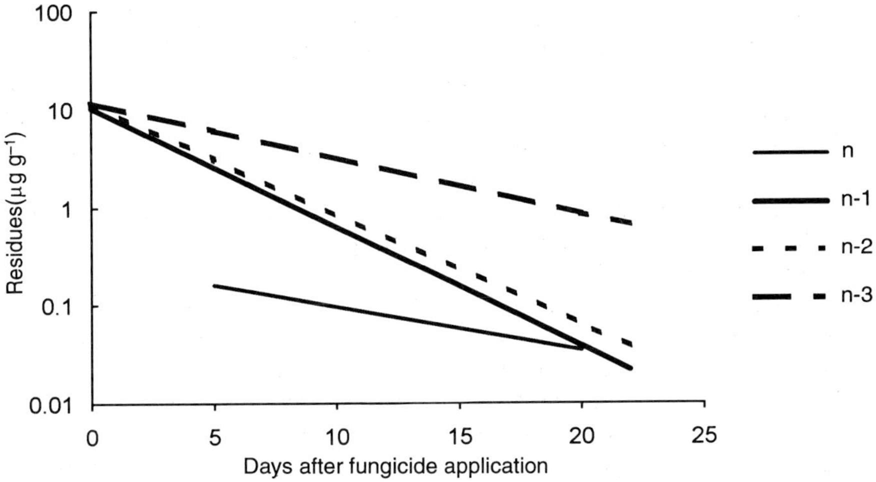

Fig. 25.2. Temporal dynamics of tricyclazole residues on leaf at each position in the field condition (Ishiguro *et al.*, 1992a). Letters, *n* and *n*-1, *n*-2, and *n*-3 mean flag leaf, second, third, and fourth leaf position. The flag leaf has not yet emerged at the fungicide spraying.

paddy fields or on nursery boxes before epidemics start. One example is probenazole. The fungicide is applied to the water surface 7–10 days before the onset of leaf blast. Growers cannot predict the initiation of an epidemic, and therefore must apply this fungicide according to the calendar. These types of fungicides are recommended for growers who prefer less labor-intensive rice cultivation, those with fields chronically affected by leaf blast epidemics, or who are risk-avoiding (conservative). Forecasts published after the onset of the epidemics, such as the ones we are trying to provide using simulation models, are not useful for timing the application of granular fungicides. Therefore, we did not include this group of fungicides in our submodel.

Fungicides that are sprayed onto plants are applied typically after epidemics have already begun. Kobayashi (1984) suggested that the first application should be done to inhibit third generation lesions. However, he realized that this was a conservative approach. Our experience suggests that this timing was more effective than others; however, no application sometimes produced the most economical results. Our objective is to find the most effective and efficient fungicide application strategies, and modelling the dynamics and mode of action of the fungicide sprays is an essential step. The fungicides are divided into those that have protection activity and those that have after-infection activity (Szkolnik, 1978). We have made efforts to model both of these activities.

A typical protectant fungicide is a melanin biosynthesis inhibitor (MBI, Table 25.1). This group mainly inhibits appresorial penetration (Woloshock and Sisler, 1982). To model the activities of the group, Ishiguro *et al.* (1988, 1992a) investigated the dynamics and effectiveness of two MBIs, fthalide and tricyclazole, sprayed on rice leaves. They estimated parameters for initial deposition, weathering and concentration–response relationship of the MBIs on rice leaves (Fig. 25.2), and found that redistribution of the fungicides occurred from fully expanded

Table 25.2. Postsymptom activities of several fungicides for blast disease[a].

Common name	Formulation	Lesion expansion rate ($mm^2\ d^{-1}$)	Sporulation capacity (spores × 10^3 $lesion^{-1}$)
Blastcidin-S	1% solution	2.33	$-0.026\ t^2 + 0.30\ t + 0.88$[c]
EDDP	30% emulsion	1.49 (6)[b]	$1.4 \exp(-0.17\ t)$
Ferimzone	30% wettable powder	1.76 (6)	$0.28 \exp(-0.35\ t)$
IBP	48% emulsion	1.17	$0.025\ t^2 - 0.34\ t + 1.4$
Isoprothiolane	40% emulsion	1.31	$-0.024\ t + 0.44$
Kasugamycin	2% solution	2.84 (4)	$-0.088\ t + 2.6$
Control		4.18 (6)	$0.031\ t + 2.9$

[a] Each fungicide was diluted ×1000 with water and sprayed 7 ml per pot just after lesion appearance.
[b] Days that lesion expansion reaches maximum size after lesion appearance.
[c] t days after fungicide spraying.

leaves to newly emerged leaves in field conditions, and that the phenomenon is critical for the longevity of the effectiveness. These parameters made it possible to model the activities of these fungicides. Some researchers claimed that tricyclazole had other modes of action. Nasuda (1977) showed that this fungicide inhibited sporulation. Okuno *et al.* (1983) suggested that tricyclazole reduced the penetration of the spore formed in lesions. However, we have not yet quantified those activities under field conditions.

Other fungicides commonly have after-infection activity, including pre- and postsymptom activity (Szkolnik, 1978), but in Japan these are often referred to as 'curative'. Some of these have both protection and after-infection activity. Ishiguro *et al.* (1992c) assessed postsymptom activity of several fungicides at the concentration that growers usually use. The fungicides suppressed sporulation as well as restricted lesion enlargement (Table 25.2). They also assessed presymptom activity of these fungicides (unpublished, data not shown). Some of those fungicides almost completely suppressed the appearance of new lesions when these were sprayed 2–3 days after infection. These fungicides have protectant activity when sprayed before infection.

The results of these experiments showed that characteristics of the fungicides were diverse. What is even more complicated is that there are several formulations that combine fungicides so that the formulation has both protectant and after-infection activity. Once the models are perfected, we will incorporate them into BLASTL and then evaluate several fungicide use strategies via trial and error simulation. The next step after this is to optimize the spraying strategies using BLASTL. There may be several proper algorithms for optimization (Reynolds, 1993). The optimal solution may identify general principles which can then be tested in field experiments.

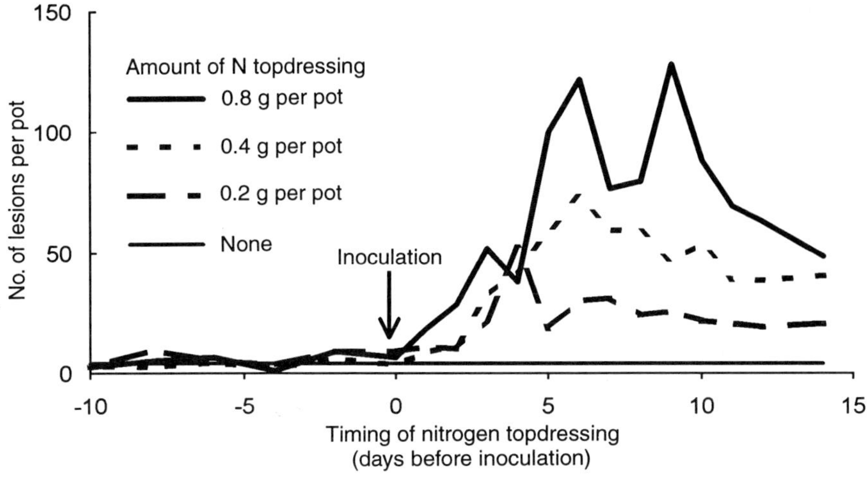

Fig. 25.3. The effect of timing nitrogen (N) topdressing application on susceptibility of rice to blast fungus. The susceptibility was evaluated by the number of primary lesions per plant (Ishiguro and Naito, 1993).

Nitrogen application submodel

There is a general agreement that excess nitrogen application to rice would induce severe outbreaks of blast disease (Hori, 1898; Kozaka, 1965). From the agronomical point of view, split application of nitrogen with careful diagnosis is recommended to achieve high yields and to prevent serious lodging (Matsushima, 1976). Also, split applications of nitrogen are generally considered to be a better strategy for rice blast management than applying nitrogen once as a basal fertilizer. Unfortunately, nitrogen topdressing sometimes enhances disease development (Kozaka, 1965). Although there are many studies concerning the relationship between timing of nitrogen topdressing and disease severity (Kürschner *et al.*, 1992), most of those unfortunately did not specify the weather conditions during the experiments. If the weather conditions were specified, we could analyze which timing of nitrogen application enhanced the infection. Because nitrogen topdressing, as well as fungicide spraying, is an important issue for rice health management and the growers have to make the decisions about these, we are trying to design a submodel for the effect of nitrogen topdressing.

We have obtained quite a simple relationship between susceptibility of rice to leaf blast and timing of nitrogen topdressing using controlled inoculation experiments (Ishiguro *et al.*, 1992b; Ishiguro and Naito, 1993; Fig. 25.3). The susceptibility increased immediately after application of nitrogen fertilizer (ammonium sulfate), reached a maximum level after a week, maintained its level for the following several days, and declined to a certain level higher than the initial one. The amount of nitrogen application influenced the maximum and final susceptibility level and the duration that the maximum susceptibility was sustained (Fig. 25.3). Nitrogen topdressing within 4 days before or after

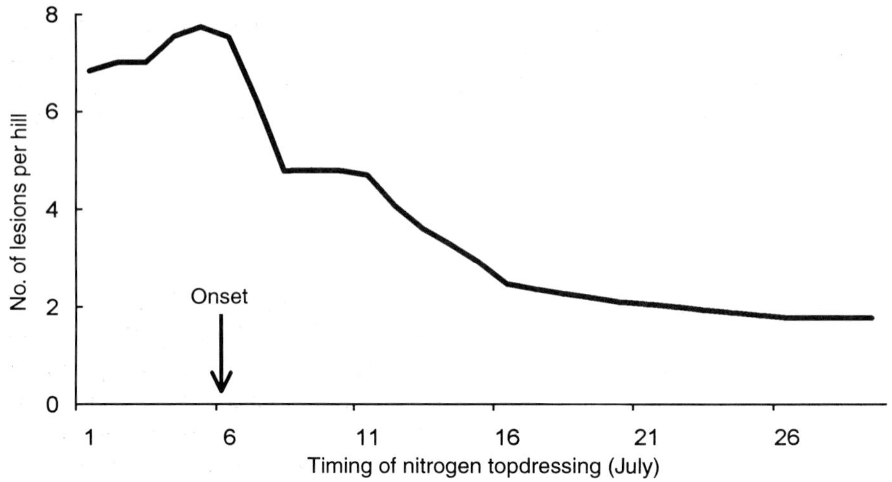

Fig. 25.4. Predicted severity of epidemics on July 31 under various timing of N application using the BLASTL model. The arrow means the onset of the leaf blast epidemic (Ishiguro and Naito, 1993).

inoculation increased sporulation capacity, although the effect was not significant (unpublished, data not shown).

These relationships were modeled and incorporated in BLASTL. The results of the simulation suggested that nitrogen topdressing at the early phase of leaf blast epidemics would cause more severe disease development than at the late phase (Fig. 25.4). The results of our field experiment agreed with this tendency. However, we have not confirmed that the relationship would be universal. Like timing fungicide application, it should be important to establish general principles for timing of nitrogen topdressing. Careful combination of the simulation and field experiments in an iterative process should enable us to approach the ultimate goal.

Panicle Blast Model, PBLAST

Although leaf blast may decrease the yield somewhat, panicle blast is the main cause of yield loss (Katsube, 1987). Leaf blast mainly plays the role of supplying inoculum (spores on leaf lesions) to panicles. So, developing a panicle blast model is more desirable than a leaf blast model. Ishiguro and Hashimoto (1988, 1990) developed a panicle blast model, PBLAST, which was written in FORTRAN77. The weather factors used in BLASTL or PBLAST are the same. PBLAST outputs the yield (or yield loss) of each rice panicle or the number of damaged grains per panicle.

PBLAST is a more ambitious model than BLASTL, because the latter is a deterministic model whereas the former is a stochastic model. Stochastic processes were incorporated in the components of spore deposition and spore penetration of PBLAST. This is because the structure of the panicle is too complex to model disease development deterministically. The model is executed using a random number generator and Monte Carlo simulation. The results obtained from each simulation run are different. Ishiguro and Hashimoto (1990) showed that at least ten iterative runs would be enough to obtain an unbiased estimate of the mean disease progress curve.

Validation of the model

Objective evaluation of PBLAST is necessary to confirm its effectiveness as a decision aid for disease management. Ishiguro and Hashimoto (1990) showed that the general behavior of the model was similar to that of a real system using 5 years of weather data. To test the model further, Hashimoto *et al.* (1993b) used 5 years of weather data obtained after the model was published, and showed that the model mimicked the real system, although fitness of the model to the real disease progression was not adequate. This might be because the model contains several tentative components that were not experimentally determined (i.e. the model still needs improvement).

Sensitivity analysis and model improvement

Compared with the leaf blast pathosystem, less quantitative knowledge has been accumulated about the components of panicle blast, possibly because experiments with panicle blast are more difficult than with leaf blast. To improve the performance of the model, many experiments for each component will be necessary. However, it is not realistic to quantify most of these components by experiments. Another strategy is to choose the critical components for performance of the model via sensitivity analysis and then make detailed experiments to quantify only those components.

The procedure is first to change the values of parameters slightly, and to assess the relative effect of each parameter on the response of the model. The relative significance of the parameters on the model's behavior should be judged by the relative sensitivity, R, as follows:

$$R = \frac{(\partial z/z)}{(\partial X_i/X_i)}$$

in which z and X_i are the values of the model's response (disease progress) and each parameter, respectively. Ishiguro and Hashimoto (1990) assessed the relative sensitivity of several parameters. The results of the sensitivity analyses showed that three processes, the process from sporulation on leaves to deposition of the spores on a panicle, the spore infection process, and the dieback process (lesion extension toward the proximal direction from a particular portion in a panicle; Hirano and Goto, 1963) are the most sensitive processes.

Sensitivity analysis showed that lesion extension from the spikelets on secondary rachilla was the most important component in the dieback phenomenon (Ishiguro and Hashimoto, 1990). The dieback phenomena were studied by

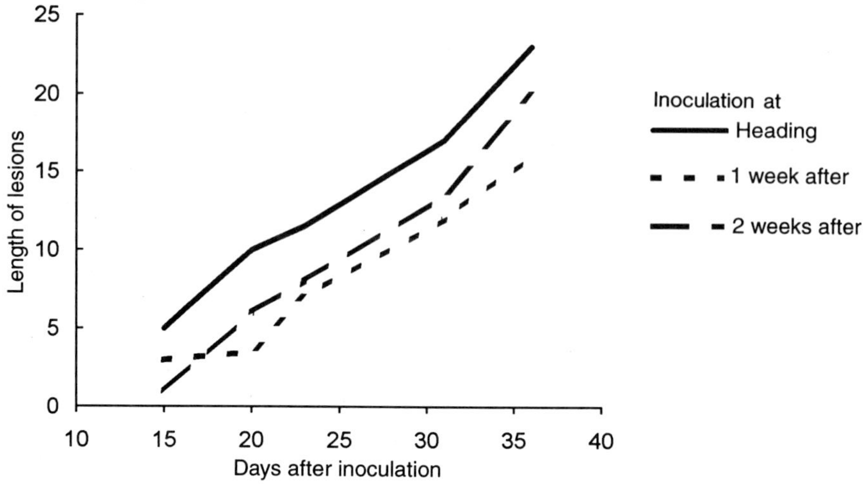

Fig. 25.5. Lesion expansion from a spikelet on secondary rachilla to proximal direction. These data were obtained by potted experiments (Harasawa *et al.*, unpublished).

Harasawa *et al.* (Fig. 25.5, unpublished), and their results allow us to determine the main cause of disease progression in the late phase of panicle blast epidemics. In the early phase of panicle blast epidemics, we know that the lesions are initiated by spores. However, it is unclear what is the main cause of disease progression during the late phase: infection by spores during the late phase or dieback. If late spore infections cause the disease to progress, then the most effective countermeasures should be to prevent late infection, and a protectant fungicide application would be effective. On the other hand, if dieback is the main cause of disease progression during the late phase, then fungicide applications would not be effective as there are no fungicides able to prevent dieback.

Issues Remaining and Future Perspective

Both BLASTL and PBLAST simulate temporal disease progress, and not spatial progress. Leaf blast is a polycyclic disease, the initial inoculum density is relatively low, and the spatial pattern of the first generation of lesions of the general epidemic in a field is usually random. This means that a component of spatial progression is not necessary in the BLASTL model, since our objective is to forecast disease severity with or without control measures. On the other hand, panicle blast has several disease cycles, the amount of initial inoculum (spores from leaf lesions) may be abundant and play an important role in the epidemics (Ishiguro and Hashimoto, 1990; Hanzawa *et al.*, 1993), and the susceptible tissues (panicles) are usually exposed for a long time against inoculum; spatial consideration may then be important for PBLASTL.

The occurrence of general epidemics of leaf blast are not explained by the BLASTL model. Additional qucstions need to be answered. For example, what is the inoculum of the general epidemics of leaf blast; how far can the inoculum disperse? Recent advances in molecular biology (Hamer *et al.*, 1989) may provide an innovative tool to answer these questions.

With the submodels for control measures, it will be easy to evaluate rice blast management strategies via simulation. However, since multiple decisions are usually necessary for management, it will not be easy to find the optimum strategy (mainly from an economic point of view). Mathematical optimization methods may be suitable for that purpose, although this has been rarely applied in disease management programs (Reynolds, 1993).

Regardless of the optimization methods employed, highly predictive simulation models are always desired. To predict future disease progression in the current season, we usually use the weather data up till the present time, followed by one of several historical weather data sets as the subsequent future weather data in our models. This procedure gives a probability of future disease progression (an uncertain future event). This probability is imperfect information about uncertain future events (we can never obtain the perfect information about future events), and is therefore sometimes not satisfactory. To introduce weather forecasting (Vincelli and Lorbeer, 1989; Bergeron and Fry, 1992; Raposo *et al.*, 1993) into our models will be an attractive way to improve the performance of the models, and should be a next step.

References

Bergeron, S.N. and Fry, W.E. (1992) Potential contribution of weather forecasts to improved fungicide spray scheduling and efficacy of potato early blight control (abstract). *Phytopathology* 82, 1091.

Campbell, C.L. and Madden, L.V. (1990) *Introduction to Plant Disease Epidemiology*. John Wiley & Sons, New York.

Fry, W.E. (1982) *Principles of Plant Disease Management*. Academic Press, New York.

Hamer, J.E., Farrall, L., Orbach, M.J., Valent, B. and Chumley, F.G. (1989) Host species-specific conservation of a family of repeated DNA sequences in the genome of a fungal plant pathogen. *Proceedings of the National Academy of Sciences of the United States of America* 86, 9981–9985.

Hanzawa, S., Ishiguro, K., Nemoto, F. and Hashimoto, A. (1993) Analysis of factors of rice panicle blast by PBLAST a computer simulation model *Annals of the Phytopathological Society of Japan* 59 (in press) [in Japanese, abstract].

Hashimoto, A., Hirano, K. and Matsumoto, K. (1984) Studies on the forecasting of rice leaf blast development by application of the computer simulation. *Special Bulletin of Fukushima Prefecture Agricultural Experiment Station* 2, 1–104 [in Japanese, English summary].

Hashimoto, A., Nemoto, F. and Hanzawa, S. (1993a) Analysis of rice blast epidemic using by simulator BLASTL. *Annals of the Phytopathological Society of Japan* 59, 62–63 [in Japanese, abstract].

Hashimoto, A., Ishiguro, K., Hanzawa, S. and Nemoto, F. (1993b) Validation of PBLAST, a simulator of rice panicle blast epidemics (abstract). *Abstracts of 6th International Congress of Plant Pathology*, 106.

Hirano, K. and Goto, K. (1963) Pathogenesis and ecology of panicle branch blast of rice plant. *Bulletin of the National Institute of Agricultural Sciences* C16, 1–66 [in Japanese, English summary].

Hori, S. (1898) Blast disease of rice plants. *Imperial Agricultural Experiment Station, Special Report* 1, 1–36 [in Japanese].

Ishiguro, K. and Hashimoto, A. (1988) Development of a simulation model for rice panicle blast epidemics. (1) Composition of the basic model. *Bulletin of Fukushima Prefecture Agricultural Experiment Station* 27, 1–19 [in Japanese, English summary].

Ishiguro, K. and Hashimoto, A. (1990) Development of a simulation model for rice panicle blast epidemics. (2) its verification and sensitive analysis. *Bulletin of Fukushima Prefecture Agricultural Experiment Station* 29, 15–70 [in Japanese, English summary].

Ishiguro, K. and Hashimoto, A. (1991) Computer-based forecasting of rice blast epidemics in Japan. In: *Rice Blast Modeling and Forecasting*. International Rice Research Institute, Los Baños, The Philippines, pp. 39–51.

Ishiguro, K. and Naito, H. (1993) Strategies for nitrogen fertilizer application to minimize risks of leaf blast epidemics of rice based on a simulation model (abstract). *Abstracts of 6th International Congress of Plant Pathology*, 105.

Ishiguro, K., Yoshitake, S., Watanabe, S., Nemoto, F. and Hashimoto, A. (1988) Temporal dynamics of fthalide on rice leaves in the field and its efficacy for preventing leaf blast. *Annual Report of the Society of Plant Protection of North Japan* 39, 63–65 [in Japanese, English summary].

Ishiguro, K., Miyagawa, S., Nemoto, F., Hashimoto, A., Ando, I., Ling, Z.Z. and Kiyosawa, S. (1989) Frequency distribution of fields versus severity of rice blast disease in a particular district (Preliminary note). *Annual Report of the Society of Plant Protection of North Japan* 40, 11–13 [in Japanese, English summary].

Ishiguro, K., Hashimoto, A. and Naito, H. (1991) Estimating parameters in BLASTL, a simulator of rice leaf blast epidemics, and its validation using the method of non-linear least squares. *Annals of the Phytopathological Society of Japan* 57, 399 [in Japanese, abstract].

Ishiguro, K., Takechi, S. and Hashimoto, A. (1992a) Behavior of tricyclazole residue on rice leaves and its efficacy for rice leaf blast control in the field. *Annals of the Phytopathological Society of Japan* 58, 259–266.

Ishiguro, K., Hayashi, N. and Naito, H. (1992b) Modeling temporal changes in susceptibility of rice leaves to blast fungus following nitrogen topdressing. *Annals of the Phytopathological Society of Japan* 58, 546 [in Japanese, abstract].

Ishiguro, K., Hayashi, N. and Naito, H. (1992c) Inhibitory effects of several fungicides on leaf lesion enlargement and spore formation on the lesions of rice blast. *Proceedings of Kanto-Tozan Plant Protection Society* 39, 13–16 [in Japanese, English summary].

Katsube, T. (1987) Relationship between timing of leaf blast infection, disease severity, and rice quality and yield. In: Yamanaka, S. and Yamaguchi, T. (eds.), *Ine Imotibyou*. Yokendo, Tokyo, pp. 76–183 [in Japanese].

Kernighan, W. and Plauger, P.J. (1978) *The Elements of Programming Style*. McGraw-Hill, New York.

Kobayashi, J. (1984) Studies on epidemic of rice leaf blast, *Pyricularia oryzae* Ca., in its early time. *Bulletin of the Akita Agricultural Experiment Station* 26, 1–84 [in Japanese, English summary].

Kobayashi, J. (1985) Method of field survey to detect general epidemics of rice blast in its starting time. *Shokubutu Boeki* 40, 429–432 [in Japanese].

Kozaka, T. (1965) Control of rice blast by cultivation practices in Japan. In: *Proceedings of the Symposium on the Rice Blast Disease*. Johns Hopkins Press, Baltimore, Maryland, pp. 421–438.

Kürschner, E., Bonman, J.M., Garrity, D.P., Tamisin, M.M., Pabale, D. and Estrada, B.A. (1992) Effects of nitrogen timing and split application on blast disease in upland rice. *Plant Disease* 76, 384–389.

Matsushima, S. (1976) *High-Yielding Rice Cultivation, a Method for Maximizing Rice Yield Through 'Ideal Plants'*. University of Tokyo Press, Tokyo.

Nasuda, K. (1977) Control effects and modes of action of tricyclazole (EL-291, Beam) on rice blast. *Bulletin of Fukui Agricultural Experiment Station* 14, 85–106.

Okuno, T., Kitamura, Y. and Matsuura, K. (1983) Mechanism of inhibitory effect of tricyclazole on secondary infection by spores of *Pyricularia oryzae*. *Journal of Pesticide Science* 8, 361–362.

Raposo, R., Wilks, D.S. and Fry, W.E. (1993) Evaluation of Potato late blight forecasts modified to include weather forecasts: A simulation analysis. *Phytopathology* 83, 103–109.

Reynolds, K.L. (1993) Simulating disease development and economic impact and the application of mathematical optimization methods for plant disease management (abstract). *Abstracts of 6th International Congress of Plant Pathology*, 25.

Szkolnik, M. (1978) Evaluation of physical modes of action of fungicides against the apple scab fungus. *New York State Agricultural Experiment Station Special Report* 28, 22–27.

Takeda, S. (1992) Seasonal occurrence of rice leaf blast at its early stage in small area of Iwate Prefecture in 1991. *Annual Report of the Society of Plant Protection of North Japan* 43, 9–12 [in Japanese].

Teng, P.S. (1981) Validation of computer models of plant disease epidemics: A review of philosophy and methodology. *Zeitschrift für Pflanzenkrankheiten Pflanzenpathologie und Pflanzenschutz* 81, 49–63.

Vincelli, P.C. and Lorbeer, J.W. (1989) BLIGHT-ALERT: A weather-based predictive system for timing fungicide application on onion before infection periods of *Botrytis squamosa*. *Phytopathology* 79, 493–498.

Waggoner, P.E. and Horsfall, J.G. (1969) EPIDEM, a simulator of plant disease written for computer. *Bulletin of the Connecticut Agricultural Experiment Station* 698, 1–80.

Woloshuk, C.P. and Sisler, H.D. (1982) Tricyclazole, pyroquilon, tetrachlorophthalide, PCBA, coumarin and related compounds inhibit melanization and epidermal penetration by *Pyricularia oryzae*. *Journal of Pesticide Science* 7, 161–166.

Zadoks, J.C. and Schein, R.D. (1979) *Epidemiology and Plant Disease Management*. Oxford University Press, New York.

26 Blast Management in High Input, High Yield Potential, Temperate Rice Ecosystems

CHANG KYU KIM

Agricultural Sciences Institute, Rural Development Administration, Suweon, Korea

INTRODUCTION

As one of the most widely cultivated food crops in the world, annual total production of rice is estimated at 520 million tonnes. Since the first description of rice blast by Soong Ying-Shin in the early 17th century (Ou, 1985), blast has become the biggest constraint to stable rice production. Rice blast still remains a serious problem where rice is cultivated regardless of rice ecosystems. The major reason for this might be the complexity of blast fungus life cycle. This chapter deals with general information on rice cultivation in Korea, the complexity of rice blast fungus, the epidemiology of rice blast, and blast management.

GENERAL INFORMATION ON RICE CULTIVATION IN KOREA

The total area of paddy rice in Korea is 1.2 million ha. As of 1990, total rice production was 5.6 million tonnes (Table 26.1; MAFF, 1991). The area in Korea changed from 1.13 million ha in 1960 to 1.24 million ha in 1990, while yield per ha increased from 2.21 tonnes in 1960 to 4.51 tonnes in 1990. During this period total fertilizer used, including on upland crops, changed from 137,000 tonnes to 1.1 million tonnes, and fungicide use from 1609 tonnes a.i. to 7778 tonnes a.i. Meanwhile, yield loss by blast was estimated as 5.6% in 1960, 4.2% in 1970, 3.9% in 1980 and 0.02% in 1990. Reasons for the recent decrease in yield loss by blast will be discussed later.

Ninety-four percent of the total rice area is machine transplanted, with 8-day-old seedlings preferred by farmers due to a shortage of rural labor. Table 26.2 shows rice area affected by outbreaks of rice disease for the period 1983–1992. Sheath blight has consistently been the most prevalent disease in the past 10 years. Rice blast has been most destructive in the temperate rice regions. However, the severity of blast epidemics in Korea has been reduced in recent years.

Table 26.1. Rice cultivation area, yield, fertilizer and fungicide consumption and yield loss caused by blast (MAFF, 1991; RDA, 1992).

Year	Acreage (Mha)	Yield ha^{-1} (Mt)	Fertilizer[a] (Mt)	Fungicide (a.i.Mt)	Yield loss caused by blast (%)
1960	1.13	2.21	137,000	1609	5.6
1970	1.21	3.25	563,000	828	4.2
1980	1.22	2.89	828,000	5448	3.9
1990	1.24	4.51	1,104,000	7778	0.02

[a] Fertilizer consumption includes total amount to all crops.

Table 26.2. Area of major rice disease outbreaks in Korea, 1983–1992 (1000 ha; RDA 1992).

Disease	1983	1984	1985	1986	1987	1988	1889	1990	1991	1992
Leaf blast	119.6	200.5	192.6	100.3	93.7	65.8	132.3	116.5	99.3	62.1
Panicle blast	7.1	5.0	8.3	9.3	12.5	4.3	11.7	3.2	4.8	7.3
Sheath blight	543.5	479.0	590.5	515.1	522.4	657.0	527.5	534.4	546.3	488.7
Bacterial blight	20.2	28.7	91.1	110.0	43.4	19.6	24.2	5.0	11.9	8.4
Stripe virus	4.6	7.6	8.2	6.2	5.4	2.5	4.7	11.3	8.3	0.5

Complexity of Blast Fungus

Rice blast fungus, *Pyricularia oryzae*, is one of the most sensitive plant pathogens to temperature changes. A series of steps are needed for conidia in a disease cycle: germination, appressorium formation, penetration, colonization and symptom development.

Germination

Germination of blast fungus conidia was shown to be best at 25–30°C, but impossible at 10–15°C (Nisikado, 1927). The optimum temperature for germination was reported to be 25–28°C by Sueda (1928). Spores of *P. oryzae* were found to germinate well at temperatures of 20, 25 and 30°C in a thin film of water (Sadasivan *et al.*, 1965). Suzuki (1969a) reported spore germination was better at higher temperatures up to 28°C. Germination rate of conidia was shown to approach 100%, 10 h after inoculation at 25 ± 1°C regardless of light or dark treatment (Kim *et al.*, 1974). These results further suggested the optimum temperature for germination was 25°C.

Appressorium formation

There appears to be divergent opinions among researchers on the effect of temperature on appressorium formation. Ito and Kuribayashi (1931) concluded that appressorium formation was better at 15–18°C than at 28°C even with reduced germination at lower temperatures. Appressoria were formed 15 h after incubation at 20–32.5 °C and 24 h after incubation at 25–35 °C (Suzuki, 1941). The optimum temperature was 28 °C for certain races (Suzuki, 1952), but different temperatures were required for appressorium formation for different races (Fujikawa *et al.*, 1954). Suzuki (1969a) reported the optimum temperature was 20 °C but that appressoria were formed well at 25, 16 and 14 °C. Yoshino (1973) concluded that 24 °C appeared to be the optimum temperature for appressorium formation. Later experiments indicated that appressorium formation began 4 h after inoculation and reached 60–80% completion 15 h after inoculation at 25 ± 1 °C (Kim *et al.*, 1974).

Penetration

In the previous studies of rice blast pathogenesis (Toyota and Suzuki, 1952; Kahn and Libby, 1958; Ohata *et al.*, 1963; Chiba *et al.*, 1968, 1973, 1977; Suzuki, 1969b; Sekiguchi and Furuta, 1970; Kato and Sasaki, 1974), penetration phase on the leaf blade was the least known parameter in spite of being the most critical factor in the development of an epidemic. Early studies dealt mostly with the mechanism and point of penetration and the factors favorable for penetration to occur. Since Sakamoto (1949, 1951) reported findings on appressorium formation, penetration and two types of cell response by leaf sheath inoculation technique, the method was intensively used (Takahashi 1951, 1956; Ohata *et al.*, 1963; Kiyosawa and Lee, 1975). A new method for observing hyphae on the leaf blade epidermis was developed by Yoshino (1971) and utilized to differentiate penetration pattern of blast fungus with regard to temperature treatment.

Colonization

Blast fungus activities were compared with seedlings by temperature treatments (Kim and Mogi, 1985). Seedlings were exposed to high temperature regimes of 29/21 °C and greenhouse conditions (33–23 °C) and low temperature regimes of 23/15 °C for 3 days prior to and 3 days after completion of inoculation. Hyphal growth in the leaf epidermis of seedlings exposed to higher temperature regimes was accelerated during 72–96 h after inoculation, while that of seedlings exposed to lower temperature regimes was apparently suppressed. The same phenomenon was found with regard to the number of cells in which hyphal growth was observed. For example, the mean number of cells in which hyphal growth was observed was 5–6 cells at the 23/15 °C regime, but reached around 30 cells for the 29/21 °C regime or greenhouse conditions at 96 h after inoculation. The suppression of blast fungus activity after penetration under a low temperature regime was directly associated with delay of symptom development.

Symptom development

The range of temperatures suitable for symptom development of rice blast disease seems to be quite variable. Kahn and Libby (1958) stated that infection occurred between 15–35 °C in dew cabinets with 16–20 h of dew exposure with the optimum at 27–29.5 °C. The minimal temperatures reported for symptom development have been 14–15 °C (Chiba *et al.*, 1968, 1977; Suzuki 1969a), 16–18 °C (Chiba *et al.*, 1973), 17 °C (Sasaki and Kato, 1971) and 18 °C (Chiba *et al.*, 1966). The optimum minimum temperatures for disease development reported are 20–23 °C (Chiba *et al.*, 1977), 20°C (Chiba *et al.*, 1973) and 22–26 °C (Sekiguchi and Furuta, 1970). The daily mean temperature during the incubation period was demonstrated to be more closely related to lesion development than was the daily minimum temperature (Chiba *et al.*, 1977).

Life cycle of blast fungus

Under temperate conditions only one rice crop is possible. Transplanting in Korea normally begins from mid-May with a transplanting density of 240,000 plants ha^{-1}. The leaf blast season starts in late June, but the severity of the epidemic depends on the impending weather conditions. Once the blast lesion appears, sporulation, release, dispersal, landing and germination successively occur in the canopy level. The fungus completes the above cycle within one week in temperate rice ecosystems and the secondary cycle repeats almost 8–11 times during the leaf blast season, resulting in a tremendous inoculum population.

A lesion will start producing conidiophores 4 h after dew formation. After a conidiophore starts sporulation, the first conidium matures in 40 min. A second conidium begins to form 20 min after the first matures. Seven to nine conidia are produced from a single conidiophore in one night. Five hours after sporulation, conidia release begins. Lesions reach their release peak 13 h after dew formation (Toyota and Suzuki, 1952).

The amount of sporulation from a single lesion during one night in the field varies from a few hundred to a few hundred thousand depending upon lesion type, age, size and weather conditions (Kim and Kim, 1993). From those sporulated conidia, 1–23% are released. It is very important to realize the complexity of the nature of the inoculum population. The average number of plants per hectare is 240,000; if there is only one lesion per plant and the daily number of conidia released is 2000, then the total number of conidia per day becomes 4.8×10^8. Sporulation from one lesion continues for more than 20 days (Kim, 1987) and there are different types of lesions in one field: fresh, old, short and long. This complexity makes it difficult to manage leaf blast properly.

Epidemiology of Rice Blast

Numerous efforts from many aspects of research have been directed to solve the rice blast problem. However, there still remain unsolved epidemiological questions. Recently, sporulation and release of conidia from naturally infected leaf blast lesions were studied under different weather conditions. In these studies, the growth of lesions and sporulation were measured using single lesions, as follows.

Table 26.3. Successive sporulation of rice blast fungus from discrete lesions in Icheon, Korea, 1991.

	Lesion length (mm)		Number of spores per lesion	
Date	Lesion 1	Lesion 2	Lesion 1	Lesion 2
Jul 1	2.9	4.9	–	–
2	3.8	6.4	240	2180
3	4.9	6.5	840	3780
4	7.0	8.4	1140	1400
5	11.2	9.8	5440	1120
6	13.7	12.9	6520	1980
7	14.8	13.9	12,540	3180
8	15.9	14.9	7260	–[a]
9	17.9	15.3	1440	1960
10	17.9	16.4	6740	2520

[a] Not detectable due to inflow of rain into the capsule.

Sporulation

A discrete lesion-bearing leaf was fixed on the KY-type spore trap (Kim and Yoshino, 1987) and sporulation amount was continuously measured for 9 days (Table 26.3). For comparison, different sized lesions of 2.9 mm and 4.9 mm were tested. For the first 3 days, sporulation amount was more on the larger lesion, but the number was greatly increased from the smaller lesion due to the rapid increment of the lesion size. The highest sporulation amount was 12,540 on the sixth day.

When sporulation amount was measured irregularly from different type and sized lesions, the amount ranged greatly from a few hundred to few hundred thousand (Table 26.4). The highest count was 220,540 from 8.8 mm-long chronic-type lesion. Sporulation amount was also different between natural lesions (pre-existing conidiophores and conidia were not removed) and from removed lesions (Table 26.5). More conidia were formed on natural lesions. In most cases, more spores were formed on the abaxial surface of leaves.

Release

Conidia release patterns can vary with weather conditions, type and age of lesions, as we have measured since 1987. Figures 26.1 and 26.2 illustrate spore catches in different devices: the KY-type spore trap (Kim and Yoshino, 1987) is used for measurement of conidia release; the rotation sampler (Suzuki, 1969b) for measurement of dynamic conidia dispersal and a horizontal spore trap for measurement of static conidia dispersal. Under clear day conditions on 9–10 July, 1987 (Fig. 26.1), conidia release measured by the KY-type spore trap and conidia dispersal measured by the other two traps exhibited one typical peak at 0600–0700 h. Suzuki (1969c) and Kato (1976) reported one peak at 0200–0300 h. However, under continuous cloudy conditions during 14–15 July, 1987, there

Table 26.4. Sporulation of *Pyricularia oryzae* from natural leaf blast lesions in Icheon, Korea, 1991.

Date	Lesion size (mm)	Lesion type[a]	No. released[b]	No. remained[c]	Total sporulation[d]
Jul 5	6.2	Acute	360	8,720	9,080
9	8.9	Chronic	3,400	24,400	27,800
	6.7	Acute	4,880	48,080	52,960
10	8.8	Chronic	14,800	205,700	220,500
	8.3	Chronic	2,080	48,760	50,840
11	15.9	Chronic	2,620	58,860	61,480
	4.4	Acute	300	1,620	1,920
12	14.2	Chronic	5,700	19,520	25,220
	6.4	Acute	1,940	13,520	15,460
24	28.0	Chronic	200	140	340

[a] Acute type: new grayish lesion without brown margin; chronic type: old lesion with yellow surroundings and distinct brown margin. Lesion type is much affected by weather conditions.
[b] Total number of conidia collected from 1800 h on day 1 until 1700 h on day 2 by the KY-type spore trap (Kim and Yoshimo, 1987).
[c] Total number of conidia remaining on both sides of lesion. Remained conidia are thoroughly scraped with a 1 ml hypodermic syringe needle and counted using a hemacytometer.
[d] Total sporulation amount derived from released and remaining conidia during 1700–0900 h. Crop growth stages during the experiment were maximum tillering to panicle initiation and the period was the rainy season. Rainfall amounts were: 5, 9 and 12 Jul: 0 mm; 10 Jul: 0.5 mm; 11 Jul: 1.4 mm and 24 Jul: 15.5 mm.

were 2–3 conidia release and dispersal peaks between 0100 and 0800 h (Fig. 26.2). Conidia release and dispersal were also observed under rainy conditions (Kim *et al.*, 1989). The data presented in Figs 26.1 and 26.2 are representative of the spore release patterns observed on many days over many years and are presented for illustrative purposes.

Blast Management

An interrelationship exists between humans and crops which has been well described by Harlan (1975): 'a fully domesticated plant cannot survive without the aid of man, but only a minute fraction of the human population could survive without cultivated plants'. This simple statement implies that agricultural scientists must exert and maintain continuous research efforts in order that human beings may survive. Blast disease is no exception and therefore varietal resistance, cultural techniques and fungicide control should be considered.

Table 26.5. Comparison of sporulation between natural leaf blast lesions and conidia- and conidiophore-removed lesions in Icheon, Korea, 1991.

	Conidia-removed lesion						Natural lesion					
Date	Size (mm)	Type	Rel[a]	Adax[b]	Abax[c]	Total[d]	Size (mm)	Type	Rel	Adax	Abax	Total
Jul 19	5.9	Acute	320	560	2,060	2,940	8.6	Chronic	8,240	2,220	63,250	73,980
	7.3	Chronic	160	540	580	1,280						
23	6.5	Acute	360	8,240	12,480	21,080	7.0	Acute	560	3,620	24,280	28,460
24	28.0	Chronic	200	100	40	340	18.3	Chronic	2,240	2,840	34,420	39,500
25	9.8	Chronic	160	840	15,520	16,520	10.5	Chronic	4,020	3,440	56,020	63,480

[a] Total number of conidia collected by the KY-type spore trap (Kim and Yoshino, 1987).
[b] Total number of conidia remaining on adaxial surface of lesion. Remaining conidia are thoroughly scraped with a 1 ml hypodermic syringe needle and counted using a hemacytometer.
[c] Total number of conidia remaining on abaxial surface of lesion. Remaining conidia are thoroughly scraped with a 1 ml hypodermic syringe needle and counted using a hemacytometer.
[d] Total sporulation amount derived from released and remaining conidia during 1700–0900 h. Crop growth stage during the experiment was panicle initiation and the period was rainy season.

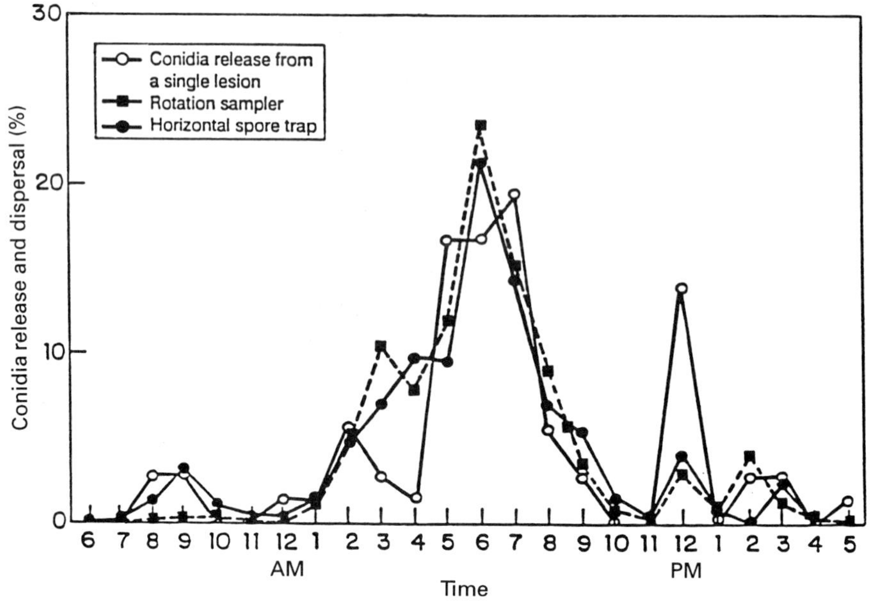

Fig. 26.1. *P. oryzae* conidia release and dispersal pattern, measured hourly by different types of spore trap under clear day conditions in Icheon, Korea, 9–10 July 1987.

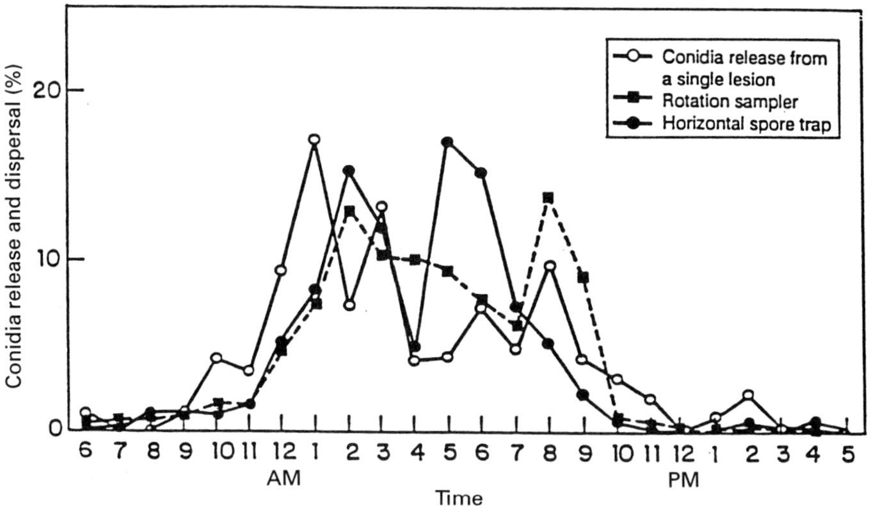

Fig. 26.2. *P. oryzae* conidia release and dispersal pattern, measured hourly by different types of spore trap under cloudy conditions in Icheon, Korea, 14–15 July 1987.

Varietal resistance

The utilization of disease resistance is not merely an *ad hoc* reaction to breakdowns in an otherwise profitable agriculture but requires long-term planning and management. Crop diseases, today, are taken very seriously and, almost without exception, disease control measures add to farming costs. Rice breeders in Korea have produced 133 rice varieties since 1931. Cultivar Tongil, the first *indica-japonica* hybrid variety released in 1971, and its sister varieties showed blast resistance which lasted at least 5 years. These varieties started expression of blast symptoms in 1976 and a severe rice blast epidemic was experienced in 1978. Coons (1953) has stated that the control of disease by use of a disease resistant variety is a painless and inexpensive method for the farmer except that he has to pay for the care and harvest of a larger crop. Our bitter experience of 1978 has shown the fallacy of this statement. There are some examples of breakdown of resistant rice cultivars to blast within a few years after release: Kwanok in 1969 (Lee, 1972); Pi No. 5 in 1962, Kusabue in 1963, Yukara, Senshuraku, Teine and Hatsuiwaimochi in 1964 (Yamada, 1965). Resistant varieties do not necessarily remain indefinitely. However, efforts to improve varietal resistance should be continued. Recently rice breeders and rice pathologists are developing quality rice varieties with resistance to blast.

Cultural techniques

Plants should be grown under proper nutrient conditions. Recommended fertilizer amount for *japonica*-type rice in Korea is 110 kg N, 70 kg P and 80 kg K ha^{-1}. Farmers used to apply 50% to 100% more nitrogen to get higher yield. This was one of the reasons for the greater blast problem in the past. In recent years less nitrogen is being used in order to reduce the high production costs for extra fertilizer, labor and chemical spray. At the same time, overdressing of nitrogen results in prolonged vegetative growth duration and delay in heading.

For machine transplanting, 130 g seeds per nursery box are needed for 35-day-old seedlings and 200 g seeds are needed for 8-day-old seedlings. Thus the latter provides more favorable conditions for leaf blast outbreak due to a larger canopy density.

Chemical control

To control blast, five applications of blasticide are recommended: one granule application to seedling box just before transplanting, two applications for leaf blast during July and two applications for neck blast just before and after heading. This schedule is normally good enough to suppress blast, but the farmers need additional applications in severe epidemic years. Blasticides widely used in Korea are Oryzamate, Kitazin, Kasugamycin, Hinosan, Fujione and Beam.

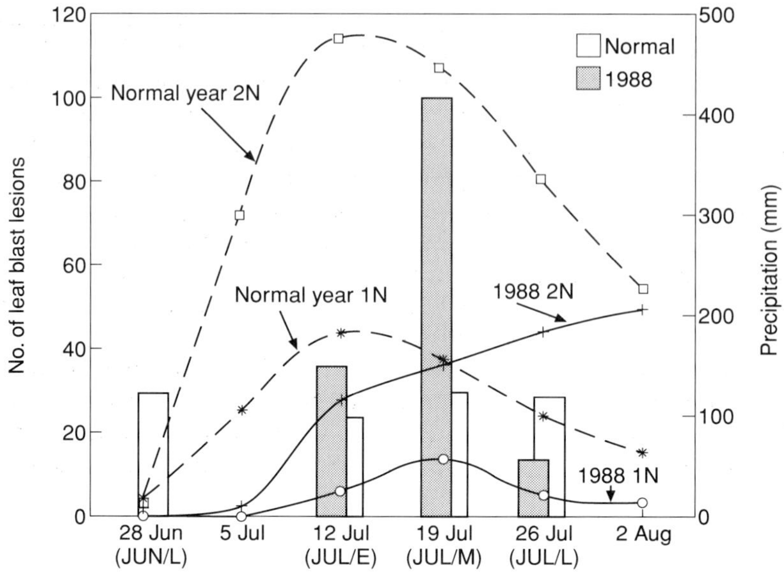

Fig. 26.3. The relationship between rainfall amount and leaf blast severity in Icheon, Korea, late June–early August, 1988.

Weather conditions

Weather conditions which cannot be artificially controlled are the most important factor governing blast epidemic. Temperature exceeding 30°C suppresses blast, but below 19°C for a few days during the reproductive stage itself gives cold damage to rice plants and also results in severe blast epidemic in temperate countries. The dew period should be continued for more than 10 h per day. Rainfall intensity over 3.5 mm h^{-1} is known to be suppressive factor against blast fungus behavior (Suzuki, 1969c). In recent years, we frequently experienced strong showers during the leaf blast season and this became one critical factor in reducing leaf blast. The data presented in Fig. 26.3 are representative of the relationship between rainfall amount and leaf blast severity in Icheon, Korea, during late June–late July, 1988. Total amount of rainfall for every 10 days in July was more or less 100 mm in a normal year, but for the first ten days in July 1988 it was 150 mm and for the middle ten days in July it was 420 mm. On the contrary, the number of leaf blast lesions per plant in the 2N plot on July 19 was 35.7 in 1988 and 107.1 in a normal year. This phenomenon demonstrates clearly that leaf blast development is greatly affected by rainfall intensity and amount during a limited period of time.

Disease forecasting system

A total of 151 forecasting stations are installed in Korea for disease and insect pest forecast. One extension worker with the aid of one assistant at each station carries out a disease and insect pest survey from 1 June to 21 September. Spore catch by rotation sampler is performed every day during 1 June–10 September. Other variables, rice plant's growth status, per cent diseased leaf area and number of leaf blast lesions per hill are scored for 10-day intervals from 20 hills until 21 August. Panicle blast severity is scored one time at 30 days after heading. Weather data are automatically collected by weather kits. The Rural Development Administration (RDA) organizes a disease forecasting conference once a week on the basis of data obtained from all stations together with rice growth status and weather data. Each province has its own meeting once a week. The forecast information announced through mass media gives farmers the opportunity to make a decision for proper control.

The recent decrease of blast outbreak in Korea can therefore be attributed to the application of chemical spray just before transplanting, which delays or suppresses initial outbreak of leaf blast, the reduced use of nitrogen fertilizer, and strong rainfall during the blast season which provides unfavorable conditions for blast epidemic.

Conclusion

Blast management is not a simple process. One or two methods alone cannot provide satisfactory blast management. The introduction of resistant varieties and proper cultural practices followed by timely application of fungicide will result in an ideal management of blast disease.

References

Chiba, S., Murakami, J., Shimada, K. and Nakagawara, T. (1966) Studies on the forecasting method of rice blast by means of sheath inoculation method. I. On the relationship between the development of rice blast and the grade of mycelial growth in the sheath cells. *Annual Report for Plant Protection, North Japan* 17, 148–166 [in Japanese, English summary].

Chiba, S., Chiba, J., Shimada, K. and Kagawa, H. (1968) Invasion of blast disease fungus and occurrence of disease under different temperature. *Annual Report for Plant Protection, North Japan* 19, 6 [in Japanese, English summary].

Chiba, S., Chiba, J., Shimada, K. and Ota, K. (1973) Daily temperature change during leaf blast infection and incubation period and number of lesion. *Annals of the Phytopathological Society of Japan* 39, 186 [Abstract in Japanese].

Chiba, S., Chiba, J., Shimada, K., Ota, K. and Kuwata, H. (1977) The influence of temperature on the lesion development by the rice blast disease. '*Pyricularia oryzae* Cavara' during the infection and latent period *Bulletin of the Aomori Agricultural Experiment Station* 22, 67–73 [in Japanese, English summary].

Coons, G.H. (1953) Breeding for resistance to disease. *Yearbook of Agriculture* 1953. USDA 174–192.

Fujikawa, T., Utsunomiya, T. and Okatome, Z. (1954) Difference in cardinal temperature for appressorium formation in conidia of *Pyricularia oryzae*. *Annals of the Phytopathological Society of Japan* 18, 161 [Abstract in Japanese].

Harlan, J.R. (1975) *Crops and Man*. American Society of Agronomy Chapter 1.

Ito, S. and Kuribayashi, K. (1931) Studies on the rice blast disease [in Japanese]. *Department of Agriculture and Forestry, Farm Bulletin, Japan* 30, 1–81.

Kahn, R.P. and Libby, J.L. (1958) The effect of environmental factors and plant age on the infection of rice by the blast fungus, *Pyricularia oryzae*. *Phytopathology* 48, 25–30.

Kato, H. (1976) Some topics in a disease cycle of rice blast and climatic factors. In: *Climate and Rice*. International Rice Research Institute, PO Box 933, Manila, The Philippines, pp. 417–425.

Kato, H. and Sasaki, T. (1974) Epidemiological studies of rice blast disease, with special reference to reproductive process in lesions on rice plants and disease forecast. *Bulletin of the National Institute of Agricultural Sciences Series C* 28, 1–61 [in Japanese, English summary].

Kim, C.K. (1987) Disease forecasting: Past and future in special reference to rice leaf blast. *Korean Journal of Plant Pathology* 3, 304–312.

Kim, C.K. and Kim, C.H. (1993) The rice leaf blast simulation model EPIBLAST. In: Penning de Vries, F.W.T. (eds), *Systems Approaches for Agricultural Development*. Kluwer Academic Publishers, The Netherlands, pp. 309–321.

Kim, C.K. and Mogi, S. (1985) Effect of temperature treatments on penetration and disease development in the leaf epidermis by the rice blast fungus, *Pyricularia oryzae* Cavara. *Bulletin of the Kyushu National Agricultural Experiment Station* 24, 53–119 [in Japanese, English summary].

Kim, C.K. and Yoshino, R. (1987) Epidemiological studies of rice blast disease caused by *Pyricularia oryzae* cavara. I. Measurement of the amount of spores released from a single lesion. *Korean Journal of Plant Pathology* 3, 120–123.

Kim, C.K., Yoshino, R. and Mogi, S. (1974) The effects of light and darkness on the infection of rice blast fungus, *Pyricularia oryzae*. *Proceedings of the Association for Plant Protection of Hokuriku* 20, 3–6 [in Japanese, English summary].

Kim, C.K., Min, H.S. and Yoshino, R. (1989) Epidemiological studies of rice blast disease caused by *Pyricularia oryzae* Cavara. IV. Conidia release under rainy conditions. *Korean Journal of Plant Pathology* 5, 60–64.

Kiyosawa, S. and Lee, E.J. (1975) Expression of genes for true and field resistance to blast disease in cells of leaf sheath of rice plant. *Japanese Journal of Breeding* 25, 145–154.

Lee, E.J. (1972) Epidemic outbreak of blast disease in the resistant variety of rice, 'Kwanok'. *Korean Journal of Plant Protection* 11, 41–43 [in Korean, English abstract].

Ministry of Agriculture, Forestry and Fishery (MAFF) (1991) *Statistical Yearbook of Agriculture, Forestry and Fisheries*. Seoul, Korea [in Korean].

Nisikado, Y. (1927) Studies on rice blast disease. *Japanese Journal of Botany* 3, 239–244.

Ohata, K., Goto, K. and Kozaka, T. (1963) Observations on the reaction of rice cells to the infection of different races of *Pyricularia oryzae*. *Annals of the Phytopathological Society of Japan* 28, 24–30 [in Japanese, English summary].

Ou, S.H. (1985) *Rice Diseases*, 2nd edn. Commonwealth Mycological Institute, Kew, UK, 380 pp.

Rural Development Administration (RDA) (1992) *Plant Protection Report for 1992*. Suweon, Korea [in Korean].

Sadasivan, T.S., Suryanarayanan, S. and Ramakrishnan, L. (1965) Influence of temperature on rice blast disease. In: *The Rice Blast Disease*. Johns Hopkins Press, Baltimore, Maryland, pp. 163–171.

Sakamoto, M. (1949) On inoculation of the leaf sheath of rice with the blast fungus. *Tohoku University Institute of Agricultural Research Bulletin* 1, 120–129 [in Japanese, English summary].

Sakamoto, M. (1951) On the new method of inoculation of rice plants with the blast fungus, *Pyricularia oryzae* Cav. *Tohoku University Institute of Agricultural Research Report* 1–2, 15–23.

Sasaki, T. and Kato, H. (1971) Climatic factors and synoptic weather charts concerning the initial stage of occurrence of leaf blast on rice plants in the paddy field. *Annual Report for Plant Protection, North Japan* 2, 38–43 [in Japanese, English summary].

Sekiguchi, Y. and Furuta, T. (1970) Ecological studies on the rice blast disease on seedling stage. 2. Environmental conditions of primary infection of the disease. *Bulletin of the Chugoku National Agricultural Experiment Station* Series E6, 81–90 [in Japanese, English summary].

Sueda, H. (1928) Studies on the rice blast disease. *Report, Government Research Institute Department of Agriculture*, Formosa 36, 1–130 [in Japanese].

Suzuki, Ha. (1941) Influence of physical and chemical factors upon the formation of appressoria in the conidia of *Pyricularia oryzae*. II. Influence of temperature. *Journal of Botany Japan* 11, 357–376.

Suzuki, Ha. (1952) Variability in germination type of conidia of *Pyricularia oryzae*. *Annals of the Phytopathological Society of Japan* 16, 36 [Abstract in Japanese].

Suzuki, Ho. (1969a) Temperature related to the spore germination and appressorium formation of rice blast fungus. *Proceedings of the Association for Plant Protection of Hokuriku* 17, 6–9 [in Japanese].

Suzuki, Ho. (1969b) Studies on the behavior of rice blast fungus spore and application to outbreak forecast of rice blast disease. *Bulletin of the Hokuriku Agricultural Experiment Station* 10, 1–118 [in Japanese, English summary].

Suzuki, Ho. (1969c) Interrelationship between the occurrence of the rice blast disease and the meteorological conditions. *Journal of Agricultural Meteorology, Japan* 24, 211–218 [in Japanese].

Takahashi, Y. (1951) Phytopathological and plant breeding investigation to determine degree of blast resistance. *Hokkaido Prefectural Agricultural Experiment Station Report* 3, 1–65 [in Japanese].

Takahashi, Y. (1956) Studies on the mechanism of resistance of rice plants to *Pyricularia oryzae*. II. Pathological changes microscopically observed in host cells in which fungus hyphae do not grow well. *Yamagata University of Agricultural Science Bulletin* 2, 37–51 [in Japanese, English summary].

Toyota, S. and Suzuki, N. (1952) Histochemical studies on the lesions of rice blast caused by *Pyricularia oryzae* Cav. I. Some observations on the sporulation on lesions of different types occurring on leaves of the same variety. *Annals of the Phytopathological Society of Japan* 17, 1–5 [in Japanese, English summary].

Yamada, M. (1965) Breakdown of highly blast resistant foreign varieties. *Plant Protection* 19, 231–234 [in Japanese].

Yoshino, R. (1971) A method for detecting hyphae of rice blast in leaf blade epidermis. *Proceedings of the Association for Plant Protection of Hokuriku* 19, 14–17 [in Japanese].

Yoshino, R. (1973) Ecological studies on the infection in rice blast epidemics. II. Inoculation temperature and progressive change of percentage of penetration. *Annals of the Phytopathological Society of Japan* 39, 186 [Abstract in Japanese].

27 Practical Approaches to Rice Blast Management in Tropical Monsoon Ecosystems, with Special Reference to Bangladesh

A.K.M. SHAHJAHAN

Plant Pathology Division, Bangladesh Rice Research Institute, Gazipur 1701, Bangladesh

INTRODUCTION

Rice blast is a cosmopolitan disease. It is prevalent in tropical forest, tropical monsoon, subtropical and temperate regions and in the continents of Africa, America (North and South), Asia, Australia and Oceania and Europe (Ou, 1985; Shahjahan *et al.*, 1987a). This suggests that the rice blast pathogen, *Pyricularia grisea*, can survive in a wide range of environments.

Blast may reduce rice yields significantly, particularly in the temperate flooded and tropical upland rice ecosystems (Ou, 1985). This loss in rice yield should be minimized in order to help the marginal and poor farmers of developing countries like Bangladesh, India, and the Philippines. To reduce losses, we need to understand the rice blast pathosystems in different ecosystems. It is known that climate influences disease incidence, distribution and severity (Reichert and Palti, 1967; Rotem, 1978). For blast development, some climatic factors such as leaf wetness period, temperature, relative humidity and rainfall are more important than others (Kahn and Libby, 1958; Hashioka, 1965; Manibhushanrao and Day, 1973; Suzuki, 1975; El-Refaei, 1977).

Knowledge of climatic requirements of a pathogen at critical stages of its development helps to predict disease occurrence provided other factors limiting disease incidence are removed (Bourke, 1970). If historical weather data and pest incidence information are available, it is possible to identify the blast endemic agroecological regions (AER) within countries as well as globally (Coakley, 1990). It would also be important to develop a strategic plan for blast management in blast endemic AERs. Development of forecasting models based on the historical weather data could serve as a tool to alert farmers of different AERs well ahead of time to take protective measures against blast outbreaks.

Epidemiology and Pathosystem Analysis as an Aid to Blast Management

The epidemiological factors conducive to blast development include:

1. The host factor – reaction to the pathogen of cultivars grown in blast endemic areas (Ou, 1985; Bonman and Mackill, 1988; Bonman *et al.*, 1989, 1992).
2. The pathogen factor – source and amount of inoculum, pathotypes/virulence spectrum (Marchetti, *et al.*, 1976; Ou, 1985).
3. The physical factors – temperature, humidity, dew or leaf wetness period, rainfall and wind velocity (Hashioka, 1965; Asai *et al.*, 1967; Manibhushanrao and Day, 1973; Kato 1974; Suzuki, 1975; El-Refaei, 1977; Ou, 1985; Shahjahan *et al.*, 1987a; Yeh *et al.*, 1989).
4. Agronomic or cultural factors – alternate hosts, soil type and moisture conditions, amount of N, P and K applied and silicon content of the soil (Volk *et al.*, 1958; Kozaka, 1965; Paik, 1975; Chung *et al.*, 1980; Lee and Lee, 1980; Bonman *et al.*, 1986; Kim, 1986; Kim *et al.*, 1986; Santos *et al.*, 1986; Mbodj *et al.*, 1987; Yamaughi and Winslow, 1987; Kim and Kim, 1990; Osuna-Conizales *et al.*, 1991; Datnoff *et al.*, 1991, 1992; Kurschner *et al.*, 1992; Winslow, 1992).

Quantitative information on the epidemiology of rice blast in the tropical regions is scanty. Quantitative information is available on the effects of dew period, leaf wetness and temperature on germination, appresorium formation, penetration and latency; of wind on spore dispersal; and of relative humidity on sporulation (Ou, 1985; Teng *et al.*, 1991).

Moisture stress, silicon content of the soil, nitrogen status of the plant, source(s) and amount of initial inoculum, and host plant age are other factors affecting infection blast development and consequently the development of epidermias. The quantitative relationships between these factors may be quite complex, and have not been adequately determined.

Pathosystem analysis is useful in management of diseases (Teng *et al.*, 1991). The blast pathosystem is known to have two major subsystems: the leaf blast pathosystem and the panicle blast pathosystem. They are treated separately because one infects the leaf during the seedling stage and the other infects the panicle usually after flowering or heading. The quantitative relationship of leaf blast to panicle blast is also not accurately known, although the pathogen (isolate) infecting leaves is also able to infect panicles (Templeton *et al.*, 1961; Ou and Nuque, 1963; Bonman *et al.*, 1989; Nahar *et al.*, 1991). Among the components of the pathosystem, germination and appressoria formation, infection, colonization, latency, sporulation, lesion enlargement, spore dispersal and spore survival are important. All the components are influenced by the physical as well as the host factors (Ou, 1985). Obviously, interruptions in the process of any of these components could be effective in managing blast.

Simulation Modeling and Forecasting as Aids to Blast Management

Efficient management is dependent on two factors: (i) being able to forecast blast occurrence well ahead of time; and (ii) to have adequate technologies for its management especially during outbreaks. Many leaf and panicle blast simulation models have been developed in both temperate and tropical regions (Hashimoto *et al.*, 1984; Koizumi and Kato, 1991; Teng and Calvero, 1991; Teng *et al.*, 1991). This reflects progress in structuring our understanding of the pathosystem which is a prerequisite for developing management strategies.

Model-generated forecasts are based on: (i) host plant criteria; (ii) disease features; (iii) spore catch; (iv) weather; (v) weather and spore catch; (vi) weather, spore catch, disease parameters and host plant criteria. Among these, forecasts based on weather appear to provide a practical approach to blast forecasting as the technique would be easier for the farmers to understand and apply for blast management. Much has been accomplished on blast forecasting in the temperate climates (Uehara *et al.*, 1988; Ishiguro and Hashimoto, 1991; Kim and Kim, 1991; Teng *et al.*, 1991). Under tropical and tropical monsoon climates the work on blast forecasting has been limited (Padmanabhan *et al.*, 1971; El-Refaei, 1977; Muralidharan and Venkata Rao, 1980; Kim *et al.*, 1988; Manibhushanrao and Krishnan, 1991; Surin *et al.*, 1991)

Blast in a Tropical Monsoon Ecosystem: Bangladesh

Bangladesh is a typical example of a region with a tropical monsoon climate, situated between 20.4°–26.4° north latitude and 88.1°–92.5° east longitude. It has two predominant seasons: hot and moist (April to October), and cool and dry (November to March). The former is known as Kharif and the latter as Rabi season. The deltaic plain land is endowed with a network of rivers and fertile soils.

Rice is the predominant crop grown throughout the year. Four types of rices are grown: Boro (November–May), Aus (April–July), Broadcast Aman or Deepwater rice (March–November), and Transplanted Aman (July–December). Vegetables, pulses, oilseeds and other cereal crops are usually grown during the Rabi (dry) season. The cropping intensity is 159% (BBS, 1992).

The country is divided into 30 agroecological regions (AER) based on the land and soil types, physical, and ecological factors (FAO, 1988; Fig. 27.1 and Table 27.2). The general pattern of rainfall, temperature and humidity during the year is shown in Fig. 27.2. The period of severe cold weather during the winter months as well as the time of onset of monsoon vary with year and region (Miah *et al.*, 1985; Shahjahan *et al.*, 1987a).

Blast is a disease of major concern in Bangladesh (Miah *et al.*, 1985). Its incidence was reduced after the introduction of the high yielding modern variety IR8 and irrigation from 1966 to 1972. There were still some sporadic occurrences on local varieties. Outbreaks began appearing again from 1973, especially on modern varieties. There is a yearly variation in blast incidence in different AERs and there is a cycle of severe outbreaks of blast in the country every

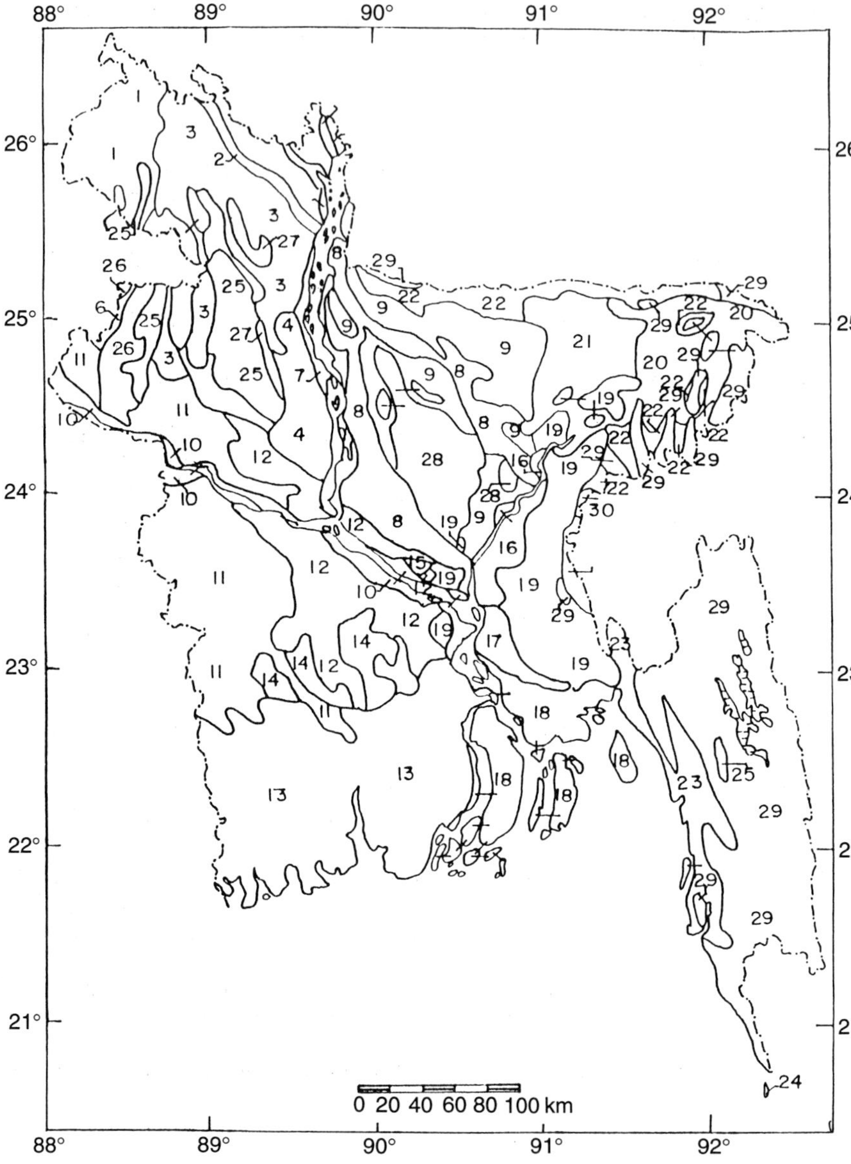

Fig. 27.1. Map of Bangladesh showing the 30 agroecological regions.

4–5 years (Table 27.1). Leaf blast is more prevalent than neck blast. The AERs where blast outbreaks usually occur are situated in the north-east, east, central, south and south-west parts of the country (Table 27.1 and Fig. 27.1). AERs in these areas vary in soil and some physical characteristics (Table 27.2).

Blast appears in Bangladesh during two periods: January–March, and October–November (Fig. 27.2). During these periods, the night temperatures are low, humidity in the morning is high, and the leaves remain wet until 9 o'clock

Table 27.1. History of blast incidence in Bangladesh for the period 1973 to 1993.

Year	Severity of incidence	Season	Type of blast	AER[a]
1973	Very low	Boro	Leaf	8, 9, 19, 20, 28
1974	Mild	Boro	Leaf	8, 9, 19, 20, 28
1975	Mild	Boro	Leaf node	8, 9, 19, 20, 28
1976	Very low	Boro	Leaf	8, 9, 19, 20, 28
1977	Very low	Boro	Leaf	8, 9, 19, 20, 28
1978	Mild	Boro	Leaf	8, 9, 19, 20, 28
1979	Mild	Boro	Leaf, neck	8, 9, 13, 17, 19, 20, 21, 22, 23, 28, 39
1980	Very severe	Boro, Aus	Leaf, leaf	8, 9, 13, 17, 19, 20, 21, 22, 23, 28, 39
1981	Mild	Boro, B. Aman	Leaf, neck Neck	8, 9, 13, 17, 19, 20, 21, 22, 23, 28, 39
1982	Mild	Boro	Leaf	8, 9, 19, 20, 28
1983	Very low	Boro	Leaf	8, 9, 19, 20, 28
1984	Low	Boro	Leaf	8, 9, 19, 20, 28
1985	Mild	Boro	Leaf	8, 9, 19, 20, 28
1986	Severe	Boro T. Aman	Leaf, neck neck	1, 3, 8, 9, 13, 17, 19, 20, 21, 22, 23, 29
1987	Mild	Boro	Leaf	1, 3, 8, 9, 13, 17, 19, 20, 21, 22, 23, 29
1988	Mild	Boro	Leaf	1, 3, 8, 9, 13, 17, 19, 20, 21, 22, 23, 29
1989	Mild	Boro	Leaf	8, 9, 13, 22, 28, 29
1990	Very severe	Boro T. Aman	Leaf, neck neck	1, 3, 8, 9, 13, 17, 19, 20, 21, 22, 23, 28, 29
1991	Mild	Boro	Leaf	1, 3, 8, 9, 13, 17, 19, 20, 21, 22, 23, 28, 29
1992	Mild	Boro, Aus	Leaf Leaf	8, 9, 13, 19, 20, 22, 23, 28
1993	Mild	Boro	Leaf	8, 9, 13, 19, 20, 22, 23, 28

[a] For AER locations see Fig. 27.1 and Table 27.2.
Source *BRRI Annual Report* 1973–1992, tour Reports of scientists of Plant Pathology Division, BRRI, and personal observations.

in the morning (personal observation). These conditions are most favorable for blast (Kahn and Libby, 1958; Hashioka, 1965; Miah *et al.*, 1981; Ou, 1985; Shahjahan *et al.*, 1991). A similar agroclimatic situation prevails in the north-eastern Indian states of West Bengal, Orissa and Assam, where blast occurrence follows the same pattern (Chakraborty, 1971; Padmanabhan *et al.*, 1971; Manibhushanrao and Day, 1973; El-Refaei, 1977).

In Bangladesh, blast is a more serious problem in dry seedbeds or in rices grown in aerobic (sandy type) soil, than in wet seedbeds or in soil that remains flooded (clayey and loamy types of soil) during the growth of the crop. This is due not only to longer periods of leaf wetness in dry land, but also to the increased

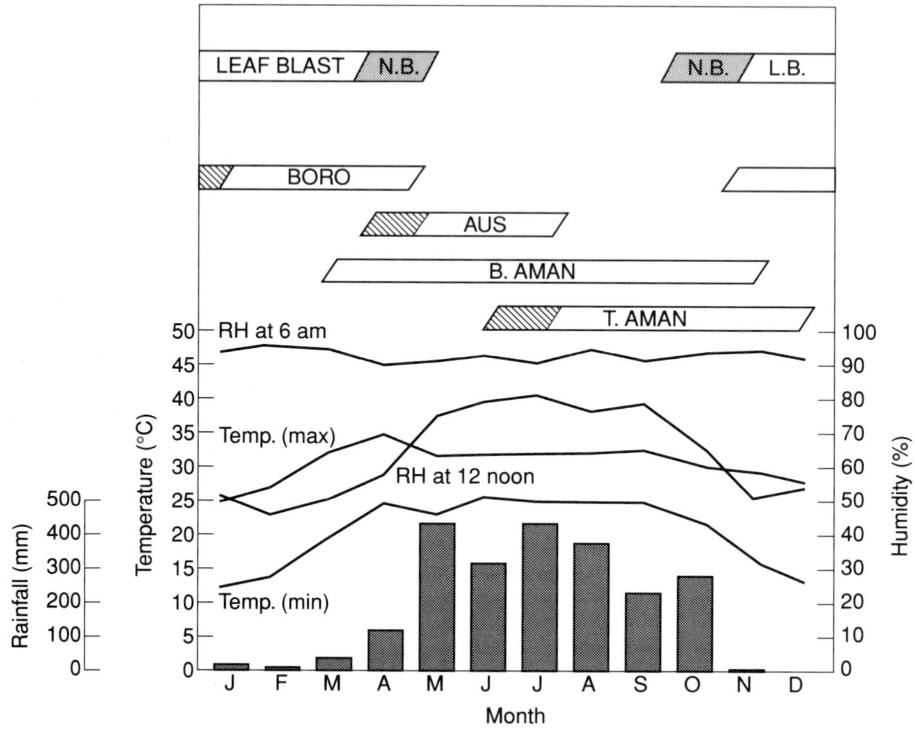

Fig. 27.2. Temperature, humidity and rainfall pattern in the rice growing seasons during an average year and occurrence of blast disease in Bangladesh.

susceptibility of the rice plant after a period of water stress (Bonman *et al.*, 1988; Gill and Bonman, 1988), a common occurrence in many upland rice environments.

Two of the most recent outbreaks occurred in 1980 and 1990 (Miah *et al.*, 1981; Shahjahan *et al.*, 1991). In both years, the humidity ranges during the outbreak period from January to March were higher at 0600 h and 1200 h, and the differences between the maximum and the minimum temperatures were less than in the non-outbreak years as shown in Fig. 27.3 (1989 outbreak year and 1990 non-outbreak year). There was also light but frequent rainfall and the foggy weather kept the rice leaves wet for longer periods in the morning. The cultivars grown, such as BR1, BR11, IR8, IR50, Pajam and some local varieties were susceptible in both the leaf and neck stages. The records clearly indicated the importance of weather as well as genotypes for blast development and hence for forecasting. However, whether the sporadic occurrence of blast in different agroecological regions and its variations in the severity from region to region within Bangladesh is also the attribute of such factors as soil and land types, moisture status of the soil, cultural practices, and cultivar genotypes, in addition to weather factors, has to be investigated through collection of adequate data from the farmers' field.

The pathogen

The blast pathogen, *P. grisea*, is known to be variable. Physiological races have been identified in many countries using international and national differentials (Matsumoto *et al.*, 1967; Marchetti *et al.*, 1976; Ou, 1985). The reports about the stability of these races are contradictory (Latterell, 1975; Ou, 1985). However, it is known that the prevalence of high populations of virulent race(s) in the presence of susceptible cultivars and favorable weather conditions leads to epidemic development. Epidemics of blast can be avoided by growing cultivars resistant to the prevailing virulent race(s) (Crill and Khush, 1979; Khush and Chaudhary, 1981; Bonman *et al.*, 1992). It is, therefore, important to identify the prevalent race(s) and their virulence spectrum so that resistant cultivars can be bred. In Bangladesh, 25 pathotypes based on the international set of differentials have been identified from 245 isolates collected from the BRRI Blast Nursery as well as from farmers' fields. Seven of these are most frequently found. Most of the pathotypes appear to be prevelant in the BRRI blast nursery and on the modern varieties or breeding lines. Among the 25 pathotypes, only six appear to be occurring both in the blast nursery and in farmers' fields, and only four were isolated from both modern and local varieties (Table 27.3). As the pathotypes were identified using international differentials, the picture could be different using local differentials. Molecular analysis of 67 of these isolates show that they belong to seven genetic lineages, most of which appear to be indigenous to Bangladesh (Levy *et al.*, 1993; Shahjahan *et al.*, 1993).

Blast caused by *P. grisea* also occurs on grass species of *Panicum, Digitaria, Leersia*, and *Eleusine* in Bangladesh (Table 27.4). The isolates of *P. grisea* from the blast lesions on these grasses belong to one rice lineage group, and they are able to infect rice producing restricted pinhead type (resistant) mostly nonsporulating lesions. On the other hand, grass lineage type (reduced copy number of the MGR probe hands) isolates of *P. grisea* have also been isolated from rice host and can re-infect rice (Shahjahan *et al.*, 1993). This makes the situation complex, but provides evidence that blast lesions in these grasses could serve as primary sources of inoculum.

The mating types of the isolates of *P. grisea* in Bangladesh have been determined. Among the isolates, 36–46% are Mat 1-1, and 20–28% are Mat 1-2; but 35–36% of the isolates mated with neither Mat 1-1 nor Mat 1-2. Mat 1-1 appears to predominate (Shahjahan and Nahar, 1993; Table 27.5). From this mating type ratio, there is the potential for sexual reproduction in nature which may drive the development of new race(s) in a locality or region.

Status of management of blast in a tropical monsoon ecosystem

Use of high inputs to obtain high yields, cultivation of susceptible cultivars, soil and soil moisture conditions, and development of new races of the pathogen favor blast development. Farmers in countries of the tropical monsoon ecosystem region are not very aware of the cultural practices that help to reduce incidence of diseases in modern cultivars due to lack of effective extension systems exacerbated by

Table 27.2. Soil and some important physical characteristics of blast endemic agroecological regions (AER) of Bangladesh.

AER[a]	Soil type	Soil texture	Land type	Average rainfall (mm)	Date when temp. fell below 20°C (average)	Cool winter period (days)	Days when temp. exceeded 40°C (average)
1	Black terai and non-calcareous brown flood plain	Loamy and sandy	Med.–high	1600–2500	26 Oct.	90–110	5–10
3	Non-calcareous gray brown flood plain	Loamy	Med.–high	1500–2300	24 Oct. 05 Nov.	90–110	5–10
8	Non-calcareous gray flood plain	Loamy	Med.–high	1500–2500	02–12 Nov.	50–70; 70–90	0.5–5
9	Non-calcareous brown gray, dark-gray flood plain	Loamy	Med.–high	2000–4000	02–12 Nov.	50–100	0.5–5
13	Non-calcareous gray flood plain (non-saline and seasonally saline)	Clayey and loamy	Med.–high	1700–3300	07–14 Nov.	30–40; 50–70	0.5–5
17	Calcareous gray flood plain and non-calcareous dark-gray flood plain	Loamy	Med. high–med. low	2000–3000	11 Nov.	50–70	0

18	Non-calcareous gray flood plain (non-saline) and calcareous alluvium	Loamy	Med.–high	2500–3000	08–11 Nov.	30–50	0
19	Non-calcareous dark-gray flood plain	Loamy	Med. high–low	2000–3000	07–15 Nov.	50–70	0
20	Non-calcareous gray flood plain	Clayey and loamy	Med. high–low	2500-5000	29 Oct–12 Nov.	70–90; 90–110	0
21	Non-calcareous gray flood plain and acid basin clays	Clayey	Med. low–very low	2500–5000	29 Oct–12 Nov.	50–70; 70–90	0
22	Gray piedmont soil and acid basin clays	Loamy and clayey	Med. high–med. low	2000–5000	24 Oct–15 Nov.	50–70; 70–90	0.5
23	Non-calcareous gray flood plain, acid sulfate and non-calcareous alluvium	Loamy, clayey and sandy	Med. high–med. low	2500–3500	10–16 Nov.	30–50; 50–70	0
25	Shallow gray and deep gray terrace soils	Clayey	High–low	1200–2000	26 Oct.–05 Nov.	70–90; 90–110	5–10

Table 27.2. *(cont.)*

AER[a]	Soil type	Soil texture	Land type	Average rainfall (mm)	Date when temp. fell below 20°C (average)	Cool winter period (days)	Days when temp. exceeded 40°C (average)
28	Deep red brown and shallow gray terrace soils	Clayey	High–med. high	2000–2300	02–05 Nov.	50–70; 70–90	0.5–5
29	Brown hill soils	Loamy and sandy loam	High	2000–>5000	24 Oct.–15 Nov.	70–90; 30–100	–

[a] 1 = Old Himalyan Piedmont Plain; 3 = Tista Meander Flood Plain; 8 = Young Brahmaputra and Jamuna Flood Plain; 9 = Old Brahmaputra Flood Plain; 13 = Ganges Tidal Flood Plain; 17 = Lower Meghna River Flood Plain; 18 = Young Meghna Estuarine Flood Plain; 19 = Old Meghna Estuarine Flood Plain; 20 = Eastern Surma–Kushiyara Flood Plain; 21 = Sylhet Basin; 22 = Northern and Eastern Piedmont Plains; 23 = Chittagong Coastal Plains; 25 = Level Barind Tract; 28 = Madhupur Tract; and 29 = Northern and Eastern Hills.

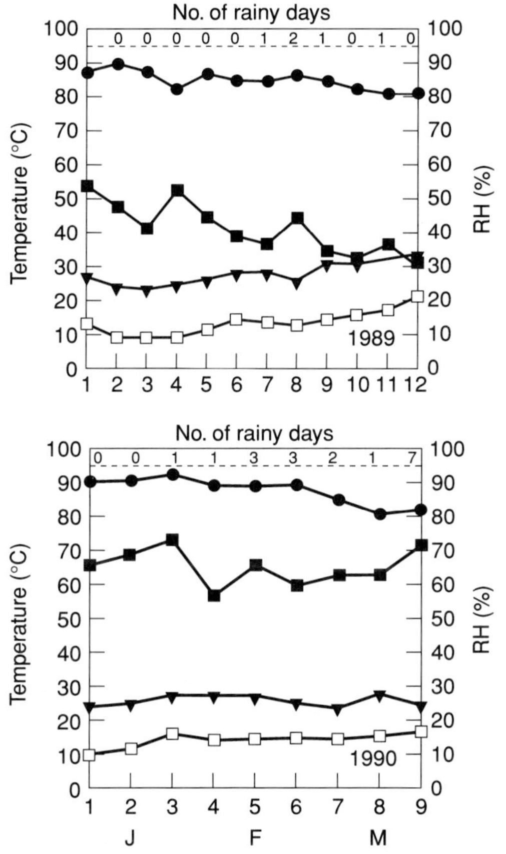

Fig. 27.3. Prevailing relative humidity (RH), temperature and number of rainy days, Jan–Mar, 1989 and 1990, respectively (J = January, F = February, and M = March). Symbols; ●, max. RH; ■, min. RH; ▼, max. temperature; □, min. temperature.

Table 27.3. Distribution of races of *P. grisea* by location and type of cultivars in Bangladesh.

Location/cultivar	No. of isolates	No. of races	Common
Blast nursery, BRRI	196	20	6
Farmers' field	49	11	
Total	245	25	
MV/lines	197	22	4
LV	48	7	
Total	245	25	
Non-rice hosts	27	2	

Table 27.4. Grass hosts of *P. grisea* in Bangladesh, and their pathogenicity to the eight international differentials and mating types.

Grass/ weed host	Places and growing seasons	No. of isolates of *P. grisea*	Pathogenicity to eight int. differentials	Mating types of *P. grisea*
Leersia hexandra (Southern cut grass/ Swamp rice grass)	Grows abundantly in wet season in swampy, shady places, drain levees, and pond ridges	12	Non-pathogenic to all	NK, Mat 1–1
Panicum repens (Torpedo grass)	Grows in both upland and lowland in the dry as well as in wet seasons	10	Pathogenic to one	NK
Digitaria setigera (Crab grass)	Grows in shady places, abundant in upland areas in both dry and wet seasons	5	Non pathogenic to all	–
Eleusine indica (Goose grass)	Grows in road-side places in uplands during both dry and wet seasons	2	Pathogenic to one	Mat 1–2, NK

NK = not known, – = not determined.

prevalent farmer illiteracy. To obtain higher yield, some farmers apply high doses of urea which is not recommended, especially in blast epidemic areas. Some fungicides are available though not in adequate quantities. Farmers in those countries used to apply pesticides frequently when the pesticides were supplied free or were subsidized. Now they are reluctant to spray even one time due to cost.

Other farmers are unable to apply even the basal dose of fertilizers which includes potassium (KCl) or irrigate the seedbed under dry or drought conditions. The choice of blast control techniques is limited for farmers of developing countries like Bangladesh. Researchers and farmers in different climatic regions and ecosystems apply various methods to manage blast. A summarized list of these methods and their effect on blast management is presented in Table 27.6. Adoption of one or more of these methods/techniques will lead to reduced incidence and/or severity of blast. However, it is now well recognized that integrated approaches are more effective and economical than the non-integrated or individual technological approaches. By integrating variety, nitrogen, fungicide treatment and water management the incidence of blast could be reduced under Bangladesh conditions. The contribution of resistant variety, however, appears to be more than the contributions of other components in blast management (Fig. 27.4). Blast severity can be minimized by avoiding high doses of nitrogen

Table 27.5. Distribution of mating types of *P. grisea* by location and type of cultivar in Bangladesh.

Location/cultivars	No. of isolates	Mating types		
		Mat 1–1	Mat 1–2	Unknown
Blast nursery, BRRI	104	42 (40.4)	28 (26.9)	34 (32.7)
Farmers' field	18	2 (11.1)	6 (33.3)	10 (55.6)
Total	122	44 (36.1)	34 (27.9)	44 (36.1)
MV/lines	90	32 (35.6)	26 (28.9)	32 (35.6)
LV	32	12 (37.5)	8 (25.0)	12 (37.5)
Total	122	44 (36.1)	34 (27.9)	44 (36.4)
Non-rice hosts	7	1 (14.3)	1 (14.3)	5 (71.4)

Figures in parentheses indicate percentages.

application and dry conditions of the field and by applying foliar fungicides to susceptible varieties (Fig. 27.4). These components could easily be incorporated into the crop management system.

The technological gap needs to be filled in developing countries by giving research emphasis on cultural and biological methods (Waterhouse, 1992; Jacobsen and Backman, 1993). The indigenous methods used by farmers in different regions should also be explored and evaluated for their effectiveness in controlling blast (Thurston, 1990, 1992; Teng *et al.*, 1991). Finally, the effectiveness and benefit of these technologies have to be demonstrated to the farmers in order to encourage adoption.

Blast can be characterized as an environment-dependent disease, though cultural factors help to aggravate the situation. Therefore, monitoring of weather is of fundamental importance in blast management. In Bangladesh as well as in other tropical countries of the monsoon region, this is not practised for the purpose of advising farmers to adopt protective or control measures at the proper time.

Although technologies for blast management in upland or cool environments in the tropical monsoon climate are limited, the following have been found to be effective especially under Bangladesh conditions:

1. Collection of seeds from healthy crops and sowing of seeds treated with fungicide such as Homai, 80 WP (Thiophanate methyl 50% + Thiram 30%).

Table 27.6. Summary of potential blast management technologies and their effect on blast.

Management technologies	Effect on blast	Environment	References
Cultivation of resistant varieties:			
(a) Vertical:			
Monogenic cvs.	Complete control	All climates and ecosystems	Crill and Khush, 1979; Crill *et al.*, 1982a,b; Ou, 1985; Ahn and Mukelar, 1986; Kim *et al.*, 1989
Gene (variety) rotation	Complete control	Temperate lowland	
(b) Horizontal: (polygenic)			
Variety mixture	Partial control	Tropical upland	Chin and Husin, 1982; Bonman *et al.*, 1986
Multilines	Partial control	Tropical upland	
Partial resistant cvs.	Partial control	All climates and ecosystems	Bonman and Mackill, 1988; Roumen, 1993; Shahjahan, 1993
(c) Transgenic plants	Complete control	All climates	
Soil amendment and deep ploughing	Reduce incidence	Temperate lowland	Kozaka, 1965; Kim and Lee, 1982; Ou, 1985
Using low seed rate in seedbed	Reduce incidence and severity	Temperate and tropical climates	Kozaka, 1965; Amin and Kalyal, 1979; Ahn and Mukelar, 1986; Prabhu and Morais, 1986; Sah and Bonman, 1992
Sowing healthy seed	Reduce incidence and severity	All climates and ecosystems	This is a general recommendation
Delay in planting	Reduce incidence and severity	Temperate and tropical climates	Chandramohan and Palaniswamy, 1963; Kozaka, 1965; Ou, 1985

Using low N_2-rate and in split	Reduce severity	All climates and ecosystems	Kozaka, 1965; Templeton *et al.*, 1970; Paik, 1975; Amin and Venkatarao, 1979; Chung *et al.*, 1980; Lee and Lee, 1980; Bernaux, 1981; Prabhu and Morais, 1986; Santos *et al.*, 1986; Mbodj *et al.*, 1987; Raid *et al.*, 1988; Correa and Zeigler, 1990; Osuna-Conizales *et al.*, 1991; Kurschner *et al.*, 1992; Shahjahan, 1993;
Using high K-rate basally or as topdress	Reduce severity	All climates and ecosystems	Kozaka, 1965; Paik, 1975; Chung *et al.*, 1980; Shahjahan, 1993
Using low P-rate in high N applied field	Reduce severity	Temperate and tropical climates	Kozaka, 1965; Paik, 1975; Ou, 1985
Maintain wet condition or flooding of the land	Reduce incidence	Temperate and tropical climates	Ahn and Mukelar, 1986; Kim, 1986; Kim *et al.*, 1986; Bonman *et al.*, 1988; Gill and Bonman, 1988; Corea and Zeigler, 1990; Kim and Kim, 1990; Shahjahan, 1993
Applying SiO_2 in low Si-content soil	Reduce severity	Temperate and tropical climates	Kozaka, 1965; Paik, 1975; Chung *et al.*, 1980; Aleshin *et al.*, 1986; Yamaughi and Winslow, 1987; Raid *et al.*, 1988; Osuna–Conizales *et al.*, 1991; Datnoff *et al.*, 1991; Winslow, 1992
Destroying weeds/alternate hosts	Reduce incidence	Temperate and tropical climates	This is a general recommendation
Wide spacing and orientation of rows	Reduce incidence	Temperate and tropical upland	Kozaka, 1965; Ahn and Mukelar, 1986; Prabhu and Morais, 1986
Fungicidal treatment (seed and foliar)	Complete/partial control	All climates and ecosystems	Ashrafuzzaman and Frederiksen, 1970; Umeda, 1973; Ahn and Rubiano, 1984; Ou, 1985; Fukunaga *et al.*, 1986; Shahjahan *et al.*, 1987a; Reddy & Satyanarayana, 1988; Correa and Zeigler, 1990; Shahjahan 1993

Table 27.6. *(cont.)*

Management technologies	Effect on blast	Environment	References
Infected stubble/residue burning	Reduce incidence	Tropical climate	This is a general recommendation
Biological methods (not yet transferable)	Reduce incidence	Temperate and tropical climates	Sy *et al.*, 1983, 1984, 1990; Iwano, 1987; Bettiol, 1988; Gnanamanickam *et al.*, 1989; Soytong and Quimio, 1989; Gnanamanickam and Mew, 1990; Fujita *et al.*, 1990; Mia *et al.*, 1990; Soytong, 1991; Rajeswari and Mariappan, 1992
Integrated management	Reduce incidence and severity	Temperate and tropical climates	Training and Extension works are needed to popularize the concept to the farmers

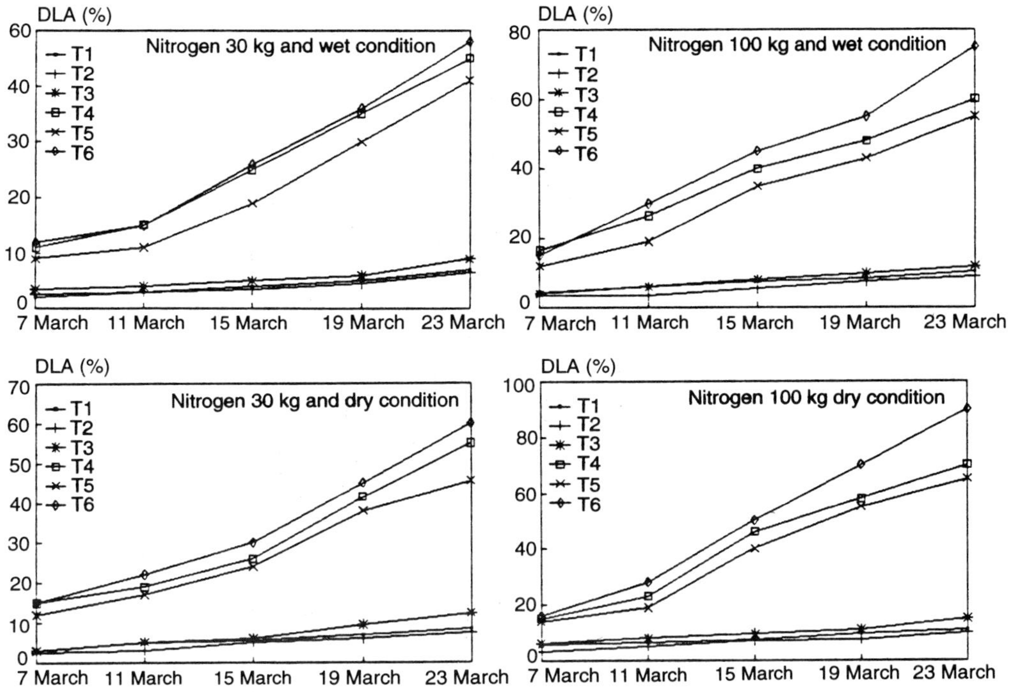

Fig. 27.4. Effect of variety, fungicide, nitrogen and water management on development of leaf blast disease of rice (BRRI, 1993). T1 = resistant variety (BR5) + seed treatment with Homai, T2 = BR5 + seed treatment + foliar spray with Homai, T3 = BR5 (control), T4 = susceptible variety (BR1) + seed treatment, T5 = BR5 + seed treatment = foliar spray with Homai, and T6 = BR1 (control).

2. Cultivation of resistant or partial resistant cultivars, such as BR5, BR7, BR10, BR12, BR14, BR16 and BR21, Kataktara, and Kalijira in the blast endemic agroecological regions.
3. Application of low dose of urea (N) – marginal and poor farmers commonly apply less fertilizer.
4. Application of KCl or ash (from the rice straw or husk and fuelwood) in the field – ash application is an indigenous method being practised by some farmers.
5. Application of irrigation water to dry fields and seedbeds especially during the seedling stage and keeping the fields flooded – during the dry season there is a scarcity of water in this region.
6. Foliar application of fungicides such as Hinosan (ediphenphos), Benlate/Fundazol (benomyl), Topsin-M (Thiophenate methyl), or Homai (Thiophenate methyl-Thiram), at the recommended rates.

Conclusion

It is clear that there is interaction between agricultural practices and blast, and such knowledge would be useful in integrating blast management systems within the overall crop management practices (Kranz, 1986).

However, knowledge of the pathosystem and development of adequate management technologies may not always bring about optimum returns. To achieve maximum benefit there should be a strong linkage between research, extension and the technology users, i.e. the farmers. Forecasting models may be useful in generating decision tools which can be used by the researchers, planners and extension workers, but not by the farmers of developing countries like Bangladesh. Thus, farmers of this region may still have to depend fully on the researchers/extension workers for information to guide their decisions.

Blast under tropical monsoon ecosystems, such as in Bangladesh, is a major constraint to rice production though it appears in epidemic form only in certain years and in certain agroecological regions. Records of blast incidence reveal that weather parameters are more important than other epidemiological factors. Although blast can be managed through either host resistance, cultural practices, chemical treatments, or biological methods (mature technologies not yet developed), the strategy for its management should be through an integrated crop management approach where the following steps are suggested:

1. Identify the blast endemic agroecological regions (AERs) through historical weather and blast incidence data and through survey data of recent years.
2. Establish a weather (temperature, humidity, rainfall, dew or leaf wetness period, wind, solar radiation, etc.) monitoring network and computerized database in all blast endemic areas.
3. Develop a surveillance system to monitor occurrence of blast, pathogen population for race changes, cultivars (genotypes) grown and the cultural practices being followed by farmers in each of those AERs.
4. Identify the technological gap that exists in order to develop new technologies for blast management.
5. Develop a control profile in relation to the agroecological factors conducive to blast development.
6. Motivate the farmers through practical training/demonstration so that they understand the usefulness of blast-reducing crop management practices.
7. Take steps to change government policy, if needed, to develop an effective technology transfer mechanism where the private sector/non-governmental organizations could be involved and to create awareness within the farming community.

References

Ahn, S.W. and Mukelar, A. (1986) Rice blast management under upland rice conditions. In: *Progress in Upland Rice Research*. IRRI, Los Baños, The Philippines, pp. 363–374.

Ahn, S.W. and Rubiano, M. (1984) Methods and timing of fungicide application to control rice blast under favourable upland condition in Colombia. *International Rice Research Newsletter* 9(5), 12.

Aleshin, N.E., Avakyan, E.R., Dyakunchak, S.A., Aleshin, E.P., Baryshok, V.P. and Voronkov, M.G. (1986) Role of Silicon in the resistance of rice to *Pyricularia*. *Doklady Akademii, Nauk SSSR* 291(2), 499–502 (*Review of Plant Pathology* 67(2), 749).

Amin, K.S. and Kalyal, I.C. (1979) Incidence of leaf blast in relation to seedling density and chemical composition of rice varieties. *Phytopathologische Zeitschrift* 96, 65–70.

Amin, K.S. and Venkatarao, G. (1979) Rice blast control by nitrogen management. *Phytopathologische Zeitschrift* 96, 140–145.

Asai, G.N. Jones, M.W. and Rorie, F.G. (1967) Influence of certain environmental factors in the field on infection of rice by *Pyricularia oryza*. *Phytopathology* 57, 237–241.

Ashrufuzzaman, H. and Frederiksen, R.A. (1970) Chemical control of rice blast caused by *Piricularia oryzae* Cav. Past II – Fungicidal effect of selected fungicides against rice blast in-vivo. *Pakistan Journal of Scientific and Industrial Research* 12, 249–256.

BBS (Bangladesh Bureau of Statstics) (1992) *Statistical Year Book of Bangladesh*. Ministry of Planning, Government of the People's Republic of Bangladesh, Dhaka.

Bettiol, W. (1988) Selection of antagonistic microorgamisms to *Pyricularia oryzae* Cav. for controlling rice (*Oryza sativa* L.) blast. PhD. Thesis, Universidade de Sao Paulo, Brazil.

Bernaux, P. (1981) Rice blast in France. In: *Proceedings of the Symposium on Rice Resistance and Blast*. Montpellier, France, pp. 23–25.

Bonman, J.M., Estrada, B.A. and Bandong, J.M. (1989) Leaf and neck blast resistance in tropical lowland cultivars. *Plant Disease* 73, 388–390.

Bonman, J.M. and Mackill, D.J. (1988) Durable resistance in rice to blast disease. *Oryza* 25, 103–110.

Bonman, J.M., Estrada, B.A. and Denton, R.I. (1986) Blast management with upland rice cultivar mixtures. In: *Progress in Upland Rice Research*. IRRI, Los Baños, The Philippines, pp. 375–382.

Bonman, J.M., Sanchez, L.M. and Mackill, A.O. (1988) Effects of water deficiency on Blast. II. Disease development in the field. *Journal of Plant Protection in the Tropics* 5, 67–73.

Bonman, J.M., Khush, G.S. and Nelson, R.J. (1992) Breeding rice for resistance to pests. *Annual Review of Phytopathology* 30, 507–528.

Bourke, P.M.A. (1970) Use of weather information in the prediction of plant disease epiphytotics. *Annual Review of Phytopathology* 8, 345–370.

Chakraborti, N.K. (1971) Relation between meteorological factors and development of foliar blast in rice. *Oryza* 8(2), 393–397.

Chandramohan, J. and Palaniswamy, S. (1963) Incidence of blast disease of rice in relation to time of planting. *Rice Newsletter* 11, 86–91.

Chin, K.M. and Husin, A.N. (1982) Rice variety mixtures in disease control. *International Conference on Plant Protection in the Tropics*. 1–4 March 1982, Kuala Lumpur, Malaysia, 6 pp.

Chung, B.K., Heo, H.Y., Cho, E.H. and Lee, J.Y. (1980) The effect of silicate, nitrogen and potassium on the incidence of rice blast disease caused by *Pyricularia oryzae* Cav. *Research Report of the Office of Rural Development, Suweon, Korea* 22, 56–62.

Coakley, S.M. (1990) Using historical weather and pest data for pest zoning. In: *Crop Loss Assessment in Rice*. IRRI, Los Baños, The Philippines, pp. 303–312.

Correa, F. and Zeigler, R.S. (1990) Components of the integrated management of the rice blast disease. A paper presentation at the International Rice Research Conference, Seoul, Korea, 27–31 Aug, 20 pp.

Crill, J.P. and Khush, G.S. (1979) Effective and stable control of rice blast with monogenic resistance. *Food and Fertilizer Technology Center Extension Bulletin* No. 128, Taiwan, 13 pp.

Crill, P., Ham, Y.S. and Beachell, H.M. (1982a) The rice blast disease in Korea and its control with race prediction and gene rotation. In: *Evolution of the Gene Rotation Concept for Rice Blast Control*. IRRI, Los Baños, The Philippines, pp. 123–130.

Crill, P., Nuque, F.L., Estrada, B.A. and Bandong J.M. (1982b) The role of varietal resistance in disease management. In: *Evolution of the Gene Rotation Concept for Rice Blast Disease*. IRRI, Los Baños, The Philippines, pp. 103–122.

Datnoff, L.E., Raid, R.N., Snyder, G.H. and Jones, B.D. (1991) Effect of calcium silicate on blast and brown spot intensities and yield of rice. *Plant Disease* 75(7), 729–732.

Datnoff, L.E., Snyder, G.H. and Daren, C.W. (1992) Influence of silicon fertilizer grades on blast and brown spot development and on rice yields. *Plant Disease* 76, 1011–1013.

El-Refaei, M.I. (1977) Epidemiology of rice blast disease in the tropics with special reference to the leaf wetness in relation to disease development. PhD Thesis, IARI, New Delhi, India.

Food and Agricultural Organization (FAO) (1988) *Land Resources Appraisal of Bangladesh for Agriculture Development*. AEZ of Bangladesh, Technical Report No. 2, 570 pp.

Fujita, Y., Sonoda, -R. and Yaegashi, H. (1990) Leaf blast suppression by pre-inoculations of some incompatible lesion-type isolates of *Pyricularia oryzae*. *Annals of the Phytopathological Society of Japan* 56, 273–275.

Fukunaga, K., Misato, T., Ishii, I., Asakawa, M. and Katagiri, M. (1986) Research and development of antibiotics for the rice blast control. *Bulletin of the National Institute of Agricultural Sciences Series C* 22, 1–94.

Gill, M.A. and Bonman, J.M. (1988) Effects of water deficit on rice blast. I. Influence of water deficit on components of resistance. *Journal of Plant Protection in the Tropics* 5, 61–66.

Gnanamanickam, S.S. and Mew, T.W. (1990) Biological control of rice disease (blast and sheath blight) with bacterial antagonists: an alternate strategy for disease management. In: Grayson, B.T., Green, M.B. and Coping L.G. (eds), *Pest Management in Rice*. Elsevier Applied Science, London, 536 pp.

Gnanamanickam, S.S., Reyes, R.C. and Mew. T.W. (1989) Biological control of rice blast (B1) with antagonistic bacteria. *International Rice Research Newsletter* 14(2), 34–35.

Hashioka, Y. (1965) Effects of environmental factors on development of causal fungus, infection, disease development and epidemiology in rice blast disease. In: *The Rice Blast Disease*. Johns Hopkins Press, Baltimore, Maryland, pp. 153–161.

Hashimoto, A., Hirano, K. and Matsumoto, K. (1984) Studies on the forecasting of rice leaf blast development by application of the computer simulation (in Japanese, English Summary) *Bulletin of the Fukushima Prefecture Agriculture Experiment Station* 2, 1–104.

Ishiguro, K. and Hashimoto, A. (1991) Computer based forecasting of rice blast epidemics in Japan. In: *Rice Blast Modeling and Forecasting*. IRRI, Los Baños, The Philippines, pp. 39–52.

Iwano, M. (1987) Suppression of rice blast infection by incompatible strain of *Pyricularia oryzae* Cav. *Bulletin of the National Agriculture Experiment Station* 75, 27–39.

Jacobsen, B.J. and Backman, P.A. (1993) Biological and cultural plant disease controls: Alternatives and supplements to chemicals in IPM Systems. *Plant Disease* 77, 311–315.

Kahn, R.P. and Libby, J.L. (1958) The effect of environmental factors on the infection of rice by the blast fungus, *Pyricularia oryzae*. *Phytopathology* 48, 25–30.

Kato, H. (1974) Epidemiology of rice blast disease. *Review of Plant Protection Research* 7, 1–20.

Khush, G.S. and Chaudhury, R.C. (1981) Role of resistant varieties in integrated pest management of rice. FFTC Extension Bulletin 162, 13 pp.

Kim, C.H. (1986) Effect of water management on the etiology and epidemiology of rice blast caused by *Pyricularia oryzae* Cav. PhD Thesis, Louisiana State University, A & M, Baton Rouge, USA. 170 pp.

Kim, C.H. and Kim, C.K. (1990) Rice blast development under different water management practices in fields in Korea. *Korean Journal of Plant Pathology* 6(1), 51–57.

Kim, C.H., Rush, M.C. and Mackenzie, D.R. (1986) Flood mediated resistance to the rice blast disease. In: Juo, A.S.R. and Lowe, J.A. (eds), *The Wetland and Rice in Subsaharan Africa*. IITA, Ibadan, Nigeria, pp. 151–60.

Kim, C.H., Mackenzie, D.R. and Rush, M.C. (1988) Field testing a computerized forecasting system for rice blast disease. *Phytopathology* 78, 931–934.

Kim, C.H., Kim, C.K. and Lee, E.J. (1989) Effect of host resistance and fungicide application on rice leaf blast management. *Korean Journal of Plant Pathology* 5(4), 377–382.

Kim, C.K. and Kim, C.H. (1991) Predicting rice blast outbreaks in Korea. In: *Rice Blast Modeling and Forecasting*. IRRI, Los Baños, The Philippines, pp. 53–68.

Kim, C.K. and Lee, S.C. (1982) Reduction of the incidence of rice neck blast by integrated soil improvement practice. *Korean Journal of Plant Protection* 21(1), 15–18.

Koizumi, S. and Kato, H. (1991) Dynamic simulation of blast epidemics using a multiple canopy spore dispersal model. In: *Rice Blast Modeling and Forecasting*. IRRI, Los Baños, The Philippines, pp. 75–88.

Kozaka, T. (1965) Control of rice blast by cultivation practices in Japan. In: *The Rice Blast Disease. Proceedings of the Symposium at IRRI, 1963*. The Johns Hopkins Press, Baltimore, Maryland, pp. 421–438.

Kranz, J. (1986) Epidemiological information as an aid in pest management. In: *Progress in Upland Rice Research*. IRRI, Los Baños, The Philippines, pp. 355–362.

Kurschner, E., Bonman, J.M., Garrity, D.P., Tamisin, M.M., Pabale, D. and Estrada, B.A. (1992) Effects of nitrogen timing and split application on blast disease in upland rice. *Plant Disease* 76, 384–389.

Latterell, F.M. (1975) Phenotypic stability of pathogenic races of *Pyricularia oryzae* and its implications for breeding of blast resistant rice varieties. In: *Horizontal Resistance to the Blast Disease of Rice*. CIAT, Series CE-9, pp. 199–234.

Lee, E.W. and Lee, B.W. (1980) Effects of split application and topdressing time of nitrogen fertilizer on rice blast disease incidence and some chemical content in the rice plant. *National University College of Agriculture Bulletin*. 5, 211–220.

Lee, Y.H., Shim, G.Y., Lee, E.J. and Mew, T.W. (1990) Evaluation of biocontrol activity of fluorescent pseudomonads against some fungal diseases in vitro and green house. *Korean Journal of Plant Pathology* 6(1), 73–80.

Levy, M., Correa-Victoria, F.J., Zeigler, R.S., Shen, Y., Shahjahan, A.K.M., Gnanamanickam, S.S., Nelson, R.E., Manry, J. and Hamer, J.E. (1993) International atlas of genetic diversity in the rice blast fungus: 1992. In: *Abstracts of the Sixth Annual Meeting of the International Program on Rice Biotechnology*. Chiang Mai, Feb. 1–5, Thailand.

Marchetti, M.A., Rush, M.C. and Hunter, W.E. (1976) Current status of rice blast in the southern United States. *Plant Disease* 60, 721–725.

Mbodj, -Y., Gaye, -S., Diaw, -S. and Sylla, M. (1987) Influence of nitrogen fertilizer on rice blast development. *Parasitica* 43, 177–185.

Manibhushanrao, K. and Day, P.R. (1973) Low night temperature and blast development on rice. *Phytopathology* 62, 1005–1007.

Manibhushanrao, K. and Krisnan, P. (1991) Epidemiology of blast (EPIBLA): a simulation model and forecasting system for tropical rice in India. In: *Rice Blast Modeling and Forecasting*. IRRI, Los Baños, The Philippines, pp. 31–38.

Matsumoto, S., Kozaka, T. and Yamada, M. (1967) Pathogenic races of *Pyricularia oryzae* cav. in Asian and some other countries. *Bulletin of the National Institute of Agriculture and Science, Tokyo*, C 23, 1–36.

Muralidharan, K. and Venkata. Rao, G. (1980) A simple method of forecasting outbreak of rice blast. *Indian Phytopathology* 33(4), 560–564.

Mia, M.A.T., Ahmed, H.U., Sharma, N.R., Ali, A. and Miah, S.A. (1990) Antifungal activity of some plant extracts. *Bangladesh Journal of Botany* 19(1), 5–10.

Miah, S.A., Shahjahan, A.K.M. and Haque, M.A. (1981) Severe blast outbreak during Boro 1979–80 season in Bangladesh. *International Rice Research Newsletter* 5(1), 10–11.

Miah, S.A., Shahjahan, A.K.M., Hossain, M.A. and Sharma, N.R. (1985) Survey of rice diseases in Bangladesh. *Tropical Pest Management* 31(3), 208–213.

Nahar, N.S., Ahmed, H.U. and Shahjahan, A.K.M. (1991) Studies on neck blast development in rice cultivars. *Bangladesh Journal of Plant Pathology* 7, 13–16.

Osuna-Canizales, F.J., De Datta, S.K. and Bonman, J.M. (1991) Nitrogen form and silicon nutrition effects on resistance to blast disease of rice. *Plant and Soil* 135, 223–231.

Ou, S.H. (1985) *Rice Diseases*, 2nd edn. Commonwealth Mycological Institute, Kew, UK, 380 pp.

Ou, S.H. and Nuque, F.L. (1963) The relation between leaf and neck resistance to the rice blast disease. *International Rice Commission Newsletter* 12(4), 30–35.

Padmanabhan, S.Y., Chakrabarty, N.K. and Row, K.V.S.R.K. (1971) Forecasting and control of rice disease. *Proceedings of the Indian National Science Academy* 37, B(6), 423–429.

Paik, S.B. (1975) Effects of silicate, nitrogen, phosphorus and potassium fertilizers on the chemical components of rice plants and on the incidence of blast disease caused by *Pyricularia oryzae* cav. *Korean Journal of Plant Protection* 14, 97–109.

Prabhu, A.S. and Morais, O.P. (1986) Blast disease management in upland rice in Brazil. In: *Progress in Upland Rice Research*. IRRI, Los Baños, The Philippines, pp. 383–392.

Raid, R.N., Jones, D.B. and Snyder, G.H. (1988) Effect of nitrogen and silicon on blast and brown spot, 1987. *Biological and Cultural Tests for Control of Plant Diseases* 3, 42.

Rajeswari, E. and Mariappan, V. (1992) Effect of plant extracts on invitro growth of rice blast (BL) pathogen *Pyricularia oryzae*. *International Rice Research Newsletter* 17(6), 24.

Reddy, A.P.K. and Satyanarayana, K. (1988) Evaluation of new organic fungicides for the control of blast disease of rice. *Pesticides* 22(9), 21–26.

Reichert, I. and Palti, J. (1967) Prediction of plant disease occurrence, a pathogeographic approach. *Mycopathologica et Mycologia Applicata* 32(4), 337–355.

Rotem, J. (1978) Climate and weather influences on epidemics. In: Horsfall, J.G. and Cowling, E.B. (eds) *Plant Disease, How Disease Develops in Populations*, Vol 2. Academic Press, New York, pp. 317–337.

Roumen, E.C. (1993) *Partial Resistance in Rice to Blast and How to Select for it*. Proefschrift Wageningen. -Met Samenvatting in het, The Netherlands, 108 pp.

Sah, D.N. and Bonman, J.M. (1992) Effects of seedbed management on blast development in susceptible and partially resistant rice cultivars. *Journal of Phytopathology* 136, 73–81.

Santos, A.B. Dos, Prabhu, A.S., Aquino, A.R.L. -de and Carvalho, J.R.P. -de. (1986) Effects of nitrogen rate, time and application methods and blast and grain yield in upland rice. *Pesquisa-Agropecuaria Brasileira* 21(7), 697–707.

Shahjahan, A.K.M. (1993) *Practical Approaches to Crop Pest and Disease Management in Bangladesh*. Bangladesh Agricultural Research Council, Dhaka, 168 pp.

Shahjahan, A.K.M. and Nahar, N.S. (1993) Determination of mating types of *Magnaporthe* populations causing blast disease of rice in Bangladesh. *International Rice Research Newsletter* 18(1), 44–45.

Shahjahan, A.K.M., Duve, T. and Bonman, J.M. (1987a) Climate and rice diseases. In: *Weather and Rice*. IRRI, Los Baños, The Philippines, pp. 125–138.

Shahjahan, A.K.M., Ahmed, H.U., Hossain, M.A. and Miah, S.A. (1987b) Evaluation of fungicides for rice blast control. *Bangladesh Journal of Plant Pathology* 3, 55–60.

Shahjahan, A.K.M., Ahmed, H.U., Miah, M.A.T., Hossain, M.A., Sharma, N.R. and Nahar, N.S. (1991) Outbreak of leaf and neck blast in Boro rice crop in Bangladesh. *International Rice Research Newsletter* 16(2), 21.

Shahjahan, A.K.M., Nahar, N.S., Levy, M., Renganathan, N. and Hamer, J.E. (1993) Genetic organization of the rice blast fungus in Bangladesh. In: *Abstracts of the Sixth Annual Meeting of the International Program on Rice Biotechnology*. Chiang Mai, Thailand, p. 185.

Soytong, K. (1991) Biological control of rice blast pathogen by coating seeds with *Chaetomium cochloides* and *C. cuniculorum*. *Khon kao Agricultural Journal* 18(2), 89–96 (undated) [*Review of Plant Pathology* 71(10), 723, 1991].

Soytong, K. and Quimio, T.H. (1989) Antagonism of *Chaetomium globosum* to the rice blast pathogen, *Pyricularia oryzae*. *Kasetsart Journal, Natural Sciences* 23, 198–203.

Surin, A., Arunyanart, P., Rojanahusdin, W., Munkong, S., Dhitikiattipong, R. and Disthaporn, S. (1991) Using empirical blast models to establish disease management recommendations in Thailand. In: *Rice Blast Modeling and Forecasting*. IRRI, Los Baños, The Philippines, pp. 69–74.

Suzuki, H. (1975) Meteorological factors in the epidemiology of rice blast. *Annual Review of Phytopathology* 13, 239–256.

Sy, A.A., Norng, K., Albertini, L. and Barrault, G. (1983) Research on biological control of *Pyricularia oryzae* Cav. III. Effect of temperature on the capacity of antagonistic microorganisums to inhibit mycelial growth of the parasite *in vitro*. *Cryptogamic Mycologie* 4(3), 245–249.

Sy, A.A., Norng, K., Albertini, L. and Petitprez, M. (1984) Research on biological control of *Pyricularia oryzae* Cav. IV. Effect of pH on the ability of antagonistic microorganisms to inhibit mycelial growth of the parasite *in vitro*. *Cryptogamic Mycologie* d5(1), 59–65.

Sy, A.A., Sarr, A., Albertini, L. and Moletti, M. (1990) [Biological control of *Pyricularia oryzae*: Parameters of stability of antogonistic activity.] *Phytopathologia Mediterranea* 29(3), 175–183.

Templeton, G.E., Johnston, T.H. and Henry, S.E. (1961) Rice blast. *Arkansas Farm Research* 10, 12.

Templeton, G.E., Wells, B.R. and Johnston, T.H. (1970) N-fertilizer applications closely related to blast at nodes and resultant lodging of rice. *Rice Journal* 73(7), 71.

Teng, P.S. and Calvero, S.B. (1991) Computer simulation of tropical rice leaf blast pathosystem using Blastim. 2. *Phytopathology* 81, 1143.

Teng, P.S., Klein-Gebbinck, H.W. and Pinnschmidt, H. (1991) An analysis of the blast pathosystem to guide modeling and forecasting. In: *Rice Blast Modeling and Forecasting*. IRRI, Los Baños, The Philippines, pp. 1–30.

Teng, P.S., Fernandez, P.G. and Garcia, R. (1991) Indigenous plant protection practices in Philippine tribal systems. *Report to GTZ Biological Plant Protection Project*, Manila, 175 pp.

Thurston, H.D. (1990) Plant disease management practices of traditional farmers. *Plant Disease* 74, 96–102.

Thurston, H.D. (1992) *Sustainable Practices for Plant Disease Management in Traditional Farming Systems*. Westview Press, Boulder, Colorado, 279 pp.

Uehara, Y., Imoto, M. and Sakai, Y. (1988) Studies on the forecasting of the rice blast development using the weather data from AMeDAS. *Bulletin of the Hiroshima Prefectural Agricultural Experiment Station* No. 51, 1–15.

Umeda, Y. (1973) Hinosan, a fungicide for control of rice blast. *Japan Pesticide Information* 17, 25–28.

Volk, R.J., Kahn, R.P. and Weintraub, R.L. (1958) Silicon control of rice plant as a factor influencing its resistance to infection by the rice plant fungus, *Pyricularia oryzae. Phytopathology* 48, 179–184.

Waterhouse, D.F. (1992) Biological control: a viable strategy for the tropics. In: Ooi, P.A.C., Lim, G.S. and Teng P.S. (eds) *Biological Control: Issues in the Tropics.* MAPPS, Kuala Lumpur, Malaysia, pp. 1–13.

Winslow, M.D. (1992) Silicon, disease resistance, and yield of rice genotypes under upland cultural conditions. *Crop Science* 32, 1208–1213.

Yamaughi, M. and Winslow, M.D. (1987) Silica reduces disease on upland rice in a high rainfall area. *International Rice Research Newsletter* 12(6), 22–23.

Yeh, W. er H., Bonman, J.M. and Lee, E.J. (1989) Effect of temperature, leaf wetness duration, and leaf age on partial resistance to rice blast. *Journal of Plant Protection in the Tropics* 6(3), 223–230.

28 Rice Breeding Programs, Blast Epidemics and Blast Management in the United States

F.N. Lee

University of Arkansas Rice Research and Extension Center, PO Box 351, Stuttgart, AR 72160, USA

Introduction

Rice production in the United States has evolved over the past 300 years from a labor intensive production system into an efficient, highly mechanized, intensively managed system where the use of airplanes, computers, lasers, and synthetic fertilizers and pesticides are routine. Rice is produced in seven states and is concentrated in three broad geographic regions; Mid-South (Arkansas, Mississippi, Missouri and North Louisiana), Gulf Coast (South Louisiana and Texas), and California. A small acreage is planted on organic soils (Histosol) in Florida. Based on the 1992 planted area of 1.24 million ha, 59.5% of the total US rice crop was planted in the Mid-South region, 27% in the Gulf Coast region and 12.7% in California. Of the total planted area, 79.2% was long grain, 20.0% was medium grain and 0.8% was short grain. Arkansas is located in the south-central US at approximately 35°N latitude and 92°W longitude and produces approximately 40% of the total US rice crop on approximately 625,000 ha.

The relative importance of rice diseases varies slightly, but generally the major diseases common to all production areas, with the exception of California, are rice blast (*Pyricularia grisea*, rice sheath blight (*Rhizoctonia solani*), stem rot (*Magnaporthe salvinii*), and kernel smut (*Tilletia barclayana*). Few diseases are present in California although stem rot and aggregate sheath spot (*Rhizoctonia oryzae-sativa*) often occur.

Cultivars utilized in the US are primarily developed by rice breeding programs associated with major state university agriculture experiment stations and the US Department of Agriculture, consisting of one or more plant breeders supported by plant pathologists and other scientists. The US programs interact freely with a unconstrained exchange of concepts, germplasm and data. Co-operative nurseries are established in each state to evaluate yield and pest reactions of advanced breeding lines. The primary objective of the rice breeding programs is to develop cultivars with high rough rice and milling yields. While new techniques utilizing anther culture, gene insertion, somaclonal variation, and other biotechniques are being incorporated into the programs, traditional plant breeding methods continue to be the primary means of cultivar development. Broad adaptability and production capabilities of cultivars is a consideration, but breeders develop and select those cultivars best adapted to the agronomic and disease requirements of their target production areas.

Cultivar resistance is the mainstay of rice disease management in the US. All breeding lines are screened in greenhouse tests for leaf reaction to those blast races most frequently found in production areas. The more advanced breeding lines are evaluated for their reaction to those races as well as others known to have adversely affected rice cultivars in the recent past. Blast race surveys normally are sufficient to detect and characterize races pathogenic to new cultivars prior to their release (Marchetti *et al.*, 1976). Field performance is a major consideration. Inoculated upland field nurseries are used to estimate cultivar susceptibility under field conditions in comparison with known susceptible cultivars. Advanced entries are also evaluated in flooded nurseries and in grower fields located throughout the production areas. Although the period required to develop new cultivars is decreasing, approximately 5–8 years of disease data are usually available at the time of cultivar release. This process has worked reasonably well, producing several high yielding cultivars with a combination of qualitative and quantitative resistance to rice blast (chapter 14, this volume).

The potential for explosive and extensive blast damage as described by Ou (1985) and as sometimes occurs in the US (Atkins, 1956; Marchetti *et al.*, 1976; Marchetti, 1983; Jones and Lee, 1988; Groth, 1992, personal communication; Whitney, 1992, personal communication) still exists because of the reliance on a few major resistance genes for rice blast control. Some cultivars have been grown for many years with minimal loss to rice blast (Table 28.1). Unfortunately, other cultivars are at risk soon after their release, often by the rapid increase of 'new' or previously infrequently occurring blast races (Table 28.2). The risk usually does not preclude their being used judiciously until better cultivars are developed.

Excellent examples of successful and 'at risk' cultivars are Lemont (Bollich *et al.*, 1985) and Newbonnet (Johnston *et al.*, 1984), which were released to growers in 1983. Both cultivars possessed substantially higher yield potential over existing cultivars, were rated resistant to then predominant races IG1 and IH1 in Arkansas and appeared to have sufficient tolerance to other known races. Lemont, as an example of a stable cultivar, seldom sustains extensive losses except in very adverse growing conditions. In contrast, Newbonnet proved to be highly susceptible to previously minor races in Arkansas, IB49 and IC17 (race determinations by Dr M.A. Marchetti), under certain field conditions. The first indication of the Newbonnet blast exposure came in 1984 when losses in a single grower seed increase field was estimated at 70% or more. Throughout Arkansas, however, Newbonnet produced record rough rice yields and by 1985 was planted to 61% of the state acreage (Table 28.3). Beginning in 1986, the volatile nature of rice blast was demonstrated by a series of annual blast epidemics that were unsurpassed in recent memory. Approximately 16,700 ha of Newbonnet in one local production area were damaged in 1986. Of that, approximately 13,000 ha sustained losses of 25% or less. Yield reductions of 26–50% were sustained over

Table 28.1. US rice cultivars grown in the Mid-South and Gulf Coast rice production areas without significant rice blast exposure.

Cultivar	Grain type	Year released	Prevalent blast race
Saturn	Medium	1964	IB45
Starbonnet	Long	1967	IG1
Brazos	Medium	1974	IG1
Mars	Medium	1977	IB1
Lemont	Long	1983	IB49, IC17
Tebonnet	Long	1984	IB49

Table 28.2. US rice cultivars grown in the Mid-South and Gulf Coast rice production areas determined to be at risk to rice blast within a short period following release.

Cultivar	Grain type	First released	Year damaged	Prevalent blast race
Gulfrose	Medium	1960	1963	IB49, IB54
Nova 66	Medium	1966	1969	IH1
Labelle	Long	1972	1975	IC17
Bonnet 73	Long	1973	1975	IB45
Lebonnet	Long	1974	1975	IC17
Newbonnet	Long	1983	1985	IB49, IC17
Texmont	Long	1990	1991	IC17
Rosemont	Long	1991	1991	IC17

Table 28.3. Percent of Arkansas rice acreage planted to selected cultivars during 1984–1993.

	Year									
	1984	1985	1986	1987	1988	1989	1990	1991	1992	1993
Starbonnet	42	3	1	1	1	–	–	–	–	–
Lebonnet	20	6	1	1	1	–	–	–	–	–
Labelle	19	4	1	1	1	–	–	–	–	–
Newbonnet	5	61	69	55	34	45	40	38	19	9
Lemont	1	12	11	17	23	23	20	15	16	12
Katy	–	–	–	–	–	1	6	13	21	19
Alan	–	–	–	–	–	–	1	4	19	29

Hectares harvested: 1984 – 465,400; 1985 – 424,930; 1986 – 428,976; 1987 – 402,671; 1988 – 453,258; 1989 – 485,633; 1990 – 485,633; 1991 – 509,915; 1992 – 534,197; 1993 – 534,197 (estimate).

Source: University of Arkansas Cooperative Extension Service.

2,350 ha. Yield reductions of 50% or greater were sustained over 1242 ha. Blast was widespread during 1987 and 1988 when an estimated 10% yield reduction occurred statewide. Regardless of the losses, a record high state yield was recorded for 1985 and state yields were considered to be exceptional in the following years. Blast continues to occur in Newbonnet and other cultivars with losses being dependent upon local environmental conditions.

Environmental conditions in Arkansas usually are not conducive to blast epidemic development especially if popular cultivars have a moderate level of horizontal (field) resistance. However, environmental variability actually was a contributing factor in the Newbonnet blast problem in Arkansas. Although races IG1 and IH1 were found from time to time on Starbonnet and other cultivars in the state during the early 1970s, losses were considered to be minor. A popular cultivar, Lebonnet, was beginning to show high levels of rice blast in Arkansas during 1978 and 1979 when a few fields sustained losses of 50% or more. Lebonnet losses were restricted to a few hectares but the disease appeared to be spreading. An unusual drought associated with the high temperature occurred in 1980 followed by extended dry periods during the 1981, 1982, and 1983 growing seasons. As a result, blast was not detected again in the state until 1984 when it was found in Newbonnet and other cultivars. Thus environmental conditions probably interfered with proper field evaluation of early population increases of Newbonnet. As blast suddenly became a major problem, many growers and agricultural advisors were slow to react because they were unfamiliar with rice blast symptoms or underestimated the potential for destruction by a disease previously considered to be a minor problem in Arkansas.

Blast Management Options

Use of blast tolerant cultivars

Blast losses have not been great on Lemont and other cultivars, but these cultivars are not particularly suited for portions of Arkansas production areas. The blast-resistant cultivar Katy (Moldenhauer *et al.*, 1990) was released for seed increase and grower use in 1989, but it lacked the desirable agronomic characteristics of Newbonnet. Also, Newbonnet sustained little damage when grown in open fields under a proper moisture regime throughout the season. As a result, growers were reluctant to switch to Katy and other less productive cultivars for economic reasons, and they often did not fully appreciate the risks involved.

Cultural manipulations

Effective alternative control measures to be used in the absence of acceptable cultivar resistance generally are not available. Blast severity may be reduced somewhat by planting in fields where local environmental or cultural factors do not promote shading or inhibit wind circulation, avoiding excessive use of nitrogenous fertilizers, and maintaining proper floodwater levels throughout the growing season. These strategies, however, often do not prevent substantial losses or are difficult for individual growers to implement. This is particularly true for the critical practice of maintaining an adequate flood over the entire field.

Fungicide applications

The fungicides benomyl and thiabendazole were available at the onset of the epidemics beginning in 1986 but adequate control data applicable to the US conditions were not available. Claims as to their effectiveness were misleading and/or instructions were unclear as to proper time and rate of application.

Fungicide tests were conducted in an effort to establish an alternative control measure and to estimate losses to rice blast. Tests were conducted in grower fields where plants already were showing blast symptoms on the leaves and collars. Uniform areas of the field were selected and marked into plots 1.8 × 4.9 m. Replicated fungicide treatments were applied with a hand-held boom at different rates and timing. Response to treatment was measured by degree of disease incidence at maturity as a visual estimate of severity on a 0–9 scale (0 = no disease, 9 = maximum losses) and/or as the percentage of diseased panicles at three predetermined points per plot. Plots were trimmed to 0.8 × 4.7 m, harvested with a small combine for rough rice, which was adjusted to a 12% moisture yield determination and % whole milled grain/% total milled grain of rough rice milled.

Benomyl proved to be the most effective registered fungicide tested and would often provide a satisfactory yield in those situations where applications were properly timed with plant growth stage, paddy floods were maintained and disease pressure was not excessive. In general, a yield response of some magnitude to benomyl application could be obtained at any growth stage from late 'boot' until panicles had been exserted for a period of 12–14 d. Multiple applications during this period provided optimal control but were neither practical nor economical.

Two critical growth stages for fungicide application appeared to be equally important in obtaining an acceptable control level. The first was a 'very late boot' stage defined as when the boot was completely expanded and the panicle was just beginning to exsert. The second stage was when the panicles were almost or had just emerged from the boot and were still susceptible to infection. Although altered by environmental conditions, this stage occurred about 7–10 days after very late boot. Two applications at these stages usually would reduce losses. Treatment efficacy was affected by initial disease pressures, growth stage uniformity in the treated area, nutrient status of the field and how well flood water depth was maintained. Data in Table 28.4 describe typical situations encountered for Newbonnet. Substantial yield reduction occurred in plots left untreated and a differential rate response to treatment was indicated by all measured parameters. Approximately 6000 kg ha^{-1} is considered to be the economic break-even point by many growers in the absence of disease. Although fungicide applications did result in increased rough rice and milling yields, additional production costs were incurred that altered the economics of the

Table 28.4. Mean response at five grower locations following very late boot and head exsertion growth stage applications of benomyl as a control measure for blast in the cultivar Newbonnet.

	Rate formulated[a] (kg ha^{-1})				
	Untreated	0.56	1.26	1.68	2.24
Location 1					
RN[b]	40	24	16	17	–
Yield[c]	3753	4889	5654	6330	–
Milling[d]	37/61	34/61	38/63	38/63	–
Location 2					
RN	67	38	33	27	27
Yield	2948	3808	4485	4740	4996
Milling	38/55	40/57	44/59	43/58	45/60
Location 3					
RN	55	30	23	18	–
Yield	4292	5081	6005	6307	–
Milling	44/60	46/62	48/63	49/64	–
Location 4					
RN	–	–	–	–	–
Yield	5936	5779	7174	7698	8083
Milling	59/66	60/67	61/67	62/67	63/69
Location 5					
RN	65	31	23	13	10
Yield	5449	6518	7204	7141	8503
Milling	44/62	48/64	48/64	52/67	–
Overall averages					
RN	46	26	21	14	–
Yield	3336	4411	5507	5824	–
Milling	38/55	48/62	49/63	50/63	–

[a] kg of benomyl formulated as 50% wetable powder ha^{-1} per application at very late boot and head exsertion growth stages.
[b] % rotten neck blast observed in plots at maturity.
[c] Rough rice yield in kg ha^{-1} adjusted to 12% moisture.
[d] Milling yield as % whole milled grain over percentage total milled grain.

situation. It is important to stress that these tests were conducted using the highly susceptible Newbonnet cultivar. In addition, the fungicide applications very seldom provided complete disease control. In extreme situations multiple applications at high rates failed to provide acceptable control.

Sources of Overwintering Inoculum

Because blast epidemics are erratic, the primary source of overwintering inoculum becomes a point of interest. As previously indicated, blast incidence was low during the early 1970s and apparently non-existent in Arkansas during

1980–1983. The prevailing belief was that inoculum was being blown into the state from down-wind diseased fields in Louisiana and Texas. In fact, the initial inoculum for diseased fields observed in the state in 1984 was probably introduced via storms that moved rapidly into Arkansas from these states. However, the distribution of seedling blast within the state and pathotype adaptation in isolated production areas suggested the spore load necessary to initiate the epidemics was originating from sources other than windborne inoculum. These observations were strengthened by the failure of blast to develop in upland trap crop plantings of susceptible indicator cultivars while fields immediately adjacent to the plantings developed blast. Therefore tests were initiated to investigate the possible overwintering sources of inoculum such as alternative hosts, stubble and seed rice.

Grass and rice stubble

In 1987 and again in 1988, we surveyed grass stubble after it had been exposed to winter environmental conditions. Stubble samples of several grasses and wild bamboo were collected in areas adjacent to rice fields throughout Arkansas. These were dried, placed in paper bags and held at room temperature until assayed. Rice stubble from areas in fields known to have been infected as well as stubble from random rice fields throughout the state were also collected. Samples were slightly moistened and placed in humidity chambers for approximately 48 h. Care was taken to have sufficient moisture for fungal growth without excessive surface moisture on the sample. Samples were examined and suspect tissue rinsed with a small volume of distilled water. The rinse water was collected and used to inoculate susceptible rice plants at the three to four leaf growth stage. Assay plants were examined for symptoms when control plants inoculated with known rice isolates began developing typical lesions.

Although *Pyricularia* spores were observed on the grass stubble and in the rinse water, lesions were not found during a careful visual examination of assay plants (Table 28.5). *Pyricularia* spores could not be found on the bamboo specimens. The fungus did overwinter on the rice stubble where approximately 46% of the samples from known diseased fields and 27.5% of samples from random fields throughout the state assayed positive for *P. grisea*. Some discolored nodes from stubble in experimental test plots were observed to be sporulating well into the growing season. We surmised from our limited effort that grasses or bamboo are not the source of overwintering inoculum for epidemics in Arkansas rice. On the other hand, rice stubble probably serves as a source of overwintering inoculum.

Literature (Ou, 1985; Mackill and Bonman, 1986) available at the time these tests were conducted indicated *P. grisea* commonly developed on other grass species and could infect rice. Itoi *et al.* (1979) suggested bamboo might serve as an overwintering host. In cross-inoculation studies, Mackill and Bonman (1986) indicated some *Pyricularia* isolates from grasses in the Philippines were pathogenic to rice. Borromeo *et al.* (1993) recently examined the DNA polymorphisms among Philippine isolates of the fungus and concluded the genetic isolation that exists between the rice and weed isolates sufficiently separates the two and that weed isolates do not play a significant role in rice blast epidemic in the

Table 28.5. Stubble types assayed as possible overwintering inoculum source of *Pyricularia grisea* following the winters of 1987 and 1988.

Stubble source	Number	
	Samples examined	Positive for *P. grisea*
Blasted rice stubble[a]	82	38
Random rice stubble[b]	58	16
Grass stubble[c]	116	0
Rice and grass mixture	3	3
Unidentified bamboo	10	0

[a] Rice stubble samples from previously identified production fields known to have been damaged with rice blast.
[b] Rice stubble samples collected from random production fields throughout Arkansas.
[c] Random grass stubble samples collected throughout Arkansas rice production areas.

Philippines. In Brazil, *Pyricularia* has developed into an important pathogen of wheat. Prabhu *et al.* (1992) and Urashima *et al.* (1993) tested isolates which were pathogenic to wheat and grasses but did not find isolates from these hosts which were pathogenic on rice. Certain rice isolates were pathogenic to wheat and several of the grasses. One can speculate that future research into the genetics of grass and rice isolates in conjunction with additional epidemiological research will demonstrate conclusively that grass hosts are not involved in rice blast epidemics and, at best, serve only as a *P. grisea* reservoir for a portion of the genetic diversity found in rice blast populations.

Seedborne inoculum

The role of seed rice as a primary overwintering source was also investigated. Cleaned Newbonnet seed samples were obtained from Arkansas processors. The Arkansas State Plant Board (ASPB) routinely certifies seed rice by various quality tests and represents a centralized collection site for much of the seed rice produced in the state. Excess seed from random 1989 and 1990 ASPB samples were obtained and assayed for *P. grisea*. Spores were extracted by placing 120 g of seed into 290 ml water and particulate matter suspended by vigorous agitation. The water was immediately passed through a sieve with 500 μm openings, collected and suspended particles pelleted using a centrifuge. The pellet was resuspended in water to inoculate susceptible plants or examined under a microscope for spore density determinations. Following the initial spore extraction, the seed was sterilized for 2 min in 0.5% sodium hypochlorite, drained to remove excess moisture, and then thinly spread in a humidity chamber for a 72 h incubation at room temperature. After incubation, the spore extraction and assay to susceptible plants was repeated. Selected seed samples were assayed an additional time immediately following

Table 28.6. Number and relative viability of *Pyricularia grisea* spores found in representative Newbonnet seed rice samples collected from Arkansas seed processors.

Sample identity	Spores per 120 g seed sample	Lesion index[a]		
		Pre-sterile[b]	Post-sterile[c]	Post-incubation[d]
ADTKS	254	0.04	0.04	0.00
WGND	127	0.19	0.00	0.33
S BEN5	382	2.00	0.10	1.21
HC1	12,723	3.00+	0.04	5.30+
SS274	13,741	2.28	0.00	0.32
BLK20	13,741	0.08	0.00	1.00

[a] Lesion index determined by dividing the total number of lesions found by the total number of inoculated plants.
[b] Viability for spores flushed from a 120 g seed sample.
[c] Viability assay for spores remaining after the initial extraction and following sample sterilization for 2 min in 0.5% sodium hypochlorite.
[d] Viability assay for spores flushed from samples incubated 72 h after sterilization for 2 min in 0.5% sodium hypochlorite.

the sterilization process. Relative spore viability was estimated using a lesion index derived by dividing the number of lesions found by the number of inoculated plants.

Desiccated, viable *Pyricularia* spores could be found in particulate dust collected when dry seed samples were placed over a screen and moved about. The number of spores flushed from the 120 g samples with water varied from none to in excess of 13.7 × 104 (Table 28.6). The average number of spores over all seed processor samples was 1.8 × 104 spores per sample. Spore viability as indicated by the lesion index also varied between samples. We found the lesion index of individual samples declined over time as samples were moved between cold storage and greenhouse for planting and concluded spore viability differences in the commercial samples resulted from differences in storage conditions. Additional spore production occurred during incubation. Of the 75 Newbonnet samples from the seed processors sixty-five were positive for *P. grisea* in the pre-sterile flush, and 45 were positive after sterilization and incubation (Table 28.7). The ASPB samples, which are more representative of total seed rice production in Arkansas, also had a high percentage of *P. grisea* positive samples (Table 28.8). A surprisingly high number of Lemont seed samples also contained *P. grisea* spores.

Considerable research since the 1920s show rice stubble and rice seed can serve as sources of *Pyricularia* inoculum (Lamey, 1970; Bernaux, 1981; Chung and Lee, 1983; Ou, 1985). Very recently, Reddy and Bastawsi (1989) concluded that spores overwintering on straw and in seed were inoculum sources for blast epidemics occurring in Egypt during 1987. Our efforts (Lee and Jones, 1989) simply reaffirmed previous research findings and provided an indication of spore numbers involved and distribution in seed throughout Arkansas. The large

Table 28.7. Newbonnet seed rice samples collected from seed processors and assayed for viable spores of *Pyricularia grisea*.

Seed lot	Number of samples assayed	Pre-sterile		Post-incubation	
		Number positive	Lesion index[a]	Number positive	Lesion index
1	16	16	1.15	11	0.11
2	23	22	0.59	15	0.21
3	16	13	0.55	9	0.09
4	16	14	0.50	11	0.23
Control	73	0	0.00	0	0.00

[a] Lesion index determined by dividing the total number of lesions found by the total number of inoculated plants.

Table 28.8. Results of assays for viable spores of *Pyricularia grisea* in random seed rice samples representative of all Arkansas rice production areas.

Cultivar	Number of samples		Percent positive[a]	
	1988	1989	1988	1989
Newbonnet	55	70	33	56
Lemont	37	28	22	54
Tebonnet	45	75	9	11
Mars	35	21	0	0
Katy	–	132	–	9

[a] Percent of the samples found to contain viable spores of *P. grisea* as determined by flushing spores from 120 g of seed and inoculating susceptible rice plants.

number of propagules found in the seed rice plus the production of new spores appear to be more than adequate to initiate seedling blast or low levels of saprophytic and pathogenic growth within the canopy.

Additional research needs to be conducted on methodology to reduce or eliminate primary overwintering blast inoculum sources particularly in temperate zones where spores are not airborne throughout the year. Modern growers certainly have the capability of completely destroying all rice stubble to prevent overwintering in localized production areas. The newer fungicides being developed appear sufficiently efficacious to totally disinfest and disinfect seed rice. Modern spore sampling could then be used to anticipate blast epidemics in highly susceptible cultivars such as Newbonnet.

Blast epidemics that recently occurred in Arkansas and other rice areas underscore the need for a coordinated multi-strategic approach to rice blast control. Cultivars utilizing a few resistance genes are and will continue as the predominant blast control method in US rice production as well in other rice production areas. However, as suggested by many, and most recently by Correa-Victoria and Zeigler (1993), the omnipresence of matching virulence within the pathogen population for host resistance factors or genes necessitate the formulation of feasible gene deployment strategies. It is equally important and even imperative that the gene deployment strategy be reinforced by effectual alternative disease control strategies for use in conjunction with cultivar resistance and when cultivar resistance is compromised by blast. This would consist of having a broad database defining the effectiveness of various agronomic practices, a thorough knowledge of the epidemiology of the blast fungus (specifically detailing the role of various inoculum sources) and includes having safe, efficacious fungicides for use in reducing losses and managing epidemics.

References

Atkins, J.G. (1956) An outbreak of *Pyricularia* on rice in 1955. *Plant Disease Reporter* 40, 372–373.

Bernaux, P. (1981) Evolution de la sensibilite des glumelles du riz a *Pyricularia oryzae* Cav. et a Drechslera oryzae (Br. de Haan) Sub et Jain: Consequences pour la transmission des maladies. *Agronomie* 1(4), 261–264 [in French, English summary].

Bollich, C.N., Webb, B.D., Marchetti, M.A. and Scott, J.E. (1985) Registration of 'Lemont' rice. *Crop Science* 25, 883–885.

Borromeo, E.S., Nelson, R.J., Bonman, J.M. and Leung, H. (1993) Genetic differentiation among isolates of *Pyricularia* infecting rice and weed hosts. *Phytopathology* 83, 393–399.

Chung, H.S. and Lee, C.U. (1983) Detection and transmission of *Pyricularia oryzae* in germinating rice seed. *Seed Science and Technology* 11, 625–637.

Correa-Victoria, F.J. and Zeigler, R.S. (1993) Pathogenic variability in *Pyricularia grisea* at a rice blast 'hot spot' breeding site in Eastern Colombia. *Plant Disease* 77, 1029–1035.

Itoi, S., Sato, F., Yamamoto, J., Uchita, Toshihisa, T. and Noda, C. (1979) Overwintering of *Pyricularia* on the living Bamboo and Bamboo grass leaves and pathogenicity of the rice blast fungus, *P. oryzae* Cavara, to bamboo and bamboo grass. *Annals of the Phytopathological Society of Japan* 45, 375–385 [in Japanese, English summary].

Johnston, T.H., Kuenzel, K.A., Lee, F.N., Wells, B.R., Henry, S.E. and Dilday, R.H. (1984) Registration of 'Newbonnet' rice. *Crop Science* 24, 209–210.

Jones, G.S. and Lee, F.N. (1988) Overwintering inoculum for rice blast *Pyricularia oryzae*. *Proceedings of the Twenty-second Rice Technical Working Group*, 154 pp.

Lamey, H.A. (1970) *Pyricularia oryzae* on rice seed in the United States. *Plant Disease Reporter* 54, 931–934.

Lee, F.N. and Jones, G.S. (1989) Primary inoculum sources of rice blast (*Pyricularia oryzae*, Cav.) in Arkansas. *Phytopathology* 79, 1141.

Mackill, A.O. and Bonman, J.M. (1986) New hosts of *Pyricularia oryzae*. *Plant Disease* 70, 125–127.

Marchetti, M.A. (1983) Dilatory resistance to rice blast in USA rice. *Phytopathology* 73, 645–649.

Marchetti, M.A., Rush, M.C. and Hunter, W.E. (1976) Current status of rice blast in the southern United States. *Plant Disease Reporter* 60, 721–725.

Moldenhauer, K.A.K., Lee, F.N., Norman, R.J., Helms, R.S., Wells, B.R., Dilday, R.H., Rohman, P.L. and Marchetti, M.A. (1990) Registration of 'Katy' rice. *Crop Science* 30, 747–748.

Ou, S.H. (1985) *Rice Diseases*, 2nd edn. Commonwealth Mycological Institute, Kew, UK 380 pp.

Prabhu, A.S., Filippi, M.C. and Castro, N. (1992) Pathogenic variation among isolates of *Pyricularia oryzae* affecting rice, wheat, and grasses in Brazil. *Tropical Pest Management* 38, 367–371.

Reddy, A.P.K. and Bastawsi, O.A. (1989) Survival of the rice blast pathogen in the Nile Delta of Egypt. *Phytopathology* 79, 1217.

Urashima, A.S., Igarashi, S. and Kato, H. (1993) Host range, mating type, and fertility of *Pyricularia grisea* from wheat in Brazil. *Plant Disease* 77, 1211–1216.

29 Strategies for the Discovery of Rice Blast Fungicides

J.D. Froyd[1] and E.H. Froeliger[2]

Fungicide Discovery[1] and Plant Biotechnology,[2] American Cyanamid Company, Agricultural Research Division, PO Box 400, Princeton, NJ 08543, USA

Introduction

Since the use of commercial fungicides became routine in the early part of this century, many different products have been used to control rice blast disease, caused by *Magnaporthe grisea* (anamorph *Pyricularia oryzae*). Rice farmers used inorganic copper-based fungicides starting around 1915, followed by organomercurials in the 1950s, antibiotics and organophosphates in the 1960s, and finally systemic fungicides of diverse chemistries in the 1970s and 1980s (Inoue, 1990).

Newer fungicides replaced older ones because they provided superior disease control, less phytotoxicity, minimal animal toxicity, lower environmental impact, no cross-resistance, unique application methods, and greater user convenience. So, what is needed now that is not being provided by today's blasticides? Or, better yet, what will be needed 8–10 years from now – the length of time it will take to commercialize a fungicide discovered today? To address this question, we should first review the current products and their usage.

Japan has the most intensive use of fungicides for the control of rice blast. Fungicide usage is high there because it is economically feasible to treat rice even when disease is not severe. Rice yields are high in Japan; losses from blast are anticipated yearly; the value of rice is 2–3 times higher than the world market; fungicide cost is only 2–3% of the crop value; and rice farmers receive technical support from many sources to control blast (Ou, 1980; Yoshino, 1988). As a result, Japan represents about one-half of the global rice fungicide market, with South Korea, Indonesia, and China having significant but smaller markets (Wood Mackenzie Agrochemical Services, 1993).

In contrast, rice fungicide usage is relatively low elsewhere in South-East Asia, India, Latin American, Africa, and the USA because yearly disease severity is often lower, chemical costs are higher in relation to the value of the crop, and/or there is less technical support for the farmer to control blast.

Table 29.1. Currently registered rice blast fungicides in Japan and some of their features (Ou, 1980; Yoshino, 1988; JPPA, 1992).

Chemical	Residual control	Systemicity	Fungal resistance	Mode of action	Other biological activities
Probenazole	+++	+++	No	Induced resistance	Bacterial leaf blight control
Isoprothiolane	++	+++	Yes	Membrane phospholipid	Planthopper control
Tricyclazole	+++	+++	No	Melanin inhibition	None
Pyroquilon	+++	+++	No	Melanin inhibition	None
Kasugamycin	+	++	Yes	Protein inhibition	Disease control on other crops
IBP	+	+++	Yes	Membrane phospholipid	Plant lodging inhibitor
Edifenphos	++	+	Yes	Membrane phospholipid	*Helminthosporium*, *Fusarium* control
Fthalide	+++	+	No	Melanin inhibition	None
Blasticidin S	+	–	Yes	Protein inhibition	None
Ferimzone	+	+	No	Membrane permeability	*Helminthosporium*, *Cercospora* control

(+) and (–) symbols designate activity ranges from none (–) to strong (+++).

Only a small portion of the rice in the USA is treated with blasticides because less than 20% of the area is threatened with blast annually (chapter 14, this volume; F. Lee, Arkansas, 1993, personal communication; Shaner, 1991). The environmental conditions favoring blast occur erratically in the USA and generally not during the major growing season (Cloud and Lee, 1992). California, the state with the highest rice yield, does not have blast. As a result, fungicide usage in the USA, and in many countries outside Japan, is economical only when damaging blast is anticipated.

Thus, rice blast fungicides developed by agrochemical companies in the past 30 years have been targeted for the large Japanese market. None of the most effective Japanese blasticides have been registered for use on rice in the USA, presumably because of the high development costs in relation to the relatively small USA market size. Products currently used in the USA are mediocre for blast control and are not even used in Japan (Cloud and Lee, 1992).

The currently registered rice blast fungicides in Japan are listed in Table 29.1 along with some significant product features. All of the products provide commercially acceptable blast control when used properly. All must be applied multiple times during the course of the rice growing season to provide complete protection from seedling, foliar, stem, and panicle blast. The market leaders in 1991, according to yen sales, were probenazole, isoprothiolane, tricyclazole, and pyroquilon (Japanese Plant Protection Association, 1992).

Several of the current products, such as probenazole, tricyclazole, pyroquilon, and fthalide, have the desirable property of providing residual blast control. Residual activity is important because blast control is achieved with fewer applications. In addition, probenazole, isoprothiolane, tricyclazole, pyroquilon, and IBP provide systemic blast control. Systemic control is a desirable property because the fungicides are absorbed by the foliage and distributed to untreated areas to provide blast control. Also, systemic fungicides are not susceptible to washoff by rain.

The antibiotics kasugamycin and blasticidin S have been promoted for curative blast control. However, performance is weak for leaf blast control and absent for stem and panicle blast control. With the introduction of ferimzone in 1992 curative blast control may become more practical (Okuno *et al.*, 1989). Fungal resistance to kasugamycin, blasticidin S, and IBP has appeared under commercial usage (Yoshino, 1988; Inoue, 1990). Results from laboratory studies suggest that the IBP resistance carries over to edifenphos and isoprothiolane. Because there are at least four different modes of action among the registered blasticides, resistance can be effectively managed by alternating product usage.

An unusual feature of modern blasticides is their specificity for rice blast control. With few exceptions, they do not provide strong control of other rice diseases or of diseases on other crops. Regardless of the diversity of chemistry and modes of action, their only commercial application is for rice blast control. Is this specificity of action coincidental or is there something unique about *M. grisea* and the rice plant that has unknowingly guided scientists over the past 30 years to repeatedly discover compounds that control only this disease? Nowhere else in the fungicide world do we encounter this degree of specificity.

With this arsenal of fungicides for rice blast control, is there a need for new blasticides? Inoue (1990) stated that 'the chance for the development of new compounds is inevitably small', and he suggested that the only opportunity for a new product was if the new blasticide had curative, preventative, and long residual properties, was fast acting, and controlled other rice diseases. These comments apply to the current, stable Japanese rice fungicide market and do not account for other problems that might arise with present products, such as the increased occurrence of pathogen resistance to existing blasticides. Also, stricter regulation could limit the future use of current products, as witnessed in recent years with fungicides in the USA and Europe. Increased environmental concerns, especially with water-applied pesticides, could place restrictions on the use of current products that are applied at active ingredient rates higher than 1 kg ha^{-1}. Many blasticides offering the best disease control are too expensive for use in countries with low value rice, regardless of the need for blast control, creating opportunities for less expensive products.

Conventional Methods for Discovering Unique Blasticides

Methods of screening for blasticides

Most modern fungicides, including those used to control rice blast, were discovered by random screening of chemical libraries and natural products. In the traditional method of screening for rice blast fungicides, milligram quantities of compounds are weighed and mixed with organic solvents or dry clays, emulsifiers, and surfactants to make formulations for application to young rice seedlings growing in small pots. The formulations are diluted with water to a desired concentration, frequently 100–400 μg a.i. ml^{-1}, and sprayed on the seedlings to runoff with a fine atomizer in a vented treatment chamber. A few hours later after the foliage has dried, the treated plants are inoculated with a spore suspension of the rice blast pathogen harvested from a nutrient medium. Inoculated plants are placed in a water saturated humidity chamber at approximately 25°C for 24–48 h to initiate infection. The plants are then placed in a greenhouse or growth chamber where disease symptoms appear about 5 days later. Treated plants are observed for disease severity and are compared with positive and negative control plants. Plants without symptoms are assumed to have been protected by the chemical treatments.

Fungicide laboratories will screen thousands of compounds each year in the search for new blasticides. Synthesized chemicals, plant extracts, and microbial fermentation broths are all capable of providing active compounds. The larger the testing capacity of the screen, the greater the chance of finding active compounds by random screening.

Lead compounds identified by screening are rarely the most active compounds in new chemistries. Generally, organic chemists prepare analogs of the lead in an attempt to improve upon the initial activity. Analog synthesis can involve the preparation and testing of hundreds of compounds over a 1–2 year period, depending on the interest in and complexity of the chemistry. Patents are filed globally to protect novel chemistries and uses, because no company will risk US$ 25–50 million and 8–10 years of effort to develop a non-proprietary blasticide.

The next step is to field test the most active compounds outdoors under conditions simulating commercial rice cultivation. Most experimental blasticides fail this evaluation and are dropped. Promising field compounds advance to full-scale development and are then entered into official registration trials, such as those conducted by the Japanese Plant Protection Association (JPPA), to be compared with other blasticides and assessed for commercial potential. In 1992, 54 candidate blasticides were evaluated in JPPA trials (Naito, 1993). Most were new formulations and combinations of known blasticides. However, four new chemicals were in the trials, including Bayer's 0301, Shionogi's SSF-126, Kureha's KNF-320, and Otsuka's OK-8801, attesting to the continued interest of companies in discovering new blasticides.

Testing for special features

Protective blast control alone is not sufficient for a candidate blasticide to meet the needs of farmers and to be competitive with existing blasticides. Additional desirable features include residual, systemic and, curative blast control, broader fungicidal spectrum, and no effects on non-target organisms.

Residual blast control

Residual blast control implies that the fungicide is resistant to biological and chemical degradation and provides disease control for approximately 15 days following application. It is measured in greenhouse studies by treating rice plants with the test compound on day 0, inoculating with the blast pathogen on subsequent days, and looking for disease control from the later inoculations. Only leaves treated with the compound are assessed. Disease formation on new leaves produced after day 0 is not a measure of residual control, since these leaves did not receive the initial treatment.

Residual control is a common feature of the leading commercial blasticides and is a necessary property for new blasticides to be competitive. Older blasticides with short residual control, such as the antibiotics blasticidin S and kasugamycin and the organophosphates IBP and edifenphos, have lost popularity in Japan to the longer residual products, such as probenazole, tricyclazole, pyroquilon, and fthalide.

Systemic blast control

A compound absorbed by the plant and translocated to untreated portions to provide disease control is called a systemic fungicide. Systemic disease control is another essential property for new blasticides, and it can be measured in several ways. In one method, the compound is added to the flood water of seedlings in pots, and the seedlings are inoculated after several days to assess for disease control on the leaves. In another method, the basal portion of an individual leaf is first treated with the compound and the entire leaf is inoculated 24h later. The distal portion of the same leaf and other leaves are subsequently assessed for disease control. Systemic blast control provides more effective disease control and the opportunity for unique application methods, such as root and seed treatments.

Several current blasticides are highly systemic in rice and provide excellent disease control from granular applications to the paddy, called submerged applications in Japan (probenazole, pyroquilon, isoprothiolane, and IBP), or to the nursery box (tricyclazole, pyroquilon, and isoprothiolane) before mechanical transplantation of the seedlings (Yoshino, 1988). Granular applications to the paddy and nursery box are convenient for the farmer and can replace multiple foliar applications. However, paddy application requires more compound per hectare than foliar and nursery box applications, treats non-target areas of the paddy, and is costly. The nursery box application, on the other hand, concentrates the blasticide on the roots of the rice plant and requires less compound.

A goal for new systemic blasticides applied to the paddy or nursery box is to provide full season blast control from a single application and at a rate less than 1 kg a.i. ha^{-1}. The advantages of such a product over current systemic usage would be fewer applications, lower compound usage, and convenience of application. Slow release granular formulations might assist with this objective by releasing the active ingredient over a longer time period, rather than the immediate release experienced from conventional formulations.

Another unique application method for blast control is seed treatment. Currently used rice seed treatments in Japan control seedborne inoculum and seedling blast in the nursery box. However, seed treatment for control of blast after transplanting is not a commercial practice. Experimentally, tricyclazole and pyroquilon can provide leaf blast control for 6–7 weeks following the direct seeding of treated seed (J. Froyd, New Jersey, personal communication). As the practice of direct seeding of rice increases in countries where early season blast is an economic problem, compounds that can provide residual control of leaf blast from a seed treatment may find a commercial application.

Curative blast control

A desirable feature lacking in most blasticides has been effective curative blast control. In spite of penetration of and systemic movement in the rice plant, current blasticides are protective and do not arrest stem and panicle blast after infections have occurred.

A new blasticide would find farmer acceptance if it could be applied at the onset of disease symptoms in stems and panicles to stop further development. In fact, the ideal blasticide should be both protective and curative. Or, a curative blasticide could be co-formulated with one having residual activity, which appears to be Takeda's strategy for developing ferimzone, which will be marketed in combination with the residual products tricyclazole and fthalide.

Curative activity can be tested for in the greenhouse by applying test compounds 1–4 days after inoculating the leaves, but before symptoms normally appear, and assessing for reduced disease or restricted symptoms.

Broader fungicidal spectrum

With few exceptions, current blasticides are specific for rice blast control and do not control other diseases (Table 29.1). Therefore, a desirable feature in a new blasticide would be control of sheath blight (*Rhizoctonia solani*), brown spot (*Helminthosporium oryzae*), and other diseases on rice. Co-formulating fungicides having different disease control spectra is currently practiced in Japan, but because cost and product usage are higher, a single broad spectrum fungicide would be more desirable. Recent disclosures on compounds from the new strobilurin fungicide chemistry suggest that broad spectrum rice disease control might be close at hand (Ammermann *et al.*, 1992; Godwin *et al.*, 1992; Masuko *et al.*, 1993).

Environmental and toxicological concerns

Currently registered blasticides have been studied extensively to define types and concentrations of residues in plants, animal toxicity, effects on non-target organisms, fate in the environment, and other properties required of regulatory agencies in user countries.

Future blasticides will have to meet these and more safety requirements before regulatory agencies will permit commercial usage. Issues of greater concern in the future may include compound load on the environment, environmental fate, and effects on additional non-target organisms.

Because blasticides are generally applied in a water environment, associated problems are soil leaching, movement, and toxicity to aquatic life. These properties should be studied early in the development of a new blasticide so that only environmentally safe blasticides are advanced to marketing. Compounds that inhibit mechanisms unique to the blast pathogen, are effective at low rates, and are applied directly to the rice plant will have fewer adverse effects on non-target organisms.

New formulations of existing blasticides that could serve as guides for the formulation of new blasticides are under investigation. For example, pyroquilon 24% jumbo granules applied at 10–13 granules per 0.1 ha achieve slower release of the active ingredient and longer disease control than from current less concentrated formulations. Probenazole 20–40% granules are applied in a water soluble film package at 10–20 packages per 0.1 ha to prevent applicator contact. Formulators of new blasticides should consider these innovations for providing greater application convenience and safety.

Novel Methods for Discovering Unique Blasticides

'Conventional' screening approaches have played an important role in the discovery and development of new blasticides in the past and will continue to play a major role in the future. However, advances in areas of plant and fungal molecular biology (Kerr, 1987) have opened new doors for fungicide discovery programs. Modern molecular genetic techniques can now be used to elucidate the biochemical and molecular basis of the interaction between *M. grisea* and its host plant (Valent and Chumley, 1991). Knowledge gained from such studies will allow the identification and characterization of appropriate target sites that can then be used for: (i) the design of chemicals that inhibit specific fungal processes or enhance the resistance of the rice plant to disease; and (ii) the development of assays to screen for novel antiblast compounds.

Choosing a target

Being mindful of the need for environmentally safe antiblast agents with new modes of action and high specificity, any process specific to the fungus or critical to the host–pathogen interaction should be considered as a desirable target site.

As mentioned earlier, the most successful antiblast agents have been those that are specific to *M. grisea* or to the interaction of the fungus with the rice plant. Thus, a promising area for target research in the search for novel blasticides would be aimed at unraveling the molecular basis of pathogenicity.

The melanin biosynthesis inhibitors (MBIs) tricyclazole, pyroquilon and fthalide are good examples of existing antiblast agents that inhibit a biochemical pathway critical to the rice blast disease cycle (reviewed by Yamaguchi and Kubo, 1992). The MBIs inhibit the reductase steps in the biosynthesis of 1,8-dihydroxynaphthalene (DHN) melanin from pentaketides. As a consequence, appressorial melanization, a process essential for successful penetration of the host plant by *M. grisea*, is blocked. Strains genetically deficient in melanin biosynthesis display a similar phenotype (Chumley and Valent, 1990).

Are there additional steps in the rice blast disease cycle that can be inhibited by new blasticides? The rice blast disease cycle begins under conditions of high humidity when airborne conidia land on the plant surface and adhere via spore tip mucilage (Hamer *et al.*, 1988). The spore germinates soon after attachment forming a short germ tube that differentiates at the tip to produce a specialized infection structure called the appressorium. Appressorium formation, triggered by contact of the germ tube with a hydrophobic surface, begins when the hyphal apex ceases forward growth and enlarges. As the appressorium matures, a single nucleus migrates into the cell and the cell wall becomes heavily melanized (Bourett and Howard, 1990). The melanin layer allows for the sufficient build up of turgor pressure within the tightly adhering appressorium (Howard *et al.*, 1991) so that a fine penetration peg can mechanically pierce the underlying cuticle and epidermal cell wall.

The penetration peg enlarges to form a short (10–20 μm) primary hypha. In a successful infection, the primary infection hyphae differentiate into much wider secondary hyphae that invade adjacent epidermal cells and the underlying mesophyll forming an extensive, branched mycelium (Heath *et al.*, 1990). Within 5–7 days, the fungus causes typical eyespot lesions within which conidiophore development is initiated. This ultimately leads to the differentiation and release of numerous new conidia for additional infections.

Identifying and cloning genes critical to the disease cycle

Each of the highly specialized steps in disease initiation and development represents a potential target for antiblast agents and is awaiting innovative research approaches. What genes or processes govern the production of adhesives or extracellular matrices involved in attachment of conidia or appressoria? What genes regulate the switch from hyphal growth into the pathway leading to formation of the specialized infection structure, the appressorium? What is the molecular basis of host and cultivar specificity? What is the molecular and cellular basis of the dimorphism of infection hyphae? How does the fungus obtain nutrients from the host? What genes direct entry into the conidial pathway and regulate steps involved in the formation of the conidiophore and mature conidia?

What genes are specifically expressed at each stage of development? The possibilities are numerous.

Useful molecular genetic techniques

Many of the modern molecular genetic techniques first developed for the yeast *Saccharomyces cerevisiae*, and soon thereafter for the filamentous fungi *Aspergillus nidulans* and *Neurospora crassa* (Timberlake, 1991), have now been extended to phytopathogenic fungi (Holden and Leong, 1989), including the rice blast fungus, *M. grisea* (Valent and Chumley, 1991). The methods of molecular biology (see Sambrook *et al.*, 1989, for detailed protocols used in recombinant DNA research), including a transformation system, are a valuable addition to classical genetic approaches to the study of rice blast disease (Valent and Chumley, 1991) and provide powerful tools for gene cloning in *M. grisea*.

Transformation of **M. grisea**

DNA-mediated transformation is an essential tool for the molecular characterization of pathogenicity and host specificity. Transformation of *M. grisea* was first achieved using the *A. nidulans argB* gene to complement an arginine auxotrophic strain deficient in ornithine carbomyltransferase activity (Parsons *et al.*, 1987). Subsequently, Leung *et al.* (1990), demonstrated transformation of *M. grisea* to hygromycin B resistance using a bacterial antibiotic resistance gene genetically engineered for expression in filamentous fungi. Drug resistance markers are particularly useful with strains that are not well characterized or when an appropriate auxotrophic strain is not available. As with other filamentous fungi (Fincham, 1989), transformation in *M. grisea* occurs by integration of the donor DNA into the genome of the recipient strain at either homologous or heterologous sites.

Several different approaches, described below, are being used to identify and clone genes critical to pathogenesis. With each of these approaches, it is crucial to be able to verify or determine whether or not a gene plays an essential role in the pathogenic process. For studies aimed at evaluating potential targets for antiblast agents, genetically engineered loss-of-function mutants can be constructed by employing a one-step gene replacement strategy (Rothstein, 1983; Fig. 29.1). This strategy uses integrative transformation as a powerful molecular genetic tool to disrupt the function of the wild-type chromosomal copy of a cloned gene.

First, the cloned gene is inactivated *in vitro*, either by inserting a transformation marker at a single restriction site within the cloned gene or by replacing an internal restriction fragment with the transformation marker. The inactivated gene is then used to transform a wild type, fully pathogenic strain of *M. grisea*. Replacement of the genomic wild-type gene with the inactivated copy occurs by a double crossover event. Generally this type of event is relatively rare in filamentous fungi. However, efficiency of this event is improved when the donor

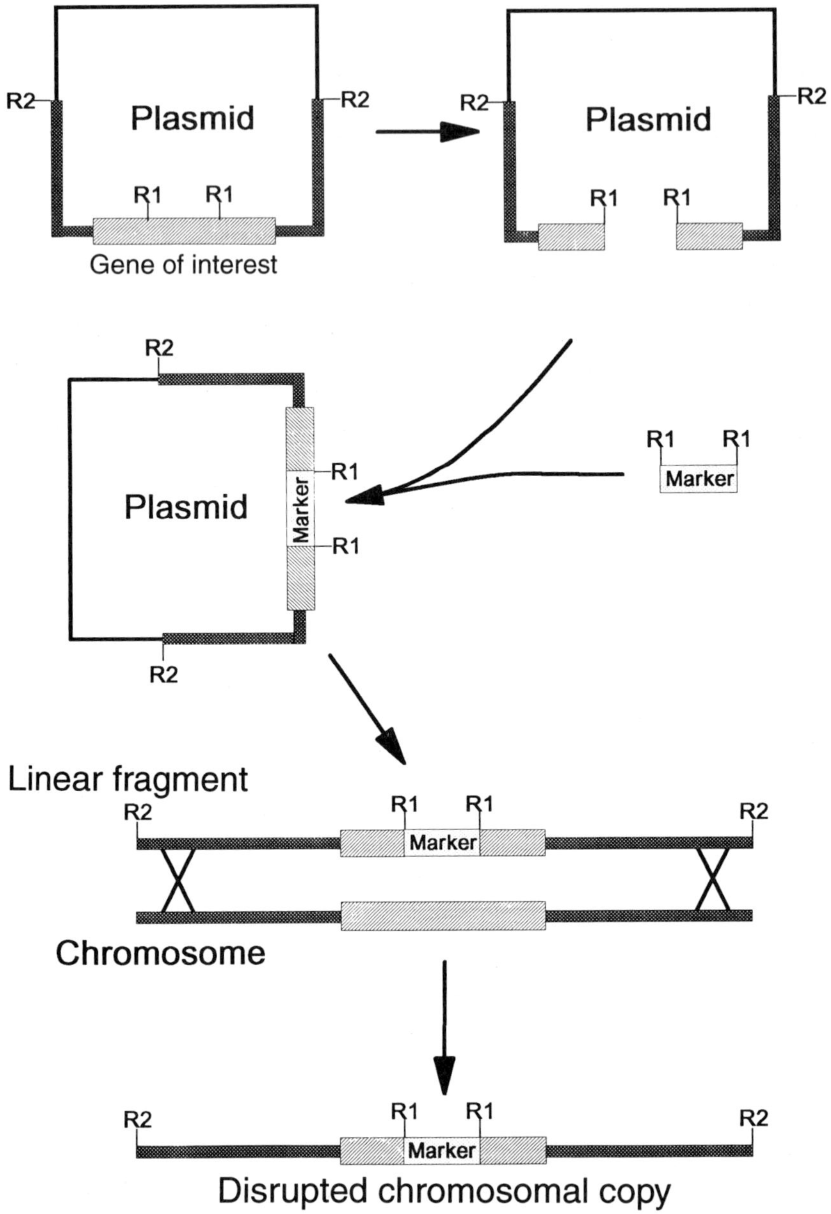

29.1. Inactivation of the wild-type copy of a cloned gene. A plasmid carrying the cloned gene of interest is digested with a restriction enzyme (R1) and the deleted fragment is replaced with a dominant selectable marker. The resulting plasmid is then digested with a second restriction enzyme (R2) to produce a linear DNA molecule. The molecule is then used to transform a wild-type recipient strain. Replacement occurs by a double crossover event. The marker provides a selection for transformation.

DNA is linear and extensive regions of homologous chromosomal DNA flank the inactivated copy.

Pathogenicity of genetically engineered loss-of-function mutants can then be determined. The ability to determine the importance of a gene product to pathogenesis is essential to the discovery of novel targets for antiblast agents.

Cloning genes by differential screening

An effective way of identifying genes playing key roles in the disease cycle involves differential screening of clone banks (Sargent, 1987). This selection strategy detects cDNA clones synthesized from mRNAs that are abundantly expressed in one cell type, tissue or developmental stage and absent or expressed at low levels in another cell type, tissue or developmental stage. For example, Talbot *et al.* (1993) used differential screening to identify *M. grisea* genes preferentially expressed during the infection process. A cDNA library was constructed using mRNA isolated from rice leaves infected with *M. grisea*. Duplicate copies of the library were screened separately with radiolabeled probes synthesized from mRNA from uninfected healthy rice leaves and from mRNA from the *M. grisea* infected rice leaves. Clones hybridizing to the infection probe only were judged to be infection-specific gene transcripts. Southern blot analysis was used to determine if these transcripts were of plant or fungal origin. Those of fungal origin were screened again with radiolabeled probes synthesized from *M. grisea* mRNA expressed in culture. In this manner, fungal transcripts preferentially expressed *in planta* were identified. Presumably, genes expressed preferentially during the infection process are likely to have important roles in pathogenesis.

Detailed characterization of one clone revealed a single gene in the *M. grisea* genome which was designated *MPG1*. The importance of *MPG1* to pathogenicity of *M. grisea* on rice was demonstrated by genetically engineering *MPG1* loss-of-function mutants using a one-step gene replacement strategy (Rothstein, 1983). Such mutants displayed a dramatic reduction in their pathogenicity and were defective in their ability to form appressoria or to conidiate. DNA sequence analysis revealed that the predicted gene product is related to fungal hydrophobins. Generally these small secreted proteins function by providing a hydrophobic shell for fungal structures such as aerial hyphae structures, conidia and fruiting bodies. In *M. grisea*, the hydrophobin encoded by *MPG1* is required for full pathogenicity.

The strategy of cloning pathogenicity related genes by differential screening can be extended to other points in the disease cycle. Lee and Dean (1993) used differential screening of a genomic library to isolate genes expressed during the development of the appressorium. Probes were prepared using mRNA from both germinated conidia induced to make appressoria structures and from non-induced germinated conidia. Several clones containing genes expressed specifically during appressoria formation have been identified. Currently, two of these genes, *mif23* and *mif29*, are being characterized by DNA sequence analysis and by gene inactivation experiments (R. Dean, Clemson University, 1993, personal communication).

The disease cycle in *M. grisea* culminates in the production of mature asexual spores, conidia. Ebbole (Texas A & M University, 1993, personal communication) is identifying genes critical to conidiophore morphogenesis and the development of mature conidia by comparing transcripts expressed during conidiation but not during mycelial growth. One cDNA clone abundantly expressed during conidiation has been characterized in detail. The predicted gene product of this clone is related to fungal hydrophobins and subsequently was shown to be identical to *MPG1* (Talbot *et al.*, 1993).

Cloning genes by function and by map position

Genes playing crucial roles in the disease cycle can be identified by the isolation and characterization of mutants defective in pathogenesis. Infection-related genes, identified through mutational analysis, can be cloned by genetic complementation of the mutations in transformation experiments. Gene inactivation experiments can be used to verify that the cloned gene carries out an essential role in the pathogenic process.

Although possible, direct selection for complementation of infection-defective mutants is not practical if each individual transformant needs to be tested for its ability to infect one or more host plants. However, it is now possible to clone genes from *M. grisea* by 'chromosome walking' (Spoerel and Kafatos, 1987) to the gene of interest from previously known linked markers. Identification of the target gene can achieved by expression of the cloned sequence in transformants.

Because of the lack of multiply marked strains of *M. grisea*, considerable effort has been made to generate genetic maps of the *M. grisea* genome using physical markers such as *M. grisea* repeat sequences (MGR; Romao and Hamer, 1992), restriction fragment length polymorphisms (RFLPs; Skinner *et al.*, 1990; Valent and Chumley, 1991) or random amplified polymorphic DNA (RAPDs; Williams *et al.*, 1991; H. Leung, Washington State University, 1993, personal communication). 'Chromosome walking' coupled with transformation experiments is being employed to clone several genes identified by mutational analysis.

Hamer *et al.* (1989b), detected the *SMO* genetic locus by screening a collection of survivors of UV light mutagenesis for individuals defective in appressoria formation. Strains carrying the *smo* mutation display abnormally shaped conidia and appressoria and are defective in the ability to infect rice (Hamer and Givan, 1990). The *SMO* locus was genetically mapped using a dispersed repeated DNA sequence family, called MGR, found in all *M. grisea* rice pathogens (Hamer *et al.*, 1989a). A clone carrying the *SMO* gene has been identified by 'walking' the chromosome from linked MGR markers. Further characterization of *SMO* is currently in progress.

Sporulation is a key factor in rice blast disease epidemics. Using chemical mutagenesis, Leung and Shi (this volume) have identified several mutations affecting conidiophore morphogenesis and development of mature conidia. Two of the mutations have been characterized in detail. Strains carrying the *con1* mutation produce single elongated conidia on each conidiophore and sporulation is reduced to 3% of wild type; strains carrying the *con2* mutation grow normally and produce normal conidiophores but no conidia are formed. RAPD-bulked segregant analysis (Michelmore *et al.*, 1991) was used to rapidly identify markers linked to the *con1* and *con2* loci. Currently, Leung *et al.* (chapter 6) are in the process of isolating the *CON1* and *CON2* genes by 'walking' the chromosome from the linked RAPD markers.

Identifying and cloning genes by insertional inactivation mutagenesis

A new strategy utilizing insertional inactivation mutagenesis has been developed by J. Sweigard and B. Valent (DuPont, 1993, personal communication) to identify genes critical to the pathogenic process. As with other filamentous fungi (Fincham, 1989), transformation in *M. grisea* occurs most frequently by nonhomologous random integration of the donor plasmid. Sweigard and colleagues used plasmids containing a dominant selectable marker to transform fully pathogenic strains of *M. grisea* and to generate a collection of mutants via insertional inactivation mutagenesis. Using a cotton swab inoculation method, individuals were screened rapidly for pathogenicity defects. Several nonpathogenic mutants were identified among 1650 insertional mutants screened. Because the mutation is associated with the dominant selectable marker, the pathogenicity gene can be cloned readily by marker rescue techniques, thus expediting molecular and biochemical analysis of the mutant phenotype. Three pathogenicity genes have already been cloned using this strategy and are currently being characterized.

Cloning genes by sequence similarity

Genes with potential importance to host–pathogen interactions can be cloned using heterologous probes. For example, the fungal enzyme cutinase may be important during the penetration of the plant cuticle and initiation of certain diseases. The single copy cutinase gene from *M. grisea*, *CUT1*, was cloned using the cutinase gene of *Colletotrichum capsici* as a heterologous probe (Sweigard *et al.*, 1992a). To test the importance of cutinase in pathogenesis, *M. grisea CUT1*-defective mutants were genetically engineered using a one-step gene disruption strategy (Rothstein, 1983). Unlike melanin biosynthesis mutants that are crucial to penetration, the *CUT1*-defective mutants displayed no obvious role in pathogenesis (Sweigard *et al.*, 1992b). Thus, cutinase would not be a good target for antiblast agents.

Numerous genes involved in conidiophore morphogenesis and the development of mature conidia have been cloned from *N. crassa* (Berlin and Yanofsky, 1985) and *A. nidulans* (Timberlake and Marshall, 1988). Ebbole (Texas A & M University, 1993, personal communication) is using cloned conidiation-specific genes from *N. crassa* as hybridization probes to isolate genes of related function from *M. grisea*. Preliminary results, using the *N. crassa con10* gene (Roberts *et al.*, 1988) as a heterologous probe, suggest that this approach will be successful. Studies directed toward the basic biology of *M. grisea* will be greatly facilitated if the wealth of information about traditional genetic organisms such as *N. crassa* and *A. nidulans* can be applied to *M. grisea*.

Cloning M. grisea *avirulence genes*

An important area in the study of plant–fungal interactions has been the molecular basis of race-specific disease resistance. Genetic analysis of the rice blast fungus suggest that there is a gene-for-gene relationship in which dominant blast resistance genes of the rice plant interact with corresponding fungal avirulence genes to produce a hypersensitive response (Leung *et al.*, 1988; Valent *et al.*, 1991; Ellingboe, 1992; Silué *et al.*, 1992). Using the tools of molecular biology, B. Valent and F.G. Chumley (chapter 7) have cloned the first avirulence gene, *Avr2-YAMO*, from the rice blast fungus. Using RFLP mapping, avirulence gene *Avr2-YAMO*, corresponding to the appropriate resistance gene in rice cultivar Yashiro-mochi, was mapped to a telomeric region. Molecular analysis of spontaneous mutations of *Avr2-YAMO* that gained virulence on rice cultivar Yashiro-mochi revealed deletions within one kilobase (kb) of the telomeric tip where *Avr2-YAMO* is linked. A 6.5 kb telomeric fragment, predicted by mutant analysis to carry the *Avr2-YAMO* gene, was subsequently cloned from the wild-type strain. To prove that *Avr2-YAMO* is responsible for Yashiro-mochi-specific resistance, the isolated clone was transformed into *M. grisea* strains virulent on Yashiro-mochi. Transformants acquiring *Avr2-YAMO* became avirulent on rice cultivar Yashiro-mochi; virulence was maintained on all other cultivars. The cloning and molecular analysis of avirulence genes from *M. grisea* is a major step toward the understanding of the molecular basis of cultivar-specific resistance to blast. Characterization of avirulence genes may reveal the function of these genes in normal cell processes of the fungus. No blast resistance genes have been cloned from rice yet, but research in this area is active (Yu *et al.*, 1991).

Induced systemic acquired resistance

The leading blasticide in Japan, probenazole, is purported to control rice blast via activation of natural defense mechanisms in susceptible rice plants (Yamaguchi, 1982). Probenazole has weak direct antifungal activity but probenazole-treated plants are able to mount a rapid and effective resistance response to infection by *M. grisea*. These responses include enhanced production of the antifungal compound α-linolenic acid, phenylalanine ammonia lyase, peroxidases, and increased lignification of rice cell walls (Koller, 1992).

Other chemicals reported to induce resistance to blast and to elicit similar biochemical responses in rice include *N*-cyanomethyl-2-chloroisonicotinamide (NCI) (Yoshida *et al.*, 1990) and dichlorocyclopropanes (Langcake and Wickens, 1975). NCI also induced resistance to bacterial leaf blight (*Xanthomonas oryzae*), suggesting activation of a broad spectrum defense mechanism. Smith and Metraux (1991) reported on the induction of systemic resistance to *M. grisea* by prior inoculation with the incompatible bacterial pathogen *Pseudomonas syringae* pv. *syringae*. However, a specific chemical elicitor was not identified.

The precise mechanisms leading to induced resistance may not be known for any of these agents. However, the discovery of other chemical elicitors for induced resistance to blast is possible and desirable, especially if the resistance includes control of other rice diseases. Toward this end, considerable effort has been directed toward understanding the molecular mechanisms of systemic acquired resistance (SAR) in plants (Enyedi *et al.*, 1992). Significant progress has been made in identifying gene families induced during the development of SAR. Of particular interest is the question of how defense-related genes are activated in plant tissue, particularly at sites distant from a localized attack by pathogen or pest. Molecular and biochemical studies have identified molecules that may serve as a signal to activate natural host defense mechanisms. One potential systemic signal, salicylic acid, appears to be specifically associated with pathogen infections and is not inducible by wounds. It is still unclear, however, whether or not salicylic acid is a primary systemic signal molecule in SAR. Identification of key components leading to SAR will provide leads for the discovery of novel blasticides that act by stimulating resistance mechanisms.

Putting newly discovered target sites to use

Once new targets have been identified and characterized, several strategies can be employed for the discovery of novel inhibitors of the infection process or for the discovery of novel inducers of host resistance.

Biorational design of antiblast agents

One approach to the discovery of new antiblast agents is through the deliberate biorational design of inhibitors directed toward predetermined and previously characterized target proteins. This strategy has been limited in the past by a general lack of ways to identify and evaluate novel antifungal target sites and inadequate computer-based tools and biophysical techniques. However, the rapid advances being made in the areas of molecular biology, computational sciences and biophysical technology makes the biorational approach to the discovery of antiblast agents feasible and cost-effective (Goosey, 1992). General strategies and principle considerations for the biorational design of antifungal inhibitors are discussed by Koller (1992).

Targeted screening

A second approach to the discovery of novel fungicides, discussed by Koller (1992), uses genetically engineered fungal organisms as target-based screening systems. In such systems, the organism is designed to selectively detect compounds that act against specific molecular targets. For example, Koller cites the use of *S. cerevisiae* expressing β-galactosidase as a screen for compounds inhibiting cell wall biosynthesis.

Kirsch and DiDomenico (1993) provide a useful introduction to target-based screening and an overview of design strategies for target based fungicide screens. Both *in vitro* screens and *in vivo* molecular genetic screens are described. An advantage of the latter is that fungal organisms are used and the response of whole organisms to a potential inhibitor can be observed. For example, Kirsch and

Lai (1986) have developed a screen for the detection of antifungal compounds acting on fungal cell wall biosynthesis using a temperature sensitive mutant of *N. crassa* that is able to divide in the absence of a cell wall. In the past, lack of knowledge about targets with critical functions in host–fungal interactions has seriously limited the use of target-based screens for the discovery of antidisease agents. Current research will enhance future opportunities in this area.

Transgenic plants

As mentioned above, a leading blasticide, probenazole, probably controls blast by inducing host resistance mechanisms. Transgenic plants or plant tissues could be used to screen for new chemicals that induce resistance mechanisms in rice to blast. Promoter regions from plant defense-related proteins fused to reporter genes such as β-glucuronidase (GUS) have been shown to be inducible in tobacco, either with natural products such as salicylic acid or with synthetic compounds (Ward *et al.*, 1993). Based on these principles, a screen for new chemical inducers of host resistance to blast could be devised (Doerner *et al.*, 1990).

Conclusions

Currently used blasticides, particularly those in Japan, are highly effective and apparently cause no adverse environmental effects to threaten their continued usage. However, in recent years farmers have switched product preference to long residual, systemic compounds that can be applied in more convenient ways than the traditional dusts and ground sprays.

The need to continue the search for new blasticides with different modes of action is important because regulatory and environmental issues and pathogen resistance unknown today could eliminate the usage of current products tomorrow. Discoverers of blasticides must be more accountable to environmental concerns of water applied pesticides, farmer needs, formulation possibilities, and new technology tools.

Targeting specific blast mechanisms for inhibition should lead to more efficient blasticides with few undesirable side-effects. Classical genetic analysis and the tools of modern molecular biology can now be combined to investigate the molecular mechanisms of fungal pathogenesis and sites for chemical inhibition. Studies on the pathways leading to systemic acquired resistance could lead to the discovery of chemicals that elicit natural resistance mechanisms in rice to blast and other diseases.

While effective and unique compounds and application methods are available to farmers at present, new strategies and technologies, integrated with traditional screening approaches, will lead to the discovery of more effective yet less toxic and more environmentally friendly fungicides.

Acknowledgments

We thank Patricia Bavoso and Candace Cook for assistance in preparing the manuscript and Laura Sarokin, Roy Chaleff, and Lynn Evans for a critical reading of the manuscript. We thank Drs J. Hamer, B. Valent, H. Leung, D. Ebbole, R. Dean, and F. Lee for communication of unpublished data.

References

Ammermann, E., Lorenz, G., Schelberger, K., Wenderoth, B., Sauter, H. and Rentzea, C. (1992) BAS 490 F, a broad spectrum fungicide with a new mode of action. *Brighton Crop Protection Conference Proceedings* 1, 403–410.

Berlin, V. and Yanofsky, C. (1985) Isolation and characterization of genes differentially expressed during conidiation of *Neurospora crassa. Molecular and Cellular Biology* 5, 849–855.

Bourett, T.M. and Howard, J.H. (1990) *In vitro* development of penetration structures in the rice blast fungus *Magnaporthe grisea. Canadian Journal of Botany* 68, 329–342.

Chumley, F.G. and Valent, B. (1990) Genetic analysis of melanin-deficient, non-pathogenic mutants of *Magnaporthe grisea. Molecular Plant-Microbe Interactions* 3, 135–143.

Cloud, G. and Lee, F. (1992) *Rice Blast: Factors Influencing Severity and Control Strategies Used to Manage the Disease*. University of Arkansas Publication MP 341.

Doerner, P.W., Stermer, B., Schmid, J., Dixon, R.A. and Lamb, C.J. (1990) Plant defense gene promoter–reporter gene fusions in transgenic plants: Tools for identification of novel inducers. *Bio/Technology* 8, 845–849.

Ellingboe, A.H. (1992) Segregation of avirulence/virulence on three rice cultivars in 16 crosses of *Magnaporthe grisea. Phytopathology* 82, 597–601.

Enyedi, A.J., Yalpani, N., Silverman, P. and Raskin, I. (1992) Signal molecules in systemic plant resistance to pathogens and pests. *Cell* 70, 879–886.

Fincham, J.R.S. (1989) Transformation in fungi. *Microbiological Reviews* 53, 148–170.

Godwin, J.R., Anthony, M.V., Clough, J.M. and Godfrey, C.R.A. (1992) ICIA5504: a novel, broad spectrum, systemic beta-methoxyacrylate fungicide. *Brighton Crop Protection Conference Proceedings* 1, 435–442.

Goosey, M.W. (1992) Towards the rational design of crop protection agents. *Pesticide Science* 34, 313–320.

Hamer, J.E., and Givan, S. (1990) Genetic mapping with dispersed repeated sequences in the rice blast fungus: mapping the *SMO* locus. *Molecular and General Genetics* 223, 487–495.

Hamer, J.E., Howard, R.J., Chumley, F.G. and Valent, B. (1988) A mechanism for surface attachment in spores of a plant pathogenic fungus. *Science* 239, 288–290.

Hamer, J.E., Farrall, L., Orbach, M.J., Valent, B. and Chumley, F.G. (1989a) Host species-specific conservation of a family of repeated DNA sequences in the genome of a fungal pathogen. *Proceedings of the National Academy of Sciences of the United States of America* 86, 9981–9985.

Hamer, J.E., Valent, B. and Chumley, F.G. (1989b) Mutations in the *SMO* genetic locus affect the shape of diverse cell types in the rice blast fungus. *Genetics* 122, 351–361.

Heath, M.C., Valent, B., Howard, R.J. and Chumley, F.G. (1990) Interactions of two strains of *Magnaporthe grisea* with rice, goose grass and weeping lovegrass. *Canadian Journal of Botany* 68, 1627–1637.

Holden, D.W. and Leong, S.A. (1989) Molecular genetic approaches to the study of fungal pathogenesis. *Annual Review of Phytopathology* 27, 463–481.

Howard, R.J., Ferrari, M.A., Roach, D.H. and Money, N.P. (1991) Penetration of hard substrates of a fungus employing enormous turgor pressures. *Proceedings of the National Academy of Sciences of the United States of America* 88, 11281–11284.

Inoue, S. (1990) Trends in the chemical control of rice diseases in Japan. *Pesticide Outlook* 1, 31–37.

Japanese Plant Protection Association (1992) Pesticide production and shipment [in Japanese]. In: *Nohyaku Yoran* (*Pesticide Yearbook*). Japanese Plant Protection Association and Ministry of Agriculture, Fisheries, and Forestry, Tokyo, Japan, pp. 108–132.

Kerr, A. (1987) The impact of molecular genetics on plant pathology. *Annual Review of Phytopathology* 25, 87–110.

Kirsch, D.R. and DiDomenico, B.J. (1993) Mechanism based screening for the discovery of novel antifungals. In: Gullo, V. (ed.), *Discovery of Natural Products with Therapeutic Potential*. Butterworth Publishers, Stoneham, Massachusetts, pp. 177–221.

Kirsch, D.R. and Lai, M.H. (1986) A modified screen for the detection of cell wall-acting antifungal compounds. *Journal of Antibiotics* 39, 1620–1622.

Koller, W. (1992) Target research in the discovery and development of antifungal inhibitors. In: Koller, W. (ed.), *Target Sites of Fungicide Action*. CRC Press, Boca Raton, Florida, pp. 255–310.

Langcake, P. and Wickens, S.G.A. (1975) Studies on the action of the dichlorocyclopropanes on host–parasite relationship in the rice blast disease. *Physiological and Molecular Plant Pathology* 7, 113–126.

Lee, Y.H. and Dean, R.A. (1993) Stage-specific gene expression during appressorium formation of *Magnaporthe grisea*. *Experimental Mycology* 17, 215–222.

Leung, H., Borromeo, E.S., Bernardo, M.A. and Nottegham, J.L. (1988) Genetic analysis of virulence in the rice blast fungus *Magnaporthe grisea*. *Phytopathology* 78, 1227–1233.

Leung, H., Lehtinen, U., Karjalainen, R., Skinner, D., Tooley, P., Leong, S. and Ellingboe, A. (1990) Transformation of the rice blast fungus *Magnaporthe grisea* to hygromycin B resistance. *Current Genetics* 17, 409–411.

Masuko, M., Niikawa, M., Kataoka, T., Ishinari, M,. Takenaka, H., Hayase, Y., Hayashi, Y. and Takeda, R. (1993) In: *6th International Congress of Plant Pathology*. International Society of Plant Pathology, Montreal, Quebec, Canada, p. 91.

Michelmore, R.W., Paran, I. and Kesseli, R.V. (1991) Identification of markers linked to disease-resistance genes by bulked segregant analysis: A rapid method to detect markers in specific genomic regions by using segregating populations. *Proceedings of the National Academy of Sciences of the United States of America* 88, 9828–9832.

Naito, H. (1993) Review of the conference of official trial results by JPPA. *Nohrin Gijutsu Shimpo* 1291, 12 [in Japanese].

Okuno, T., Furusawa, I., Matsuura, K. and Shishiyama, J. (1989) Mode of action of ferimzone, a novel systemic fungicide for rice diseases: Biological properties against *Pyricularia oryzae in vitro*. *Phytopathology* 79, 827–832.

Ou, S.H. (1980) A look at worldwide rice blast disease control. *Plant Disease* 64, 439–445.

Parsons, K.A., Chumley, F.G. and Valent, B. (1987) Genetic transformation of the fungal pathogen responsible for rice blast disease. *Proceedings of the National Academy of Sciences of the United States of America* 84, 41161–4165.

Roberts, A.N., Berlin, V., Hager, K.M. and Yanofsky, C. (1988) Molecular analysis of a *Neurospora crassa* gene expressed during conidiation. *Molecular and Cellular Biology* 8, 2411–2418.

Romao, J. and Hamer, J.E. (1992) Genetic organization of a repeated DNA sequence family in the rice blast fungus. *Proceedings of the National Academy of Sciences of the United States of America* 89, 5316–5320.

Rothstein, R.J. (1983) One-step gene disruption in yeast. *Methods in Enzymology* 101, 202–211.

Sambrook, J., Fritsch, E.F. and Maniatis, T. (1989) In: Nolan, C. (ed.), *Molecular Cloning: A Laboratory Manual*. Cold Spring Harbor Laboratory, New York.

Sargent, T.D. (1987) Isolation of differentially expressed genes. *Methods in Enzymology* 152, 423–432.

Shane, G. (1991) Fungicide benefits assessment: Cereal crops. *NAPIAP Fungicide Assessment Report*, USDA, Washington DC.

Silué, D., Tharreau, D. and Notteghem, J.L. (1992) Identification of *Magnaporthe grisea* avirulence genes to seven rice cultivars. *Phytopathology* 82, 1462–1467.

Skinner, D.Z., Leung, H. and Leong, S.A. (1990) Genetic map of the blast fungus, *Magnaporthe grisea*. In: O'Brian, S.J. (ed.), *Genetic Maps*. Cold Spring Harbor Laboratory, New York, pp. 3.82–3.83.

Smith, J.A. and Metraux, J.P. (1991) *Pseudomonas syringae* pv. *syringae* induces systemic resistance to *Pyricularia oryzae* in rice. *Physiological and Molecular Plant Pathology* 39, 451–461.

Spoerel, N.A. and Kafatos, F.C. (1987) Isolation of full length genes: walking the chromosome. *Methods in Enzymology* 152, 598–603.

Sweigard, J.A., Chumley, F.G. and Valent, B. (1992a) Cloning and analysis of *CUT1*, a cutinase gene from *Magnaporthe grisea*. *Molecular and General Genetics* 232, 174–182.

Sweigard, J.A., Chumley, F.G. and Valent, B. (1992b) Disruption of a *Magnaporthe* cutinase gene. *Molecular and General Genetics* 232, 183–190.

Talbot, N.J., Ebbole, D.J. and Hamer, J.E. (1993) Identification and characterization of *MPG1*, a gene involved in pathogenicity from the rice blast fungus *Magnaporthe grisea*. *The Plant Cell* 5, 1575–1590.

Timberlake, W.E. (1991) Cloning and analysis of fungal genes. In: Bennett, J.W. and L.L. Lasure (ed.), *More Gene Manipulations in Fungi*. Academic Press, San Diego, California, pp. 51–85.

Timberlake, W.E. and Marshall, M.A. (1988) Genetic regulation of development in *Aspergillus nidulans*. *Trends in Genetics* 4, 162–169.

Valent, B. and Chumley, F.G. (1991) Molecular genetic analysis of the rice blast fungus, *Magnaporthe grisea*. *Annual Review of Phytopathology* 29, 443–467.

Valent, B., Farrall, L. and Chumley, F.G. (1991) *Magnaporthe grisea* genes for pathogenicity and virulence identified through a series of back-crosses. *Genetics* 127, 87–101.

Ward, E.R., Ryals, J.A. and Miflin, B.J. (1993) Chemical regulation of transgene expression in plants. *Plant Molecular Biology* 22, 361–366.

Williams, J.G.H., Kubelik, A.R., Rafalski, J.A. and Tingey, S.V. (1991) Genetic analysis with RAPD markers. In: Bennett, J.W. and Lasure, L.L. (eds), *More Gene Manipulations in Fungi*. Academic Press, New York, pp. 431–439.

Wood Mackenzie Agrochemical Services (1993) Agrochemical product groups. *Agrochemical Service* 1, Rice-15.

Yamaguchi, I. (1982) Fungicides for control of rice blast disease. *Journal of Pesticide Science* 7, 307–316.

Yamaguchi, I. and Kubo, Y. (1992) Target sites of melanin biosynthesis inhibitors. In: Koller, W. (ed.), *Target Sites of Fungicide Action*. CRC Press, Boca Raton, Florida, pp. 101–115.

Yoshida, H., Konishi, K., Koike, K., Nakagawa, T., Sekido, S. and Yamaguchi, I. (1990) Effect of N-cyanomethyl-2-chloroisonicotinamide for control of rice blast. *Pesticide Science* 15, 413–417.

Yoshino, R. (1988) Present status of occurrence and control of blast disease in Japan. *Japan Pesticide Information* 52, 3–8.

Yu, Z.H., Mackill, D.J., Bonman, J.M. and Tanksley, S.D. (1991) Tagging genes for blast resistance in rice via linkage to RFLP markers. *Theoretical and Applied Genetics* 81, 471–476.

30 Biological Control of Rice Leaf Blast

A.A. Sy[1], L. Albertini[2] and M. Moletti[3]

[1]*WARDA/ADRAO, 01 BP 2551, Bouaké Côte d'Ivoire;* [2]*ENSAT/INP, Laboratoire d'Ingénierie Agronomique, 145 av. de Muret, 31076 Toulouse, Cedex, France;* [3]*ENTE NAZIONALE RISI, Department of Plant Pathology, Strada per Ceretto, 4 27030 Castello D'Agogna (PV), Italy*

Introduction

Initial screening of a range of microorganisms for the suppression of *Pyricularia oryzae* growth *in vitro* indicated that many species were promising candidates for use as biological control agents for rice blast disease (Sy *et al.*, 1983a). The objectives of this chapter are: (i) to assess the *in vivo* performance level of antagonists that exhibited high *in vitro* effectiveness; (ii) to determine the optimum dose and the limits to biotreatment use; and (iii) to assess the efficiency spectrum of *Chaetomium globosum* towards three combinations of rice cultivars and *Pyricularia oryzae* isolates. Results of three levels of experimentation are considered:

1. In vivo *preliminary screening*. Twenty antagonistic microorganisms were tested for their ability to inhibit *in vivo* leaf blast development. The test microorganisms were selected from a population of 120 organisms previously characterized for their reaction to temperature, pH, media, light, *in vivo* innocuity in relation to rice, and their ability to inhibit mycelium growth and/or sporulation and/or conidia germination and early growth of *P. oryzae* (Sy, 1976, 1987; Sy *et al.*, 1983b, 1984).
2. In vivo *advanced screening*. Eight microorganisms with the best performance in the first level of screening were selected and used with three doses (three antagonist/parasite concentration ratios). The objective was to determine the level of intrinsic effectiveness (minimum dose producing the maximum efficiency) as well as the limits of each biotreatment (minimum dose below which no significant protection is noted and maximum dose above which a toxic effect is triggered).
3. *Spectrum of actions*. Following the second set of experiments, the best biotreatment (*C. globosum*, which totally suppressed the leaf blast expression with a relatively low application rate) was selected and its activity spectrum determined on three differential strains of *P. oryzae* associated with their respective hosts (Sy, 1987).

Materials and Methods

The antagonistic microorganisms involved in these experiments were isolated from (i) rice, (ii) barley straw, (iii) soil samples, or (iv) from various international collections (The Netherlands, Czechoslovakia, Sweden).

For the first and second screening levels, the host and parasite combination was as follows: Se302G rice cultivar and Se3 isolate of *P. oryzae* native of Casamance, Senegal (Sy, 1987). For testing of the action spectrum of *C. globosum*, the host and parasite combinations were as follows (Sy, 1987): the Maratelli cultivar combined with the MAR-L5 isolate of *P. oryzae* (native of Po Valley, Italy); the Ikong-Pao cultivar combined with the IKP-N7 isolate of *P. oryzae* (native of the Senegal Valley); and the 6044 cultivar combined with the SIX-L1 of *P. oryzae* (native of Sine Saloum, Senegal).

All experiments were conducted under controlled conditions: 14 hours light at 28°C and 10 hours dark at 24°C; minimal relative humidity of 76°C, raised to 96–100% for 4 days following the application of treatments.

The pathogen and antagonist were inoculated simultaneously (by fine spraying) on to hosts in the presence of a 0.25% gelatin solution. All treatments (including appropriate checks i.e. distilled water or suspension of the pathogen conidia or suspension of the antagonist particles) were set out according to a randomized complete bloc design with four replications, each single plot having 30 vigorous 2-week-old plants at the time of treatment application.

The application dose for each biotreatment was estimated by the number of the antagonist particles (number of conidia or ascospores or vegetative particles in the case of *Micromonospora*) per unit volume (i.e. antagonist concentration) divided by the number of parasite conidia per unit volume (i.e. parasite concentration). Thus, the estimated quotient represents the antagonist/parasite concentration ratio.

Two complementary variables were assessed (assessement was based on a rating scale of categories 1–5, indicating degrees of severity: 1 = no lesion and 5 = more than 10 lesions per unit surface) to evaluate the impact of the disease following the biotreatment application. The first set of observations were for incidence (I) = percentage of affected plants. The second set of observations were for severity (S) (Sy, 1987) calculated as follows:

$$S(\%) = \frac{\Sigma_i(i-1)}{E(i-1)} \times \frac{Y_i}{N} \times 100$$

where i = category i.e. 1, 2, . . ., 5; Y_i = number of plants per category i; and N = total number of observed plants.

The effectiveness of the biological treatment was estimated by (i) the rate of incidence inhibition [INH.(I)] as compared to the control and (ii) the rate of severity inhibition or [INH.(S)]. After their transformation (arc sin %), data were analyzed through two-way analysis of variance, cross-classification, fixed model and multiple means comparisons were made using the Student–Newman–Keuls test.

Disease assessment was measured starting from day 6 after treatment application and continued for 7 days thereafter. Data presented here refer to the last scoring date (27 DAS).

Results and Discussion

In vivo preliminary screening

Data on the suppression of the incidence and severity of leaf blast are shown in Table 30.1. The table shows each biotreatment, the antagonist/parasite concentration ratios, average inhibition rates (actual) and statistical differences. *Trichoderma harzianum* (CPO-80) and *Micromonospora* sp. (PS6-2) were the most effective in terms of incidence (91.6% and 90.8% inhibition respectively) and severity (97.4% and 97.3% inhibition respectively). Despite its low concentrations ratio, *C. globosum* (N 76–1) was the third most effective treatment (87.5% incidence inhibition and 95.8% severity inhibition).

The antagonistic potential of a microorganism species *vis-à-vis* a given pathogenic agent depends on its characteristics, such that related species of the same genus often respond quite differently. For example, *T. harzianum* exhibits 91.6 and 97.4% inhibition of incidence and severity in comparison to *Trichoderma pseudokoningii* and *Trichoderma viride* which exhibit 0% inhibition levels. On the other hand, very distant species such as *T. harzianum* with 91.6% inhibition of incidence and 97.4% inhibition of severity in relation to *Micromonospora* sp. with 90.8 and 97.3% respectively can manifest the same efficiency potential *vis-à-vis* the same pathogenic agent.

Similar differential responses can be observed between strains of the same species. For example, in contrast to the exceptional performance of the *T. harzianum* (CPO-80) strain, the *T. harzianum* (N 68) strain appears to be ineffective. The same is true of *Trichothecium roseum* (T 372), (inhibition rates of 77.5% incidence and 90.6% severity) as compared to *T. roseum* (N 57) (0% and 13.1%). Therefore, crucial determinants of effectiveness may be based on the biological and physiological characteristics of different strains.

With reference to the relative importance of the antagonist/parasite concentration ratios, experimental data also show important differences in the effects of different concentration ratios on performance (Table 30.1), although these tend to be species specific.

With reference to the direct or indirect character of the antagonist and parasite interactions, we also examined whether the antagonist acts directly through physical or chemical means on the pathogen or if the host must intervene (elicitation phenomenon). In particular, we sought to answer the following two questions: How can the complete lack of *in vivo* performance of *T. viride* (N 76–5) and *T. pseudokoningii* (N 39) be explained while they inhibited 100% of mycelium growth of Se3 at pH 4.10 in *in vitro* tests? What explains the *in vitro* and *in vivo* performance differences of *C. globosum* (N 76–1) and of *T. harzianum* (CPO-80) especially as compared with mycelium growth? In the first case, the relative efficiency of *T. viride* and *T. pseudokoningii* is due to their rapid growth which results in early colonization followed by the exhaustion of

Table 30.1. Mean suppression of incidence and severity by 20 potential biological control agents of the development of rice blast disease on the rice cultivar Se302G when co-inoculated with isolate Se3.

Treatment[a]	[Antagonist] / [Parasite]	Mean inhibition incidence (%)[b]	Treatment[a]	[Antagonist] / [Parasite]	Mean inhibition severity (%)[b]
Check (Se3)	0	0.0 a	Check (Se3)	0	0.0 a
T. roseum (N 57)	21	0.0 a	*S. koorchalomoides* (C 302G)	20	0.0 a
Trichoderma sp. (BRL 5)	993	0.0 a	*T. pseudokoningii* (N 69)	1050	0.0 a
S. koorchalomoides (C302G)	20	0.0 a	*T. viride* (N-76–5)	96	0.0 a
T. pseudokoningii (N 69)	1050	0.0 a	*Trichoderma* sp. (IKP 2)	175	0.0 a
T. viride (N 76–5)	796	0.0 a	*T. roseum* (N 57)	21	13.1 b
Trichoderma sp. (IKP 2)	175	0.0 a	*T. roseum* (N 39)	36	26.4 c
T. koningii	100	10.8 b	*Trichoderma* sp. (BRL 5)	993	28.8 c
Fusarium sp. (C302G)	40	12.7 b	*Fusarium* sp. (C302G)	40	39.4 d
T. harzianum (N 68)	997	18.9 c	*T. koningii* (N 70)	100	42.1 d
Bacteria PTZ	1341	37.5 d	*T. harzianum* (N 68)	997	50.9 e
Trichoderma sp. (CLY 3)	153	49.2 e	*T. harzianum* (N 72)	294	77.9 f
T. koningii (T 341)	313	50.8 e	Bacteria PTZ	1341	78.9 f
T. harzianum (N 72)	294	55.0 e	*Trichoderma* sp. (CLY 3)	153	79.8 f
Micromonospora sp. (PN7–6)	878	66.6 f	*Micromonospora* sp. (PN7–6)	878	80.8 f
T. roseum (N 39)	36	71.5 f	*T. koningii* (T 341)	313	81.3 f
Trichoderma sp. (IKP-6)	1084	76.7 g	*Trichoderma* sp. (IKP 6)	1084	90.3 g
T. roseum (T 372)	984	77.5 g	*T. roseum* (T 372)	984	90.6 g
C. globosum (N 76–1)	16	87.5 h	*C. globosum* (N 76–1)	16	95.8 h
Micromonospora sp. (PS6–2)	584	90.8 i	*Micromonospora* sp. (PS6–2)	584	97.3 hi
T. harzianum (CPO-80)	1156	91.6 i	*T. harzianum* (CPO-80)	1156	97.4 i

[a] Figures in parentheses indicate accession number.
[b] Means followed by the same letter do not differ ($P = 0.05$) according to multiple comparison of arc-sin transfunction of percentage inhibition or severity using the Student–Newman–Keuls test.

nutrient compounds (Sy *et al.*, 1978a,b). In the second case, our preliminary investigations (Sy, 1987) suggested the involvement of biochemical mechanisms (antibiotics, enzymes, toxins). Metabolites generated by liquid cultures or ascospores inhibited conidia germination of *P. oryzae* up to 100%. In addition, the acetonic extract and the chloroformic, etheric and aqueous phases inhibit the germination of Se3 and CRR-450 isolates by up to 100% whereas the proteic precipitate inhibits only 0% and 27.2% respectively. When the aqueous phase was subjected to column chromatography (Trisacryl GF 05 to separate from 300 nm to 2500 nm; empty volume assessed with Dextran Blue; wavelength from 180 nm to 200 nm; absorption peak of 189 nm), the maximum inhibition rate was 77.8% (against CRR-450 isolate) for the fraction corresponding to K_{av} equal to 185.1/1000. Further experimentation is required to verify the *in vivo* efficiency of the various treatments used (crude filtrate, acetonic extract, organic and aqueous phases, fraction which K_{av} is equal to 185.1/1000).

In vivo advanced screening

In order to determine the optimum dose and the limits to biotreatment use, we compared the Se302G × *P. oryzae* (Se3) combination with each of eight antagonistic microorganisms selected at three concentration levels. The means of responses to each treatment were compared statistically and ranked (Table 30.2). It was observed that *C. globosum* (N 76–1) can induce 100% inhibition at a concentration ratio of only 36. Moreover, this treatment guaranteed an inhibition level of 47.5% for incidence and 80.5% for severity at the lowest concentration ratios (equal to 2) used in this experiment. The antagonistic potential of each microorganism seems to be the determining factor even though optimization of potential depends on the dosage used. Furthermore, it should be pointed out that with *C. globosum*, an *in vivo* concentration ratio of 36 and an *in vitro* ratio of 50 both resulted 100% inhibition.

Performance spectrum of *C. globosum* (N 76–1)

In tests to assess the performance spectrum of *C. globosum* (N 76–1), it was found that *C. globosum* (N 76–1) provides full protection of these rice cultivars against the associated strain of *P. oryzae* with 100% inhibition of the development of leaf blast.

This type of investigation into host and parasite interactions should be expanded because it may hold the key to determining the performance limits of biological treatments interacting with a highly variable parasite. This approach will also help to determine the degree of biotreatment reliability.

Conclusions

From the 120 microorganisms that were initially tested *in vitro* and *in vivo* (Sy *et al.*, 1983a,b; Sy, 1987), only eight were selected based on their sound

Table 30.2. Mean suppression of incidence and severity by eight selected strains of potential biological control agents at different [antagonist]:[parasite] ratios on rice cultivar Se302G when co-inoculated with isolate Se3.

Treatment[a]	[Antagonist]/[Parasite]	Mean inhibition incidence (%)[b]	Treatment[a]	[Antagonist]/[Parasite]	Mean inhibition severity (%)[b]
Check (Se3)	0	0.0 a	Check (Se3)	0	0.0 a
T. roseum (T 372)	10	0.0 a	*Trichoderma* sp. (IKP 6)	35	0.0 a
Trichoderma sp. (IKP 6)	35	0.0 a	*Trichoderma* sp. (IKP 6)	85	0.0 a
Trichoderma sp. (IKP 6)	85	0.0 a	*T. roseum* (T 372)	10	2.1 b
T. roseum (T 372)	99	9.2 b	*T. koningii* (N 70)	1126	40.1 c
T. koningii (N 70)	100	10.8 b	*T. koningii* (N 70)	100	42.1 c
T. koningii (N 70)	1126	23.4 c	*T. roseum* (T 372)	99	54.2 d
Bacteria PTZ	8092	36.7 d	*T. koningii* (N 70)	5000	58.5 d
Bacteria PTZ	1341	37.5 d	Bacteria PTZ	8092	78.6 de
Bacteria PTZ	16144	40.8 de	Bacteria PTZ	1341	78.9 e
T. koningii (N 70)	5000	43.4 def	*Trichoderma* sp. (CLY 3)	1531	79.4 e
C. globosum (N 76–1)	2	47.5 efg	Bacteria PTZ	16144	79.8 e
Trichoderma sp. (CLY 3)	1531	48.4 fgh	*T. koningii* (T 341)	1697	79.8 e
T. koningii (T 341)	312	50.8 gh	*C. globosum* (N 76–1)	2	80.5 e
Trichoderma sp. (CLY 3)	153	51.3 ghi	*Trichoderma* sp. (CLY 3)	153	81.3 e
T. koningii (T 341)	1697	52.7 ghi	*T. koningii* (T 341)	312	81.3 e
Trichoderma sp. (CLY 3)	3473	54.4 hi	*Trichoderma* sp. (CLY 3)	3473	81.5 e
T. koningii (T 341)	11279	57.5 i	*T. koningii* (T 341)	11279	81.7 e
Trichoderma sp. (IKP 6)	1086	76.7 j	*Trichoderma* sp. (IKP 6)	1086	90.0 f
T. roseum (T 372)	984	77.5 j	*T. roseum* (T 372)	984	90.6 f
C. globosum (N 76–1)	16	87.5 k	*C. globosum* (N 76–1)	16	95.8 g
T. harzianum (CPO-80)	1158	91.5 l	*T. harzianum* (CPO-80)	1158	97.3 h
T. harzianum (CPO-80)	1354	100 m	*T. harzianum* (CPO-80)	1354	100 i
T. harzianum (CPO-80)	13125	100 m	*T. harzianum* (CPO-80)	13125	100 i
C. globosum (N76–1)	36	100 m	*C. globosum* (N 76–1)	36	100 i

[a] Figures in parentheses indicate accession number.
[b] Means followed by the same letter do not differ ($P = 0.05$) according to multiple comparison of arc-sin transfunction of percentage inhibition or severity using the Student–Newman–Keuls test.

level of efficiency. The highest rates of disease incidence reduction varied from 77.5% to 91.6% following foliar treatment by *T. roseum, C. globosum, Micromonospora* sp. *and T. harzianum* (CP0–80). Similarly, leaf blast severity inhibition rates varied from 90.3% to 97.4% with biotreatments from *Trichoderma* sp. (str-IKP-6), *T. roseum, C. globosum, Micromonospora* sp. (PS6–2) and *T. harzianum*. When the same concentrations were used, *C. globosum* was the most effective of all the 120 microorganisms tested (100% inhibition of leaf blast in controlled conditions with a concentration ratio as low as 36%).

In order to address the large variability of the blast fungus, it is important to determine the efficiency spectrum of the best biotreatments selected towards a range of differential *P. oryzae* strains.

Among the key issues for a control treatment regime are: the time of biotreatment application, the reaction of selected biotreatments against pesticides, biotreatment efficiency compared to the best chemical treatments, residual biofungicide effect and cost. All these represent priority topics for future research on biological control against blast.

References

Sy, A.A. (1976) Contribution à l'étude de *P. oryzae* CAV: Morphologie, biologie, physiologie. Recherche *in vitro* d'antagonistes dans une perspective de lutte biologique. PhD Thesis, U.P.S. de Toulouse, pp. 1–190.

Sy, A.A. (1987) Lutte biologique contre la pyriculariose du riz (*P. oryzae* CAV.): Stratégie et application. PhD Thesis, Institut National Polytechnique INP (Université de Toulouse IV), No. d'ordre 120, 1–157.

Sy, A.A., Albertini, L. and Hammant, C.L. (1978a) Recherches sur la lutte biologique contre *P. oryzae* cav. I. Action *in vitro* d'antagonistes fongiques sur la croissance mycélienne du parasite. *Revue Générale de Botanique*, T.85, 1004–1005–1006, janvier, février, mars 1978, 63–81.

Sy, A.A., Albertini, L. and Hammant, C.L. (1978b) Recherches sur la lutte biologique contre *P. oryzae* cav. II. Action *in vitro* d'antagonistes fongiques sur la germination des conidies du parasite. *Revue Générale de Botanique*, T.85, 1004–1005–1006, janvier, février, mars 1978, 83–90.

Sy, A.A., Norgn, K., Albertini, L. and Barrault, G. (1983a) Recherches sur la lutte biologique contre *P. oryzae* CAV. III. Influence de la température sur l'aptitude des germes antagonistes à inhiber *in vitro* la croissance mycélienne du parasite. *Cryptogamie, Mycologie* T.4, 245 – 249.

Sy, A.A., Albertini, L., Zohouri, P., Norgn, K. and Petitprez, M. (1983b) Biological control of *P. oryzae* Cav: *In vivo* assays. *Fourth International Congress of Plant Pathology*, Melbourne, Australia, Abstracts of papers (No. 931).

Sy, A.A., Norgn, K., Albertini, L. and Petitprez, M. (1984) Recherches sur la lutte biologique contre *P. oryzae* Cav. IV. Influence du pH sur l'aptitude des germes antagonistes à inhiber *in vitro* la croissance mycélienne du parasite. *Cryptogamie, Mycologie* T.5, 59–65.

V Farmers' Perspectives

31 Crop–Livestock Interactions: Implications for Crop Improvement in Sustainable Agriculture

J.D. Reed

Department of Meat and Animal Science, University of Wisconsin-Madison, 1675 Observatory Drive, Room 256, Madison, WI 53706–1284, USA

Introduction

Crop and livestock production may interact in ways which either complement or compete for natural and human resources. The nature of these interactions is an important component of resource-poor farms in mixed farming systems in tropical developing countries (Reed, 1988; McIntire *et al.*, 1992). A typical resource-poor farm can be described as 'a unit operated by a single family with little or no outside labor and with returns that provide subsistence, but little for savings, capital investment or for the purchase of external inputs, and with significant dependence on livestock' (McDowell, 1977a). Income from animal related products is typically 40% of the gross farm income, but may be as high as 70% (Gryseels and Asamenew, 1985). Ruminant livestock contribute a large diversity of non-food products, while milk and meat are often secondary (McDowell, 1977b).

The most important animal products that complement crop production are animal traction for cultivation and manure for increasing soil fertility. However, other uses may also be important, especially the role of livestock as a source of capital for purchasing seeds and food when crops fail. Manure is used for increasing soil fertility because fertilizers are unavailable or too expensive. Animal traction is used for cultivation, weeding, carting, water lifting and processing food crops. Ruminant livestock (cattle, buffalo, sheep, goats, and camelids) and equines increase overall farm productivity by grazing uncultivated land and through the utilization of crop residues for feed. Swine, poultry and a large diversity of other species complement food production by using kitchen wastes and the by-products of food processing, therefore increasing the overall productivity of the crop.

However, livestock may also compete with crop production for the use of resources (land, labor and capital) in mixed farming systems. Increases in the area of land cultivated leads to restriction of grazing areas and the potential of damage to crops from grazing animals. Additional labor for herding livestock during the cropping season may not be available because of emigration to urban areas and demands for cultivation. Cattle, sheep and goats may be tethered for extended periods of time during the cropping season. Tethered livestock have low productivity because they have insufficient feed and increased exposure to disease (especially internal parasites). Development of rice cultivation and vegetable gardening in seasonally flooded pasture land is another form of competition between livestock and crop production.

Livestock and Sustainable Agriculture

'Sustainable agriculture' has been defined as 'an integrated system of plant and animal production practices having a site-specific application that will, over the long term: (i) satisfy human food and fiber needs; (ii) enhance environmental quality and the natural resource base upon which the agricultural economy depends; (iii) make the most efficient use of nonrenewable resources and on-farm resources and integrate, where appropriate, natural biological cycles and controls; (iv) sustain the economic viability of farm operations; and (v) enhance the quality of life for farmers and society as a whole' (National Agricultural Research, Extension and Training Policy Act of 1977, as amended by Section 1603 of the Fact Act). Almost all definitions of sustainable agriculture include the need for long-term maintenance of natural resources and agricultural productivity, minimal adverse environmental impacts, adequate economic returns to farmers, optimal crop production with minimized chemical inputs, satisfaction of human needs for food and income, and provision for the social needs of farm families and communities (NRC, 1991).

The role of livestock in promoting the sustainability of mixed farming systems in developing countries is implicit in these definitions. Livestock production has a direct impact on four of the eight elements of sustainable agriculture as listed by the NRC (1991): conservation of soil and water resources and the impact on fertility of the soil's physical and biological characteristics, processes and cycles; cultural practices for improving soil fertility, controlling erosion, and maximizing biological production potential (for example, tillage methods, crop residue management, irrigation, alley cropping, and agroforestry); indigenous practices and uses of germplasm and the economic and cultural consequences of biodiversity loss and preservation; and, the consequence of converting forest and savannah lands into range for cattle production. Livestock production systems also affect the other elements of sustainable agriculture (integrated pest management, institutional arrangements, resource management, land tenure, public policy and production of food crops and cash crops) (NRC, 1991).

Technology for the development of sustainable agriculture will need to strengthen the complementarity of crop and livestock production in order to be acceptable to most resource-poor farmers. Biological and social scientists need to focus on practices that are already being used by farmers in order to understand how farming systems have evolved and to determine the 'best bet' research and development areas to pursue. Important research issues in sustainable agriculture and natural resource management that involve the interaction of soils, crops, forests, rangelands and livestock in fragile ecosystems are: the role of livestock in nutrient cycles and the maintenance of soil fertility; the use of livestock to optimize the use of on-farm biological resources and the potential conflict between livestock grazing crop residues and their use for soil conservation, organic matter and other effects on soil characteristics; the importance of livestock in the management of forest resources and the introduction of agroforestry systems for improved livestock production.

Crop/Livestock Interactions and Rice Blast

Crop/livestock interactions may have both a direct and indirect effect on the incidence of crop diseases such as rice blast. The hyphae and conidia from affected plant material which remains in the field is a major cause of primary infection of rice in the subsequent crop (Kozaka, 1965). Hyphae and conidia on straw and stubble remain viable for up to one year under dry conditions but are viable for shorter periods of time under wet conditions (Kozaka, 1965). Livestock may have a direct effect on the incidence of rice blast because they consume the hyphae and conidia present in rice straw which is used for feed. Removing affected plant material from the field is important to the control of primary infection. The removal of straw for feeding domestic ruminants and the grazing of stubble removes affected plant material. It is unlikely that the hyphae and conidia remain viable after passage through the ruminant digestive tract. The practice of removing straw for feeding and grazing stubble is common in drier regions where upland rice is grown and may be important in controlling sources of primary infection.

However, *Pyricularia oryzae* produces a mycotoxin, tenuazonic acid (Lebrun *et al.*, 1990), which may have a direct effect on livestock consuming the contaminated straw. Tenuazonic acid has been shown to be toxic to poultry (Giambrone *et al.*, 1978), young mice (Davis *et al.*, 1977), and has been implicated as a contributing factor in human disease (Steyn and Rabie, 1976). Tenuazonic acid caused depressed growth rate and feed efficiency when fed to young chickens at a rate of 1.25 or 2.5 mg kg^{-1} of body weight (Giambrone *et al.*, 1978). No morbidity was observed in this research but the chicks showed several lesions which indicated toxic effects. Although, the author is not aware of studies of the effects of tenuazonic acid on ruminants there are other fungi which grow on rice straw and produce mycotoxins which cause diseases in ruminants. Degnala disease in cattle and buffalo is caused by the consumption of rice straw which is contaminated with *Fusarium* species (Marasas *et al.*, 1984). Degnala disease is characterized by edema of the legs and followed by necrosis, gangrene and sloughing of the extremities. The disease occurs in India and Pakistan in lowland rice growing regions and has also been reported in China (Marasas *et al.*, 1984).

The use of manure from livestock may have an indirect effect on the incidence of rice blast. Animal manure is used to improve and manage soil fertility in most resource-poor farming systems. Manure from grazing ruminants may contain high amounts of biogenic silica (up to 20% of dry matter) because the silicon accumulated by grasses is indigestible (Jones and Handreck, 1965; Reed, 1983). As discussed below, silicon deficiency in rice leads to a high incidence of rice blast. Inorganic fertilizers do not correct this problem unless they contain silicate for the specific purpose of supplying available Si (Datnoff *et al.*, 1991). Manure from grazing ruminants would supply available Si in addition to plant nutrients and organic matter.

The shift from animal draft to power tillers for cultivating rice may also have an indirect effect on the incidence of rice blast. Time of planting is important because the interaction between humidity, temperature, and stage of maturity influence the susceptibility of the plant to blast infection. Time of planting depends on availability of labor and draft power in resource-poor farming systems (Orr, 1991). Technologies which improve the efficiency of allocation of labor and draft power resources may assist resource-poor farmers to develop cultural practices to prevent rice blast. The use of power tillers in combination with animal traction could improve the ability of farmers to control time of planting in relation to incidence of rice blast infection.

Interdisciplinary Research on Cereal Crop Residues

The use of cereal crop residues as a feed for ruminant livestock is an example of interdisciplinary research on crop/livestock interactions which is relevant to assisting resource-poor farmers develop sustainable production systems. At least 50% of annual metabolizable energy intake by domestic ruminants in developing countries is derived from fibrous crop residues (McDowell, 1977a). However, there are large differences among farming systems and countries in the efficiency of utilization of crop residues. In the highlands of Ethiopia, farmers efficiently utilize the crop residues from teff, barley and wheat. These cereals are harvested by sickle at a height of less than 10 cm and the entire crop is threshed by trampling with oxen and equines. After threshing the straw is tightly stacked and protected from rain by covering with long straw or grass (Goe, 1987). When crop residues are grazed in the field there are large losses of 40–50% of the dry matter from trampling and spoilage (Chandler, 1983; Tessema, 1983). In Mexico and India resource poor farmers harvest, store and sell the crop residues from rice, maize, sorghum, and millet for livestock feeding (Reed, 1980; Parthasarathy Rao, 1985). In other countries such as The Gambia cereal crop residues are often left in the field although more efficient use could increase livestock production (Njie, 1986; Russo, 1986).

Low digestibility and voluntary intake are major constraints to increasing the productivity of livestock fed crop residues. However, research has shown that there is over a 20 unit range in digestibility of crop residues among different varieties of several food crops (Reed *et al.*, 1988a). Level of feed offered also has a large influence on the intake and growth rate of ruminants fed cereal crop residues (Owen and Aboud, 1988; Aboud *et al.*, 1990). Although much research has been devoted to upgrading crop residues through various chemical treatments (Jackson, 1978), little attention has been given to natural variation in the nutritive value of untreated crop residues as influenced by species, variety and environment. Chemical treatment of crop residues has limited applicability to animal production in tropical and sub-tropical countries. Emphasis should be transferred to exploit genetic variation in crop residue quality (Reed *et al.*, 1988a).

Differences in growing conditions due to season, elevation or latitude cause crop residues from the same variety to vary widely in nutritive value (Blummel, 1988). High temperature during growth increases cell wall and crude lignin content and decreases digestibility (Deinum, 1976). High humidity and rain during and after grain harvest cause deterioration of nutritive value. Loss of leaves through wind or trampling of crop residues left in the field also causes deterioration. The losses can be eliminated by improved conservation practices.

Quantity and quality of crop residue are important criteria in a farmer's decision to grow a particular variety (McDowell, 1988). Resource-poor farmers have rejected new varieties that were proven successful at plant breeding centers when they realized that yield and nutritive value of the crop residue was unacceptable (Capper *et al.*, 1988; Nordblom, 1988; Ceccarelli, 1993).

Crop improvement programs can have a large effect on the nutritive value and utilization of crop residues. In areas where resource-poor farmers rely on crop residues for feeding livestock on farm or for sale to specialized livestock operations, the amount and nutritive value of crop residues should be considered in crop improvement programs. The effect of breeding for increased yield and disease resistance on nutritive value of crop residues needs to be determined. Improvements in the nutritive value of crop residues could substantially increase ruminant productivity on resource-poor farms in developing countries.

Nutritive Value of Cereal Crop Residues

Cereal crop residues are deficient in protein but supplementation with non-protein nitrogen and protein does not always lead to improved intake and digestibility because other factors limit nutritive value. These factors need to be determined because, within the range of energy intake of cereal crop residues, large increases in animal productivity can be achieved by relatively small increases in digestibility and intake.

Cell wall carbohydrates (cellulose and hemicellulose) account for 80% of the dry matter in cereal crop residues and represent a large source of energy for ruminant feeding. However, the ability of rumen microorganisms to digest these carbohydrates is limited by the presence of lignin and other phenolic compounds (Hartley, 1981). The digestibility of cereal crop residues is correlated with the nature and amount of lignin associated with their cell walls. Environment also affects the nature and amount of lignin and related cell wall phenolics leading to substantial variation in digestibility. In sorghum, high levels of phenolic pigmentation are associated with higher levels of lignin and lower digestibility of cell wall carbohydrates (Reed *et al.*, 1987; Reed *et al.*, 1988b; Mueller-Harvey and Reed, 1992).

Biogenic silica deposited in the cell wall of grasses also has a negative effect on the digestibility of cell wall carbohydrates. Silica is absorbed from the soil solution of silicic acid and deposited in the cell wall. Most grasses will accumulate silica, but it is uncommon in legumes and other dicots. However, *Aspilia mossambicensis* (Compositae) accumulates over 10% of the dry matter of leaves as biogenic silica (Reed, 1983). The silica is deposited in stylus-like trichomes covering the leaf surface and may be an important deterrent to herbivores. Cell wall silica is in the opaline form. Silica increases insect and disease resistance of cereals such as rice, oats, wheat and corn (Yoshida *et al.*, 1962; Jones and Handreck, 1967; Rojanaridpiched *et al.*, 1982). Silica in grasses may also be a deterrent to herbivores (Reed, 1983; McNaughton *et al.*, 1984; Gali-Muhtasib *et al.*, 1992). Silica lowers the digestibility of grasses (Van Soest and Jones, 1968). Biogenic silica is indigestible and dilutes the energy value of grasses (Jones and Handreck, 1965). Silica is also involved in the formation of urinary calculi (Jones and Handreck, 1965).

Rice Straw

Rice is the best known example of a cereal crop that accumulates silica (Yoshida *et al.*, 1962). Rice straw has 8–15% silica on a dry matter basis. Rice hulls have over 20% silica making them indigestible and potentially harmful if fed at a high level (Van Soest, 1970).

Rice grown on soils which are deficient in Si were highly susceptible to rice blast and the incidence of blast decreased by 30% when the rice was fertilized with available Si (Datnoff *et al.*, 1991). There was a corresponding increase in the silica content of the rice plant tissue as fertilizer Si increased (Datnoff *et al.*, 1991). Resistance to blast in goosegrass (*Eleusine indica*) is associated with the deposition of biogenic silica which was triggered in response to fungal attack (Heath *et al.*, 1992). These results indicate that resistance to fungal diseases in rice is associated with increased silica content which would potentially decrease the feeding value of the straw.

However, research conducted at IRRI and the University of the Philippines indicates that selection for resistance to insects, diseases and lodging may improve the nutritive value of rice straw (Khush *et al.*, 1988). These researchers also showed that there was no relationship between *in vitro* organic matter digestibility and silica content of rice straw from plants grown by hydroponics in solutions ranging from 0 to 400 p.p.m. of SiO_2. These results suggest that silica accumulation in rice straw does not have a negative effect on digestibility and contradicts previous research which indicated that biogenic silica is a major factor in limiting the digestibility of rice straw (Van Soest, 1982).

However, there are several problems with the interpretation of the results reported by Khush *et al.* (1988). Techniques of *in vitro* estimates of digestibility can often be misleading because microbial residues and ash confound differences in true digestibility of plant organic matter (Reed and Goe, 1989). Genetic and environmental effects on digestibility of cereal crop residues is expressed by differences in the digestion of the cell wall carbohydrates (Reed *et al.*, 1987), which can easily be estimated by the extraction of plant samples and *in vitro* residues with neutral detergent by the methods of Goering and Van Soest (1970). Results can then be expressed as *in vitro* digestibility of the neutral-detergent fiber fraction (Reed and Goe, 1989).

Another problem in estimating genetic and environmental effects on the digestibility of straw is that plant parts have different extents of digestion. In most cereals the leaf blade has a higher digestibility than the leaf sheath which is higher than the internodes. However, in the rice plant the internodes often have higher digestibility than the leaves (Khush *et al.*, 1988). Since most of the biogenic silica is deposited in the leaf blade fraction, the higher digestibility of the internodes suggests that silica is having a negative effect on digestion of cell wall carbohydrates in the blades. Studies on the effect of Si availability to the plant on the digestibility of the crop residue should be made based on plant fractions (leaf blade, leaf sheath and internodes) and not the whole plant because of possible effects on the ratio of plant parts.

The mechanism by which biogenic silica may inhibit the digestion of cell wall carbohydrates by rumen microorganisms and prevent fungal infection is very relevant to the problem of determining the role of silica in the rice plant. If deposition of silica is a specific response to fungal attack as suggested by Heath *et al.* (1992), then correlations of resistance with total plant silica may be very misleading. This same logic would also apply to correlations of total plant silica and digestibility. Genetic differences are likely to be tissue specific. In this case, low levels of biogenic silica may impart disease resistance without having a negative effect on straw digestibility. Obviously more research of an interdisciplinary nature is required to investigate the mechanisms by which biogenic silica affects microbial and fungal invasion and digestion of the cell walls of the rice plant.

Conclusion

Crop/livestock interactions in sustainable agriculture is a research area which has a large potential for assisting resource-poor farmers in improving production. However, most of the research themes are complex in that they require interdisciplinary teams of scientists and do not necessarily lead to a commodity-based production system or a disciplinary focus. This complexity also presents difficulties for national and international agricultural research institutes which have adopted a commodity-oriented focus to their programs and crop and livestock research are often conducted at separate institutions. Research on the nutritive value and other uses of cereal crop residues is one of these themes. The author coordinated a workshop titled 'Plant breeding and the nutritive value of crop residues' at the International Livestock Center for Africa in 1987 (Reed *et al.*, 1988a). The general recommendation was that: 'The international agricultural research centers (IARCs) and the national agricultural research systems (NARs) should recognize that farmers grow crops not only to feed themselves and their families but also to feed their animals. The rejection by farmers of high-yielding varieties because of their low straw yield or poor quality of straw or stover shows that attention must be paid to residue yield and quality in cereal crops'. The barley program at the International Center for Agricultural Research in the Dry Areas has conducted some of the pioneering research which led to the development of this recommendation (Ceccarelli, 1993). However, institutional and disciplinary boundaries are a major obstacle to overcome in both NARs and IARCs.

References

Aboud, A.O.A., Reed, J.D., Owen, E. and McAllan, A. (1990) Feeding sorghum stover to Ethiopian sheep: effect of stover variety and amount offered on growth, intake and selection. Summaries, Meeting of the British Society of Animal Production, March 19–21 1990.

Blummel, M. (1988) The nutritive value of teff straw and methods of its assessment. MSc Thesis, University of Aberdeen, UK.

Capper, B.S., Thomsen, E.F. and Herbert, F. (1988) Genetic variation in the feeding value of barley and wheat straw. In: Reed, J.D., Capper, B.S. and Neate, P.J.H. (eds.), *Plant Breeding and the Nutritive Value of Crop Residues*. Proceedings of a Workshop held at ILCA, Addis Ababa, 7–10 December 1987. International Livestock Centre for Africa, Addis Ababa, Ethiopia, pp. 177–193.

Ceccarelli, S. (1993) Plant breeding technologies relevant to developing countries. In: Gill, M., Owen, E., Pollott, G.E. and Lawrence, T.L.J. (eds), *Animal Production in Developing Countries*. BSAP Occasional Publication Number 16. The British Society of Animal Production. pp. 37–46.

Chandler, D.R. (1983) The grazing animal and land pressures in small farm systems: must ruminant nutrition create inevitably a confrontation between arable lands and grazing requirements. In: *Draft Power and Animal Feeding, CIMMYT East and Southern Africa Workshop, Swaziland*, 4–6 October 1983.

Datnoff, L.E., Raid, R.N. and Snyder, G.H. (1991) Effect of calcium silicate on blast and brown spot intensities and yields of rice. *Plant Disease* 75, 729–732.

Davis, N.D., Diener, U.L. and Morgan-Jones, G. (1977) Tenuazonic acid production by *Alternaria alternata* and *A. tenuissima* isolated from cotton. *Microbiology* 34, 155–157.

Deinum, B. (1976) Effect of age, leaf number and temperature on cell wall and digestibility of maize. In: *Carbohydrate Research in Plants and Animals*. Miscellaneous Papers, Agricultural University, Wageningen, No. 12, 29–41.

Gali-Muhtasib, H.U., Smith, C.C. and Higgins, J.J. (1992) The effect of silica in grasses on the feeding behavior of the prairie vole, *Microtus ochrogaster*. *Ecology* 73, 1724–1729.

Giambrone, J.J., Davis, J. and Diener, U.L. (1978) Effect of tenuazonic acid on young chickens. *Poultry Science* 57, 1554–1558.

Goe, M.R. (1987) Animal traction on smallholder farms in the Ethiopian highlands. PhD Thesis, Cornell University, Ithaca, New York, 380 pp.

Goering, H.K. and Van Soest, P.J. (1970) Forage fiber analysis. *USDA Agricultural Handbook*, No. 379, pp. 1–20.

Gryseels, G. and Asamenew, G. (1985) Links between livestock and crop production in the Ethiopian highlands. *International Livestock Center for Africa Newsletter* 4(2), 5–6.

Hartley, R.D. (1981) Chemical constitution, properties and processing of lignocellulosic wastes in relation to nutritional quality for animals. *Agriculture and Environment* 6, 91–113.

Heath, M.C., Howard, R.J., Valent, B. and Chumley, F.G. (1992) Ultrastructural interactions of one strain of *Magnaporthe grisea* with goosegrass and weeping lovegrass. *Canadian Journal of Botany* 70, 779–787.

Jackson, M.G. (1978) Treating straw for animal feeding. *FAO Animal Production and Health Paper 10*. Food and Agricultural Organization of the United Nations, Rome.

Jones, L.H.P. and Handreck, K.A. (1965) The relation between the silica content of the diet and the excretion of silica by sheep. *Journal of Agricultural Science* 65, 129–134.

Jones, L.H.P. and Handreck, K.A. (1967) Silica in soils, plants, and animals. *Advances in Agronomy* 19, 107–149.

Khush, G.S., Juliano, B.O. and Roxas, D.B. (1988) Genetic selection for improved nutritional quality of rice straw – a plant breeder's viewpoint. In: Reed, J.D., Capper, B.S. and Neate, P.J.H. (eds), *Plant Breeding and the Nutritive Value of Crop Residues*. Proceedings of a Workshop held at ILCA, Addis Ababa, 7–10 December 1987. International Livestock Centre for Africa, Addis Ababa, Ethiopia, pp. 261–279.

Kozaka, T. (1965) Control of rice blast by cultivation practices in Japan. In: *The Rice Blast Disease*. Johns Hopkins Press, Baltimore, Maryland, pp. 421–438.

Lebrun, M.H., Dutfoy, F., Gaudemer, F., Kunesch, G. and Gaudemer, A. (1990) Detection and quantification of the fungal phytotoxin tenuazonic acid produced by *Pyricularia oryzae*. *Phytochemistry* 29, 3777–3783.

Marasas, W.F.O., Nelson, P.E. and Toussoun, T.A. (1984) *Toxigenic Fusarium Species*. The Pennsylvania State University Press, University Park, Pennsylvania.

McDowell, R.E. (1988) Importance of crop residues for feeding livestock in smallholder farming systems. In: Reed, J.D., Capper, B.S. and Neate, P.J.H. (eds), *Plant Breeding and the Nutritive Value of Crop Residues*. Proceedings of a Workshop held at ILCA, Addis Ababa, 7–10 December 1987. International Livestock Centre for Africa, Addis Ababa, Ethiopia, pp. 3–27.

McDowell, R.E. (1977a) Are US animal scientists prepared to help small farmers in developing countries. *Cornell International Agricultural Mimeograph 58*. Cornell University, Ithaca, New York.

McDowell, R.E. (1977b) *Ruminant Products: More than Meat and Milk*. Winrock Report. Winrock International Livestock Research and Training Center, Petit Jean Mountain, Morrilton, Arkansas.

McIntire, J., Bourzat, D. and Pingali, P. (1992) *Crop-Livestock Interaction in Sub-Saharan Africa*. The World Bank, Washington DC.

McNaughton, S.J., Tarrants, J.L., McNaughton, M.M. and Davis, R.H. (1984) Silica as a defense against herbivory and a growth promoter in African grasses. *Ecology* 66, 528–535.

Mueller-Harvey, I. and Reed, J.D. (1992) Identification of phenolic compounds and their relationship to *in vitro* digestibility of sorghum leaves from bird-resistant and non-bird-resistant varieties. *Journal of the Science of Food and Agriculture* 60, 179–196.

Njie, M. (1986) A survey of crop residues and agro-industrial by-products in The Gambia and their nutritive value for ruminants. MSc Thesis, The Swedish University of Agricultural Sciences, Uppsala, Sweden.

Nordblom, T.L. (1988) The importance of crop residues as feed resources in West Asia and North Africa. In: Reed, J.D., Capper, B.S. and Neate, P.J.H. (eds), *Plant Breeding and the Nutritive Value of Crop Residues*. Proceedings of a Workshop held at ILCA, Addis Ababa, 7–10 December 1987. International Livestock Centre for Africa, Addis Ababa, Ethiopia, pp. 41–64.

NRC (1991) *Toward Sustainability*. National Academy Press, Washington DC, 147 pp.

Orr, A. (1991) A generalized model of economic constraints on crop turnaround time. *IRRI Research Paper Series* No. 145.

Owen, E. and Aboud, A.A.O. (1988) Practical problems of feeding crop residues. In:

Reed, J.D., Capper, B.S. and Neate, P.J.H. (eds), *Plant Breeding and the Nutritive Value of Crop Residues*. Proceedings of a Workshop held at ILCA, Addis Ababa, 7–10 December 1987. International Livestock Centre for Africa, Addis Ababa, Ethiopia, pp. 133–156.

Parthasarathy Rao, P. (1985) Marketing of fodder in rural and urban areas of India. In: Agricultural markets in the semi-arid tropics. *Proceedings of the International Workshop*, 24–28 October 1983, ICRISAT Center, Patancheru, India.

Reed, J.D. (1980) The alkali treatment of cereal crop residues. MSc Thesis, Cornell University, Ithaca, New York.

Reed, J.D. (1983) The nutritional ecology of game and cattle on a Kenyan ranch. PhD Thesis, Cornell University, Ithaca, New York.

Reed, J.D. (1988) Improving livestock productivity by strengthening crop/livestock interactions in smallholder farming systems in Africa. In: Stone, J.C. (ed.), *The Exploitation of Animals in Africa*. Aberdeen University African Studies Group, Aberdeen, pp. 71–79.

Reed, J.D. and Goe, M.R. (1989) Estimating the nutritive value of cereal crop residues: implications for developing feeding standards for draught animals. *ILCA Bulletin* 34, 2–9.

Reed, J.D., Tedla, A. and Kebede, Y. (1987) Phenolics, fibre and fibre digestibility in the crop residue from bird resistant and non-bird resistant sorghum varieties. *Journal of the Science of Food and Agriculture* 39, 113–121.

Reed, J.D., Capper, B.S. and Neate, P.J.H. (1988a) *Plant Breeding and the Nutritive Value of Crop Residues*. Proceedings of the Workshop held at ILCA, Addis Ababa, 7–10 December 1987. International Livestock Centre for Africa, Addis Ababa, Ethiopia.

Reed, J.D., Kebede, Y. and Fussell, L.K. (1988b) Factors affecting the nutritive value of sorghum and millet crop residues. In: Reed, J.D., Capper, B.S. and Neate, P.J.H. (eds), *Plant Breeding and the Nutritive Value of Crop Residues*. Proceedings of a Workshop held at ILCA, Addis Ababa, 7–10 December 1987. International Livestock Centre for Africa, Addis Ababa, Ethiopia, pp. 41–64. and 233–251.

Rojanaridpiched, C., Gracen, V.E. and Coors, J. (1982) Mechanism of second brood corn bored resistance in corn. Paper presented at 37th North East Corn Improvement Conference, New York.

Russo, S. (1986) The use of crop residues for animal feed in The Gambia. *Mixed Farming and Resource Management Project, Technical Report No. 11*. Ministry of Agriculture, Government of The Gambia.

Steyn, P.S. and Rabie, C.J. (1976) Characterization of magnesium and calcium tenuazonate from *Phoma sorghina*. *Phytochemistry* 15, 1977–1979.

Tessema, S. (1983) Animal feeding in small farm systems. In: *Draft Power and Animal Feeding, CIMMYT East and Southern Africa Workshop*, Swaziland, 4–6 October 1983.

Van Soest, P.J. (1970) The role of silicon in the nutrition of plants and animals. Cornell Nutrition Conference for Feed Manufacturers, Buffalo, New York.

Van Soest, P.J. (1982) *Nutritional Ecology of the Ruminant*. O & B Books, Corvallis, Oregon.

Van Soest, P.J. and Jones, L.H.P. (1968) Effect of silica in forages upon digestibility. *Journal of Dairy Science* 51, 1644–1648.

Yoshida, S., Ohnishi, Y. and Kitagishi, K. (1962) Histochemistry of silicon in rice plant. I. a new method for determining the localization of silicon within plant tissues. *Soil Science and Plant Nutrition* 8, 30–41.

32 Assessing Indigenous and Traditional Knowledge in Farming Systems

H.D. THURSTON

Department of Plant Pathology, Cornell University, Ithaca, NY 14853, USA

INTRODUCTION

Local growers near Montilla, Spain, use a curious practice to control a grape disorder known as esca, caused by the fungi *Stereum hirsutum* and *Phellinus igniarius*. The disease complex has been a serious problem of grapes in the Mediterranean area, apparently for millennia. The pathogens cause a rot of the trunk and branches. Today in Andalusia, Spain, a practice for control of the disease is to split the trunk of the grape vine with a hatchet, to clean out any infected tissue, and to insert a stone to keep the trunk separated. I visited grape vineyards near Montilla where the practice is still being used; the visit stimulated my interest in the practice, and I searched for literature on the subject. Aristotle, who lived from 384–323 BC, wrote: 'A tree that has a hard bark and has become barren, if its roots be split and a stone inserted in the cleft, it will become fruitful again' (Ross, 1913, as cited by Orlob, 1973). Columela (1988), a Roman who lived in Southern Spain in 60 AD, described splitting the trunk of the vine down the middle, and inserting soil. Some references suggested that the method has been used since 'time immemorial' in Greece and Asia Minor. Palladius, a Roman who lived in the fourth or fifth century, suggested driving a wedge into the roots to cure unhealthy grapes, olives, figs, palms, and peaches (Orlob, 1973). A similar surgical practice for chestnuts and other woody plants was described by Ibn Al-Awam (1988), a Spanish Moor writing in the 1200s. About 1350 AD the German Gottfried van Franken related that in Greece, when grape vines began to deteriorate, the roots were split, a rock inserted, and they were then covered with soil and manure (Orlob, 1973).

It is highly possible that these practices for control of the esca disease were adaptations of Aristotle's original recommendation. The practice is still common today in Andalusia, Spain, and I was informed that the alternative to the practice is the use of an arsenical fungicide. Without understanding the many centuries of experience behind the above practice, one might ridicule it and consider it ludicrous, whereas in fact the practice appears to be effective and useful. We scientists should be cautious about dismissing what appear to be ridiculous farmer practices. Traditional and indigenous knowledge can be overvalued or romanticized, but that is surely better than deriding, despising, or ignoring it.

In this chapter I will relate some of the strategies used by indigenous farmers for plant disease management, and, where appropriate, describe practices they use for the management of the rice blast disease caused by *Pyricularia oryzae*. In the context of this chapter, indigenous farmers are peoples native to an area who continue to use the accumulated knowledge of traditional agriculture transmitted by their ancestors. The term 'traditional' is usually associated with primitive agricultural systems or preindustrial peasant agriculture. Traditional farming usually is based on agriculture that has been practised for many generations. The accumulated knowledge on agriculture is considerable, as humans began crop production perhaps 10,000 years ago.

It is important to understand thoroughly the indigenous farming systems and practices used in developing countries if agricultural scientists are to be effective in ameliorating the problems of food production in the developing world. This is essential so that researchers address appropriate problems in the context of farmers' systems, and so that efficient, proven practices can be disseminated to other farmers. Far too many giant projects in developing countries costing huge sums have failed dismally and have caused serious environmental problems due to a lack of sufficient understanding of indigenous agriculture. Outside intervention is often misguided and irrelevant when it is undertaken without a thorough understanding of local, indigenous agricultural practices and systems.

Much can be found in the literature on indigenous knowledge and traditional agriculture, but little can be found on these subjects in the literature of agricultural science. Most practices for disease management used by indigenous farmers in developing countries are cultural practices, but little information is available in an easily accessible form on the cultural practices used in indigenous systems. Many of the indigenous farmer practices are given in Table 32.1. The table indicates that most, but not all, of these practices are sustainable in the long term. Traditional cultivars or landraces selected over millennia are also most important, especially as regards disease resistance. Landraces are usually genetically diverse and in balance with their environment and endemic pathogens. They are dependable and stable in that, although not necessarily high yielding, they yield some harvest under all but the worst conditions. Although some serious overuse and misuse of chemical controls by indigenous farmers occurs, generally the high cost of chemicals inhibits their use.

The extensive agricultural knowledge of indigenous farmers often is not appreciated by todays' scientists. Conklin (1954) described the knowledge of the Hanunóo, a mountain tribe in the Philippines:

> On some aspects of agriculture their knowledge is amazingly wide, accurate, and practical. They distinguish 10 basic and 30 derivative soil and mineral categories and understand the suitability of each for various crops as well as the effects of erosion, exposure, and over-farming. They distinguish over 1,500 useful plant types, including 430 cultigens, and they discern minute differences in vegetative structure.

Table 32.1. Sustainability, external inputs needed, and labor requirements of selected plant disease management practices of traditional farmers.

Practice	Sustainable?	External inputs	Labor
Adjusting crop density	Yes	Low	Low
Adjusting depth of planting	Yes	Low	Low
Adjusting time of planting	Yes	Low	Low
Altering of plant and crop architecture	Yes	Low	High
Biological control (soilborne pathogens)	Yes	High	High
Burning	Yes[a]	Low	High
Fallowing	Yes	Low	Low
Flooding	Yes	Low	High
Manipulating shade	Yes	Low	Low
Mulching	Yes	High	High
Multistory cropping	Yes	Low	Low
Multiple cropping	Yes	Low	High
Planting diverse crops	Yes	Low	Low
Planting in raised beds	Yes	High	High
Rotation	Yes	Low	Low
Site selection	Yes	Low	Low
Tillage	No	Low	High
Using organic amendments	Yes	High	High
Weed control	No	Low	High

[a] Under high population pressure the slash and burn system is neither stable nor sustainable. (Source: Thurston, 1992).

The plant classification system of the Mayan (Tzeltal) Indians of Mexico provides another example of indigenous knowledge. Berlin *et al.* (1974) stated: '471 mutually exclusive generic taxa were established as legitimate Tzeltal plant groupings'.

Much of the literature on indigenous and traditional agriculture is anecdotal rather than experimental, much to the distress of scientists who believe that only information obtained by scientific methods is of real value. Also, indigenous agriculture often includes a mixture of superstitious, religious, and magical beliefs.

Chinese writings 5000 years ago mentioned rice as an important food (Grist, 1986). For millennia the major disease management practices for rice used by traditional farmers have been cultural control practices and the use of landraces. Although indigenous farmers continue such practices today, the use of cultural controls for managing the rice blast disease has been largely neglected in the scientific literature. For example, in the book *The Rice Blast Disease*, published for the International Rice Research Institute (IRRI) in 1965, there were only 17 pages out of 507 on cultural controls (IRRI, 1965). I attended CIAT's (Centro Internacional de Agricultura Tropical) rice blast symposia in 1971 (CIAT, 1975) and IRRI's symposia in 1979 (IRRI, 1979). Cultural controls were never mentioned in either symposium.

Studies over the last two decades indicate that 'modern' agriculture has lost many of the building blocks of sustainable systems as found in traditional agriculture systems, including those pertaining to disease management. To develop more sustainable disease management practices the basic principles from traditional systems should be reincorporated into agricultural systems without reducing crop productivity. Obviously, efforts should concentrate on understanding and re-establishing genetic variability in crops, but efforts should also concentrate on the use of organic soil amendments to quantitatively and qualitatively increase disease suppressive microflora and soil fauna, and on evaluating the multitude of cultural disease management practices in traditional systems. To improve indigenous systems efforts should concentrate on understanding both the systems and the farmers and maintaining existing sustainable components while building productivity components and new management strategies into the system. An extended philosophy of integrated pest management (IPM) should be developed with an ecological and systems perspective to problem solving and the use of tools such as agroecological databases, epidemiology, microbial ecology, anthropology, and biotechnology. I do not believe that recombinant DNA technology products will become the major component of new sustainable systems, but rather they may provide valuable tools for pest identification, for a better understanding of the mechanisms of action at a molecular level, and provide technologies for systems which already contain the basic building blocks needed for sustainability. Selected chemicals are likely to have a role in disease management in sustainable systems if used prudently and in small quantities.

Most scientists and many of the world's farmers have abandoned traditional farming practices and systems in an effort to increase food production and income and to improve the efficiency of their land and labor use. However, the resulting 'modern' systems largely ignore many of the sustainable practices, including crop protection practices, that have evolved among farmers to cope with the multitude of environmental stresses under which crops are grown. Slowly, we have come to realize that many of the modern systems are not sustainable, and if we do not devise better agricultural systems we will weaken our ability to produce food in the future. Today there are serious concerns about so-called 'modern' agriculture; that it is extremely energy intensive, the genetic base is narrow, and stress on increasingly high yields and efficiency leads to monoculture, and sometimes to serious erosion, pollution, and excessive pesticide residues. A historical perspective on the practices and genetic materials used by traditional farmers through the centuries may help us to develop a truly sustainable agriculture. Perhaps we can learn not only from the ongoing successes of traditional farmers, but also analyze and understand the reasons for past collapses of some of their agricultural systems and societies.

Much of what follows in this chapter comes from the writings of anthropologists, archaeologists, ecologists, geographers, and ethnobotanists who record information on traditional agriculture practices, rather than from plant pathologists or other agricultural scientists. Space allows only a sampling of the many of cultural practices used by indigenous farmers. Cultural practices for control of rice diseases, especially blast, will be emphasized, but examples from other crops and diseases will also be given.

Adjusting Density and Spacing

Farmers have been concerned with the appropriate distance of planting for millennia. Columela's book was republished in 1988 (Columela, 1988). A Roman who lived in Spain in about 50 AD, he discussed at length the various factors affecting the quantity of cereal seed to be planted in a given field, and their effect on crop density. The Moor, Ibn Al-Awam (1988), writing during the 12th century, recommended the correct distance of planting for numerous crops grown in Islamic Spain. Plant pathologists have made relatively few studies of the effect of plant or crop density on disease, but it is recognized that dense populations usually contribute to disease epidemics.

Close planting of rice, especially in nurseries, generally favors blast development. For example, Amin and Katyal (1979) reported that as the seeding rate in rice was increased, the incidence of *P. oryzae* increased. Kozaka (1965) reported that close spacing tended to increase the severity of rice blast in Japan, but that in less fertile soil this was not always the case. The opposite effect was noted with bacterial blight (*Xanthomonas campestris* pv. *oryzae*) which was more severe with wider spacing (Have and Kaufmann, 1972). They postulated that at the late stages of development wider spaced plants had a higher N content which predisposed them to disease. The highest yields were obtained in the closer spacings, as positive effects of high nitrogen levels tended to mask differences in bacterial blight due to spacing.

Sheath blight, caused by *Thanatephorus sasakii*, has become a major problem of the high yielding varieties (HYVs) of rice in Asia, and is most destructive under conditions of high humidity and high temperature. Dense plant stands significantly change the plant microenvironment. The high plant populations generally used for HYVs have greatly increased the incidence of sheath blight in Asia during the last few decades.

Adjusting Depth of Planting

The uniform and perfect stands of many crops grown by indigenous farmers in Asia is impressive. These perfect stands are testimony to their knowledge and skill in their use of this cultural practice. Indigenous farmers in Colombia knew the proper depth to plant crops and were surprised to find that everyone did not possess such basic knowledge. Kozaka (1965) noted that in Japan deep transplanting reduced plant growth, delayed maturity and resulted in a higher incidence of infection by *P. oryzae*. Thus, depth of planting is also an important consideration for blast management.

Adjusting Time of Planting

The date that a farmer chooses to plant a crop may be influenced by past experience, group traditions, superstitions, phase of the moon, magic, existing and expected weather, and advice from family and neighbors. Detailed instructions on what to plant for each month of the year was given in the writings of the Arabs, the Romans, the Incas, and many other traditional societies. Ibn Al-Awam (1988) gave the recommended time for planting for hundreds of different crops. Numerous Japanese references indicate that early planting of rice reduces damage from rice blast. Kozaka (1965) suggested that in Japan there is less *P. oryzae* on early transplanted rice because of low air temperatures at tillering time when rice plants are most susceptible, as well as high temperatures at heading, the critical time for neck rot infection.

Using Clean Seed

Clean seed and propagating material is basic to a thriving agriculture. The major plant pathogens – bacteria, fungi, nematodes, viruses and virus-like pathogens – are sometimes transmitted by seed or propagating material, but fungi and viruses are the most commonly transmitted. Ancient farmers selected the best, flawless seed for planting. In 60 AD Columela (1988), cited Virgil (70–19 BC) as suggesting that cereals would degenerate if the largest seed was not selected, one by one, for planting each year. Ibn Al-Awam (1988) devoted an entire section in his book on agriculture to careful seed selection and to the standards that should be used when selecting wheat seed. Good weight, brightness, hardness, plumpness, and color, were criteria he described. Such seed selection criteria assist in the production of healthy crops. In Japan in the 1930s, Ito, cited by Padwick (1950), recommended avoiding the use of diseased seed for blast management 'especially in cases where the infection appeared in blackening of the ears'.

Ancient farmers usually prepared seed beds with great care. Today, especially in Asia, many indigenous farmers still grow rice seedlings in small, carefully tended seed beds before transplanting, and select the healthiest seedlings for transplanting. Seedlings can be checked for freedom from disease at the time of selection for transplanting, and this constitutes an important indigenous practice for disease management.

Fallowing

Fallowing has been practiced for millennia. The ancient Sumerians, one of the world's oldest civilizations, practiced fallowing (La Placa and Powell, 1990). Fallowing was also mandated in ancient Jewish religious law. Exodus 23:10–11 reads: 'For six years you shall sow your land and gather in its yield; but the seventh year you shall let it rest and lie fallow'. Fallowing is a highly effective cultural practice for managing many pathogens, especially soilborne fungi and nematodes, and, although many examples could be cited, only one especially interesting example will be given.

Fallowing is often more effective in reducing pathogen populations in combination with crop rotations as shown in the following case. Potato cyst nematodes (*Globodera rostochiensis* and *G. pallida*) originated in the South American Andes. In Peru, before the arrival of the Spanish, Inca farmers used both fallow and rotations for potatoes, according to Garcilaso de la Vega, El Inca (de la Vega, 1966) and Felipe Guaman Poma de Ayala (Poma de Ayala, 1987). Six to eight year rotations are still used today by some indigenous communities in the Andes. Brush (1977) describes a typical rotation/fallow as follows:

> A third stratagem used by Uchucmarcan peasants to assure a potato harvest is to cultivate fields for only one to three years before returning them to a long fallow of eight or more years. Farmers usually sow potatoes in the first year and other Andean tubers – oca (*Oxalis tuberosa*), mashua (*Tropaeolum tuberosum*), and ullucu (*Ullucus tuberosum*) – for one or two subsequent years. The long fallow period lowers subsistence risk in two ways: by reducing the amount of erosion and soil loss and by killing disease vectors such as nematodes and fungi, which remain in the soil and depend on the continued potato planting to survive.

The non-host crops grown during the long rotation also controlled nematodes. In New York, Brodie's (1984) experiments indicated that nematode densities in the soil decline 30–50% annually when a non-host crop was grown. Thus, the Peruvian farmer's rotation with non-hosts of the potato cyst nematode was also a sound nematode management practice.

Through centuries of trial and error Andean Indian civilizations probably learned that a 7- or 8-year rotation/fallow gave the highest potato yields. Studies in Rothamsted, UK, demonstrated that a 7-year fallow reduced potato cyst nematode populations to below their economic threshold (Jones, 1972). Today, some of the highest population levels in the world of potato cyst nematodes are present in much of the Peruvian Andes, as the traditional long fallow and rotation period is practised only by indigenous farmers in remote areas. To the Spanish, the Inca system of fallow and rotation probably seemed senseless, and thus long fallow periods were abandoned in Spanish-dominated Peru, and serious losses due to the potato cyst nematodes have been there ever since.

FIRE

Fire has been deliberately used for managing agriculture for millennia; probably since agriculture began (Hardison, 1976). Between 2 and 5% of the globe is burned annually, mainly for agriculture (Monastersky, 1988).

Slash-and-burn (also called shifting or swidden) agriculture is estimated to support from 200 to 500 million farmers (Myers, 1988). The system has been frequently condemned as destructive, but in its defense Greenland (1975) wrote: 'In the majority of lowland areas of the humid tropics it has been a stable system, providing a limited number of people living on sufficient land as a continuing method of food production, requiring little in terms of inputs'. The similarity of indigenous slash-and-burn practices in Asia, Africa, and Latin America is striking. The effect of burning on plant pathogens in slash-and-burn systems has rarely been studied. Caveness (1972) is one of the few researchers who has made such studies. He found that burning a 10 cm leaf litter layer destroyed root-knot nematodes (*Meloidogyne incognita*) to a depth of 9 cm and sheath nematodes (*Hemicycliophora* spp.) to a depth of 19 cm. Considering the magnitude of slash-and-burn agriculture in the developing world, it is strange that from the millions spent to help indigenous farmers in developing countries, only a small number of articles can be found examining the value of fire as a pest management measure.

Fire has been used for the control of rice pathogens. Padwick (1950) recommended 'burning straw and stubble from a diseased crop' in order to reduce overseasoning inoculum of *P. oryzae*. Burning of rice stubble and straw was common in Vietnam according to Hardison (1976) and was believed to reduce inoculum of *P. oryzae* (rice blast), *Gibberella fujikuroi* (foot rot), and *Magnaporthe salvinii* (stem rot). In California, burning of rice straw was considered the most effective method of reducing inoculum of stem rot of rice (*M. salvinii*). Burning of rice stubble has been recommended for sheath blight management (*Thanatephorus sasakii*) in the Philippines (Hardison, 1976). Burning rice crop residues has been recommended for management of the rice nematode (*Ditylenchus angustus*) according to Luc *et al.* (1990).

FLOODING

Flooding can be disastrous or of immense value. Flooding has periodically enriched soils in many of the deltas of the world, such as those of the Tigris and Euphrates, the Nile in Egypt, the Amazon, and the immense river deltas in Asia. During the period 2111 BC to 2003 BC, in what is now Iraq, ancient Sumerians flooded their fields and subsequently puddled them with oxen, a practice still used in rice cultivation in Southern Asia (Maekawa, 1990). Deep-water rice varieties, which can elongate as floodwaters rise, are commonly found in delta areas such as those in Bangladesh, India, Thailand, and Vietnam. Flooding eradicates or lowers soilborne insect and pathogen populations by rendering soils anaerobic.

The paddy rice growing system is one of the oldest uses of flooding for plant disease management, as traditional farmers have flooded their rice paddies for thousands of years. The system is perhaps our oldest sustainable agricultural system. Historically, without high inputs of fertilizer or pesticides, rice grown in this system annually produced 1–2 tonnes of paddy per hectare. Some yields in Asia today exceed 6 tonnes ha^{-1}. Ponnamperuma (1972) describes paddy soils as follows:

> Paddy soils are soils that are managed in a special way for the wet cultivation of rice. The management practices include: a) leveling of the land and construction of levees to impound water; b) puddling (plowing and harrowing the water-saturated soil); c) maintenance of 5–10 cm of standing water during the 4–5 months the crop is on the land; d) draining and drying the fields at harvest; and e) reflooding after an interval which varies from a few weeks to as long as 8 months.

Some authorities suggest that weed control is probably the major benefit of the rice paddy system; however, some have also noted the value of flooding for the management of pathogens, especially soilborne pathogens (Kelman and Cook, 1977; Cook and Baker, 1983). The anaerobic or near-anaerobic conditions produced by flooding reduces the number of fungi, insects, and nematodes in the soil and, by controlling many weeds, which may harbor rice pathogens and insects, reduces disease and insect damage.

Rice blast is less severe on flooded paddy rice than on upland rice. One possible reason why blast is less severe is that fewer hours of dew occur in flooded than in upland rice, and thus *P. oryzae*, which requires free moisture for spore penetration, is less serious due to the shorter infection periods which occur under flooded conditions. Bonman (cited by Gupta and O'Toole, 1986) suggested that upland rice plants have a lower silica content than flooded plants. The shorter periods of intense rainfall in tropical rice growing areas, in comparison to the 'drizzles' that often occur in temperate areas, may also explain the severity of the disease in some temperate regions. Heavy rains may wash *P. oryzae* spores off rice leaves. Also, it is well known that rice plants are more susceptible when grown under 'dry' conditions and more resistant under moist or flooded conditions (Kahn and Libby, 1958; Ou, 1972). A water deficit during the vegetative stage of dry-season upland rice increased the severity of both leaf and neck blast of rice according to Bonman *et al.* (1988). Thus, flooding may operate in several ways to reduce the severity of rice blast.

Kelman and Cook (1977) stated: 'The practice of flooding fields for paddy rice and the use of organic material as fertilizers are apparently key factors in the general absence of soilborne diseases in China'. Cook (1986) wrote: 'The use of paddy rice in a rotation is among the most effective natural treatments ever discovered for control of pathogens in the soil'. *Fusarium oxysporum* f. sp. *vasinfectum* (fusarium wilt of cotton) was managed in China by growing paddy rice every 3 or 4 years in rotation with cotton, and this flooding reduced fungal inoculum in the soil (Cook, 1981). Credit for the development of these practices obviously belongs to ancient traditional farmers.

Cook and Baker (1983) noted that flooding is not always completely effective, and suggested that pathogens, such as those that have propagules able to float (for example, the sclerotia or resting bodies of *Rhizoctonia solani*, which causes sheath blight of rice), may survive in rice fields. Algae in flooded rice may also produce nitrogen, which in turn may affect rice diseases (e.g. high nitrogen causes rice to be more susceptible to *P. oryzae*). Flooding of rice fields also appears to control nematodes (Castillo and Litsinger, 1978; IRRI, 1978).

Mulching

Mulching means different things to different people. It has been simply defined as 'application of a covering layer of material to the soil surface' (Rowe-Dutton, 1957) or 'any covering placed over the soil surface to modify soil physical properties, create favorable environments for root development and nutrient uptake, and reduce soil erosion and degradation' (Wilson and Akapa, 1983). Mulches contribute to disease management in various ways. Reduction or prevention of soil splashing is an important function of mulches in the management of some plant pathogens. In addition, the organic material produced by mulches plays an important role in producing suppressive soils that reduce or eliminate soilborne pathogens. Traditional farmers, especially the Chinese (Youtai, 1987), have used mulches in their agriculture for millennia. It is often difficult to distinguish between mulches, crop residues, organic fertilizer, and green manures, because in much of the literature authors often do not specify whether these organic amendments were used as a covering layer on the soil or incorporated into the soil.

Space does not allow a description of the innumerable use of mulches in the past and the present by farmers, but the *frijol tapado* (covered beans) system used on steep slopes in Central and South America will be used as an example. Bean seeds are broadcast into selected weeds, then the weeds are slashed with a machete so the broadcast seeds are covered with a mulch of weeds (Galindo *et al.*, 1982). The beans grow through the mulch and subsequently the foliage of the bean plants and the mulch together effectively smother weed growth. The weed mulch also prevents soil splashing, which was found in a Costa Rican study by Galindo *et al.* (1983a,b) to be the major source of inoculum of *Thanatephorus cucumeris* (causal agent of web blight). The disease is effectively managed today by thousands of indigenous farmers who use the traditional practice of *frijol tapado*, even in areas where climate is optimal for web blight development. Although yields from *frijol tapado* are somewhat low, the system is stable, sustainable, environmentally sound, and economic returns to labor are high. Even on steep mountain slopes, the mulch effectively prevents serious erosion.

Organic Soil Amendments

Anyone seriously interested in the long history of rice production, going back perhaps to 5000 BC, should study the book *Farmers of Forty Centuries* by F.H. King (1926). Dr King was a professor of soils at the University of Wisconsin and once worked with the USDA. During a trip to China, Japan, and Korea in 1907 he took some excellent photographs that illustrate his book and provide a glimpse of what indigenous agriculture looked like before European 'modern' science had a significant impact on Asian agriculture. Stable and often surprisingly productive systems are found in the many long established traditional and indigenous Asian systems. One of the best examples of such systems is found in the intensive and highly productive rice growing systems of mainland China. There is still much to learn about the mechanisms which kept these systems stable, sustainable, and productive for millennia. Dr King's observations on agriculture in Asian countries still make fascinating reading, and, in today's world, where energy shortages and concern about the environment are increasing, his observations are highly pertinent:

> It could not be other than a matter of the highest industrial, educational and social importance to any nation if it could be furnished with a full and accurate account of all those conditions which have made it possible for such dense populations to be maintained upon the products of Chinese, Korean, and Japanese soils. Many of the steps, phases and

> practices through which this evolution has passed are irrevocably buried in the past, but such remarkable maintenance efficiency attained centuries ago and projected into the present with little apparent decadence merits the most profound study.

Ancient rice growing systems incorporated large quantities of organic matter into soil. The value of manures was established in China before the 5th century BC according to Youtai (1987). Numerous authorities in the discipline of plant pathology suggest that the incorporation of large quantities of organic matter into soil generally results in reduced soilborne disease in addition to other important agronomic benefits. King (1926) observed:

> In China enormous quantities of canal mud are applied to the fields, sometimes at the rate of even 70 and more tons per acre. So, too, where there are no canals, both soil and subsoil are carried in to the village and there they are, at the expense of great labor, composted with organic refuse, then dried and pulverized, and finally carried back to the fields to be used as home-made fertilizers. Manure of all kinds, human and animal, is religiously saved and applied to the fields in a manner which secures an efficiency far above our own practices.

Cook and Baker (1983) in their book *The Nature and Practice of Biological Control of Plant Pathogens* wrote that about 80% of the People's Republic of China's fertilizer needs are realized with organic sources such as composted crop residues, green manures, and human and animal wastes. Over 100 tonnes of compost is often used on a single hectare of land. They stated:

> Perhaps the best large-scale demonstration of effective biological control by cultural practices is the widespread multiple-cropping organic system used in the People's Republic of China. The agriculture of that country, which feeds nearly one fourth of the earth's population, clearly demonstrates that farming can be both intensive and sustainable and, if stabilized for years or perhaps for centuries, can provide a biological balance and disease suppression similar in effect to the disease suppression that can occur with prolonged monoculture of some crops.

There are important relationships between the above and indigenous knowledge systems for the control of rice blast. For example, according to Kozaka (1965): 'In Japan, compost is the most popular fertilizer as a source of silica'. Kozaka (1965), Datnoff *et al.* (1991), Grist (1986), and Ou (1972) cite references on the value of silica for increasing the resistance of rice plants to blast, and Kozaka notes 'In volcanic ash soil, or degraded paddy soil, a heavy silica supply results in less infection by rice blast and Helminthosporium leaf spot as well as better growth and high yield'. The heavy applications of compost used on rice in Asia appear to have contributed not only to plant nutrition, but also significantly to disease resistance. Gupta and O'Toole (1986) cited Matsuo (1954) as writing that compost and farmyard manure were effective in increasing resistance to blast by supplying them with silicic acid.

Raised Beds

Raised beds, ridges, and mounds are commonly used in Asia for agriculture, especially in high rainfall and swampy areas. Herklots (1972) emphasizes the importance of raised beds in traditional Asian agriculture as follows:

> Throughout the monsoon countries of South-East Asia this is the almost universal technique adopted where water is abundant or too abundant. In the vast silted plains of the river valleys of Thailand, of the Mekong delta of Cambodia and of the Pearl River of South China and even of the smaller rivers of Taiwan raised vegetable beds are a feature of the countryside, especially near the large cities.

Over 2000 years ago ridge farming (*tai-t'ien*) was introduced to the Han Court in China (Hsu, 1980). The 'dike-pond system' found in the Pearl River Delta of Southern China is described by Ruddle and Zhong (1988). The system has been evolving for 2000 years and includes agriculture and fish ponds. The system includes an area of 800 square kilometers and supports a dense population of 1.2 million. Luo and Han (1990) call this raised bed system the 'deep ditch-high bed system' and note that the ditches produce rice, water hyacinth, snails, and fish, while the 48,000 ha of such raised beds are intercropped with various vegetables, flowers, and fruit. The system provides an outstanding example of long-term agricultural sustainability.

One of the most studied raised field system is the *chinampas* or 'floating gardens' of the Valley of Mexico, which the Spanish conquistadores erroneously thought floated. The Spanish who arrived in Mexico in 1521 were amazed by the immense areas in *chinampas* cultivated by the Aztecs. Some are still farmed near Mexico City at Xochimilco. The great productivity of the *chinampas* has been cited as a reason why the Aztecs grew from a small group to a powerful tribe that controlled most of Mexico.

The *chinampas* are generally rectangular in shape and divided by canals with the surface of the *chinampas* about a meter above the canal water level (Coe, 1964). Mud rich in nutrients from the bottom of the canals is dredged up and spread on the *chinampas*. Also, aquatic weeds and animal manure are spread on the surface. During the time of the Aztecs a variety of crops were grown on the *chinampas*, and numerous crops are still grown on the remaining *chinampas* today (Muñoz, 1986). Continuous cropping was possible in the *chinampa* system because of sophisticated water control, multiple cropping, high levels of organic material and nutrients periodically added to the system, and the transplanting of healthy, selected seedlings with strong root systems. These practices contribute towards good disease management. The *chinampas* were found to have soils suppressive to both fungi (Lumsden *et al.*, 1987) and plant parasitic nematodes (Zuckerman *et al.*, 1989). Known antagonists such as *Trichoderma* spp., *Pseudomonas* spp., and *Fusarium* spp. suppressed the activity of *Pythium aphanidermatum* and other *Pythium* spp. in studies of *chinampa* soils.

For centuries in Southern China, after two or three crops of rice are grown in a field, the soil is shaped into raised beds for growing various vegetables (King, 1926; Williams, 1979). The same system has been described in Taiwan (Su, 1979).

A similar system is described in West and Central Africa where there are 85 million ha of inland valleys (International Institute for Tropical Agriculture, 1988). On over 80% of the inland valley fields in this region rice is grown in the wet season and then the soil is shaped into mounds or raised beds for vegetables, cassava, or sweet potatoes during the dry season. Organic matter consisting of crop residues and weeds is incorporated into the mounds, and through flooding, many insects and pathogens in the soil are destroyed. Thus, traditional farmers from both Asia and Africa developed similar systems that included flooding and raised beds, and these systems, which appear to be effective against soilborne organisms, are still in common use by indigenous farmers today.

Rotations

Crop rotation is an ancient agronomic practice important to the management of plant pathogens, especially soilborne pathogens. The history of rotations in China goes back thousands of years (Food and Agriculture Organization, 1980). Legumes such as alfalfa, broad beans, peas, lupines and vetches were used by the Romans in their crop rotations. The Roman Virgil (70–19 BC), as translated by Lewis (1941), recommended the following:

> See, too, that your arable land lies fallow in due rotation, and leave the idle field alone to recoup its strength: or else, changing the seasons, put down to yellow spelt, a field where before you raised the bean with its rattling pods, or the small seeded vetch, or the brittle stalks and rustling haulm of the bitter lupine, so too are the fields rested by a rotation of crops, and unploughed land promises to pay you.

Although it is important to emphasize rotation for plant disease management, it should be noted that the much-maligned monocultures can be sustainable and productive with some crops and in some areas. Many tropical crops, especially perennials such as sugarcane, coffee, tea, cacao, and bananas, are grown in monocultures for extended periods.

Terraces

The agricultural terraces found worldwide in mountainous areas are used primarily by indigenous farmers. The organic matter that is commonly incorporated into terrace soils is essential for their productivity and contributes in a major way to assuring their continual use for long periods of time. Terraces reduce erosion, manage irrigation water, and provide a level surface on which to cultivate crops.

According to Ruthenberg (1980), rice has been grown annually by Ifugao farmers near Bontoc in the Philippines on terraces in the highlands of Luzon for over 2000 years, apparently without major disease problems. However, no record of disease problems during this time was found in the literature. The Ifugao also grow sweet potatoes in their irrigated terraces in rotation with rice. Yen (1974) described the incorporation of organic matter into these terraces:

> In the spring preparation of fields for rice planting, large parties cut and bring in selected shrubby second growth from nearby gullies for incorporation into the soil by puddling with hands and feet. The soil has been made ready for this by early precrop inundations. After rice harvest, and the drainage of any residual water, preparation for the sweet potato planting entails the incorporation of the cereal stubble.

The incorporation of large quantities of organic matter into the soils of terraces helps to explain the sustainability of these systems over centuries.

Multiple Cropping

Multiple cropping, a common practice in tropical indigenous agriculture systems today, is an ancient practice. Dalrymple (1971) wrote that the practice may have been used before the time of Christ in Egypt and Babylon and was used about 1000 BC in India. During the Ming Dynasty (1644–1840) multiple cropping was found all over China (Youtai, 1987).

The combination of maize, beans, and squash is commonly used by many indigenous farmers throughout North and South America. Kirkby *et al.* (1980) found an astonishing 100 different crop associations grown by indigenous farmers in a 30,000 ha area near Riobamba, Ecuador. The Chinese may have more land in multiple cropping than all developing countries combined (Dalrymple, 1971). Kass (1978) reported that in Brazil 99% of the maize and rice were grown in mixed stands in the Amazon. In Taiwan two crops of rice are commonly followed by two to three upland crops of vegetables (Anonymous, 1974). De Datta (1987) noted that in Indonesia upland rice was often grown in combination with crops such as cassava.

It is difficult to generalize with any degree of accuracy about disease management through the use of multiple cropping, as disease may increase or decrease depending on numerous factors affecting the crop, the pathogen, and the environment. Recommendations on multiple cropping need to be thoroughly tested and often site-specific recommendations will be required.

Shade

The use of shade is not common in rice disease management, but for some crops, especially tropical plantation tree crops, management of shade or light is important and affects yields and disease severity. Shade was almost universally used for coffee until the latter half of this century, and some tea was also grown under shade. In the fifteenth century Fernandez de Oviedo (1986) related that Nicaraguan Indians associated shade trees with cacao, and noted that they pruned the trees for appropriate shade. The same author wrote the following regarding the use of shade for managing weeds in cassava: 'And thus the cassava continues to grow, and the weeds are controlled, until such time as the cassava dominates (shades) the weeds'. One of the reasons for the popularity of cassava among indigenous farmers in the hot, humid tropics is its ability to produce a heavy canopy that effectively shades weeds.

Weeds and Wild Plants

Many indigenous farmers spend more time in weed control than any other single activity. Kasasian (1971) even suggests that more human effort is spent on weed control than on any other human activity. Herbicides are the major means of combating weeds in developing countries and account for 42% of the world's pesticide sales (Edwards *et al.*, 1991). However, weed control practices used by indigenous farmers consist of many of the same cultural practices used to control pathogens. Flooding, burning, adjusting crop density, multiple cropping, clean seed, and the use of cover crops and mulches that smother weeds are some examples. Weed management is especially important in rice culture. Without weed control rice production could be diminished by up to 80% (De Datta and Lacsina, 1972).

Some vector-borne viruses spread through weed hosts and may cause serious losses to crops of indigenous farmers. Examples are barley yellow dwarf virus, cowpea mosaic virus, various maize viruses, hoja blanca virus and rice tungro virus. In addition to viruses, weeds (*Leersia* spp.) growing near paddy fields were an important source of bacterial leaf blight of rice (*Xanthomonas campestris* pv. *oryzae*) inoculum (Mizukami and Wakimoto, 1969). Padmanabhan (1977) related that *Leersia hexandra* and *Echinochloa colona* are sources of inoculum for *Cochliobolus miyabeanus* (causal agent of brown spot of rice) and recommend their removal.

Conclusion

A historical perspective on the practices used by traditional rice farmers through the centuries may help us to learn to conserve energy better, maintain natural resources, reduce chemical use and develop a truly sustainable agriculture. Systems that have lasted for thousands of years certainly justify serious study, and perhaps we can learn not only from the successes of ancient farmers, but also their failures. One should be cautious about dismissing indigenous practices that may seem ridiculous at first glance.

References

Al-Awam, Ibn (Abu Zacaria Iahia) (1988) *Libro de Agricultura*. 2 Vols. J.A. Banqueri. (trans.). Clasicos Agrarios, Ministry of Agriculture, Pesca y Alimentación, Madrid, 1454 pp.

Amin, K.S. and Katyal, J.C. (1979) Incidence of leaf blast in relation to seedling density and chemical composition of rice varieties. *Phytopathology Zeitschrift* 96, 65–70.

Anonymous (1974) *Multiple Cropping Systems in Taiwan*. Food and Fertilizer Technology Center for the Asian and Pacific Region, Taipei, Taiwan. 77 pp.

Berlin, B.D.E., Breedlove, D.E. and Raven, P.H. (1974) *Principles of Tzeltal Plant Classification. An Introduction to the Botanical Ethnography of a Mayan-Speaking People of Highland Chiapas*. Academic Press, New York. 660 pp.

Bonman, J.M., Sanchez, L.M. and Mackill, A.O. (1988) Effects of water deficit on rice blast II. Disease-development in the field. *Journal of Plant Protection in the Tropics* 5, 76–73.

Brodie, B.B. (1984) Nematode parasites of potato. In: Nickle, W. (ed.), *Plant and Insect Nematodes*. Marcel Dekker, Basel, pp. 167–212.

Brush, S.B. (1977) Farming the edge of the Andes. *Natural History* 86(5), 32–40.

Castillo, M. and Litsinger, J.A. (1978) Plant parasitic nematodes of mung bean in Philippines. In: *Proceedings of the 1st International Mungbean Symposium*. AVRDC, Taiwan, pp. 195–200.

Caveness, F.E. (1972) Changes in plant parasitic nematode populations on newly cleared land. *Nematropica* 2(1), 1–2.

Centro Internacional de Agricultura Tropical (1975) *Horizontal Resistance to the Blast Disease of Rice*. Series CE-9. CIAT, Cali, Colombia, 246 pp.

Coe, M.D. (1964) The chinampas of Mexico. *Scientific American* 211, 90–98.

Columela, L.J.M. (1988) De Los Trabajos de Campo. In: Holgado, R.A. (ed.), *Ministry of Agriculture, Pesca y Alimentación*. Siglo XXI de España Editores, Madrid, 339 pp.

Conklin, H.C. (1954) An ethnoecological approach to shifting agriculture. *Transactions of the New York Academy of Sciences* (Ser. 2) 17, 133–142.

Cook, R.J. (1981) Fusarium diseases in the People's Republic of China. In: Nelson, P.E., Toussoun, T.A. and Cook, R.J. (eds), *Fusarium Diseases, Biology and Taxonomy*. Pennsylvania State University Press, University Park, Pennsylvania, pp. 53–55.

Cook, R.J. (1986) Plant health and the sustainability of agriculture, with special reference to disease control by beneficial microorganisms. *Biological Agriculture Horticulture* 3, 211–232.

Cook, R.J. and Baker, K.F. (1983) *The Nature and Practice of Biological Control of Plant Pathogens*. American Phytopathology Society, St Paul, Minnesota, 539 pp.

Dalrymple, D.G. (1971) *Survey of Multiple Cropping in Less Developed Nations*. Foreign Economic Development Report No. 12. Foreign Economic Research Services, USDA/USAID, Washington DC, 108 pp.

Datnoff, L.E., Raid, R.N., Snyder, G.H. and Jones, D.B. (1991) Effect of calcium silicate on blast and brown spot intensities and yields of rice. *Plant Disease* 75, 729–732.

De Datta, S.K. and Lacsina, R.Q. (1972) Weed control in flooded rice in tropical Asia. In: *Proceedings of the 11th British Weed Control Conference*, Brighton, UK, pp. 472–478.

De Datta, S.K. (1987) *Principles and Practices of Rice Production*. Krieger, Malabar, Florida, 618 pp.

De la Vega, Garcilaso, El Inca (1966) *Royal Commentaries of the Incas and General History of Peru. Part 1*. H.V. Livermore (trans.). University of Texas Press, Austin, 627 pp.

Edwards, C.A., Thurston, H.D. and Janke, R. (1991) Integrated pest management for sustainability in developing countries. In: *National Research Council; Toward Sustainability; A Plan for Collaborative Research on Agriculture and Natural Resource Management*. National Academy Press, Washington DC, pp. 109–133.

Fernandez de Oviedo, G. (1986) In: Ballesteros, M. (ed.) *Sumario de la Natural Historia de las Indias*. Historia 16, Madrid, 181 pp.

Food and Agriculture Organization (1980) *China: Multiple Cropping and Related Crop Production Technology*. Plant Production and Protection. Paper 22. FAO, Rome, 57 pp.

Galindo, J.J., Abawi, G.S., Thurston, H.D. and Galvez, G. (1982) 'Tapado', controlling web blight of beans on small farms in Central America. *New York Food and Life Sciences* 14, 21–25.

Galindo, J.J., Abawi, G.S., Thurston, H.D. and Galvez, G. (1983a) Source of inoculum and development of bean web blight in Costa Rica. *Plant Disease* 67, 1016–1021.

Galindo, J.J., Abawi, G.S., Thurston, H.D. and Galvez, G. (1983b) Effect of mulching on web blight of beans in Costa Rica. *Phytopathology* 73, 610–615.

Greenland, D.J. (1975) Bringing the green revolution to the shifting cultivator. *Science* 190, 841–844.

Grist, D.H. (1986) *Rice*. Longman, London, 599 pp.

Gupta, P.C. and O'Toole, J.C. (1986) *Upland Rice. A Global Perspective*. International Rice Research Institute, Los Baños, The Philippines, 360 pp.

Hardison, J.R. (1976) Fire and flame for plant disease control. *Annual Review of Phytopathology* 14, 355–379.

Have, H. and Kaufmann, H.E. (1972) Effect of nitrogen and spacing on bacterial leaf blight of rice. *Indian Farming* 22(10), 8–13.

Herklots, G.A.C. (1972) *Vegetables in South-east Asia*. Hafner, New York, 525 pp.

Hsu, Cho-yun (1980) *Han Agriculture*. University of Washington Press, Seattle, Washington, 377 pp.

International Institute for Tropical Agriculture (1988) *Annual Report for 1987*. IITA, Ibadan, Nigeria.

International Rice Research Institute (1965) *The Rice Blast Disease*. The Johns Hopkins Press, Baltimore, Maryland, 507 pp.

International Rice Research Institute (1978) *Annual Report for 1977*. International Rice Research Institute, Los Baños, The Philippines.

International Rice Research Institute (1979) *Rice Blast Workshop*. International Rice Research Institute, Los Baños, The Philippines, 222 pp.

Jones, F.G.W. (1972) Management of nematode populations in Great Britain. In: *Proceedings of the Tall Timbers Conference*, February 24–25 1972. Tall Timbers Research Station, Tallahassee, Florida, pp. 81–107.

Kahn, R.P. and Libby, J.L. (1958) The effect of environmental factors and plant age on the infection of rice by the blast fungus *Pyricularia oryzae*. *Phytopathology* 48, 25–30.

Kasasian, L. (1971) *Weed Control in the Tropics*. Leonard Hill, London, 307 pp.

Kass, D.L. (1978) *Polyculture Cropping Systems: Review and Analysis. Cornell International Agricultural Bulletin*, 32. Cornell University, Ithaca, New York, 69 pp.

Kelman, A. and Cook, R.J. (1977) Plant pathology in the People's Republic of China. *Annual Review of Phytopathology* 15, 409–429.

King, F.H. (1926) *Farmers of Forty Centuries*. Harcourt Brace, New York, 379 pp.

Kirkby, R., Gallegos, P. and Cornick, T. (1980) *Metodologia para el desarrollo de technologia agricola apropiadaparapequeños productores experiencias del Proyecto Quimiag-Penipe, Ecuador*. Ministerio de Agriculture y Ganaderia, Ecuador, 38 pp.

Kozaka, T. (1965) Control of rice blast by cultivation practices in Japan. In: *The Rice Blast Disease*. The Johns Hopkins Press, Baltimore, Maryland, pp. 421–438.

La Placa, P.J. and Powell, M.A. (1990) The agricultural cycle and the calendar at Pre-Sargonic Givsu. *Bulletion of Sumerian Agriculture* 5, 75–104.

Lewis, C.D. (1941) *The Georgics of Virgil: A New Translation. Book 1*. Jonathan Cape, London.

Luc, M., Sikora, R.A. and Bridge, J. (eds) (1990) *Plant Parasitic Nematodes in Subtropical and Tropical Agriculture*. CAB International, Wallingford, UK, 629 pp.

Lumsden, R.D., García-E., R., Lewis, J.A. and Frias-T., G.A. (1987) Suppression of damping-off caused by *Pythium* spp. in soil from the indigenous Mexican chinampa agricultural system. *Soil Biology and Biochemistry* 19, 501–508.

Luo, S.M. and Han, C.R. (1990) Ecological agriculture in China. In: Edwards, C.A. Lal, R., Madden, P., Miller, R.H. and House, G. (eds), *Sustainable Agricultural Systems*. Soil, Water Conservation Society, Ankeny, Iowa, pp. 299–322.

Maekawa, K. (1990) Cultivation methods in the Ur III period. *Bulletin of Sumerian Agriculture* 5, 115–145.

Matsuo, T. (1954) *Rice Culture in Japan*. Ministry of Agriculture and Forestry, Japanese Government, Tokyo, 119 pp.

Mizukami, T. and Wakimoto, S. (1969) Epidemiology and control of bacterial leaf blight of rice. *Annual Review of Phytopathology* 7, 51–72.

Monastersky, R. (1988) Legacy of fire: the soil strikes back. *Science Newsletter* 133, 231.

Muñoz, E. (1986) Producción de maíz, frijol, y calabaza en un sistema hidráulico de chinampa. *Turrialba* 36, 369–373.

Myers, N. (1988) News and comment: tropical forests: shifting cultivators in the Philippines. *Plants Today*, January–February, 1.

Orlob, G.B. (1973) Ancient and medieval plant pathology. *Pflanzenschutz-Nachrichten Bayer* 26, 65–294.

Ou, S.H. (1972) *Rice Diseases*, 2nd edn. Commonwealth Mycological Institute, Kew, UK, 368 pp.

Padmanabhan, S.Y. (1977) Cochliobolus miyabeanus. In: Kranz, J., Schmutterer, H. and Koch, W. (eds), *Diseases, Pests, and Weeds in Tropical Crops*. Verlag Paul Parey, Berlin, pp. 106–108.

Padwick, G.W. (1950) *Manual of Rice Diseases*. Commonwealth Mycological Institute, Kew, UK, 198 pp.

Poma de Ayala, F.G. (1987) In: Murra, J.V., Adorno, R. and Urioste, J.L. (eds), *Nueva Cronica y Buen Gobierno*. Historia 16, Madrid, 1384 pp.

Ponnamperuma, F.N. (1972) The chemistry of submerged soils. *Advances in Agronomy* 24, 29–96.

Rowe-Dutton, P. (1957) *The Mulching of Vegetables*. Tech. Comm., No. 24. Commonwealth Agricultural Bureau, Farnham Royal, UK, 169 pp.

Ruddle, K. and Zhong, G. (1988) *Integrated Agriculture-Aquaculture in South China. The Dike-Pond System of the Zhujiang Delta*. University Press, Cambridge, 164 pp.

Ruthenberg, H. (1980) *Farming Systems in the Tropics*. Clarendon Press, Oxford, 424 pp.

Su, Kuang-chi (1979) Role of tomato in multiple cropping in Asian Vegetable and Development Center. In: *Proceedings of the 1st International Symposium on Tropical Tomato*, Oct. 23–27, 1978. Shanhua, Taiwan, Republic of China.

Thurston, H.D. (1992) *Sustainable Practices for Plant Disease Management in Traditional Farming Systems*. Westview, Boulder, Colorado, 279 pp.

Yen, D.E. (1974) *The Sweet Potato and Oceania. An Essay in Ethnobotany*. Bishop Museum Press, Honolulu, Hawaii, 389 pp.

Youtai, Y. (1987) Agricultural history over seven thousand years. In: Wittwer, S., Youtai, Y., Han, S. and Lianzheng, W. (eds), *Feeding a Billion: Frontiers of Chinese Agriculture*. Michigan State University Press, East Lansing, pp. 19–33.

Williams, P.H. (1979) Vegetable crop production in the People's Republic of China. *Annual Review of Phytopathology* 17, 311–324.

Wilson, G.F. and Akapa, K.L. (1983) Providing mulches for no-tillage in the tropics. In: Akobundu, I.O. and Deutsch, A.E. (eds), *No-tillage Crop Production in the Tropics*. Monrovia, Liberia, August 1981. IPPC, Oregon State University, Corvallis, pp. 51–65.

Zuckerman, B.M., Dicklow, M.B., Coles, G.C., Garcia-Espinosa, R. and Marban-Mendoza, N. (1989) Suppression of plant parasitic nematodes in the chinampa agricultural soils. *Journal of Chemical Ecology* 15, 1947–1955.

33 Rice, Reason, and Resistance: A Comparative Study of Farmers' vs. Scientists' Perception and Strategies

V.D. NAZAREA-SANDOVAL AND R.E. RHOADES

University of California, Santa Cruz, CA 95064 and Department of Anthropology, University of Georgia, Athens, GA 30602, USA

INTRODUCTION

This chapter is based on two premises. The first one is that farming is not just a random combination of soil, plants and water manipulated by a person called a 'farmer'. Farming is performance (the phrase 'agriculture as performance' has been used by Richards, 1989b, in a slightly different manner and context) in the same way that science is performance. Both performances require a 'script' or program as a means of orchestrating how things are done and how they are communicated. For science, it is the scientific method – a systematic, rigorous methodology of proof and disproof that is well understood by its disciples. In farming, the script is based on the farmers' own understanding of what they consider as significant factors and relationships in their environment – in other words, their ethnoscience – and how these can be combined to form a workable whole. Second, while guided by different scripts, both scientific performance and farming performance require innovation and experimentation. Without doubt, scientists are innovators and tireless experimenters. What is less frequently recognized, however, is that farmers are innovators, too [Recently, a growing body of literature has called attention to the vitality and significance of farmer experimentation (Gupta, 1987; Richards, 1989a; Rhoades and Bebbington, 1994).]. More importantly, they are innovators under a tighter set of constraints embedded in a system of constant flux over which they have little control. As such, their 'reality checks' are unsparing, their margin of error tight. As Richards (1989a) perceptively put it, 'farmers have to innovate, even to stay in place'.

Our purpose in writing this chapter is not only to explore the divergence between scientists' scripts and farmers' script (for interesting theoretical reasons) – and to argue for the value of the latter (for urgent practical ones) – but also to analyze the underlying foundations of curiosity, rationality, and adaptation common to both. We believe this is important because few if any agricultural problems are unidimensional, yielding to formulas and prescriptions, however scientifically robust these may be. In the case of rice blast management, for example, both scientists and farmers are confronted with a complex, dynamic disease, highly sensitive to environment, management, and rice cultivars. The fact that farmers and rice are still around attests to some success in the management of blast but does not diminish the threat it poses. We can accelerate our understanding of the problem – and increase the relevance and acceptability of potential solutions – by appreciating the keen observation and strong motivation behind the strategies employed by farmers in crop selection and disease management.

Farmers' versus Scientists' Perspectives and Strategies

Literature on farmer experimentation and decision making strongly points to the internal coherence and adaptive significance of agriculture technologies evolved *in situ* (Chacon and Gliessman, 1982; Rhoades and Booth, 1982; Nazarea-Sandoval, 1991). A closer examination provides interesting evidence that in dealing with similar problems and working towards nearly identical goals – basically, greater productivity and sustainability of agriculture – farmers and scientists none the less utilized different criteria and 'scripts' for evaluating options, managing resources, and formulating strategies. In simple terms, the difference can be illustrated by the fact that rice scientists are interested in 'how rice grows' whereas rice farmers are interested in 'how to grow rice' (Rhoades and Bebbington, 1983). This is roughly equivalent to Levi-Strauss' (1966) distinction between 'science of the abstract' and 'science of the concrete', respectively. From these parallel frameworks, the output of scientists may be seen to consist of findings expressed in neat tables, statistics, and graphs which lead to solid research publications. On the other hand, the output of the farmers may be viewed as consisting of energy, food, and, ultimately, survival.

Clearly, there is a difference in orientation and motivation, not to mention constraints, between farmers and scientists. But while the above dichotomies are fundamentally true, they dramatize just a little bit more than they illuminate – as dichotomies are wont to do. Table 33.1 summarizes the definitions given by local farmers in Kabaritan, The Philippines, for some rice-associated plants and arthropods. (The research was conducted in an irrigated rice-farming village in the province of Laguna. Farmer definitions were elicited from a small, representative sample consisting of male and female farmers belonging to different socioeconomic groups; Nazarea-Sandoval, 1987.) Side by side it provides scientific descriptions of the organisms, permitting a comparison of farmer definitions with scientific definitions. Examining the two columns, we cannot fail to notice the utilitarian character of farmer definition; there is emphasis on use, effect, and management. For example, farmers observe that azolla (*Azolla* sp.) controls grass in rice fields and that it can be fed to 'tilapia' (*Tilapia nolotica*) in fish ponds. Likewise, 'bayakibok' (*Echinochloa crus-galli*), while considered a weed, is also valued as livestock feed. For all their utilitarian slant, however, farmer definitions also demonstrate considerable intellectual curiosity and effort. Farmers make thorough morphological observations and provide detailed descriptions of growth habits way beyond the requisites of purely agronomic, or even market, considerations. The level of sophistication of their observations is also notable from their recognition of the existence of different *types* of an organism (e.g. different types of *Azolla* and different types of spiders). Moreover, like true systematists and ecologists, they pay attention to prevailing associations between different organisms, as in the case of *Azolla* and 'kiyapo' (*Pistia stratoites*) as well as between 'bayakibok' and rice.

An even more subtle evidence of farmers' power of observation and inference is the attention given to thresholds and points of diminishing returns. Note how *Azolla*, which is promoted by extension agents as an inexpensive biofertilizer, is described by farmers as being 'hot' or deleterious for rice plants beyond a certain level of abundance, a point that we will return to later in this chapter. In the same vein, spiders, which are much touted as 'friendly insects', have been observed by farmers to cause some harm to rice plants when they spin so much web that they bind up the leaves and make them turn yellow. Even so, there is a certain degree of parallelism between local farmers' observations and agricultural scientists' prescriptions. For instance, farmers make the observations that mole crickets (*Gryllotalpa africana*) fly around any light source and scientists recommend light trapping as a means of control.

In relation to rice blast, a much-needed feedback which may be input to research and development is the level at which crop loss due to disease is acceptable. If science cannot reduce losses to zero, as seems likely, what levels of loss are acceptable and, as a corollary, what control measures are adequate? Farmer input with regards to thresholds is particularly essential if we are to apply the partial resistance approach. Before a certain level of resistance is targeted, it will be useful to analyze what level of risk is acceptable. In other words, up to what point are farmers (and scientists) willing to tolerate the appearance of the disease in the fields given the explosive nature of blast epidemics? Insights into these questions can only be gained by paying close attention to farmers' perspectives and strategies.

Comparison of Indigenous Beliefs and Scientific Beliefs

Column 1 in Table 33.2 presents a list of indigenous beliefs and practices associated with sweet potato cultural management in Bukidnon, The Philippines. In a recent annual meeting of the International Potato Center in South-East Asia, scientists from various disciplines were asked to comment on the indigenous beliefs and practices. The other columns of Table 33.2 summarize scientific evaluations and comments by different scientists (one set of responses was chosen for each discipline). This is by no means exhaustive or representative of 'the scientific stand' but it reveals interesting differences in perspectives between farmers and scientists, and even among scientists themselves, with respect to cultural management practices.

One thing that is obvious from an examination of column 1 is the adaptive value of indigenous knowledge in terms of the total agricultural system. For example, ginger, which has quite a pungent odor, is being used effectively as a weevil

Table 33.1. Comparison between farmers' definitions and scientists' definitions of rice-associated plants and anthropods.

Name		Definitions		
Common	Scientific	Farmers'	Scientists'	Source
Azolla	*Azolla* sp.	There are two kinds of azolla, a green and a red one; both have small, round leaves. These plants never grow tall, they just float on the surface of the water along with the 'kiyapo' (*Pistia stratoites*). They multiply and spread fast in rice fields and fish ponds, and can be fed to 'tilapia' (*Tilapia nibotica*) in fish ponds. They are helpful in keeping grasses from growing in rice fields but 'hot' for rice plants when grown in abundance	Free-floating, heterosporous aquatic pteridophytes, considered to contain six extant species in two sections; all species normally contain an N-fixing cyanobacterium. If other nutritional factors are adequate, *Azolla–Anabaesa* associates can exhibit rapid vegetative growth in areas where combined N levels are limiting to other aquatic plants, thus providing N to the environment. Traditionally grown as biofertilizer or green manure crop for lowland rice	Peters (1984)
Bayakibok	*Echinochloa crus-galli*	Have tall stems with elongated leaves. They grow in between rice plants in the fields. They look like rice plants but are taller, with sharper stems and leaves, and they flower. They can be fed to livestock. Cause damage to rice plants, hence should be removed from rice fields	An erect annual up to 2 m high, with thick roots and stout spongy stems. Leaves up to 40 cm long and 5–15 mm wide. Inflorescence pinkish to purplish, occasionally green, soft nodding panicle, 6–20 cm long with dense spikelets. Lowest branches the longest, occasionally 10 cm long, often rebranched and spread at maturity. Propagation by seed	Moody *et al.* (1984)

Gagamba (spider)	*Argiope catenulata* *Araneus inustus*	Rounded bodies and fat abdomens, with many legs; creeps on different surfaces and leaves webs wherever they pass. There are field and house spiders; the field spider spins webs around plants usually causing no harm, but when the webs get too thick, the leaves turn yellow and the rice plants become weak. The house spider is a nuisance because it makes too much cobweb; children play with these	Make circular webs in the rice canopy, capturing flying prey as large as butterflies and grasshoppers. They live 2–3 months and lay 600–800 eggs. They are late colonizers of rice fields, but are found in all rice environments; *Agiope* has yellow and grayish white markings on the abdomen; *Araneus* has a black ovoid band on the abdomen. The female lays eggs within folded leaves and covers them with white silken threads; preys on small insects – leafhoppers, planthoppers, and flies	Shepard *et al.* (1987)
Suhong (mole cricket)	*Gryllotalpha africana*	Elongated bodies, brownish in color; they have two hands which they use in the soil; they have wings, too. They live under the ground. They fly at night and fly in circles around any light source. They destroy the roots of rice plants and eat the stems; these insects are eaten by Ilocanos	Destroys the roots and base of young plants and can cause damage by burrowing under lowland rice in which no standing water can be maintained. When there is abundant water, the pest is confined to the dikes; do little damage to the plants. Control measure consists of flooding of entire field; light trapping as well as spreading poisoned rice bran bait in infested paddies have also been recommended	Cendana and Calora (1964)

Table 33.2. Scientific opinions on indigenous agricultural knowledge pertaining to sweet potato cultivation in Bukidnon, The Philippines.

Indigenous cultural management beliefs and practices	Scientific evaluations and comments by discipline				
	Plant pathologist	Entomologist	Agronomist	Plant breeder	Plant physiologist
Grasses and leaves are burned and the ashes are mixed with the cuttings so there will be many roots	Ashes provide phosphorus which enhances tuber formation	Organic fertilizer	Ashes probably provide some nutrients at planting	Ashes contain minerals	No comment
Ginger and peanuts are mixed with the planting materials for weevil resistance and for good harvest, respectively	Ginger may be weevil-repellent and intercropping is the method of pest control	Peanut can fix N, thus increase yield; peanut and ginger act as insect repellent	No comment	Needs to be verified	No comment
Holy water is mixed with some cuttings so God will watch over and protect the crop	Purely fanatic	No comment	No comment	Water will be good for cuttings	No comment
Male should plant and he should be naked so the harvest will be good	Foolish!	No comment	No comment	No comment	No comment

Chemicals are not used, especially for sweet potatoes	Depends!	Chemicals are expensive considering the low price of sweet potatoes	Depends on the kind; if indigenous fertilizers are used, plants grow more vines and leaves	Chemical fertilizers have long-term negative effects on the environment	Sweet potatoes are not as sensitive as other crops to pests and diseases; hence no need for chemicals
Grasses and weeds are cut using scythe or long-bladed knife to maintain cleanliness of the area	Sound	Weeds compete for light, nutrients, and water	This minimizes competition for light, nutrients, and water	The farmer has little choice	If farmer has enough resources to clear the area then he should; if not, the plants will not be terribly affected
While weeding, trim off tops of plants so these will have more roots and will not die easily	Trimming is a kind of stress which induces the plant to produce fruit/tuber to perpetuate its species	Pruning sweet potato plants will increase or decrease yield depending on the volume of leaves removed	Depends on the timing; probably good when done early and when growth is too luxuriant	No comment	No comment
Rats are left alone because they cannot eat everything anyway	Sweet potato is a low-income crop so use of pesticide will just increase cost of production and reduce return on investment	No comment	Chemical control is expensive and dangerous	No comment	No comment

Table 33.2. *(cont.)*

Indigenous cultural management beliefs and practices	Scientific evaluations and comments by discipline				
	Plant pathologist	Entomologist	Agronomist	Plant breeder	Plant physiologist
Allow insects and other pests to eat as much as they can because if they are 'touched' they will become angry and damage everything in the field	Nonsense!	Allow insects to 'be' because some of them serve as biological control	No comment	No comment	No comment
Prayers are said to the gods (e.g. the god of rats) to implore them to spare/protect one's fields	Nonsense!	No comment	No comment	Won't do any harm	No comment

Data were collected using participant observation, life histories, and ethnoscientific methods. This was part of the Memory Banking Project sponsored by the Users' Perspective with Agricultural Research and Development (UPWARD) based at the International Potato Center (CIP). A working paper, 'Of Memories and Varieties', and a Memory Banking Protocol are currently in press.

repellent; groundnut, with its nitrogen-fixing property, is utilized for intercropping; and clearing the field of grasses, which shelter competitors, is practiced as a means of weed and pest control. Several scientists agreed with the soundness of these practices, a few of them going to the extent of elaborating the scientific bases of such practices.

It is also notable that *some* scientists have the ability to see beyond differences in terminology and phraseology and to understand the reason behind practices. One local practice of leaving insects alone because to 'touch' them would only make things worse may appear on the surface as irrational or superstitious. However, the entomologist explained that such a belief could prevent people from inadvertently killing 'friendly insects' along with the not-so-friendly ones, thus upsetting the ecological balance of the farming system and doing more harm than good. Permitting the survival of beneficial insects is, of course, one of the basic principles of integrated pest management (IPM). Likewise, the local belief that grasses and leaves should be included with planting materials has been justified by many of the scientists interviewed as a way of increasing the organic matter content of the soil. This is not to say that all scientists have the perceptiveness to appreciate the practical value of farmer knowledge. The comments, 'scientifically unsound', 'purely fanatic', 'nonsense', are interspersed with scientific explanations validating farmers' agricultural beliefs and practices.

To return to the thesis that was made earlier, there are several points of convergence between scientific knowledge and indigenous knowledge. The former, earlier referred to as the 'science of the abstract' following Levi-Strauss, in fact hopes for concrete applications and must meet very concrete tests. On the other hand, indigenous knowledge or the 'science of the concrete' is in fact based on quite sophisticated abstractions. Moreover, in both, redundancy is used to secure critical relationships or important outcomes. For instance, looking at Table 33.2, it will be noted that many of the practices are working towards the same goal, be it soil enrichment or pest control. In the same way, scientific solutions are designed with backup systems such that if one line of defence fails there is another approach – or, in the case of resistance, another gene – to buffer the system from collapse (chapter 16, this volume). A third point of convergence is the perceived necessity for, and the actual practice of, replication in both scientific trials and farmer trials. While this may not be so obvious in the latter case, farmers actually set aside small portions of their field for trying out innovations and compare the observed results with performance in 'untreated controls'. They also compare notes with other farmers who have also tried the innovation or the treatment as well as with those who have not. Lastly, within their own fields, farmers make detailed observations over a span of time, sometimes even intergenerationally, and therefore approximate 'vertical' replication.

Need and Basis for Complementation: 'Beyond Political Correctness'

Much has been written about the value of local farming techniques for agriculture research and development (Brokensha *et al.*, 1980; Fujisaka, 1992; Sperling,

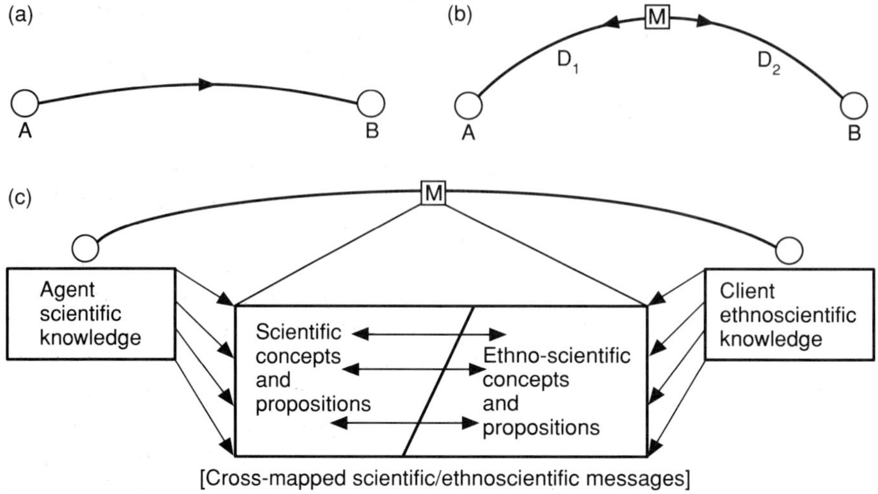

Fig. 33.1 (a) Source–receiver, 'arc' model of communication; (b) Demiarc model of communication; (c) Adaptation of the demiarc model to agricultural knowledge utilization (from Meehan, 1980).

1992). It is in fact currently fashionable – in other words, politically correct – to criticize science for its ostensibly objective, overly empirical approach (Chambers, 1990; Bentley and Andrews, 1991). In particular, agricultural science has come under fire for neglecting adaptations which have fostered long-term survival of farm populations despite limited resources. We believe the next constructive step is to seriously analyze the interface between agricultural science and indigenous knowledge and to study how local perspectives and strategies can be effectively and fruitfully integrated into ongoing research and development. Central to the operationalization of indigenous knowledge systems in formal agricultural projects would be the search for a common ground between western science and ethnoscience.

Meehan (1980) contrasted the conventional extension model with the demiarc model of communication (Fig. 33.1). In the former, a message from a supposedly more knowledgeable agent is presumed to go through a smooth channel and be passively received by a supposedly less-knowledgeable recipient. New agricultural technologies, for example, are generated in agricultural research centers and farm stations and transmitted to the local farming population by extension agents. The transmission or diffusion has been assumed to proceed uneventfully by virtue of the inherent superiority of the new technologies over traditional ones. Thus, most non-adoption cases have been conventionally attributed to farmers' lack of access to credit and infrastructure (Rogers, 1962), if not to their ignorance and traditionalism. On the other hand, the demiarc model stresses that the channel is far from smooth and non-resistant – that, in fact, transmission across the medium requires a negotiation between the agent's scientific knowledge and the local client's understanding of significant factors and relationships within his or her farm, also referred to as client ethnoscientific knowledge. This transmission would require a cross-translation between Western scientific concepts and propositions and equally valid ethnoscientific concepts and propositions. Needless to say, such negotiation can only proceed in a climate of mutual respect and appreciation of the need for complementation.

Such complementation is crucial to the understanding and management of a complex disease like rice blast. Scientists working on the problem have investigated lines of defense that are in fact quite dynamic, either in the development stage (e.g. lineage exclusion proposed by Zeigler *et al.*, chapter 16, this volume, or partial resistance proposed by Roumen, chapter 15, this volume) or in the utilization/adoption stage (e.g. cultivar mixtures proposed by Mundt, chapter 17, this volume). For all these projects, farmers' input in shaping the research and their cooperation in executing the recommendations are indispensable given the investment required for development in the first case and the intense farmer involvement needed in managing mixtures properly in the second case. It cannot be overstressed that the earlier farmer input is sought, the more likely the technologies generated by the research process will be relevant and successful. In short, the terms of reference of the research generating the technology must take into account the users of the technology.

The complementation needs to be fluid and iterative and can be found at the interface between Western science and ethnoscience. Attempts have already been underway for a number of years to bridge the gap between the two. Perhaps the best known case is that of the farmer-back-to-farmer model as proposed by Rhoades and Booth (1982). This model (Fig. 33.2) was generated as a result of a post-harvest experience with potatoes in which scientists, both social and technical, were initially attempting to reduced storage 'losses'. The concept turned out to be alien to the farmers who were more attuned to storage of different varieties of *seed* potatoes. Realizing their oversight, the post-harvest team then approached the research process from the clients' perspective by implementing a series of participatory technology adaptations based on local knowledge. The result was a highly successful and low-cost seed storage technology which was accepted by thousands of farmers around the world (Rhoades *et al.*, 1983).

A more recent attempt at integrating scientific – particularly biological – concepts and methods with indigenous beliefs and practices is the Memory Banking Project on sweet potatoes (Nazarea-Sandoval, 1992). Based on the observation that attrition of cultural knowledge is accompanied by corresponding genetic erosion, the research sought indigenous knowledge associated with traditional varieties of crops and methods for documenting this (Fig. 33.3). The project demonstrated the complementarity between cultural and genetic diversity, and underscored the need for supplementing passport data and genetic evaluation, as routinely collected for accessions in gene banks, with cultural information in the form of a memory bank (Fig. 33.4). Production or genetic scientists outside the farmers' cultural context may not appreciate the strengths (e.g. resistances, tolerances, yield) and, therefore, may under estimate or miss totally the material's potential uses as well as opportunities for problem identification. The full strengths of natural germplasm may not be appreciated outside their cultural content and, therefore, potential uses and opportunities for problem identification may be lost. The pay-off will be the documentation and preservation of

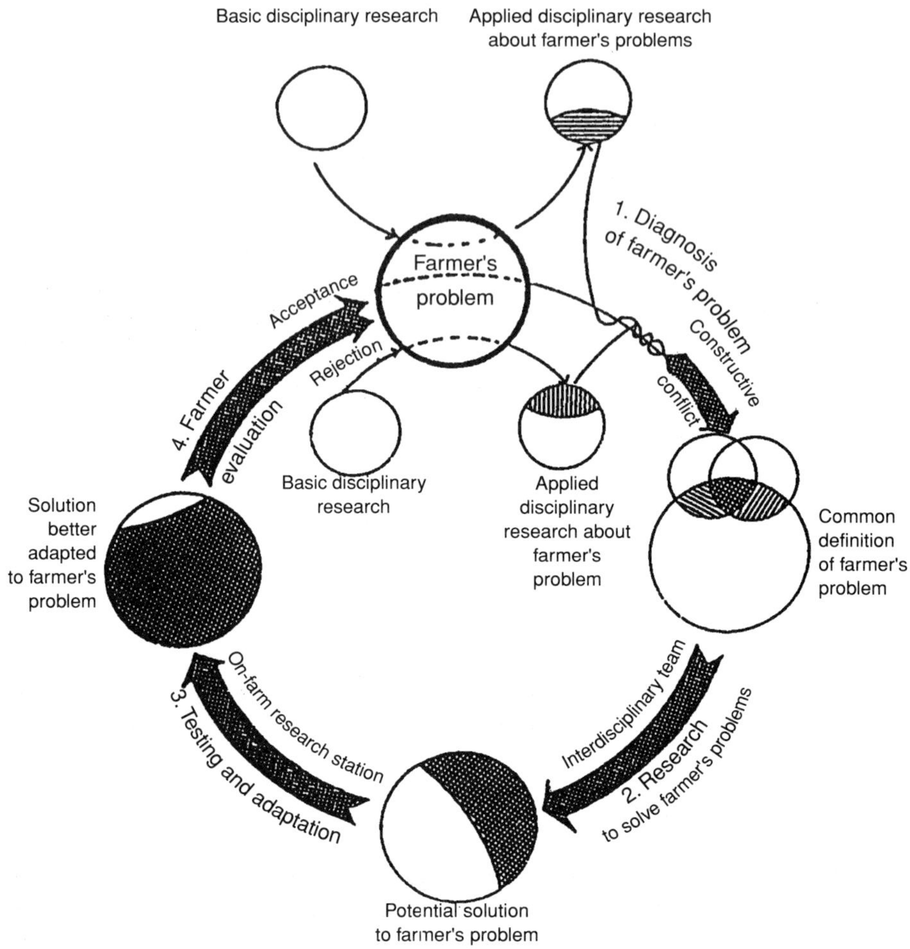

Fig. 33.2. 'Farmer-back-to-farmer' – a model generating acceptable technology (from Rhoades and Booth, 1982).

indigenous technologies which will provide us with the flexibility and span of options needed to deal with an environment that is impossible to completely predict or control. As a case in point, farmer practices in the management of crop diseases, such as the use of multiple varieties and resistant landraces, crop rotation and inter-cropping, fallowing and clearing – mostly low external input and user-friendly technologies – may provide (and in some cases, already have provided) valuable leads for scientific investigation.

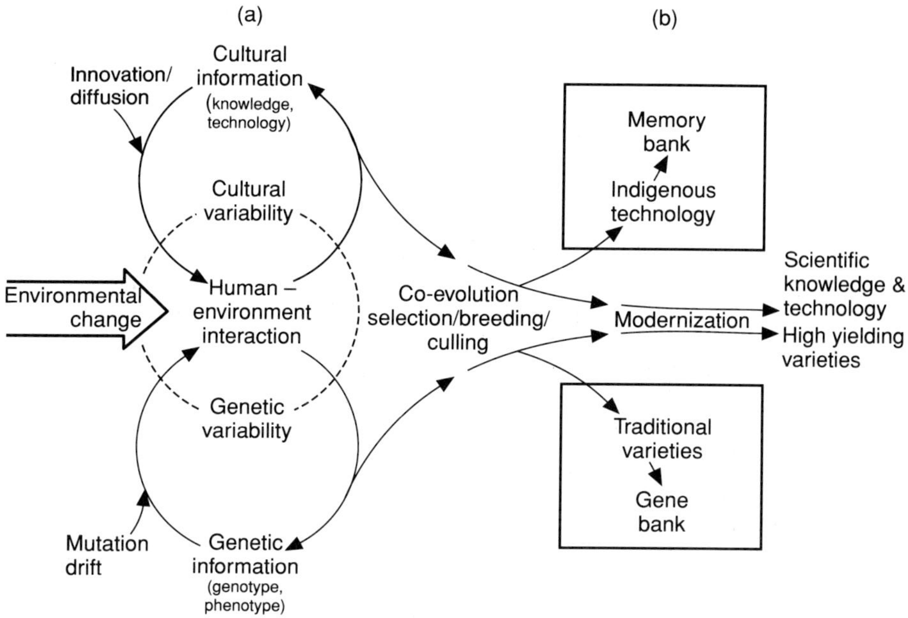

Fig. 33.3. (a) Co-evolution of cultural and genetic information systems (see also Bajema, C.J. (1972). Transmission of information about the environment in human species. *Social Biology* 19, 224–226). (b) Parallelism and complementation between gene bank and memory bank. (from Nazarea-Sandoval, 1992).

Resistance on Two Fronts

This volume has focused on understanding and controlling rice blast disease on an international scale. There has been much discussion on the desired development of resistance of rice plants to blast as well as the undesired development of resistance of rice blast to 'blasticides'. One uncomfortable but promising thought that has not been explored is the concept of farmer resistance to 'farmercides'. This refers to resistance among farmers, particularly those in the developing world, to the overpowering idea of the superiority of Western science – a hegemony that undermines their confidence in their own system of indigenous knowledge and modes of experimentation. If not for this resistance, the reification of the supremacy of 'modern' technologies over local adaptations could irretrievably push farmers over the brink toward instability and dependence in the pursuit of intensification.

Close inspection of why farmers maintain some traditional methods and technologies instead of indiscriminately adopting 'improved' ones often reveals

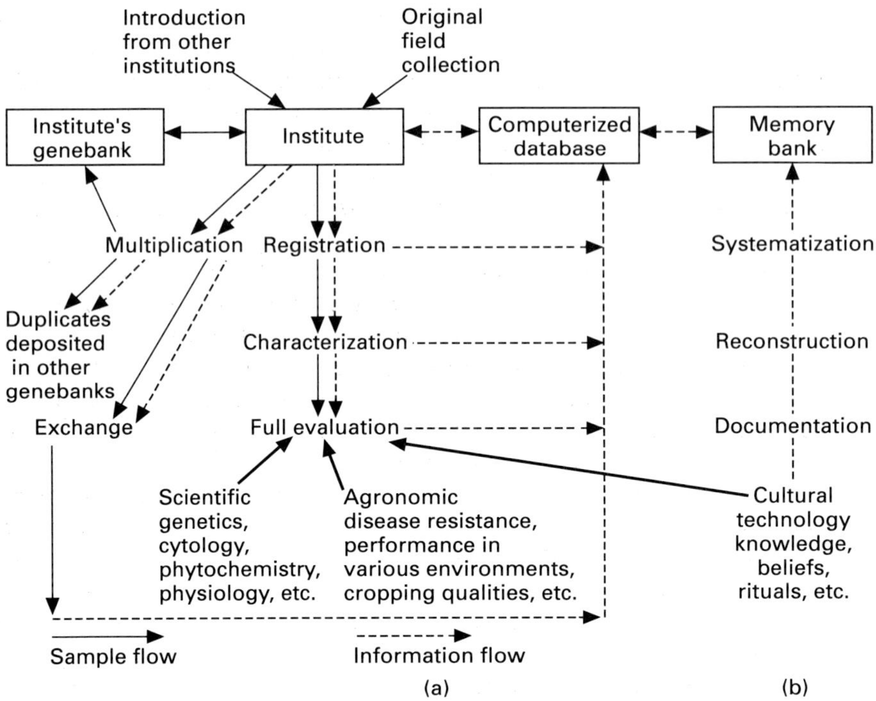

Fig. 33.4. Proposed integration of memory banking with gene banking procedures. (a) Procedures followed in processing a gene-bank accession (from Plucknett D.L. *et al.* (1987) *Gene Banks and the World Food,* p. 74. (b) Procedures followed in memory banking.

the wisdom of their resistance. For example, during the 1970s and 1980s Andean potato farmers consistently rejected large, centralized government stores despite the structures' technical superiority and a host of economic studies predicting that farmers who utilized them would increase their profits. A follow-up study on why this was so revealed that farmers in the Andes preferred the ancient system of storing near their homes for better security and closer management (Rhoades, 1983). Likewise, in Kabaritan, a farming village in the Philippines, it was observed that some farmers actually scoop out *Azolla*, a much-touted nitrogen-fixer, from their fields arguing that when it reaches a certain density it actually becomes 'hot' for the rice plants. According to their perspective, over-abundant *Azolla* 'mats' form a hospitable habitat for 'enemies' of rice plants. Could this observation be accounted for, at least partially, by the fact that too much *Azolla* may produce too high a level of nitrogen and thus increase the susceptibility of rice to blast? Also contributing to the perception of 'hotness' may be the lodging of rice plants associated with excess nitrogen or, conversely, the fact that nitrogen may be tied up in the *Azolla* biomass thus effecting a net reduction in nitrogen available to rice.

Based on countless observations and more macro-analysis (Altieri and Anderson, 1986; Bebbington, 1991) farmers have their reasons (a fact that should not be too surprising) and, therefore, the faith of farmers in their own observational skills should be encouraged rather than undermined. The same radical transformation can be accomplished by stimulating a favorable climate for vigorous experimentation and actually supporting the ability of farmers to reflect and, when deemed necessary, to resist. In short, farmers should be supported to *adapt* rather than adopt 'the message'. This framework is based on the negotiation of the *science* rather than the technology of agriculture, optimally a science that takes into account farmers' constraints and opportunities and respects their ethnoscience.

We have attempted in this chapter to show the basic commonality between farmers' and scientists' observations and experimentation underlying the differences in their 'scripts'. Applied to the problem of rice blast, farmers make sharp judgments about thresholds and their perceptions on acceptable level of risk and adequate level of control need to be taken into account in developing strategies based on partial resistance. Moreover, farmers' terms of reference must guide investments in research instead of simply providing *post hoc* explanations of failure or non-acceptance of agricultural technologies. If we ignore these factors, Western science could end up being a 'farmercide' by not recognizing the fragile equilibrium of the rice–farmer–blast relationship and disrupting the system by pushing it to ever-higher, *and* less stable, levels of intensification.

But resistance is not a monopoly of farmers. Scientists through centuries have been known to resist both religious and outmoded scientific dogma. In the Kuhnian paradigm of scientific revolutions, it is this very resistance that leads to significant shifts from dominant theories of 'normal' science and, ultimately, to significant contributions in science or scientific revolutions. There is a need to stretch and challenge 'habits of the mind' (Margolis, 1993) that create set ways of seeing and doing things. The conservative tendency that characterizes conventional science has been succinctly captured by Mundt in his comment that scientists have the propensity to ask the same questions over and over again and to be pleasantly surprised each time at getting the same answers (chapter 17, this volume). There is an urgent need on the part of agricultural scientists to be more creative, even playful, and to pay attention to the out-of-the-ordinary or *resistant* activities of farmers. Why are they doing (or not doing) certain things when scientific logic dictates otherwise? As Gupta (1987) pointed out, scientists should be stimulated by the odd and the unexpected in farmers' fields. Then, using methods and instruments not readily accessible to the farming population, scientists may be able to approach problems – both old and new – with fresh insight and direction.

What this will accomplish is a liberation from perceptual sclerosis or the hardening of perception that makes us think only in terms of tried and tested and therefore safe paradigms of normal science. This is particularly critical when scientific research interfaces with development issues on which the quality of life (including the quality of the environment) of millions of people may depend, as in the control of important diseases of food crops. As Illich (1987) pointed out:

> The only way to reverse the disastrous trend of underdevelopment, hard as it is, is to learn to laugh at expected solutions in order to change the demands which make them necessary. Only free men can change their

minds and be surprised and while no men are completely free, some are freer than others.

References

Altieri, M.A. and Anderson, M.K. (1986) An ecological basis for the development of alternative agricultural systems for small farms in the Third World. *American Journal of Alternative Agriculture* 1, 30–38.

Bebbington, A. (1991) Indigenous agricultural knowledge systems, human interests, and critical analysis: reflections on farmer organization in Ecuador. *Agriculture and Human Values, Winter and Spring* 14, 23.

Bentley, J. and Andrews, K.L. (1991) Pest, peasants, and publications. Anthropological and entomological views of an integrated pest management program for small-scale Honduran farmers. *Human Organization* 50(2), 113–124.

Brokensha, D.D.M.W. and Werner, O. (eds) (1980) *Indigenous Knowledge Systems* and *Development*. University Press of America, Boston.

Cendana, S.M. and Calora, F.B. (1964) Insect pests of rice plants in the Philippines. In: *The Major Insect Pests of the Rice Plant*. Symposium Proceedings, International Rice Research Institute, Los Baños, The Philippines, September 1964.

Chacon, J.C. and Gleissman, S.R. (1982) Use of the "Non-Weed" concept in traditional tropical agro-ecosystems of South-Eastern Mexico. *Ecosystems* 8, 1–11.

Chambers, R. (1990) Farmer first: a practical paradigm for Third World agriculture. In: Altieri, M.A. and Hecht, S. (eds), *Agroecology and Small-farm Development*. CRC Press, Boca Raton, Florida.

Fujisaka, S. (1992) Farmer knowledge and sustainability in rice-farming systems: blending science and indigenous innovation. In: Moock, J.L. and Rhoades, R.E. (eds), *Diversity, Farmer Knowledge, and Sustainability*. Cornell University Press, Ithaca, New York.

Gupta, A. (1987) Organizing the poor client responsive system: Can the tail wag the dog? Paper presented in Workshop on Farmers and Agriculture Research: Complementary Methods. Institute of Development Studies, University of Sussex, July 16–31 1987.

Illich, I. (1987) *Towards a History of Needs*. Pantheon Books, New York.

Levi-Strauss, C. (1966) *The Savage Mind*. University of Chicago Press, Chicago.

Margolis, H. (1993) *Paradigms and Barriers: How Habits of Mind Govern Scientific Beliefs*. University of Chicago Press, Chicago.

Meehan, P. (1980) Science, ethnoscience, and agricultural knowledge utilization. In: Brokensha, D., Warren, D.M. and Werner, O. (eds), *Indigenous Knowledge Systems and Development*. University Press of America, Washington DC.

Moody, K., Munroe, C.E., Lubigan, R.T. and Paller, E.G., Jr (1984) *Major Weeds of the Philippines*. Weed Society of the Philippines and University of the Philippines at Los Baños, The Philippines.

Nazarea-Sandoval, V. (1987) Philippine rural cultivators in transition: Operational reality and cognized models in agricultural decision making. PhD Thesis, University of Kentucky, USA.

Nazarea-Sandoval, V. (1991) Ethnoagronomy and ethnogastronomy: On indigenous typology and use of biological resources. *Agricultural and Human Values* 8, 121–131.

Nazarea-Sandoval, V. (1992) Memory banking of indigenous technologies associated with traditional crop varieties: A focus on sweet potatoes. In: Rhoades, R., Sandoval, V.N.

and Bagalano, C.P. (eds), *Sweet Potato Cultures in Asia and the South Pacific*. User's Perspective with Agricultural Research and Development, Los Baños, The Philippines.

Peters, G.A. (1984) *Azolla-Anabaena* symbioses: Basic biology, use and prospects for the future. In: *Practical Applications of Azolla for Rice Production*. Proceedings of the International Workshop, Mayaguez, Puerto Rico, November 17–19 1992.

Plucknett, D.L., Smith, N.J.H., Williams, J.T. and Murthi Anishetty, N.M. (1987) *Gene Banks and the World's Food.* Princetown University Press, Princetown, USA.

Richards, P. (1989a) Farmers also experiment: A neglected intellectual resource in African science. *Discovery and Innovation* 1(1), 19–25.

Richards, P. (1989b) Agriculture as performance. In: Chambers, R., Pacey, A. and Thrupp, L.A. (eds), *Farmer First: Farmer Innovation and Agricultural Research*. International Technology Publications, London.

Rhoades, R. (1983) *Breaking New Ground: Agricultural Anthropology*. International Potato Center, Lima, Peru.

Rhoades, R. and Bebbington, A. (1994) Farmers who experiment: an untapped resource. In: Warren, D.M., Brokensha, D. and Slikkerveer, L.J. (eds), *Indigenous Knowledge Systems: The Cultural Dimensions of Development*. Kegan Paul, London (in press).

Rhoades, R. and Booth, R. (1982) Farmer-back-to-farmer. *Agricultural Administration* 11, 127–137.

Rhoades, R., Booth, R.H. and Potts, M. (1983) Farmer acceptance of improved potato storage practices in developing countries. *Outlook on Agriculture* 12(1), 12–16.

Rogers, E. (1962) *The Diffusion of Innovations.* Hutchinson, London.

Shepard, B.M., Barrion, A.T. and Litsinger, J.A. (1987) *Friends of the Rice Farmer: Helpful Insects, Spiders, and Pathogens*. International Rice Research Institute, Los Baños, The Philippines.

Sperling, L. (1992) Farmer participation and the development of bean varieties in Rwanda. In: Mooch, J.L. and Rhoades, R.E. (eds), *Diversity, Farmer Knowledge, and Sustainability*. Cornell University Press, Ithaca, New York.

VI Summary of Discussion Group Findings

Summary of Discussion Group Findings

The following observations and ideas were generated during the break-out sessions when conferees were organized into small groups (10–12 people per group) to discuss topics presented at the symposium and to share working experience with rice blast disease. These sessions provided an informal opportunity to foster communication among researchers. The summary below does not reflect a consensus of opinions among conferees; rather, it serves to highlight the views and thoughts of researchers from very diverse disciplines.

Regional Assessment of Blast Problems

Many conferees raised the question of how serious blast is in different countries and rice growing environments. In general, the disease is most serious in intensive production areas (e.g. Korea and Japan) and less so in areas with low fertilizer use (e.g. most production areas in Africa). However, it is recognized that the disease is a potentially explosive problem in currently low-production countries as soon as fertilizer use is increased (e.g. rainfed lowlands). In order to guide management efforts, there is a need to assess closely crop loss due to blast in different rice production environments.

Control and Management Strategies

Host plant resistance and breeding

Host plant resistance remains the core of blast control strategies; however, there are different experiences and views on the kinds of resistance most appropriate to ensure durability. Due to the short-term effectiveness of single-gene, race-specific resistance, increasingly emphasis is being placed on partial or incomplete resistance. New methods to identify and select for partial resistance are being developed. Specifically, the use of molecular markers to track qualitative and quantitative resistance factors appears immediately useful. Also, suggestions were made to intensify the exploitation of interspecific crosses for blast resistance. New methods in wide-crosses and in tagging resistance genes are expected to widen the germplasm available for blast control.

Some conferees noted that there is little incentive to breed for durable resistance in countries where chemical control is available and cost-effective. Also, breeding for blast resistance may not be a top priority in areas where the occurrence of blast is sporadic and chemical control can be used as a back-up measure. Whether this situation will change is likely to be determined by the economics of rice production, such as marketing and price support policies for rice production and pesticide use.

Management tactics

Population biology and epidemiology

Traditional virulence analysis is considered highly essential for effective management of blast but there is a feeling that this approach has been diluted in recent years. Some conferees advocated a sustained and systematic effort to monitor the virulence spectra of the pathogen in different geographic regions. The successful management of wheat rusts in Australia and North America through long-term virulence surveys was cited as a potential model for blast control.

New excitement has been generated by the use of molecular markers to understand the population biology and evolutionary history of the blast fungus. DNA markers can be used to identify pathogen lineages found worldwide. To make practical use of the molecular data, knowledge of the relationship between genetic lineages and virulence of blast isolates will be critical. Some conferees pointed out the need and opportunity to merge epidemiological and population genetics data to develop an integrated management plan.

Promoting diversity with cultivar mixtures

Although practical and experimental work have suggested the usefulness of cultivar mixtures for disease control, it remains an under-utilized strategy in blast management. Some conferees felt that cultivar deployment (in mixtures or in rotation) can be the fastest means to achieve production stability against blast epidemics because it makes use of existing genetic resources and production systems. New tools to introduce resistance into rice (e.g. transgenic technology, marker-assisted selection), if used appropriately, can enhance the use of cultivar mixtures. The impact of using cultivar mixtures on the pathogen population can also be evaluated by monitoring the pathogen with genetic markers.

Biocontrol

Some promising results of blast control have been obtained using antagonistic organisms to suppress disease development. Biocontrol is expected to be a significant component in the integrated management of blast.

The 'best-mix' approach

Because of the diversity of rice production environments, a sound knowledge of the cultural practices, farming system, and ecology at the local level is necessary to make effective management choices. Valuable lessons can be learned from alternative practices and indigenous knowledge at a local level. The future challenge seems to be in selecting the best combination of approaches to solve a local problem. Stronger extension support is needed in this regard.

Basic Biology of Host–Pathogen Interactions

Many conferees felt that an understanding of the basic biology of the pathogen will play a pivotal role in developing future strategies for blast control. The rapid progress in understanding the genetic, molecular, and morphogenetic bases of virulence and pathogenesis is very encouraging. There is a critical mass of researchers involved in the molecular analysis of the fungus that multi-lab collaboration will be possible as well as beneficial. For example, effort is being made to integrate molecular maps of *M. grisea* and to share markers among laboratories. Some conferees pointed out that, in sharp contrast to the studies of the pathogen, progress has been slow in understanding the mechanisms of resistance at the biochemical and molecular level in the rice plant.

Biotechnology for Blast Control

Although most conferees were encouraged by the progress made in the basic studies of the pathogen, they did not have an unrealistic expectation of the application of biotechnology to solving the blast problem. The complexity of rice blast disease will make short-term solutions unlikely but most conferees have long-term hopes for biotechnology and basic research.

The most immediate benefit is likely to come from the use of molecular markers in gene tagging. Some expressed concern that biotechnology will further promote genetic uniformity whereas others argued that biotechnology can be used to enhance genetic diversity in rice production (e.g. in multilines or cultivar mixtures). Some conferees pointed out that benefits from biotechnology go beyond transgenic plants. Break-throughs may come from biocontrol and the development of environmentally safe fungicides. To make biotechnology and basic research relevant to rice production, there is a need to form good linkages between breeders and biotechnologists.

Communication Network

All conferees agreed to the need to strengthen communication between basic and applied scientists. Scientific jargon should be reduced to improve communication. As a step to foster communication, conferees supported the proposal of creating 'Blast Corner' in the International Rice Research Newsletter (IRRN) to post rapid communications and announcements concerning blast research. This idea will be explored with IRRN editors. To encourage the use of electronic mails in communication, an e-mail address list will be compiled and distributed to blast researchers. For those without e-mail, some proposed the CG centers to provide the necessary linkages.

It was proposed that an International Blast Symposium be held every 5 years. The symposium will most likely be held in conjunction with a major scientific conference to reduce travel costs. A committee was formed to follow up on this proposal.

Database

Many researchers felt a need to develop a reliable database. The database will include GIS data, yield loss data, virulence surveys, rice cultivar deployment data, etc. Some suggested ways to standardize data entry. The database can be maintained in a centralized facility, modeled upon the USDA Cereal Rust Laboratory.

Additional Questions

Many more interesting research issues were raised but there was insufficient time to discuss them in the break-out sessions. Some popular questions were: How do we produce durable blast resistance in various agronomically desirable backgrounds? Is it possible to engineer polygenic resistance? How can we best integrate the use of markers to select for polygenic resistance in a breeding program?

VII Abstracts of Poster Presentations

High Resolution Mapping of Blast-Resistance Genes – a Forestep for Positional Cloning of the Genes

S. Kawasaki[1,2], M. Satoh[1,3], M. Miyamoto[1,4] and I. Ando[5]
[1]National Institute of Agrobiological Resources (NIAR); [2]PRESTO; [3]Science and Technology Agency; [4]Faculty of Biological Sciences, Ibaraki University; [5]National Agriculture Research Center, Kan'non-dai 2-1-2, Tsukuba, Ibaraki 305, Japan

To date 14 resistance genes on 8 loci have been identified in Japan based on their characteristic resistance spectra to the standard test fungus strains. To avoid breakdown by specific fungus strains, cyclic short-term use of these genes in NILs (nearly isogenic lines) will provide an environment friendly measure for crop protection.

From a biological viewpoint, rice blast resistance genes provide an excellent model system for studying the molecular biology of resistance genes: (i) the genome size of rice may be the smallest in major crops ($\sim$420 Mb); (ii) correspondingly physical length for a unit genetic distance is small ($\sim$210 kb cM^{-1}); (iii) rice has the most well-established transformation system in monocotyledons; (iv) abundant RFLP markers developed in NIAR ($>$1000) and Cornell University (590) are available; (v) genetic and physiological information about resistance genes and assay methods are established.

Three *indica* or *javanica*-derived resistance genes (*Pi-z*^{t}, *b*, *ta*^{2}) have been introgressed to *japonica* cultivars, respectively, in 3–7 independent lines. This is especially fitted for NIAR probes developed for *indica* × *japonica* crossings. RFLP markers of more than 1300, and 700 RAPD primers were screened with these NILs, and the loci of the genes were marked as the consensus *indica* type region on *japonica* background, and finally determined by F_2 analysis of $\sim$100 susceptible homozygote individuals. *Pi-z*^{t}, *Pi-b*, and *Pi-ta*^{2} were mapped on the chromosome 6 ($\sim$100 cM), chromosome 2 ($\sim$0 cM), and chromosome 12 ($\sim$70 cM), respectively, on NIAR RFLP map. The map positions of *Pi-z*^{t} and *Pi-ta*^{2} seem to correspond to those of *Pi-2(t)* and *Pi-4(t)*, respectively. In each case of *Pi-b* and *Pi-ta*^{2} genes, one marker was found to completely cosegregate with these genes, corresponding to $<$ 0.5 cM of distance ($<$ 200 kb). One and two markers were found at 0.5 cM and 0.6 cM from respective genes.

Using these markers, positional cloning of the genes is being attempted. A library of rice genome is being constructed from green leaves to avoid genetic mobilization in cultured cells.

Comparison of Rice Lines Derived through Anther Culture and the Pedigree Method in Relation to Blast (*Pyricularia grisea*) Resistance

C.P. Martinez, F. Correa-Victoria, M.C. Amezquita and R.S. Zeigler[1]
Centro Internacional de Agricultura Tropical (CIAT), AA 6713, Cali, Colombia; [1]Present address: Entomology and Plant Pathology Division, International Rice Research Institute, PO Box 933, Manila, The Philippines

Rice blast (*Pyricularia grisea*) is considered the most important disease of rice worldwide. Development of resistant cultivars has been the most effective method to control it; the pedigree method (PM) has been used in this regard but recently, the production of double-haploids through anther culture (AC) has been proposed as an effective, and economic breeding tool. The objectives of this study were to generate rice lines by PM and AC and to compare their efficiency in relation to blast resistance.

Crosses were made between the rice cultivar Fanny (highly susceptible to blast) and 11 cultivars with different degrees of resistance/susceptibility to blast. Using PM, the F_2, F_3, F_4, F_5, and F_6 generations were evaluated and selected for blast resistance through subsequent semesters. Via AC, AC-R_2 (from F_1 plants), and AC-F_2 (from F_2-blast susceptible plants) lines were obtained using a liquid potato-medium for callus induction and an MS medium for plant regeneration. Three sets of lines (F_6, AC-R_2 and AC-F_2)

were planted together and evaluated for blast resistance under upland conditions in a blast prone environment. Rice blast reactions were evaluated on a 0–9 scale and analyzed as a discrete variable following the multinomial distribution. Three response variables, leaf and neck blast, and general reaction to blast (maximum score between the first two variables), were considered. Stratified analysis using the Cochran–Mantel–Haenszel statistic was performed to test the association between methods and blast resistance, and resistance stability across crosses; all data processing and analysis was performed through SAS version 6.07.

The percentage of resistant lines as well as stability of resistance was significantly ($P = 0.01$) greater in PM-derived lines, with a lower mean blast score, suggesting that PM is more efficient than AC in relation to selection for blast resistance. Nevertheless, in Fanny/Ceysvonni, and Fanny/LAC 23 crosses the percentage of stable-resistant lines was greater in the case of AC.

Crosses between susceptible cultivars (Fanny/CICA 4, and Fanny/Colombia 1) yielded resistant lines by either method (PM or AC). This excludes somaclonal variation as a possible mechanism and indicates that recombination of possibly minor genes took place and was fixed through either method. However, the stability of the resistance was greater in PM-derived lines.

Mapping of a Cluster of Resistant Genes to Colombian Rice Blast MGR Lineages

J. Tohme[1], M.V. Montenegro[1], A. Badillo[1], G. Gallego[1] and F. Correa[2]
[1]Biotechnology Research Unit; [2]Rice Program, CIAT, AA 6713, Cali, Colombia

Genetic resistance to *Pyricularia grisea* in Colombia is short-lived, rarely being effective for more than 3 years due to the presumed diversity of the fungus and the difficulties in selecting for rice genotypes with specific gene combinations. Using MGR fingerprinting the genetic structure of the fungus populations has been determined. Fourteen statistically distinct lineages have been identified; each is associated with a specific subset of cultivars and a specific subset of pathotypes. Resistant cultivars with unknown resistance genes to one or several lineages have been identified. We are tagging specific genes and gene combinations that provide resistance against part or entire MGR lineage rather than just against individual races.

Different populations of double haploid lines have been generated from crosses between Fanny, a susceptible cultivar to most MGR lineages in Colombia, and resistant cultivars to several MGR lineages (IRAT 13, Carreon, Colombia 1 and LAC 23). The double haploid lines were screened with blast isolates belonging to six MGR lineages and RFLP markers from the Cornell rice molecular map. RAPD primers were screened using the bulk segregant analysis method. A cluster of resistance genes to isolates from two lineages, SRL-1 and SRL-6, was identified on chromosome 4 in the cross IRAT 13 × Fanny. RAPD fragments of the tightly linked primers have been cloned and sequenced to develop sequence characterized amplified regions (SCARs) to facilitate future PCR based screening. The relationship between the tagged genes and MGR lineages resistance and its implication for durable rice blast breeding programs are discussed.

Population Structure and Virulence Diversity of the Rice Blast Fungus in Colombia

J. Manry[1], F. Correa-Victoria[2], J. Hamer[1] and M. Levy[1]
[1]Department of Biological Sciences, Purdue University, West Lafayette, IN 47907, USA; [2]Rice Program, Centro Internacional de Agricultura Tropical, Cali, Colombia

More than 700 field isolates of the rice blast fungus in Colombia have been characterized for MGR-DNA fingerprint (a profile of 50 60 *Eco*RI RFLPs for the dispersed middle-repetitive sequence, MGR586) and for pathotype (virulence spectrum) on international differential and other rice cultivars. The isolates were collected over a multi-year period (1988–1992) and represented a total sampling of more than 75 pathotypes and more than 90 cultivars growing under a variety of disease pressure regimes. The sampling provided the opportunity to study the structure of rice blast fungus populations under the chronic epidemic conditions of a disease nursery (Santa Rosa), in 'trap nurseries' placed in a newly cultivated region (Altillanura), and in typical commercial plantings. The fingerprints separate the isolates into only 14 distinct haplotype groups, each of which was diagnostic over time and across broad geographic distributions. This pattern implied that each group represented a clonal genetic lineage and that recombination between lineages was not an important feature for generating virulence diversity. Generally, each lineage was compatible with only a limited number of cultivars and most cultivars were susceptible to only one or a few lineages. Cultivar–lineage associations were generally maintained throughout Colombia, although several lineages were detected only in the sampling of the Altillanura region. Each lineage expressed multiple, but generally lineage-specific, pathotypes. The lineage-specific pathotypes were often organized into networks of virulence forms related by single differences in compatibility on the international differential cultivars. These features indicated that lineages have definable and, at least over several years, limited virulence spectra. Pathotype diversity per lineage was highest in the disease nursery epidemic populations and lowest in commercial fields, indicating that resistance breeding under epidemic nursery conditions is efficient but may amplify the rate of evolution of new pathotypes and resistance breakdown. These results show that, in spite of the huge apparent virulence diversity of the rice blast fungus in Colombia, the genetic structure of the pathogen population is relatively simple. Furthermore, since a lineage defines a historical virulence range, it is possible to design relatively simple breeding strategies for developing rice cultivars that are resistant to an entire complement of lineages and their summary pathotype diversity.

DNA Fingerprinting and Pathotype Characterization of *Pyricularia oryzae* Isolates

H. Zhu[1], W. Jiang[2], J. Wang[2], C. Lei[2], Z. Ling[2] and L. Zhu[1]
[1]*Institute of Genetics, Academia Sinica, Beijing 100101, PRC;* [2]*Institute of Crop Breeding and Cultivation, CAAS, Beijing, China*

By selecting from a genomic library of rice blast fungus strain Hokul, we have cloned a repeated sequence named POR6 and verified that it is a genome specific and dispersed repeated sequence, which can be used to fingerprint various kinds of rice blast pathotypes. It was sequenced and the copy number has been determined. The electrophoresis karyotype of Hokul reveals 4 chromosomes ranging from about 2 to 6 Mb in length. POR6 has been proven to be distributed on all of them by Southern analysis. Some blast isolates in our laboratory were proven variable from generation to generation in their pathotypes identified by the traditional method. By using POR6 as a probe, we can detect 3 to 4 bands of variation in DNA fingerprints across generations.

Mitotic Stability of and Low Hybridization Potential between Two DNA Fingerprint Groups of *Magnaporthe grisea*

J.Q. Xia[1], J.C. Correll[1] and D.D. Rhoads[2]
[1]*Department of Plant Pathology;* [2]*Biological Science, University of Arkansas, Fayetteville, AR 72701, USA*

MGR586-DNA fingerprinting was used to examine the mitotic stability and hybridization potential of two DNA-fingerprint groups of *Magnaporthe grisea* under laboratory conditions. Two haplotypes, representing fingerprint groups A and D commonly found on rice in Arkansas, were grown singly or in co-culture on solid medium, liquid medium, or on rice tissue. A total of 355 monoconidial cultures were recovered at various periods of time, and examined for their DNA fingerprints. Overall, the two MGR586 DNA haplotypes remained stable over the 162–171 day study period. Sixteen isolates recovered belonged to one of seven non-parental haplotypes identified; the DNA fingerprints of these haplotypes differed by 1–5% from the parental haplotypes. Of 97 isolates recovered from solid culture, a single non-parental haplotype was identified from the co-culture treatment after 171 days. Of the 200 isolates recovered from liquid culture, 15 were non-parental types and represented seven haplotypes. A single non-parental isolate was recovered from the D group control isolate grown singly after 67 days. The other 14 non-parental isolates were recovered from the co-culture treatment; three were recovered after 38 days, four after 67 days, and seven after 162 days. The appearance of non-parental haplotypes was more likely due to minor mutations during incubation than hybridization between the two groups. All isolates recovered from a lesion from co-inoculated rice leaves were one haplotype. The data indicate that there was competition in both artificial media and host tissue between the two MGR586 haplotypes.

Characterization of a Rice cDNA Exhibiting Elevated Expression During Infection with *Magnaporthe grisea*

J.D. McGee[1], N.J. Talbot[2], J.E. Hamer[2] and T.K. Hodges[1]
[1]Department of Botany and Plant Pathology; [2]Department of Biology, Purdue University, West Lafayette, IN 47907, USA

Rice blast, caused by *Magnaporthe grisea*, is one of the most important diseases of rice throughout the world, and it would be desirable to genetically engineer rice to be resistant to this disease. One approach to this problem would be to isolate a gene from rice containing a promoter that is specifically induced during infection by the pathogenic fungus, and then to use this promoter to drive genes encoding known antifungal enzymes such as chitinase, peroxidase, and β-1,3-glucanase in transgenic plants. Over-production of these enzymes at the site of infection may suppress the development of the fungus. Rice cDNA clones corresponding to genes exhibiting elevated expression during rice blast infection have been isolated by differential screening. One of these cDNA clones, MAG-7, was chosen for characterization.

Analysis of the MAG-7 clone indicated that it contained a 550 bp cDNA insert of an estimated 850 bp full-length mRNA. Southern hybridization of genomic DNA isolated from rice varieties CO39, IR36, and IR54 with the MAG-7 probe showed that there are one to two copies of the gene per haploid genome in each of the three varieties. Northern blot analysis of RNA isolated from CO39 seedling leaves inoculated with *M. grisea* demonstrated that the MAG-7 transcript was first detectable at 24 hours after infection and at higher levels after 36, 48, and 96 hours. Sequence studies revealed that MAG-7 cDNA shares strong homology to two pathogenesis-related proteins from potato (33% identity at the amino acid level). Genomic clones of MAG-7 have been isolated from a rice genomic library and a detailed restriction map of the MAG-7 gene has been constructed. The genomic clone is approximately 4.5 kb in length. Current work is focused on determining the DNA sequence of the MAG-7 gene and its promoter region.

Restoration of Appressorium Deficient Mutants of *Magnaporthe grisea* by Cyclic AMP

D.S. Whitehead, Y.H. Lee and R.A. Dean
Department of Plant Pathology and Physiology, Clemson University, Clemson, SC 29634, USA

Successful infection by *Magnaporthe grisea* depends upon the formation of a dome-shaped, highly melanized infection structure: the appressorium. The differentiation of this unique, specialized cell from the tip of the emerging germ tube is a response to environmental stimuli. The fate of a germ tube tip depends on the hydrophobicity of the surface on which it is growing. On hydrophobic surfaces, we have observed that the germ tube ceases polar extension and the tip swells to form an appressorium. By contrast, on hydrophilic surfaces polar growth continues and the hypha branch to form vegetative mycelium. The exogenous and the endogenous signalling mechanisms involved in surface recognition and the transfer of this information into the cell leading to infection-related morphogenetic events remain to be elucidated. We have recently shown that cyclic AMP (cAMP) is important to the regulation of this crucial developmental process. To further elucidate the mechanisms involved in infection structure formation, several appressorium deficient mutants were generated by UV mutagenesis. Following mutagenesis, surviving conidia underwent one asexual reproductive cycle before being evaluated for appressorium formation. From approximately 800 survivors, four were unable to initiate appressorium formation on hydrophobic surfaces and two formed abnormal infection structures. An additional mutant, although able to conidiate, was unable to germinate. All mutants were essentially normal in other aspects of growth and sporulation. We now report that several of these mutants can be restored to form normal appressoria by cAMP, its analogs and IBMX (a phosphodiesterase inhibitor). These data provide further evidence that cAMP plays a central role in appressorium formation and suggests several appressorium deficient mutants have defects in pathways involved in the synthesis or degradation of cAMP.

Studies on Pathogenicity Factors of the Blast Fungus in the Host Species Specificity

S. Mayama and H. Kato
Department of Plant Protection, Kobe University, Kobe 657, Japan

The blast fungus, *Magnaporthe grisea*, parasitizes not only rice but also many other grasses, sedges and gingeberous plants. Molecular studies demonstrated that they are phylogenetically related with a common ancestry; however, the host range of the fungus is generally restricted in each cultivated plant species. The specificity in parasitism of the blast fungus in rice cultivar and other plant species is an important phenomenon for seeking strategies of disease control. It is interesting to study the involvement of biologically-active compounds as pathogenicity factor, such as host-specific toxins, in the host species specificity. The infectivity of the blast fungus isolated from *Digitaria sanguinalis*, *Setaria italica*, *Eleusine coracana*, *Panicum miliaceum*, *Triticum aestivum* and *Oryza sativa* was examined with over 24 species in seven families. Microscopical study of infection sites was done in the pathogenic and non-pathogenic blast-infected plant leaves. The types of cellular response were clarified and the restricted host range was confirmed. The germination fluids of the blast fungus isolated from the host plants were obtained and fractionated into the ethylacetate-soluble and aqueous solutions. The fractions of germination fluids were mixed with conidia of the blast fungus non-pathogenic to the plants to test whether the fractions show any infection-inducing activity. The ethylacetate fractions of the blast fungus isolates from *S. italica*, *P. miliaceum* and *T. aestivum* showed infection-inducing activity to the blast fungus nonpathogenic to their host plants. The infection-inducing activity to nonpathogenic blast fungus in *S. italica* was clearly detected with the fractions of germination fluid of the isolates from *S. italica* in the host plant, but not in the non-host plant. Also, the germination fluids of the high virulent type of isolates from *D. sanguinalis* could promote the growth of the weak-virulent isolate. The results suggest the involvement of pathogenicity expressing compounds in germination fluids of the blast fungus.

Genome Mapping of *Magnaporthe grisea*

N. Hayashi and H. Naito
National Agriculture Research Center, Tsukuba, Ibaraki 305, Japan

A restriction fragment length polymorphism (RFLP) map has been constructed for the nuclear genome of the plant pathogenic ascomycete *Magnaporthe grisea* (anamorph = *Pyricularia oryzae* and *Pyricularia grisea*). The segregation of 61 RFLP markers, five host-specificity gene loci and the mating type locus was analyzed among 82 randomly selected colonies from a cross between parental strains CHNOS 37-1-1 and CHNOS 32-2-4, which are field isolates and pathogenic on rice (126.0 and 017.5 in the race code based on Japanese differential rice varieties, respectively). The present map consists of thirteen linkage groups. These linkages totalled 372 centimorgans (cM). One RFLP marker appears to be closely linked to the putative avirulence gene *Avr1-TA* which functions specifically to prevent strains of rice blast pathogen from infecting rice cultivar Tsuyuake. Whole chromosomes from the parental fungus strains were separated using contour-clamped homogeneous electric field (CHEF) electrophoresis. Six chromosomes were identified; one of the bands appears to be a doublet band. The estimated size of these bands, based on migration rates relative to those of chromosomal DNA of *Schizosaccharomyces pombe*, are from 1.8 to more than 10 megabases (Mb). The total genome size was estimated to be roughly from 34 to 38 Mb.

Magnaporthe grisea Pathogenicity Genes Cloned by Insertional Inactivation Mutagenesis

J. Sweigard, A. Walter and B. Valent
DuPont Science and Engineering Laboratories, Wilmington, DE 19880-0402, USA

We have developed methods for isolation of non-pathogenic mutants of the rice blast fungus using integration of transforming DNA to tag genes involved in pathogenicity. Seven non-pathogenic mutants were obtained in a screen of over 1600 hygromycin-resistant transformants. The pathogenicity defect in five of these mutants segregated as a single gene linked to the hygromycin resistance marker. Plasmid sequences along with flanking fungal genomic sequences have been recovered from four of the transformants. These rescued plasmids have been used as hybridization probes to identify homologous cosmids from a library of wild-type pathogen genomic DNA. In three cases so far, corresponding cosmids transformed into the nonpathogenic mutants restore these mutants to full pathogenicity. The identified pathogenicity genes are now being subcloned.

Microconidium Formation in *Magnaporthe grisea*

H. Kato, S. Mayama, R. Sekine and A.S. Urashima
Faculty of Agriculture, Kobe University, Kobe 657, Japan

Magnaporthe grisea produces pyriform conidium as an anamorph and spindle-shaped ascospore as a teleomorph. An additional conidium state was found.

The nucleus staining with Giemsa solution was tried in the area of mycelium confluence of both mating types to detect the process of nucleus migration. Some isolates possessing mating ability in the strains of finger millet, rice and wheat produced phialides and microconidia. Microconidia are hyaline, cylindrical, rounded at the end first, then lunate with thin cell wall, no septum, 5–8 (average 6) μm long, 0.5–0.8 (0.7) μm wide, and with one nucleus. The first rod-shaped microconidium is formed blastically from the conidiogenous cell apex within the collarette. At least two successive rod-shaped microconidia are formed in blastic from alternate side of the conidiogenous cell. Each conidium continues the growth and secedes. Subsequent microconidia are produced from the same locus. Accumulated microconidia curved form a globose mass with slime in most at the tip of a phialide. The shape of microconidia changes to crescent.

The phialide is darkly pigmented vase-shaped, spherical to obclavate, with a tapered apex, terminal collarette, basal septum, thick-walled, 5.9–12.5 (average 8.9) μm long, 3.3–7.2 (4.5) μm thick in the broadest part, arising solitary or sympodially from aerial hyphae.

Cytological Analyses of Leaf Lesions Induced by *Pyricularia oryzae* and Tenuazonic Acid

S. Ranomenjanahary, M.H. Lebrun, C. Duchartre and J. Orcival
Laboratory of Cryptogamy, University of Paris XI, 91405 Orsay, France

Leaf lesions induced by *Pyricularia oryzae* and by tenuazonic acid (TA) on one resistant and one susceptible rice varieties were examined cytologically. TA is a phytotoxic metabolite produced by *P. oryzae* during its infection of rice leaves. It was observed that the mode of spreading of brown margin lesions caused by both the pathogen and TA was similar either on compatible rice varieties (slow-appearing light browning, after 96 h if it occurs) or on some varieties with high level of partial resistance (rapid and intense browning, in 48 h).

Leaf sections in lesions of the susceptible variety Maratelli and the resistant variety IRAT 13 were examined four days after inoculation with a compatible isolate CI41 or with TA. It was observed that the hyphae appeared in epidermal or in parenchyma cells, in intercellular spaces in the mesophyll and in vascular tissues. In the resistant variety hyphae were rare and located within the brown area which was formed by a deposit of pigments in the cell walls and the cells themselves constituting an impervious barrier which prevented the diffusion of hydrophobic molecules and the progression of hyphae. Also, it was shown that most of peroxidase activity induced by *P. oryzae* and TA, mainly parietal, was located in the brown area. It was greater in the resistant variety than in the susceptible one and more intense when induced by *P. oryzae*.

Examination of ultrathin sections of infected leaves revealed that leaf cells of Maratelli did not present any structural change except for the presence of osmophilic inclusions in a few cells. In IRAT 13 inoculated with TA some damaged cells were observed due to bursting of chloroplasts and swelling of thylakoids.

Intracellular green lipid globules were present in healthy tissues but did not seem to exist in younger leaves of Maratelli. These lipid globules were abundant, brownish and of a larger size in vacuole of the brown cells surrounding the lesions caused by *P. oryzae* and TA. These globules were fewer in number and of lighter color in Maratelli. The presence of these globular structures in the cytoplasm and the vacuole suggested that they were first produced in the cytoplasm and then accumulated in the vacuole followed by lysis of the cytoplasm membrane. All these phenomena might participate in the defense reaction of the infected rice leaves. The brown pigmentation surrounding the lesions might contribute to check *P. oryzae* infection of the leaf and thus determine lesion size depending on the timing of appearance of browning.

Studies on the Perfect Stage of *Pyricularia oryzae* in China

Y. Shen[1], P.L. Zhu[1], X.P. Yuan[1], M.Z. Jin[2], R.Y. Chai[2] and Q.S. Zhang[2]
[1]*Department of Plant Protection, China National Rice Research Institute, Hangzhou 310006, China;* [2]*Zhejiang Academy of Agricultural Sciences, Hangzhou 310021, China*

The formative ability and mating type of 1235 isolates of *Magnaporthe grisea* (*Pyricularia oryzae*) from 24 provinces and cities (autonomous regions) in China have been tested using four standard isolates from IRRI. Most of the isolates can produce perithecia on oatmeal agar. The formative fertility of the perfect stage of *P. oryzae* differed greatly among isolates. The amount of perithecium production was related to the mating type of standard isolates. Isolates from rice and crabgrass belonged to two mating types, designed as A and a. All of them belonged to type A except four isolates from Yunnan and Zhejiang provinces. Several hermaphroditic isolates have also been found in China. The electrophoretic profiles of soluble protein and lactate dehydrogenase (*Ldh*) have been studied, and no polymorphism in *Ldh* isozyme was found. The study results will lay the foundation for genetics of *M. grisea* and the disease ecology research.

Genetic Organization of Rice Blast Fungus in China

P.L Zhu[1], Y. Shen[1], X.P. Yuan[1], M. Levy[2], J. Manry[2] and J. Hamer[2]
[1]*Department of Plant Protection, China National Rice Research Institute, Hangzhou 310006, China;* [2]*Department of Biological Sciences, Purdue University, West Lafayette, IN 47907, USA*

From April 1991 to August 1992, about 140 field isolates of *Pyricularia grisea* were obtained from rice blast nurseries and popular cultivars in different locations in each province. Sixteen of these came from various host species. The MGR-DNA fingerprints of *c.* 100 isolates have been studied with an MGR586 probe/EcoRI digest combination. The preliminary results are as follows:

The rice blast fungus in China is characterized by a high level of genomic RFLP polymorphism. MGR-DNA fingerprints demonstrate that, as elsewhere, rice blast populations are composed of a small number of easily identified clonal lineages, each with defined pathotype diversity. Although more than 30 pathotypes are present in the samples, only 15 distinct lineages were identified; lineage diversity may increase as some yet unassigned isolates are characterized. Isolates within a lineage express more than 90% fingerprint similarity while different lineages are distinguished by having less than 80% similarity. Each lineage contains one to many pathotypes. The same pathotype may be expressed in more than one lineage. The results parallel those found in the USA and Colombia. Thus far, the two provinces appear to have moderately different populations of the pathogen.

We also found that one of the lineages common to both provinces, PRC-VIII, is the same (more than 95% fingerprint similarity) as one found in Colombia, lineage SRL-6. SRL-6 is the dominant lineage in Colombia commercial fields. Pathogenicity tests, using both Chinese and international differentials, are underway to further assess the comparative features of these widely separated clones which share a recent common ancestor. We expect to find more cases of international 'migration' of lineages. This information will be extremely valuable for targeting useful versus non-durably resistant germplasm in different rice growing regions.

The MGR-DNA fingerprints of 16 weed pathogen isolates gave results very similar to earlier reports. Very few copies of MGR586-hybridizing sequences exist in the genomic DNA of grass pathogens, making diagnosis of host range straightforward. We did find two grass blast isolates (host species not identified) from Hunan province with a rice pathogen MGR-DNA profile, but no isolates from rice with grass pathogen profiles. Consequently, in the southern rice growing region of China it appears that grass pathogens are not likely to be infection sources of rice blast disease, while some rice pathogens can infect both cultivated and feral hosts.

We will continue to determine the pathotype variation within lineages in China and the lineage distribution on specific Chinese commercial cultivars in the near future. We expect that this information will be the basis from which new, durable resistance breeding strategies and more effective cultivar deployment schemes will emerge.

An Attempt to Construct a Physical Map of *Magnaporthe grisea*

Yi-Jung Chen[1] and Hsin-Kan Wu[2]
[1]*Department of Agronomy, National Taiwan University;* [2]*Institute of Botany, Academia Sinica, Taipei, Taiwan*

Rice blast fungus, *Magnaporthe grisea*, is one of the filamentous, heterothellic ascomycetes. However, the construction of a physical map for this fungus is accessible only recently after the successful isolation of GUY11 that was isolated by Notteghem in France. In this study, hermaphroditic and fertile GUY11 was crossed to four local races (II-1, IH-1, IF-4 and IG-2) and one isolate MAT1 that is pathogenic to grass. Young assci that were dissected from fertile perithecia were stained with latic-orcein for counting chromosomes at meiosis. We found that the chromosome number in *M. grisea* is nine. When the intact chromosome DNAs of the isolates were separated in pulsed field gel electrophoresis (PFGE) apparatus (CHEFII), nine bands could be resolved at most. Chromosomal polymorphisms for both number and length were observed among isolates. It revealed that chromosomes among isolates may have structural rearrangement. To find chromosome-specific segments, RAPD markers were labeled with ^{32}P and used as probes in Southern blot analysis of chromosomal DNA bands of the isolates separated by PFGE. Some individual RAPD markers showed positive hybridization to a single band of certain size in all the isolates used. It showed that those markers are chromosome specific. These results suggest the possibility of constructing a physical map of *M. grisea* in the near future.

Molecular Analysis of Spontaneous Mutations in *Magnaporthe grisea*

M.-H. Lebrun[1], F. Chumley[2] and B. Valent[2]
[1]*Institute de Genetique et Microbiologie, Orsay, France;* [2]*E.I. Dupont de Nemours, USA*

We have used the nitrate reductase gene (*nial*) as a tool to survey spontaneous deletion and insertion events in *M. grisea*. We obtained spontaneous mutants in *nial* by selecting for resistance to chlorate. An internal fragment of *nial* was obtained by PCR using degenerate oligonucleotides corresponding to domains conserved among fungal nitrate reductases. The wild-type *nial* gene was cloned by hybridization of cosmid libraries with this PCR probe. Spontaneous mutants from 5 isolates were analyzed by hybridization with *nial*. We identified only one insertion (+10 kb) and two deletions (−1 kb and −10 kb) among 120 mutations. We also obtained spontaneous mutations in the *Aspergillus nidulans* gene for nitrate reductase (*niaD*), introduced into *M. grisea* by transformation. Almost all 40 spontaneous mutants, obtained from two independent single copy transformants, had undergone a deletion of all the plasmid at high frequency (1 or 0.1%). We also obtained one mutant with a 5.6 kb insertion in the promoter regions of *niaD*. The element inserted in *niaD* belongs to a family of *M. grisea* LTR-retrotransposons, *Fos1* and MAGGY. Sequences of LTR and reverse transcriptase domains differed from those of MAGGY by only a few substitutions. The retroelement target site (ATATT) was duplicated. All revertants from this mutant had one copy of the LTR left at the point of insertion. All these characteristics showed that this element belongs to a family of active LTR retrotransposons. This mutation is the first example of insertional mutagenesis due to a transponson in *M. grisea*.

Phylogenetic Relationships among *Magnaporthe grisea* Isolates from Different Host Plants

M.P. Capy[1], M.-H. Lebrun[1], J.L. Notteghem[2] and Y. Brygoo[1]
[1]*Institute de Genetique et Microbiologie, Orsay;* [2]*CIRAD-CA, Montpellier, France*

Genetic diversity of *Magnaporthe grisea* isolates from different host plants was evaluated with ribosomal DNA polymorphisms (RFLP, ITS sequences). Seventy-five isolates from all possible geographic origins for each host plant were grouped in only seven distinct rDNA haplotypes (RFLP). We sequenced ITS (350 bp long) from a few isolates from each rDNA haplotype. The ITS sequence phylogenetic tree showed high relatedness between isolates from rice, *Eleusine* spp. and wheat (rDNA haplotypes 1, 2, and 3: 0.3–1% nucleotide divergence). Such differences are expected within a polymorphic species. Other rDNA RFLP haplotypes (4, 5, and 6) were monophyletic and as divergent (7 and 9% nucleotide divergence) as related species in other fungal genera (*Fusarium*, *Aspergillus* or *Colletotrichum*). Therefore, these haplotypes should be considered as other *Magnaporthe* species. Using RFLP from random single copy DNA (scDNA), we obtained the same grouping and phylogenetic tree as with rDNA; scDNA allowed subdivisions of each rDNA haplotypes in a few related groups, composed of isolates from only one host plant, whatever their geographic origin. When comparing all isolates from one host plant, we found two situations. Isolates from rice, wheat and *Digitaria* spp., were monophyletic, but those from *Eleusine* spp. and *Pennisetum* spp. come from two distinct phylae. This biphyletic origin might be due to independent acquisition of pathogenicity for the same host during the evolution of *Magnaporthe* species or might be a host specificity shared with a common ancestor. Rice isolates are monophyletic and genetically distinct from all others. Such an accumulation of specific polymorphisms could be explained by a strong genetic drift during specialization on a particular host followed by a strong genetic isolation of these isolates from other *M. grisea* populations.

Genetic and Physical Mapping of a Gene Affecting Virulence in *Magnaporthe grisea*

Mark L. Farman[1] and Sally A. Leong[1,2]
[1]*Department of Plant Pathology;* [2]*USDA-ARS Plant Disease Research Unit, 1630 Linden Drive, Madison, WI 53706, USA*

We are interested in understanding the molecular basis of host specificity. Towards this end, we are attempting to clone a gene affecting the ability of *Magnaporthe grisea* to grow on rice cultivar CO39. Genetic analysis of virulence in progeny of a cross between GUY11 and 2539 indicated that a single locus was responsible for the expression of avirulence/virulence on CO39. This locus (avrCO39) was mapped to one end of chromosome 1 in a previously constructed RFLP map of *M. grisea*.

Unfortunately, chromosome 1 was poorly represented by RFLP markers. This chromosome was excised from a CHEF gel and used as a hybridization probe to identify cosmids containing DNA from this chromosome. This resulted in the identification of eleven more markers which map to chromosome 1.

In order to determine the physical distance between markers flanking avrCO39, we are employing novel chromosome cleavage techniques. By making a single, site-specific cleavage of the genome at the nearest marker, CH5-120H, we liberated a 1.2 Mb restriction fragment. Hybridization of other surrounding markers to this fragment indicated that it contains the avrCO39 gene. If we assume that this fragment is telomeric and recombination is evenly distributed along this region of chromosome 1, we calculate that CH5-120H should lie approximately 290 kb from avrCO39. A chromosome walk from 1.2H so far has proceeded 110 kb towards avrCO39. A recombination point was identified after approximately 30 kb.

Rice Blast Evaluation in Populations Developed Through Anther Culture and Pedigree Selection

I. Perez, Z. Lentini, E.P. Guimaraes and F. Correa-Victoria
Centro Internacional de Agricultura Tropical (CIAT), AA 6713, Cali, Colombia

Rice blast, caused by the fungus *Pyricularia grisea*, is the most important disease in most rice growing areas. Breeders have developed resistant lines to control the disease, but the resistance breaks down shortly after release. This lack of durability is believed to be related, in part, to difficulties in evaluating and selecting resistant materials in segregating populations. Anther culture (AC) quickly generates homozygous genotypes from segregating lines, which may facilitate the breeder to assess the potential of a line. This study compares leaf (Bl) and panicle (NBl) blast in populations developed through AC and Pedigree (P), to determine if AC can be used as a tool to develop resistant materials. Eleven double crosses were processed through AC and the same combinations were advanced to the S_2 generation, using the P method. Ten R_2 and ten S_2 lines from each cross were randomly chosen. The 220 lines were planted under rainfed upland conditions, following a split-plot design with three replications, in 1992. The site used was Villavicencio, Colombia, a location of high disease pressure. The Bl and NBl were measured three times in the leaf and two times in the panicle, using the 'Standard Evaluation System for Rice'. Only the highest score was considered for analysis. Ten plants per line for P and five for AC were evaluated to represent the variability within line. The data were analyzed by using the line mean as base. The Bl average was the same for both AC and P lines (2.82). The AC materials presented a wider range of disease reactions (1.14–5.47) compared with P lines (1.85–4.74). These results suggest that the development of lines through AC does not skew the population toward either resistant or susceptible reactions, but provides a better chance for selecting extremes. The heritability for AC (narrow sense) was 0.54 and for P (broad sense) was 0.67. For NBl, a significant difference existed between the two means. The AC lines were more susceptible (5.53) than the P materials (4.57). The range of genetic variability for disease reactions was wider for AC (2.90–7.14) than for P (3.63–6.73). These findings suggest that the AC produced extremes not found in the P lines. Therefore through selection, more resistant lines could be obtained through AC than through P. AC showed 0.66 for narrow sense heritability, whereas P had 0.58 for the broad sense. These results indicate that AC is helpful in selecting high levels of Bl and NBl resistance. The heritability values suggest that genetic progress could be made without major effort.

Studies on Rice Cultivar Resistant Spectrum to *Magnaporthe grisea* and its Application

Peicun Xiao[1] and Yunliang Peng[2]
[1]*Neijiang Institute of Agricultural Sciences, Nei Jiang, Sichuan Province;* [2]*Sichuan Academy of Agricultural Sciences, Chengdu City, 64100, China*

'Resistance spectrum' analysis of rice cultivars to *M. grisea* was conducted on the basis of 'gene to gene' theory and intraorganismal genetics. Physiologic races of *M. grisea* isolates, resistance frequencies and resistant types of rice cultivars were obtained from this analysis. The results of a 15-year research indicated that rice cultivar resistance frequencies were highly negatively correlated to the disease epidemic areas and disease loss and so the rice cultivar change could be predicted by resistance frequencies. No relationships between the predominant physiologic race frequencies and prevailing rice cultivar resistance frequencies were found. Thirty-five resistant cultivar types different from all of 13 Japanese monogene varieties and seven Chinese physiologic race-discriminating host varieties were obtained from the analysis of the reaction type data of 50 cultivars to 331 *M. grisea* isolates from different rice areas of Sichuan province, which indicated the complicated genetic background of the resistance. The cultivars with high resistance frequencies were selected to be the resistance endowers of the multiline variety breeding programs in Sichuan.

Screening of Resistance Resources and Gene Analysis of Resistance for Blast Disease

Zhongzhuan Ling[1], Jiulin Wang[1], Cailin Lei[1] and Wanru Jiang[2]
[1]Rice Breeding Lab; [2]Department of Genetics, Institute of Crop Breeding and Cultivation, CAAS, Beijing, W1 100081, China

Zhongzhua No. 8 and Zhonghua No. 9 were developed by using another culture method. These two varieties have resistance gene *Pi-z*^{t} which derived from Japanese variety Toride 2. Before releasing Zhongzhua No. 8 and Zhonghua No. 9, *Pi-z*^{t} gene was not infected by any pathogenic races in rice-cropping regions of north China. After releasing these two varieties, many new pathogenic races that can invade *Pi-z*^{t} gene occurred in rice-cropping regions of north China. These new pathogenic races invaded a lot of cultivated varieties and many breeding materials which include developing lines and resistance donors.

In order to provide new resistance resources with broader resistance spectrums for rice breeding for resistance to blast disease, many local *japonica* varieties from Yunnan province in China were identified for their blast resistance by using seven Japanese representative strains and 15 Chinese representative strains and new strains with broad virulence spectrum. Through a long-term test, 80 highly resistant varieties were selected from 596 tested local *japonica* varieties.

Twelve Japanese differential varieties, which have a single resistance gene, respectively, were introduced to China in the 1980s. Some of the above highly resistant varieties were analyzed for their resistance genes by seven Japanese representative strains and 12 differential varieties. Analyzed resistant varieties include Hongliandaogu-1, Mengwanggu-1, Dabainuo, Haobian-1, Zhamianni-1 and Buzha-1. The results indicated that Haobian-1 and Zhamianni-1 have one dominant resistance gene and Mengwanggu-1, Hongliandaogu-1 and Dabainuo-1 have two dominant resistance genes, respectively. An allelism test was conducted between these genes and name genes carried in Japanese differential varieties. On the basis of results of the allelism test, we can conclude that: (i) analyzed local *japonica* varieties from Yunnan province have new resistance genes different from named genes in Japanese differential varieties; (ii) the resistance spectrum of new genes in these resistant varieties is broader than that of named genes. Therefore, these new resistance genes have more highly available value in rice breeding for resistance to blast disease in north China.

Recurrent Selection for Rice Blast (*Pyricularia grisea*) Resistance

E.P. Guimaraes and F. Correa-Victoria
Centro Internacional de Agricultura Tropical (CIAT), AA 6713, Cali, Colombia

Blast (*Pyricularia grisea*) is the most widespread disease affecting rice. Breeders have tried several approaches in developing varieties with stable resistance, but without success. The literature reports that many genes with major and minor effects control disease reaction. An untested, but possibly useful, method in dealing with the problem may therefore be recurrent selection. This report shows data obtained after one cycle of selection for leaf (Bl) and panicle blast (NBl) in a population that is being developed through recurrent selection. Thirty parents, with different levels of resistance and/or resistant to different pathotypes of the blast fungus were chosen to create a basic population (GC-91), in 1990. In the first intermating, each parent was combined with three others; in the second, crosses were made between F_1s (one to five others). The generation S_0 was planted in an area free of blast, for seed increase and selection for highly heritable traits. The S_1 generation, formed for approximately 25 plants per cross, was conducted under high blast pressure in Villavicencio, Colombia. Individual plant selection was practiced to generate the S_2 lines. Ten plants were scored and averaged, to represent a reaction of a S_2 line. Three populations were derived from the GC-91, using leaf and panicle blast as criteria for selection. Population 1 was composed of lines with Bl and NBl scores of $\leqslant 3$; population 2 had plants with Bl and NBl scores equal to 4 or 5; and population 3 included Bl and NBl scores of $\leqslant 5$. Not necessarily all plants within a selected line met the criteria. To begin the next cycle, the 58 (C1P1), 51 (C1P2), and 58 (C1P3) best lines, respectively, to the three populations, were selected as parents for crossing. Comparing the results with the GC-91 population (Bl = 2.43 and NBl = 3.02), the following were observed: (a) in population 1, the selected C1P1 parents averaged 0.62 for Bl and 1.28 for NBl, suggesting that rapid progress can be made for resistance to the disease. This may be a result of selection of genes with major effects; (b) in population 2, the mean of the 51 C1P2 lines showed the highest scores for Bl (3.68) and NBl (2.24). This was expected because of the selection criterion used. Perhaps the genes with major effects were defeated and the lines have genes with minor effects, which can be accumulated through recurrent selection; and (c) for population 3, the aim was to accumulate genes with major and minor effects. After one cycle of selection, the average data for the 58 C1P3 parents (Bl = 1.48 and NBl = 1.49) suggested that progress was made.

Resistance to Rice Blast Disease in Egypt

M.R. Sehly and M.S. Balal
Rice Research and Training Center, Sakha, Kafr El-Sheikh 33717, Egypt

Rice blast disease caused by *Pyricularia oryzae* is the most important rice disease in Egypt. Host plant resistance is considered as the primary means of controlling blast. However, *P. oryzae* is known to be highly variable. Several tests were used for the screening of different groups of breeding material for blast resistance. Promising lines, observation lines and IRBN lines were evaluated under blast nursery test conditions in the period of 1988–1992. More than 60% of the material tested was resistant to or moderately resistant to the blast races prevalent in Egypt with scores of 1–2 and 3 respectively. Multilocation test (trap varieties) were used to evaluate the important promising lines only with commercial cvs and some Japanese differential varieties under natural infection conditions in farmers' fields. IR28, Giza 181, GZ 1368 S-5-4 and ECIA 31-104 were resistant at all locations distributed over the six rice growing governorates in Egypt during the period of 1988–1992. Greenhouse tests were used to evaluate the promising lines as well as international and Japanese differential varieties, against single isolates of *P. oryzae* collected from different locations of the six rice growing governorates. The most common race groups identified in the period of 1988–1992 were IC, ID, IG, IH and II, in addition to IE5 and IF1 in the 1992 season. Common races in the IC race group were IC13, IC17, IC25, IC29 and IC31, while ID13 and ID15 were common in the ID race group. The mean percentage level of resistance of the single genes in the Japanese differential varieties against single isolates of *P. oryzae* as a mean of 5 years was evaluated. The most effective genes were *Pi-b*, *Pi-k*, $Pi\text{-}z^t$, *Pi-z* and $Pi\text{-}k^m$ in the varieties BL-1, kanto 51, Toride 1, Fukunishiki and Tsuyake respectively.

Silicon and Benomyl Reduce Blast Incidence and Increase Rice Yields

L.E. Datnoff[1] and G.H. Snyder[2]
[1]*Department of Plant Pathology;* [2]*Soil and Water Science, University of Florida-IFAS, Everglades Research and Education Center, Belle Glade, FL 33430, USA*

Rice grown in Florida is rotated with sugarcane and vegetables on Histosols with organic matter contents in excess of 80%. Silicon (Si) concentrations are very low in these soils. Recent research has demonstrated that blast caused by *Pyricularia grisea* can be reduced and yields increased by amending with a plant-available Si source. Based on this information, it was hypothesized that Si will enhance host plant resistance and control blast to the same degree as a fungicide while increasing yields. Experiments were conducted at two locations. Treatments consisted of plots nonamended and amended with calcium silicate slag (Si source) at 10 Mg ha^{-1}, and nonsprayed and sprayed with benomyl at 1.68 kg product ha^{-1}. Fungicide sprays were applied with a hand-held boom at panicle differentiation, boot, heading and heading + 14 days. Blast incidence was determined by examining 25 heads of rice per experimental unit and calculating the percentage of those heads exhibiting symptoms. Rough rice yields were calculated as threshed grain weight adjusted to 12% moisture. Disease incidence in the controls averaged 73% whereas Si-amended, benomyl and the combination of both treatments averaged 36%, 27% and 13%, respectively, across both locations. Average rough rice yields increased over the controls 26%, 16% and 62% for Si-amended, benomyl and the combination of both treatments, respectively, across both locations. These data suggest that Si may reduce or eliminate fungicide inputs while providing a sustainable method for managing rice blast and improving yields.

Identification of Stable Resistance to *Pyricularia oryzae* in Upland Rice Cultivars

A.S. Prabhu[1], O.P. Morais[2], M.C. Fillippi[1], N.S. Souza[2], L.G. Barros[2] and D.M.P. Azevedo[3]
[1]*CNPAF/EMBRAPA, C.P. 179, 74001-970 Goiania, GO;* [2]*EMPAER/Mato Groso, C.P. 225, 78070-000 Cuiba, Mato Groso;* [3]*CNPAF, C.P. 406, 78900-000, Porto Velho, Rondonia*

The stability of resistance to leaf blast was studied in field experiments conducted with six upland rice cultivars across 17 different environments. Three locations, three years and two levels of fertilizer (N_{10} and N_{60}) were used to obtain different environments. The area-under-disease-progress-curve (AUDPC) was used to measure leaf blast severity over time. Log transformation was chosen to reduce the heterogeneity of error mean square. The stability analysis was made using transformed values of AUDPC. The stability of different cultivars was analyzed using the following bisegmented model of linear regression: $Y_{ij} = \beta_{oi} + \beta_{1i}x_j + \beta_{2i}T(x_j)$ where $T(x_j) = 0$, if $x_j \leqslant 0$; $T(x_j) = x_j - x_p$ when $x_j > 0$; $\bar{x}_p$ = mean of $\bar{x}_p$ positives. This multiple regression model with two independent variables detect the differential performance of cultivars in favorable and unfavorable environments. The ideal cultivar is the one with low values of AUDPC and stable performance in both favorable and unfavorable environments. A cultivar is considered stable when unit regression coefficient $\beta_i = 0$, $\beta_1 + \beta_2 = 0$, deviation from the regression as small as possible ($s^2d = 0$) and the AUDPC values are low and close to 0. Cultivars with values $\beta_1 = 0$ under unfavorable environments and with values $\beta_1 + \beta_2 \geqslant 1$ under favorable environments are suited only under conditions unfavorable to leaf blast development. Among the cultivars tested, the improved ones Centro America and Guarani exhibited relatively high stability both under favorable and unfavorable environments whereas the cv. Rio Paranaiba showed stability only under unfavorable conditions. Even though the cv. Cuiabana had low mean values of AUDPC, it exhibited low stability in both environments. The widely cultivated commercial cvs. IAC 47 and IAC 165, utilized as controls for comparing with the improved rice cultivars, were inferior in performance both in relation to disease severity and stability. The analysis of physiologic races showed that the race IB-9 is predominant and virulent in all test cultivars and locations.

Research of Durable Resistance to Rice Blast Disease in *Oryza*

Peng Shaoqiu, Liu Erming and Huan Feiyan
Plant Protection Institute, Human Academy of Agricultural Science Changsha, 410125, China

From 1972 to 1992, 35,000 materials, including upland, very old traditional varieties, wild *Oryza* and cultivars had been collected. Five varieties with durable resistance, Mowanggu, Xiangzi 3150, Gumei No. 2, Tetep and *Oryza minuta*, were screened by resistant blast differentiation from latitude disease nurseries of different ecological types in China, and analysis of several resistant components and in comparison them with susceptible varieties, Xiangaizao No. 7 and B40. The study showed that their scales of leaf-panicle blast were 0–4, grades of spore layer scales of lesions were 1.667–19.667, yielding spore numbers were 0–4.83 spores on average per field of vision at 400 times under microscope, percentages of leaf lesion expansion per day were 0.027–0.040, and their frequencies of resistance to 122 strains from Chinese different ecological rice zones and Japan were 95.88–100%.

Mowanggu, Xiangzi 3150 and Tetep are old local traditional varieties, Gumei No. 2 has been cultivated for about ten years in south-west China. *Oryza minuta* is a wild *Oryza* from IRRI. The research puts forward four points: (i) the more the variety resists numbers of strains from different ecological disease zones, the more stable and durable its resistance is; (ii) some varieties' resistance with hypersensitivity, such as Xiangzi 3150, Tetep, etc. are a type of durable blast resistance; some varieties with moderate resistance, for example *Oryza minuta*, may be another type of durable blast resistance; (iii) durable blast resistance of a variety is specific in time and space.

SSF-126: A Novel Antifungal Alkoxyiminoacetamide Derivative for Control of Rice Blast

M. Masuko, N. Niikawa, T. Kataoka, M. Ichinari, T. Takenaka, Y. Hayase, Y. Hayashi and R. Takeda
Aburahi Laboratories, Shionogi & Co. Ltd., Shiga, 520-34, Japan

New alkoxyiminoacetamide derivatives were synthesized and their biological characteristics as a new class of fungicide were assessed.

Since Becker *et al.* first reported the fungicidal properties of strobilurins and other natural products, structural modification of those leads has progressed aggressively, and β-methoxyacrylates have been introduced as promising candidates for a new class of fungicide which has a broad spectrum and a new mode of action. Our alkoxyiminoacetamides were derived from carbamoylisoxazole derivatives which were previously synthesized and assessed as a new blasticide, by cleavaging the isoxazole ring, and optimized. The alkoxyiminoacetamide derivatives exhibited similar biological properties to β-methoxyacrylates. From the structure–activity relationship, it was demonstrated that when oxyimino moiety is methyl ether (R1 = Me), and acetamide moiety is mono-methyl amine (either R2 or R3 is H and the rest is Me), the fungicidal activity was the best. The phenoxy group should be attached to the ortho-position of the phenyl ring, and the other substituents on the phenyl ring (R4) of phenylacetamide moiety besides the ortho-phenoxy group reduced the fungicidal activity. Between two geometrical isomers, the (*E*)-form was stronger than the (*Z*)-form.

Consequently, (*E*)-2-methixyimino-*N*-methyl-1-(2-phenoxyphenyl)acetamide (SSF-126) exhibited excellent fungicidal activity. Although SSF-126 is active against a wide range of diseases, it shows excellent activity against rice blast by leaf spray as well as submerged application into paddy water. On the other hand, the compound showed weak fungicidal activity *in vitro*. It did not inhibit spore germination or mycelial growth of *Pyricularia oryzae* at less than 200 p.p.m. In pot tests, SSF-126 showed good preventive and curative properties by foliar application; however, the controlling activity of SSF-126 against rice blast was excellent when it was applied into the paddy water. SSF-126 also exhibited good control against rice sheath blight and *Helminthosporium* leaf spot as well. Under field conditions, SSF-126 showed excellent control against ear blast as well as leaf blast by one-shot application into the paddy water by granular formulation just before or even after the outbreak of symptom on the leaf at the dose rate of 1800 g a.i. ha^{-1}.

Effect of Two Octadecadienoic Acids on Rice Resistance to Blast at Seedling Stage

F.-M. Song[1], X.-C. Ge[1], Z. Zheng[1], W.-L. Wu[2] and Y.-L. Wu[2]
[1]Department of Plant Protection, Zhejiang Agricultural University, Hangzhou 310029; [2]Shanghai Institute of Organic Chemistry, Academia Sinica, Shanghai 200032, China

Two kinds of oxygenated unsaturated fatty acids, i.e. 11, 12, 13-trihydroxy-9*Z*, 15*Z*-octadecadienoic acid and 11-hydroxy-12,13-epoxy-(9*Z*,15*Z*)-octadecadienoic acid were synthesized and tested for their biological activities in the resistance of rice seedlings to blast disease. Both compounds had weak suppression on the mycelial growth but strong inhibition on the conidia germination of *Pyricularia grisea in vitro*. Foliar application with either of these compounds at three-to-four-leaf stage of rice seedlings, followed in 2 or 5 days by inoculation with the compatible strain, markedly reduced the disease severity, the percentage of diseased leaves and the lesion number and size. For the best expression of the induced resistance, the lag between induction and challenge inoculation for the hydroxy C_{18} fatty acid was 5 days but for the epoxy C_{18} fatty acid it was 2 days. The effective concentration of either of the chemicals for foliar application was 5 $\mu g\ ml^{-1}$, resulting in over 50% reduction in the disease severity. Preinoculation of the pathogen followed in 5 or 10 days by foliar spraying of either compound at 5 or 10 $\mu g\ ml^{-1}$ also suppressed the disease with significantly reduced lesion numbers and size. It appears that treatment with these unsaturated fatty acids before or after infection may enhance the rice resistance to blast disease.

Response of Drought-Tolerant, Submergence-Tolerant and Aromatic Rice Genotypes to Leaf and Neck Blast Disease

Shailaja Hittalmani[1], H.E. Shashidhar[2] and S.C. Chandrashekhar
University of Agricultural Sciences, Bangalore-65, India; [1]Present address: Department of Plant Breeding, International Rice Research Institute, PO Box 933, Manila, The Philippines; [2]Department of Biotechnology, Texas Technical University, Mail Stop 2122, Lubbock, TX 79409, USA

Seventeen rice genotypes including the locals and the improved lines known to have shown considerable tolerance to low moisture stress, submergence tolerance and possess grain aroma were scored for leaf blast at seedling stage and for neck blast at harvest to Indian races of *Pyricularia oryzae*. This experiment was undertaken in the wet season in the open fields when the natural fungal inoculum is very high in the atmosphere. The blast symptoms on young leaves and the older stem of the panicle was scored based on the number of lesions on the young leaves and the extent on the neck of the panicle at harvest and converted to percentage respectively (IRRI Standard Evaluation System).

The blast score results indicated the differential reaction of the rice genotype to the blast fungus. Low leaf blast and neckblast scores were recorded for genotypes that were tolerant to drought as well as excess moisture condition while most of the aromatic genotypes showed higher blast scores indicating their possible susceptibility to different indigenous races of blast pathogen. The high yielding varieties recorded higher blast scores but were much lower than for the susceptible check variety HR-12.

Among the genotypes observed, the submergence-tolerant strain DWR-4107 had a leaf blast score of 1 and Salumpikit, a drought-tolerant strain, also had leaf blast score 1 and the lowest percentage of neck blast (1.82%) disease. DWR-4107 is in the final stages in the field trials for release for cultivation.

Trap Plant Method to Predict the Occurrence of Rice Blast

A. Surin, U. Pliansinchai and D. Sophanocora
Rice Pathology Research Group, Department of Agriculture, Bangkok, Thailand

Managing rice blast requires a forecasting system which is simple and easy to use. To develop and test such a system, we conducted experiments at the Supanburi Field Crops Research Centre (SFCRC) during August 1992 to March 1993. Experimental plots were planted monthly under irrigated and rainfed conditions with a variety, RD23, susceptible to blast disease. The disease severity and climatic conditions were recorded monthly. Leaf blast was first seen on 7 September 1992. Lesions appeared over the whole plot (1 m^2) but severity was low. During September 1992, average daily temperature showed a general decreasing trend while dew incidence increased. The disease severity was higher in the following month. The climatic conditions and disease severity through the crop were statistically analyzed and results indicated a strong correlation between disease severity, dew, temperature and humidity. We can conclude that as average temperatures declined with the dry 'winter' season, dew incidence increased, relative humidities exceeded 60% and the disease incidence and severity exceeded our local economic threshold level.

In the same field, nine rice varieties with different reactions to *Pyricularia oryzae* were also planted. Each row was 50 cm long; SPR60 was used as border rows. At the end of September 1992, all varieties showed symptoms of rice leaf blast (LB). Slides coated with gelatin were attached to lesions with size and symptoms typical to each variety, and approximately the same across varieties. The number of conidia on each slide from each of the nine varieties was determined. There was no significant difference in lesion size among the nine varieties but the number of conidia/lesions was highly significant ($P < 0.01$). Though the resistant and moderately resistant varieties showed typical blast lesions, the number of conidia per lesion was not as high as in the susceptible variety. This indicated that the occurrence of rice blast requires not only favorable microclimatic conditions but also a susceptible rice variety.

Antibiotic Production Mediates the Biological Control of Rice Blast by *Pseudomonas fluorescens*

S.S. Gnanamanickam[1], R. Valasubramanian[1], A. Chatterjee[2], A.K. Chatterjee[2] and T.W. Mew[3]
[1]*University of Madras, India 600 25;* [2]*University of Missouri-Columbia, USA;* [3]*International Rice Research Institute, PO Box 933, Manila, The Philippines*

We collected a total of 1085 strains of *Pseudomonas fluorescens* from rice tissue samples gathered from different rice growing regions of India and screened them in the laboratory for their antagonism towards the rice blast fungus, *Pyricularia grisea*. Two hundred and seven strains (19.1%) induced growth-inhibition and eight of the strains inhibited *P. grisea* up to a maximum of >35 mm diameter on PDA plates. Further evaluations for blast suppression due to bacteria treatments were made in the greenhouse and field plots.

Isolation of antifungal antibiotic(s) (*Afa*) from the efficient strains of *P. fluorescens* were made by growing them in shake cultures and processing the culture fluids further by extraction with methylene chloride and by purification of the biologically active material by thin layer and column chromatography. Characterization of the antiblast substances has led us to identify the production of phenazine(s).

A genetic analysis of *Afa* was initiated for *P. fluorescens* strain 7-14. This strain does not harbor a plasmid indicating that the gene(s) specifying its antibiotic production are located within the chromosome. By using the plasmid vehicle, pUT/km, we performed random mini-Tn5 km mutagenesis and have isolated mutants which produce little or no antibiotic (0.12%)(*Afa*). Southern hybridization with the mini-Tn5 probe revealed that insertions within 3.3 kb, 3.9 kb, 7.2 kb or 10.6 kb *Sst1* chromosomal fragments affected *Afa* production. Complementary analysis suggested that the *Afa* genes may be clustered in the cosmid pAKC 901.

In a field test for biological suppression of rice blast, strain 7-14 afforded 79 and 82% reduction of leaf and neck blasts while its *Afa* mutants afforded 34 and 12% control of leaf and neck blasts suggesting that *Afa* production mediates the biological control of rice blast by *P. fluorescens*.

Biosynthetic Study of Diterpene and Flavone Phytoalexins in Rice Plants

O. Kodama, H. Kato and T. Akatsuka
School of Agriculture, Ibaraki University, Ami, Ibaraki 300-03, Japan

We isolated and identified several diterpene phytoalexins, oryzalexins A, B, C, D, E and S, and the flavone phytoalexin, sakuranetin, from blast-infected or UV-irradiated rice leaves. Among these oryzalexins, oryzalexin D (3,7-dihydroxy-sandaracopimaradiene) and oryzalexin E (3,9-dihydroxy-sandaracopimaradiene) show the highest antifungal activity and were produced from 3-hydroxy-sandaracopimaradiene by inducible cytochrome P-450 dependent monooxygenase(s). The sakuranetin synthesizing enzyme, naringenin 7-*O*-methyltransferase, was induced in the irradiated rice leaves. Consequently, sakuranetin did accumulate in the rice leaves. Because of the low antifungal activity of the precursors of these phytoalexins, naringenin and 3-hydroxy-sandaracopimaradiene, the conversion of these compounds to active phytoalexins by either cytochrome P-450 dependent monooxygenase(s) or naringenin 7-*O*-methyltransferase is very important for the dynamic defense mechanism of the rice plant.

Risk Analysis of Rice Leaf Blast Epidemics on Asian Countries Associated with Global Climate Change by Simulation Studies

Y. Luo[1] and P.S. Teng
Entomology and Plant Pathology Division, International Rice Research Institute, PO Box 933, Manila, The Philippines; Present address: [1]Department of Plant Pathology, University of Arkansas, 217 Plant Science, Fayetteville, AR 72701, USA

The two simulation models, CERES-RICE (a rice growth simulation model) and BLASTSIM (a rice leaf blast epidemic simulation model), were coupled by considering the effects of leaf blast on rice leaf photosynthesis and biomass production. At least 5 years of historical daily weather data for a given location of an Asian country were utilized to produce estimated daily weather data of each location by means of two weather generators, WGEN and WMAK, from DSSAT. Effects of global climate change on blast epidemics in terms of temperature and rainfall changes were considered in simulation studies. Thirty years' daily weather data produced by one of the above two generators for each location under each condition of global climate change were used for driving the combined model to simulate blast epidemics. Distributions of AUDPCs, maximum blast severities and estimated yield losses caused by leaf blast resulting from 30 simulation years were analyzed statistically. Using this simulation approach, 61 locations in seven Asian rice growing countries were studied for risk analysis of blast epidemics associated with global climate changes.

Changes in rainfall levels had no significant effect on variation of disease epidemics in all cases because that rainfall changes led only to very slight changes in dew period in the BLASTSIM model. In most locations, changing temperature had significant differences on disease development. However, these effects vary at different agroecological zones. In cool subtropical zones like Japan and north of China, climate changes by elevation of temperature up to +3°C will initially be associated with higher risk of blast epidemics. Situations in humid tropics and warm humid subtropics are opposite; lower temperature would lead higher risk of blast epidemics. In some situations, a large change in AUDPC did not result in more yield losses. Same AUDPCs probably do not cause same yield losses in different growing seasons under different weather conditions.

Foliar Fungicides for Management of Blast in Louisiana

D.E. Groth[1] and M.C. Rush[2]
[1]*Rice Research Station, Louisiana Agricultural Experiment Station, LSU Agricultural Center, PO Box 1429, Crowley, LA 70527-1429;* [2]*Department of Plant Pathology and Crop Physiology, Baton Rouge, LA 70803, USA*

Louisiana provides a warm, humid climate favorable for endemic and epidemic development of blast caused by the fungus *Pyricularia grisea*. Blast is the second most important rice disease in the United States after sheath blight (*Rhizoctonia solani*). The long-grain cultivars grown in the southern United States tend to have higher levels of partial resistance to blast than medium grain cultivars. Preventive fungicidal sprays at boot and heading can suppress rotten-neck symptoms, although the primary management tool used is disease resistance.

Field tests of foliar fungicides were conducted at the Rice Research Station in Crowley, Louisiana. The susceptible cultivar M201 was used. Fungicides were applied in water at 186 l ha^{-1} with a CO_2 pressurized, backpack sprayer. Fungicides were applied at the boot (B) (panicle 3–6 cm), and heading (H) (80–90% of panicles emerging) growth stages. Benlate (benomyl) was applied at 0.55 and 1.1 kg a.i. ha^{-1} and Beam (tricyclazole) was applied at 0.28 kg a.i. ha^{-1}. Blast severity was recorded as percent panicles infected at or near maturity. The center four rows of seven row plots were harvested with a small plot combine. Rough rice grain yields were calculated as kg ha^{-1} at 12% moisture for each treatment.

Benlate treatments consistently increased yields, resulting in 911 and 1328 kg ha^{-1} mean yield increases over the unsprayed check, respectively, for the low and high rates over 4 years. Beam treatments resulted in a mean 1214 kg ha^{-1} yield increase. Base yield levels (3174 kg ha^{-1}) were low in these trials due to high levels of disease pressure from blast, which was not completely controlled by these fungicides. The experimental fungicide RO43-2664, tested in 1991, completely controlled blast and yielded over 4070 kg ha^{-1} higher than the unsprayed check. This provided a good indication of the potential yield loss from blast. Benlate was the most effective against rice blast of the fungicides labeled in the United States in these tests. This fungicide suppressed blast significantly, but did not control the disease completely. Benlate was also effective for leaf blast control, but re-establishing the flood on a drained field often was as effective as fungicide applications for reducing disease development. Two 1.1 kg a.i. ha^{-1} applications of Benlate were more effective for controlling rotten-neck blast than the 0.55 kg ha^{-1} application, but the cost may prohibit using this rate. The fungicide Beam has performed very well against blast in most trials, but this fungicide is not labeled for use in the United States. Other fungicides have been tested for blast control with varying levels of activity, but none are near registration at this time.

Index

Note – page numbers in *italics* refer to figures and tables